Signal Processing for Autonomous Systems

Yong Ching Lim • Paulo S. R. Diniz •
Yih-Fang Huang

Signal Processing for Autonomous Systems

Springer

Yong Ching Lim
Singapore, Singapore

Yih-Fang Huang
Dept of Electrical Engineering
University of Notre Dame
Notre Dame, IN, USA

Paulo S. R. Diniz
Federal University of Rio de Janeiro
Rio de Janeiro, Rio de Janeiro, Brazil

ISBN 978-3-032-00674-5 ISBN 978-3-032-00675-2 (eBook)
https://doi.org/10.1007/978-3-032-00675-2

This Springer imprint is published by the registered company Springer Nature Switzerland AG
The registered company address is: Gewerbestrasse 11, 6330 Cham, Switzerland

Contents

Acronyms

4-QAM	Quadrature amplitude modulation
A/D	Analog-to-digital
ADAS	Advanced driver assistance systems
ADC	Analog-to-digital converter
AESA	Active electronically scanned array
AF	Application function
AI	Artificial intelligence
AMF	Mobility management function
AMPS	Advanced mobile system
APD	Avalanche photodiode
ATM	Asynchronous transfer system
AUSF	Authentication server function
AWGN	Additive white Gaussian noise
balun	Bal-to-unbalanced
BDS	BeiDou system
BLUE	Best linear unbiased estimator
BPF	Bandpass filters
BSC	Base station controller
BTS	Base transceiver station
BW	Bandwidth
CA	Coarse acquisition
CDF	Cumulative distribution function
CDMA	Code division multiple access
CP	Cyclic prefix
CRLB	Cramér-Rao lower bound
CTE	Coefficient of thermal expansion
D/A	Digital-to-analog
DAB	Digital radio broadcasting or digital audio broadcasting
DAC	Digital-to-analog converter
dBi	dB-isotropic
dBd	dB-dipole

DFT	Discrete Fourier transform
DL	Diode lasers
DMT	Discrete multitone
DOA	Direction of arrival
DPSSL	Diode pumped solid state laser
DS	Direct sequence
DS-CDMA	Direct sequence CDMA system
DWMT	Discrete wavelet multitone
ECEF	Earth centered, earth fixed
EEL	Edge-emitting laser
EFoD	Exponential family of distributions
EM	Electromagnetic
eMBB	Enhanced mobile broadband
ESPRIT	Estimation of signal parameters using rotational invariance technique
ETSI	European Standardization Union
FDD	Frequency division duplex
FDMA	Frequency division multiple access
FET	Field effect transistor
FFT	Fast Fourier transform
FH	Frequency hopping
FH-CDMA	Frequency-hopping CDMA
FHSS/LBT	Frequency hopping spread spectrum/listen before talk
FL	Fiber laser
FMCW	Frequency-modulated continuous wave
GGSN	Gateway GPRS support node
GHz	Gigahertz
GLONASS	Globalnaya Navigationnaya Sputnikovaya Sistema
GMSK	Gaussian minimum shift keying
gNB	Generalized NodeB
GNSS	Global Navigation Satellite System
GPRS	Generalized packet radio system
GPS	Global Positioning System
GSM	Global system for mobility
GSO	Geostationary orbit
HEMT	High electron mobility transistors
HPBW	Half-power beamwidth
IEEE	Institute of Electrical and Electronics Engineering
IF	Intermediate frequency
IGSO	Inclined geostationary orbit
IMT	International Mobile Communications
IP	Internet protocol
IRNSS	Indian Regional Navigation Satellite System
ISI	Intersymbol interference
ITU	International Telecommunication Union

KKT	Karush-Kuhn-Tucker
LAN	Local area networks
LED	Light-emitting diode
LHCP or LHEP	Left-hand circularly or elliptically polarized
Lidar	Light detection and ranging
LMMSE	Linear minimum mean square error estimator
LNA	Low-noise amplifier
LPF	Lowpass filters
LSE	Least squares estimator
LTE	Long-term evolution
MA	Multiple access
MAN	Metropolitan area networks
MAP	Maximum a posteriori
MC	Multicarrier
MEMS	Microelectromechanical system
MEO	Medium earth orbit
MIC	Microwave integrated circuit
MIMO	Multiple-input multiple-output
MISO	Multiple-input single-output
MLE	Maximum-likelihood estimate
mMTC	Massive machine type of communications
mmWave	Millimeter wave
MMI	Multimode interference
MMIC	Monolithic microwave integrated circuit
MMPC	Multipixel photon counter
MMSE	Minimum mean squared error
MSC	Mobile switching center
MSE	Mean-squared error
MT	Multitone
MUSIC	Multiple signal classifier
MVU	Minimum variance unbiased estimator
MVUE	Minimum variance unbiased estimator
NEF	Network exposure function
NF	Noise figure
NG-RAN	Next-generation radio access network
NR	New radio
NRF	Network repository function
NSSF	Network slice selection function
OFDM	Orthogonal frequency-division multiplex
OFDMA	Orthogonal frequency division multiple access
OPA	Optical phased array
OTH	Over-the-horizon
PA	Power amplifier
PAN	Personal area networks
PAPR	Peak to average power

PCF	Policy control function
Pcode	PR code
PDF	Probability density function
PESA	Passive electronically scanned array
PIC	Parallel interference canceler
PIN	p-type, intrinsic, n-type
PLF	Polarization loss factor
PLL	Phase-locked loop
PMT	Photomultiplier tube
PNT	Positioning, navigation, and timing
PR	Pseudo-random
PRI	Pulse repetition interval
PSK	Phase shift keying
PSTN	Public switched telephone network
PZT	Lead-zirconate-titanate
QoS	Quality of services
QZSS	Quasi-Zenith Satellite System
Radar	Radio detection and ranging
RFID	RF IDentification
RHCP or RHEP	Right-hand circularly or elliptically polarized
RNC	Radio network controller
RSSI	Radio strength indications
SATNAV	Satellite navigation
SCP	Service communication proxy
SDMA	Space-division multiple access
SGSN	Serving GPRS support node
S-IC	Self-interference cancellation
SIC	Serial interference canceler
SI-DS	Self-interference to the desired-signal ratio
SINR	Signal-to-interference-noise ratio
SiPM	Silicon photomultiplier
SIR	Signal-to-interference ratio
SNR	Signal-to-noise ratio
SOI	Silicon-on-insulator transistor
SPAD	Single-photon avalanche diode
SR	Shift registers
SSL	Solid-state lasers
SVD	Singular value decomposition
SWIR	Shortwave infrared
TDD	Time division duplex
TDM	Time-division multiplex
TDMA	Time division multiple access
TH	Time hopping
TMUX	Transmultiplexer
TOA	Time of arrival

TRM	Transmitter/receiver module
UDM	Unified data management
UDR	Unified data repository
ULA	Uniform linear array
UMTS	Universal mobile telecommunication system
UMVE	Uniformly minimum variance estimator
UPF	User plane function
URLLC	Ultra-reliable and low latency
UV	Ultraviolet
VCO	Voltage-controlled oscillator
VCSEL	Vertical-cavity surface-emitting laser
VGA	Variable gain amplifiers
VLR	Visitor location register
VSWR	Voltage standing wave ratio
WAN	Wide area networks
WCDMA	Wideband CDMA
WGS-84	World geodetic system 1984
WiMAX	Worldwide interoperability for microwave access
ZP	Zero prefix

Chapter 1
Summary

There are many types of autonomous vehicles. They may be operating on land, underground, on water surface, underwater, in air, or in space. A self-driving car is an example of autonomous vehicles. Most modern vehicles have some automatic functions such as speed control. Many passenger cars have automatic parking function. An aircraft may have auto pilot. At the two extremes are the fully manual with no automatic feature and the fully autonomous requiring no human intervention in its operation. In between the two extremes are the increasingly autonomous vehicles. Certain basic automatic functions are not considered "automatic." For example, a vehicle flashlight that flashes automatically instead of requiring a human to turn it on and off is considered "manual"; otherwise, there would be no "fully manual" vehicle in these modern days. Fully autonomous vehicles are not expected to have the intelligence of a human. For example, a fully autonomous passenger car is not required to have the ability to recognize a malfunctioning traffic light and negotiate its way crossing a congested malfunctioning traffic junction; otherwise, a fully autonomous passenger car will not be expected any time soon. To what degree of automation will a vehicle be considered an autonomous vehicle? Different individual, different institution, and different jurisdiction may have different perception and definition.

The 2021 SAE (Society of Automotive Engineers) standard J3016_202104 of April 2021 classifies on-road autonomous vehicles into the following six levels.

Level 0 – No driving automation.

Level 1 – Driver assistance such as cruise control and lane keep assist.

Level 2 – Vehicle has ADAS (advanced driver assistance systems) such as steering, throttling, accelerating, and braking capabilities. A driver must sit in the driver's seat and actively monitor the surroundings and must be ready to take over control of the vehicle at any time.

Level 3 – Vehicle has full environmental detection capabilities. Driver monitors surroundings passively but must be ready to take over control of the vehicle at any time. The technological difference between Level 2 and

Y. C. Lim et al., *Signal Processing for Autonomous Systems*,
https://doi.org/10.1007/978-3-032-00675-2_1

Level 3 is substantial but not significant from a human perspective; a driver must still be in the driver's seat.

Level 4 – Vehicle requires no driver interaction but still include functional driving equipment such as steering wheel and pedals.

Level 5 – Vehicle is completely autonomous and does not have driving equipment such as steering wheel and pedals.

Autonomous vehicles at various levels of automation make use of sensors such as radar, sonar, lidar, camera, Global Navigation Satellite System (GNSS), accelerometer, speedometer, odometer, gyro, altimeter, and pressure meter to perceive the condition of its surroundings. Signage must be correctly interpreted. Signage information and sensor information are fed to a sophisticated AI (artificial intelligence) algorithm interpreting the ever-changing surrounding and making decisions on a real-time basis.

An autonomous vehicle may also communicate with other autonomous vehicles directly. The 5G system has very short latency. One of the objectives of very short latency is for supporting communication between autonomous vehicles.

Signals from the antenna are converted into digital form and processed digitally. Digital filters are used at various stages of signal processing to enhance essential information embedded in the signal and reject noise that corrupts the signal. Digital filters are also employed in various stages of communication for extracting essential information.

AI is an important aspect of autonomous vehicles and warrants a complete volume by itself and so is not covered in this book. This book focuses on signal processing techniques for autonomous vehicles. It is divided into three main parts. Part I is devoted to discussing the fundamentals of signal processing for better understanding of Parts II and III. Part II discusses the various techniques used in determining the direction, velocity, and location of an autonomous vehicle and other vehicles in the vicinity of the autonomous vehicle. Part III covers the various techniques used in an autonomous vehicle's communication system. A brief summary of each of these three parts is presented below.

1.1 Part I: An Introduction to Statistical Estimation Theory

Part I presents the fundamental concepts of statistical estimation theory in Chaps. 2–3. We shall present two main methodologies for statistical estimation, namely, Fisher and Bayes. The Fisher estimation methodology relies critically on the notion of likelihood functions and is usually used to estimate unknown deterministic parameters. On the other hand, the Bayesian estimation methodology is derived with knowledge of the a posteriori distribution and is usually used to estimate random signals or parameters.

Statistical estimation theory is the basic working principle for most communication and signal processing systems. In wireless communication systems, for example, the problem of interference mitigation is usually solved with estimation methods based on some statistical modeling of signals and noise. Statistical estimation methods are also used in timing acquisition and carrier frequency offset estimation in most communication systems. They are also important tools for location systems and navigation systems for autonomous vehicles.

The objective of Part I is to present essential fundamental concepts of statistical signal processing, with a focus on estimation. It should be noted that these statistical concepts and the tools are applicable to a wide variety of areas where statistical models are used. Our coverage here should provide an easy-to-access reference to the essential concepts and useful tools of mathematical statistics underpinning the working principles of the systems to be discussed throughout this book.

Chapter 2 presents the basic principles of Fisher statistics, which are built upon the notion of likelihood functions. The concepts of sufficient statistics and that of exponential family of distributions form an essential part of this discussion. Then, the notion of minimum variance unbiased estimator will be introduced. Lehmann-Scheffè theorem will be presented with the aid of the notion of *complete sufficient statistics*, which is a very useful tool of finding minimum variance unbiased estimators. Information inequality, including Cramèr-Rao bound, will then be presented. Discussion on Fisher statistics will conclude with an outline of the important properties of maximum-likelihood estimates (MLEs), as well as some application examples in signal processing and communications.

Bayesian estimation methodology is presented in Chap. 3 primarily for estimating random signals or parameters with a known a posteriori distribution. The key objective of Bayesian estimation is to minimize the *Bayes risk*, which is defined to be the expected value of the *cost*. Within this framework, depending on how the cost function is defined, different estimation methods can be derived. For example, if the cost function is defined to be the mean-squared error, the resulting Bayesian estimation method is the conditional mean estimator, which is the minimum mean-squared error (MMSE) estimator. Another example of a Bayesian estimator is the maximum a posteriori (MAP) estimator, obtained by maximizing the a posteriori probability density function.

1.2 Part II: Direction and Location Estimation

Satellite navigation (SATNAV) system is commonly used for position estimation. Currently, several satellite navigation systems are in operation or in the process of construction. They include the GPS (USA), GLONASS (RUSSIA), BeiDou (China), Galileo (EU and ESA), IRNSS (India), and QZSS (Japan) and are briefly discussed in Chap. 4. Included in Chap. 4 are brief discussions on some commonly used terminologies on satellite orbits, WGS-84 reference ellipsoid, and methods for location estimation.

Lidar and radar are very commonly used for distance and velocity estimations. Radar uses electromagnetic (EM) waves at RF frequencies, which are tens of GHz, whereas lidar uses laser. Laser is also a form of EM wave but at much higher frequencies. Lidar and radar hardware are discussed in Chaps. 5 and 6, respectively.

Various types of common laser sources and detectors are discussed in Sects. 5.3 and 5.4, respectively. A laser beam must be steered in order for it to perceive the conditions of a reasonable area of its surrounding instead of only one sport. Commonly used methods for steering a laser beam include mechanical spinning, MEMS, and optical phased array. They are discussed in Sect. 5.5.

A radar system involves the transmission and reception of RF EM waves. The properties of EM waves are discussed in Sects. 6.2 and 6.3.1. The structure involves in the transmission and reception of EM waves is called the antenna. There are many types of antenna; they are discussed in Sects. 6.3.3–6.3.9. The direction steering of RF EM wave may be achieved using phased array. The implementation of RF phased array requires RF switches, phase shifters, and attenuators. RF switches, phased shifters, and attenuators are discussed in Sect. 6.4, 6.5, and 6.6, respectively. RF EM wave steering may also be achieved using microwave lens as discussed in Sect. 9.19.

After a brief discussion on the hardware essentials for lidar and radar presented in Chaps. 5 and 6, respectively, we proceed to discuss the signal processing techniques for velocity, distance, and direction estimations using lidar/radar in Chaps. 7–10.

The velocity of a moving object may be estimated from either (1) the first principle defined as the distance moved per unit time or (2) Doppler shift when the object is illuminated by a sinusoidal signal. The illuminant may be a laser, radar, or an acoustic source. Distance estimation may be achieved by using a pulse or a chirp. These are discussed in Sect. 7.4. When a chirp is used, the estimation problem is essentially a problem of estimating the frequency of a sinusoid. Frequency estimation may be accomplished using the discrete Fourier transform (DFT) or the fast Fourier transform (FFT) as discussed in Sect. 7.5.

Frequency estimation has wide applications. Many signal processing techniques have been developed for frequency estimation with resolution higher than that can be achieved using the Fourier transform method. Three of the methods, namely, (a) linear prediction, (b) maximum entropy method, and (c) maximum likelihood method, are discussed in Chap. 8.

A wave, regardless of whether it is laser, RF, or acoustic, can be transmitted towards a given direction using an array of omnidirectional transmitters with the proper selection of the phase and amplitude of the signal in each transmitter. An array of omnidirectional receivers can also be made to be directionally selective in reception by summing properly phase-shifted and amplitude-scaled versions of the signal received from each receiver. The process of phase shifting and amplitude scaling of the signal in each transmitter or receiver for directivity is called array beamforming. This is discussed in Chap. 9.

The analogue signals of a system in the form of voltage or current are signals with real values. However, in many digital signal processing algorithms, signals are assumed to have complex values. Chapter 10 is devoted to the exploration of the methods for converting signals in real form to signals in complex form.

1.3 Part III: Introduction to Wireless Communications—MIMO Systems and Radar

In real life, there is a growing need to communicate large amounts of information through a propagation medium. A key player in most transmissions is the propagation of electromagnetic energy through the air or wired channels. In particular, we are interested in wireless communication systems benefiting from widespread digital technology. In wireless communications, the channel is the medium through which the electromagnetic waveform, as well as other means, energy propagates. Part III of the book discusses several concepts related to wireless communications that apply beyond the domains of the radio spectrum domain. Chapters 11–17 cover the fundamental aspects of cellular communications to the use and design of multiple-input and multiple-output (MIMO) systems so crucial for detection, location, and estimation purposes.

Part III starts with briefly describing the classical digital communication systems in Chap. 11. Next, the main strategies to allow multiple access to wireless channels in a shared environment are discussed in Chap. 12, including describing some classes of wireless network services currently utilized.

Chapter 13 explains the basic principles of spectrum reutilization akin to cellular communications. The chapter also discusses the evolution of cellular communication standards. Wireless channel modeling is a crucial ingredient to deploy the network infrastructure successfully; as such, Chap. 14 discusses the subject to equip the reader with the basic ideas.

Some wireless networks use the code division multiple access (CDMA) systems and still find application in some localization systems. As a result, Chap. 15 discusses some properties and variations of the CDMA systems, highlighting the generations of the spreading codes, which are valuable tools in many building blocks of communication systems. The chapter also discusses some receiver solutions that can be connected to the methods mentioned in Part II, as well as some concepts of channel capacity to evaluate the efficacy of spectral use.

Chapter 16 presents in detail the orthogonal frequency-division multiplex (OFDM) system, along with its multiple access version, which is the primary transmission system to cope with frequency-selective channels. This chapter presents several computationally efficient transceiver configurations along with the concepts of bit loading and the optimal equalization approach.

Chapter 17 presents some features and configurations of the MIMO systems. The MIMO model encompasses several communication systems, including CDMA and multicarrier systems. The chapter discusses the use of statistical signal processing tools to design optimized MIMO systems. This chapter also presents some mathematical properties of the data rate capacity in MIMO systems and an important filter-bank description of the transceivers. This so-called transmultiplex description provides many insights into the block transmission solutions, as described in Chap. 17. The signal processing aspects of the analysis and implementation of automotive radar systems are presented in Chap. 18, discussing the key issues related to range-Doppler estimation. The chapter includes details related to data organization and sampling and presents a joint radar wireless communication system employing multicarrier transmission. Chapter 18 also introduces some tools related to MIMO radar, such as virtual arrays, and details some examples, including a direction-of-arrival case. Several previously discussed tools, such as MIMO and OFDM systems, are adapted to the typical implementation of these systems.

In summary, this book covers a wide range of topics pertaining to the domains of direction and localization of objects, with autonomous vehicles as the primary motivation. However, we divided the text into three somewhat self-contained parts to provide an overall view of the tools taking part in the most practical solutions. Part I provides an overview of the statistical theory of estimation, from which many solutions discussed in Parts II and III rely. Part II discusses the main concepts, devices, standard solutions, and some signal processing tools required to estimate the location of subjects successfully. In practice, the methods employed to detect and estimate the localization of vehicles, for instance, should be well understood in the sense of the quality and robustness of the solutions. Part III delves into explaining the basics of wireless communications, an area that is crucial in current and future solutions to the problems addressed in this book. The coming solutions to widespread high-data-rate and low-latency wireless communications will be essential to reliable and accurate localization solutions. Many forthcoming solutions will be deployed with the aid of wireless networks such as the 5G, WiFi, and low-earth orbit satellites, along with the increasing computational power of autonomous vehicles and the growing availability of cloud computing and federated learning methods. This book covers many important techniques taking part in ongoing and future solutions for accurate and reliable estimation of direction and localization.

Part I
An Introduction to Statistical Estimation Theory

Chapter 2
Fisher Statistics and Maximum Likelihood Estimation

2.1 Introduction

In this chapter, we present the basic principles of statistical estimation as they relate to the other parts of this book. Statistical estimation theory has a wide range of applications in solving real-world problems, some of which are presented in this book. For example, determining the location of a moving object can be formulated as an estimation problem. Estimation of the frequency of sinusoidal signals is also a common practical problem, such as those required in estimating the speed of a vehicle. Designing precoders and equalizers in the transceivers of multi-input multi-output (MIMO) communication systems also calls for estimation methods.

In developing statistical estimation algorithms, the underlying statistical distribution of the signals is usually assumed to be known, except for certain parameters of the distribution (e.g., mean and variance). Those parameters need to be estimated in order to formulate implementable solutions. Furthermore, those unknown signals or parameters can be either deterministic or random. Depending on the characteristics of the signals and parameters to be estimated, and depending on the measure of estimation quality (which is often defined in terms of estimation error), many different estimation methods have been derived. In general, the objective of estimation is to optimize (minimize) some function of the estimation errors.

Generally speaking, the studies of statistical estimation theory can be broadly classified into two major schools of thoughts: Fisher school of statistics and Bayesian school of statistics. Both of these methodologies are derived based on the presumed knowledge of the statistical distributions of the underlying random variables. However, there are fundamental differences between the two approaches. Fisher school of statistics assumes knowledge of the a priori distribution, while the Bayesian school of statistics requires knowledge of the a posteriori distribution. Fisher's approach to estimation problems is commonly applied to estimating deterministic signals/parameters, whereas the Bayesian approach is more often used to estimate random signals/parameters. It is worth noting that there are other

Y. C. Lim et al., *Signal Processing for Autonomous Systems*,
https://doi.org/10.1007/978-3-032-00675-2_2

estimation methods like those of least squares that do not require the knowledge of distribution like Fisher or Bayesian methods do. However, they are also a popular tool that is used in many applications. Some discussions on the least squares methods can be seen in Sect. 17.3.2 of Chap. 17.

The focus of this chapter will be on presenting some key elements in Fisher estimation theory and employing MLE for estimating unknown deterministic parameters. To begin, we first define *sufficient statistic* and introduce the notion of *exponential family of distributions* (EFoD), which constitute an essential part of this discussion. The EFoD has many useful properties in estimation, particularly because it offers an easy way to identify sufficient statistic. The optimality theory in the framework of Fisher statistics and comparison of estimators will be considered in the latter part of the chapter. For that, we shall introduce the notion of minimum variance unbiased estimator (MVUE), and present Lehmann-Scheffè theorem, facilitated by the concept of *complete sufficient statistic*, which is a very useful tool of finding MVUE. Information inequality and Cramér-Rao lower bound (CRLB) will be shown as a metric of optimality of estimators measured in terms of the variance of estimation errors. The discussion of Fisher statistics will conclude with a summary of the important properties of MLE, as well as an application example in signal processing. Materials presented in this chapter also lay the foundation for later chapters of this book, particularly Chap. 8, *Frequency Estimation*, which presents MLE and CRLB for the associated problem, and Chap. 17, which considers MLE as one of the tools for designing MIMO communication systems.

2.2 Likelihood Function

The fundamental concept upon which the Fisher school of statistics is built is that of *likelihood function* (Fisher 1922). The likelihood function has the identical functional form as the a priori probability density function (PDF) of the observed random variables. In Fisher statistics, the likelihood function is the objective function to be optimized. In essence, the likelihood function characterizes the *likelihood*, which is defined in terms of the probability, that a particular parameter (e.g., mean or variance of the signal distribution) could have produced the set of outcomes. Highlighting the Fisher statistical approach to estimation is the method of maximum likelihood estimation (MLE), which has found a wide range of applications in signal processing, communications, and control systems. Some of those examples are seen in, e.g., Chaps. 8 and 17 of this book.

Though the likelihood function is exactly the same as the a priori PDF in its functional form, its independent variable is the parameter to be estimated. On the other hand, the independent variable of the PDF is the value of the random variable. Stated in another way, the PDF is a function defined on the sample space of the random variable, while the likelihood function is a function defined on the parameter space. Thus, the value of the PDF varies with respect to the value of the random variable given a fixed parameter, while the value of the likelihood function varies

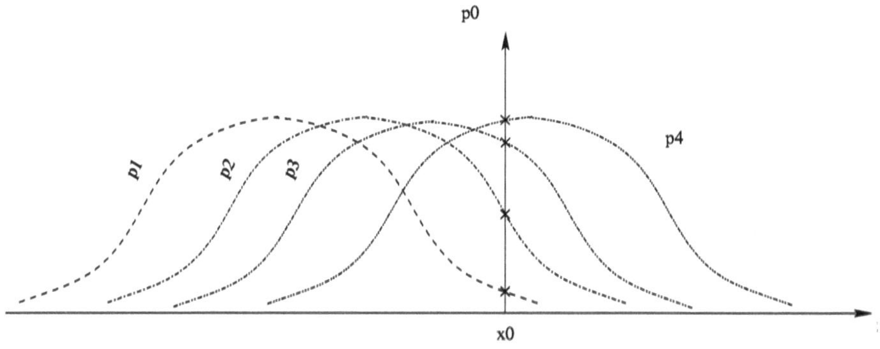

Fig. 2.1 Likelihood function, $L(x_0; \theta)$ and probability density functions $f_X(x; \theta_1)$, $f_X(x; \theta_2)$, $f_X(x; \theta_3)$, $f_X(x; \theta_4)$

with respect to the value of the parameter given a particular realization of the random variable.

Consider a random variable X that has a cumulative distribution function (CDF) $F_X(x)$ whose PDF is parameterized by an *unknown* parameter θ, denoted by $f_X(x; \theta)$. The likelihood function (with respect to the parameter θ) is defined to be

$$L(x; \theta) = f_X(x; \theta) \tag{2.1}$$

As noted above, the distinction between $L(x; \theta)$ and $f_X(x; \theta)$ is that the value of $L(x; \theta)$ varies with respect to θ while x is fixed, while the value of $f_X(x; \theta)$ varies with x for a given value of θ. In other words, for a given realization of the random variable $X = x$, $L(x; \theta)$ is a function of only θ. Figure 2.1 illustrates the distinction between a likelihood function and the PDFs.

Thus, the likelihood function provides a quantitative measure of how *likely* that a particular realization (observation) of the random variable would have been produced from a particular parameter value of the distribution. The higher the value of the likelihood function is, the more likely the particular value of the parameter would have produced that realization of X. Hence, the cost function in Fisher's estimation paradigm is the likelihood function, and the objective is to find a value of the parameter that maximizes the likelihood function, thus the term *maximum likelihood estimate* (MLE).

The MLE can be derived as follows: Let X be a random variable and $f_X(x; \theta)$ be its PDF parameterized by θ that is to be estimated. Accordingly, the likelihood function is given by (2.1), and the MLE for θ is defined by

$$\hat{\theta}_{MLE}(x) = \text{Arg}\{\max_{\theta} L(x; \theta)\} \tag{2.2}$$

which is the value of θ that yields the maximum value of $L(x;\theta)$ over the entire parameter space. If the likelihood function $L(x;\theta)$ is differentiable with respect to θ, then a necessary condition for $L(x;\theta)$ to be maximized is that

$$\frac{\partial}{\partial\theta}L(x;\theta)\bigg|_{\theta=\hat{\theta}_{MLE}}=0. \tag{2.3}$$

The above equation is sometimes referred to as the *likelihood equation.* In general, it is more convenient to consider the *log-likelihood function,* which is defined as

$$l(x;\theta)=\ln L(x;\theta) \tag{2.4}$$

Accordingly, we can define the *log-likelihood equation* as

$$\frac{\partial}{\partial\theta}l(x;\theta)\bigg|_{\theta=\hat{\theta}_{MLE}}=0. \tag{2.5}$$

Since $l(x;\theta)=\ln L(x;\theta)$ is a monotone nondecreasing function of $L(x;\theta)$, maximizing $l(x;\theta)$ is equivalent to maximizing $L(x;\theta)$. If the likelihood function (thus the log-likelihood function) is differentiable to the second order, a sufficient condition for the likelihood function to be maximized is that $L(x;\theta)$ is concave downwards, i.e.,

$$\frac{\partial^2}{\partial\theta^2}L(x;\theta)\bigg|_{\theta=\hat{\theta}_{MLE}}\leq 0. \tag{2.6}$$

or, equivalently

$$\frac{\partial^2}{\partial\theta^2}l(x;\theta)\bigg|_{\theta=\hat{\theta}_{MLE}}\leq 0. \tag{2.7}$$

The following simple examples illustrate some practical applications of the method of MLE.

Example 2.1 To facilitate better planning of power distribution, a utility company needs to find out the fraction of households that own electric vehicles (EVs) in residential communities. They survey a total of K households in a residential community and characterize their survey results in terms of a set of independent and identically distributed (*i.i.d.*) Bernoulli random variables. Specifically, they use $X_1, X_2, \cdots, X_K$ to represent the survey results of a residential community, where $X_i = 0$ if the i-th household does not own an EV, and $X_i = 1$ if it does.

The problem is formulated with the marginal probability mass function (PMF) of X_i, defined as

$$P_X(x_i;\theta) = \begin{cases} 1-\theta & \text{if } x_i = 0 \\ \theta & \text{if } x_i = 1 \end{cases}$$

where $0 < \theta < 1$ is the probability that a household owns an EV. Since X_i's, $i = 1, 2, \cdots, K$, are *i.i.d.* random variables, their joint PMF, thus the corresponding likelihood function, is the product of the marginal ones and is given by

$$L(\mathbf{x}, \theta) = \prod_{i=1}^{K} \mathrm{p}(x_i;\theta) = \theta^n (1-\theta)^{K-n} \tag{2.8}$$

where $n = \sum_{i=1}^{K} x_i$ is the total number of households that own an EV among those K households surveyed. The goal here is to find an MLE for θ. As usual, it is easier to work with the log-likelihood function, which is

$$l(\mathbf{x}, \theta) = n\ln(\theta) + (K-n)\ln(1-\theta)$$

Employing (2.5), i.e., differentiating the above equation with respect to θ and setting it equal to zero, yields

$$\left.\left(\frac{n}{\theta} - \frac{(K-n)}{(1-\theta)}\right)\right|_{\theta=\hat{\theta}_{\mathrm{MLE}}} = 0$$

Thus,

$$\hat{\theta}_{\mathrm{MLE}} = \frac{n}{K} \tag{2.9}$$

Thus, the MLE for the probability that a household owns an EV in a residential community is simply the ratio of the number of households that own EVs to the total number of households surveyed, which is consistent with our intuition. This example illustrates the well-known result that the MLE of the probability of success in a Bernoulli trial is the ratio of the number of successes to the total number of trials, which is sometimes referred to as *relative frequency*.

Example 2.2 Consider a problem of estimating a deterministic DC signal embedded in additive white Gaussian noise, using a set of observations. This problem can be formulated as

$$X_i = \theta + N_i$$

where N_i, $i = 1, 2, \cdots, K$, is a set of *i.i.d.* Gaussian random variables with zero mean and known variance σ^2, θ is the DC signal to be estimated, and X_i is the sequence of observations. Since N_i's are *i.i.d.* Gaussian random variables and θ is deterministic (even though unknown), X_i's are *i.i.d.* Gaussian random variables with mean θ, thus the likelihood function as defined in (2.1) is given by

$$L(\mathbf{x};\theta) = \left(\frac{1}{\sqrt{2\pi\sigma^2}}\right)^K \exp\left[-\frac{1}{2\sigma^2}\sum_{i=1}^{K}(x_i-\theta)^2\right] \tag{2.10}$$

and the log-likelihood function is

$$l(\mathbf{x};\theta) = -K\ln(\sqrt{2\pi\sigma^2}) - \frac{1}{2\sigma^2}\sum_{i=1}^{K}(x_i-\theta)^2 \tag{2.11}$$

Solving the *log-likelihood equation*, (2.5), yields

$$\hat{\theta}_{MLE}(\mathbf{x}) = \frac{1}{K}\sum_{i=1}^{K}x_i \tag{2.12}$$

This example shows another well-known result that the MLE for a DC signal embedded in a sequence of *i.i.d.* Gaussian random variables (which is commonly referred to as *additive white Gaussian noise*) is the *sample mean* of the observations, which is also consistent with our intuition. There are more to say about relative frequency and sample mean of these two examples as MLE.

According to the *strong law of large numbers* (Papoulis and Pillai 2002), the relative frequency given in (2.9) in Example 2.1 converges with probability one (w.p.1.) to θ, i.e., the probability of success in the Bernoulli trials. Likewise, the sample mean (2.12) converges to the DC signal θ w.p.1. The convergence w.p.1. can be understood intuitively as follows: The estimate would be very close to the true parameter value when the number of observations is large enough, and it would be almost surely equal to the true parameter value when the number of observations approaches infinity.

Another important property of the MLE in the above two examples is that they both are directly related to *sufficient statistic* for the parameter to be estimated. Specifically, the relative frequency (2.9) in Example 2.1 is a sufficient statistic for estimating the probability of success θ. Also, the sample mean (2.12) in Example 2.2 is a sufficient statistic for estimating the DC signal θ.

The concept of sufficient statistic is critical to the optimum properties of MLE and, in general, to Fisher statistics. Generally speaking, the sufficient statistic can be extracted from the likelihood function, and the MLE is almost always directly proportional to the sufficient statistic, which is why MLE has many optimal properties as an estimate. This is the topic to be discussed in the next section.

2.3 Sufficient Statistics

A sufficient statistic is defined in reference to a particular parameter (or signal) to be estimated. Assume that $\mathcal{X}$[1] is a random vector representing a set of observed variables whose marginal distribution is parameterized by an unknown variable θ. A function of the random vector $\mathcal{X} = \mathbf{x}$, denoted by $T(\mathbf{x})$, is said to be a *sufficient statistic* if the probability distribution of $\mathcal{X}$ given $T(\mathbf{x}) = t$ is independent of θ. In other words, if $T(\mathbf{x})$ is a sufficient statistic, then all the information regarding the estimation of θ that can be extracted from the observation is contained in $T(\mathbf{x})$. Another way to explain this is that a sufficient statistic is a function of the observed data whose use involves no loss of information relevant to the parameter that is to be estimated. Thus, if an estimator is derived based on a sufficient statistic, e.g., MLE, it is derived from the most relevant information, thus it would be a *good* estimate.

The *Fisher factorization theorem* stated below provides a formal analytical characterization for the sufficient statistic and is sometimes taken as a definition for the sufficient statistic.

Fisher Factorization Theorem (Bickel and Doksum 1977) Given a random vector $\mathcal{X}$ whose statistical distribution is characterized by a parameter θ. A function, $T(\mathbf{x})$, of the observation of the random vector $\mathcal{X} = \mathbf{x}$ is a sufficient statistic if the likelihood function (associated with the distribution of $\mathcal{X}$) can be expressed as a product of two functions, with one being a function of both $T(\mathbf{x})$ and θ, while the other one being a function of $\mathbf{x}$ only. Specifically, if the likelihood function $L(\mathbf{x}; \theta)$ can be expressed as

$$L(\mathbf{x}; \theta) = h(T(\mathbf{x}), \theta)g(\mathbf{x}), \tag{2.13}$$

then $T(\mathbf{x})$ is the sufficient statistic.

Observe that the likelihood function (2.8) in Example 2.1 can be expressed as

$$L(\mathbf{x}, \theta) = \theta^n(1-\theta)^{K-n} = \left(\frac{\theta}{(1-\theta)}\right)^n (1-\theta)^K \tag{2.14}$$

We can define $h(T(\mathbf{x}), \theta) = \left(\frac{\theta}{(1-\theta)}\right)^n (1-\theta)^K$ and $g(\mathbf{x}) = 1$. Accordingly, $T(\mathbf{x}) = n$ is a sufficient statistic for estimating θ for the Bernoulli distribution in Example 2.1, and the MLE for θ is $\hat{\theta}_{\text{MLE}} = \frac{n}{K}$ which is simply the sufficient statistic $T(\mathbf{x}) = n$ normalized by the total number of observations. In fact, it should be noted that sufficient statistic can usually be scaled by a constant to suit the underlying problem. Thus, while $T(\mathbf{x}) = n$ is a sufficient statistic in Example 2.1, $T(\mathbf{x}) = \frac{n}{K}$ is also a sufficient statistic.

---

[1] In this chapter, we use $\mathcal{X}$ to denote a random vector and $\mathbf{x}$ to denote its realization. Additionally, the notation $\mathbf{x}$ is also adopted to represent deterministic vectors in the remaining parts of the book, but the context leads to no confusion.

Similar to Example 2.1, the likelihood function (2.10) in Example 2.2 can be expressed as

$$L(\mathbf{x};\theta) = \exp\left[\frac{\theta}{\sigma^2}\left(\sum_{i=1}^{K} x_i\right) - \frac{K\theta^2}{2\sigma^2}\right]\left(\frac{1}{\sqrt{2\pi\sigma^2}}\right)^K \exp\left[-\frac{1}{2\sigma^2}\sum_{i=1}^{K} x_i^2\right] \tag{2.15}$$

Matching the terms of (2.15) to those of (2.13), it can be seen easily that if we set

$$h(T(\mathbf{x}),\theta) = \exp\left[\frac{\theta}{\sigma^2}\left(\sum_{i=1}^{K} x_i\right) - \frac{K\theta^2}{2\sigma^2}\right]$$

and

$$g(\mathbf{x}) = \left(\frac{1}{\sqrt{2\pi\sigma^2}}\right)^K \exp\left[-\frac{1}{2\sigma^2}\sum_{i=1}^{K} x_i^2\right],$$

then $T(\mathbf{x}) = \sum_{i=1}^{K} x_i$ is clearly a sufficient statistic for estimating θ. Again, the sufficient statistic is not unique, for it can be scaled by a constant. Thus, the sample mean, (2.12), is a sufficient statistic, just as $T(\mathbf{x}) = \sum_{i=1}^{K} x_i$ is. A convenient way to identify sufficient statistic is through the notion of *exponential family of distributions* (EFoD).

2.4 Exponential Family of Distributions

The exponential family of distribution (EFoD) is a family of distributions that is parameterized by a parameter of finite value and that has some well-defined exponential form. The specific form of the EFoD offers many convenient tools in the study of statistics, including identification of sufficient statistics.

If the PDF of a distribution, parameterized by a parameter θ, can be expressed a

$$f_{\mathcal{X}}(\mathbf{x}|\theta) = \{\exp[c(\theta)T(\mathbf{x}) + d(\theta) + S(\mathbf{x})]\}\, I_A(\mathbf{x}) \tag{2.16}$$

where the functions $c(\theta)$ and $d(\theta)$ depend only on the parameter θ, while $T(\mathbf{x})$ and $S(\mathbf{x})$ depend only on the realization of the random vector $\mathcal{X}$, i.e., $\mathbf{x}$, and $I_A(\mathbf{x})$ is an indicator function defined as

$$I_A(\mathbf{x}) = \begin{cases} 1 & \text{if } \mathbf{x} \in A \\ 0 & \text{otherwise} \end{cases} \tag{2.17}$$

where set A is the range space of $\mathcal{X}$. It is important to note that set A in the above definition is independent of the parameter θ. A distribution whose PDF can be expressed in the form given in (2.16) is said to belong to the *exponential family of distributions* (EFoD) (Bickel and Doksum 1977). The EFoD has many common features that are important in theory and applications of statistics.

It can be verified easily that many commonly used distributions such as Gaussian, exponential, Laplace (i.e., two-sided exponential), binomial, and Poisson distributions all belong to the EFoD. When a distribution belongs to the EFoD, its PDF can be expressed in the form of (2.16) and its associated likelihood function is given by

$$L(\mathbf{x};\theta) = \{\exp\left[c(\theta)T(\mathbf{x}) + d(\theta) + S(\mathbf{x})\right]\} I_A(\mathbf{x}) \tag{2.18}$$

Employing the Fisher factorization theorem, (2.13), we can identify $h(T(\mathbf{x}), \theta)$ and $g(\mathbf{x})$ as follows:

$$h(T(\mathbf{x}), \theta) = \exp\left[c(\theta)T(\mathbf{x}) + d(\theta)\right] \tag{2.19}$$

and

$$g(\mathbf{x}) = \{\exp\left[S(\mathbf{x})\right]\} I_A(\mathbf{x}) \tag{2.20}$$

Thus, we can see clearly that $T(\mathbf{x})$ in the definition of EFoD, as in (2.18), is a sufficient statistic for the estimation of the parameter θ, and it is sometimes referred to as *natural sufficient statistic*.

As a simple example, consider the Gaussian likelihood function, (2.10), in Example 2.2, and use the expression given in (2.15), and we have

$$\begin{aligned} L(\mathbf{x};\theta) &= \left(\frac{1}{\sqrt{2\pi\sigma^2}}\right)^K \exp\left[-\frac{1}{2\sigma^2}\sum_{i=1}^K (x_i-\theta)^2\right] \\ &= \{\exp\left[c(\theta)T(\mathbf{x}) + d(\theta) + S(\mathbf{x})\right]\} I_A(\mathbf{x}) \end{aligned} \tag{2.21}$$

where $c(\theta) = \frac{\theta}{\sigma^2}$, $T(\mathbf{x}) = \sum_{i=1}^K x_i$, $d(\theta) = -\frac{K\theta^2}{2\sigma^2}$, $S(\mathbf{x}) = -\frac{K}{2}\ln(2\pi\sigma^2) - \frac{1}{2\sigma^2}\sum_{i=1}^K x_i^2$, and $I_A(\mathbf{x})$ is the indicator function for which A is the set of real-valued vectors. Accordingly, $T(\mathbf{x}) = \sum_{i=1}^K x_i$ is a sufficient statistic for estimating θ, and we have seen from Example 2.2 that the MLE for θ is just $T(\mathbf{x})$ normalized by the sample size K, i.e., which is the sample mean.

As another example of EFoD, let us consider *Poisson* distribution.

Example 2.3 Given a set of *i.i.d. Poisson* random variables $X_i, i = 1, 2, \cdots, K$, whose marginal PMF, parameterized by the mean arrival rate λ, is given by

$$P_{X_i}(x_i) = \begin{cases} \frac{\lambda^{x_i}}{x_i!}\exp[-\lambda] & \text{if } x_i = 0, 1, 2, \cdots \\ 0 & \text{otherwise} \end{cases} \tag{2.22}$$

Since X_i's are *i.i.d.*, the joint PMF is given by

$$\begin{aligned} P_{\mathcal{X}}(\mathbf{x}) &= \prod_{i=1}^{K} P_{X_i}(x_i) \\ &= \prod_{i=1}^{K} \frac{\lambda^{x_i}}{x_i!} \exp[-\lambda] \\ &= \prod_{i=1}^{K} \exp\left[x_i \ln(\lambda) - \ln(x_i!) - \lambda\right] \\ &= \exp\left[(\ln(\lambda))\left(\sum_{i=1}^{K} x_i\right) - K\lambda - \sum_{i=1}^{K} \ln(x_i!)\right] I_A(\mathbf{x}) \end{aligned} \tag{2.23}$$

where $I_A(\mathbf{x})$ is the indicator function whose value is 1 if all components of $\mathbf{x}$ are nonnegative integers and is 0 otherwise. Note that, as we have done and will do throughout this chapter and the next, we use the boldface $\mathcal{X}$ to denote the k-dimensional random vector whose components are $X_1, X_2, \cdots, X_K$, and $\mathbf{x}$ to denote the realization of $\mathcal{X}$ whose components are $x_1, x_2, \cdots, x_K$. Matching the terms in the last equality of (2.23) with (2.16), i.e.,

$$P_{\mathcal{X}}(\mathbf{x}) = \{\exp\left[c(\lambda)T(\mathbf{x}) + d(\lambda) + S(\mathbf{x})\right]\} I_A(\mathbf{x}),$$

we see that $c(\lambda) = \ln(\lambda)$, $T(\mathbf{x}) = \sum_{i=1}^{K} x_i$, $d(\lambda) = -K\lambda$, and $S(\mathbf{x}) = -\sum_{i=1}^{K} \ln(x_i!)$.

Example 2.3 shows that *Poisson* distribution also belongs to EFoD. The sufficient statistic for estimating the mean arrival rate λ is $T(\mathbf{x}) = \sum_{i=1}^{K} x_i$. Furthermore, by differentiating the last equality in (2.23) with respect to λ and setting it equal to zero, we can obtain the MLE for λ as

$$\hat{\lambda}(\mathbf{x})|_{\text{MLE}} = \frac{T(\mathbf{x})}{K} = \frac{\sum_{i=1}^{K} x_i}{K}$$

Thus, the MLE for the mean arrival rate is also the sample mean of the number of event points.

We can now explore EFoD further by noting that the log-likelihood function corresponding to any member of the EFoD is given by

$$l(\mathbf{x}; \theta) = c(\theta)T(\mathbf{x}) + d(\theta) + S(\mathbf{x}) \tag{2.24}$$

Differentiating (2.24) with respect to θ and setting it equal to zero clearly shows that the MLE for θ is always a function of $T(\mathbf{x})$. The following theorem further illustrates this useful property that MLE is a function of the sufficient statistic when the PDF of the underlying distribution belongs to the EFOC.

Theorem (Bickel and Doksum 1977) Let $L(\mathbf{x}; \theta)$ belong to the EFoD defined in (2.16). Assume that c is a one-to-one function of θ and C is the interior of the range space of the function $c(\theta)$. If the equation

$$\mathbb{E}_{\theta}\{T(\mathcal{X})\} = T(\mathbf{x}) \tag{2.25}$$

has a solution $\hat{\theta}(\mathbf{x})$ for which $c(\hat{\theta}(\mathbf{x})) \in C$, then $\hat{\theta}(\mathbf{x})$ is the unique MLE for θ. Note that the expectation defined in the above equation is a function of the parameter θ, which is to be estimated. The above theorem can be illustrated by revisiting Example 2.2, in which the sufficient statistic is $T(\mathbf{x}) = \sum_{i=1}^{K} x_i$. Thus, $\mathbb{E}_\theta\{T(\mathcal{X})\} = K\theta$. Employing (2.25) yields $\hat{\theta}_{MLE} = \frac{1}{K}\sum_{i=1}^{K} x_i$. Similarly, we can apply the above theorem to Example 2.3 to verify that the MLE for the mean arrival rate, λ, in a *Poisson* distribution is $\frac{T(\mathbf{x})}{K} = \frac{\sum_{i=1}^{K} x_i}{K}$.

The definition of EFoD given in (2.16) is for the case when the parameter θ is a scalar. This can be extended directly to the case of vector parameters. In particular, we can define a k-parameter exponential family of distributions (k-EFoD) as

$$f_{\mathcal{X}}(\mathbf{x}; \underline{\theta}) = \left\{\exp\left[\sum_{i=1}^{k} c_i(\underline{\theta})T_i(\mathbf{x}) + d(\underline{\theta}) + S(\mathbf{x})\right]\right\} I_A(\mathbf{x}) \tag{2.26}$$

where $\underline{\theta}$ is a k-dimensional parameter vector. Employing the Fisher factorization theorem with a k-dimensional vector parameter $\underline{\theta}$ in (2.13), we can see from (2.26) that, for each i, $T_i(\mathbf{x})$ is a sufficient statistic for θ_i, because $h(T(\mathbf{x}, \underline{\theta})) = \exp\left[\sum_{i=1}^{k} c_i(\underline{\theta})T_i(\mathbf{x}) + d(\underline{\theta})\right]$, and $g(\mathbf{x}) = \{\exp[S(\mathbf{x})]\}\, I_A(\mathbf{x})$.

Example 2.4 Given a set of *i.i.d.* real-valued Gaussian random variables $\{X_i\}$, $i = 1, 2, \cdots, K$, with unknown mean μ and variance σ^2 for each X_i. The joint PDF of $\mathcal{X}$, where $\mathcal{X} = [X_1 X_2 \cdots X_K]^{\mathrm{T}}$, is given by

$$\begin{aligned} f_{\mathcal{X}}(\mathbf{x}) &= \left(\frac{1}{\sqrt{2\pi\sigma^2}}\right)^K \exp\left[-\frac{1}{2\sigma^2}\sum_{i=1}^{K}(x_i-\mu)^2\right] \\ &= \left\{\exp\left[\frac{\mu}{\sigma^2}\sum_{i=1}^{K} x_i - \frac{1}{2\sigma^2}\sum_{i=1}^{K} x_i^2 - K\frac{\mu^2}{2\sigma^2} - \frac{K}{2}\ln(2\pi\sigma^2)\right]\right\} I_A(\mathbf{x}) \end{aligned} \tag{2.27}$$

where $I_A(\mathbf{x})$ is 1 if $x_i \in R$ for all i, and is 0 otherwise. If we now define the parameter vector $\underline{\theta} = [\mu, \sigma^2]^{\mathrm{T}}$. We can match the last equation in (2.27) with (2.26) term by term as follows: $c_1(\underline{\theta}) = \frac{\mu}{\sigma^2}$, $T_1(\mathbf{x}) = \sum_{i=1}^{K} x_i$, $c_2(\underline{\theta}) = \frac{1}{2\sigma^2}$, $T_2(\mathbf{x}) = \sum_{i=1}^{K} x_i^2$, $d(\underline{\theta}) = -K\frac{\mu^2}{2\sigma^2} - \frac{K}{2}\ln(2\pi\sigma^2)$, and $S(\mathbf{x}) = 0$. Thus, the joint Gaussian PDF belongs to the two-parameter EFoD.

Some may question if the correspondence between $T_i(\mathbf{x})$ and θ_i for each i may be somewhat arbitrary. In most cases, such a correspondence can be determined quite unambiguously by examining the order of statistics.

The fact that the MLE is a function of the sufficient statistic helps to ensure its quality as an estimator, for it is obtained based on the most relevant information. This will be discussed in more detail in the subsequent sections.

2.5 Measuring the Quality of Estimators

Once an estimator is derived, a natural (and an important) question to ask is *how good this estimator is*. Given an estimator, $\hat{\theta}(\boldsymbol{\mathcal{X}})$, for the parameter θ based on the observation $\boldsymbol{\mathcal{X}}$, three criteria are typically used to evaluate its quality:

1. Bias: This is a measure of how far, on its average value, the estimate is from the true parameter value.
2. Variance: This provides a measure of dispersion of the estimates that the estimator renders. For the estimation of deterministic parameters and signals, variance of the estimates is the same as variance of estimation error, which is a commonly used performance measure.
3. Consistency: This is a statistical property that is examined when the sample size (i.e., the dimension of the vector $\boldsymbol{\mathcal{X}}$) approaches infinity. If $\hat{\theta}(\boldsymbol{\mathcal{X}})$ *converges with probability one* to θ, it is said to have *strong consistency*. If it *converges in probability*, the estimator has *weak consistency*.

Our discussion in this section will focus on the first two criteria. We begin by defining *unbiased estimator*.

Definition (Unbiased Estimator) If an estimator, denoted by $\hat{\theta}(\boldsymbol{\mathcal{X}})$, for a parameter θ based on the observation $\boldsymbol{\mathcal{X}}$ satisfies $\mathbb{E}\{\hat{\theta}(\boldsymbol{\mathcal{X}})\} = \theta$, then $\hat{\theta}(\boldsymbol{\mathcal{X}})$ is said to be an *unbiased* estimator.

Specifically, if the expected value of the estimator is equal to the true value of the parameter to be estimated, then it is an unbiased estimator. Unbiasedness is a desirable property for estimation because it assures that, on average, the estimate is equal to the value of the true parameter.

In general, we should be interested in finding estimators that are unbiased and have minimum variance for its estimates. To reduce the extent of variation in the values of the estimate, we would like to have an estimator with *uniformly* minimum variance, i.e., with minimum variance regardless of the true value of the parameter. Practically, some uniformly minimum variance estimator (UMVE) may be of little practical value. To see this, consider a simple estimator that assumes a constant value regardless of the observed data. This estimator obviously has the minimum variance, which is zero. However, it is useless as an estimator for any parameter to be estimated, because it renders estimates totally independent of the observation. Also, comparing any other estimator's error variance to this estimator's (which is zero) offers no insights at all, for all variances are nonnegative. Thus, a more sensible idea is to include bias in the consideration and restrict our attention to the class of *unbiased estimators*. The following definition epitomizes this concept.

Definition (Uniformly Minimum Variance Unbiased Estimator) A *uniformly minimum variance unbiased estimator* (UMVUE) is an unbiased estimator whose variance is minimum among the class of all unbiased estimators, regardless of the true value of the parameter.

Understandably, it is generally difficult to find UMVUE. However, with the concept of *complete sufficient statistic* (Bickel and Doksum 1977), this can be made easier.

Definition (Complete Sufficient Statistic) A sufficient statistic $T(\mathbf{x})$ is said to be *complete* if the only function defined on the range space of $T(\mathbf{x})$ that satisfies

$$\mathbb{E}_\theta\{f(T(\mathcal{X}))\} = 0 \quad \text{for all } \theta$$

is the function $f(T(\mathbf{x})) \equiv 0$.

The above definition implies that *if a sufficient statistic is complete, there is only one unbiased estimator that is a function of this sufficient statistic.* This statement can be elucidated as follows:

Let $T(\mathbf{x})$ be a complete sufficient statistic for estimating a parameter θ and $\hat{\theta}(T(\mathbf{x}))$ be an unbiased estimator. Suppose there exists another unbiased estimator that is also a function of $T(\mathbf{x})$, say $\hat{\theta}'(T(\mathbf{x}))$. Obviously, $\mathbb{E}_\theta\{\hat{\theta}(T(\mathcal{X}))\} = \mathbb{E}_\theta\{\hat{\theta}'(T(\mathcal{X}))\} = \theta$ for all θ; thus, $\mathbb{E}_\theta\{\hat{\theta}(T(\mathcal{X})) - \hat{\theta}'(T(\mathcal{X}))\} = 0$. Since $T(\mathbf{x})$ is a complete sufficient statistic, it must be true that $\hat{\theta}(T(\mathcal{X})) - \hat{\theta}'(T(\mathcal{X})) \equiv 0$. Thus, $\hat{\theta}(T(\mathbf{x})) \equiv \hat{\theta}'(T(\mathbf{x}))$. Hence, there is only one unbiased estimator that is a function of $T(\mathbf{x})$.

In general, it is not easy to determine if a sufficient statistic is complete by simply employing the definition above. Fortunately, the sufficient statistic identified for any distribution that belongs to the k-EFoD is a *complete sufficient statistic.*

Example 2.5 Given a set of observations $X_1, X_2, \cdots, X_K$ that are *i.i.d. Poisson* random variables, each of which has the mean arrival rate λ. We have shown, in Example 2.3, that Poisson distribution belongs to EFoD, and the sufficient statistic for estimating λ is given by $\sum_{i=k}^{K} x_k$. We now show that $T(\mathbf{x}) = \sum_{i=k}^{K} x_k$ is a *complete sufficient statistic*. Since the sum of K *i.i.d. Poisson* random variables with a mean arrival rate of λ is also a *Poisson* random variable and has a mean arrival rate of $K\lambda$, the PMF of $T(\mathcal{X})$ is given by

$$P_T(i) = e^{-K\lambda}\frac{(K\lambda)^i}{i!},$$

where $i = \sum_{k=1}^{K} x_k = T(\mathbf{x})$. Thus, for any given function $f(T(\mathcal{X}))$,

$$\mathbb{E}_\lambda\{f(T(\mathcal{X}))\} = \mathbb{E}_\lambda\left\{f\left(\sum_{k=1}^{K} x_k\right)\right\} = \sum_{i=0}^{\infty} f(i)P_T(i) = e^{-K\lambda}\sum_{i=0}^{\infty} f(i)\frac{(K\lambda)^i}{i!}$$

Since $\lambda > 0$ and $i \geq 0$, $f(T(\mathcal{X})) \equiv 0$ must hold if $\mathbb{E}_\lambda\{f(T(\mathcal{X}))\} = 0$. Therefore, $T(\mathcal{X})$ is a complete sufficient statistic.

The above example further illustrates that the sufficient statistic for a parameter of a distribution that belongs to EFoD is a complete sufficient statistic.

Now that we understand what a complete sufficient statistic is, we shall learn how to find an (unbiased) estimator as a function of the complete sufficient statistic. The following theorem by Lehmann-Scheffé is a useful tool for finding an unbiased estimator from complete sufficient statistic. In fact, it will render a uniformly minimum variance unbiased estimator, i.e., UMVUE.

Theorem ((Lehmann-Scheffé) Bickel and Doksum 1977) Consider the estimation of a parameter vector $\underline{\theta}$. Let $\underline{T}(\mathbf{x})$ be a complete sufficient statistic and $\hat{\underline{\theta}}(\mathbf{x})$ be an arbitrary unbiased estimator for the parameter $\underline{\theta}$. Then, $\hat{\underline{\theta}}^*(\mathbf{x}) = \mathbb{E}\{\hat{\underline{\theta}}(\boldsymbol{X})|\underline{T}(\mathbf{x})\}$ is a *uniformly minimum variance unbiased estimator* (UMVUE) for $\underline{\theta}$. Furthermore, if all components of $\text{Var}_{\underline{\theta}}[\hat{\underline{\theta}}^*(\boldsymbol{X})]$ are finite for all $\underline{\theta}$, then $\hat{\underline{\theta}}^*(\mathbf{x})$ is the *unique* UMVUE. (UMVUE).

Here, we use the notation $\underline{T}(\mathbf{x})$ to represent the vector of sufficient statistics. Specifically, the i-th component of $\underline{T}(\mathbf{x})$, i.e., $T_i(\mathbf{x})$, is the sufficient statistic for the i-th component of the parameter vector $\underline{\theta}$, namely, θ_i. With the Lehmann-Scheffé theorem, we can see that, given a complete sufficient statistic $\underline{T}(\mathbf{x})$, there are two ways to find a UMVUE: (1) Find an arbitrary unbiased estimator $\hat{\underline{\theta}}(\mathbf{x})$ and then $\mathbb{E}\{\hat{\underline{\theta}}(\boldsymbol{X})|\underline{T}(\mathbf{x})\}$ is an UMVUE, according to the Lehmann-Scheffé Theorem; and (2) find a function of $\underline{T}(\mathbf{x})$, say $f(\underline{T}(\mathbf{x}))$, such that $f(\underline{T}(\mathbf{x}))$ is an unbiased estimator for θ. Then, since $\mathbb{E}\{f(\underline{T}(\boldsymbol{X}))|T(\mathbf{x})\} = f(\underline{T}(\mathbf{x}))$, $f(\underline{T}(\mathbf{x}))$ is a UMVUE for $\underline{\theta}$. To prove the Lehmann-Scheffé theorem, we need the following lemma and the Rao-Blackwell theorem (Bickel and Doksum 1977).

Lemma 2.1 *Let X and Y be two arbitrary random variables with $\mathbb{E}\{|Y|\} < \infty$. Then*

$$\text{Var}[\mathbb{E}\{Y|X\}] \leq \text{Var}[Y]$$

Furthermore, if $\text{Var}[Y] < \infty$, *the strict inequality holds unless $Y = \mathbb{E}\{Y|X\}$ or, equivalently, unless Y is a function of X.*

Theorem (Rao-Blackwell) Let $\underline{T}(\mathbf{x})$ be a sufficient statistic vector and $\hat{\underline{\theta}}(\mathbf{x})$ be an arbitrary estimator for the parameter vector $\underline{\theta} \in \boldsymbol{\Theta}$, with each component in $\mathbb{E}_{\underline{\theta}}\{|\hat{\underline{\theta}}(\boldsymbol{X})|\}$ being finite for all $\underline{\theta}$, where $|\hat{\underline{\theta}}(\boldsymbol{X})|$ is a vector whose components are the absolute values of the corresponding components of the parameter vector $\hat{\underline{\theta}}(\boldsymbol{X})$. Define $\hat{\underline{\theta}}^*(\mathbf{x}) = \mathbb{E}\{\hat{\underline{\theta}}(\boldsymbol{X})|\underline{T}(\mathbf{x})\}$. Then, for all $\underline{\theta} \in \boldsymbol{\Theta}$,

$$\mathbb{E}_{\underline{\theta}}\{||\hat{\underline{\theta}}^*(\boldsymbol{X}) - \underline{\theta}||^2\} \leq \mathbb{E}_{\underline{\theta}}\{||\hat{\underline{\theta}}(\boldsymbol{X}) - \underline{\theta}||^2\}$$

Furthermore, if all components of $\text{Var}_{\underline{\theta}}[\hat{\underline{\theta}}^*(\boldsymbol{X})]$ are finite, the strict inequality holds unless $\hat{\underline{\theta}}^*(\mathbf{x}) = \hat{\underline{\theta}}(\mathbf{x})$.

The proof for the Rao-Blackwell theorem is quite straightforward by employing Lemma 2.1. Specifically, for any estimator $\hat{\underline{\theta}}(\boldsymbol{X})$,

$$\mathbb{E}_{\underline{\theta}}\{||\hat{\underline{\theta}}(\boldsymbol{X}) - \underline{\theta}||^2\} = \text{Var}_{\underline{\theta}}[\hat{\underline{\theta}}(\boldsymbol{X})] + b^2(\underline{\theta}, \hat{\underline{\theta}}(\boldsymbol{X})) ,$$

where

$$b(\underline{\theta}, \hat{\underline{\theta}}(\mathcal{X})) := \mathbb{E}\{(\hat{\underline{\theta}}(\mathcal{X}) - \underline{\theta})\} = \mathbb{E}\{\hat{\underline{\theta}}(\mathcal{X})\} - \underline{\theta}$$

is the bias of the estimator. Also, by the *law of iterated expectation*, we see that

$$\mathbb{E}_{\underline{\theta}}\{\hat{\underline{\theta}}^*(\mathcal{X})\} = \mathbb{E}_{\underline{\theta}}\{\mathbb{E}\{\hat{\underline{\theta}}(\mathcal{X})|\underline{T}(\mathbf{x})\}\} = \mathbb{E}_{\underline{\theta}}\{\hat{\underline{\theta}}(\mathcal{X})\};$$

thus,

$$b^2(\underline{\theta}, \hat{\underline{\theta}}(\mathcal{X})) = b^2(\underline{\theta}, \hat{\underline{\theta}}^*(\mathcal{X}))$$

In the last equality, we used the fact that $b^2(\underline{\theta}, \hat{\underline{\theta}}(\mathcal{X})) := b^{\mathrm{T}}(\underline{\theta}, \hat{\underline{\theta}}(\mathcal{X}))b(\underline{\theta}, \hat{\underline{\theta}}(\mathcal{X}))$. With Lemma 2.1, the theorem can now be easily established, and the equality holds if and only if $\hat{\underline{\theta}}(\mathbf{x}) = \hat{\underline{\theta}}^*(\mathbf{x})$.

Once we have the Rao-Blackwell theorem, the Lehmann-Scheffé theorem can be proved quite similarly with the fact that $\underline{T}(\mathbf{x})$ is a complete sufficient statistic.

2.6 Information Inequality and Cramér-Rao Lower Bound

In this section, the discussion will be focused on the variance of the estimates. As stated before, if we focus only on estimating deterministic, as opposed to random parameters, then the variance of the estimates is the same as the variance of the estimation error. The estimation error variance is one of the most commonly used performance measures in statistical signal processing. In the estimation of deterministic parameters, *Fisher's information* is closely related to the variance of the estimators, and the information inequality gives us the lowest error variance that can be achieved by any estimators.

To present Fisher's information, we first need to have the following two regularity conditions:

(I) The set $A = \{\mathbf{x} : f_{\mathcal{X}}(\mathbf{x}; \theta) > 0\}$ does not depend on the value of the parameter θ, $\theta \in \Theta$ (where Θ is the parameter space). Also, for all $\mathbf{x} \in A$ and for all $\theta \in \Theta$, $\frac{\partial}{\partial\theta} f_{\mathcal{X}}(\mathbf{x}; \theta)$ exists and is finite.

(II) If $T(\mathbf{x})$ is any statistic such that $\mathbb{E}_{\theta}(|T(\mathbf{x})|) < \infty$ for all $\theta \in \Theta$. Then

$$\frac{\partial}{\partial\theta}\left[\int_{-\infty}^{\infty} \cdots \int_{-\infty}^{\infty} T(\mathbf{x}) f_{\mathcal{X}}(\mathbf{x}; \theta) d\mathbf{x}\right] = \int_{-\infty}^{\infty} \cdots \int_{-\infty}^{\infty} T(\mathbf{x}) \frac{\partial}{\partial\theta} [f_{\mathcal{X}}(\mathbf{x}; \theta)]\, d\mathbf{x} \tag{2.28}$$

Remarks

1. The regularity condition defined in (2.28) guarantees that the order of expectation operation and differentiation is interchangeable and it implies that

$$\mathbb{E}_\theta\left\{\frac{\partial}{\partial\theta}l(\mathbf{x};\theta)\right\}=0 \quad \text{for all } \theta\in\Theta \tag{2.29}$$

where $l(x;\theta)$ is the log-likelihood function as defined in (2.4). This can be seen easily by realizing

$$\begin{aligned}\mathbb{E}_\theta\left\{\frac{\partial}{\partial\theta}l(\mathbf{x};\theta)\right\} &= \int\cdots\int\left(\frac{1}{f_X(\mathbf{x};\theta)}\frac{\partial}{\partial\theta}f_X(\mathbf{x};\theta)\right)f_X(\mathbf{x};\theta)d\mathbf{x}\\ &= \int\cdots\int\frac{\partial}{\partial\theta}f_X(\mathbf{x};\theta)d\mathbf{x}\\ &= \frac{\partial}{\partial\theta}\int\cdots\int f_X(\mathbf{x};\theta)d\mathbf{x}\\ &= \frac{\partial}{\partial\theta}(1)\\ &= 0\end{aligned}$$

2. Any PDF that belongs to the k-EFoD satisfies the regularity conditions defined above, provided that $c(\theta)$ has nonvanishing derivatives.

Definition (Fisher's Information) Assuming that the regularity conditions are satisfied, Fisher's information is defined as

$$I(\theta):=\mathbb{E}_\theta\left\{\left[\frac{\partial}{\partial\theta}l(\mathbf{x};\theta)\right]^2\right\} \tag{2.30}$$

Furthermore, if the regularity condition (II), (2.28), can be extended to hold for the second-order derivatives, namely,

$$\frac{\partial^2}{\partial\theta^2}\left[\int_{-\infty}^{\infty}\cdots\int_{-\infty}^{\infty}T(\mathbf{x})f_X(\mathbf{x};\theta)d\mathbf{x}\right]=\int_{-\infty}^{\infty}\cdots\int_{-\infty}^{\infty}T(\mathbf{x})\frac{\partial^2}{\partial\theta^2}\left[f_X(\mathbf{x};\theta)\right]d\mathbf{x} \tag{2.31}$$

for any statistic $T(\mathbf{x})$ such that $\mathbb{E}_\theta(|T(\mathbf{x})|)<\infty$ for all $\theta\in\Theta$. Then,

$$I(\theta)=\mathbb{E}_\theta\left\{\left[\frac{\partial}{\partial\theta}l(\mathbf{x};\theta)\right]^2\right\}=-\mathbb{E}_\theta\left\{\frac{\partial^2}{\partial\theta^2}l(\mathbf{x};\theta)\right\} \tag{2.32}$$

To show this, note that

$$\begin{aligned}\frac{\partial^2}{\partial\theta^2}l(\mathbf{x};\theta) =& \frac{\partial}{\partial\theta}\left[\frac{1}{f_{\mathcal{X}}(\mathbf{x};\theta)}\left(\frac{\partial}{\partial\theta}f_{\mathcal{X}}(\mathbf{x};\theta)\right)\right]\\ =& \frac{1}{f_{\mathcal{X}}(\mathbf{x};\theta)}\left[\frac{\partial^2}{\partial\theta^2}f_{\mathcal{X}}(\mathbf{x};\theta)\right] - \frac{1}{f^2_{\mathcal{X}}(\mathbf{x};\theta)}\left[\frac{\partial}{\partial\theta}f_{\mathcal{X}}(\mathbf{x};\theta)\right]^2\end{aligned}$$

The expected value of the first term of the last equality vanishes due to the regularity condition, (2.31), and due to the fact that $f_{\mathcal{X}}(\mathbf{x};\theta)$ is a PDF; thus, its integration over the entire sample space is equal to 1. Specifically,

$$\begin{aligned}\mathbb{E}_\theta\left\{\frac{1}{f_{\mathcal{X}}(\mathbf{x};\theta)}\left[\frac{\partial^2}{\partial\theta^2}f_{\mathcal{X}}(\mathbf{x};\theta)\right]\right\} =& \int_{-\infty}^{\infty}\cdots\int_{-\infty}^{\infty}\frac{\partial^2}{\partial\theta^2}\left[f_{\mathcal{X}}(\mathbf{x};\theta)\right]d\mathbf{x}\\ =& \frac{\partial^2}{\partial\theta^2}\left[\int_{-\infty}^{\infty}\cdots\int_{-\infty}^{\infty}f_{\mathcal{X}}(\mathbf{x};\theta)d\mathbf{x}\right]\\ =& \frac{\partial^2}{\partial\theta^2}[1] = 0\end{aligned}$$

Also, as we have seen previously,

$$\frac{1}{f^2_{\mathcal{X}}(\mathbf{x};\theta)}\left[\frac{\partial}{\partial\theta}f_{\mathcal{X}}(\mathbf{x};\theta)\right]^2 = \left[\frac{\frac{\partial}{\partial\theta}f_{\mathcal{X}}(\mathbf{x};\theta)}{f_{\mathcal{X}}(\mathbf{x};\theta)}\right]^2 = \left[\frac{\partial}{\partial\theta}l(\mathbf{x};\theta)\right]^2$$

Taking the expectation of the above yields

$$\mathbb{E}_\theta\left\{\left[\frac{\partial}{\partial\theta}l(\mathbf{x};\theta)\right]^2\right\} = \mathbb{E}_\theta\left\{\frac{1}{f^2_{\mathcal{X}}(\mathbf{x};\theta)}\left[\frac{\partial}{\partial\theta}f_{\mathcal{X}}(\mathbf{x};\theta)\right]^2\right\} = -\mathbb{E}_\theta\left\{\frac{\partial^2}{\partial\theta^2}l(\mathbf{x};\theta)\right\} \tag{2.33}$$

which proves (2.32).

Theorem (Information Inequality) Let $\hat{\theta}(\mathcal{X})$ be an estimator such that its variance $\mathrm{Var}_\theta[|\hat{\theta}(\mathcal{X})|] < \infty$, for all θ in the parameter space Θ. Define $\psi(\theta) = \mathbb{E}_\theta\{\hat{\theta}(\mathcal{X})\}$. Assume that the regularity conditions hold:

Then

$$\mathrm{Var}_\theta[\hat{\theta}(\mathcal{X})] \geq \frac{[\psi'(\theta)]^2}{I(\theta)}, \tag{2.34}$$

where $\psi'(\theta) := \frac{\partial\psi(\theta)}{\partial\theta}$.

Proof The proof for the information inequality is straightforward with the use of Schwartz inequality (Bickel and Doksum 1977). By definition,

$$\psi(\theta) = \mathbb{E}_\theta\{\hat{\theta}(\mathcal{X})\} = \int \cdots \int \hat{\theta}(\mathbf{x}) f_{\mathcal{X}}(\mathbf{x};\theta) d\mathbf{x}$$

Thus,

$$\frac{\partial}{\partial\theta}\left(\mathbb{E}_\theta\{\hat{\theta}(\mathcal{X})\}\right) = \frac{\partial}{\partial\theta}\left(\int \cdots \int \hat{\theta}(\mathbf{x}) f_{\mathcal{X}}(\mathbf{x};\theta) d\mathbf{x}\right)$$

The regularity condition (II), (2.28), implies

$$\frac{\partial}{\partial\theta}\left(\mathbb{E}_\theta\{\hat{\theta}(\mathcal{X})\}\right) = \int \cdots \int \hat{\theta}(\mathbf{x}) \left[\frac{\partial}{\partial\theta}\{f_{\mathcal{X}}(\mathbf{x};\theta)\}\right] d\mathbf{x} \tag{2.35}$$

and

$$\frac{\partial}{\partial\theta}\left(\int \cdots \int f_{\mathcal{X}}(\mathbf{x};\theta) d\mathbf{x}\right) = \int \cdots \int \frac{\partial}{\partial\theta}\{f_{\mathcal{X}}(\mathbf{x};\theta)\} d\mathbf{x} = 0$$

Since $\mathbb{E}_\theta\{\hat{\theta}(\mathcal{X})\}$ is a variable independent of $\mathbf{x}$, the above equality implies that

$$\int \cdots \int \mathbb{E}_\theta\{\hat{\theta}(\mathcal{X})\} \left(\frac{\partial}{\partial\theta}\{f_{\mathcal{X}}(\mathbf{x};\theta)\}\right) d\mathbf{x} = 0 \tag{2.36}$$

Combining (2.35) and (2.36) yields

$$\begin{aligned}\frac{\partial}{\partial\theta}\left(\mathbb{E}_\theta\{\hat{\theta}(\mathcal{X})\}\right) &= \int \cdots \int \left(\hat{\theta}(\mathbf{x}) - \mathbb{E}_\theta\{\hat{\theta}(\mathcal{X})\}\right)\left(\frac{\partial}{\partial\theta}\{f_{\mathcal{X}}(\mathbf{x};\theta)\}\right) d\mathbf{x} \\ &= \int \cdots \int \left(\hat{\theta}(\mathbf{x}) - \mathbb{E}_\theta\{\hat{\theta}(\mathcal{X})\}\right)\left(\frac{\partial}{\partial\theta}\{\ln f_{\mathcal{X}}(\mathbf{x};\theta)\}\right) f_{\mathcal{X}}(\mathbf{x};\theta) d\mathbf{x}\end{aligned}$$

Define

$$h(\mathbf{x}) = \left(\hat{\theta}(\mathbf{x}) - \mathbb{E}_\theta\{\hat{\theta}(\mathcal{X})\}\right)(f_{\mathcal{X}}(\mathbf{x};\theta))^{1/2}$$

and

$$g(\mathbf{x}) = \left(\frac{\partial}{\partial\theta}\{\ln f_{\mathcal{X}}(\mathbf{x};\theta)\}\right)(f_{\mathcal{X}}(\mathbf{x};\theta))^{1/2}$$

Then, by Schwartz inequality, namely,

$$\left(\int\cdots\int h(\mathbf{x})g(\mathbf{x})d\mathbf{x}\right)^2 \leq \left(\int\cdots\int h^2(\mathbf{x})d\mathbf{x}\right)\left(\int\cdots\int g^2(\mathbf{x})d\mathbf{x}\right),$$

we have

$$\left(\frac{\partial}{\partial\theta}\left\{\mathbb{E}_\theta\{\hat{\theta}(\boldsymbol{\mathcal{X}})\}\right\}\right)^2 \leq \text{Var}[\hat{\theta}(\boldsymbol{\mathcal{X}})]I(\theta)$$

Thus,

$$\text{Var}[\hat{\theta}(\boldsymbol{\mathcal{X}})] \geq \frac{\left[\frac{\partial}{\partial\theta}\left(\mathbb{E}_\theta\{\hat{\theta}(\boldsymbol{\mathcal{X}})\}\right)\right]^2}{I(\theta)}$$

and the information inequality is now proven.

Remarks

1. Fisher's information is nonnegative and it is additive if the set of random variables are independent. Thus, if a set of observations are independent, their individual information adds up to aid in the estimation.
2. The expression of (2.32) suggests that Fisher's information can be viewed as the *average* of the distribution's sensitivity to the parameter that is being estimated. Furthermore, the information inequality gives a lower bound on the variance that any estimator can achieve. Thus, it is intuitively reasonable that the information inequality characterizes the best that an estimator can do, as measured by the error variance.
3. If $\hat{\theta}(\boldsymbol{\mathcal{X}})$ is an unbiased estimator for θ, i.e., $\mathbb{E}_\theta\{\hat{\theta}(\boldsymbol{\mathcal{X}})\} = \theta$, then the information inequality, (2.34), reduces to

$$\text{Var}_\theta[\hat{\theta}(\boldsymbol{\mathcal{X}})] \geq \frac{1}{I(\theta)}, \tag{2.37}$$

   which is the well-known *Cramér-Rao lower bound* (CRLB).
4. In statistical signal processing, closeness to the CRLB is often considered as a measure of *efficiency* of an (unbiased) estimator. In particular, one may define the *Fisherian efficiency* of an unbiased estimator $\hat{\theta}(\boldsymbol{\mathcal{X}})$ as

$$\mathcal{E}(\hat{\theta}(\boldsymbol{\mathcal{X}})) = \frac{I^{-1}(\theta)}{\text{Var}_\theta[\hat{\theta}(\boldsymbol{\mathcal{X}})]} \tag{2.38}$$

   Note that $\mathcal{E}$ is always less than one, and the larger $\mathcal{E}$ is, the more efficient that estimator is. In fact, when $\mathcal{E} = 1$, the estimator achieves the CRLB and is said to be an efficient estimator in the Fisherian sense.

5. *If an unbiased estimator has a variance that achieves the CRLB for all $\theta \in \Theta$, it coincides with the MLE and it must be a UMVUE.* Furthermore, it is known that UMVUEs are consistent estimators.

The last remark is an important property of MLE. To illustrate this point, consider again Example 2.2, in which

$$\frac{\partial}{\partial \theta} l_X(\mathbf{x}; \theta) = \frac{1}{\sigma^2} \sum_{i=1}^{K} x_i - \frac{K\theta}{\sigma^2}$$

and

$$\frac{\partial^2}{\partial \theta^2} l_X(\mathbf{x}; \theta) = -\frac{K}{\sigma^2}$$

Hence, $I(\theta) = K/\sigma^2$. Furthermore, since $\hat{\theta}_{MLE} = (1/K) \sum_{i=1}^{K} x_i$, this MLE is unbiased and its error variance achieves the CRLB for all θ. Thus, $\hat{\theta}_{MLE}$ in this case is the UMVUE. However, one must keep in mind that this is a special case, for the MLE is often not unbiased and its error variance does not always achieve the information inequality bound.

In general, it can be shown that an unbiased estimator, $\hat{\theta}(\mathbf{x})$, can be found to achieve the CRLB for all θ if, and only if,

$$\frac{\partial}{\partial \theta} \ln f_X(\mathbf{x}; \theta) = g(\theta)(\hat{\theta}(\mathbf{x}) - \theta) \tag{2.39}$$

for some deterministic function $g(\theta)$. To prove this statement, one only needs to show that $g(\theta) = I(\theta)$. Differentiating both sides of the equality, (2.39), gives

$$\frac{\partial^2}{\partial \theta^2} \ln f_X(\mathbf{x}; \theta) = \left(\frac{\partial}{\partial \theta} g(\theta) \right) (\hat{\theta}(\mathbf{x}) - \theta) - g(\theta)$$

Taking expectation on the right side of the equality of the above equation yields

$$\mathbb{E}_\theta \left\{ \left(\frac{\partial}{\partial \theta} g(\theta) \right) (\hat{\theta}(\boldsymbol{X}) - \theta) - g(\theta) \right\} = \left(\frac{\partial}{\partial \theta} g(\theta) \right) \mathbb{E}_\theta \{ \hat{\theta}(\boldsymbol{X}) - \theta \} - g(\theta),$$

The above equality holds because $g(\theta)$ is a deterministic function of a deterministic variable θ; thus, its derivative can be moved outside of the expectation operation. Now, given that $\hat{\theta}(\boldsymbol{X})$ is an unbiased estimator, i.e., $\mathbb{E}_\theta\{\hat{\theta}(\boldsymbol{X}) - \theta\} = 0$, we have

$$\mathbb{E}_\theta \left\{ \left(\frac{\partial}{\partial \theta} g(\theta) \right) (\hat{\theta}(\boldsymbol{X}) - \theta) - g(\theta) \right\} = -g(\theta)$$

Recall that Fisher's information number is $I(\theta) = -\mathbb{E}\{\frac{\partial^2}{\partial\theta^2}(\ln f_X(\mathbf{x};\theta))\}$. Thus, $g(\theta) = I(\theta)$. Realizing that (2.39) equal to zero when $\theta = \hat{\theta}_{MLE}$ (the long-likelihood equation, (2.5)) shows that the unbiased estimator $\hat{\theta}(\mathbf{x})$ is indeed the MLE of θ, and it is the UMVUE whose error variance achieves the CRLB.

The above results on CRLB can be extended to the vector parameter case quite straightforwardly. Let $\underline{\theta} := [\theta_1, \theta_2, \cdots, \theta_k]^{\mathrm{T}}$ be a k-dimensional parameter vector. Subject to the regularity conditions, we define Fisher's information matrix $\underline{I}(\underline{\theta})$ as a $k \times k$ matrix whose i–jth component is given by

$$\mathbb{E}_{\underline{\theta}}\left\{\left(\frac{\partial}{\partial\theta_i}\ln f_X(\mathbf{x};\underline{\theta})\right)\left(\frac{\partial}{\partial\theta_j}\ln f_X(\mathbf{x};\underline{\theta})\right)\right\} = -\mathbb{E}_{\underline{\theta}}\left\{\frac{\partial^2}{\partial\theta_i\partial\theta_j}\ln f_X(\mathbf{x};\underline{\theta})\right\} \tag{2.40}$$

Thus, the CRLB for the vector parameter case can be stated as follows:

If $\hat{\underline{\theta}}(\mathbf{x})$ is an unbiased estimator for a k-dimensional vector $\underline{\theta}$ and its error covariance matrix is given by

$$\mathrm{Var}_{\underline{\theta}}[\hat{\underline{\theta}}(\boldsymbol{X})] := \mathbb{E}_{\underline{\theta}}\left\{(\hat{\underline{\theta}}(\boldsymbol{X}) - \underline{\theta})(\hat{\underline{\theta}}(\boldsymbol{X}) - \underline{\theta})^H\right\} \tag{2.41}$$

Then, $\left(\mathrm{Var}_{\underline{\theta}}[\hat{\underline{\theta}}(\boldsymbol{X})] - \underline{I}^{-1}(\underline{\theta})\right)$ is a positive semi-definite matrix.

For the i-th element in the parameter vector $\underline{\theta}$, i.e., θ_i, the above result can be stated more specifically:

$$\mathrm{Var}_{\theta_i}[\hat{\theta}_i(\boldsymbol{X})] \geq [I^{-1}(\theta)]_{ii} \tag{2.42}$$

which is similar to (2.37).

Before we leave these topics on information inequality and CRLB, it is worth noting that a counterexample regarding the regularity conditions is that of the uniform distribution. Consider, for example, $\{X_i\}, i = 1, 2, \cdots, K$ is a set of *i.i.d.* random variables whose marginal distribution is uniform in $[0, \theta], \theta > 0$. The PDF this uniform distribution does not belong to the EFoD, for the indicator function varies with θ, the parameter to be estimated. Furthermore, its log-likelihood function violates the regularity condition (2.29). In particular,

$$\mathbb{E}_{\theta}\left\{\frac{\partial}{\partial\theta}\ln f_X(\mathbf{x};\theta)\right\} = \mathbb{E}_{\theta}\left\{\frac{\partial}{\partial\theta}\ln\left(\frac{1}{\theta^K}\right)\right\} = -\frac{K}{\theta} \neq 0$$

However, a sufficient statistic can still be found for estimating θ. It is $\max\{X_i\}, i = 1, 2, \cdots, K$, and that is the $\hat{\theta}_{MLE}$.

Section 8.4 presents some further discussions related to the CRLB applied to a problem of sinusoidal frequency estimation.

2.7 Properties of MLE

The MLE has many interesting properties, and it is the purpose of this section to list some of those properties. Silvey (Billingsley 1975) provides more detailed discussions on those properties.

1. It is easy to see that, in general, the MLE is not *unbiased*. This is because "unbiasedness" was not part of the objective in seeking MLE. However, it is always asymptotically unbiased.
2. As shown in the previous sections, MLE is a function of the sufficient statistic. This can also be seen from the Fisher factorization theorem.
3. MLE is asymptotically a minimum variance unbiased estimator.
4. MLE is consistent. In particular, it converges to the parameter with probability one (or in probability).
5. Under the *regularity conditions*, if there exists an unbiased estimator whose variance attains the CRLB, it is the MLE.
6. Generally speaking, the MLE of a function of a parameter (or signal) is the function of the MLE of that parameter (or signal). This is commonly known as the *invariance property* of the MLE.

The fact that MLE is a function of sufficient statistic means that it depends on *relevant information*. This does not necessarily mean that it will always be the *best* estimator (in the sense of, say, minimum variance), for it may not make the best use of the information. However, when the sample size is large, all the information relevant to the unknown parameter is essentially used by the MLE estimator. This explains why MLE is *asymptotically* unbiased. It is consistent, and its variance achieves asymptotically the CRLB.

We now present a signal processing example for MLE from Kay (1993) to conclude this section.

Example 2.6 (Kay 1993) Consider the problem of estimating the phase of a sinusoidal signal received with additive Gaussian noise. The problem is formulated as

$$X_i = A\cos(\omega_o i + \theta) + N_i \quad i = 0, 1, \cdots, K-1$$

where $\{N_i\}$ is an *i.i.d.* sequence of zero-mean Gaussian random variables with variance σ^2. Employing (2.5), MLE can be obtained by minimizing

$$\sum_{i=0}^{K-1}(x_i - A\cos(\omega_o i + \theta))^2$$

Differentiating the above expression with respect to θ and setting it equal to zero yields

$$\sum_{i=0}^{K-1} x_i \sin(\omega_o i + \hat{\theta}_{MLE}) = A \sum_{i=0}^{K-1} \sin(\omega_o i + \hat{\theta}_{MLE}) \cos(\omega_o i + \hat{\theta}_{MLE})$$

For ω_o not near 0 and π, we have

$$\frac{1}{K} \sum_{i=0}^{K-1} \sin(\omega_o i + \hat{\theta}_{MLE}) \cos(\omega_o i + \hat{\theta}_{MLE}) = \frac{1}{2K} \sum_{i=0}^{K-1} \sin(2\omega_o i + 2\hat{\theta}_{MLE}) \approx 0$$

Therefore,

$$\sum_{i=0}^{K-1} x_i \sin(\omega_o i + \hat{\theta}_{MLE}) \approx 0$$

so that

$$\sum_{i=0}^{K-1} x_i \sin(\omega_o i) \cos(\hat{\theta}_{MLE}) \approx - \sum_{i=0}^{K-1} x_i \cos(\omega_o i) \sin(\hat{\theta}_{MLE})$$

Then the MLE for the phase can be approximated as

$$\hat{\theta}_{MLE} \approx - \arctan \frac{\sum_{i=0}^{K-1} x_i \sin(\omega_o i)}{\sum_{i=0}^{K-1} x_i \cos(\omega_o i)}$$

It can be further shown that the CRLB for estimating θ can be approximated by $\frac{2\sigma^2}{KA^2}$. Interested readers should probe further to evaluate the error variance of this MLE and compare it to the CRLB.

It is easy to see that implementation of MLE and the evaluation of the CRLB can become complex and numerically difficult, especially when the underlying distribution is non-Gaussian. If the parameter to be estimated is a simple scalar and its admissible values are confined within a finite interval, one can usually employ search algorithms to obtain satisfactory results. If such is not the case, more sophisticated numerical optimization algorithms will be needed to render good estimation results. Among others, iterative algorithms such as Newton-Raphson method and the expectation-maximization (EM) algorithm are often employed.

2.8 Conclusion

Maximum likelihood estimation has been used successfully in many different applications. Its wide adoption is attributable primarily from the many nice properties discussed in this chapter. Most of those properties are direct outcomes of the fact that MLEs are derived from the sufficient statistic. This chapter presented foundational materials to help readers better understand the underlying principles for the methods used in many parts of the book relevant to estimation of signal parameters, including frequency estimation in Chap. 8 and precoder and equalizer design for MIMO systems in Chap. 17.

References

P. Bickel, K. Doksum, *Mathematical Statistics—Basic Ideas and Selected Topics* (Holden-Day, New York, 1977)
P. Billingsley, *Statistical Inference* (Chapman and Hall, London, 1975)
P. Billingsley, *Probability and Measure* (Wiley, New York, 1979)
R. Fisher, On the mathematical foundations of theoretical statistics, in *Contributions to Mathematical Statistics*, ed. by R.A. Fisher (1950). Probability and Measure (Wiley, New York, 1922)
R. Gallager, *Low-Density Parity-Check Codes* (MIT Press, Cambridge, MA, 1963)
S. Haykin, *Communication Systems*, 4th edn. (Wiley, New York, 2001)
S. Kassam, *Signal Detection in Non-Gaussian Noise*, 1st edn. (Springer, New York, 1988)
S.M. Kay, *Fundamentals of Statistical Signal Processing: Estimation Theory* (Prentice Hall, Upper Saddle River, NJ, 1993)
S.M. Kay, *Fundamentals of Statistical Signal Processing: Detection Theory* (Prentice Hall, Upper Saddle River, NJ, 1998)
M.G. Kendall, A. Stuart, *The Advanced Theory of Statistics*, vol. II, 2nd edn. (Macmillan Publishing Co., New York, 1977)
A. Papoulis, S.U. Pillai, *Probability, Random Variables and Stochastic Processes*, 4th edn. (McGraw-Hill, New York, 2002)
H.V. Poor, *An Introduction to Signal Detection and Estimation*, 2nd edn. (Springer, New York, 1994)
L.L. Scharf, *Statistical Signal Processing: Detection, Estimation, and Time Series Analysis* (Addison-Wesley, Reading, MA, 1991)
H. Stark, J.W. Woods, *Probability, Random Processes and Estimation Theory*, 3rd edn. (Prentice-Hall, Upper Saddle River, 2002)
J.B. Thomas, *An Introduction to Statistical Communication Theory* (Wiley, New York, 1969)
H.L. Van Trees, *Detection, Estimation, and Modulation Theory* (Wiley, New York, 1968)
C. Weber, *Elements of Detection and Signal Design* (Springer (a reprint of the McGraw-Hill Edition, 1968), New York, 1987)

Chapter 3
Bayesian Estimation

3.1 Introduction

This chapter presents the basic principles of Bayesian estimation. Bayesian estimation methods have been used extensively in many different fields, including communication systems (see, e.g., Chap. 17 of this book) and machine learning for the derivation of inference rules. In coding/decoding of communication theory, *belief propagation* was derived from the Bayesian statistical paradigm and was actually a new name for *message passing decoder* (Gallager 1963); see, e.g., Kschischang et al. (2001); MacKay (2003).

In principle, the objective of Bayesian estimation is to minimize the *Bayes risk*, which is defined to be the expected value of the *cost*. Within this framework, different cost functions would lead to different estimation methods. For example, if the cost function is defined to be the mean-squared error, the Bayesian estimation method leads to the minimum mean-squared error (MMSE) estimator, which turns out to be the conditional mean estimator. On the other hand, if the cost function is the a posteriori PDF, the resulting Bayesian estimator is the maximum a posteriori (MAP) estimator, which is obtained by maximizing the a posteriori PDF.

3.2 Bayesian Estimation

Derivations of Bayesian estimation algorithms require the knowledge of the a posteriori distribution of the signals (or signal parameters) that are to be estimated. The a posteriori distributions can be obtained by employing Bayes rules, thus the name Bayesian estimation. This is in contrast to the *likelihood function* needed in Fisher estimation method, namely, the maximum likelihood estimation. The likelihood function, as we have shown in Chap. 2, is essentially the a priori probability density function (PDF).

Y. C. Lim et al., *Signal Processing for Autonomous Systems*,
https://doi.org/10.1007/978-3-032-00675-2_3

The derivations shown in this chapter follow closely those given in Van Trees (1968). Consider a random variable X that is a function of another random variable S. In practice, X is the observed variable and S is the *signal* (or signal parameter) that is to be estimated. Denote the estimate as $\hat{S}(X)$ and the estimation error as $\mathcal{E} \triangleq S - \hat{S}(X)$. One can then define the cost function based on the estimation error, which, of course, depends on both X and S, namely, $C(\mathcal{E}) = C(S, X)$.

As stated previously, the objective of Bayesian estimation is to minimize the *Bayes risk* $\mathcal{R}_\mathrm{B}$, which is defined as the expected value of the cost. Specifically, the Bayes risk is defined by

$$\mathcal{R}_\mathrm{B} := \mathbb{E}\{C(\mathcal{E})\} = \int_{-\infty}^{\infty}\int_{-\infty}^{\infty} C(s, x) f_{S,X}(s, x) dx ds \tag{3.1}$$

where $f_{S,X}(s, x)$ is the joint PDF of the two random variables S and X. In practice, the joint PDF is usually not directly obtainable. However, by employing the definition of conditional PDF, we have

$$f_{S,X}(s, x) = f_{S|X}(s|x) f_X(x) \tag{3.2}$$

where $f_{S|X}(s|x)$ is the a posteriori PDF of S given the observation X. Substituting (3.2) into (3.1), the Bayes risk can be expressed as

$$\mathcal{R}_\mathrm{B} = \int_{-\infty}^{\infty}\left\{\int_{-\infty}^{\infty} C(s, x) f_{S|X}(s|x) ds\right\} f_X(x) dx \tag{3.3}$$

Note that the expression (3.3) does not specify the relation between S and X; thus, it is suitable for any functional relationship between S and X. In Bayesian estimation, the relationship between S and X is characterized through the a posteriori distribution, which is assumed known, or can be obtained.

Since the cost function is always nonnegative and so is the PDF, minimizing the Bayes risk is equivalent to minimizing

$$\int_{-\infty}^{\infty} C(s, x) f_{S|X}(s|x) ds \tag{3.4}$$

Depending on how the cost function is defined, the Bayesian estimation principle leads to different algorithms. Two of the most commonly used cost functions are

$$C_{ms}(\epsilon) = |\epsilon|^2 \tag{3.5}$$

$$C_{MAP}(\epsilon) = \begin{cases} 0 & \text{if } |\epsilon| \le \frac{\Delta}{2} \\ 1 & \text{if } |\epsilon| > \frac{\Delta}{2} \end{cases} \quad \text{where} \quad \Delta \ll 1 \tag{3.6}$$

where ϵ is a realization of the random variable $\mathcal{E}$, i.e., the estimation error. These example cost functions result in two popular estimators, namely, the minimum mean-squared error (MMSE) estimator and the maximum a posteriori (MAP) estimator. These two estimators will be discussed in more details in the following sections.

3.3 Minimum Mean-Squared Error Estimation

When the cost function is defined to be the mean-squared error (MSE), (3.5), the Bayesian estimate can be derived by substituting $C(\epsilon) = |\epsilon|^2 = (s - \hat{s}(x))^2$ into (3.3). Hence,

$$\mathcal{R}_{\mathrm{Bms}} = \int_{-\infty}^{\infty} \left\{ \int_{-\infty}^{\infty} (s - \hat{s}(x))^2 f_{S|X}(s|x) ds \right\} f_X(x) dx \tag{3.7}$$

Denote the resulting estimate as $\hat{s}_{ms}(x)$, and use the same argument that leads to (3.4); then a necessary condition for minimizing $\mathcal{R}_{\mathrm{Bms}}$ is

$$\frac{d}{d\hat{s}} \int_{-\infty}^{\infty} (s - \hat{s}(x))^2 f_{S|X}(s|x) ds = 0 \tag{3.8}$$

when the differentiation is evaluated at $\hat{s} = \hat{s}_{ms}(x)$. Consequently,

$$\int_{-\infty}^{\infty} (s - \hat{s}_{ms}(x)) f_{S|X}(s|x) ds = 0 \tag{3.9}$$

and

$$\hat{s}_{ms}(x) = \int_{-\infty}^{\infty} s f_{S|X}(s|x) ds \tag{3.10}$$

In obtaining (3.10), we have used the facts that $\hat{s}_{ms}(x)$ is a constant with respect to s in carrying out the integration in (3.9) and that integration of the a posteriori PDF $f_{S|X}(s|x)$ over the entire sample space of s is equal to 1.

The estimate defined by (3.10) is precisely the mean value of the a posteriori distribution, namely, the *conditional mean* of the signal given the observation. In other words, *the minimum mean-squared error (MMSE) estimate is the conditional-mean estimate*. If the a posteriori distribution is Gaussian, then the conditional-mean estimate is a linear function of X, regardless of the functional relation between S and X. The following theorem is one way to present this very useful result.

Theorem (Stark and Woods 2002) Let $\boldsymbol{X} = [X_1, X_2, \cdots, X_K]^T$ be a random vector and S be a signal to be estimated. Assume that $\boldsymbol{X}$ and S are jointly Gaussian with zero means and are uncorrelated, i.e., X_i and S are jointly Gaussian and are uncorrelated for all i. The MMSE estimate of S based on $\mathcal{X}$ is $\mathbb{E}\{S|\mathcal{X} = \mathbf{x}\} = \sum_{i=1}^{K} a_i x_i$ if a_i is chosen such that

$$\mathbb{E}\left\{\left(S - \sum_{i=1}^{K} a_i X_i\right) X_j\right\} = 0 \tag{3.11}$$

for any $j = 1, 2, \cdots, K$.

The above theorem can be proven simply as follows: Let $\hat{S} = \sum_{i=1}^{K} a_i X_i$ be an arbitrary estimate for S. Clearly, $\hat{S}$ is a linear combination of the observations $X_i = x_i, i = 1, 2, \cdots, K$. We shall show first that $\hat{S}$ is an unbiased estimate for S, i.e., $\mathbb{E}\{\hat{S}\} = \mathbb{E}\{S\}$, if $\mathcal{X} = [X_1, X_2, \cdots, X_K]^T$ and S are jointly Gaussian with zero means.

Define the error signal $\mathcal{E} = S - \hat{S} = S - \sum_{i=1}^{K} a_i X_i$. Then,

$$\mathbb{E}\{\mathcal{E}\} = \mathbb{E}\{S - \hat{S}\} = \mathbb{E}\{S\} - \sum_{i=1}^{K} a_i \mathbb{E}\{X_i\} \tag{3.12}$$

Since S and $\mathcal{X}$ are both zero mean, $\mathbb{E}\{\mathcal{E}\} = 0$; thus, $\hat{S}$ is an unbiased estimate for S. Note that, in addition to having zero mean, $\mathcal{E}$ is also a Gaussian random variable. It is worth mentioning that $\hat{S}$ being an unbiased estimate for S does not require that S and $\mathcal{X}$ be uncorrelated, or jointly Gaussian.

Now, since $\mathcal{E} = (S - \sum_{i=1}^{K} a_i X_i)$, (3.11) implies

$$\mathbb{E}\{\mathcal{E} X_j\} = 0 = \mathbb{E}\{\mathcal{E}\}\mathbb{E}\{X_j\} \tag{3.13}$$

for all $j = 1, 2, \cdots, K$. The last equality holds because both X_j and $\mathcal{E}$ are zero-mean Gaussian random variables. With (3.13), we have just shown that $\mathcal{E}$ and X_j are uncorrelated Gaussian random variables; thus, they are independent. Therefore,

$$\mathbb{E}\{\mathcal{E}|\mathcal{X} = \mathbf{x}\} = \mathbb{E}\{\mathcal{E}\} = \mathbb{E}\{S - \hat{S}\} = \mathbb{E}\{S\} - \mathbb{E}\left\{\sum_{i=1}^{K} a_i X_i\right\} = 0 \tag{3.14}$$

since $\hat{S}$ is an unbiased estimate for S. Furthermore,

$$\mathbb{E}\{\mathcal{E}|\mathcal{X} = \mathbf{x}\} = \mathbb{E}\{(S - \hat{S})|\mathcal{X} = \mathbf{x}\} \tag{3.15}$$

$$= \mathbb{E}\{S|\mathcal{X} = \mathbf{x}\} - \mathbb{E}\{\hat{S}|\mathcal{X} = \mathbf{x}\} \tag{3.16}$$

$$= \mathbb{E}\{S|\mathcal{X} = \mathbf{x}\} - \mathbb{E}\{\sum_{i=1}^{K} a_i X_i|\mathcal{X} = \mathbf{x}\} \tag{3.17}$$

$$= \mathbb{E}\{S|\mathcal{X} = \mathbf{x}\} - \sum_{i=1}^{K} a_i x_i = 0 \tag{3.18}$$

Thus, $\sum_{i=1}^{K} a_i x_i = \mathbb{E}\{S|\mathcal{X} = \mathbf{x}\}$, i.e., the conditional mean of S given $\mathcal{X}$, which is the MMSE estimate for S. In particular,

$$\hat{s}_{ms} = \mathbb{E}\{S|\mathcal{X} = \mathbf{x}\} = \sum_{i=1}^{K} a_i x_i \ , \tag{3.19}$$

if a_i is chosen such that $\mathbb{E}\{(S - \sum_{i=1}^{K} a_i X_i)X_j\} = 0$, for all $j = 1, 2, \cdots, K$.

The above theorem is a version of the well-known *orthogonality principle* in estimation theory. In essence, the orthogonality principle states that the MMSE is achieved if the error signal is *orthogonal* to the observation. In probability theory, the term *orthogonality* is implied by the *uncorrelatedness* with the zero-mean assumption. In particular, the two random variables $\mathcal{E}$ and $\mathcal{X}$ are said to be *orthogonal* to each other if their cross-correlation is zero, i.e., if $\mathbb{E}\{\mathcal{E}\mathcal{X}\} = 0$. If at least one of them has zero mean, they are uncorrelated. Furthermore, when the underlying distribution is Gaussian, the *uncorrelatedness* is equivalent to *independence*, which offers many convenient analytical tools, e.g., the joint distribution is the product of the marginal distributions.

The orthogonality principle can also be understood in terms of the concept of vector space, i.e., the MMSE estimate for S is the orthogonal projection of S into the space of $\mathcal{X}$, since the orthogonal projection has the minimum error.

It is important to note that, in general, if the a posteriori distribution is not Gaussian, the conditional mean, i.e., MMSE, estimate is usually not a linear function of the observation. However, one can still consider *linear* MMSE (LMMSE) estimate using orthogonality principle, even though the resulting estimate is not the conditional mean estimate. Thus, the LMMSE estimation methodology can be derived without the knowledge of the a posteriori distribution, unlike the typical Bayesian estimation methods. Chapter 17 (Sect. 17.3.5) of this book offers more insights into this.

3.4 Maximum a Posteriori Estimation

If the cost function is defined as in (3.6), the Bayes risk is

$$\mathcal{R}_{\text{BMAP}} = \int_{-\infty}^{\infty} f_X(x) \left\{ 1 - \int_{\hat{s}_{MAP}-\frac{\Delta}{2}}^{\hat{s}_{MAP}+\frac{\Delta}{2}} f_{S|X}(s|x)ds \right\} dx$$

To minimize $\mathcal{R}_{\text{BMAP}}$, we maximize $\int_{\hat{s}_{MAP}-\frac{\Delta}{2}}^{\hat{s}_{MAP}+\frac{\Delta}{2}} f_{S|X}(s|x)ds$. When Δ is extremely small (as is required), this is equivalent to maximizing $f_{S|X}(s|x)$. Thus, $\hat{s}_{MAP}(x)$ is the value of s that maximizes $f_{S|X}(s|x)$. Normally, it is often more convenient (for algebraic manipulation) to consider $\ln f_{S|X}(s|x)$, since $\ln(x)$ is a nondecreasing function of x, and the $\ln(\cdot)$ operation would remove the exponential term that appears in many PDFs. A necessary condition for maximizing $\ln f_{S|X}(s|x)$ is

$$\frac{\partial}{\partial s} \ln f_{S|X}(s|x)\Big|_{s=\hat{s}_{MAP}(x)} = 0. \tag{3.20}$$

The above equation is often referred to as the MAP equation.

Note that the a posteriori PDF, $f_{S|X}(s|x)$, can be derived from the a priori PDF, i.e., $f_{X|S}(x|s)$, with the help of Bayes rule, i.e.,

$$f_{S|X}(s|x) = \frac{f_{S,X}(s,x)}{f_X(x)} = \frac{f_{X|S}(x|s)f_S(s)}{f_X(x)} \tag{3.21}$$

Since $f_X(x)$ is independent of S, the MAP equation can also be written as

$$\frac{\partial}{\partial s} \ln f_{X|S}(x|s) + \frac{\partial}{\partial s} \ln f_S(s)\Big|_{s=\hat{s}_{MAP}(x)} = 0. \tag{3.22}$$

In summary, the MAP estimate for s is the value of s that maximizes the a posteriori PDF, or, equivalently, the log function of the a posteriori PDF.

3.5 Examples

This section presents some simple examples to illustrate basic procedures for obtaining MMSE estimate (which is the conditional mean estimate) and MAP estimate.

Example 3.1 (Van Trees 1968) Let $X_i, i = 1, 2, \cdots, K$ be a set of random variables modeled as follows:

$$X_i = S + N_i \quad i = 1, 2, \cdots, K$$

where S is a zero-mean Gaussian random variable with variance σ_s^2 and $N_i, i = 1, 2, \cdots, K$ is a set of independent and identically distributed (*i.i.d.*) zero-mean Gaussian random variables with variance σ_n^2. The goal here is to find the conditional mean estimate and MAP estimate for S, given the observations $X_i, i = 1, 2, \cdots, K$.

Denote $\boldsymbol{X} = [X_1 X_2 \cdots X_K]^{\mathrm{T}}$. We can obtain the conditional mean estimate and the MAP estimate by first deriving the a posteriori PDF as follows: Due to the *i.i.d.* assumption, the joint PDF of $\boldsymbol{X}$ given $S = s$ is the product of the marginal PDFs. Thus, the above problem statement is formulated by

$$f_{\boldsymbol{X}|S}(\mathbf{x}|s) = \prod_{i=1}^{K} \frac{1}{\sqrt{2\pi}\sigma_n} \exp\left[-\frac{(x_i - s)^2}{2\sigma_n^2}\right]$$

and

$$f_S(s) = \frac{1}{\sqrt{2\pi}\sigma_s} \exp\left[-\frac{1}{2}\frac{s^2}{\sigma_s^2}\right]$$

Further, from (3.21), we have

$$\begin{aligned} f_{S|\boldsymbol{X}}(s|\mathbf{x}) =& \frac{f_{\boldsymbol{X}|S}(\mathbf{x}|s) f_S(s)}{f_{\boldsymbol{X}}(\mathbf{x})} \\ =& \frac{1}{f_{\boldsymbol{X}}(\mathbf{x})}\left[\left(\frac{1}{\sqrt{2\pi}\sigma_n}\right)^K \frac{1}{\sqrt{2\pi}\sigma_s}\right] \exp\left[-\frac{1}{2}\left(\frac{\sum_{i=1}^K (x_i - s)^2}{\sigma_n^2} + \frac{s^2}{\sigma_s^2}\right)\right] \\ =& C(\mathbf{x}) \exp\left[-\frac{1}{2\sigma_p^2}(s - \frac{\sigma_s^2}{\sigma_s^2 + \sigma_n^2/K}(\frac{1}{K}\sum_{i=1}^K x_i))^2\right] \end{aligned} \quad (3.23)$$

where $\sigma_p^2 \triangleq \left(\frac{1}{\sigma_s^2} + \frac{K}{\sigma_n^2}\right)^{-1} = \frac{\sigma_s^2\sigma_n^2}{K\sigma_s^2 + \sigma_n^2}$, and $C(\mathbf{x})$ is a constant with respect to s.

The above expression shows that the a posteriori distribution is Gaussian, for (3.23) is clearly a Gaussian PDF. For Gaussian distribution, the mean is the value that maximizes its PDF. Thus, we can conclude that the conditional mean estimate and the MAP estimate are equivalent in this example. In particular,

$$\hat{s}_{ms}(\mathbf{x}) = \hat{s}_{MAP}(\mathbf{x}) = \frac{\sigma_s^2}{\sigma_s^2 + \sigma_n^2/K}\left(\frac{1}{K}\sum_{i=1}^K x_i\right) \quad (3.24)$$

Some useful insights into Bayesian estimation can be gained through this example.

Remarks

1. This example looks similar to Example 2.2 in that it is also estimating the mean of the distribution of the observed variables. The difference is that, in Example 2.2, the mean is an unknown deterministic variable, whereas the mean in this example is an unknown random variable with known a priori distribution. These two examples also help elucidate the difference between the Fisher approach and the Bayesian approach to estimation. The former is derived from the *likelihood*

function, while the latter is derived from the a posteriori distribution, which can be derived from the a priori distribution of the signal to be estimated.
2. Examining (2.12) and (3.24) reveals that the difference between the Fisherian estimate (i.e., MLE) and the Bayesian estimate (i.e., MMSE and MAP in this case) is the scaling factor, i.e., $\frac{\sigma_s^2}{\sigma_s^2+\sigma_n^2/K}$. This scaling factor appears as a result of the randomness of S, i.e., the signal to be estimated. Note that if $\sigma_s^2 \ll \frac{\sigma_n^2}{K}$, which is like when the signal-to-noise ratio (SNR) is small, the a priori knowledge is more useful than the observed data and the estimate is very close to the a priori mean, i.e., 0. In this case, the a posteriori distribution almost has no effect on the value of the estimate. On the other hand, if $\sigma_s^2 \gg \frac{\sigma_n^2}{K}$ (i.e., the SNR is large), the estimate is directly related to the observed data as it is the sample mean, while the a priori knowledge is of little value.
3. The equivalence of $\hat{s}_{MAP}$ to $\hat{s}_{ms}(\mathbf{x})$ is not restricted to the case of Gaussian a posteriori PDF. The following are two example conditions that warrant such equivalence Van Trees (1968):

   a. If the a posteriori PDF is symmetric with respect to its mean, and if the cost function is symmetric and strictly convex, i.e.,

   $$C(\lambda x_1 + (1-\lambda)x_2) \leq \lambda C(x_1) + (1-\lambda)C(x_2)$$

   where the inequality is strict unless $x_1 = x_2$, then the MAP estimate is equivalent to the MMSE estimate.
   b. If the a posteriori PDF is symmetric with respect to its mean, and is unimodal and satisfies

   $$\lim_{s\to\infty} C(s) f_{S|\mathbf{X}}(s|\mathbf{x}) = 0 \; ,$$

   and if the cost function is symmetric and nondecreasing (in terms of the distance from the symmetry point), then the MAP estimate is equivalent to the MMSE estimate.

We now use the following examples to show that the MMSE estimate and the MAP estimate are, in general, not equal.

Example 3.2 In this example, we show how the different a priori distributions of the signal to be estimated could lead to different kinds of MMSE and MAP estimates. The objective here is to estimate a random signal embedded in a set of *i.i.d.* Gaussian random variables. However, the signal has an asymmetric PDF, which results in an asymmetric a posteriori PDF. Specifically, we consider an exponentially distributed random signal embedded in *i.i.d.* Gaussian noise. This problem is formulated as observing a set of random variables $X_i = S + N_i, i = 1, \ldots, K$.

The unknown signal S is assumed to have an exponential a priori PDF:

$$f_S(s) = \begin{cases} \lambda \exp(-\lambda s) & \text{if } s > 0 \\ 0 & \text{if } s < 0 \end{cases} \tag{3.25}$$

where $\lambda > 0$ and $N_i, i = 1, \cdots, K$ is a set of *i.i.d.* zero-mean Gaussian random variables with variance σ^2, which are independent of S. We can find the MAP estimator of S as follows: Recall that the a posteriori PDF can be expressed as

$$f_{S|\mathcal{X}}(s|\mathbf{x}) = \frac{f_{\mathcal{X}|S}(\mathbf{x}|s) f_S(s)}{f_{\mathcal{X}}(\mathbf{x})} \tag{3.26}$$

and since $N_i, i = 1, 2, \cdots, K$ is an *i.i.d.* set of random variables,

$$f_{\mathcal{X}|S}(\mathbf{x}|s) = \prod_{i=1}^{K} f_{X_i|S}(x_i|s) \tag{3.27}$$

$$= \left(\frac{1}{\sqrt{2\pi\sigma^2}}\right)^K e^{-\frac{1}{2\sigma^2}\sum_{i=1}^{K}(x_i-s)^2} \tag{3.28}$$

Thus,

$$f_{S|\mathcal{X}}(s|\mathbf{x}) = \begin{cases} \frac{1}{f_{\mathcal{X}}(\mathbf{x})} \lambda e^{-\lambda s} \left(\frac{1}{\sqrt{2\pi\sigma^2}}\right)^K e^{-\frac{1}{2\sigma^2}\sum_{i=1}^{K}(x_i-s)^2} & \text{if } \lambda > 0 \\ 0 & \text{otherwise} \end{cases} \tag{3.29}$$

To find the MAP estimate for S, denoted by $\hat{s}_{\text{MAP}}$, we seek to maximize the a posteriori PDF given in (3.29). This can be achieved by differentiating (3.29) with respect to s. Since $f_{\mathcal{X}}(\mathbf{x})$ is independent of s, we can find the maximum by differentiating $\lambda e^{-\lambda s} \left(\frac{1}{\sqrt{2\pi\sigma^2}}\right)^K e^{-\frac{1}{2\sigma^2}\sum_{i=1}^{K}(x_i-s)^2}$ with respect to s. This process can be simplified by taking the logarithm of (3.29), i.e.,

$$\ln\left[\lambda e^{-\lambda s} \left(\frac{1}{\sqrt{2\pi\sigma^2}}\right)^K e^{-\frac{1}{2\sigma^2}\sum_{i=1}^{K}(x_i-s)^2}\right] \tag{3.30}$$

$$= \ln\lambda - \frac{K}{2}\ln(2\pi\sigma^2) - \left[\lambda s + \frac{1}{2\sigma^2}\sum_{i=1}^{K}(x_i - s)^2\right] \tag{3.31}$$

Differentiating (3.31) with respect to s and setting it equal to zero yields

$$\hat{s}_{\text{MAP}} = \frac{1}{K}\left(\sum_{i=1}^{K} x_i\right) - \frac{\sigma^2\lambda}{K} \tag{3.32}$$

Note that the above solution for $\hat{s}_{\text{MAP}}$ has a constraint that $\hat{s}_{\text{MAP}} > 0$, for the true value $s > 0$ as defined in (3.25). Thus, it is required that $\frac{1}{K}\left(\sum_{i=1}^{K} x_i\right) - \frac{\sigma^2\lambda}{K} > 0$. Otherwise, we would have no solution for $\hat{s}_{\text{MAP}}$.

Recall in Example 2.2, (2.12) shows that the MLE of the deterministic DC signal embedded in *i.i.d.* Gaussian random sequence (i.e., additive white Gaussian noise) is the sample mean of the observations. In contrast, in this chapter, Examples 3.1 and 3.2 show that the MAP estimate of a random signal, though also embedded in additive white Gaussian noise, is not quite the same as the sample mean as in Example 2.2. If the random signal is Gaussian, as in Example 3.1, the difference is a scaling factor to the sample mean. If the random signal has an exponential distribution, as in Example 3.2, the difference is an additional term to the sample mean. Those differences are due to the randomness of the signal S, and they depend on the a priori distribution of S.

Finding the conditional mean estimate for Example 3.2 can be done similarly to what was done in Example 3.1. However, the analytical expressions are rather complicated; thus, it is omitted here.

Example 3.3 In wireless communication systems, the received signals are often subject to the so-called Rayleigh fading. As such, the magnitude of the received signal can be formulated as a random variable with a Rayleigh distribution. A simplified form of the problem can be formulated as follows: Given a Rayleigh random variable X with a PDF given by

$$f_{X|A}(x|\alpha) = \begin{cases} 2\alpha x e^{-\alpha x^2} & \text{if } x > 0 \\ 0 & \text{otherwise} \end{cases} \tag{3.33}$$

where $\alpha > 0$ is a realization of an exponential random variable A whose PDF is

$$f_A(\alpha) = \frac{1}{2} e^{-\alpha/2} \tag{3.34}$$

Given an observation of X, denoted by x, find the MAP estimate of A and the MMSE estimate of A.

We shall first find the MAP estimate for α, denoted by $\hat{\alpha}_{\text{MAP}}$. To find $\hat{\alpha}_{\text{MAP}}$, we seek to find a value of α that maximizes the a posteriori PDF, i.e.,

$$\hat{\alpha}_{\text{MAP}} = \text{Arg}\{\max f_{A|X}(\alpha|x)\}$$

From Bayes' rule,

$$f_{A|X}(\alpha|x) = \frac{f_{X|A}(x|\alpha) f_A(\alpha)}{f_X(x)} \tag{3.35}$$

Note that $f_X(x)$ is considered a constant while maximizing $f_{A|X}$ with respect to α. Thus, maximizing $f_{A|X}(\alpha|x)$ is equivalent to maximizing $f_{X|A}(x|\alpha)f_A(\alpha)$, which is

$$f_{X|A}(x|\alpha)f_A(\alpha) = \alpha x e^{-\alpha(x^2+\frac{1}{2})} \tag{3.36}$$

Differentiating the above with respect to α,

$$\frac{d}{d\alpha}\left\{\alpha x e^{-\alpha(x^2+\frac{1}{2})}\right\} = x e^{-\alpha(x^2+\frac{1}{2})} - \alpha x\left(x^2+\frac{1}{2}\right)e^{-\alpha(x^2+\frac{1}{2})}$$

and setting it equal to zero yields

$$x e^{-\alpha(x^2+\frac{1}{2})}\left(1-\alpha\left(x^2+\frac{1}{2}\right)\right) = 0$$

Since $x > 0$ and $e^{-\alpha(x^2+\frac{1}{2})} > 0$ for any real-valued α, the MAP estimate for α is obtained by setting $(1-\alpha(x^2+\frac{1}{2})) = 0$; thus, the MAP estimate for α is given by

$$\hat{\alpha}_{\mathrm{MAP}} = \frac{1}{x^2+\frac{1}{2}} \tag{3.37}$$

Next we shall find $\hat{\alpha}_{ms}$, which is the conditional mean of the a posteriori distribution, i.e., $\hat{\alpha}_{ms} = \mathbb{E}\{A|X\}$. To find this conditional mean, we need to find the a posteriori PDF first. This can be achieved by using the Bayes rule, i.e., (3.35) and combining (3.33) and (3.34) to obtain

$$f_{A|X}(\alpha|x) = \frac{f_{X|A}(x|\alpha)f_A(\alpha)}{f_X(x)} = \frac{1}{f_X(x)}\alpha x e^{-\alpha(x^2+\frac{1}{2})} = \frac{f_{X,A}(x,\alpha)}{f_X(x)} \tag{3.38}$$

The PDF $f_X(x)$ is obtained by

$$f_X(x) = \int_{-\infty}^{\infty} f_{X,A}(x,\alpha)d\alpha = \int_0^{\infty} \alpha x e^{-\alpha(x^2+\frac{1}{2})}d\alpha$$

The above integration can be calculated by performing integration by parts to obtain

$$f_X(x) = \frac{x}{(x^2+\frac{1}{2})^2} \tag{3.39}$$

Substituting (3.39) into (3.38) gives the a posteriori PDF of A given X, i.e.,

$$f_{A|X}(\alpha|x) = \alpha\left(x^2+\frac{1}{2}\right)^2 e^{-\alpha(x^2+\frac{1}{2})} \tag{3.40}$$

We can now find the mean of the a posteriori distribution by multiplying α to (3.40) and integrating, i.e.,

$$\mathbb{E}\{A|X\} = \int_{-\infty}^{\infty} \alpha f_{A|X}(\alpha|x) d\alpha \tag{3.41}$$

$$= \left(x^2 + \frac{1}{2}\right)^2 \int_0^{\infty} \alpha^2 e^{-\alpha(x^2+\frac{1}{2})} d\alpha \tag{3.42}$$

The above integration can be calculated by performing integration by parts twice, similar to how (3.39) was derived, to obtain the conditional mean estimate, i.e.,

$$\hat{\alpha}_{ms} = \mathbb{E}\{A|X\} = \frac{2}{(x^2 + \frac{1}{2})} \tag{3.43}$$

Comparing (3.43) with (3.37), we see that, when the a posteriori is not symmetric (see, e.g., Remark 3 following Example 3.1), the MAP estimate and the conditional mean estimate may be different.

Example 3.4 A young woman is looking to set up a coffee shop in the center of the business district of a city, and she needs to estimate the number of customers that may come to the shop in the morning. She understands that the number of customers that come to the shop follows a *Poisson* distribution, with a mean arrival rate A, which is a random variable with an exponential distribution. Specifically, the number of customers arriving at the shop follows a *Poisson* distribution specified by

$$\text{Prob}\{N = n|A = a\} = \frac{a^n}{n!} \exp[-a] , \tag{3.44}$$

and the a priori PDF of A is

$$f_A(a) = \lambda \exp[-\lambda a] \quad \text{if} \quad a > 0 \tag{3.45}$$

with $f_A(a) = 0$ for any $a \leq 0$, and λ is a known deterministic parameter. In particular, the shop owner is trying to estimate the mean arrival rate A.

We can find the MAP estimate of A and the conditional mean estimate of A as follows.

MAP Estimate

From the problem statement, we know that the number of customers that come into the coffee shop is formulated by (3.44), with parameter A modeled as an exponential random variable with known λ, whose PDF specified by (3.45). Accordingly, the

MAP equation, (3.22), is given by

$$\frac{\partial}{\partial a}\left\{\ln\left(\frac{a^n}{n!}e^{-a}\right)+\ln\left(\lambda e^{-\lambda a}\right)\right\}\bigg|_{a=\hat{a}_{MAP}}=0 \tag{3.46}$$

Note that

$$\ln\left(\frac{a^n}{n!}e^{-a}\right)+\ln\left(\lambda e^{-\lambda a}\right) \tag{3.47}$$

$$=n\ln a-\ln n!-a+\ln\lambda-\lambda a \tag{3.48}$$

$$=n\ln a-a(1+\lambda)+\ln\lambda-\ln n! \tag{3.49}$$

Taking the derivative of the last equation above with respect to a and setting it to zero yields

$$\hat{a}_{\text{MAP}}=\frac{n}{\lambda+1} \tag{3.50}$$

MMSE Estimator

To begin, we note that the marginal probability mass function, $p_N(n)$, of N is given by

$$p_N(n)=\int_0^\infty \Pr[N=n|A=a]f_A(a)da \tag{3.51}$$

$$=\int_0^\infty e^{-a}\frac{a^n}{n!}\lambda e^{-\lambda a}da \tag{3.52}$$

$$=\frac{\lambda}{n!}\int_0^\infty e^{-a(\lambda+1)}a^n da \tag{3.53}$$

Employing Euler exponential (3.54)

$$\int e^{-a(\lambda+1)}da=-\frac{1}{\lambda+1}e^{-a(\lambda+1)}+K \tag{3.54}$$

and integrating by parts yields

$$\int e^{-a(\lambda+1)}a^n da=\frac{-e^{-a(\lambda+1)}}{\lambda+1}a^n+\frac{n}{\lambda+1}\int e^{-a(\lambda+1)}a^{n-1}da \tag{3.55}$$

$$=\frac{-e^{-a(\lambda+1)}}{\lambda+1}a^n+\frac{-ne^{-a(\lambda+1)}}{(\lambda+1)^2}a^{n-1} \tag{3.56}$$

$$+\frac{n(n-1)}{(\lambda+1)^2}\int e^{-a(\lambda+1)}a^{n-2}da \tag{3.57}$$

$$= \frac{-e^{-a(\lambda+1)}}{\lambda+1}a^n - \frac{ne^{-a(\lambda+1)}}{(\lambda+1)^2}a^{n-1} \tag{3.58}$$

$$- \frac{n(n-1)e^{-a(\lambda+1)}}{(\lambda+1)^3}a^{n-2} - \dots \tag{3.59}$$

The above recursion leads to

$$\int e^{-a(\lambda+1)}a^n da = -\sum_{i=0}^{n} \frac{n!/(n-i)!}{(\lambda+1)^{i+1}} e^{-a(\lambda+1)}a^{n-i} \tag{3.60}$$

Marginalization related to the integral limits leads to

$$p_N(n) = \frac{\lambda}{n!}\left(0 - \left(-\frac{n!}{(\lambda+1)^{n+1}}\right)\right) \tag{3.61}$$

$$= \frac{\lambda}{(\lambda+1)^{n+1}} \tag{3.62}$$

From (3.62), we get

$$f_A(a|n) = \frac{\frac{\lambda a^n}{n!}e^{-a(\lambda+1)}}{\frac{\lambda}{(\lambda+1)^{n+1}}} \tag{3.63}$$

$$= \frac{(\lambda+1)^{n+1}e^{-a(\lambda+1)}a^n}{n!} \tag{3.64}$$

The MMSE is obtained by taking the conditional expectation:

$$\hat{a} = \mathbb{E}[A|n] \tag{3.65}$$

$$= \int_0^{\infty} a \frac{(\lambda+1)^{n+1}e^{-a(\lambda+1)}a^n}{n!} da \tag{3.66}$$

$$= \frac{(\lambda+1)^{n+1}}{n!} \int_0^{\infty} e^{-a(\lambda+1)}a^{n+1} da \tag{3.67}$$

$$= \frac{(\lambda+1)^{n+1}}{n!} \frac{(n+1)!}{(\lambda+1)^{n+2}} \tag{3.68}$$

$$\hat{a}_{ms} = \frac{n+1}{\lambda+1} \tag{3.69}$$

This example shows another case that MMSE estimate and MAP estiamte are not equal.

3.6 Conclusion

This chapter presents two main Bayesian estimation methods, namely, the MMSE estimator and the MAP estimator. The MMSE estimator is the conditional mean estimator, i.e., the mean of the a posteriori distribution, whereas the MAP estimate is the value that maximizes the a posteriori PDF. We showed that when the a posteriori distribution is Gaussian, those two estimates coincide, and they are linear functions of the observation. When the a posteriori distribution is not Gaussian, however, those properties may not hold, and, in general, both the MMSE estimate and MAP estimate are nonlinear functions of the observation. Apart from these we can restrict the estimates to be linear functions of the observation and seek to minimize the mean-square error of the estimates, thereby leading to the so-called linear MMSE, or LMMSE estimates. If the underlying processes are assumed to have zero mean, which is often the case, orthogonality principle can be used to derive the LMMSE estimate. Discussions on some applications of these estimation methods (MAP, MMSE, and LMMSE) to multi-input multi-output communications will be seen in Chap. 17 of this book.

References

R. Gallager, *Low-Density Parity-Check Codes* (MIT Press, Cambridge, MA, 1963)

S.M. Kay, *Fundamentals of Statistical Signal Processing: Estimation Theory* (Prentice Hall, Upper Saddle River, NJ, 1993)

F.R. Kschischang, B.J. Frey, H.-A. Loeliger, Factor graphs and the sum-product algorithm. IEEE Trans. Inf. Theory **47**(2), 498–519 (2001)

D.J.C. MacKay, *Information Theory, Inference and Learning Algorithms* (Cambridge University Press (chapters 16), Cambridge, 2003), pp. 24–26

H. Stark, T.W. Woods, *Probability, Random Processes and Estimation Theory*, 3rd edn. (Prentice Hall, Upper Saddle River, 2002)

H.L. Van Trees, *Detection, Estimation, and Modulation Theory* (Wiley, New York, 1968)

Part II
Detection and Location Estimation

Chapter 4
Navigation Satellite System

4.1 Introduction

Consider the scenario shown in Fig. 4.1 consisting of an object and a few reference points R1, R2, R3, R4, and R5. Suppose that the coordinates of the reference points are known. It is possible to estimate the coordinates of the object if the distances d_1, d_2, d_3, d_4, and d_5 between the object and the reference points are also known. This is the basis of a satellite navigation (or satnav) system. In a satellite navigation system, the reference points are the satellites. The distance between the object and a satellite is determined from the duration the electromagnetic wave takes to travel from the satellite to the object. Each navigation satellite carries one or more atomic clocks keeping very accurate time reference; the time between satellites is also very accurately synchronized. In a three-dimensional space, if the object also carries an atomic clock synchronized to those of the satellites, three satellites are sufficient for geospatial positioning. However, in practice, the object does not have an accurate time reference and so at least four satellites are needed for geospatial positioning.

Currently, several satellite navigation systems are in operation or in the process of construction (United nations office for outer space affairs Online; International committee on global navigation satellite systems ICG; Bonnor 2012; Zaidi and Suddle 2006; Teunissen and Montenbruck 2017; Lechner and Baumann 2000; Dow et al. 2009; Li et al. 2015; Yang et al. 2019; Li et al. 2020; Benedicto et al. 2000; Ivanov and Salischev 2009; Xu and Xu 2016; Feng and Wang 2008; Langley 1998; Takasu and Yasuda 2009; Teunissen et al. 2014). If the system provides global geospatial positioning, it is called a GNSS (Global Navigation Satellite System). If the system provides geospatial positioning only in a region but does not provide global coverage, it is called an RNSS (Regional Navigation Satellite System). Today, the GNSSs are GPS (USA), GLONASS (Russia), BeiDou (China), and Galileo (EU). India's IRNSS (Indian Regional Navigation Satellite System) comprising seven satellites provides geospatial positioning in a region around India. Japan's QZSS (Quasi-Zenith Satellite System) comprising four satellites is a satellite-based

Y. C. Lim et al., *Signal Processing for Autonomous Systems*,
https://doi.org/10.1007/978-3-032-00675-2_4

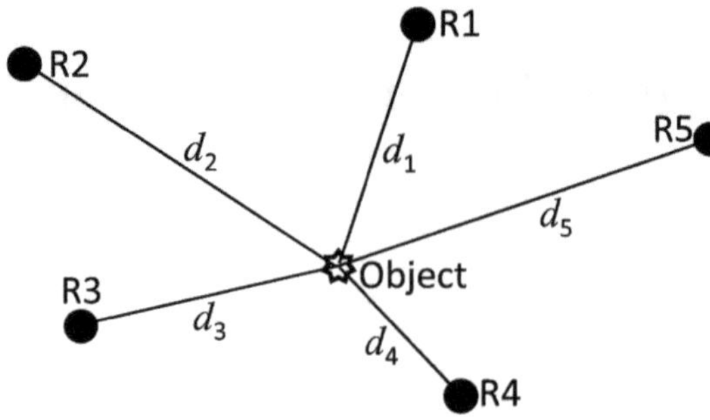

Fig. 4.1 An object and a few reference points

augmentation system providing GPS augmentation information for civil aircraft under the Fukuoka Flight Information Region.

4.2 Satellite Orbits

The trajectory of a satellite round the Earth is called the satellite's orbit. In general, a satellite's orbit is elliptical. Three of the commonly encountered terms of the elliptical orbit of a satellite are apogee, perigee, and inclination.

1. **Apogee**: The point of the orbit furthest from the Earth is called the apogee.
2. **Perigee**: The point of the orbit nearest the Earth is called the perigee. The orbit is a circular orbit if its perigee and apogee are equal.
3. **Inclination**: The inclination is the tilt of orbit relative to the equatorial plane. If the inclination is 0°, the orbit is in the same plane as the equator. If the inclination is 90°, the orbit passes over North and South Poles.

There are three orbits commonly used by navigation satellite systems.

1. **Medium Earth Orbit** (MEO): MEO is sometimes called intermediate circular orbit (ICO). The orbit's altitude is approximately 20,000 km above sea level. MEO satellites travel round the Earth in about 12 hours.
2. **Geostationary Orbit** (GSO): GSO is 35,786 km above sea level on the equator. A GSO satellite is stationary with respect to an observer on Earth.
3. **Inclined Geostationary Orbit** (IGSO): IGSO is 35,786 km above sea level with an inclination to the equator. To an observer on Earth, an IGSO satellite moves in the figure "8." The top and bottom loops of the figure "8" may or may not be symmetrical. See Fig. 4.2. IGSO is the basis for the implementation of RNSS.

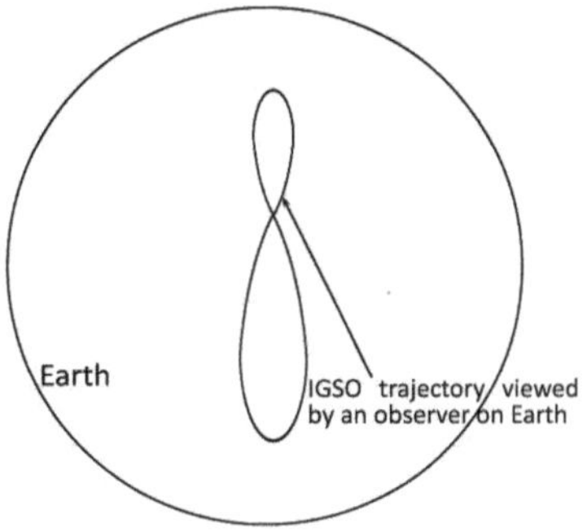

Fig. 4.2 IGSO trajectory viewed from Earth

4.3 Existing Satnav Systems

Several satnav systems are either fully operational or in the process of construction:

1. **GPS** (Global Positioning System). GPS is fully operational. It maintains at least 24 operational MEO satellites. The MEO satellites are in six orbital planes with at least four in each plane. Each of the plane has five slots. The system is capable of supporting up to 30 ($5 \times 6 = 30$) satellites on orbit. The orbital inclination is approximately 55° at an altitude of 20,200 km. Its nominal period is 11 hours 58 minutes and 2 seconds.
2. **GLONASS** (Globalnaya Navigationnaya Sputnikovaya Sistema). GLONASS is fully operational. It contains 21 MEO satellites in three orbital planes, with three on-orbit spares. The orbital inclination is approximately 64.8° at an altitude of 19,130 km. Its nominal period is 11 hours 15 minutes.
3. **BDS** (BeiDou System). BDS-3 is fully operational since July 2020 offering both passive and subscribers-only active services. BDS GNSS is BDS-3. BDS-1 (decommissioned in 2012) and BDS-2 are RNSS commissioned in the years 2000 and 2011, respectively.
   BDS-3 has 30 satellites including 24 MEO, three GSO, and three IGSO satellites. The MEO orbital inclination is approximately 55° in three planes at an altitude of 21,528 km. Its nominal period is 12 hours 53 minutes and 24 seconds. These 24 MEO satellites are sufficient to provide global services. The IGSO satellites cross the equator at 135° East, 148° East, and 106° West, respectively, with an inclination of 55° ±7°. The GSO are at 48° East, 149° East, and 88° West.
4. **Galileo**. Galileo consists of 30 MEO satellites at an altitude of 23,222 km with 24 of them in full service; the remaining six are spares. The satellites are in three orbital planes with approximately 56° inclination. Each orbital plane has eight operational satellites and two spares. Its nominal period is 14 hours 5 minutes.
5. **IRNSS** (Indian Regional Navigation Satellite System). The operational name is NAVIC (NAVigation with Indian Constellation). IRNSS consists of seven satellites including three GSO satellites placed at 33° East, 83° East, and 131.5°

East, and two IGSO satellites that cross the equator at 55° East and 111.75° East, respectively, all with inclination 29°.

6. **QZSS** (Quasi-Zenith Satellite System). It is also known as Michibiki. Currently, it is a four-satellite system augmenting GPS. It has one GSO and three IGSO satellites. The GSO satellite is placed at 127° East above the equator. The IGSO crosses the equator at 135° East ±5° with inclination 43° ±4°. The IGSO satellites spend approximately 13 hours in the northern hemisphere and roughly 11 hours in the southern hemisphere. The number of satellites will be increased to seven in the year 2025 to become independent of GPS.

4.4 WGS-84 Reference Ellipsoid

An ellipsoid is defined by its (1) semi-major axis a and (2) semi-minor axis b. See Fig. 4.3. The flattening factor f of the ellipsoid is defined as

$$f = \frac{a - b}{a}. \tag{4.1}$$

In the Worldwide Reference Ellipsoid WGS-84 (World Geodetic System 1984), the ellipsoid is geocentrically positioned with respect to the center of the Earth. Such a system is called ECEF (Earth Centered, Earth Fixed). The positive X-axis lies on the equatorial plane and passes through the intersection between the equator and the Greenwich meridian (0° meridian). The Y-axis lies on the equatorial plane and is offset 90° to the east of the X-axis. The Z-axis coincides with the Earth's axis of rotation and extends through the geographical North Pole. See Fig. 4.4. Statistics for

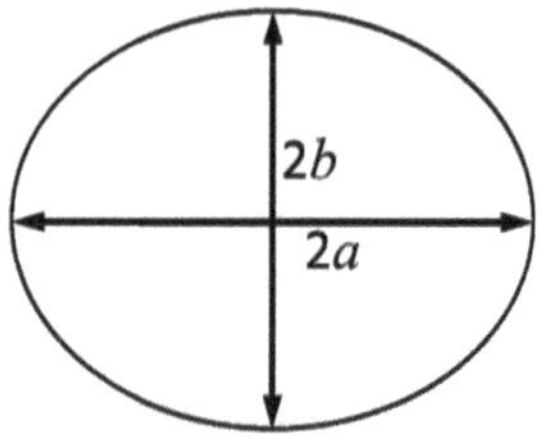

Fig. 4.3 An ellipsoid

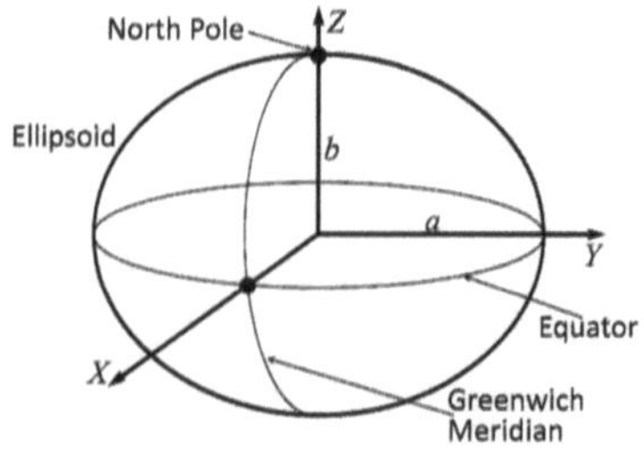

Fig. 4.4 WGS-84 Cartesian coordinates

Fig. 4.5 Earth's axis length and reciprocal of flattening factor

WGS-84 Reference Ellipsoid		
Semi-major axis, a (m) Equatorial radius (Definition)	Semi-minor axis, b (m) Polar radius (Derived from a and $\frac{1}{f}$)	Reciprocal of flattening, $\frac{1}{f}$ (Definition)
6,378,137.0	6,356,752.314,2	298.257,223,563

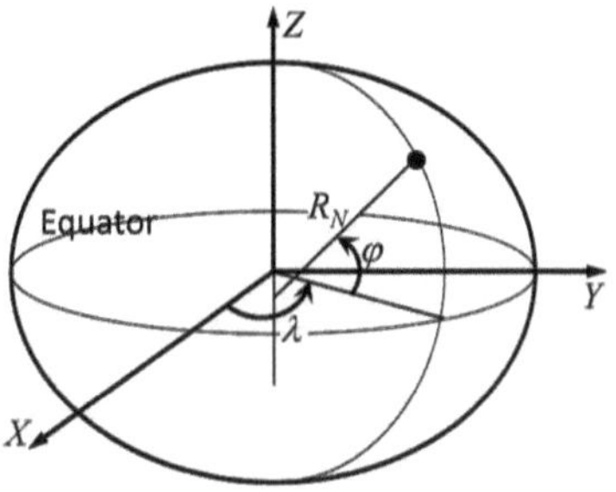

Fig. 4.6 Latitude φ of a point on the ellipsoid

the Earth's axis length and the reciprocal of flattening factor are shown in Fig. 4.5. The Earth's surface is uneven and the length of a and b depends on the actual locations the measurements are taken. WGS-84 defines the Earth's equatorial radius a to be exactly 6,378,137.0 meters and the reciprocal of flattening $\frac{1}{f}$ to be exactly 298.257,223,563. The value of the polar radius b is derived from the values of a and $\frac{1}{f}$. The eccentricity e is given by

$$e^2 = \frac{a^2 - b^2}{a^2} = 2f - f^2, \tag{4.2a}$$

$$b = a\sqrt{1 - e^2}. \tag{4.2b}$$

In WGS-84, the square of the eccentricity e^2 for the Earth is a derived quantity and is 0.006,694,379,990,14. The definition of the geodetic latitude φ of a point on the ellipsoid is the angle between the normal of the point and the equatorial plane. See Fig. 4.6. Except for points on the equator or at the poles, the normal at a point on the surface of an ellipsoid does not pass through the center of the ellipsoid. The radius of curvature $R_N(\varphi)$ of the prime vertical of the ellipsoid is given by

$$R_N(\varphi) = \frac{a^2}{\sqrt{a^2\cos^2(\varphi) + b^2\sin^2(\varphi)}} = \frac{a}{\sqrt{1 - e^2\sin^2(\varphi)}}. \tag{4.3}$$

The prime vertical is the vertical circle passing East and West through the zenith of a specific location; the end of the radius $R_N(\varphi)$ is always on the polar axis. At the pole, $R_N(90°) = \frac{a^2}{b}$ = 6,399.594 km. At the equator, $R_N(0°) = a$ = 6,378.137,0 km.

The Cartesian coordinates (x, y, z) in terms of the ellipsoid parameters are

$$x = R_N(\varphi)\cos(\varphi)\cos(\lambda), \tag{4.4a}$$

$$y = R_N(\varphi)\cos(\varphi)\sin(\lambda), \tag{4.4b}$$

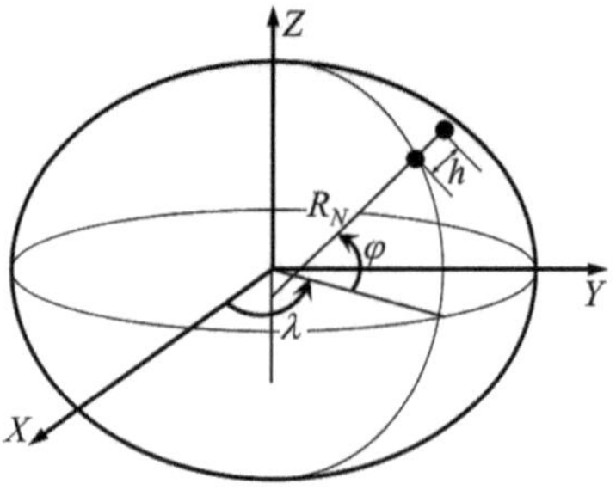

Fig. 4.7 A point at height h above (x, y, z)

$$z = \left(\frac{b^2}{a^2}\right) R_N(\varphi) \sin(\varphi). \tag{4.4c}$$

For a point with coordinates (x, y, z) on the Earth's surface, its longitude λ is given by

$$\lambda = \tan^{-1}\left(\frac{y}{x}\right). \tag{4.5}$$

For a point at height h above (x, y, z) as shown in Fig. 4.7, its coordinates (x', y', z') are given by

$$x' = (R_N(\varphi) + h) \cos(\varphi) \cos(\lambda), \tag{4.6a}$$

$$y' = (R_N(\varphi) + h) \cos(\varphi) \sin(\lambda), \tag{4.6b}$$

$$z' = \left(\frac{b^2}{a^2} R_N(\varphi) + h\right) \sin(\varphi). \tag{4.6c}$$

4.5 Radius of the Earth

The Earth is not only not a perfect sphere, but its surface is uneven. The density of the material inside the Earth is also not homogeneous. The radius of the Earth depends on the measurement method and the actual locations the measurements are taken. We have discussed some of the definitions for the radius of the Earth above. The Earth's equatorial radius a is defined to be 6,378.137,0 km in WGS-84. The Earth's polar radius b is 6,356.752,314,2 km. The radius of curvature $R_N(\varphi)$ of the prime vertical is defined in (4.3).

There are many other definitions for the Earth's radius. The radius of curvature on the Earth's surface in the (North-South) meridian at geodetic latitude φ denoted by $R_M(\varphi)$ is given by $R_M(\varphi) = \frac{a^2b^2}{((a\cos(\varphi))^2 + (b\sin(\varphi))^2)^{\frac{3}{2}}}$. While the end of the radius of the prime vertical $R_N(\varphi)$ is always on the polar axis, and the end of the meridian radius $R_M(\varphi)$ is not on the polar axis; it is in the quadrant on the other side of the equator. At the pole, $R_M(90°) = \frac{a^2}{b}$ = 6,399.594 km. At the equator, $R_M(0°) = \frac{b^2}{a}$ = 6,335.439 km. The meridional mean, denoted by M_r, is given by $M_r = \frac{2}{\pi}\int_0^{\frac{\pi}{2}} \sqrt{a^2\cos^2(\varphi) + b^2\sin^2(\varphi)}d\varphi$. The approximation $M_r \approx \left(\frac{a^{\frac{3}{2}}+b^{\frac{3}{2}}}{2}\right)^{\frac{2}{3}}$ = 6,367.449,1 km differs from the exact result by less than 1 micrometer. The approximation $M_r \approx \frac{a+b}{2}$ = 6,367.445 km differs from the exact result by about 4 meters.

The distance on the Earth's surface at geodetic latitude φ to the Earth's center denoted by $R(\varphi)$ is called the Earth's geocentric radius and is given by $R(\varphi) = \sqrt{\frac{(a^2\cos(\varphi))^2+(b^2\cos(\varphi))^2}{(a\cos(\varphi))^2+(b\cos(\varphi))^2}}$. The geocentric latitude θ of a point on the Earth is the angle between the geocentric radius of the point and the equatorial plane.

The volumetric radius of the Earth denoted by R_3 is the radius of a sphere whose volume is equal to the volume of the Earth and is given by $R_3 = a^3b^{\frac{1}{3}}$. The Earth's volumetric radius is 6,371.000,8 km. The authalic radius of the Earth denoted by R_2 is the radius of a sphere that has the same surface area as the reference ellipsoid and is given by $R_2 = \sqrt{\frac{a^2+\frac{ab^2}{\sqrt{a^2-b^2}}\ln\left(\frac{a+\sqrt{a^2-b^2}}{b}\right)}{2}} = \sqrt{\frac{a^2}{2}+\frac{b^2}{2}\frac{\tanh^{-1}e}{e}} = \sqrt{\frac{A}{4\pi}}$, where A is the surface area of the reference ellipsoid. The Earth's authalic radius is 6,371.007,2 km. The International Union of Geodesy and Geophysics (IUGG) defines the mean radius of the Earth denoted by R_1 to be $R_1 = \frac{2a+b}{3}$ and is 6,371.008,8 km. Hence, $R_M(90°) = R_N(90°) > R_N(0°) = a > R_1 > R_2 > R_3 > M_r > b = R_M(0°)$.

The average distance from the center of the Earth to the surface is 6,371.230 km. It is larger than R_1 but smaller than a. The above are some of the commonly encountered definitions for the radius of the Earth. There are many other definitions.

4.6 Location Estimation

Consider the estimation of the coordinate (x_0, y_0) of a point on a two-dimensional plane as shown in Fig. 4.8. Suppose that a reference point at location (x_1, y_1) transmits a pulse to location (x_0, y_0). Encoded in the pulse is the time t_1 of the transmission. The pulse is received at location (x_0, y_0) at time $t_{0,1}$. Assume that both t_1 and $t_{0,1}$ are atomic clock time. If the velocity of the pulse is c, the distance d_1 between (x_1, y_1) and (x_0, y_0) is given by

$$d_1 = c(t_{0,1} - t_1) = \sqrt{(x_1 - x_0)^2 + (y_1 - y_0)^2}. \tag{4.7}$$

Suppose that a second reference point at location (x_2, y_2) also transmits a pulse to location (x_0, y_0). The time t_2 of the transmission is also encoded in the pulse. The pulse is received at location (x_0, y_0) at time $t_{0,2}$. Assume that both t_2 and $t_{0,2}$ are

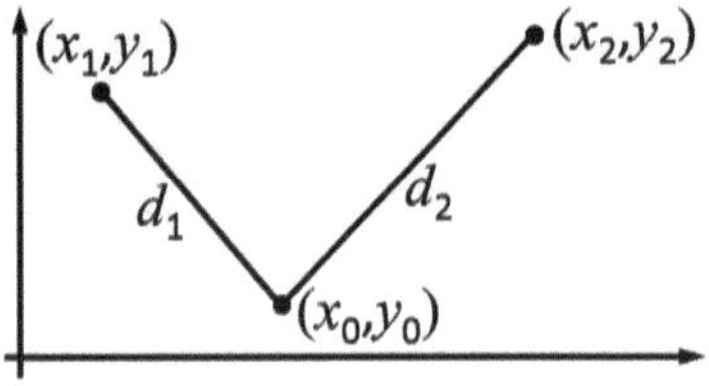

Fig. 4.8 Location estimation on a two-dimensional plane

also atomic clock time. Let the distance between (x_2, y_2) and (x_0, y_0) be d_2. The distance d_2 is given by

$$d_2 = c(t_{0,2} - t_2) = \sqrt{(x_2 - x_0)^2 + (y_2 - y_0)^2}. \tag{4.8}$$

Define

$$\Delta_k = t_{0,k} - t_k. \tag{4.9}$$

From (4.7), (4.8), and (4.9), we have

$$d_k = \sqrt{(x_k - x_0)^2 + (y_k - y_0)^2} = c\Delta_k \quad , \quad k = 1, 2. \tag{4.10}$$

The coordinate (x_0, y_0) can be obtained by solving the simultaneous Eqs. (4.10).

In a GNSS, the clocks of the reference points (i.e., the satellites) are atomic clocks and so are very accurate and very well synchronized. The clock at location (x_0, y_0) is not an atomic clock and is not synchronized to that of the reference points. The atomic clock time $t_{0,1}$ and $t_{0,2}$ are not known in location (x_0, y_0). Instead of the atomic clock time $t_{0,k}$, the corresponding time recorded in location (x_0, y_0) is $\hat{t}_{0,k}$, i.e., the time of the clock at location (x_0, y_0). There is a difference between $t_{0,k}$ and $\hat{t}_{0,k}$. The clock at location (x_0, y_0) is a very stable crystal clock, and so the difference between $t_{0,k}$ and $\hat{t}_{0,k}$ for all k is constant over a short duration sufficiently long for collecting all the relevant satellite data. Let the time displacement between $t_{0,k}$ and $\hat{t}_{0,k}$ be Δ; Δ is unknown. Let

$$\hat{t}_{0,k} = t_{0,k} + \Delta, \tag{4.11}$$

and let

$$\hat{\Delta}_k = \hat{t}_{0,k} - t_k. \tag{4.12}$$

From (4.9), (4.11), and (4.12), we have

$$\Delta_k = \hat{\Delta}_k - \Delta. \tag{4.13}$$

Substituting (4.13) into (4.10),

$$d_k = \sqrt{(x_k - x_0)^2 + (y_k - y_0)^2} = c(\hat{\Delta}_k - \Delta) = c\hat{\Delta}_k - c\Delta \quad , \quad k = 1, 2. \tag{4.14}$$

There are three unknowns, x_0, y_0, and Δ, in (4.14), but there are only two equations for $k = 1, 2$. To solve for (x_0, y_0), we need another equation. We need a third reference point at location (x_3, y_3). Let the distance between (x_3, y_3) and (x_0, y_0) be d_3. The distance d_3 is given by (4.10) with $k = 3$. Given (4.10) with $k = 1, 2, 3$, there are three unknowns and three equations, and so the three unknowns x_0, y_0, and Δ can be solved. In a two-dimensional space, we need three reference points. In a three-dimensional space, we need four reference points. The reference points are the navigational satellites. Let the four reference points in the three-dimensional space be (x_1, y_1, z_1), (x_2, y_2, z_2), (x_3, y_3, z_3), (x_4, y_4, z_4), respectively. Hence, we have

$$d_k = \sqrt{(x_k - x_0)^2 + (y_k - y_0)^2 + (z_k - z_0)^2} = c\hat{\Delta}_k - c\Delta \quad , \quad k = 1, 2, 3, 4. \tag{4.15}$$

(x_0, y_0, z_0) may be obtained by solving (4.15).

4.7 Linearization of Distance Function

The expression (4.15) is nonlinear. It can be linearized over a small region by its Taylor series expansion. An iterative method for solving (4.15) is shown below. It starts with assuming an initial approximation for the location (x_0, y_0, z_0). The expression (4.15) is linearized in the vicinity of (x_0, y_0, z_0) using its Taylor series expansion. The linearized (4.15) is solved and a better approximation for the location of (x_0, y_0, z_0) is obtained. This better approximation is then used as initial solution to obtain further improvement on the location of (x_0, y_0, z_0). The iteration is repeated until a criterion to stop the iteration is met. Following (4.15), define the distance function $d_k(x, y, z)$ given by

$$\begin{aligned} d_k(x, y, z) &= \sqrt{(x_k - x)^2 + (y_k - y)^2 + (z_k - z)^2}, \\ &= c\hat{\Delta}_k - c\Delta. \end{aligned} \tag{4.16}$$

Let the partial differential of $d_k(x, y, z)$ with respect to x, y, z be $g_{k,x}(x, y, z)$, $g_{k,y}(x, y, z)$, and $g_{k,z}(x, y, z)$, respectively. Partial differentiation of (4.16) with respect to x, y, z gives

$$g_{k,x}(x, y, z) = \frac{\partial d_k(x, y, z)}{\partial x} = \frac{x - x_k}{d_k(x, y, z)}, \tag{4.17a}$$

$$g_{k,y}(x, y, z) = \frac{\partial d_k(x, y, z)}{\partial y} = \frac{y - y_k}{d_k(x, y, z)}, \tag{4.17b}$$

$$g_{k,z}(x, y, z) = \frac{\partial d_k(x, y, z)}{\partial z} = \frac{z - z_k}{d_k(x, y, z)}. \tag{4.17c}$$

The first-order Taylor series expansion of (4.16) around the coordinate (x_0, y_0, z_0) is given by

$$\begin{aligned} d_k(x, y, z) \approx\ & d_k(x_0, y_0, z_0)+ \\ & g_{k,x}(x_0, y_0, z_0)\delta x + g_{k,y}(x_0, y_0, z_0)\delta y + g_{k,z}(x_0, y_0, z_0)\delta z, \end{aligned} \tag{4.18a}$$

$$\begin{aligned} \approx\ & d_k(x_0, y_0, z_0)+ \\ & \frac{x_0 - x_k}{d_k(x_0, y_0, z_0)}\delta x + \frac{y_0 - y_k}{d_k(x_0, y_0, z_0)}\delta y + \frac{z_0 - z_k}{d_k(x_0, y_0, z_0)}\delta z, \end{aligned} \tag{4.18b}$$

where $\delta x = x - x_0 \ , \ \delta y = y - y_0 \ , \ \delta z = z - z_0 \ .$

In (4.18), $|\delta x|$, $|\delta y|$, and $|\delta z|$ are small. From (4.16) and (4.18), we have

$$\begin{aligned} & c\hat{\Delta}_k - d_k(x_0, y_0, z_0) \\ & \quad \approx g_{k,x}(x_0, y_0, z_0)\delta x + g_{k,y}(x_0, y_0, z_0)\delta y + g_{k,z}(x_0, y_0, z_0)\delta z + c\Delta, \end{aligned} \tag{4.19a}$$

$$\approx \frac{x_0 - x_k}{d_k(x_0, y_0, z_0)}\delta x + \frac{y_0 - y_k}{d_k(x_0, y_0, z_0)}\delta y + \frac{z_0 - z_k}{d_k(x_0, y_0, z_0)}\delta z + c\Delta. \tag{4.19b}$$

Define

$$\boldsymbol{g}_k(x_0, y_0, z_0, c) = [g_{k,x}(x_0, y_0, z_0) \ , \ g_{k,y}(x_0, y_0, z_0) \ , \ g_{k,z}(x_0, y_0, z_0) \ , \ c]^{\mathrm{T}}, \tag{4.20a}$$

$$= \left[\frac{x_0 - x_k}{d_k(x_0, y_0, z_0)} \ , \ \frac{y_0 - y_k}{d_k(x_0, y_0, z_0)} \ , \ \frac{z_0 - z_k}{d_k(x_0, y_0, z_0)} \ , \ c\right]^{\mathrm{T}}, \tag{4.20b}$$

$$\boldsymbol{u} = [\delta x \ , \ \delta y \ , \ \delta z \ , \ \Delta]^{\mathrm{T}}, \tag{4.21}$$

and

$$p_k(x_0, y_0, z_0) = c\hat{\Delta}_k - d_k(x_0, y_0, z_0). \tag{4.22}$$

From (4.19), (4.20), (4.21), and (4.22), we have

$$p_k(x_0, y_0, z_0) \approx \boldsymbol{g}_k(x_0, y_0, z_0)^{\mathrm{T}}\boldsymbol{u}. \tag{4.23}$$

The vector $\boldsymbol{u}$ contains four unknown elements. If there are four satellites, there will be four sets of data for (4.23). Replacing the approximation sign of (4.23) by equality sign, $\boldsymbol{u}$ can be solved. If there are more than four satellites, an optimization method can be used to obtain the "best fit" solution. Suppose that there are K satellites where $K \geq 4$. We have K sets of data for (4.23). Define

$$\boldsymbol{p} = \begin{bmatrix} p_1(x_0, y_0, z_0) \\ p_2(x_0, y_0, z_0) \\ \vdots \\ p_K(x_0, y_0, z_0) \end{bmatrix}, \quad \mathbf{G} = \begin{bmatrix} \boldsymbol{g}_1(x_0, y_0, z_0, c)^{\mathrm{T}} \\ \boldsymbol{g}_2(x_0, y_0, z_0, c)^{\mathrm{T}} \\ \vdots \\ \boldsymbol{g}_K(x_0, y_0, z_0, c)^{\mathrm{T}} \end{bmatrix}, \quad \boldsymbol{u} = \begin{bmatrix} \delta x \\ \delta y \\ \delta z \\ \Delta \end{bmatrix}. \tag{4.24}$$

The least squares solution for K sets of data for (4.23) is

$$\boldsymbol{u} = [\mathbf{G}^{\mathrm{T}}\mathbf{G}]^{-1}\mathbf{G}^{\mathrm{T}}\boldsymbol{p}. \tag{4.25}$$

After $\boldsymbol{u}$ is obtained, the location (x_0, y_0, z_0) is updated. Let $(x_{\text{updated}}, y_{\text{updated}}, z_{\text{updated}})$ denote the updated (x_0, y_0, z_0). The updated location is given by

$$(x_{\text{updated}}, y_{\text{updated}}, z_{\text{updated}}) = (x_0 + \delta x, y_0 + \delta y, z_0 + \delta z). \tag{4.26}$$

The updated location $(x_{\text{updated}}, y_{\text{updated}}, z_{\text{updated}})$ replaces (x_0, y_0, z_0) in the next iteration.

4.8 RTK (Real-Time Kinematics)

In a navigational satellite system, each satellite transmits a unique code of itself with a time-stamp. A mobile receiver matches the code to its internally stored code. The difference between the received time and the transmit time is the time taken by the satellite signal to travel from the satellite to the mobile receiver. The time difference is multiplied by the speed of light ($299,792,458\,\mathrm{m/s}$) to obtain the distance. Every nanosecond error in estimating the delay time will cause an error of 0.3 meter in the estimated distance.

The satellite signal passes through the ionosphere and the troposphere before reaching the mobile receiver. The speed of light in vacuum, the ionosphere, and the troposphere are different. This causes a few meters or a few tens of meters of error in the estimated horizontal location of the mobile receiver.

A more accurate estimation may be obtained from the phase of the carrier signal from the satellites. Suppose that the kth satellite is transmitting with carrier frequency f_c with phase θ_k and has a distance $d_{mob,k}$ from a mobile receiver. The wavelength λ of the carrier is c/f_c where c is the speed of light. Let its carrier signal received by the mobile receiver at time t be $s_{mob,k}(t)$ and is given by

$$s_{mob,k}(t) = A\cos\left(2\pi f_c t + \theta_k - \frac{2\pi d_{mob,k}}{\lambda} + e_{mob,k}\right), \tag{4.27}$$

where $e_{mob,k}$ is the phase error due to the change in the speed of light in the ionosphere and troposphere.

The mobile receiver generates a local sinusoid $s_{mob,loc}(t)$ with phase $\theta_{mob,loc}$ at the same frequency as the satellite and is given by

$$s_{mob,loc}(t) = B\cos(2\pi f_c t + \theta_{mob,loc}). \tag{4.28}$$

The product $s_{mob,k}(t) \times s_{mob,loc}(t)$ is given by

$$\begin{aligned} s_{mob,k}(t) \times s_{mob,loc}(t) =& AB\cos\left(2\pi f_c t + \theta_k - \frac{2\pi d_{mob,k}}{\lambda} + e_{mob,k}\right) \\ & \times \cos(2\pi f_c t + \theta_{mob,loc}) \\ =& \frac{AB}{2}\left[\cos\left(4\pi f_c t + \theta_k - \frac{2\pi d_{mob,k}}{\lambda} + \theta_{mob,loc} + e_{mob,k}\right)\right. \\ & \left. + \cos\left(\frac{2\pi d_{mob,k}}{\lambda} + \theta_{mob,loc} - \theta_k - e_{mob,k}\right)\right]. \end{aligned} \tag{4.29}$$

The $\cos\left(4\pi f_c t + \theta_k - \frac{2\pi d_{mob,k}}{\lambda} + \theta_{mob,loc} + e_{mob,k}\right)$ term in (4.29) is a high-frequency term with frequency $2f_c$ and can be removed by lowpass filtering. The $\cos\left(\frac{2\pi d_{mob,k}}{\lambda} + \theta_{mob,loc} - \theta_k - e_{mob,k}\right)$ term is a constant. After lowpass filtering of $s_{mob,k}(t) \times s_{mob,loc}(t)$, we have

$$\begin{aligned} & \text{lowpassed}\left\{s_{mob,k}(t) \times s_{mob,loc}(t)\right\} \\ &= \frac{AB}{2}\cos\left(\frac{2\pi d_{mob,k}}{\lambda} + \theta_{mob,loc} - \theta_k - e_{mob,k}\right). \end{aligned} \tag{4.30}$$

The values of A and B can be easily estimated from $s_{\text{mob},k}(t)$ and $s_{\text{mob,loc}}(t)$. To simplify notation, let $\frac{2\times \text{lowpassed}\{s_{\text{mob},k}(t)\times s_{\text{mob,loc}}(t)\}}{AB}$ be denoted by $\xi_{\text{mob},k}$.

$$\frac{2 \times \text{lowpassed}\left\{s_{\text{mob},k} \times s_{\text{mob,loc}}\right\}}{AB} = \xi_{\text{mob},k}. \tag{4.31}$$

From (4.30) and (4.31),

$$\cos\left(\frac{2\pi d_{\text{mob},k}}{\lambda} + \theta_{\text{mob,loc}} - \theta_k - e_{\text{mob},k}\right) = \xi_{\text{mob},k}. \tag{4.32}$$

The distance $d_{\text{mob},k}$ can be written in terms of the number of wavelength λ. Let $d_{\text{mob},k}$ be equal to an integer number of wavelength plus a fractional wavelength where the integer is $N_{\text{mob},k}$ and the fraction is $\frac{\phi_{\text{mob},k}}{2\pi}$. Hence,

$$d_{\text{mob},k} = \left(N_{\text{mob},k} + \frac{\phi_{\text{mob},k}}{2\pi}\right)\lambda = N_{\text{mob},k}\lambda + \frac{\phi_{\text{mob},k}}{2\pi}\lambda. \tag{4.33}$$

From (4.32) and (4.33),

$$\begin{aligned}\xi_{\text{mob},k} &= \cos\left(\frac{2\pi d_{\text{mob,k}}}{\lambda} + \theta_{\text{mob,loc}} - \theta_k - e_{\text{mob},k}\right)\\ &= \cos(2\pi N_{\text{mob},k} + \phi_{\text{mob},k} + \theta_{\text{mob,loc}} - \theta_k - e_{\text{mob},k}). \end{aligned} \tag{4.34a}$$

$$2\pi N_{\text{mob},k} + \phi_{\text{mob},k} + \theta_{\text{mob,loc}} - \theta_k - e_{\text{mob},k} = \cos^{-1}(\xi_{\text{mob},k}). \tag{4.34b}$$

The error due to the change in the speed of light in the ionosphere and the troposphere can be compensated by using a reference station with a known location if the receiver is near ($\lesssim$20 km) the reference station. The changes in the condition of the ionosphere and the troposphere are slow enough to allow correction information to be sent from the reference station to the receiver through Internet link.

Consider a reference station in the neighborhood ($\lesssim$20 km away) of the mobile receiver. Let the corresponding $d_{\text{mob},k}$, $N_{\text{mob},k}$, $\phi_{\text{mob},k}$, $\theta_{\text{mob,loc}}$, $\xi_{\text{mob},k}$, and $e_{\text{mob},k}$ be $d_{\text{ref},k}$, $N_{\text{ref},k}$, $\phi_{\text{ref},k}$, $\theta_{\text{ref,loc}}$, $\xi_{\text{ref},k}$, and $e_{\text{ref},k}$, respectively. Since the location of the reference station is known, $d_{\text{ref},k}$ can be estimated and so is known. Replacing "mob" in (4.34) by "ref," we have

$$\frac{2\pi d_{\text{ref},k}}{\lambda} + \theta_{\text{ref,loc}} - \theta_k - e_{\text{ref},k} = \cos^{-1}(\xi_{\text{ref},k}). \tag{4.35}$$

If the receiver is near ($\lesssim$20 km) the reference station, $e_{\text{ref},k} \approx e_{\text{mob},k}$; we shall assume that $e_{\text{ref},k} = e_{\text{mob},k}$. Given this assumption, rearranging (4.35), we have

$$\theta_k + e_{\text{mob},k} = \theta_k + e_{\text{ref},k} = \frac{2\pi d_{\text{ref},k}}{\lambda} + \theta_{\text{ref,loc}} - \cos^{-1}(\xi_{\text{ref},k}). \tag{4.36}$$

Eliminate $\theta_k + e_{\mathrm{mob},k}$ in (4.34b) using (4.36), and we have

$$2\pi N_{\mathrm{mob},k} + \phi_{\mathrm{mob},k} + \theta_{\mathrm{mob,loc}} = \cos^{-1}(\xi_{\mathrm{mob},k}) - \cos^{-1}(\xi_{\mathrm{ref},k}) + \frac{2\pi d_{\mathrm{ref},k}}{\lambda} + \theta_{\mathrm{ref,loc}}. \tag{4.37}$$

For two satellites $k = k1$ and $k = k2$, respectively, we have

$$2\pi N_{\mathrm{mob},k1} + \phi_{\mathrm{mob},k1} + \theta_{\mathrm{mob,loc}} = \cos^{-1}(\xi_{\mathrm{mob},k1}) - \cos^{-1}(\xi_{\mathrm{ref},k1}) + \frac{2\pi d_{\mathrm{ref},k1}}{\lambda} + \theta_{\mathrm{ref,loc}}. \tag{4.38a}$$

$$2\pi N_{\mathrm{mob},k2} + \phi_{\mathrm{mob},k2} + \theta_{\mathrm{mob,loc}} = \cos^{-1}(\xi_{\mathrm{mob},k2}) - \cos^{-1}(\xi_{\mathrm{ref},k2}) + \frac{2\pi d_{\mathrm{ref},k2}}{\lambda} + \theta_{\mathrm{ref,loc}}. \tag{4.38b}$$

Eliminate $\theta_{\mathrm{mob,loc}}$ and $\theta_{\mathrm{ref,loc}}$ by subtracting (4.38a) from (4.38b):

$$\begin{aligned}\phi_{\mathrm{mob},k2} - \phi_{\mathrm{mob},k1} &= \cos^{-1}(\xi_{\mathrm{mob},k2}) - \cos^{-1}(\xi_{\mathrm{mob},k1}) - \cos^{-1}(\xi_{\mathrm{ref},k2}) + \cos^{-1}(\xi_{\mathrm{ref},k1}) \\ &+ \frac{2\pi d_{\mathrm{ref},k2}}{\lambda} - \frac{2\pi d_{\mathrm{ref},k1}}{\lambda} - 2\pi \left(N_{\mathrm{mob},k2} - N_{\mathrm{mob},k1}\right).\end{aligned} \tag{4.39}$$

Replacing “mob” in (4.31) by “ref,” we have

$$\xi_{\mathrm{ref},k} = \frac{2 \times \mathrm{lowpassed}\left\{s_{\mathrm{ref},k} \times s_{\mathrm{ref,loc}}\right\}}{AB}. \tag{4.40}$$

In (4.39), $\xi_{\mathrm{mob},k}$ and $\xi_{\mathrm{ref},k}$ are obtained from (4.31) and (4.40), respectively. Since the location of the reference station is known, $\frac{2\pi d_{\mathrm{ref},k2}}{\lambda}$ and $\frac{2\pi d_{\mathrm{ref},k1}}{\lambda}$ are also known. The unknown quantities are $N_{\mathrm{mob},k2}$ and $N_{\mathrm{mob},k1}$. There are many methods for estimating $N_{\mathrm{mob},k2}$ and $N_{\mathrm{mob},k1}$. One may use several of them to estimate $N_{\mathrm{mob},k2}$ and $N_{\mathrm{mob},k1}$.

References

J. Benedicto, S.E. Dinwiddy, G. Gatti, R. Lucas, M. Lugert, *GALILEO: Satellite System Design and Technology Developments* (European Space Agency, Germany, 2000)

N. Bonnor, A brief history of global navigation satellite systems. J. Navigation **65**(1), 1–14 (2012)

J.W. Dow, R.E. Neilan, C. Rizos, The international GNSS service in a changing landscape of global navigation satellite systems. J. Geodesy **83**, 191–198 (2009)

Y. Feng, J. Wang, GPS RTK performance characteristics and analysis. J. Global Positioning Systems **1**(13), 1–8 (2008)

International committee on global navigation satellite systems (ICG). https://www.unoosa.org/oosa/en/ourwork/icg/icg.html

N. Ivanov, V. Salischev, The GLONASS system—an overview. J. Navigation **45**(2), 175–182 (2009)

R.B. Langley, GPS RTK. GPS World **9**(9), 70–76 (1998)

W. Lechner, S. Baumann, Global navigation satellite systems. Comput. Electron. Agric. **25**(1-2), 67–85 (2000)

X. Li, X. Zhang, X. Ren, M. Fritsche, J. Wickert, H. Schuh, Precise positioning with current multi-constellation global navigation satellite systems: GPS, GLONASS, Galileo and BeiDou. Sci. Rep. **66**(1), 8328 (2015)

R. Li, S. Zheng, E. Wang, J. Chen, S. Feng, D. Wang, L. Dai, Advances in BeiDou navigation satellite system (BDS) and satellite navigation augmentation technologies. Satell. Navig. **1**(1), 12 (2020)

T. Takasu, A. Yasuda, Development of the low-cost RTK-GPS receiver with an open source program package RTKLIB, in *Proceedings of the 2009 International Symposium on GPS/GNSS, Jeju, Korea* (2009), pp. 1–6

P.J.G. Teunissen, O. Montenbruck, *Springer Handbook of Global Navigation Satellite Systems* (Springer, Berlin, 2017)

P. Teunissen, R. Odolinski, D. Odijk, Instantaneous BeiDou+ GPS RTK positioning with high cut-off elevation angles. J. Geodesy **88**, 335–350 (2014)

United nations office for outer space affairs. https://www.unoosa.org/oosa/index.html

G. Xu, Y. Xu, *GPS: Theory, Algorithms and Applications* (Springer, Berlin, 2016)

Y. Yang, W. Gao, S. Guo, Y. Mao, Y. Yang, Introduction to BeiDou-3 navigation satellite system. Navigation **66**(1), 7–18 (2019)

A.S. Zaidi, M.R. Suddle, Global navigation satellite systems: a survey, in *Proceedings of the 2006 International Conference on Advances in Space Technologies* (2006), pp. 84–87

Chapter 5
Introduction to Lidar

5.1 Introduction

Lidar (**Li**ght **d**etection **a**nd **r**anging) is a sensing method that uses light (usually laser) to detect objects and determine their ranges from reflected or backscattered light. It is also called laser radar (Lidar base specification Online; Wandinger 2005; Behroozpour et al. 2017; Li and Ibanez-Guzman 2020; Filgueira et al. 2017; Warren 2019; Roriz et al. 2022; Csanyi and Toth 2007; Chow 2012; Agrawal 2013; Dupuis 1987; Bowers 1987; Adams et al. 1985; Lan et al. 2021; Liu et al. 2021; Ryvkin et al. 2020; Samuel and Turnbull 2007; Blanchard et al. 2008; Maximov et al. 2008; Avrutin and Ryvkin 2019; Cosendey et al. 2012; Shau et al. 2001; Podoskin et al. 2020; Zhdanov et al. 2004; Boulon 2012; Fang et al. 2014; Godard 2007; Jauregui et al. 2013; Canning 2006; Fermann and Hartl 2013; Costela et al. 2003; Oehme et al. 2006; Su et al. 2011; Cong et al. 2016; Campbell 2016; Woodson et al. 2006; Liu et al. 2007; Jiang et al. 2007; Yamamoto et al. 2007; Nagano et al. 2011; Zambra et al. 2004; Saldanha et al. 2017; Wang et al. 2020; Zhou et al. 2008; Tuantranont et al. 2001; Brandl et al. 2013; Poulton et al. 2018; Gozzard et al. 2020; McManamon et al. 1996, 2009; Wesstrom et al. 2004; Yam and Hassan 2005). It will not be surprising if all future highly autonomous vehicles have lidar systems installed. Lidar determines the range and speed of an object by illuminating the object with an optical signal and measures the time of flight and the frequency shift of the optical signal reflected or backscattered from the object.

Figure 5.1 shows the atmospheric opacity for various wavelengths of electromagnetic wave with wavelength ranging from 100 nm to 1 mm. There are two windows where the atmosphere is transparent to light—one is at the visible light region and the other is centered around 10 μm. The sun's surface temperature is about 5,800°K. Black body at 5,800°K has the strongest radiation in the visible light region. Solar radiation is significantly weaker at 10 μm wavelength. Hence, solar radiation is strongest on the Earth's surface at the visible light region. There is no surprise that creatures on Earth develop eyesight to see visible light. The visible

Y. C. Lim et al., *Signal Processing for Autonomous Systems*,
https://doi.org/10.1007/978-3-032-00675-2_5

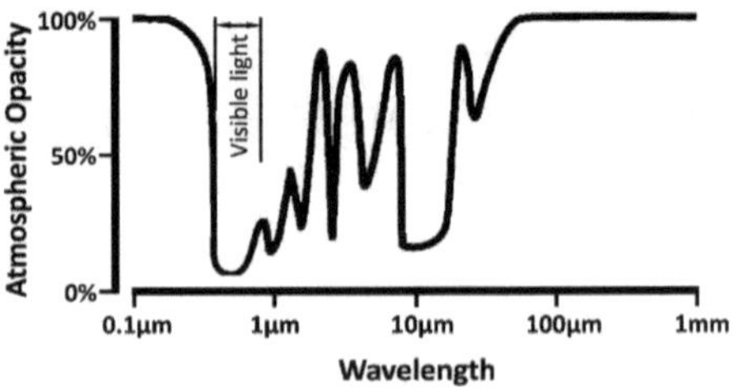

Fig. 5.1 Atmospheric opacity versus wavelength

light that we refer to are the wavelength visible to human. Birds are able to see part of the ultraviolet (UV) spectrum and are tetrachromats; they see four colors—UV, blue, green, and red. The birds' world is more colorful than human's! Mosquitoes, bullfrog, and some snake species are able to see infrared. Goldfish's and zebrafish's vision extends from the infrared spectrum all the way into the ultraviolet spectrum; they are tetrachromats.

There are many types of lidar system being used in, as well as being developed for, autonomous vehicles. Each of them has its advantages and disadvantages. Different lidar type has different operation range, spatial and range resolutions, implementation cost, safety issue, size, robustness against vibration, resistance to abuse, and etc. It is too early to predict which types of lidar system will be the prevailing types in the future. In this chapter, we shall described some of the lidar hardware components being used in vehicles or in the process of being developed in the laboratories.

The hardware components common to all lidar systems include a (a) laser source, (b) laser detector, and (c) laser scanner.

5.2 Laser Wavelength, Safety, and Cost

The 0.85–0.95 μm NIR (**n**ear **i**nfra**r**ed) and 1.55 μm (1.55 μm = 1550 nm) SWIR (**s**hort **w**ave **i**nfra**r**ed) are the most commonly used lasers in industry. For comparison, the wavelength of visible light spectrum is 0.38–0.7 μm, and the ITU (International Telecommunication Union) THF band is 100–1000 μm.

The retina is sensitive to light and can be damaged if the light energy impinging on it exceeds a certain safety limit. The cornea, lens, aqueous humor, and vitreous humor in the eye strongly absorb laser wavelengths longer than 1.4 μm. Hence, for laser wavelength longer than 1.4 μm, the laser energy finally impinging on the retina is greatly reduced. The threshold of laser energy required to cause damage to the cornea, lens, aqueous humor, and vitreous humor is also much higher than that required to cause damage to the retina. Thus, the 1.55 μm SWIR laser is safer than the 0.85 − 0.95 μm NIR laser to the eyes and so a stronger 1.55 μm laser pulse may be used. Nevertheless, a small fraction of the laser energy will still impinge on the retina. Furthermore, the 1.55 μm SWIR laser can still cause damage to the cornea, lens, aqueous humor, and vitreous humor if the power is too high.

The 1.55 μm SWIR laser needs an InGaAs (indium gallium arsenide) detector, which is significantly more expensive than the silicon-based detector needed by the 0.85–0.95 μm NIR laser. Furthermore, the efficiency of the InGaAs detector is lower than that of the silicon detector, and the atmospheric water absorption for 1.55 μm laser is stronger than that for the 0.85–0.95 μm laser. Hence, 0.85–0.95 μm NIR laser is currently the most commonly used wavelength for autonomous vehicles; this may change in the future.

5.3 Laser Sources

Laser, the acronym for **l**ight **a**mplification by **s**timulated **e**mission of **r**adiation, is a coherent light source. Its spatial coherence allows it to be focused and stay as a narrow beam over great distances. However, it should be noted that a laser beam must be focused by means of a lens for it to stay a narrow beam. This is because laser will be diffracted as it passes through the aperture of the device that emits the laser. The laser's temporal (or longitudinal) coherence allows it to be emitted with very narrow spectrum or to produce very short pulse (with femtosecond duration) with broad spectrum. (The shorter is the pulse, the wider is the spectrum.) The temporal coherence of a light wave is the ability of the light wave to remain at a single frequency with constant phase. The distance that the laser light travels while remaining coherent is called its coherence length. The coherence length is given approximately by

$$\text{Coherence length} \approx \frac{c}{n\Delta f} = \frac{\lambda^2}{n\Delta\lambda}, \tag{5.1}$$

where $c(\approx 299,792,458\,\text{ms}^{-1})$ is the speed of light in a vacuum, n is the refractive index of the medium, Δf is the bandwidth of the source, λ is the signal wavelength, and $\Delta\lambda$ is the width of the range of wavelengths. The bandwidth is the full width at half maximum linewidth. (If the peak of the spectrum is $Peak_{max}$, there will be a $\frac{Peak_{max}}{2}$ on each side of $Peak_{max}$. The bandwidth is the separation between these two $\frac{Peak_{max}}{2}$). If the spectrum distribution is Gaussian,

$$\text{Coherence length} \approx \frac{2\ln(2)}{\pi}\frac{\lambda^2}{n\Delta\lambda}. \tag{5.2}$$

The coherence length divided by the speed of light is the duration that the laser stays coherent.

Figure 5.2 shows the structure of a typical laser-emitting device. It consists of an active region (the cavity or the gain medium) sandwiched between two reflective facets. One facet is nontransparent and highly reflective reflecting almost all the light. The other facet is reflective but is partially transparent allowing some of the light to pass through; this is the facet where the laser beam emits.

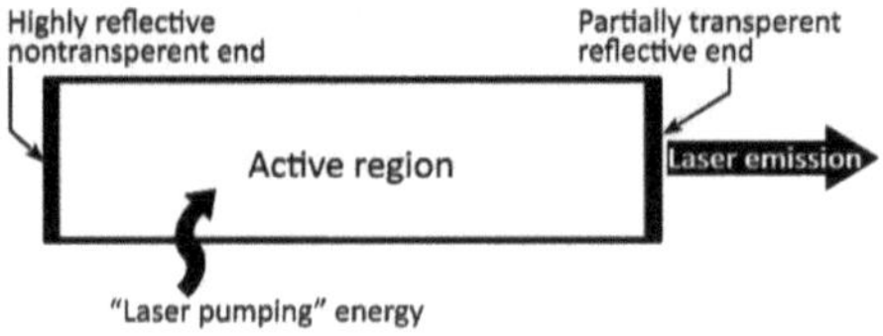

Fig. 5.2 Structure of a typical laser-emitting device

An electron may transit from a higher energy level to a lower energy level on its own causing the emission of a photon with wavelength λ. Let the difference between the energy levels where the electron transits be E_T and let the frequency of the photon be ν. E_T is proportional to ν; the constant of proportionality is Planck's constant h:

$$E_T = h\nu. \tag{5.3}$$

Planck's constant $h \approx 6.626, 070, 15 \times 10^{-34}$ joule-second (or J Hz^{-1}). Since $c = \lambda\nu$, λ is given by

$$\lambda = \frac{hc}{E_T}. \tag{5.4}$$

An electron may also be stimulated to transit from a higher energy level to a lower energy level by a photon of the correct wavelength; this is called stimulated emission. A photon emitted by stimulated emission matches exactly in phase, wavelength, and direction of travel, the original photon causing the emission.

A photon, whether emitted by spontaneous emission or stimulated emission, will be reflected back and forth many times between the two reflective facets shown in Fig. 5.2. This will cause more electrons to emit photons by stimulated emission producing optical gain.

A wave with wavelength λ will reflect back and forth with constructive interference in a cavity with length L given by

$$L = k \times \frac{\lambda}{2}, \tag{5.5}$$

where k is an integer called the mode order. Other wavelengths will die out eventually by destructive interference. The different wavelengths that a cavity can support are the longitudinal modes of the cavity. The frequency difference $\Delta\nu$ between any two consecutive modes k and $k + 1$ is given by

$$\Delta\nu = \frac{c}{2nL}, \tag{5.6}$$

where n is the refractive index of the cavity material.

Energy is supplied into the active region in a process called pumping. When a laser device is in laser action, it is called lasing.

Table 5.1 Common semiconductor laser materials

Materials	Wavelengths
AlGaInP/GaAs	620–680 nm
GaAs/AlGaAs	720–850 nm
GaAs/InGaAs	900–1100 nm
InP/InGaAsP	1000–1650 nm

The most popular laser sources are (1) diode lasers (DL), (2) solid-state lasers (SSL), (3) fiber lasers, and (4) dye lasers. A diode laser is a laser emitted by a semiconductor laser diode, which is electrically a PIN (***p***-type, **i**ntrinsic, ***n***-type) diode. A PIN diode is a diode with a wide undoped intrinsic region sandwiched between a p-type region and an n-type region. Diode lasers can be classified into two categories—edge-emitting laser (EEL) and vertical-cavity surface-emitting laser (VCSEL). Some common semiconductor materials and their laser wavelengths are shown in Table 5.1. A solid-state laser is a laser whose active region is solid state other than semiconductor. A fiber laser is a solid-state laser whose active region is an optical fiber.

5.3.1 Edge-Emitting Laser (EEL) Diode

In an edge-emitting semiconductor laser (also called in-plane laser or Fabry-Perot diode laser), the laser light propagates in a direction along the wafer surface of the semiconductor chip and emits out at an edge. There are many different designs for edge-emitting laser diode. The most popular being the double heterostructure laser diode, a simplified structure of which is shown in Fig. 5.3. In the double heterostructure laser diode, more than one type of semiconductor materials is used. The active region is the thin middle layer where the electrons and holes exist simultaneously. When the PIN junction is forward biased with large applied voltage, there will be a high population density (typically several orders of magnitude higher than the intrinsic carrier concentration) of electrons in the higher energy conduction band and holes in the lower energy valence band; this is a condition of population inversion. When electrons and holes are present in the same region, an electron

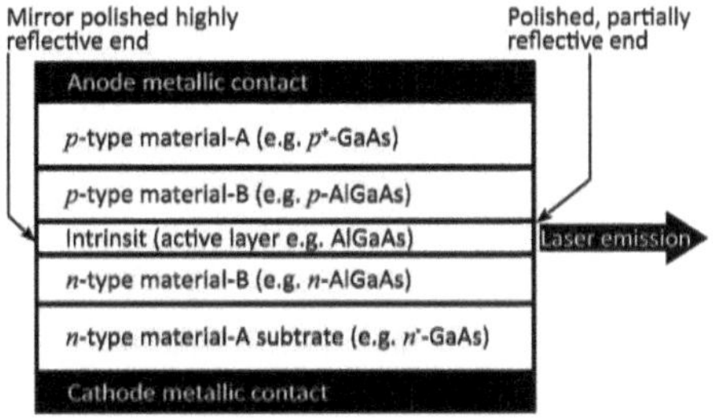

Fig. 5.3 Double heterostructure laser diode

may move from a higher energy conduction band to combine with a hole at the lower energy valence band. The energy difference between the electron's original state and the hole's state may be emitted in the form of a photon in the process called spontaneous (or recombination) emission. Electrons and holes may coexist in proximity without recombining for a certain time called the recombination time. A nearby photon with energy equal to the recombination energy can stimulate a recombination producing another photon in the process called stimulated emission. A resonant cavity in the active region between the highly reflective end and partially reflective end is created for the light to reflect back and forth many times. The large bandgap semiconductor has a lower refractive index than that in the active region. Light is reflected within the heterojunction forming an optical waveguide in the transverse direction. Hence, light is confined to the region where light amplification takes place.

As light is reflected back and forth in the optical cavity, it is amplified by stimulated emission. Some light is lost due to absorption, and some light is lost due to incomplete reflection from the end facets. If the amplification is larger than lost, lasing occurs. Reflection occurs at both end facets. Since significantly more light emits from the partially reflective end than from the very highly polished reflective end, the partially reflective end is the laser emission end. An EEL diode usually transmits an elliptical laser beam. It requires additional beam-shaping optics to convert the elliptical beam into a circular beam. Since its aperture is very small, it is highly divergent. A lens is needed for focusing. As the laser beam is emitted in a direction parallel to the substrate, a 2D laser array on a single chip cannot be made. Nevertheless, an EEL diode bar can be easily made. Such diode bars can be stacked to form a 2D array. See Fig. 5.4.

5.3.2 *Vertical-Cavity Surface-Emitting Laser (VCSEL) Diode*

While the optical cavity axis of the EEL diode is perpendicular to the direction of current flow, the optical axis of the vertical-cavity surface-emitting semiconductor laser (VCSEL) diode is along the direction of current flow. In contrast to the EEL diode where the active region length is the horizontal length of the active region between the two reflective ends, the active region length of the VCSEL diode is the

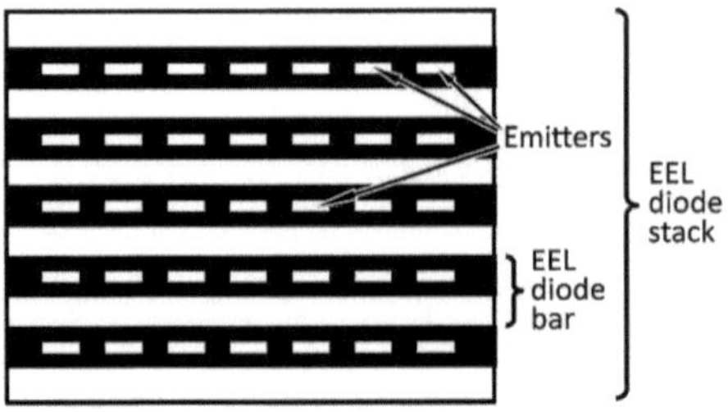

Fig. 5.4 An EEL diode stack

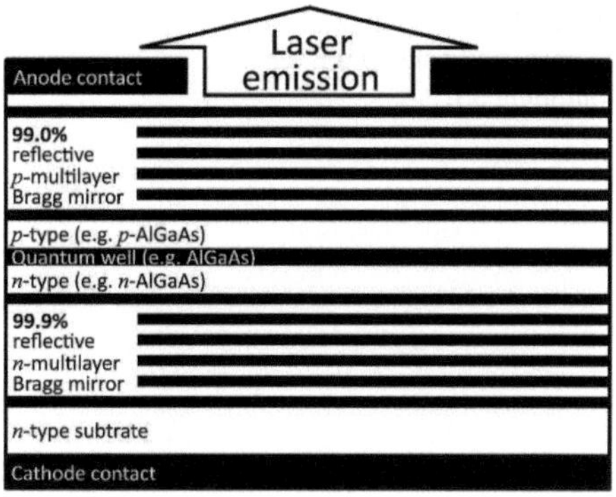

Fig. 5.5 A simplified structure of a VCSEL diode

thickness of the active region. The active region length of the VCSEL diode is very short compared to the EEL diode. A simplified structure of a VCSEL diode is shown in Fig. 5.5.

In the VCSEL diode, the thickness of the active region is made thin enough so that it becomes a quantum well. The reflectors are the multilayer Bragg mirror. A Bragg mirror is a periodically structured diffracting grating made from alternating high and low refractive index quarter-wave thick dielectric multilayer; it is frequency selective reflecting light with a particular wavelength. If the thicknesses of alternating layers are d_1 and d_2 with refractive indices n_1 and n_2, the reflected light will have constructive interference if its wavelength λ satisfies

$$\lambda = 2n_1d_1 + 2n_2d_2. \tag{5.7}$$

The bottom Bragg mirror is about 99.9% reflective. The top Bragg mirror is about 99.0% reflective. Laser emits through the aperture cut through the top anode contact.

VCSEL is usually made to output a circular beam. As the laser beam is emitted in a direction perpendicular to the substrate, it is easy to make a 2D laser array on a single chip with a large number of VCSEL diodes. The emitting aperture is typically 5–25 um in diameter, which is much bigger than EEL and so the divergence of the resulting beam is much lower. Compared to the EEL diode, the VCSEL has lower power limit.

5.3.3 Solid-State Laser (SSL)

A solid-state laser is a laser whose active region (or gain media or cavity) is solid state as opposed to gaseous state or liquid state. Although semiconductor is also solid state, a semiconductor laser is not referred to as a solid-state laser; it is referred to as semiconductor laser or diode laser. Figure 5.6 shows the structures of two SSLs. Diode lasers are used as pump light. There are two main SSL configurations—edge pumping and side pumping. In edge pumping, light enters

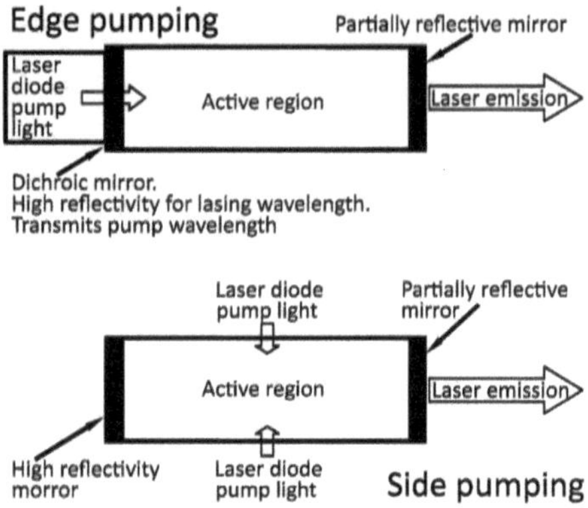

Fig. 5.6 Edge (longitudinal) pumping and side (transverse) pumping SSL

the active region where stimulated emission takes place through a dichroic mirror, which is highly reflective for the lasing wavelength but transparent for the pump wavelength. For side pumping, light enters the active region from the sides. For both configurations, laser emits from the active region through the partially reflective mirror. Diode laser-pumped SSL is also called DPSSL (Diode Pumped Solid State Laser.) The active region may be ruby (of historical importance only) or crystal or glass doped with rare earth.

Ruby is crystalline Al_2O_3 with a small fraction of the aluminum Al^{3+} ions replaced by chromium Cr^{3+} ions. The characteristic pink or red color of ruby is due to the Cr^{3+} ions. In a ruby laser, photons at the absorption wavelength of ruby (centered on 400 nm and 550 nm) are pumped into the active region. After being pumped into the active region, the photons are absorbed by the chromium ions raising the energy level of the electrons of the chromium ions from a ground energy level to a high energy level. The electrons at the high energy level rapidly undergo non-radiative transitions (a transition that does not result in the emission of light) to two metastable levels with long lifetime. In a non-radiative transition, the energy released in the transition is dissipated as heat. The metastable energy levels have a long lifetime (about 4 milliseconds). This causes chromium ions to build up in the metastable levels resulting in a population inversion—the condition required for stimulated emission. Some chromium ions will decay spontaneously to the ground-state level emitting 694.3 nm red light (with a line-width of 0.53 nm); the light will be reflected back and forth between the mirrors producing stimulated emission.

Ruby laser is a three-level system. The three levels are (1) the ground energy level, (2) the high level as a result of pumping, and (3) the metastable level. See Fig. 5.7

While ruby laser is a three-level system, neodymium-doped lasers are four-level system. Examples of neodymium-doped lasers are Nd:YAG (neodymium-doped yttrium aluminum garnet; Nd:$Y_3Al_5O_{12}$) and neodymium glass (Nd:Glass). The most common glass materials are the oxide glasses such as SiO_2 and P_2O_5. Most practical lasers are four-level system.

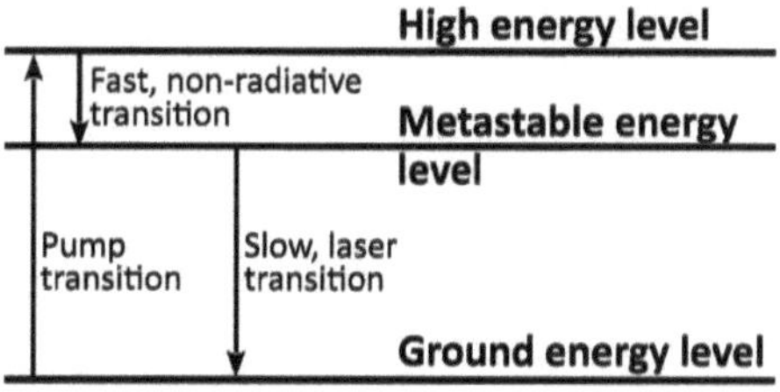

Fig. 5.7 The three energy levels of a ruby laser

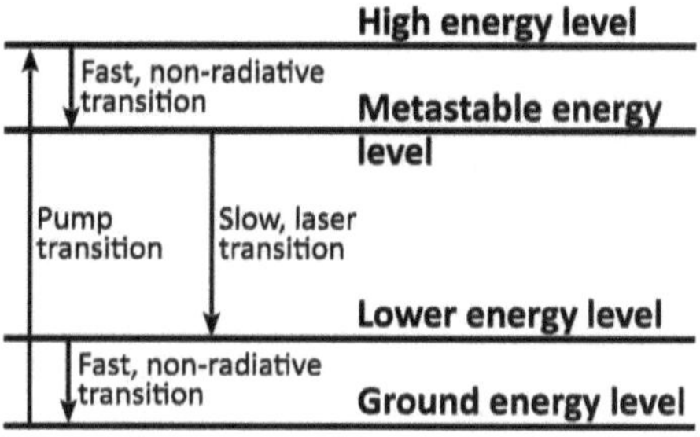

Fig. 5.8 Four-energy level system

In a four-level system, photons at the absorption wavelength of the active region are pumped into the active region raising the electrons in the active region from a ground energy level to a high energy level. Electrons at the high energy level rapidly undergo non-radiative transitions to a metastable level with long lifetime resulting in population inversion. Some electrons at the metastable energy level will decay spontaneously to a lower energy level emitting light. The light is reflected back and forth between the mirrors causing stimulated emission. Electrons at the lower energy level will rapidly undergo non-radiative transitions to the ground energy level. See Fig. 5.8.

In addition to chromium (Cr) and neodymium (Nd), many other transition metals and rare-earth elements have been used as dopant. Some of the examples are ytterbium (Yb), erbium (Er), praseodymium (Pr), thulium (Tm), holmium (Ho), titanium (Ti), vanadium (V), nickel (Ni), and cobalt (Co). Er:glass has attracted particular attention in eye-safe laser.

5.3.4 Fiber Laser (FL)

A fiber laser is essentially a solid-state laser with an optical fiber active region. A fiber laser consists of typically a few meters of cylindrical dielectric optical fiber cable with an active core doped with transitional metal or rare earth ions embedded in an outer cladding material of slightly lower refractive index. Silica glass is the most widely used core material for operation in the visible and near-infrared operation. Zirconium (Zr) fluoride materials are used for operations at longer wavelengths. The choice of the dopant is decided by the wavelength requirement. At both ends are mirrors. An appropriate laser diode pumps photons into the active

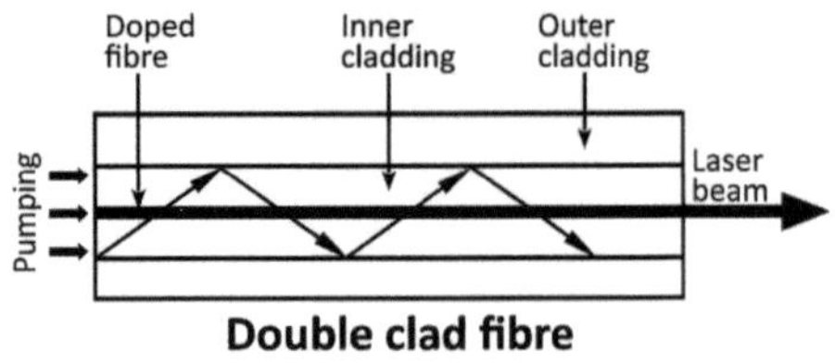

Fig. 5.9 Light travels in a double-clad fiber

region. A fiber laser can be end pumped or side pumped. Light rays incident at the core at angles lower than the critical angle undergo total internal reflection and are guided through the core. Light rays with any other inclination are not guided and lose their energy in the cladding. In the case of solid-state lasers, the photons move to and fro in a straight line between resonator mirrors, whereas in the case of a fiber-optic laser, the photons move in a zigzag path guided by total internal reflection phenomenon. See Fig. 5.9. Other aspects of the fiber-optic laser are the same as those of the solid-state laser.

5.3.5 *Dye Laser*

A dye laser is a laser based on a dye as the laser gain medium. It has a broad wavelength tunability and ultrashort pulse generation. Unfortunately, the dye degrades and must be replaced after a given period of operation. There are some successes in the development of rotating dye disk with dyes in plastic fiber; further improvement is needed before it can truly be conveniently used in autonomous vehicles.

Figure 5.10 shows the structure of a wavelength-tunable dye laser. The dye is usually liquid in the form of a thin dye jet, or pumped through a thin cuvette, or in plastic fiber. Many of the dyes and solvents are toxic and carcinogenic. The mirrors are highly reflective, while the output coupler is a partially transparent mirror. The birefringent tuner is an optical filter whose refractive index depends on the polarization direction.

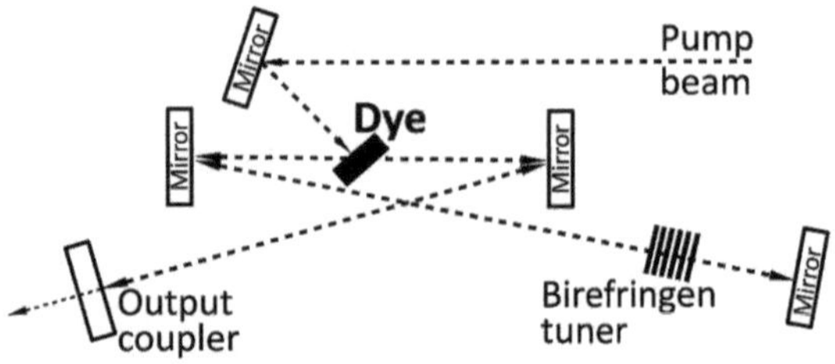

Fig. 5.10 A wavelength-tunable dye laser

5.4 Photodetector

A photodetector converts light energy impinges on it into electrical form. Silicon-based detectors are used for wavelength between 0.3 μm and 1.1 μm. InGaAs detectors are used for wavelength longer than 1.1 μm. There are many types of photodetector. The most popular detectors are (1) PIN photodiode, (2) avalanche photodiode (APD), (3) single-photon avalanche diode (SPAD), and silicon photomultiplier (SiPM).

5.4.1 PIN Photodiode

A reversed-biased PIN diode is a photodetector. Photons absorbed in the intrinsic region create electron-hole pairs. The reversed bias field sweeps the electron-hole pairs out of the region creating photocurrent. The photocurrent is proportional to the number of photons entering the intrinsic region of the diode but is relatively constant over a range of applied reversed biasing voltage. See Fig. 5.11 within the PIN region.

PIN diode has very large bandwidth but their sensitivity may not be sufficient for low light applications.

5.4.2 Avalanche Photodiode (APD)

The reversed bias of an avalanche photo detector is much higher than that of a PIN photo detector. Just as a PIN photo detector, photons absorbed in the intrinsic region creates electron-hole pairs. The reversed bias field sweeps the electron-hole pairs out of the region creating photo current. However, the high reversed bias voltage accelerates the carriers generating additional secondary electrons by impact ionization. This creates an avalanche resulting in a gain factor of photo current. The gain factor may be a few hundreds. Hence, the photo current and the gain factor increase with increasing reversed voltage. See Fig. 5.11 within the APD region.

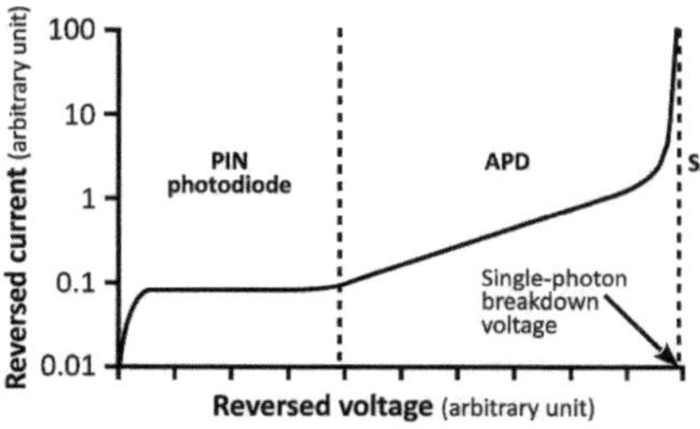

Fig. 5.11 Diode current vs reversed voltage plot for a given light intensity

Unless the reverse bias is too close to the breakdown voltage, in the avalanche mode, the APD photocurrent is proportional to the number of incident photons. An APD has similar bandwidth as a PIN photodiode but its sensitivity is much higher than that of a PIN photodiode.

5.4.3 Single-Photon Avalanche Diode (SPAD)

When the reversed voltage is slightly above the single-photon breakdown voltage, a single electron-hole pair in the intrinsic region will cause a self-sustained avalanche. The photocurrent is a fixed maximum that is independent on the number of photons; the photocurrent is not linearly proportional to the number of photons and so cannot be used to indicate the light intensity. See Fig. 5.11 at the SAPD region.

To detect the arrival of subsequent photons, the SAPD's avalanche process must first be stopped. This can be achieved by a quenching circuit, which reduces the reversed bias voltage to below a critical value for a short time to stop the avalanche. After the recovery time, the photodiode is ready to detect the next photon. Higher reverse voltage produces higher electric field inside the device leading to higher photon detection efficiency and, hence, higher gain. However, higher reversed voltage has a higher chance of false trigger due to thermal noise.

SPAD is so sensitive that it can detect the presence of a single photon. However, its bandwidth is low due to the quenching activity.

5.4.4 Multipixel Photon Counter (MMPC)

Multipixel photon counter is also called silicon photomultiplier (SiPM). It is essentially an array of SPAD microcells each with its own quenching circuit. Each microcell either conducts its own maximum photocurrent or does not conduct; there is no intermediate value. Depending on the design, the microcells may be of the same size or of different sizes. If they have different sizes, a larger cell will have a higher chance of receiving a photon. The output of the cells may or may not be weighted and then summed to form the final output.

MMPCs are used in scenarios of very dim light. If the light intensity is so high that all the microcells conduct its own maximum photocurrent, the MMPC saturates at its maximum output. If not all the SPADs conduct photocurrent, the output of the MMPC is an analogue value between its maximum and zero. The number of photon is assumed to be proportional to the MMPC output, but this may not be true since some microcells may receive more than one photon.

The bandwidth of MMPC is the same as that of SAPD. The advantage of MMPC is its very high gain; a gain of a million is achievable. It is particularly suitable for low light scenario.

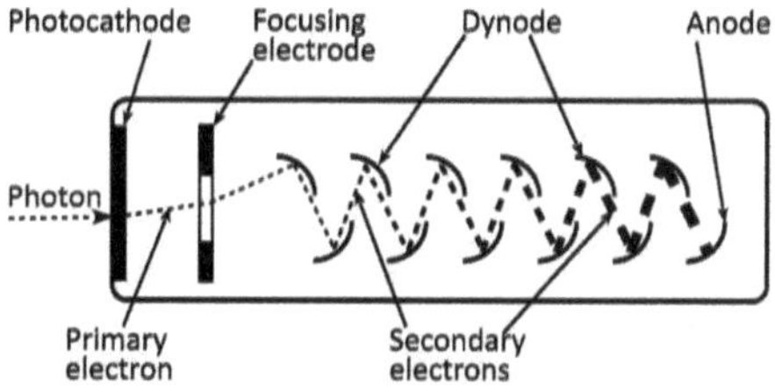

Fig. 5.12 Schematic of a photomultiplier tube

5.4.5 *Photomultiplier Tube (PMT)*

A photomultiplier tube is a vacuum phototube. A schematic of the photomultiplier tube is shown in Fig. 5.12. When a photon strikes the photocathode, an electron is released. This electron is referred to as the primary electron. The primary electron is accelerated by the positive potential of a series of electrodes called dynodes. The impact of the electrons on the dynodes generates more secondary electrons. This results in very high gain (such as 10^8). Single-photon detection with very wide bandwidth (>1 GHz) is possible. However, being a vacuum tube, it is bulky and fragile. Furthermore, it requires high voltage supply.

5.5 Laser Beam Steering

A lidar system must be able to perceive its surrounding, detecting and localizing the presence of pedestrians and other relevant stationary or moving objects with an acceptable resolution. A laser emitter emits a narrow beam of light shining onto a spot on an object. If the spot is reflective, it is possible to estimate the spot's distance and speed from the reflected light. It is necessary to move the light spot from one location to another to estimate the distance and speed of the new location. The moving of the laser beam from one direction to another is called scanning or beam steering. Many methods for beam steering exist. Some of them are currently being used or have great potential for being used in the future. The selection of the most appropriate scanning strategy is determined by many factors such as cost, eye-safety, robustness, and physical form factor.

5.5.1 *Mechanical Spinning System*

A mechanical spinning system consists of a rotating assembly of mirrors and prisms driven by an electric motor. A galvanometric nodding mirror as shown in Fig. 5.13 is usually used for vertical scan. The nodding mirror tilts the laser to generate a vertical scan. Horizontal scan may be achieved by spinning the entire system around a vertical axis. Another commonly used horizontal scan mechanism is a spinning

Fig. 5.13 Schematic of a mechanical spinning beam steering system

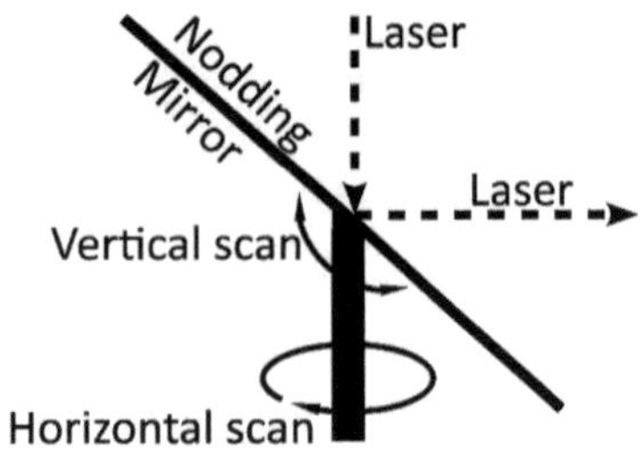

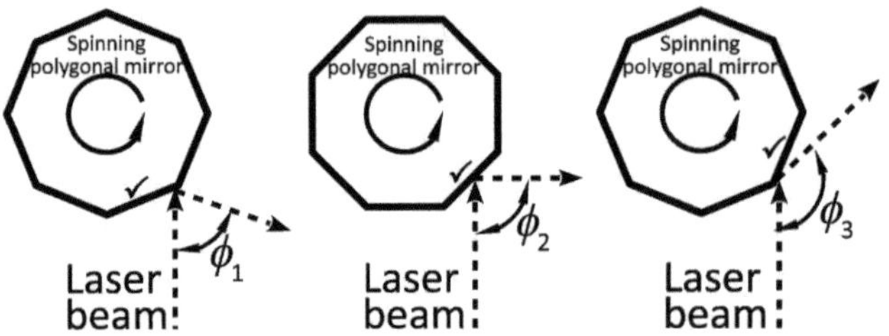

Fig. 5.14 Schematic of a spinning polygonal-mirror system

polygonal mirror system as shown in Fig. 5.14. To facilitate understanding the scanning operation of Fig. 5.14, a tick mark is placed next to the mirror that reflects the laser. When the reflecting surface enters the laser zone, the laser is reflected at an optical angle of ϕ_1. As the mirror rotates, the angle widens. Just before the reflecting surface exits the laser zone, the laser is reflected at an optical angle of ϕ_3. The two extreme optical angles of the mirror are ϕ_3 and ϕ_1. The maximum optical angular steering range is $\phi_3 - \phi_1$.

The optical angular steering range is different from the mechanical angular range the mirror rotates. The optical angular steering range is the angular change in direction of the reflected light. The mechanical angular range the mirror rotates is the physical angular rotation of the mirror. To determine the relationship between the two, note from Fig. 5.15 that

$$\phi + 2\theta = 180^\circ. \tag{5.8}$$

Differentiating (5.8) with respect to θ, we have

$$\frac{\mathrm{d}\phi}{\mathrm{d}\theta} = -2. \tag{5.9}$$

The negative sign in (5.9) indicates that ϕ increases as θ decreases. As the mirror rotates, θ changes and, hence, ϕ will also change. The mechanical angular range the mirror rotates is the change in θ denoted by $\Delta\theta$. It can be seen from (5.9) that the mirror's mechanical angular change of $\Delta\theta$ will produce a change in ϕ denoted by

Fig. 5.15 Relation between relevant angles of an incident ray

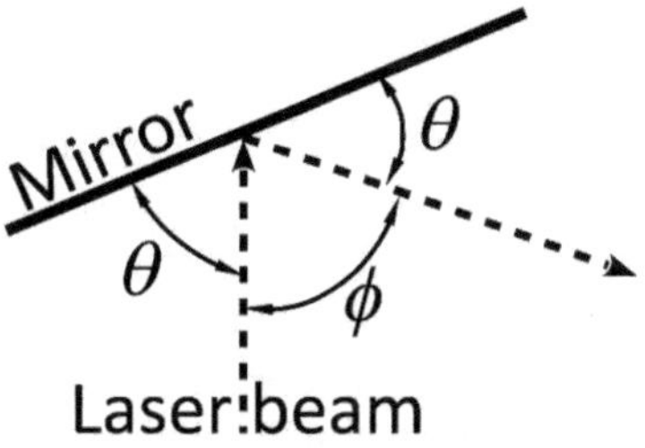

Fig. 5.16 Angle between neighboring sides of a hexagonal mirror and that for an octagonal mirror

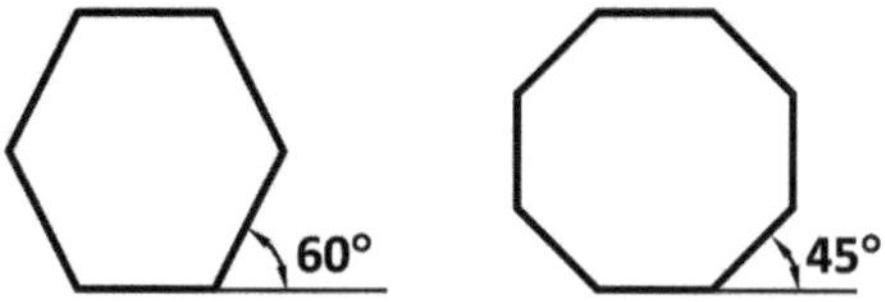

$\Delta\phi$ by $2\Delta\theta$ (ignoring the negative sign).

$$\Delta\phi = -2\Delta\theta. \tag{5.10}$$

Let $\Delta\phi_{\max}$ and $\Delta\theta_{\max}$ denote the maximum possible change in ϕ and the maximum possible change in θ, respectively. From Fig. 5.14, it can be seen that $\Delta\theta_{\max}$ is equal to the angle between two consecutive sides of the polygonal mirror. The angle between the two consecutive sides for an octagonal mirror is 45° and that for a hexagonal mirror is 60°. See Fig. 5.16. Hence, the maximum optical angular steering ranges for an octagonal mirror and a hexagonal mirror are 90° and 120°, respectively. In general, the angle between two neighboring sides of an N-sided mirror is $\frac{360°}{N}$. Thus, the maximum optical angular steering range of an N-sided mirror is

$$|\Delta\phi_{\max}| = |2\Delta\theta_{\max}| = \frac{720°}{N}. \tag{5.11}$$

Mechanical spinning systems are bulky, fragile, power-hungry, and misalignment-prone under shock and vibration. Owing to the large inertia of the rotating module, mechanical spinning system has low frame rate. Multiple beams are usually used to improve frame rate.

5.5.2 Microelectromechanical System (MEMS)

MEMS (**m**icro-**e**lectro-**m**echanical **s**ystem or **m**icro**e**lectro**m**echanical **s**ystem) is a miniature integrated device with both mechanical and electrical components. Its size may range from a few micrometers to a few millimeters. It is fabricated using integrated circuit fabrication techniques.

5.5.2.1 Effects of Dimension Scaling

The physical size of a MEMS micro-object may be several orders of magnitude smaller than its large-size cousin. The very large difference in sizes results in tremendous difference on many aspects on the physical properties and behaviors between the micro-object and large-size object. We shall examine some of the properties when the linear dimensions of an object are scaled by a factor s. Both its volume and mass will be scaled by the factor s^3, and both its surface area and cross-sectional area will be scaled by the factor s^2.

$$\text{Linear dimension} \propto s, \tag{5.12a}$$

$$\text{Area} \propto s^2, \tag{5.12b}$$

$$\text{Volume, mass, weight} \propto s^3. \tag{5.12c}$$

We shall see how the rotational kinetic energy E_R stored in a uniform disk with radius r and mass m spinning with angular velocity $\dot{\theta}$ about an axis perpendicular to the disk passing through the center of the disk is affected by scaling. The radius of gyration k of the disk is given by

$$k = \frac{r}{\sqrt{2}}. \tag{5.13}$$

The rotational kinetic energy E_R of an object is given by

$$E_\mathrm{R} = \frac{1}{2} m k^2 \dot{\theta}^2. \tag{5.14}$$

Substituting (5.13) into (5.14), the rotational kinetic energy of the disk is

$$E_\mathrm{R} = \frac{1}{4} m r^2 \dot{\theta}^2 \propto s^3 \times s^2 = s^5. \tag{5.15}$$

Since $m \propto s^3$ and $r^2 \propto s^2$, $E_\mathrm{R} \propto s^5$. As s scales down, the rotational kinetic energy E_R scales down with the fifth power of s and so it diminishes rapidly. Hence, a flywheel or a rotating disk is a good energy storage media in large size object but a bad one in MEMS.

The area of a solar panel mounted at the top of an object and, hence, its power generation capability, are proportional to s^2.

The electrostatic force F_e exerted on a charge q placed in an electric field E_o is

$$F_\mathrm{e} = E_\mathrm{o} q. \tag{5.16}$$

For a parallel plate capacitor with potential difference V between the two plates and plate separation d, the total electric field E_{T} between the two plates is

$$E_{\mathrm{T}} = \frac{V}{d}. \tag{5.17}$$

For a given voltage, the electric field $\propto s^{-1}$. As s becomes very small, the electric field may become very large. Care should be taken to ensure the insulation does not break down.

The total electric field E_{T} is contributed by both plates with each plate contributing $\frac{E_{\mathrm{T}}}{2} = \frac{V}{2d}$. The charge q on each plate of the capacitor is

$$q = CV, \tag{5.18}$$

where C is the capacitance. The force exerted on each plate is due to the electric field of the other plate; the electric field due to a charge does exert a force on itself. The electric field E_{o} that exerts a force F_{e} on a plate is $\frac{E_{\mathrm{T}}}{2} = \frac{V}{2d}$. The force is given by

$$F_{\mathrm{e}} = E_{\mathrm{o}}q = \frac{CV^2}{2d}. \tag{5.19}$$

Let the area of each of the parallel plate be A.

$$C = \frac{\varepsilon A}{d}, \tag{5.20}$$

where ε is the permittivity of the dielectric between the two plates. From (5.19) and (5.20),

$$F_{\mathrm{e}} = E_{\mathrm{o}}q = \frac{\varepsilon AV^2}{2d^2} \propto \frac{s^2}{s^2} = \text{constant}. \tag{5.21}$$

Under linear scaling of s, both A and d^2 are scaled by s^2. Hence, the electrostatic force is not scaled. This is very significant because other forces such as the weight of the various structures of the MEMS device are scaled by higher powers of s. For example, gravitational force mg where g is the gravitational acceleration is scaled by s^3. Electrostatic force, which is negligible for large-size objects, becomes significant in micro-devices.

The resistance R of a conductor is given by

$$R = \frac{\rho L}{A} \propto \frac{s}{s^2} = s^{-1}, \tag{5.22}$$

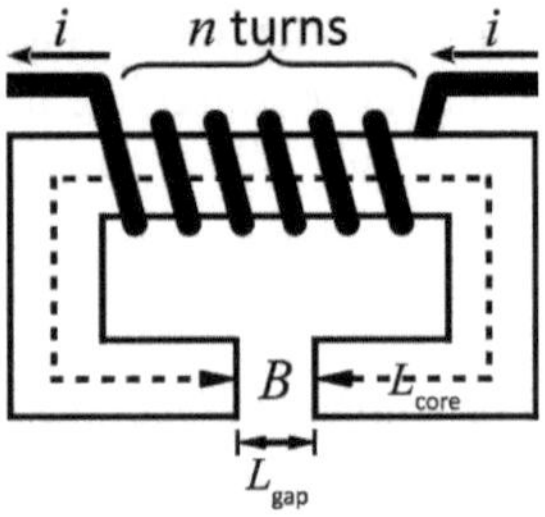

Fig. 5.17 An electromagnet

where ρ, L, and A are its resistivity, length, and cross-sectional area, respectively. Hence, the resistance scales with s^{-1} and for a given voltage, current scales with s.

$$i \propto s \qquad \text{for a given voltage.} \tag{5.23}$$

Figure 5.17 shows an electromagnet with n turns carrying a current i. The core has permeability μ_{core}, magnetic field strength H_{core}, and length L_{core}. The gap has permeability μ_{gap}, magnetic field strength H_{gap}, and length L_{gap}. The magnetic flux density in the gap is B. According to Ampère's circuital law,

$$\text{Magnetomotive force} = ni = H_{\text{core}}L_{\text{core}} + H_{\text{gap}}L_{\text{gap}} = B\left(\frac{L_{\text{core}}}{\mu_{\text{core}}} + \frac{L_{\text{gap}}}{\mu_{\text{gap}}}\right). \tag{5.24}$$

From (5.24),

$$B = \frac{ni}{\frac{L_{\text{core}}}{\mu_{\text{core}}} + \frac{L_{\text{gap}}}{\mu_{\text{gap}}}} \propto i \times s^{-1} = \text{constant for } i \propto s, \tag{5.25a}$$

$$\propto s^{-1} \text{ for } i = \text{constant.} \tag{5.25b}$$

If the structure providing B is an electromagnet within the MEMS structure, its physical dimension will be scaled with the MEMS and so its current $i \propto s$. In this case, B is independent of s, i.e., B remains unchanged. If the electromagnet is external to the MEMS device, i does not scale with s and so $B \propto s^{-1}$, i.e., B increases when the linear dimension is scaled down.

The thick dotted line of Fig. 5.18 shows a straight cantilever beam with negligible mass, length L, width W, and thickness t without the application of the force F. The

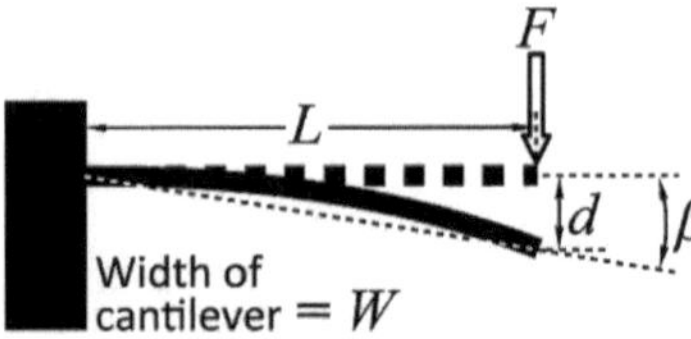

Fig. 5.18 A cantilever with negligible mass subject to a force, F

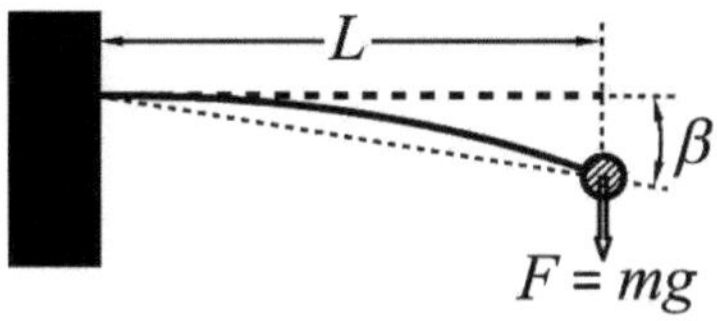

Fig. 5.19 A cantilever with negligible mass loaded with a load of mass m

Table 5.2 Young's modulus of some materials

Materials	Young's modulus/GPa	Materials	Young's modulus/GPa)
Magnesium	45	Steel	200
Aluminum	69	Tungsten	400
Bronze	100–125	Silicon carbide	450
Copper	117	Tungsten carbide	400–650
Silicon/SiO_2	140–180	Diamond	1220

applied force F bends the cantilever by a small deflection d as shown by the thick solid line. If the angle β is small, $\beta = \tan(\beta)$ and so β is given by

$$\beta = \frac{d}{L} = \frac{4FL^2}{\Xi W t^3} \propto F s^{-2}, \tag{5.26}$$

where Ξ is the elastic modulus (also known as Young's modulus or modulus of elasticity or tensile modulus) of the cantilever material. Suppose that the cantilever is loaded with a load of mass m, see Fig. 5.19. Let g be the gravitational acceleration. Substituting $F = mg$ into (5.26), and taking note of (5.12c), we have

$$\beta = \frac{d}{L} = \frac{4mgL^2}{\Xi W t^3} \propto s. \tag{5.27}$$

As can be seen from (5.27), for the same material, as the linear dimension scales down, β decreases with s; as a consequence, MEMS structures are "relatively very stiff." A list of the Young's modulus for some materials are presented in Table 5.2. The unit GPa in Table 5.2 is gigapascals; one pascal is one newton per square meter ($1\ \text{Pa} = 1\ \text{Nm}^{-2}$).

In MEMS, light is steered using mirrors. While a large-size mirror system is misalignment-prone under shock and vibration, MEMS structure is relatively very stiff, and misalignment under shock and vibration is negligible as can be seen from (5.27) since s is very small.

All devices have their own natural resonance frequencies. A device's natural resonance frequency should be far away from any vibration frequency of the application. For automotive application, the natural resonance frequency should be higher than 800 Hz for MEMS mirror.

The beam divergence angle of a laser beam is the derivative of the beam radius with respect to the axial position in the far field. See Fig. 5.20. For a diffraction-limited laser beam with Gaussian profile and moderately strong divergence, the half

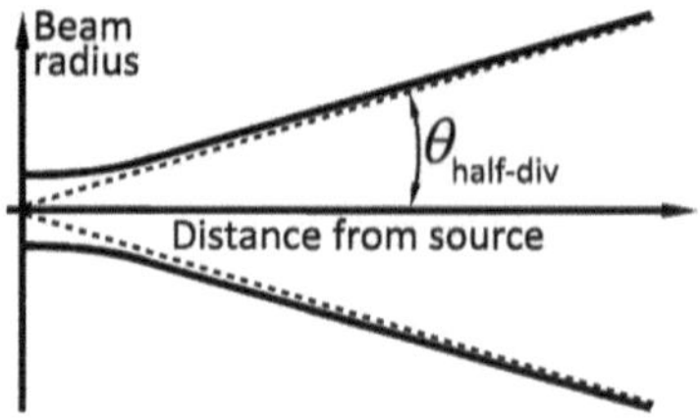

Fig. 5.20 Beam divergence angle

divergence angle $\theta_{\text{half-div}}$ of the laser beam is given in terms of its wavelength λ and beam radius r by

$$\theta_{\text{half-div}} = \frac{\lambda}{\pi r}. \tag{5.28}$$

The beam radius for a Gaussian profile beam is the radius where the beam intensity is $\frac{1}{e^2}$ its maximum intensity where $e \approx 2.718281828459$ is the base of the natural logarithm; it is 1.18 times that of the half-power radius. For $\theta_{\text{half-div}} = 0.5$ milliradian and $\lambda = 1550$ nm, $r \approx 1$ mm. The expression (5.28) is for perfect beam quality. For lower quality laser such as that emitted from an edge-emitting laser diode, $\theta_{\text{half-div}}$ is roughly 1.5 times larger. The half-beam radius of the laser is limited by the MEMS mirror dimension and so the MEMS mirror must be sufficiently large. For lower quality laser such as that emitted from an edge-emitting laser diode, the MEMS mirror dimension is usually about 3×3 mm^2.

A MEMS mirror is driven by an actuator to an appropriate angle to reflect light to a desired direction. There are several drive methods for driving MEMS mirror. The most commonly used methods are electrostatic, magnetic, piezoelectric, and electrothermal.

There are two modes of operation for MEMS mirror–linear mode and resonant mode (also called nonlinear mode). In linear mode, the mirror's optical deflection is accurately controlled by the driving voltage or current. The optical deflection may not be proportional to the drive voltage or current but the angle reproducibility relative to the drive voltage or current is very high. Linear mode is not suitable for high-speed operation. In linear mode, the mirror's operation frequency must be at least an order of magnitude lower than its resonant frequency.

In resonant mode operation, the MEMS mirror is driven by a periodic drive voltage or current at the resonance frequency of the MEMS structure. The resonance frequency of a MEMS structure depends on its structure, mass, and spring constant. At resonance, it is possible to obtain large optical angle at reduced drive voltage or current, but it is difficult to control its optical deflection with high accuracy. Whether the drive voltage or current is a sine wave or a square wave, the mirror swings with a sine wave in resonant mode.

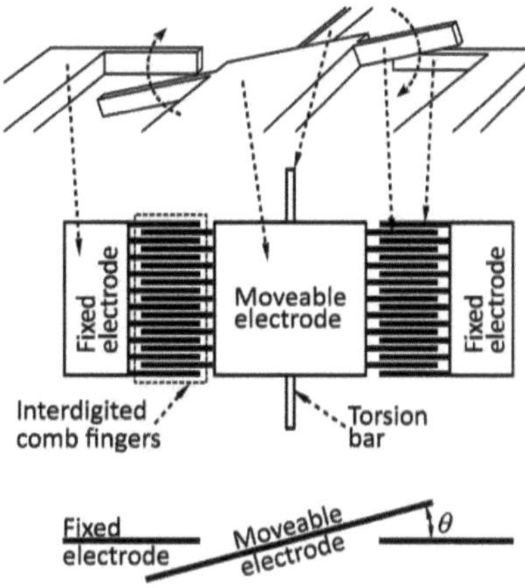

Fig. 5.21 An electrostatic actuator

5.5.2.2 Electrostatic Actuator

An electrostatic actuator consumes no static current. Nevertheless, during the instance when circuit capacitance changes or circuit voltage changes, current will flow momentarily to charge or discharge circuit capacitances. Thus, an electrostatic actuator consumes dynamic current. An electrostatic actuator requires high voltage (50–150 V). Its optical deflection angle is usually about $\pm$ 10°in nonlinear mode and about $\pm$ 5°in linear mode.

Figure 5.21 shows an electrostatic comb actuator. It consists of a moving electrode with a row of comb fingers on each side and two fixed electrodes with a row of comb fingers on one side of each electrode. Figure 5.21a shows a comb finger on each side of the electrodes. Suppose that when the angle between the moving electrode and the fixed electrode is θ, the capacitance formed between the moving electrode and the fixed electrodes is $C(\theta)$, and the torque exerted on the moving electrode by the torsion spring is $\Gamma(\theta)$. See Fig. 5.21c. Assume that the potential difference between the moving electrode and the fixed electrode is V and is constant. The charge and energy stored in $C(\theta)$ are $C(\theta)V$ and $0.5C(\theta)V^2$, respectively. Suppose that changing θ by an infinitesimal amount $\mathrm{d}\theta$ causes $C(\theta)$ to change by $\mathrm{d}C(\theta)$. The charge and energy stored in $C(\theta) + \mathrm{d}C(\theta)$ are $C(\theta)V + V\mathrm{d}C(\theta)$ and $0.5C(\theta)V^2 + 0.5V^2\mathrm{d}C(\theta)$, respectively. The charge stored in the capacitor is increased by $V\mathrm{d}C(\theta)$; this increased charge is supplied by the voltage source. The energy supplied by the voltage source associated with the increased charge is $voltage \times charge$ and is $V \times V\mathrm{d}C(\theta) = V^2\mathrm{d}C(\theta)$. The energy supplied by the torsion bar for this infinitesimal change in θ is $\Gamma(\theta)\mathrm{d}(\theta)$. The increase in the stored energy in the capacitor is derived from the voltage source and the torsion bar. Hence,

$$V^2\mathrm{d}C(\theta) + \Gamma(\theta)\mathrm{d}(\theta) = 0.5V^2\mathrm{d}C(\theta).$$

$$\therefore \quad \Gamma(\theta) = -0.5V^2\frac{\mathrm{d}C(\theta)}{\mathrm{d}(\theta)}. \tag{5.29}$$

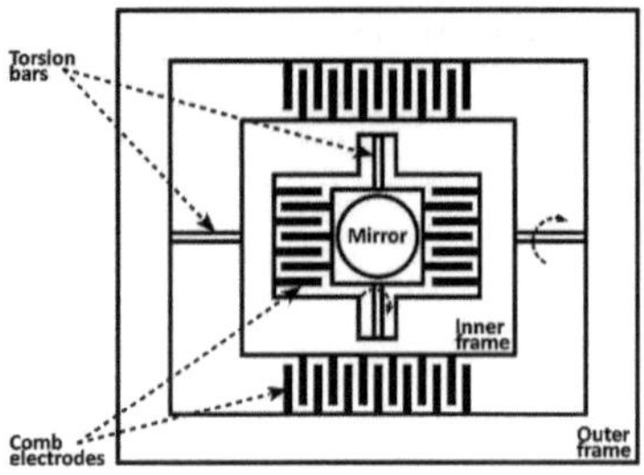

Fig. 5.22 Two-dimensional scanning using two comb actuators

The negative sign in (5.29) indicates that the torque is in the direction that opposes the increase in $C(\theta)$. If the yield point or elastic limit is not exceeded, $\Gamma(\theta)$ is proportional to θ, but $C(\theta)$ is not a linear function of θ.

Two comb actuators can be arranged as shown in Fig. 5.22 to provide two-dimensional scanning.

5.5.2.3 Magnetic Actuator

The magnetic actuator offers a wider optical deflection angle than an electrostatic actuator. Its optical deflection angle is typically about $\pm$ 25°in nonlinear mode and about $\pm$ 15°in linear mode. Its drive voltage is typically 5 V and its current consumption is about 20 mA.

Figure 5.23 shows the structure of a magnetic actuator. The magnet is external to the MEMS. Suppose that the coil has n turns and has an area $A = W \times L$. Suppose also that the coil carries a current i and the plane of the coil is at an angle of θ to the magnetic flux B. See Fig. 5.24. The torque of the coil $\Gamma(\theta)$ is given by

$$\Gamma(\theta) = BAni\cos(\theta). \tag{5.30}$$

The torque is maximum at $\theta = 0$. A two-dimensional scan system can be constructed from two sets of magnetic actuators or one set of magnetic actuator and one set of electrostatic comb actuator.

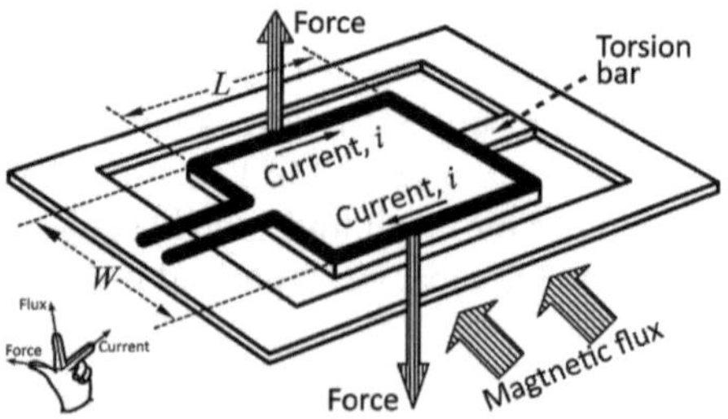

Fig. 5.23 A magnetic actuator

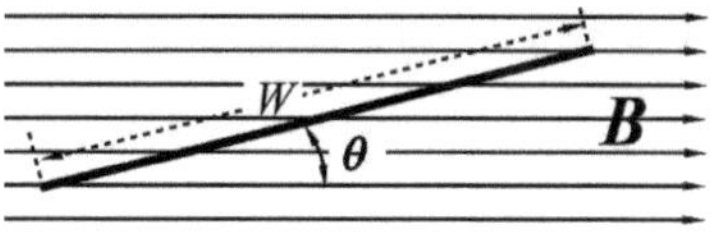

Fig. 5.24 The plane of the coil is at an angle of θ to the magnetic field

5.5.2.4 Piezoelectric Actuator

A piezoelectric material produces a physical displacement as a result of an applied electric field. The material consists of many tiny polar domains with similar dipole arrangement. If most of the polar domains are aligned, the material exhibits piezoelectric property. If the polar domains are randomly arranged, it does not exhibit piezoelectric behavior. If the polar domains are randomly arranged, it can be induced to exhibit macroscopic polarization by subjecting it to a very strong electric field (with field strength in the order of megavolt per meter). During the process of inducing the material to polarize, the material elongates in the direction of the poling field and contracts in the lateral direction. See Fig. 5.25. The shape deformation and polarization is permanent until it is depoled by a very large opposite field or by an elevated temperature. A material that can be induced to possess piezoelectric property is called ferroelectric. If a material is naturally electrically polarized, it is called pyroelectric.

If a voltage is applied to the polarized piezoelectric material in the same direction as the poling voltage, the material elongates further along the polar axis and contracts laterally. The material returns to its original poled dimension when the applied voltage is removed. See Fig. 5.26. If the applied voltage is opposite to the direction of the poling voltage, the material contracts along the polar axis and

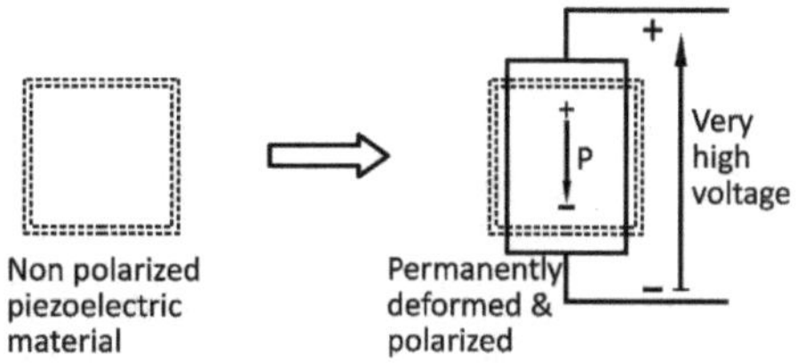

Fig. 5.25 Deformation is exaggerated in the drawing

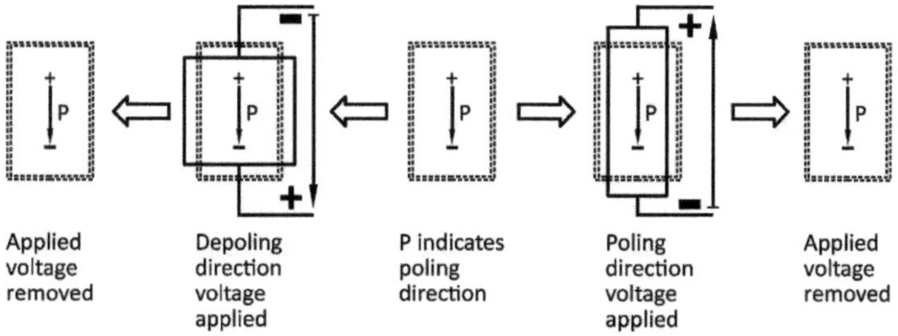

Fig. 5.26 Deformation of piezoelectric material subject to applied voltage

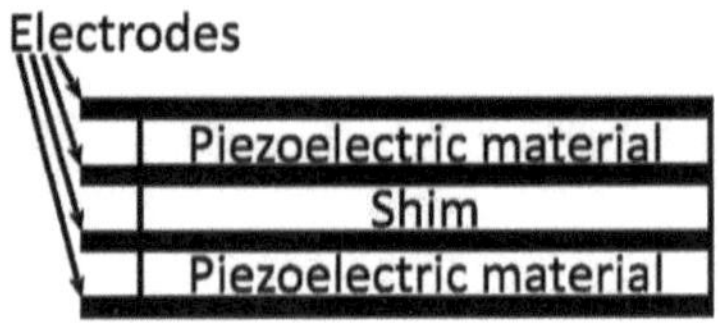

Fig. 5.27 Bending actuation

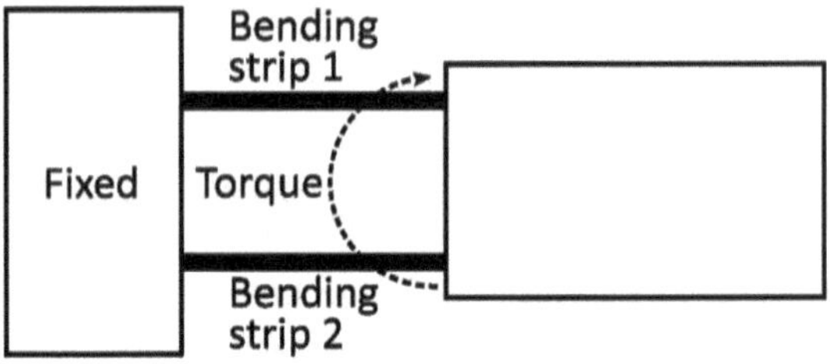

Fig. 5.28 Torsion actuation

expands laterally. The material returns to its original pole dimension when the applied voltage is removed. If the applied voltage opposite to the direction of the poling voltage is too large, it may either depol the material or reverse the direction of polarization.

Most of the piezoelectric actuators available on the market are made from the ceramic material lead-zirconate-titanate (PZT). PZT is both pyroelectric and ferroelectric. There are a large number of piezoelectric MEMS mirror actuator designs. Figure 5.27 shows a bimorph actuator design. In the design, very high voltages are first applied to the electrodes to pole the dielectric. Subsequently, during operation, suitable voltages are applied to the electrodes such that one of the piezoelectric expands laterally and the other contracts laterally; the strip bends. Torsion actuation can be achieved by applying the voltages such that the two bending strips bend in opposite directions. See Fig. 5.28.

5.5.2.5 Electrothermal Actuator

An electrothermal actuator uses the asymmetric thermal expansion of its parts to produce actuation. In general, it has a slow switching speed because of the large time constants of the thermal processes. Nevertheless, fast switching speed designs exist. There are many different designs of an electrothermal actuator. Figure 5.29 shows a bimorph design and a hot-arm-cold-arm design.

Both bimorph and hot-arm-cold-arm designs consist of two layers. In a bimorph design, the two layers are made from different materials. One layer is made from material with high CTE (coefficient of thermal expansion) and another layer is made from material with low CTE. Upon actuation, heat will generate more expansion in the high-CTE layer.

For the hot-arm-cold-arm design, the two layers are made from the same material. The hot arm has smaller cross-sectional area, so it heats up more and expands more

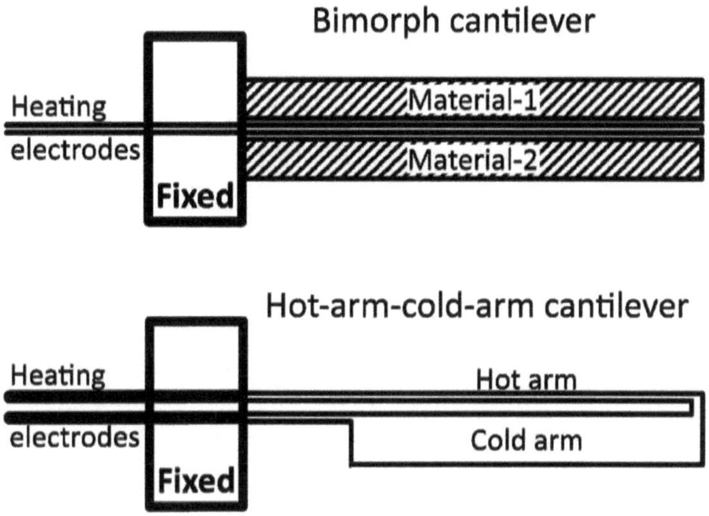

Fig. 5.29 Electrothermal actuators

when an electric current flows through the two layers. The unequal expansion in the two layers causes the cantilever to bend.

5.5.3 Optical Phased Array

Although MEMS is classified as "solid state," it is essentially still mechanical; it has mechanical moving parts. OPA (optical phased array) is a true solid-state device without mechanical moving parts and so in principle, its scanning speed is not limited by mechanical inertia, and it is free from mechanical wear and tear, robust, insensitive to acceleration and vibration, and highly compact. It is a developing technology with high potential and will not be surprising if it becomes the dominant approach for laser scanning. A very important aspect of OPA is its compatibility with modern digital signal processing techniques; this allows the application of the entire package of already well-developed knowledge of digital phased array signal processing techniques. In this section, we shall present a brief description of OPA's fundamental operating principle and the components essential to its implementation and leave the digital signal processing parts to Chaps. 7–9.

Consider a system where a laser beam is split into two equal beams and emitted through two identical apertures as shown in Fig. 5.30. Suppose that position P is equidistant from the two laser-emitting apertures. The light received at P from both apertures are the same. Suppose that the light received at P from each aperture is $a_p s(t - \theta)$. The value a_p represents the attenuation from the light-emitting aperture

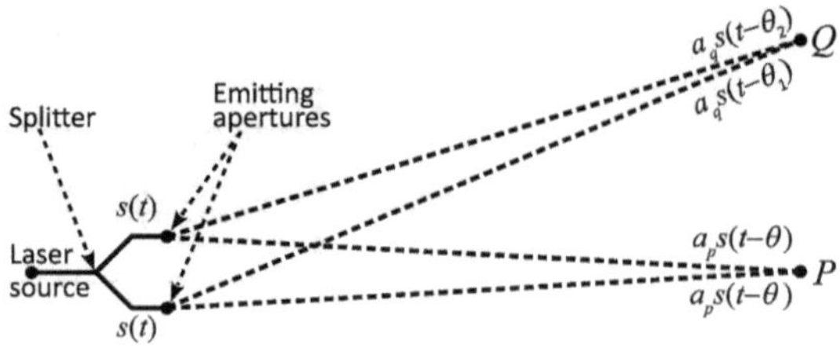

Fig. 5.30 Directional selectivity of a two-element OPA

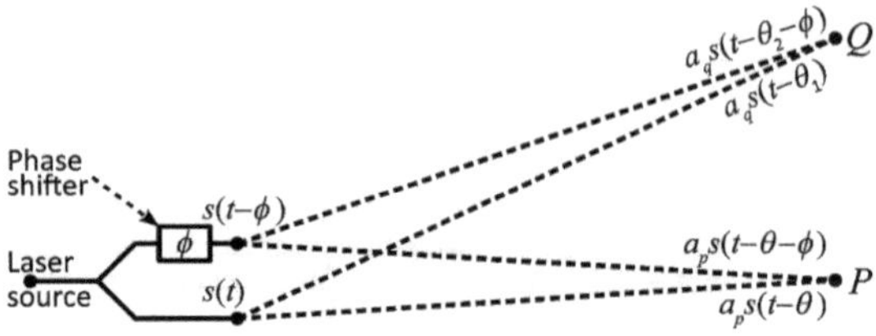

Fig. 5.31 Steering a two-element OPA

to P. Hence, the total light received at P is $2a_p s(t-\theta)$; its magnitude $= 2a_p$ times that of an emitting aperture.

Consider another location Q. Location Q has different distances from the two light-emitting apertures but the difference is small and so the amplitude attenuation is the same. Let a_q represents the attenuation and so the total light received at Q is $a_q s(t-\theta_2) + a_q s(t-\theta_1)$; its magnitude $< 2a_q$ times that of an emitting aperture because $a_q s(t-\theta_2)$ and $a_q s(t-\theta_1)$ are not in phase.

If the distance between P and the emitting apertures and that between Q and the emitting apertures are roughly the same, $a_p \approx a_q$. Hence, the light magnitude received at P is larger than that at Q; the transmission is directionally selective. If there are many emitting apertures, it is possible to make the light received at Q to be arbitrarily small compared to that at P; this means that the light can be beamed towards a given direction.

The light can also be steered to beam towards any desired direction. Suppose that a phase shifter is added into one of the light paths as shown in Fig. 5.31. If $\phi = \theta_1 - \theta_2$, the light magnitude received at Q will be $2a_q$ times that of an emitting aperture and the light received at P will be $a_p s(t-\theta-\theta_1+\theta_2) + a_p s(t-\theta)$; the magnitude is $< 2a_q$ times that of an emitting aperture, if $a_p \approx a_q$. The beam has been steered from P to Q!

A typical configuration of an OPA is shown in Fig. 5.32. Commonly used beam splitters are (1) multimode interference (MMI) beam splitters, (2) star couplers, and (3) directional couplers. The MMI splitters distribute input power into each output channel equally. The star coupler is smaller but may produce phase misalignments due to path-length mismatch; the phase misalignments must be subsequently compensated. The directional coupler transfers power between adjacent parallel optical waveguides placed closed together; the amount of power transferred depends on the coupling length.

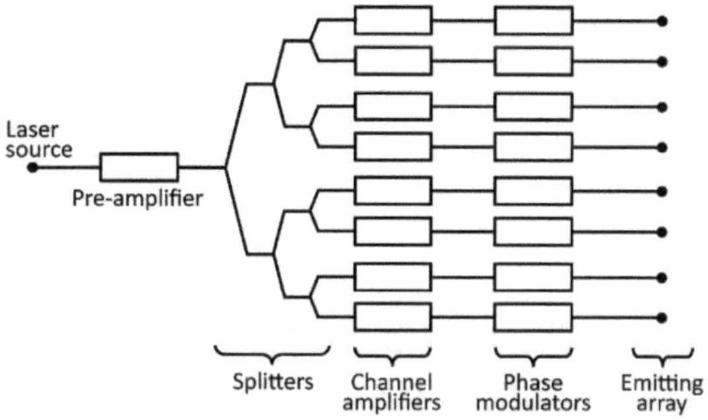

Fig. 5.32 A typical configuration of an OPA

There are two commonly used emitting array. One is the waveguide grating antenna, which is a multi-slit diffraction grating. The other one is the end-fire array with power radiated from the facets of waveguides.

The most commonly used method for phase modulation is by varying the refractive index n of a section of a waveguide. If the light waveform with frequency f at the entrance of a section of the waveguide with length d and reflective index n is $\sin(2\pi f t + \alpha)$, at the exit of that section, the waveform will be $\sin(2\pi f t + \alpha - \theta)$ where θ is given by

$$\theta = \frac{2\pi f d n}{c}, \tag{5.31}$$

where c is the speed of light in a vacuum. The refractive index is changed by either the thermo-optic (TO) approach of changing the temperature of the section or the electro-optic (EO) approach of carrier injection into a PIN (see Sect. 5.4.1) structure exhibiting Pockels effect. The EO approach is faster and more energy efficient than the TO approach.

5.6 Tunable Laser

A tunable laser is a laser whose wavelength can be adjusted (i.e., tuned) during operation. The wavelength of some tunable lasers can be tuned over hundreds of nanometer, whereas some can only be tuned over a fraction of a nano-meter.

Most laser diodes can be tuned over a few nanometers by varying the junction temperature or drive current. Some external-cavity diode lasers and distributed Bragg reflector lasers can be tuned over a few tens of nanometer; the tuning range can be increased by adding intracavity diffraction grating. Figure 5.33 shows the structures of two external cavity lasers. In Fig. 5.33a, the emission wavelength is tuned by rotating the diffraction grating; the disadvantage is that the direction of emission rotates as the diffraction grating rotates. In Fig. 5.33b, the emission

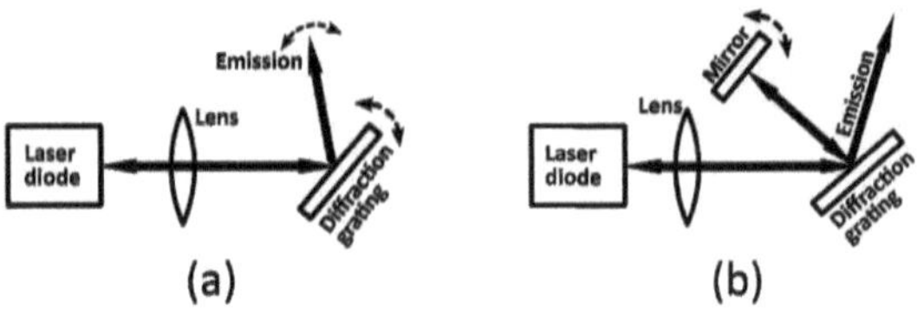

Fig. 5.33 Two structures of an external-cavity diode laser

wavelength is tuned by rotating the mirror; the direction of emission is not affected by tuning.

Dye lasers have very broad tunability. Some solid-state lasers such as the titanium-sapphire and chromium-doped Cr:ZnSe (zinc selenide) and Cr:ZnS (zinc sulphide) lasers can be tuned over hundreds of nanometer in the near- and mid-infrared spectrum.

Exotic laser type such as the free electron laser fed with high energy electrons from an electron accelerator can be tuned over enormous wavelength ranges.

5.7 Light-Emitting Diode (LED)

Figure 5.34 shows the structure of a semiconductor LED. LED is not a laser diode and operates at lower drive current; see Fig. 5.35. LED light comes from spontaneous emission without stimulated emission and is not coherent. For short-range applications, LED may be used as a light source instead of a laser. The advantage of an LED is its lower cost. Table 5.3 shows a list of commonly used LED materials. The emission color (i.e., the center wavelength) of an LED is determined

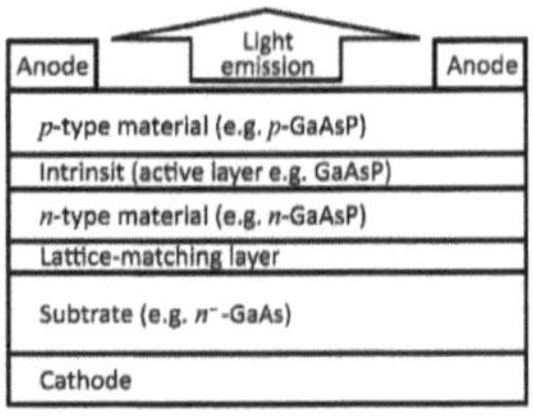

Fig. 5.34 Structure of an LED

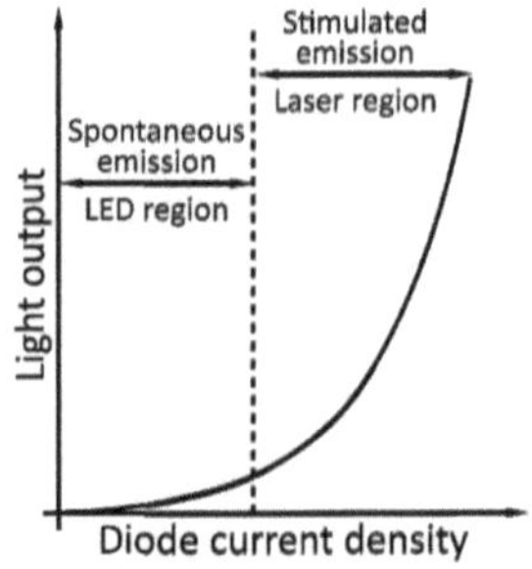

Fig. 5.35 LED region and laser region of a diode

Table 5.3 Some common LED materials

Materials	Wavelength	Voltage		Color
		min	Max	
Diamond, BN, AlN	< 400 nm	3.1 V	4.4 V	Ultraviolet
InGaN	400–450 nm	2.8 V	4.0 V	Violet
ZnSe, InGaN, SiC, Si	450–500 nm	2.5 V	3.7 V	Blue
GaInP, AlGaInP, AlGaP, InGaN	500–570 nm	1.9 V	4.0 V	Green
GaAsP, AlGaInP, GaP	570–590 nm	2.1 V	2.2 V	Yellow
GaAsP, AlGaInP, GaP	590–610 nm	2.0 V	2.1 V	Orange
AlGaAs, GaAsP, AlGaInP, GaP	610–760 nm	1.6 V	2.0 V	Red
GaAs, AlGaAs	>760 nm		1.9 V	Infrared

by the bandgap energy of the semiconductor material used. The achievable internal quantum efficiency depends on the emission wavelength. The internal quantum efficiency for GaAsP at 650 nm (red) is close to 100% but is lower at 620 nm (red) where the human eye is more sensitive. The internal quantum efficiency for GaAsP decreases further in the green and yellow regions. As temperature increases, the emission wavelength increases and the quantum efficiency decreases.

Internal quantum efficiencies above 90% are achievable at the blue and ultraviolet regions using InGaN. Green and yellow can be obtained by increasing the indium content but the efficiency drops sharply as the wavelength is increased.

White light can be obtained either by appropriate mixing of red, green, and blue or by using a blue LED and phosphor. Phosphor converts part of the blue light into light with longer wavelength in the appropriate composition to form white light.

OLED (organic LED) is constructed using methods completely different from semiconductor LED. OLED is constructed from thin elastic layers of organic materials and electrodes.

References

M.J. Adams, J.V. Collins, I.D. Henning, Analysis of semiconductor laser optical amplifiers. IEE Proc. J. (Optoelectronics) **132**(1), 58–63 (1985)

G.P. Agrawal, N.K. Dutta, *Semiconductor Laser* (Springer, Berlin, 2013)

E.A. Avrutin, B.S. Ryvkin, Effect of spatial hole burning on output characteristics of high power edge emitting semiconductor lasers: A universal analytical estimate and numerical analysis. J. Applied Phys. **125**(2), 023108 (2019)

B. Behroozpour, P.A.M. Sandborn, M.C. Wu, and B.E. Boser, Lidar system architectures and circuits. IEEE Comm. Mag. **55**(10), 135–142 (2017)

N. Yu, R. Blanchard, J. Fan, F. Capasso, T. Edamura, M. Yamanishi, H. Kan, Small divergence edge-emitting semiconductor lasers with two-dimensional plasmonic collimators. Applied Phys. Lett. **93**(18), 181101 (2008)

G. Boulon, Fifty years of advances in solid-state laser materials. Opt. Mater. **34**(3), 499–512 (2012)

J.E. Bowers, High speed semiconductor laser design and performance. Solid-State Electron. **30**(1), 1–11 (1987)

P. Brandl, S. Schidl, A. Polzer, W. Gaberl, H. Zimmermann, Optical wireless communication with adaptive focus and MEMS-based beam steering. IEEE Photon. Tech. Lett. **25**(15), 1428–1431 (2013)

J.C. Campbell, Recent advances in avalanche photodiodes. J. Lightwave Tech. **34**(2), 278–285 (2016)

J. Canning, Fibre lasers and related technologies. Optics Lasers Eng. **44**(7), 647–676 (2006)

W.W. Chow, S.W. Koch, M. Sargent, *Semiconductor-Laser Physics* (Springer, Berlin, 2012)

H. Cong, C. Xue, J. Zhang, F. Yang, K. Yu, Z. Liu, X. Zhang, B. Cheng, Q. Wang, Silicon based GeSn pin photodetector for SWIR detection. IEEE Photon. J. **8**(5), 1–6 (2016)

G. Cosendey, A. Castiglia, G. Rossbach, J.F. Carlin, N. Grandjean, Blue monolithic AlInN-based vertical cavity surface emitting laser diode on free-standing GaN substrate. Applied Phys. Lett. **101**(15), 151113 (2012)

A. Costela, I. García-Moreno, R. Sastre, Polymeric solid-state dye lasers: recent developments. Phys. Chem. Chem. Phys. **5**(21), 4745–4763 (2003)

N. Csanyi, C.K. Toth, Improvement of lidar data accuracy using lidar-specific ground targets. Photogrammetric Engin. Remote Sens. **73**(4), 385–396 (2007)

R. Dupuis, An introduction to the development of the semiconductor laser. IEEE J. Quantum Electron. **23**(6), 651–657 (1987)

H.H. Fang, J. Yang, J. Feng, T. Yamao, S. Hotta, H.B. Sun, Functional organic single crystals for solid-state laser applications. Laser Photonics Rev. **8**(5), 687–715 (2014)

M.E. Fermann, I. Hartl, Ultrafast fibre lasers. Nat. Photonics **7**(11), 868–874 (2013)

A. Filgueira, H. González-Jorge, S. Lagüela, L. Díaz-Vilariño, P. Arias, Quantifying the influence of rain in lidar performance. Measurement **95**, 143–148 (2017)

A. Godard, Infrared (2-12um) solid-state laser sources: a review. C. R. Phys. **8**(10), 1100–1128 (2007)

R.D. Gozzard, L.E. Roberts, J.T. Spollard, P.G. Sibley, D.A. Shaddock, Fast beam steering with an optical phased array. Optics Lett. **45**(13), 3793–376 (2020)

C. Jauregui, J. Limpert, A. Tünnermann, High-power fibre lasers. Nat. Photonics **7**(11), 861–867 (2013)

X. Jiang, M.A. Itzler, R. Ben-Michael, K. Slomkowski, InGaAsP–InP avalanche photodiodes for single photon detection. IEEE J. Select. Topics Quantum Electron. **13**(4), 895–905 (2007)

Y. Lan, G. Yang, Y. Liu, Y. Zhao, Z. Wang, T. Li, A. Demir, 808 nm broad-area laser diodes designed for high efficiency at high-temperature operation. Semicon. Sci. and Tech. **36**(10), 105012 (2021)

Y. Li, J. Ibanez-Guzman, Lidar for autonomous driving: the principles, challenges, and trends for automotive lidar and perception systems. IEEE Sig. Proc. Mag **37**(4), 50–61 (2020)

Lidar base specification. https://pubs.er.usgs.gov/publication/tm11B4

M. Liu, C. Xu, X. Bai, X. Guo, J.C. Campbell, Z. Pan, M.M. Tashima, High-performance InGaAs/InP single-photon avalanche photodiode. IEEE J. Select. Topics Quantum Electron. **13**(4), 887–894 (2007)

Y. Liu, G. Yang, Z. Wang, T. Li, S. Tang, Y. Zhao, Y. Lan, A. Demir, High-power operation and lateral divergence angle reduction of broad-area laser diodes at 976 nm. Optics Laser Tech. **141**(10), 107145 (2021)

M.V. Maximov, Y.M. Shernyakov, I.I. Novikov, Y.L. Karachinsky, N.Y. Gordeev, U. Ben-Ami, D. Bortman-Arbiv, A. Sharon, V.A. Shchukin, N.N. Ledentsov, T. Kettler, High-power low-beam divergence edge-emitting semiconductor lasers with 1-and 2-D photonic bandgap crystal waveguide. IEEE J. Select. Top. Quantum Electron. **14**(4), 1113–1122 (2008)

P. McManamon, T. Dorschner, D. Corkum, L. Friedman, D. Hobbs, M. Holz, S. Liberman, H. Nguyen, D. Resler, R.C. Sharp, E.A. Watson, Optical phased array technology. Proc. IEEE **84**(2), 268–298 (1996)

P. McManamon, P. Bos, M. Escuti, J. Heikenfeld, S. Serati, H. Xie, E.A. Watson, A review of phased array steering for narrow-band electrooptical systems. Proc. IEEE **97**(6), 1078–1096 (2009)

T. Nagano, K. Yamamoto, K. Sato, N. Hosokawa, A. Ishida, T. Baba, Development of multi-pixel photon counter (MPPC), in *2011 IEEE Nuclear Science Symposium Conference Record* (2011), pp. 1657–1659

M. Oehme, J. Werner, E. Kasper, M. Jutzi, M. Berroth, High bandwidth Ge p-i-n photodetector integrated on Si. App. Phys. Lett. **89**(7), 071117 (2006)

A.A. Podoskin, V.S. Golovin, P.S. Gavrina, D.A. Veselov, V.V. Zolotarev, V.V. Shamakhov, D.N. Nikolaev, I.S. Shashkin, S.O. Slipchenko, N.A. Pikhtin, P.S. Kopèv, Properties of external-cavity high-power semiconductor lasers based on a single InGaAs quantum well at high pulsed current pump. JOSA B **37**(3), 784–788 (2020)

C.V. Poulton, M.J. Byrd, E. Timurdogan, P. Russo, D. Vermeulen, M.R. Watts, Optical phased arrays for integrated beam steering, in *Proceedings of the 2018 IEEE 15th International Conference on Group IV Photonics (GFP)* (2018), pp. 1–2

R. Roriz, J. Cabral, T. Gomes, Automotive lidar technology: a survey. IEEE Trans. Intel. Transport. Syst. **23**(7), 6282–6297 (2022)

B.S. Ryvkin, E.A. Avrutin, J.T. Kostamovaara, Asymmetric-waveguide, short cavity designs with a bulk active layer for high pulsed power eye-safe spectral range laser diodes. Semicon. Sci. Tech. **35**(8), 085008 (2020)

R. Saldanha, L. Grandi, Y. Guardincerri, T. Wester, Model independent approach to the single photoelectron calibration of photomultiplier tubes. Nucl. Instr. Meth. Phys. Research Sec. A: Acce., Spectro., Detect. Ass. Equip. **863**, 35–46 (2017)

I.D.W. Samuel, G.A. Turnbull, Organic semiconductor lasers. Chem. Rev. **107**(4), 1272–1295 (2007)

R. Shau, M. Ortsiefer, J. Rosskopf, G. Böhm, F. Köhler, M.C. Amann, Vertical-cavity surface-emitting laser diodes at 1.55 μm with large output power and high operation temperature. Electron. Lett. **37**(21), 1295–1296 (2001)

S. Su, B. Cheng, C. Xue, W. Wang, Q. Cao, H. Xue, G. Zhang, Y. Zuo, Q. Wang, GeSn pin photodetector for all telecommunication bands detection. App. Phys. Lett. **19**(7), 6400–6405 (2011)

A. Tuantranont, V.M. Bright, J. Zhang, W. Zhang, J.A. Neff, Y.C. Lee, Optical beam steering using MEMS-controllable microlens array. Micromachines **91**(3), 363–372 (2001)

U. Wandinger, *Introduction to Lidar* (Springer, Berlin, 2005)

D. Wang, C. Watkins, H. Xie, MEMS mirrors for LiDAR: a review. Micromachines **11**(5), 456 (2020)

M.E. Warren, Automotive lidar technology, in *Proceedings of the 2019 Symposium on VLSI Circuits* (2019), C254–C255

J.O. Wesstrom, G. Sarlet, S. Hammerfeldt, L. Lundqvist, P. Szabo, P.J. Rigole, State-of-the-art performance of widely tunable modulated grating Y-branch lasers, in *Proceedings of the 2004 Optical Fiber Communication Conference*, vol. 1 (2004), p. 389

M.E. Woodson, M. Ren, S.J. Maddox, Y. Chen, S.R. Bank, C. Campbell, Low-noise AlInAsSb avalanche photodiode. App. Phys. Lett. **108**(8), 081102 (2016)

F.K. Yam, Z. Hassan, Innovative advances in LED technology. Microelectron. J. **36**(2), 129–137 (2005)

K. Yamamoto, K. Yamamura, K. Sato, S. Kamakura, T. Ota, H. Suzuki, S. Ohsuka, Development of multi-pixel photon counter (MPPC), in *2007 IEEE Nuclear Science Symposium Conference Record* (2007), pp. 895–905

G. Zambra, M. Bondani, A.S. Spinelli, F. Paleari, A. Andreoni, Counting photoelectrons in the response of a photomultiplier tube to single picosecond light pulses. Rev. Sci. Instruments **75**(8), 2762–2765 (2004)

B.V. Zhdanov, T. Ehrenreich, R.J. Knize, Narrowband external cavity laser diode array. Electron. Lett. **29**(4), 361–363 (2004)

Q. Zhou, P. Ben-Tzvi, D. Fan, A.A. Goldenberg, Design of fast steering mirror systems for precision laser beams steering, in *Proceedings of the 2008 International Workshop on Robotic and Sensors Environments* (2008), pp. 144–149

Chapter 6
Introduction to Radar

6.1 Introduction

While lidar uses laser to detect and determine the range and speed of an object, radar (**ra**dio **d**etection **a**nd **r**anging) uses radio wave to achieve the same function (Balanis 1992, 2011; Milligan 2005; Stutzman and Thiele 2012; Khan et al. 2015; Waterhouse 1995; James et al. 1986; Herd and Conway 2015; Hajimiri et al. 2005; Parker and Zimmermann 2002a,b; Gloeckler 1990; Mailloux 1982, 2017; Visser 2006; Hansen 2009). Both laser and radio wave are electromagnetic (EM) waves but they belong to different wavelength domains. Lasers operates at the infrared, visible light, and ultraviolet domains of the EM spectrum, whereas wavelengths of radar waves are longer than 0.1mm.

A radar system consists of (1) a transmitter producing the desired electrical signals in the radio frequency domain, (2) a transmitting antenna converting the electrical signals from the transmitter into radio waves and radiating the radio waves into free space, (3) a receiving antenna receiving radio waves from free space and converting them into electrical signals, (4) a receiver that selects the desired electrical signals received by the antenna, and (5) a processor to determine the properties of the objects.

Radar technology is a mature technology. It was first developed for military use during World War II. Today, it is being widely used in many application areas including military, spacecraft, aircraft, ships, motor vehicles, traffic-law enforcement, weather prediction, terrain mapping, etc.

In this chapter, we shall described some properties of EM wave such as (1) radio frequency bands, (2) atmospheric absorption, and (3) EM wave polarization as well as radar hardware components including (1) antenna, (2) RF switch, (3) phase shifter, and (4) attenuator. These form part of the background knowledge for the algorithms presented in Chaps. 7–9.

Y. C. Lim et al., *Signal Processing for Autonomous Systems*,
https://doi.org/10.1007/978-3-032-00675-2_6

6.2 Radio Spectrum

Radio spectrum is the part of EM spectrum used in modern technology such as radio communication. Radio waves are used in radar systems to determine the direction and distance of a target from an observer. The target may be within the line-of-sight or may be over-the-horizon (OTH). For line-of-sight operation, the lower bound of the frequency used is limited by the sizes of the antenna and the object; the upper bound is limited by atmospheric absorption of radio wave.

Radio wave spectrum is divided into bands by various incompatible naming systems. The International Telecommunication Union (ITU) radio bands are listed in Table 6.1, the IEEE radar bands are listed in Table 6.2, and radar-frequency bands as defined by EU, NATO, and USA for ECM systems are listed in Table 6.3.

The use of low frequencies for radio communication is limited by the size of the antenna. Very few systems operate at below 30 kHz because the antenna will be too large and the information rate is too slow. Nevertheless, it is used for communication with submerged submarines. The VLF (3–30 KHz) band is used for near-surface communication. The ELF and SLF (3–300 Hz) bands are used for operating depth

Table 6.1 ITU bands

ITU band number	Abbreviation	Band name	Frequency	Wavelength
0	TLF	Tremendously low frequency	<3 Hz	>100,000 km
1	ELF	Extremely low frequency	3–30 Hz	100,000 – 10,000 km
2	SLF	Super low frequency	30–300 Hz	10,000 – 1000 km
3	ULF	Ultralow frequency	0.3–3 kHz	1,000 – 100 km
4	VLF	Very low frequency	3–30 kHz	100–10 km
5	LF	Low frequency	30–300 kHz	10–1 km
6	MF	Medium frequency	0.3–3 MHz	1–0.1 km
7	HF	High frequency	3–30 MHz	100–10 m
8	VHF	Very high frequency	30–300 MHz	10–1 m
9	UHF	Ultrahigh frequency	0.3–3 GHz	1–0.1 m
10	SHF	Super high frequency	3–30 GHz	100–10 mm
11	EHF	Extremely high frequency	30–300 GHz	10–1 mm
12	THF (THz)	Tremendously high frequency (Terahertz)	0.3–3 THz	1–0.1 mm

Table 6.2 IEEE radar bands

Abbreviation	Band name	Frequency
HF	High frequency	3–30 MHz
VHF	Very high frequency	30–300 MHz
UHF	Ultrahigh frequency	0.3–1 GHz
L	Long wave	1–2 GHz
S	Short wave	2–4 GHz
C	Compromise between S & X	4–8 GHz
X		8–12 GHz
K_u	Kurz-under	12–18 GHz
K	Kurz	18–27 GHz
K_a	Kurz-above	27–40 GHz
V		40–75 GHz
W		75–110 GHz
mm or G	millimeter	110–300 GHz

Table 6.3 Radar-frequency bands as defined by EU, NATO, and USA for ECM systems

Band name	Frequency
A	< 250 MHz
B	250–500 MHz
C	500–1000 MHz
D	1–2 GHz
E	2–3 GHz
F	3–4 GHz
G	4–6 GHz
H	6–8 GHz
I	8–10 GHz
J	10–20 GHz
K	20–40 GHz
L	40–60 GHz
M	60–100 GHz

communication. Recently, there are attempts to use the blue lasers for submerged communications.

Atmospheric absorption limits the useful range for communication at high frequencies. Figure 6.1 shows the atmospheric absorption of EM waves at altitudes less than 1 km. The first peak occurs at 22 GHz due to water vapor. The second peak occurs at 63 GHz due to oxygen. The total atmospheric absorption is the sum of those due to both of them. At 30 GHz, the useful range for communication is a few km. At 300 GHz, the useful range for communication reduces to a few tens of meters. The 0.3–3 THz band is regarded as a radio frequency band by ITU but regarded as part of the far infrared band by several other naming systems. The atmosphere is almost opaque to the terahertz band.

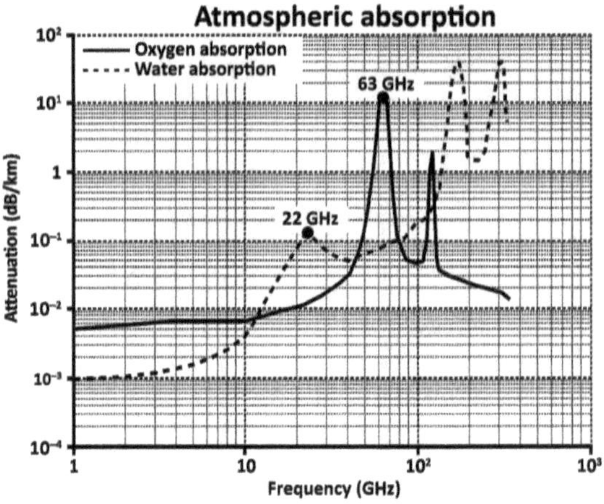

Fig. 6.1 Atmospheric absorption of EM waves at < 1 km altitudes

Radar speed guns used in traffic-law enforcement normally operate at the IEEE K_u, K, and K_a bands. Millimeter wave (mmWave) radar operates in the 76−81 GHz band with corresponding wavelength 3.95−3.70 mm.

In general, the use of radio spectrum is governed by licensing regulation and differs from country to country. Certain industrial, scientific, and medical (ISM) devices emit powerful radio wave. Such devices include microwave ovens, medical diathermy machines, etc. ISM devices are allocated certain frequency bands for their operations. The use of ISM bands does not require government licensing and communication devices using these ISM bands must tolerate interference generated by ISM devices. Since the use of ISM bands does not require government licensing, many short-range low-power wireless communication systems use these bands. ISM bands are tabulated in Table 6.4. Home microwave ovens, Bluetooth, and many radio control equipment operate at the 2.45 GHz ISM band. Laptops, tablet computers, printers, and smartphones talk to each other via the IEEE 802.11 protocol using the 2.45 GHz and 5.8 GHz ISM bands. Personal area networks implementing IEEE 802.15.4 use the 915 MHz and 2.45 GHz ISM bands. Contactless smart cards use the 13.56 MHz and 27.12 MHz ISM bands. The most commonly used band for RFID is the 13.56 MHz ISM band.

6.3 Antenna

There are two angles commonly referred to in an antenna. They are the azimuth and elevation angles. Consider a vector and a horizontal plane as shown in Fig. 6.2. Suppose that the vector is not on the plane. Project the vector perpendicularly onto the plane. Consider a reference direction on the horizontal plane. The azimuth angle is the angle of the perpendicular projection of the vector on the horizontal plane

Table 6.4 Industrial, scientific, and medical (ISM) bands

Frequency range		Center frequency	Bandwidth	Type
6.765 MHz	6.795 MHz	6.78 MHz	30 kHz	A
13.553 MHz	13.567 MHz	13.56 MHz	14 kHz	B
26.957 MHz	27.283 MHz	27.12 MHz	326 kHz	B
40.66 MHz	40.70 MHz	40.68 MHz	40 kHz	B
902 MHz	928 MHz	915 MHz	26 MHz	B
2.4 GHz	2.5 GHz	2.45 GHz	100 MHz	B
5.725 GHz	5.875 GHz	5.8 GHz	150 MHz	B
24.0 GHz	24.25 GHz	24.125 GHz	250 MHz	B
61.0 GHz	61.5 GHz	61.25 GHz	500 MHz	A
122 GHz	123 GHz	122.5 GHz	1 GHz	A
244 GHz	246 GHz	245 GHz	2 GHz	A

Type A: Availability subject to local acceptance

Type B: Worldwide availability

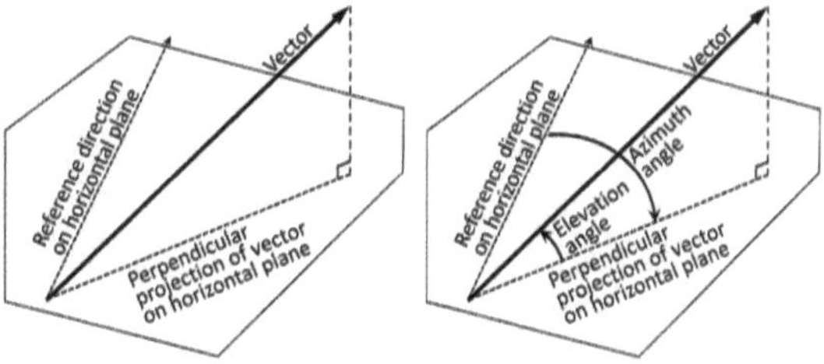

Fig. 6.2 Azimuth and elevation angles

measured clockwise with respect to the reference. The elevation angle is the angle between the vector and the horizontal plane measured vertically.

There are two types of antenna—the transmitting antenna and the receiving antenna. A transmitting antenna converts voltage/current produced by the transmitter into EM wave and radiate it into free space. A receiving antenna converts EM wave into voltage/current for the receiver to operate.

As the name suggests, the far field of an antenna is the region far from the antenna. For an antenna, distances further away than the Fraunhofer distance given by $\frac{2D^2}{\lambda}$ are regarded as far field where D is the largest dimension of the radiator (or the diameter of the antenna). In the far field, radiation from an antenna resembles that of a plane wave; its electric field vector $\boldsymbol{E}$ is orthogonal to its magnetizing field vector $\boldsymbol{H}$. In the far field, the electric field strength and magnetizing field strength are inversely proportional to the distance and so the power is inversely proportional to the square of the distance.

The Poynting vector $\boldsymbol{S}$ is the power flow per unit area of the EM wave and is given by the cross product of the electric field vector $\boldsymbol{E}$ and magnetizing field vector $\boldsymbol{H}$.

$$\boldsymbol{S} = \boldsymbol{E} \times \boldsymbol{H}. \tag{6.1}$$

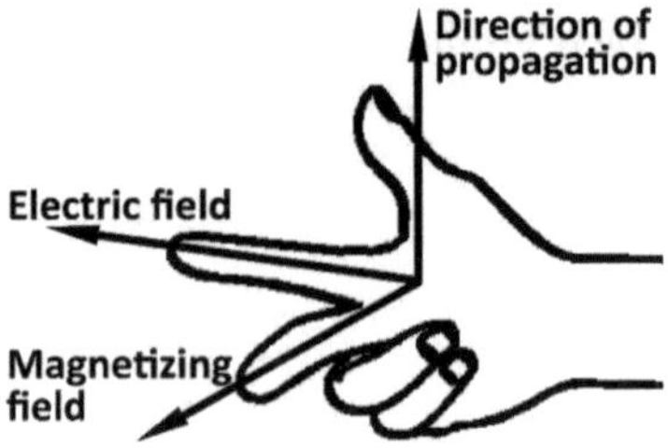

Fig. 6.3 Right-hand rule for $\boldsymbol{S} = \boldsymbol{E} \times \boldsymbol{H}$

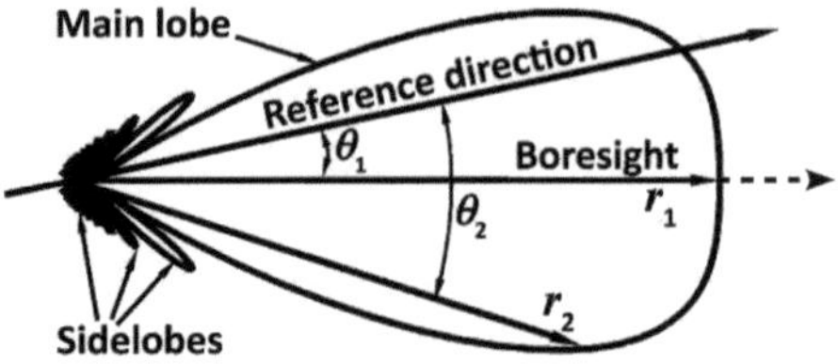

Fig. 6.4 Radiation pattern of an arbitrary antenna

The cross product right-hand rule for the direction of $\boldsymbol{S}$ in (6.1) is shown in Fig. 6.3. When an EM wave is reflected from a good electric conducting surface, $\boldsymbol{E}$ and $\boldsymbol{S}$ reverse in directions, and the direction of $\boldsymbol{H}$ remains unchanged.

The radiation pattern of an antenna is the variation of the radiation strength (power or electric field strength or magnetizing field strength) radiated by an antenna as a function of the angular direction measured from a reference in the far field. Figure 6.4 shows an example of a radiation pattern of an arbitrary antenna. The radiation strength at direction θ_1 and θ_2 is r_1 and r_2, respectively.

A radiation pattern is "isotropic" if the radiation pattern is the same in all directions. An isotropic antenna is a conceptual antenna and does not exist in practice. The isotropic antenna is used as a reference for comparing the performance of a real antenna. The gain of an antenna in dBi (dB-isotropic) is the ratio of its gain to that of an isotropic antenna expressed in dB. The gain of an antenna in dBd (dB-dipole) is the ratio of its gain to that of a half-wave dipole antenna expressed in dB. The value in dBi is the value in dBd plus 2.15. A radiation pattern is "omnidirectional" if it has no azimuth angle preference but may vary with elevation angle.

The radiation pattern of a practical antenna has directional preference. The response lobe with the largest magnitude (maximum gain) is called the main lobe. Response lobes other than the main lobe are called the sidelobes. The axis of the maximum gain is called the boresight of the antenna. In Fig. 6.4, the boresight is at an angle of θ_1, and the radiation strength at the boresight is r_1.

The radiation efficiency $E_{\text{radiation}}$ of an antenna is defined as

$$E_{\text{radiation}} = \frac{P_{\text{radiated}}}{P_{\text{input}}}, \tag{6.2}$$

where P_{radiated} is the power radiated by the antenna and P_{input} is the power input to the antenna. When an antenna is being supplied by a transmission line or a waveguide, some of the power arriving from the transmission line or waveguide will be reflected back due to impedance mismatch. The total radiation efficiency E_{total} of an antenna is defined as

$$E_{\text{total}} = M_{\text{L}} \times E_{\text{radiation}}, \tag{6.3}$$

where M_{L} is the ratio of the power absorbed by the antenna to the total power arriving at the antenna; it is a measure of the impedance matching. Since $0 \leq M_{\text{L}} \leq 1$, $E_{\text{total}} \leq E_{\text{radiation}}$. The antenna efficiency usually refers to E_{total}. The antenna efficiency of a well-designed dish antenna, horn antenna, or half-wavelength dipole is almost 100% or 0dB. The efficiency of a mobile phone antenna is between 0.2 and 0.7 (−7dB to −1.5dB). At AM radio frequencies, a car radio antenna is very much shorter than a half-wavelength, and so the efficiency is usually around 1% (−20dB). The efficiency of an antenna when used as a radiating antenna is the same as that when used as a receiving antenna; this is called antenna reciprocity. When used as a receiving antenna, an efficiency of 70% means its electrical output is 70% that of an antenna with an efficiency of 100%.

The antenna gain G_{antenna} is the ratio of the power transmitted in the direction of peak radiation of an antenna to that of an isotropic antenna. It is defined as

$$G_{\text{antenna}} = \frac{P_{\text{peak}}}{P_{\text{isotropic}}}, \tag{6.4}$$

where P_{peak} is the power transmitted in the direction of peak radiation and $P_{\text{isotropic}}$ is the power transmitted if the antenna is replaced by a conceptual isotropic antenna. It should be noted high P_{peak} is achieved by reducing the radiation in directions other than the direction of peak radiation. This means that high gain antenna is very directional. If the direction of transmission or reception is known, high gain is desirable; otherwise, high gain is undesirable.

The antenna directivity D_{antenna} is related to the antenna gain G_{antenna} and antenna radiation efficiency $E_{\text{radiation}}$ by

$$G_{\text{antenna}} = E_{\text{radiation}} \times D_{\text{antenna}} \tag{6.5}$$

For efficient transmission or reception, the impedance of the transmission line must be well matched to the antenna's impedance. The parameter VSWR (voltage standing wave ratio) describes how well the antenna is impedance matched to the transmission line it is connected to. The voltage standing wave ratio is given in terms of the reflection coefficient s_{11} by

$$\text{VSWR} = \frac{1 + s_{11}}{1 - s_{11}} \tag{6.6}$$

The smaller is VSWR or s_{11}, the better is the impedance match. The impedance matching is said to be good if VSWR ≤ 2 or $s_{11} \leq \frac{1}{3}$.

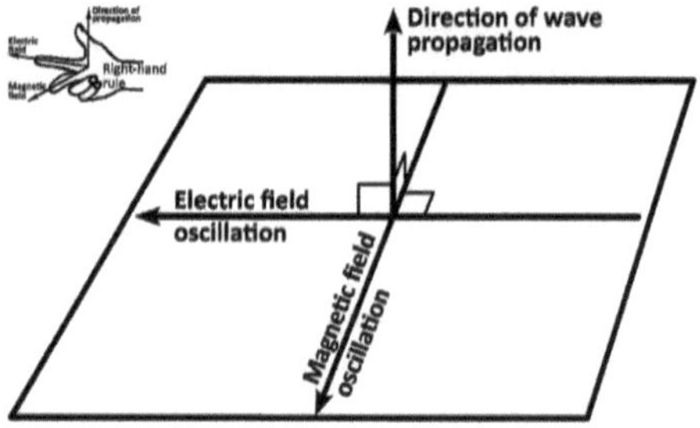

Fig. 6.5 Electric field, magnetizing field, and direction of propagation are perpendicular to each other

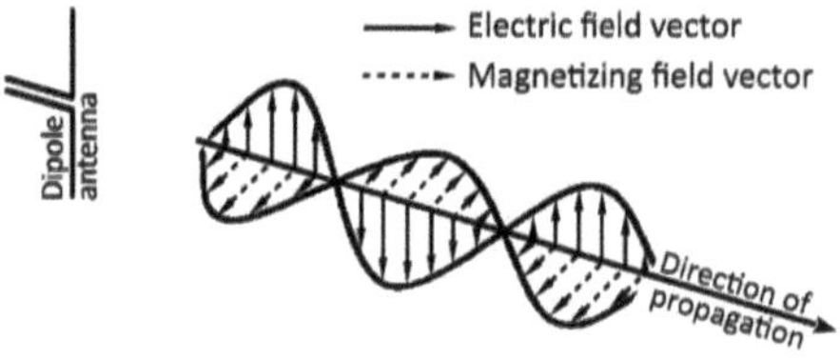

Fig. 6.6 A linearly polarized wave

6.3.1 Polarization

The electric field and magnetizing field of an EM wave are perpendicular to each other and oscillate in a plane perpendicular to the direction of travel of the wave as shown in Fig. 6.5. The figure that the electric field traces out while propagating is called the polarization. If the electric field of an EM wave stays along a single line, the EM wave is said to be linearly polarized. In addition, if the electric field is vertical with respect to the ground, it is called vertically polarized. If it is horizontal, it is called horizontally polarized. The EM wave in Fig. 6.6 is vertically polarized.

If the electric field rotates in a circle, the EM wave is said to be circularly polarized. If the electric field rotates in an ellipse, the EM wave is said to be elliptically polarized. If you are facing an incoming circularly or elliptically polarized EM wave and if the electric field rotates counterclockwise, the EM wave is said to be right-hand circularly or elliptically polarized (RHCP or (RHEP). If the electric field rotates clockwise, the EM wave is said to be left-hand circularly or elliptically polarized (LHCP or LHEP). An RHCP or RHEP EM wave reflected off a surface becomes LHCP or LHEP and vice versa.

The polarization of an antenna is the polarization of the EM wave it radiates. An antenna that radiates a linearly polarized EM wave is a linearly polarized antenna. An antenna can only communicate with another antenna of the same polarization. A vertically polarized antenna will not be able to receive radiation from a horizontally polarized antenna. An RHCP antenna will not be able to receive an LHCP EM wave. The RHCP EM wave transmitted by an RHCP antenna becomes LHCP when it is reflected by an object and so an LHCP antenna is needed to receive the reflected signal.

The power loss due to polarization mismatch is called polarization loss factor (PLF). For two linearly polarized antennas that are rotated from each other by an

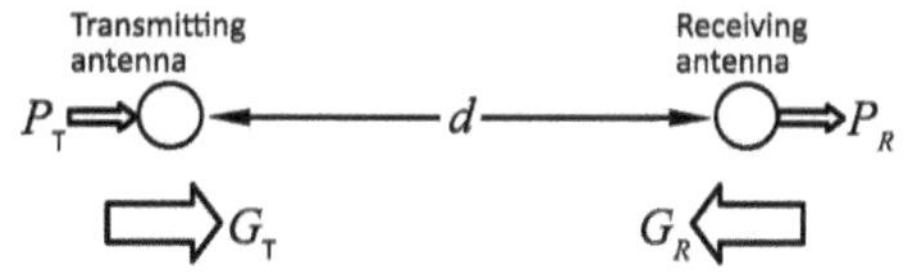

Fig. 6.7 Transmission from one antenna to another

angle ϕ, the PLF is given by

$$\text{PLF} = \cos^2(\phi). \tag{6.7}$$

A circularly polarized antenna may be regarded as two antennas separated by a rotation angle of 90°. As a result, the linearly polarized antenna will have a polarization mismatch loss of 0.5 (−3dB) when receiving radiation from a circularly polarized antenna.

6.3.2 *Friis Transmission Formula*

Consider a transmitting antenna with input power $P_{\rm T}$ and antenna gain $G_{\rm T}$ radiating towards a receiving antenna at distance d in the far field. Let the antenna gain of the receiving antenna be $G_{\rm R}$ and let its electrical power output be $P_{\rm R}$. See Fig. 6.7. The various losses of the two antennas have already been absorbed into $G_{\rm T}$ and $G_{\rm R}$. Assume that the polarizations of the transmitting antenna and receiving antenna are the same. Friis transmission formula states that

$$P_{\rm R} = \frac{P_{\rm T} G_{\rm T} G_{\rm R} \lambda^2}{(4\pi d)^2}. \tag{6.8}$$

It can be seen from (6.8) that the received power for a given transmitted power is proportional to the square of the wavelength. Hence, for efficient power transfer, longer wavelength transmission is preferred.

6.3.3 *Dipole Antenna*

The dipole antenna is an open-circuited wire fed at its center as shown in Fig. 6.8. We shall assume that the diameter of the dipole is small relative to the wavelength. The current $i(t, l)$ at time t and distance l from the center of the dipole is a standing wave and is given by

$$i(t, l) = I(l) \cos(\omega t), \tag{6.9}$$

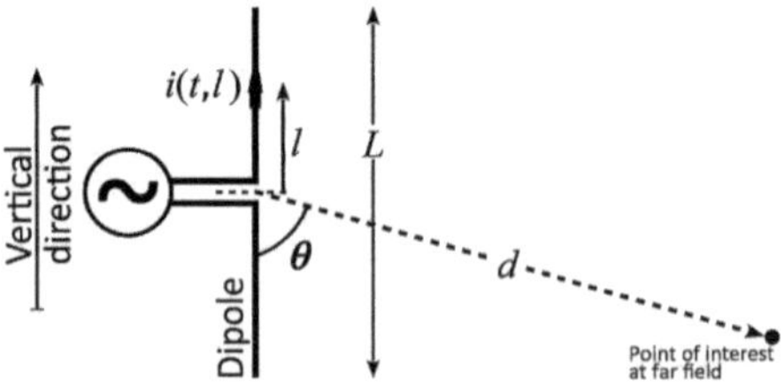

Fig. 6.8 A dipole antenna

where ω is the frequency and $I(l)$ is given by (6.10).

$$I(l) = I_0 \sin\left(k_0\left(\frac{L}{2} - |l|\right)\right) \quad , \quad -\frac{L}{2} \le l \le \frac{L}{2}, \tag{6.10}$$

where I_0 is a constant, L is the length of the dipole, and $k_0 = \frac{2\pi}{\lambda}$ is the wave number. If $L = \frac{\lambda}{2}$, $I = I_0$ at $l = 0$.

Its far-field electric field strength $E(\theta, d)$ and magnetizing field strength $H(\theta, d)$ are given by

$$E(\theta, d) = \frac{\eta_0 I_0 \cos(\omega t - k_0 d)}{2\pi d} \bullet \frac{\cos\left(\frac{k_0 L}{2}\cos(\theta)\right) - \cos\left(\frac{k_0 L}{2}\right)}{\sin(\theta)}, \tag{6.11a}$$

$$H(\theta, d) = \frac{E(\theta, d)}{\eta_0}, \tag{6.11b}$$

where d is the distance between the dipole and the point of interest in the far field and η_0 is the intrinsic impedance of free space ($= 120\pi \approx 377\Omega$). The intrinsic impedance η of any medium where a wave travel is defined as the ratio of the wave's electric field strength E to its magnetizing field strength H, i.e., $\eta = \frac{E}{H}$.

The wave radiating from a dipole is linearly polarized in parallel to the length of the dipole. In Fig. 6.8, it is vertically polarized if the length of the dipole is positioned vertically.

If the dipole is fed at its center, and if the current distribution along the dipole antenna is sinusoidal, the driving point impedance ($R_{\text{dipole}} + jX_{\text{dipole}}$) at feedpoint is given by

$$\begin{aligned} R_{\text{dipole}} = \frac{\eta_0}{2\pi \sin^2\left(\frac{k_0 L}{2}\right)} \Bigg(& \gamma + \ln(k_0 L) - C_i(k_0 L) \\ & + \frac{\sin(k_0 L)\left(S_i(2k_0 L) - 2S_i(k_0 L)\right)}{2} \\ & + \frac{\cos(k_0 L)\left(C_i(2k_0 L) - 2C_i(k_0 L)\right) + \gamma + \ln(\frac{k_0 L}{2})}{2} \Bigg), \end{aligned} \tag{6.12a}$$

$$X_{\text{dipole}} = \frac{\eta_0}{2\pi \sin^2\left(\frac{k_0 L}{2}\right)} \left(S_i(k_0 L) + \frac{\cos(k_0 L)\left(-S_i(2k_0 L) + 2S_i(k_0 L)\right)}{2} \right.$$
$$\left. + \frac{\sin(k_0 L)\left(C_i(2k_0 L) - 2C_i(k_0 L)\right) + C_i\left(\frac{2k_0 a^2}{L}\right)}{2} \right), \qquad (6.12\text{b})$$

where $\gamma = \lim_{n\to\infty}\left(-\ln(n) + \sum_{k=1}^{n}\frac{1}{k}\right) \approx 0.5772156649$ is Euler's constant, a is the radius of the conductor, $S_i(x) = \int_0^x \frac{\sin(t)}{t}\mathrm{d}t = \sum_{k=1}^{\infty}(-1)^{k-1}\frac{x^{2k-1}}{(2k-1)(2k-1)!}$ is the sine integral function, and $C_i(x) = -\int_x^{\infty}\frac{\cos(t)}{t}\mathrm{d}t = \gamma + \ln(x) + \sum_{k=1}^{\infty}\frac{(-x^2)^k}{2k(2k)!}$ is the cosine integral function.

The bandwidth of a dipole antenna can be widened by increasing the radius of the dipole.

A dipole antenna is omnidirectional. In application where it is necessary to isolate the antenna from what is behind it, the radiation to the back hemisphere can be eliminated in two ways. One way is to back the antenna by an absorbing material. Another way is to reflect the radiation to the back hemisphere to the front by using a conducting back plane. EM wave reflected by a conducting plane appears to be radiated from a mirror image of the original antenna on the conducting back plane by a source opposite in electrical polarity. If the back plane is placed very near to (or at integer multiple of half-wavelength from) the antenna, the field generated by the mirror image will cancel the field generated by the antenna. Hence, the conducting plane should be placed at quarter-wavelength (or integer multiple of half-wavelength plus quarter-wavelength).

6.3.3.1 Short Dipole Antenna

If $L \ll \lambda$, say $L < 0.1\lambda$, the dipole is called a short dipole. Since $\sin(x) \approx x$ for small values of $|x|$, (6.9) becomes

$$I(l) \approx I_S\left(1 - \frac{2|l|}{L}\right), \quad \text{where } I_S = \frac{I_0 \pi L}{\lambda}. \qquad (6.13)$$

The current distribution depicted in (6.13) is triangular as shown in Fig. 6.9. Note that $\cos\left(\frac{k_0 L}{2}\cos(\theta)\right) - \cos\left(\frac{k_0 L}{2}\right) \approx \frac{k_0^2 L^2 \sin^2(\theta)}{8}$ for small values of $|\frac{k_0 L}{2}|$. Hence, (6.11) becomes

$$E(\theta, d) = \frac{\eta_0 k_0 I_S L \cos(\omega t - k_0 d)}{8\pi d}\sin(\theta), \qquad (6.14\text{a})$$

$$H(\theta, d) = \frac{E(\theta, d)}{\eta_0}. \qquad (6.14\text{b})$$

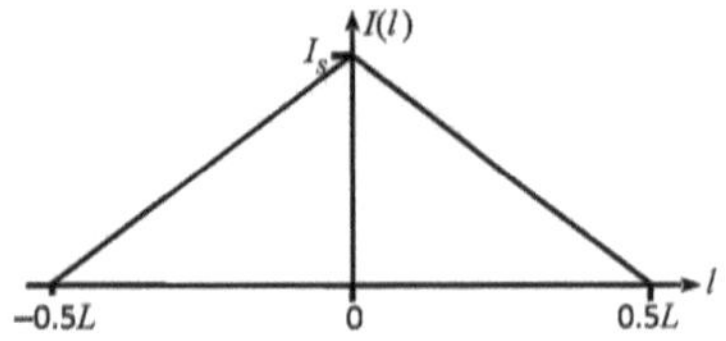

Fig. 6.9 Current distribution along a short dipole antenna

From (6.14), it can be seen that $E(\theta, d) \propto L$ indicating that longer short dipole will radiate more power. The power at a point of interest is proportional to the mean value of $|E(\theta, d)|^2$.

The real part of the feedpoint impedance of the short dipole consists of $R_{\text{radiation}}$ representing the energy radiated and R_{loss} representing loss due to the finite conductivity of the antenna. The energy stored in the fields is represented by the imaginary component jX of the driving point impedance. Hence, the feedpoint impedance is $(R_{\text{radiation}} + R_{\text{loss}} + jX)$ where

$$R_{\text{radiation}} \approx 20\pi^2 \left(\frac{L}{\lambda}\right)^2 \approx 197 \left(\frac{L}{\lambda}\right)^2, \tag{6.15}$$

and

$$R_{\text{loss}} \approx \frac{L}{6\pi a}\sqrt{\frac{\omega\mu_0}{4\sigma}}, \tag{6.16}$$

where a is the radius of the dipole, σ is the conductivity of the dipole, and $\mu_0 = 4\pi \times 10^{-7}$ is the permeability of free space. For $L \ll \lambda$, $R_{\text{radiation}}$ may become less than R_{loss} and the antenna becomes very inefficient. The reactive component is given by

$$X \approx -\frac{120\lambda}{\pi L}\left(\ln\left(\frac{L}{2a}\right) - 1\right). \tag{6.17}$$

The negative sign in (6.17) indicates that the impedance is capacitive. For $L \ll \lambda$, $\frac{\lambda}{L}$ becomes large and so $|jX|$ becomes much larger than $(R_{\text{radiation}} + R_{\text{loss}})$; this makes impedance matching difficult.

6.3.3.2 Half-Wavelength Dipole Antenna

The half-wavelength dipole antenna is an important special case of the dipole antenna where $L = 0.5\lambda$. In this case, (6.10) becomes

$$I(l) = I_0 \cos(k_0 l). \tag{6.18}$$

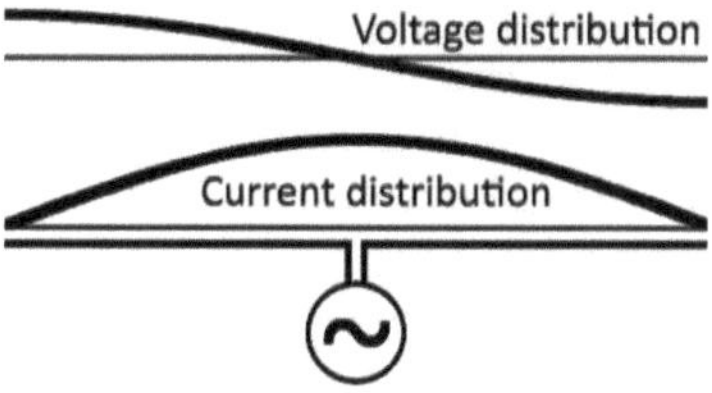

Fig. 6.10 Current and voltage distributions along a half-wave dipole antenna

For $L = \frac{\lambda}{2}$, $\frac{k_0 L}{2} = \frac{\pi}{2}$ and (6.11) becomes

$$E(\theta, d) = \frac{\eta_0 I_0 \cos(\omega t - k_0 d)}{2\pi d} \bullet \frac{\cos\left(\frac{\pi \cos(\theta)}{2}\right)}{\sin(\theta)}, \tag{6.19a}$$

$$H(\theta, d) = \frac{E(\theta, d)}{\eta_0}. \tag{6.19b}$$

Its current and voltage distributions along the half-wave antenna are shown in Fig. 6.10. The current peaks at the feedpoint and is zero at the extreme ends of the antenna. The voltage is zero at the feedpoint and peaks at the extreme ends. Hence, it has low impedance at the feedpoint. The currents at the two antenna arms are in phase and radiate. The voltages at the two antenna arms are opposite in phase and cancel each other. Hence, the antenna is a "current radiator." Wire antennas such as the dipole, monopole, and spiral antennas are current radiators.

Its directivity is $\frac{4}{\int_0^{2\pi} \frac{1-\cos(x)}{x} dx} = 1.64$ (2.15 dBi $= 10 \times \log(1.64)$), and its half-power beamwidth is 78°. Its feedpoint impedance is $(R_{\text{radiation}} + jX)$, where $R_{\text{radiation}} = \frac{\eta_0}{4\pi} \int_0^{2\pi} \frac{1-\cos(x)}{x} dx = 73.1$ Ohms and $X = 42.5$. The jX term of the feedpoint impedance can be removed by slightly decreasing the dipole length. The length reduction is a function of the diameter of the dipole. For thin dipole with diameter $= 10^{-5}\lambda$, the required length reduction is 2%. For thick dipole with diameter $= 0.008\lambda$, the required length reduction is 6%.

6.3.3.3 Half-Wavelength Folded Dipole Antenna

If the ends of two half-wavelength dipoles are connected to each other forming a loop as shown in Fig. 6.11, the resulting antenna is called a folded dipole antenna. The feedpoint impedance of the half-wavelength folded dipole antenna is four times that of the half-wavelength dipole antenna. The radiation pattern of the half-wavelength folded dipoles has the same form as that of half-wavelength dipoles.

For a three-wire half-wavelength folded dipole antenna as shown in Fig. 6.12, the feedpoint impedance is nine times that of the half-wavelength dipole antenna.

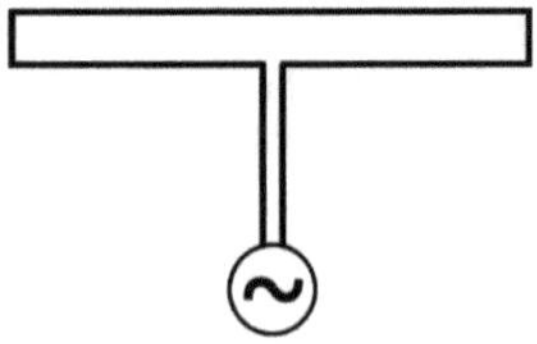

Fig. 6.11 A 0.5λ folded dipole antenna. Feedpoint impedance is four times that of a 0.5λ dipole

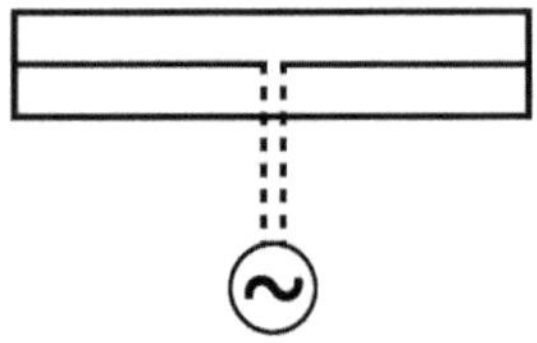

Fig. 6.12 A three-wire 0.5λ folded dipole antenna. Feedpoint impedance is nine times that of a 0.5λ dipole

Table 6.5 Dipole directivity for various L values

L	Directivity (dBi)	Comment
$\ll 0.5\lambda$	1.76	Poor efficiency
0.5λ	2.15	Most common
λ	4.0	Fat dipole only
1.25λ	5.2	Best directivity
1.5λ	3.5	
2λ	4.3	

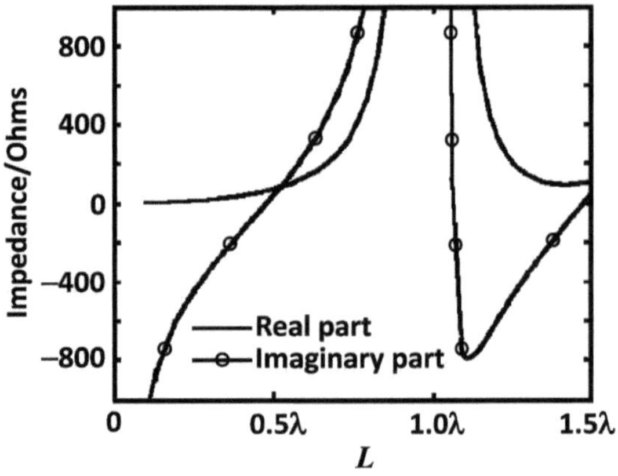

Fig. 6.13 Feedpoint impedance versus L plot for a dipole antenna fed at the center

6.3.3.4 Dipole Antenna with $L > 0.5\lambda$

Except for the case where $L = 1.25\lambda$, dipole antennas with $L > 0.5\lambda$ are seldom used. The special case where $L = 1.25\lambda$ has the highest directivity for L around that values; however, its feedpoint impedance is very reactive and requires impedance matching network. Table 6.5 lists the directivity for several L values. Figure 6.13 shows a feedpoint impedance versus L plot.

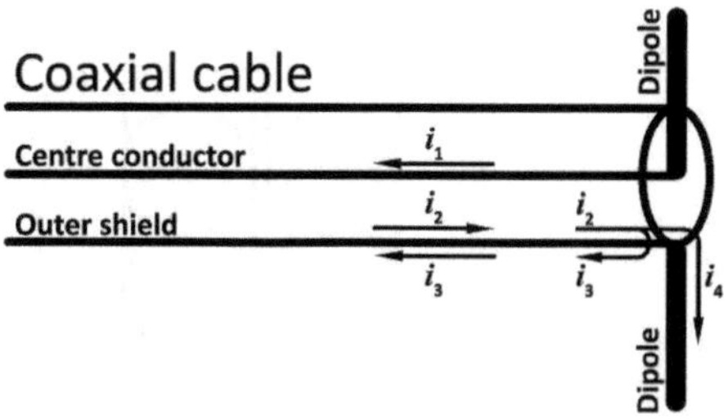

Fig. 6.14 A coaxial cable feeding a dipole antenna

Fig. 6.15 A balun interfacing a dipole to a coaxial cable

6.3.3.5 Feeding a Dipole Antenna

Figure 6.14 shows a coaxial cable feeding a dipole antenna. Current i_1 of the coaxial cable's center conductor flows away from the dipole. Current i_2 is equal in magnitude but opposite in direction to current i_1 flows in the inner part of the coaxial cable's outer shield and splits into currents i_3 and i_4 as shown in Fig. 6.14. Current i_4 flows into the dipole. Current i_3 returns along the outer part of the coaxial cable's outer shield. The net current of the coaxial cable is $i_1 + i_3 - i_2 \neq 0$; the coaxial cable radiates.

A dipole antenna expects a balanced feed. A twin lead is a balanced transmission line but has the problem that it can be electrically disturbed by a nearby conductor. A coaxial cable has one conductor grounded; it is not a balanced line. A balun (**bal**-to-**un**balanced) is needed if a dipole is fed using a coaxial cable. A balun is an adaptor for interfacing a balanced port such as a dipole antenna to an unbalanced port such as a coaxial cable as shown in Fig. 6.15.

There are two general categories of baluns—the voltage balun and the current balun. The voltage balun uses transformer action to establish equal voltage magnitude across the two sides of a balanced load. A current balun allows differential currents to pass but chokes common mode currents; there is no transformer action.

Figure 6.16 shows two voltage baluns. Figure 6.16a uses a transformer. The impedance ratio on the two sides of the transformer is equal to the square of the turns ratio on the two sides. The turns ratio of the transformer can be selected for impedance matching. Figure 6.16b uses an autotransformer with turns ratio 1:2; the impedance ratio will be 1:4.

Figure 6.17 shows three current baluns. Figure 6.17a uses a common mode choke. The impedance of the common mode choke is very high for common mode current and, hence, blocking current i_3 shown in Fig. 6.14.

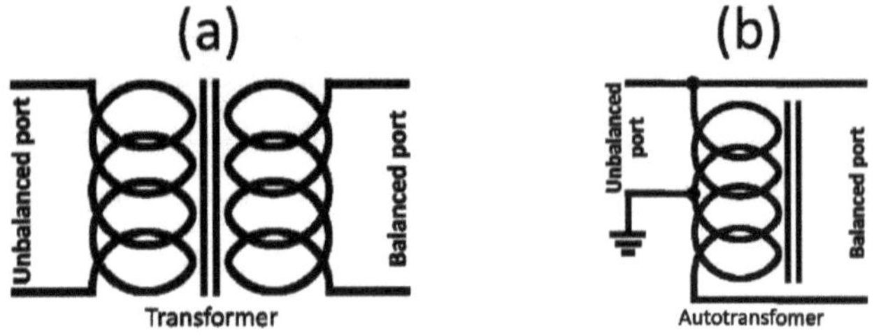

Fig. 6.16 Voltage baluns

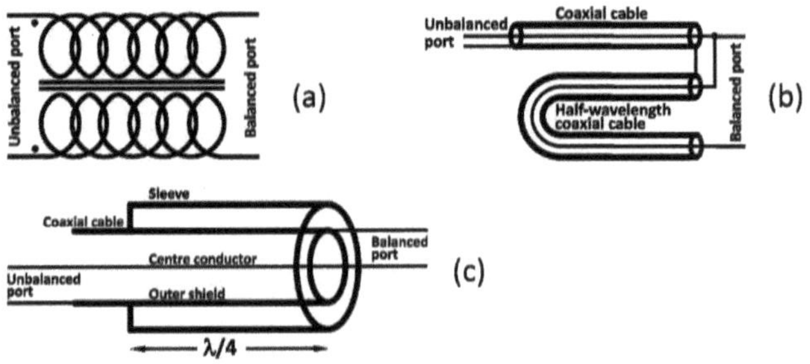

Fig. 6.17 Current baluns

Figure 6.17b is a half-wavelength loop balun. The delay caused by the half-wavelength coaxial cable inverts the signal for the other part of the dipole. This is the best design if a narrow bandwidth is desired.

Figure 6.17c is a sleeve balun. The length of the sleeve is quarter-wavelength and is shorted to the outer shield of the coaxial cable at the unbalanced side. The sleeve and the outer shield of the coaxial cable form a short circuited quarter-wavelength transmission line, which has infinite impedance for signal traveling from the balanced side, and hence, blocking current i_3 shown in Fig. 6.14.

6.3.4 Monopole Antenna

Figure 6.18a shows a monopole antenna of length L fed at the extreme end nearest to the ground plane. A monopole antenna is one-half of a dipole antenna. It is always mounted above a ground plane, which can be the Earth's surface or a metallic plate (which can be the casing of a device such as a mobile phone) connected to the ground. Its radiation pattern is the same as a dipole antenna with length $2L$. The part of the dipole equivalent below the ground is the mirror image (but reversed in electrical polarity) of the monopole on the ground plane. See Fig. 6.18b. The radiation pattern is affected by the size of the ground plane. For a quarter-wavelength monopole $L = 0.25\lambda$, the monopole's feedpoint impedance z_{monople} is half of that

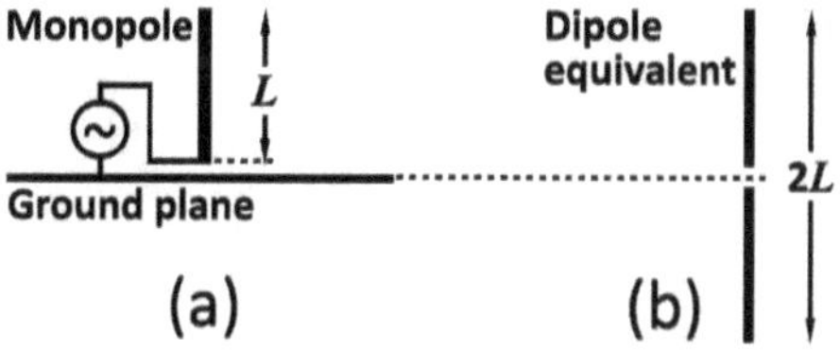

Fig. 6.18 (**a**) Monopole antenna. (**b**) Dipole equivalent

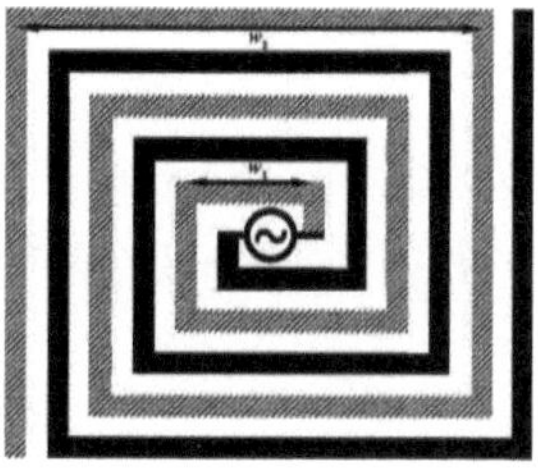

Fig. 6.19 A square spiral antenna fed at the center

of a half-wavelength dipole and so $z_{\text{monople}} = 36.5 + j21.25$ Ohms. The directivity of a monopole with length L is twice that of a dipole with length $2L$. This is because no radiation occurs below the ground plane; hence, the antenna is effectively twice as "directive."

6.3.5 Spiral Antenna

Spiral antennas are circularly polarized, with wide bandwidth and low directivity. Its radiation pattern peaks at the direction perpendicular to the plane of the spiral with half-power beamwidth (HPBW) $\approx 70°\text{–}90°$. It is fed at the center as shown in Fig. 6.19 for a square spiral antenna.

A spiral antenna radiates from a region called the active region. The active region is the circumference of the spiral, which is equal to one wavelength. As the spiral radiates, energy is removed from the spiral. As a result, the magnitude of the current dies off in the spiral arms with distance from the center of the spiral. As a consequence, there is little reflection of current from the end of the spiral antenna.

A spiral antenna transmits circularly polarized EM waves. It receives circularly polarized EM waves of the same polarization orientation (right-hand or left-hand oriented) and linearly polarized waves in any polarization orientation. The spiral antennas in Figs. 6.19 and 6.20 radiate right-hand circularly polarized (RHCP) EM wave out of the diagram and left-hand circularly polarized (LHCP) EM wave into the diagram.

There are many types of spiral antenna but the most commonly used are Archimedean spiral, logarithmic spiral, and square spiral. The logarithmic spiral antenna is usually regarded as frequency independent because its fractional band-

Fig. 6.20 Round spiral antennas

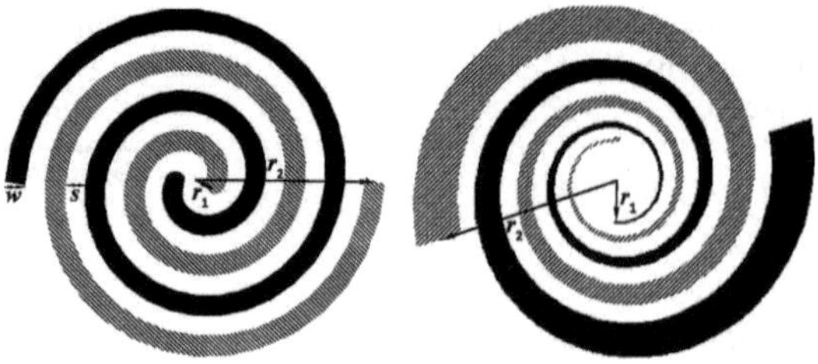

width $f_{\text{bandwidth}}$ given by

$$f_{\text{bandwidth}} = \frac{f_{\text{higher cut-off}} - f_{\text{lower cut-off}}}{f_{\text{centre}}}, \tag{6.20}$$

may be as high as 30. The Archimedean spiral antenna is the most commonly used spiral antenna. A two-arm Archimedean spiral is shown in Fig. 6.20a. Let the two arms be arm $-a$ and arm $-b$. Let r_a and r_b be the distances of arm $-a$ and arm $-b$ from the center of the spiral, respectively. Let w be the width of each arm and s be the spacing between each turn. Let r_1 and r_2 be the inner radius and outer radius of the spiral, respectively. Expressions for r_a and r_b are given by

$$r_a = r_0\phi + r_1 \ , \text{ and} \tag{6.21a}$$

$$r_b = r_0(\phi - \pi) + r_1, \phi \geq \pi, \tag{6.21b}$$

where

$$r_0 = \frac{s + w}{\pi}. \tag{6.22}$$

Suppose that the spiral has N turns.

$$w + s = \frac{r_2 - r_1}{2N}. \tag{6.23}$$

The low-frequency operating point f_{low} is determined by the outer radius r_2 and is given by

$$f_{\text{low}} = \frac{c}{2\pi r_2}, \tag{6.24}$$

where c is the speed of light. The high-frequency operating point f_{high} is determined by the inner radius r_1 and is given by

$$f_{\text{high}} = \frac{c}{2\pi r_1}. \tag{6.25}$$

A spiral antenna is self-complementary if the metal and air regions of the antenna are equal. For the case of the Archimedean spiral antenna, this corresponds to $s = w$. The feedpoint impedance $z_{\text{self-comp}}$ of a self-complementary antenna is given by

$$z_{\text{self-comp}} = \frac{\eta_0}{2} = \frac{120\pi}{2} \approx 188.5\Omega. \tag{6.26}$$

Expression (6.26) applies to logarithmic spiral and square spiral as long as they are self-complementary.

For the logarithmic spiral antenna shown in Fig. 6.20b, expressions for r_a and r_b are given by

$$r_a = r_0 e^{\alpha\phi} \quad \text{, and} \tag{6.27a}$$

$$r_b = r_0 e^{\alpha(\phi-\pi)}, \phi \geq \pi. \tag{6.27b}$$

In (6.27), α is called the flare rate. It is the rate at which the spiral grows with angle. If α is too small, the spiral is tightly wrapped around itself giving poor radiation. If α is too large, the spiral does not wrap around itself and behaves like a dipole. A commonly used value is $\alpha = 0.22$. The low-frequency operating point f_{low} is given by

$$f_{\text{low}} = \frac{c}{2\pi r_2}. \tag{6.28}$$

The high-frequency operating point f_{high} is given by

$$f_{\text{high}} = \frac{c}{4r_1}. \tag{6.29}$$

The spiral antenna is a balanced antenna. It needs a balun if it is fed from a coaxial cable. A solution is to use the coaxial cable's outer shield as one arm of the spiral antenna as shown in Fig. 6.21. This balun is called the infinite balun. In this way, the coaxial cable serves as one arm of the spiral antenna as well as the transmission line. Since the impedance of the coaxial cable is 50Ω, there will be significant mismatch in impedance without an impedance matching network.

6.3.6 Microstrip (Patch) Antenna

Microstrip antenna is also called patch antenna or PCB antenna. It can be easily fabricated on PCB and is very widely used in mobile phones. Its radiation pattern peaks in the direction coming out and perpendicular to the center of the microstrip antenna. It is linearly polarized in the horizontal direction of Fig. 6.22a. A microstrip antenna consists of a high conductivity patch separated from a high conductivity

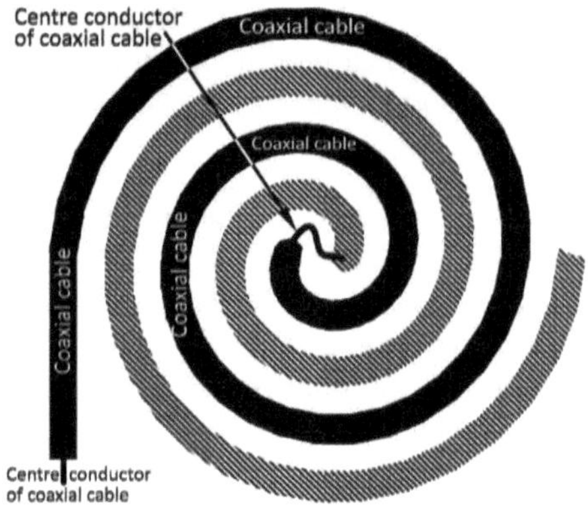

Fig. 6.21 The coaxial cable's outer shield is one arm of the spiral antenna

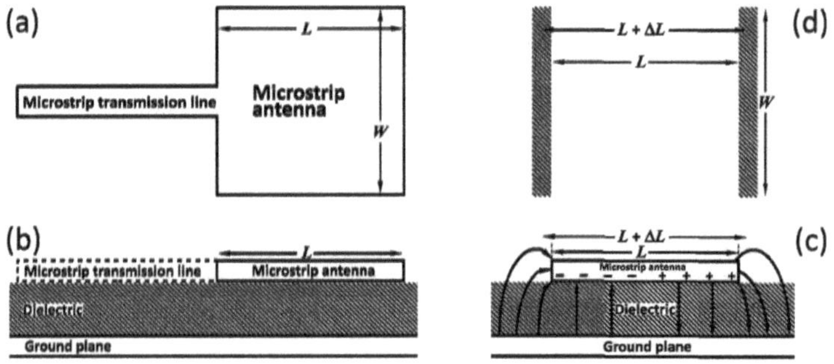

Fig. 6.22 A microstrip antenna. (**a**) Top view. (**b**) Side view. (**c**) Fringing fields. (**d**) Equivalent radiator bars

ground plane by a dielectric layer and fed by a microstrip transmission line. The thickness of the dielectric is very much smaller than λ but should not be smaller than 0.02λ; otherwise, the antenna efficiency will be seriously degraded. The thickness of the patch is very much smaller than λ.

Figure 6.22a shows the top view and Fig. 6.22b shows the side view of a patch antenna. Figure 6.22c shows the electric field in the dielectric under the patch and the fringing electric fields near the edges of the patch. The electric field bulges out of the patch's edges. As a consequence of the bulging electric field, $L + \Delta L = 0.5\lambda$. The extent of the bulge is determined by the thickness and the permittivity $\varepsilon = \varepsilon_r \varepsilon_0$ of the dielectric where ε_r and ε_0 are the relative permittivity and permittivity of free space, respectively. The bulge extends further out from the edges if the dielectric thickness is increased or ε_r is decreased. The fringing fields at the two edges and near the surface of the dielectric are both in the same direction. In Fig. 6.22c, they are both in the direction from left to right. The radiation of the microstrip antenna is produced by these fringing electric fields on the edges of the patch. The currents on the patch and on the ground plane are equal in magnitude but opposite in direction; their radiation effects cancel each other and so the currents do not cause radiation. Hence, the microstrip antenna is a "voltage radiator" as opposed to the wire antennas such as the dipole antennas that are "current radiators." Since a thicker dielectric with a smaller permittivity will result in the fringing electric field extending further out, a thicker dielectric with a smaller permittivity yields better radiation. For the microstrip transmission line where radiation should be avoided, a thinner dielectric with a larger permittivity is desirable. This represents a design trade-off in selecting the dielectric's thickness and permittivity.

The frequency of operation of the microstrip antenna of Fig. 6.22a is given by

$$\begin{aligned} f_0 &= \frac{c}{2L\sqrt{\varepsilon_r}}, \\ &= \frac{1}{2L\sqrt{\varepsilon_r \varepsilon_0 \mu_0}}, \end{aligned} \tag{6.30}$$

where c, ε_0, and μ_0 are the speed of light in, permittivity of, and permeability of vacuum, respectively, and $c^2\varepsilon_0\mu_0 = 1$.

The radiation pattern of the microstrip antenna shown in Fig. 6.22 is that of two bar radiators shown in Fig. 6.22d. First, we shall analyze the radiation pattern of one of the bar radiators. We shall replace each bar radiator by an equivalent $2N + 1$ isotropic point radiators placed on a straight line as shown in Fig. 6.23a where $N \to \infty$. We shall assume that two neighboring point radiators are separated by a distance d and so $2Nd = W$. As $N \to \infty$, $d \to 0$. Suppose that the electric field of the bar radiator is $\hat{E}_0$. The electric field of each point radiator will be $\frac{\hat{E}_0}{2N+1}$. Let the electric field at time t of the point radiator at the center (at point **O** in Figs. 6.23b and c) be $\frac{\hat{E}_0}{2N+1}e^{j\omega t}$ where ω is the frequency of radiation. Let the bar radiator/point radiators be lying on a reference plane. Let the axis $\theta = 0°$ be perpendicular to the reference plane and passing through the point **O**. Let the axis $\phi = 0°$ be lying on the reference plane, perpendicular to the line of point radiators, and passing through the point **O**. (See Fig. 6.23b.)

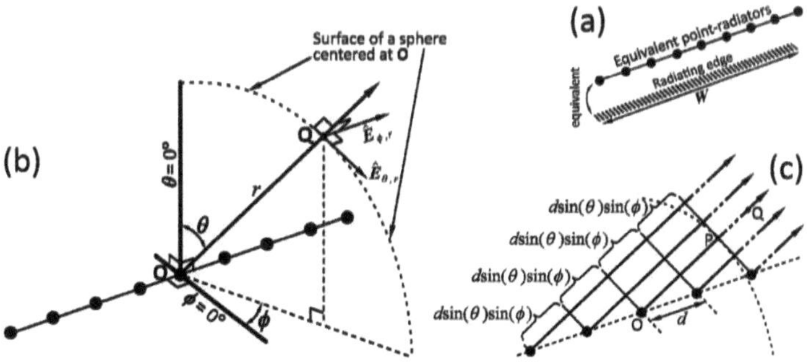

Fig. 6.23 (**a**) A bar radiator is modeled by $2N + 1$ point radiators. (**b**) Definitions for the angles θ and ϕ. (**c**) Path difference from the point radiators to far field

Consider a point **Q** in the far field at a distance r from **O**. Consider a radiation in the direction $\overrightarrow{\mathrm{OQ}}$ as shown in Fig. 6.23b. Let θ be the angle between the $\theta = 0°$ axis and the line $\overrightarrow{\mathrm{OQ}}$. Let ϕ be the angle between the $\phi = 0°$ axis and the perpendicular projection of the line $\overrightarrow{\mathrm{OQ}}$ on the reference plane. Let (θ, ϕ) denote the radiation direction of the radiation along $\overrightarrow{\mathrm{OQ}}$. The electric field $\hat{E}_Q$ at the point **Q** must lie in the plane perpendicular to $\overrightarrow{\mathrm{OQ}}$. Resolve $\hat{E}_Q$ into two orthogonal components $\hat{E}_{\theta,r}$ and $\hat{E}_{\phi,r}$. The vector $\hat{E}_{\theta,r}$ lies on the same plane as the circle centered at **O**, passing through **Q**, and perpendicular to the reference plane. The vector $\hat{E}_{\theta,r}$ touches this circle at **Q**.

Consider planes perpendicular to $\overrightarrow{\mathrm{OQ}}$ passing through each point radiator as shown in Fig. 6.23c. It can be shown that the separation between consecutive plane is $d\sin(\theta)\sin(\phi)$. As $r \to \infty$, the EM wave at **Q** becomes plane wave. The path difference between consecutive point radiators to **Q** is $d\sin(\theta)\sin(\phi)$. The electric field at **Q** due to the point radiator at **O** is $\frac{1}{r}\frac{\hat{E}_0}{2N+1}e^{j\omega(t-\tau)}$ where τ represents the phase shift due to the distance between **O** and **Q**. Ignoring polarization, the electric field at **Q** due to the mth point radiator will be $\frac{1}{r}\frac{\hat{E}_0}{2N+1}e^{j(\omega t-\tau+mk_0 d\sin(\theta)\sin(\phi))}$ where $k_0 = \frac{2\pi}{\lambda}$. The polarization of the bar radiator is in the plane perpendicular to the bar radiator. Taking polarization into consideration, $\hat{E}_{\theta,r}$ is given by

$$\begin{aligned}
\hat{E}_{\theta,r} &= \frac{\hat{E}_0}{r(2N+1)}e^{j\omega(t-\tau)}\cos(\phi)\sum_{m=-N}^{N} e^{jmk_0 d\sin(\theta)\sin(\phi)}, \\
&= \frac{\hat{E}_0}{r(2N+1)}e^{j\omega(t-\tau)}\cos(\phi) \\
&\quad\times \sum_{m=-N}^{N} \frac{e^{-j(N+0.5)k_0 d\sin(\theta)\sin(\phi)} - e^{j(N+0.5)d\sin(\theta)\sin(\phi)}}{e^{-0.5jd\sin(\theta)\sin(\phi)} - e^{0.5jk_0 d\sin(\theta)\sin(\phi)}}, \\
&= \frac{\hat{E}_0}{2r(N+0.5)}e^{j\omega(t-\tau)}\cos(\phi)\frac{\sin((N+0.5)k_0 d\sin(\theta)\sin(\phi))}{\sin(0.5k_0 d\sin(\theta)\sin(\phi))}. \qquad (6.31)
\end{aligned}$$

As $N \to \infty$, $(N+0.5)d \to 0.5W$ and $\sin(0.5k_0 d\sin(\theta)\sin(\phi)) \to 0.5k_0 d\sin(\theta)\sin(\phi)$. Hence, as $N \to \infty$, (6.31) becomes

$$\begin{aligned}
\hat{E}_{\theta,r} &= \frac{\hat{E}_0\, e^{j\omega(t-\tau)}}{r}\frac{\sin(0.5k_0 W\sin(\theta)\sin(\phi))}{0.5k_0 W\sin(\theta)\sin(\phi)}\cos(\phi), \\
&= \frac{\hat{E}_0\, e^{j\omega(t-\tau)}}{r}A(\theta,\phi)\cos(\phi)\ , \quad \text{where} \qquad (6.32a) \\
A(\theta,\phi) &= \frac{\sin(0.5k_0 W\sin(\theta)\sin(\phi))}{0.5k_0 W\sin(\theta)\sin(\phi)}. \qquad (6.32b)
\end{aligned}$$

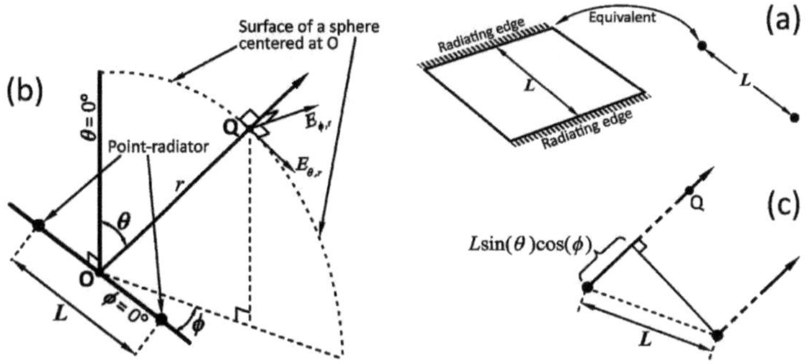

Fig. 6.24 (**a**) Radiating edges modeled by two point radiators. (**b**) Definitions for the angles θ and ϕ. (**c**) Path difference from the point radiators to far field

Similarly, it can be shown that

$$\hat{E}_{\phi,r} = \frac{\hat{E}_0\, e^{j\omega(t-\tau)}}{r} A(\theta,\phi)\cos(\theta)\sin(\phi). \tag{6.33}$$

The far-field radiation pattern of the bar radiator is equivalent to that of a point radiator placed at **O** whose $\hat{E}_{\theta,r}$ and $\hat{E}_{\phi,r}$ are given by (6.32) and (6.33), respectively.

The patch antenna shown in Fig. 6.22 has two radiating edges. In the derivation of its far-field radiation pattern, each radiating edge is replaced by a point radiator as shown in Fig. 6.24a. Using the same definition for θ, ϕ, and (θ,ϕ) as shown in Figs. 6.23b, 6.24b is obtained. The path difference between the two point radiators to the far-field point **Q** is $L\sin(\theta)\cos(\phi)$ as shown in Fig. 6.24c. Hence, the contributions to $E_{\theta,r}$ at **Q** from the two point radiators are $\frac{\hat{E}_0\, e^{j\omega(t-\tau)}}{r} A(\theta,\phi)\cos(\phi)e^{-0.5jk_0L\sin(\theta)\cos(\phi)}$ and $\frac{\hat{E}_0\, e^{j\omega(t-\tau)}}{r} A(\theta,\phi)\cos(\phi)e^{0.5jk_0L\sin(\theta)\cos(\phi)}$, respectively. Thus,

$$E_{\theta,r} = \frac{\hat{E}_0\, e^{j\omega(t-\tau)}}{r} A(\theta,\phi)\cos(\phi)\left(e^{-0.5jk_0L\sin(\theta)\cos(\phi)} + e^{0.5jk_0L\sin(\theta)\cos(\phi)}\right), \tag{6.34a}$$

$$= \frac{2\hat{E}_0\, e^{j\omega(t-\tau)}}{r} A(\theta,\phi)B(\theta,\phi)\cos(\phi)\ , \quad \text{where} \tag{6.34b}$$

$$B(\theta,\phi) = \cos(0.5k_0L\sin(\theta)\cos(\phi)). \tag{6.34c}$$

Similarly, it can be shown that

$$E_{\phi,r} = \frac{2\hat{E}_0\, e^{j\omega(t-\tau)}}{r} A(\theta,\phi)B(\theta,\phi)\cos(\theta)\sin(\phi). \tag{6.35}$$

The factor $e^{j\omega(t-\tau)}$ in (6.34) and (6.35) is a complex sinusoid with magnitude equal to 1. Let $\bar{E}_0 = |2\hat{E}_0\, e^{j\omega(t-\tau)}| = 2\hat{E}_0$ and the corresponding $E_{\phi,r}$ and $E_{\theta,r}$ be $\bar{E}_{\phi,r}$ and $\bar{E}_{\theta,r}$, respectively. In this case, (6.34) and (6.35) becomes

$$\bar{E}_{\theta,r} = \frac{\bar{E}_0}{r} A(\theta,\phi) B(\theta,\phi)\cos(\phi), \tag{6.36a}$$

$$\bar{E}_{\phi,r} = \frac{\bar{E}_0}{r} A(\theta,\phi) B(\theta,\phi)\cos(\theta)\sin(\phi), \tag{6.36b}$$

$$A(\theta,\phi) = \frac{\sin(0.5k_0 W \sin(\theta)\sin(\phi))}{0.5k_0 W \sin(\theta)\sin(\phi)}, \tag{6.36c}$$

$$B(\theta,\phi) = \cos(0.5k_0 L \sin(\theta)\cos(\phi)). \tag{6.36d}$$

We shall consider the special case where $L = W = 0.5\lambda$. For $\phi = 0$, $\bar{E}_{\theta,r} = \frac{\bar{E}_0}{r}\cos(0.5\pi\sin(\theta))$ and $\bar{E}_{\phi,r} = 0$. The electric field $\bar{E}_{\theta,r}$ is maximum at $\theta = 0$ and decreases to 0 when $|\theta|$ increases to 0.5π. For $\phi = 0.5\pi$, $\bar{E}_{\phi,r} = \frac{\bar{E}_0}{r}\frac{\sin(0.5\pi\sin(\theta))}{0.5\pi\sin(\theta)}\cos(\theta)$ and $\bar{E}_{\theta,r} = 0$. The electric field $\bar{E}_{\phi,r}$ is maximum at $\theta = 0$ and decreases to 0 when $|\theta|$ increases to 0.5π.

For a half-wave ($L = 0.5\lambda$) patch antenna, the feedpoint impedance depends on the width W and the location of the feedpoint. In Fig. 6.22a, the feedpoint is at the edge. At the edge, the voltage is maximum and the current is minimum and so the feedpoint impedance is high. As the current increases and the voltage decreases towards the center, the feedpoint impedance is decreased if the antenna is fed nearer the center. This can be achieved by using an inset feed as shown in Fig. 6.25. If R_0 is the feedpoint impedance at the edge, the feedpoint impedance R_v for a patch with an inset feed is given by

$$R_v = R_0\cos^2\left(\frac{\pi v}{L}\right) = R_0\cos^2\left(\frac{2\pi v}{\lambda}\right), \quad L = \frac{\lambda}{2}. \tag{6.37}$$

Patch antenna may be designed to radiate circularly polarized radiation. There are many ways for achieving this. Figure 6.26 shows two of these methods. In Fig. 6.26a, the patch is a square patch fed at two edges with 90° out of face signals. Circular polarization is obtained from two orthogonal linearly polarized waves in phase quadrature in the time domain. Figure 6.26b is a single-feed design. The patch is almost square (but not strictly square) and is corner-fed.

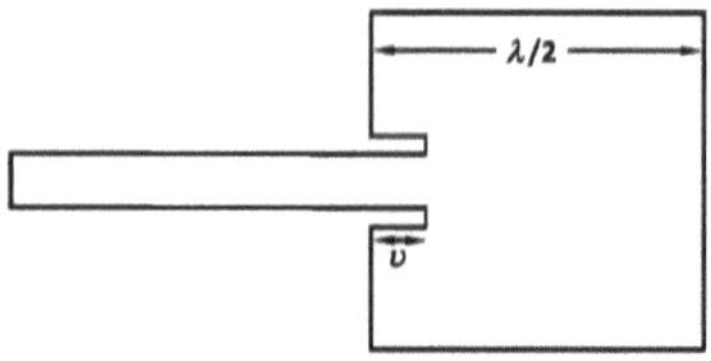

Fig. 6.25 An inset feed

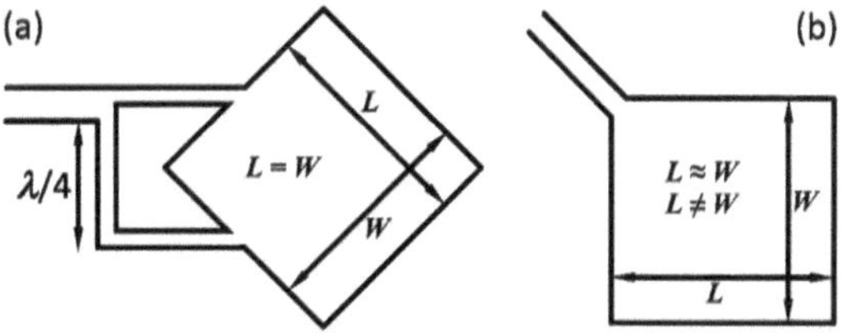

Fig. 6.26 Circularly polarized rectangular patch antenna. (**a**) Separate orthogonal feeds. (**b**) Single-corner feed

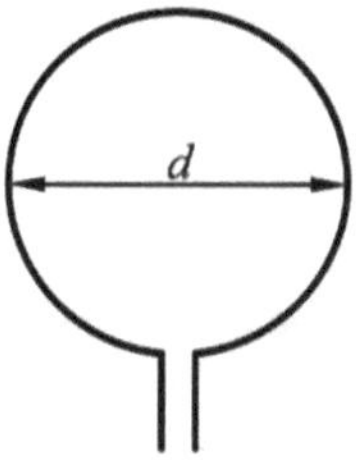

Fig. 6.27 Loop antenna

6.3.7 Loop Antenna

Figure 6.27 shows a circular loop antenna but the loop can be of any shape. A loop antenna is fed from a balanced source. There are two types of loop antenna, namely, resonant loop (or large loop) and small loop.

6.3.7.1 Resonant Loop Antenna

The resonant loop antenna is also called the large loop antenna. At its operating frequency, the wavelength is slightly less than the circumference πd for a circular antenna. Its radiation pattern peaks in the direction perpendicular to the plane of the antenna. Its polarization depends on the feedpoint. For the feedpoint as shown in Fig. 6.27, it is horizontally polarized. Its impedance is purely real. An advantage of the loop antenna is its compactness; its diameter is $\approx \frac{\lambda}{\pi}$ for a circular loop, and its width is $\approx \frac{\lambda}{4}$ for a square loop.

6.3.7.2 Small Loop Antenna

A small loop antenna is a loop antenna whose circumference is very much smaller than a wavelength. It has low radiation resistance and high inductive reactants. It is not a good radiator and so it is often used as a receiving antenna because of its compactness. Let N be the number of turns of the conductor around the loop and let

A be the area of the loop antenna. The radiation resistance of the small loop is given by

$$R_{\text{small loop}} = \frac{8\eta_0\pi^3}{3}\left(\frac{NA}{\lambda^2}\right)^2 \approx \left(\frac{177NA}{\lambda^2}\right)^2, \tag{6.38}$$

where $\eta_0 = 120\pi \approx 377\Omega$ is the intrinsic impedance of free space. The inductive component of the impedance at frequency f_0 Hz is given by

$$X_{\text{small loop}} = \pi f_0 \mu_0 d\left[\ln\left(\frac{4d}{b}\right) - 1.75\right], \tag{6.39}$$

where b is the radius of the wire.

While the radiation pattern of the resonance loop peaks along the axis perpendicular to the loop, the radiation pattern of the small loop antenna peaks in the direction parallel to the plane of the loop. The current in the small loop is approximately constant; let the current be I_0. The electric field strength E at distance r from the loop and at an angle θ from the axis perpendicular to the loop is given by

$$E = \frac{\eta_0 k_0^2 d^2 I_0}{d}\sin(\theta), \tag{6.40}$$

where $k_0 = \frac{2\pi}{\lambda}$ is the wave number. The electric field strength is insensitive to the orientation of the feed position. The polarization is in the plane of the loop.

The response of a small loop antenna is proportional to the rate of change of the magnetic flux through the loop and so it is also known as magnetic loop antenna. Including a material with high magnetic permeability into the core will increase its radiation resistance by orders of magnitude, so its performance improves by orders of magnitude. As the human body does not significantly affect the magnetic fields, the human body does not significantly affect the small loop antenna.

6.3.7.3 Halo Antenna

Figure 6.28 shows a halo antenna where the circumference of the antenna is approximately 0.5λ. Superficially, a halo antenna looks like a loop antenna but it

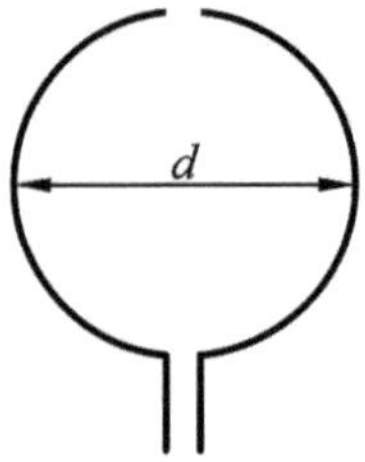

Fig. 6.28 Halo antenna

is actually not. A halo antenna is actually a half-wavelength dipole antenna bent into a circle with an electrical break directly opposite the feedpoint. For the same operating frequency, the halo antenna is half the size of the resonant loop antenna. A halo antenna is a self-resonant antenna with purely resistive feedpoint impedance at its designed operating frequency. The current distribution along the two arms of a halo antenna is similar to that along the two arms of a half-wave dipole. The radiation pattern and polarization of the halo antenna are similar to that of the small loop antenna with the exception that the halo radiates slightly more towards the split in the loop and a small amount in the direction perpendicular to the loop plane.

6.3.8 Horn Antenna

Figure 6.29 shows several horn antennas. Horn antennas are widely used in frequencies above 300 MHz. They are used in speed enforcement cameras. A horn antenna has no resonant elements and so has wide bandwidth; its higher cutoff frequency is typically ten times higher than its lower cutoff frequency. Its gain is typically 10 dBi–20 dBi. A horn antenna has very little loss and so its directivity is roughly equal to its gain. It is fed by a section of a waveguide, and the wave guide is usually fed with a short dipole or monopole attached to one side of the waveguide. It is commonly used as a feed antenna feeding larger reflector antenna.

The most commonly used horn shapes are the pyramidal and conical horns shown in Fig. 6.29 because they have straight sides and are easy to design and fabricate. The horn shape that gives the minimum reflected power is the exponential horn; it does not have straight sides and is used only in special application such as big satellite antenna and radio telescope. A horn antenna in the shape of a four-sided pyramid as in Fig. 6.29a and flares in both the directions of the electric and magnetizing fields is called pyramidal horn. A horn that flares only in the direction of the magnetizing field as in Fig. 6.29b is called a H-plane horn. A horn that flares only in the direction

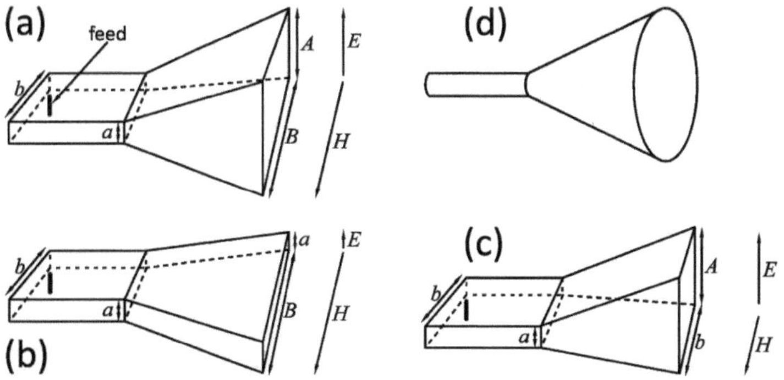

Fig. 6.29 (**a**) Pyramidal horn, (**b**) H-plane horn, (**c**) E-plane horn, and (**d**) conical horn

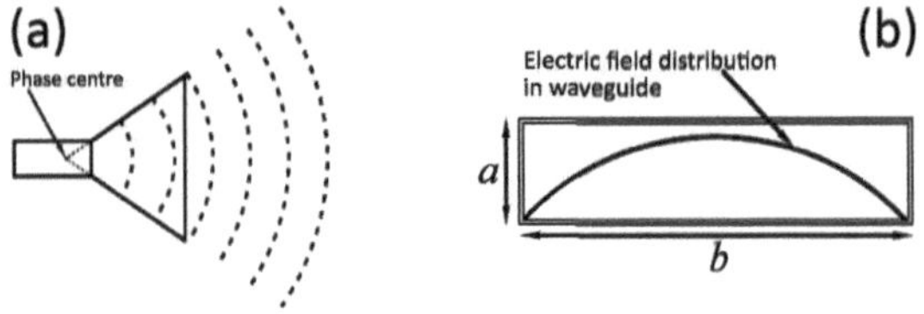

Fig. 6.30 EM waves travel down a horn as spherical wavefront

of the electric field as in Fig. 6.29c is called an E-plane horn. A horn in the shape of a cone with a circular cross section as in Fig. 6.29d is called a conical horn. E-plane horn and H-plane horn are collectively called sectoral horn. A sectoral horn produces a fan-shaped beam. The H-plane horn produces a beam narrow in the plane of the magnetizing field but wide in the plane of the electric field. The E-plane horn produces a beam narrow in the plane of the electric field but wide in the plane of the magnetizing field.

EM waves travel down a horn as spherical wavefront originating from a point called the phase center, which is the apex of the horn. See Fig. 6.30a. The electric field distribution in the waveguide is shown in Fig. 6.30b; the electric field strength peaks at the center and decreases to zero at the walls

6.3.9 Parabolic Reflector Antenna

Figure 6.31 shows a parabolic dish antenna. Parabolic antennas belong to a class of reflector antennas where the reflectors are parabolic. The reflector may be a solid metal surface or a metal screen or metal wire grill as long as the holes are smaller than 0.1λ. A parallel wire grill reflector only reflects the component of EM wave whose polarization is parallel to the grill elements. If the incident wave is circularly polarized, the reflected wave will be linearly polarized in parallel to the grill elements. If the incident wave is linearly polarized with polarization perpendicular to the grill elements, it will not be reflected. Note that a shiny reflector also reflects sunlight and the focused sunlight may generate enough heat to cause overheating to the receiver. A parabolic antenna is most often fed by a horn at the focal point of the antenna.

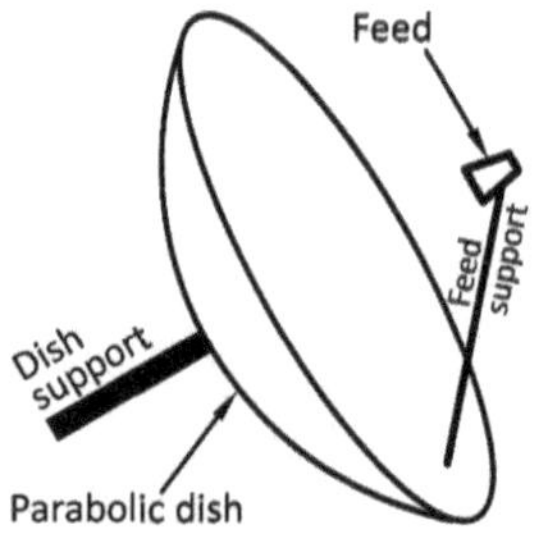

Fig. 6.31 Parabolic antenna

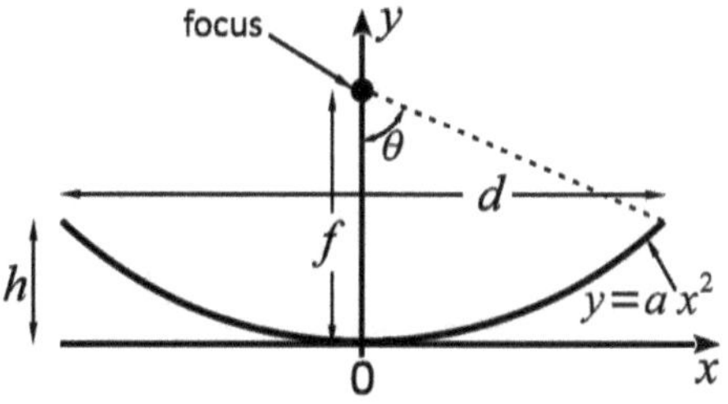

Fig. 6.32 Parameters of a parabola

The geometry of a parabolic dish antenna is best understood by considering the parabola $y = ax^2$ with focal length f and aperture d as shown in Fig. 6.32. The parameters of the parabola are related to each other by

$$y = ax^2, \tag{6.41a}$$

$$f = \frac{1}{4a}, \tag{6.41b}$$

$$h = \frac{ad^2}{4}, \tag{6.41c}$$

$$f\text{-number} = \frac{f}{d} = \frac{1}{4ad} = \frac{1}{4\tan(0.5\theta)}. \tag{6.41d}$$

Since the equivalent area of an isotropic antenna is $\frac{\lambda^2}{4\pi}$ and the area of the antenna aperture is $\pi\frac{d^2}{4}$, the maximum gain $G_{\max}$ of the dish antenna is given by

$$G_{\max} = \frac{\text{Area of aperture}}{\text{Area of isotropic antenna}} = \frac{4\pi}{\lambda^2} \times \frac{\pi d^2}{4} = \frac{\pi^2 d^2}{\lambda^2}. \tag{6.42}$$

In practice, there are various factors such as feed spillover (where some of the radiation from the feed falls outside the edge of the dish), feed illumination taper (where the outer part of the dish is illuminated with a lower intensity, which effectively reduces the aperture area), aperture blockage (blockage due to the feed support structure in front of the dish), dish shape error, etc. causing reduction in the radiation efficiency. The reduction in radiation efficiency is represented by aperture efficiency ρ. For parabolic antenna, typically, $0.5 < \rho < 0.7$. Taking the aperture efficiency into consideration, the antenna gain G is given by

$$G = \rho\left(\frac{\pi d}{\lambda}\right)^2 = \rho\pi^2\left(\frac{d}{\lambda}\right)^2. \tag{6.43}$$

It can be seen from (6.43) that the antenna gain is proportional to $\left(\frac{d}{\lambda}\right)^2$.

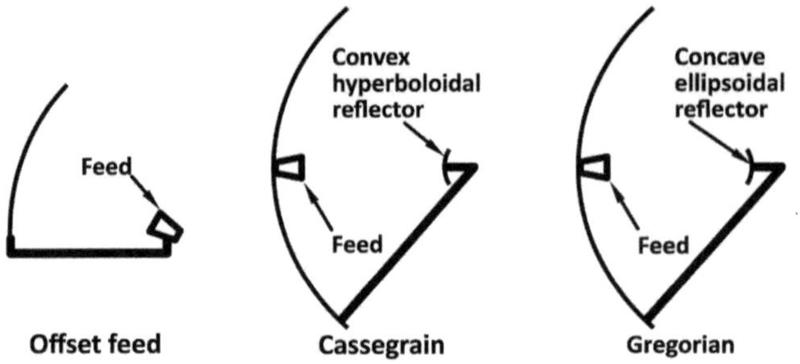

Fig. 6.33 Feed structures of parabolic antenna

The half-power beamwidth ϕ is given by

$$\phi = \frac{\lambda}{d} \text{ radian} = \frac{360}{2\pi} \times \frac{\lambda}{d} \text{ degree.} \tag{6.44}$$

Combining (6.43) and (6.44), we have

$$G = \rho\left(\frac{\pi}{\phi}\right)^2, \quad \phi \text{ in radian.} \tag{6.45}$$

Front feed as shown in Fig. 6.31 is the most common feed method for the parabolic antenna. However, there are many other commonly used feed structures. Some of them are shown in Fig. 6.33. In the offset feed antenna, the feed structure is moved out of the beam path so that it does not block the beam. In Cassegrain and Gregorian antennas, the feed is located on or behind the dish. A convex hyperboloidal secondary reflector is used in the case of the Cassegrain antenna, whereas a concave ellipsoid secondary reflector is used in the case of the Gregorian antenna.

6.4 RF Switch

The implementation of variable phase shifters and variable attenuators often requires RF switches. There are two main types of RF switches, namely, semiconductor RF switch and MEMS RF switch.

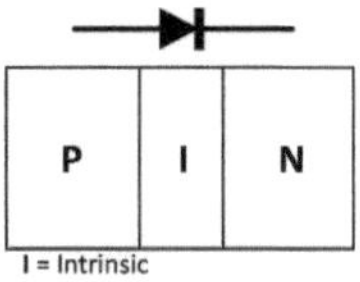

Fig. 6.34 Structure of a PIN diode

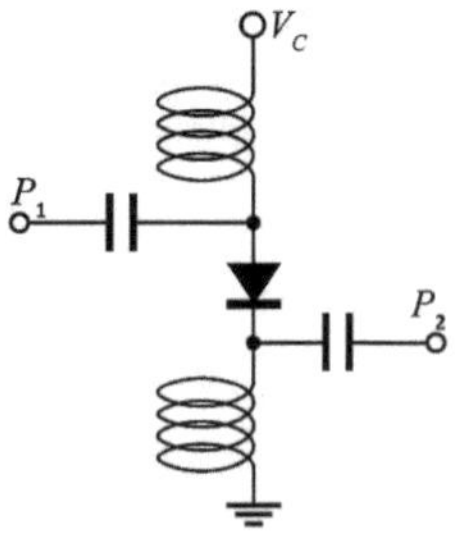

Fig. 6.35 PIN diode RF switch

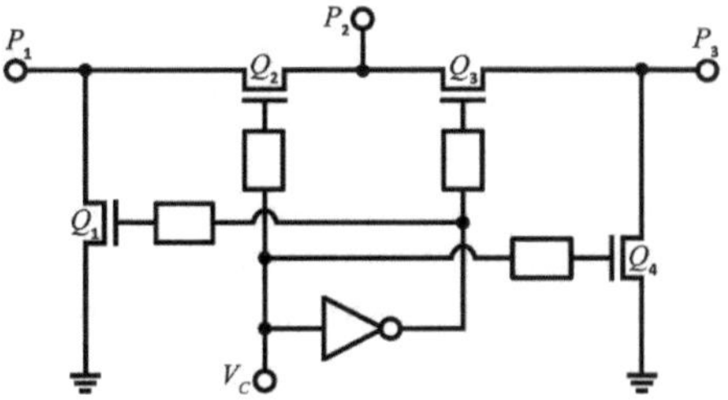

Fig. 6.36 RF FET switch

6.4.1 Semiconductor RF Switch

Commonly used semiconductor RF switches are PIN diode switch and FET (field effect transistor) switch. A PIN diode is a diode and differs from the conventional diode in that it has an intrinsic (I) region between the conventional positively doped (P) and negatively doped (N) regions as shown in Fig. 6.34. Figure 6.35 shows an implementation of an RF switch using a PIN diode. When the control voltage V_C is positive, the PIN diode is forward biased and behaves like a short circuit; port P_1 is shorted to port P_2. When the control voltage V_C is negative, the PIN diode is reverse biased and becomes an open circuit; port P_1 is disconnected from port P_2.

PIN diode switches are often used in "chip and wire" MIC (**m**icrowave **i**ntegrated **c**ircuit) implementation. MIC should not be confused with MMIC, which is monolithic. A PIN diode has a better frequency response at high frequencies than an FET when used as a switch, but the wirebond inductance associated with MIC construction limits its application at high frequencies.

Historically, FET RF switches were implemented using GaAs high electron-mobility transistors (HEMTs). In recent years, HEMT has been replaced by silicon-on-insulator (SOI) transistor. Figure 6.36 shows an implementation of an RF switch using FET. When the control voltage V_C is high, transistors Q_2 and Q_4 are turned on but Q_1 and Q_3 are turned off; port P_1 is shorted to port P_2; port P_3 is shorted to ground. When the control voltage V_C is low, transistors Q_1 and

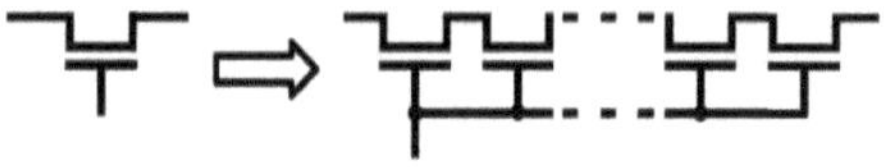

Fig. 6.37 Each FET in Fig. 6.36 is implemented using a series of cascaded FETs

Fig. 6.38 Each inductor and its parallel stray capacitor form a parallel resonant circuit

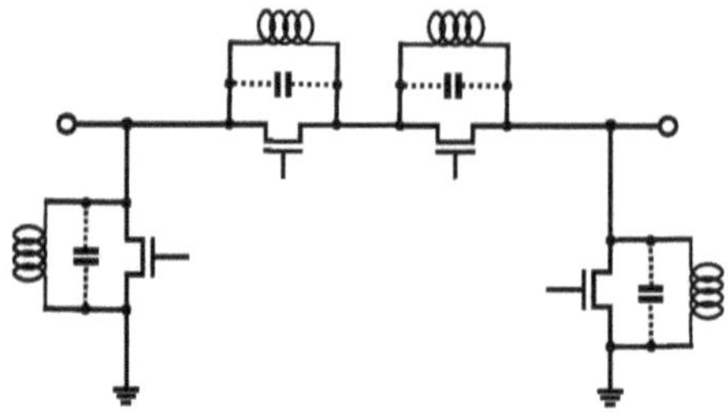

Q_3 are turned on but Q_2 and Q_4 are turned off; port P_3 is shorted to port P_2; port P_1 is shorted to ground. In actual implementation, each FET in Fig. 6.36 is usually implemented using a series of cascaded FETs as shown in Fig. 6.37 for the dual purposes of increasing the breakdown voltage and improving the isolation. The drain-source stray capacitance will degrade the performance of the FET. This problem can be overcome by connecting an inductor in parallel with the stray capacitance as shown in Fig. 6.38. In Fig. 6.38, the dotted line capacitors represent the stray capacitances. The value of the inductors is selected such that each inductor and its parallel stray capacitor form a parallel resonant circuit at the frequency of operation. This parallel tune circuit method may be used in other scenarios to nullify stray capacitance. FET switches are often used in MMIC (**m**onolithic **mi**crowave **i**ntegrated **c**ircuit) implementation at frequency up to millimeter waves.

6.4.2 MEMS RF Switch

Figure 6.39 shows an electrostatic actuated MEMS switch. It consists of a cantilever suspended over a substrate. The cantilever may be a gold membrane. There is a conductive actuation electrode deposited on the substrate below the cantilever forming the bottom actuation electrode. If the cantilever is not conductive, there will also be a conductive electrode deposited underneath the cantilever and directly above the bottom actuation electrode. When a large potential difference (a few tens

Fig. 6.39 Electrostatic actuated MEMS switch

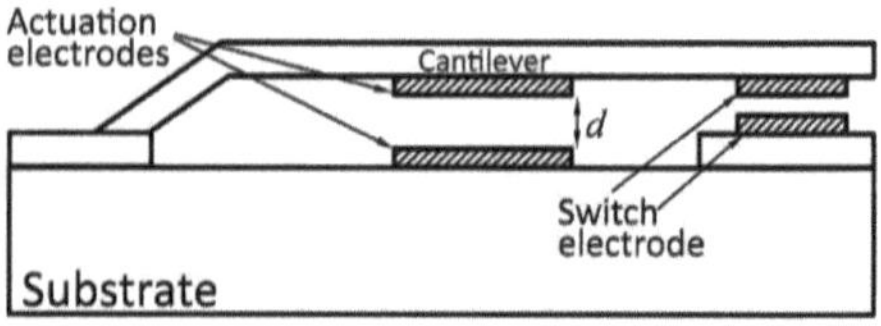

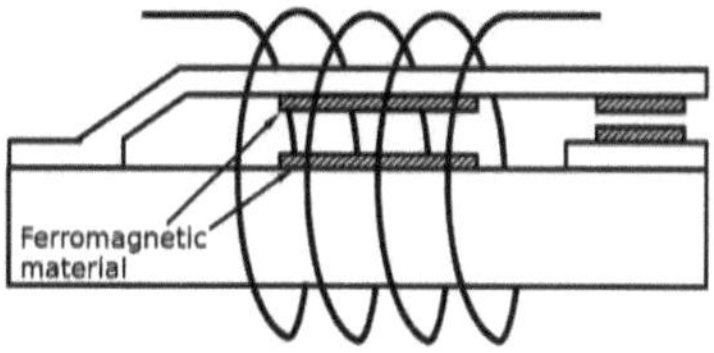

Fig. 6.40 Electromagnetic actuated MEMS switch

of volts) is applied to the two actuation electrodes, an electrostatic force will pull the cantilever down and the cantilever will make contact with the switch electrode. The actuation electrodes will not make contact with each other and so there will be no static current flowing through the actuation electrodes. Electrostatic actuated MEMS consumes no static power. If the area of the actuation electrode is A and if the separation between the two actuation electrode is d, the electrostatic force of attraction F_{attract} exerted on the actuation electrodes is given by

$$F_{\text{attract}} = \frac{\varepsilon_0 A V^2}{2d^2}, \tag{6.46}$$

where V is the potential difference between the actuation electrodes and ε_0 is the permittivity of air. There are many actuation methods for MEMS switch. If the actuation electrodes are replaced by ferromagnetic material, and if a coil is wound on the structure as shown in Fig. 6.40, the MEMS is electromagnetically actuated. There are other forms of actuation. Piezoelectric actuation (Chap. 5, Sect. 5.5.2.4) and electrothermal actuation (Chap. 5, Sect. 5.5.2.5) are another two possibilities.

6.5 Phase Shifter

A phase shifter is a very useful RF component that provides a change in the phase of the signal. Phase shifter is critical in the implementation of phased arrays. An ideal phase shifter only shifts the phase of the signal; it is lossless and impedance matched to both the source and load. Phase shifting may be performed at either the high frequency RF stage or the IF (intermediate frequency) stage. Performing phase shifting at the IF stage is easier and more accurate than performing it at the RF stage, but for a given bandwidth, the relative bandwidth (i.e., bandwidth divided by center frequency) at the IF stage is much larger.

Two signals with the same frequency but with different phases may be digitally synthesized and frequency translated to RF frequency by modulation to produce two RF signals with the same frequency but with different phases assuming that the modulation process does not introduce phase mismatch.

Before proceeding further, we shall examine the amount of phase shift required in the implementation of a phase array. Consider an N-element linear array with separation d between consecutive array elements as shown in Fig. 6.41. Suppose that the array is steered with an arbitrary broadside angle θ where $\theta \leq \theta_{\max}$. In the

Fig. 6.41 Linear phase array

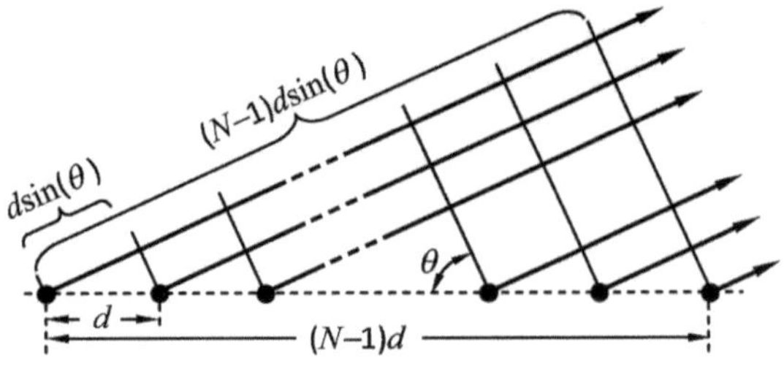

far field, the path difference between consecutive array elements is $d\sin(\theta)$, and the maximum path difference is $(N-1)d\sin(\theta_{\max})$. Let $\phi_{N,\max}$ be the maximum phase shift of the Nth array element. We have

$$\phi_{N,\max} = (N-1)k_0 d\sin(\theta_{\max}), \tag{6.47}$$

where $k_0 = \frac{2\pi}{\lambda}$. The phase shift of other array elements will be less than $(N-1)kd\sin(\theta_{\max})$. If $\phi_{N,\max} < 2\pi$, the maximum required phase shift is $(N-1)k_0 d\sin(\theta_{\max})$. If $\phi_{N,\max} \geq 2\pi$, the maximum required phase shift is 2π since any phase shift larger than 2π can be reduced to less than 2π by subtracting an integer number of 2π from it. Consider an example with $d = 0.5\lambda$, $N = 32$, and $\theta_{\max} = 30°$. This will lead to the result $\phi_{N,\max} = 15.5\pi > 2\pi$; in this case, the required phase shift may be any arbitrary angle between 0 and 2π.

Phase shifter may be classified into two broad categories, namely, (a) discrete step phase shifter, which is also called digital phase shifter, and (b) continuously tunable phase shifter, which is also called analogue phase shifter.

Discrete step phase shifter consists of a bank of fixed phase shifters that can be switched in and out of a circuit. Switches produce insertion loss and must be compensated.

6.5.1 *Switched Line Phase Shifter*

Signal takes time to travel through a transmission line. If the length of the transmission line is l and if the wavelength of the signal traveling in the transmission line is λ_l, the phase difference between the signal at the two ends of the transmission line will be $\frac{2\pi l}{\lambda_l}$ radians $\frac{l}{\lambda_l} \times \frac{360}{2\pi}{}^{\circ}$. For example, 180° and 90°phase difference can be achieved with $l = 0.5\lambda$ and $l = 0.25\lambda$, respectively. Thus, a section of a transmission line may be used as a phase shifter as shown in Fig. 6.42. Switched line phase shifters are usually used only for coarse angular resolution such as 180°, 90°, and 45°. The filter-type phase shifters described below are used in better resolutions.

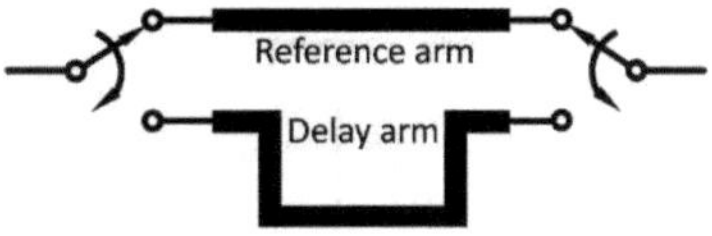

Fig. 6.42 Switched line phase shifter

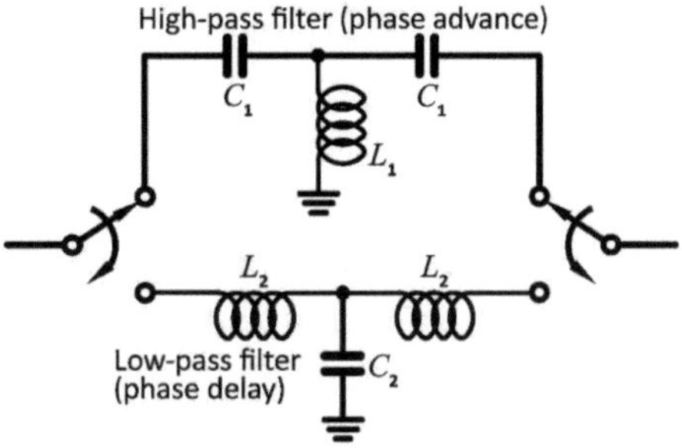

Fig. 6.43 High-pass low-pass phase shifter

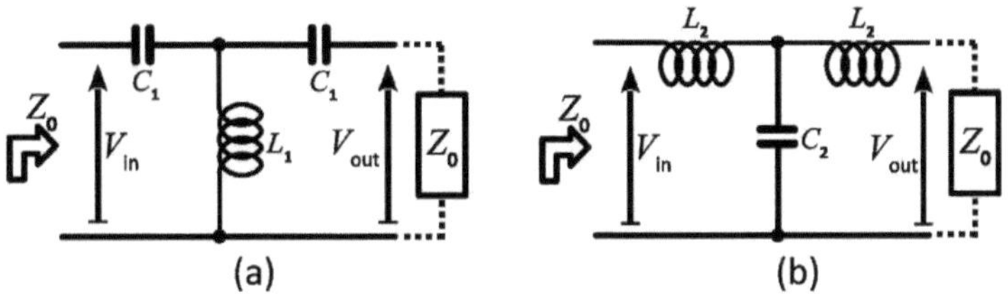

Fig. 6.44 T-section

6.5.2 *High-Pass Low-Pass Phase Shifter*

The high-pass low-pass phase shifter shown in Fig. 6.43 consists of a high-pass filter and a low-pass filter. The high-pass filter produces phase advance and the low-pass filter produces phase delay.

Suppose that the characteristic impedance of the line is Z_0 and so the load of the high-pass low-pass phase shifter is Z_0. See Fig. 6.44. In order for the driving point impedance at the input end to be also Z_0, (6.48) must be satisfied:

$$2\omega^2 L_1 C_1 = 1 + \omega^2 C_1^2 Z_0^2, \tag{6.48a}$$

$$2L_2 C_2 = C_2^2 Z_0^2 + \omega^2 L_2^2 C_2^2. \tag{6.48b}$$

It can be shown that $\frac{V_{\text{out}}(e^{j\omega})}{V_{\text{in}}(e^{j\omega})}$ is given by

$$\frac{V_{\text{out}}(e^{j\omega})}{V_{\text{in}}(e^{j\omega})} = \frac{\omega^2 L_1 C_1}{\omega^2 L_1 C_1 - 1 - j\omega C_1 Z_0} = e^{j\phi}, \tag{6.49a}$$

$$\frac{V_{\text{out}}(e^{j\omega})}{V_{\text{in}}(e^{j\omega})} = \frac{1}{1 - \omega^2 L_2 C_2 + j\omega C_2 Z_0} = e^{j\phi}, \tag{6.49b}$$

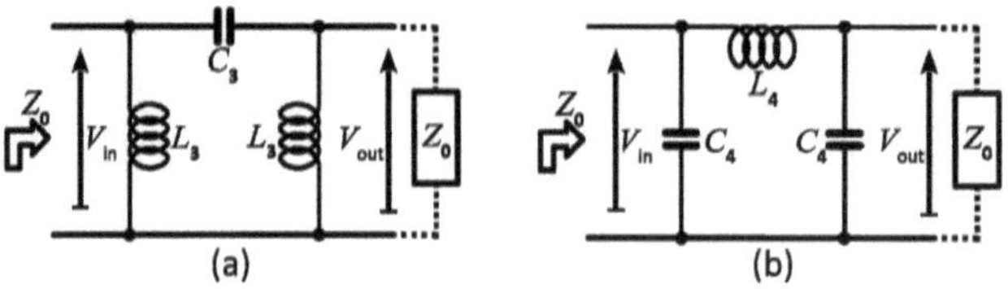

Fig. 6.45 π-section

where ϕ is the phase shift of the phase shifter. In (6.49), $\frac{V_{out}(e^{j\omega})}{V_{in}(e^{j\omega})} = e^{j\phi}$ since a phase shifter is lossless and only shifts the phase of the signal. From (6.48) and (6.49), we have

$$C_1 = \frac{\sin(\phi)}{\omega Z_0(1 - \cos(\phi))}, \tag{6.50a}$$

$$L_1 = \frac{Z_0}{\omega \sin(\phi)}, \tag{6.50b}$$

$$C_2 = \frac{\sin(-\phi)}{\omega Z_0}, \tag{6.50c}$$

$$L_2 = Z_0 \frac{1 - \cos(\phi)}{\omega \sin(-\phi)}. \tag{6.50d}$$

Note that in (6.50c) and (6.50d), ϕ is negative because low-pass filter produces phase delay.

The T-section in Figs. 6.43 and 6.44 may be substituted by the π-section in Fig. 6.45. For the circuits shown in Fig. 6.45, in order for the driving point impedance at the input end to be also Z_0, (6.51) must be satisfied:

$$\omega^2 L_3^2 = (2\omega^2 L_3 C_3 - 1) Z_0^2, \tag{6.51a}$$

$$\omega^2 L_4 C_4 = (2 - \omega^2 L_4 C_4)\omega^2 C_4^2 Z_0^2. \tag{6.51b}$$

It can be shown that $\frac{V_{out}(e^{j\omega})}{V_{in}(e^{j\omega})}$ is given by

$$\frac{V_{out}(e^{j\omega})}{V_{in}(e^{j\omega})} = \frac{\omega^2 L_3 C_3 Z_0}{(1 - \omega^2 L_3 C_3) Z_0 + j\omega L_3} = e^{j\phi}, \tag{6.52a}$$

$$\frac{V_{out}(e^{j\omega})}{V_{in}(e^{j\omega})} = \frac{Z_0}{(1 - \omega^2 L_4 C_4) Z_0 + j\omega L_4} = e^{j\phi}. \tag{6.52b}$$

From (6.51) and (6.52), we have

$$C_3 = \frac{1}{\omega Z_0 \sin(\phi)}, \tag{6.53a}$$

$$L_3 = \frac{Z_0 \sin(\phi)}{\omega(1 - \cos(\phi))}, \tag{6.53b}$$

$$C_4 = \frac{1 - \cos(\phi)}{\omega Z_0 \sin(-\phi)}, \tag{6.53c}$$

$$L_4 = \frac{Z_0 \sin(-\phi)}{\omega}. \tag{6.53d}$$

Note that in (6.53c) and (6.53d), ϕ is negative because low-pass filter produces phase delay.

6.5.3 Switched Filter Phase Shifter

A phase shifter can be implemented by using only the high-pass or low-pass section. Figure 6.46 shows such configurations of switched filter phase shifter. Switches S_1 and S_2 are in antiphase, i.e., when one is closed, the other is opened. When S_1 is closed, port P_1 is shorted to port P_2. When switch S_2 is closed, port P_1 and port P_2 are the two ends of a π configured filter. Figure 6.46a is high-pass and Fig. 6.46b is low-pass. The phase shift is that of the filter.

Figure 6.47 shows FET implementations for the switched filter phase shifter shown in Fig. 6.46. The value of L_2 is selected such that L_2 and C_2 resonate at the

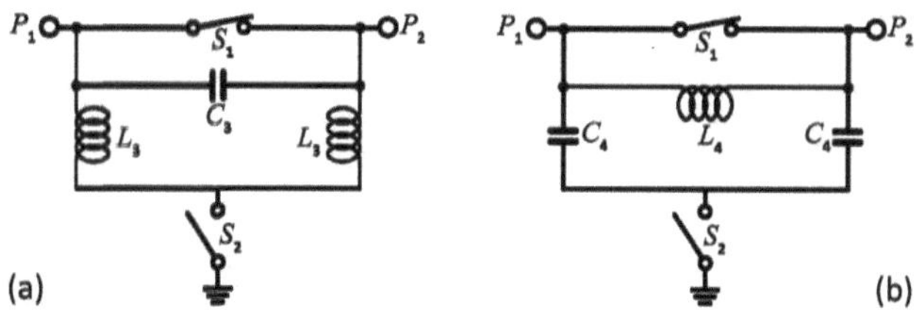

Fig. 6.46 Switched π-filter phase shifter

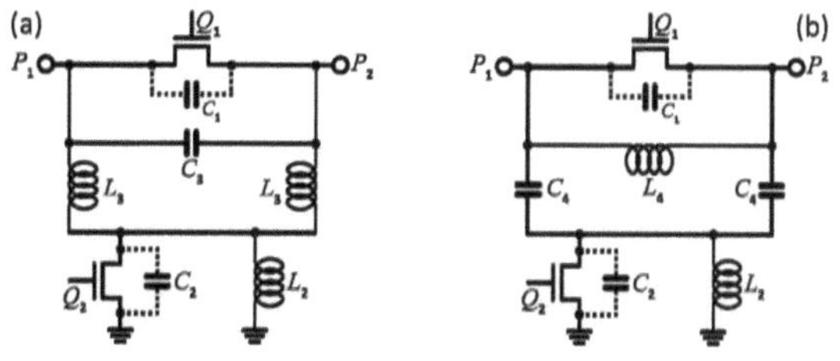

Fig. 6.47 FET implementation for Fig. 6.46

operating frequency as already discussed in Sect. 6.4.1 under RF FET switch. The drain-source capacitance of Q_1 is in parallel with C_3 in Fig. 6.47a and in parallel with L_4 in Fig. 6.47b. Hence, the effective capacitance $C_{3,\text{effective}}$ and inductance $L_{4,\text{effective}}$ at frequency ω become

$$C_{3,\text{effective}} = C_1 + C_3, \tag{6.54a}$$

$$L_{4,\text{effective}} = \frac{L_4}{1 - \omega^2 L_4 C_1}. \tag{6.54b}$$

6.5.4 Loaded-Line Phase Shifter

Figure 6.48 shows a loaded-line phase shifter. The loaded-line phase shifter consists of a transmission line loaded with a reactant jX_i at each end. The length of the transmission line is usually equal to a quarter-wavelength. The phase shift is determined by the value of jX_i.

6.5.5 Continuously Varying Phase Shifter

A continuously varying phase shifter may be realized using any one of the circuits in Figs. 6.46, 6.47, and 6.48 by replacing the fixed value capacitor by a variable capacitance diode. A variable capacitance diode is commonly called varactor. A varactor is always reversed biased in a circuit; its capacitance is controlled by the reversed biasing voltage. Its electrical symbol is the same as that of a diode with the single bar replaced by a double bar.

In Fig. 6.49a, the varactor D is reversed biased by a control voltage V_C through two large resistors R. The two capacitors C are for blocking the biasing voltage from the rest of the circuit. Two inductors L may be inserted to improve isolation as shown in Fig. 6.49b.

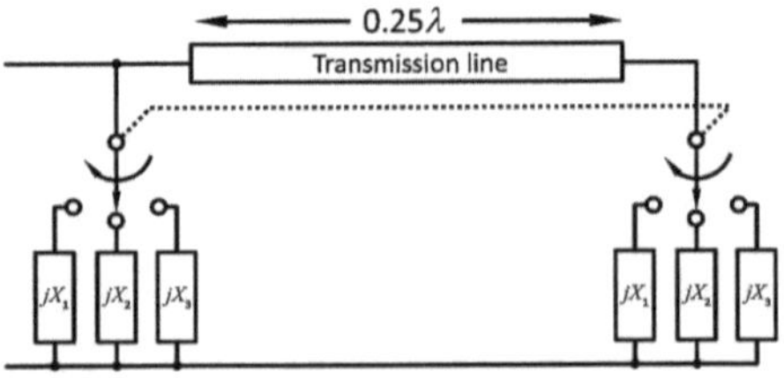

Fig. 6.48 Loaded-line phase shifter

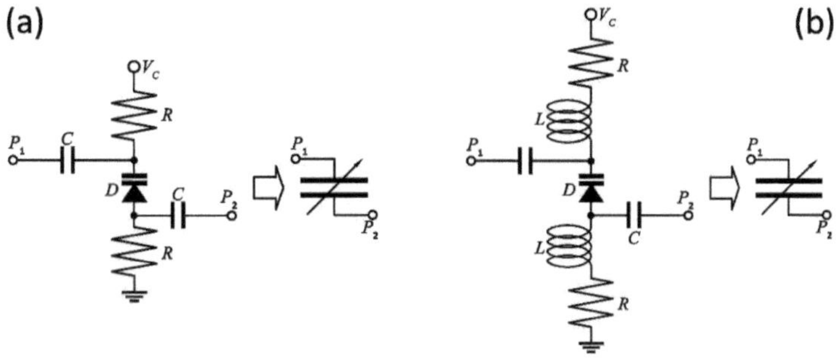

Fig. 6.49 Biasing a varactor

6.6 Attenuator

As the name suggests, an attenuator is used for attenuating signal. Figure 6.50 shows the circuit symbol of an attenuator. There are many ways for implementing an attenuator. Figure 6.51a shows a resistive T-network implementation. Let R_0 be the characteristic impedance of the transmission lines connected to both ends of the attenuator. In order for the driving point impedance at the input end to be R_0, (6.55) must be satisfied.

$$2R_1R_2 = R_0^2 - R_1^2 \ , \quad R_0 > R_1 > 0. \tag{6.55}$$

Let $A = \frac{V_{in}}{V_{out}} > 1$. It can be easily shown that

$$A = \frac{V_{in}}{V_{out}} = \frac{R_0 + R_1 + R_2}{R_2} > 1. \tag{6.56}$$

Fig. 6.50 Circuit symbol of an attenuator

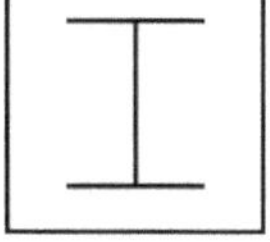

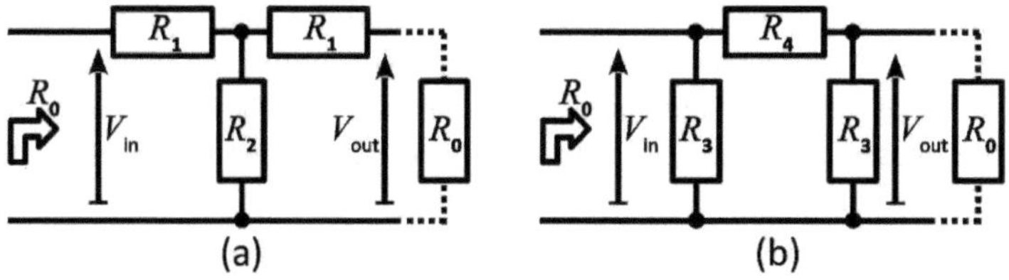

Fig. 6.51 Resistive attenuator

From (6.55) and (6.56), we have

$$R_1 = R_0 \frac{A-1}{A+1} \text{ , and} \tag{6.57a}$$

$$R_2 = R_0 \frac{2A}{A^2-1}. \tag{6.57b}$$

Figure 6.51b shows a resistive π-network implementation. In order for the driving point impedance at the input end to be R_0, (6.58) must be satisfied:

$$R_4\left(R_3^2 - R_0^2\right) = 2R_0^2 R_3 \text{ , } \quad R_3 > R_0 > 0. \tag{6.58}$$

It can be easily shown that

$$A = \frac{V_{\text{in}}}{V_{\text{out}}} = \frac{R_0 R_3 + R_0 R_4 + R_3 R_4}{R_0 R_3} > 1. \tag{6.59}$$

From (6.58) and (6.59), we have

$$R_3 = R_0 \frac{A+1}{A-1} \text{ , and} \tag{6.60a}$$

$$R_4 = R_0 \frac{A^2-1}{2A}. \tag{6.60b}$$

If the resistors in Fig. 6.51 are tunable, the attenuator will become a tunable attenuator. The high-frequency resistance of a forward-biased PIN diode is inversely proportional to its biasing current and so is often used as a tunable resistor.

6.7 Phased Array

The antenna we discussed in Sect. 6.3 consists of a single radiator. The antenna gain, directivity, and total power radiated can be increased by using two or more radiators. An antenna consisting of two or more radiators is called a phased array or antenna array. Figure 6.52a shows the schematic of a radiating phased array, and Fig. 6.52b shows the schematic of a receiving phased array. By controlling the phase and power of the EM wave radiated from each individual radiator, the direction of maximum gain of the phased array can be arbitrarily steered without physically altering the position or orientation of the individual radiating antenna. It is also possible to null the response of the phased array to signals from any arbitrary direction; this provides an excellent means for rejecting interference from particular directions. We shall discuss this in detail in Chap. 9.

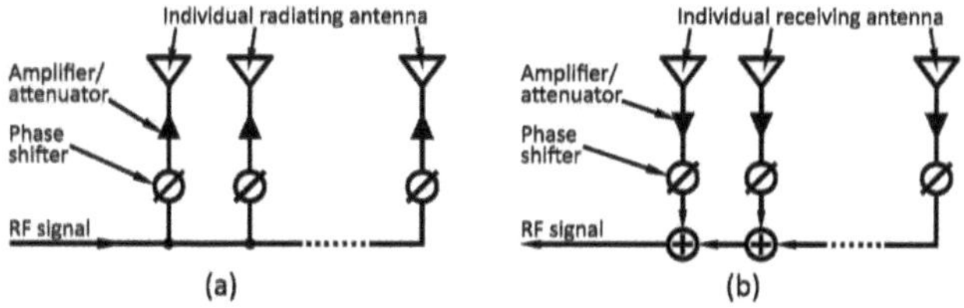

Fig. 6.52 (**a**) Radiating phased array. (**b**) Receiving phased array

Signal processing techniques have been developed to provide super-resolution in determining the direction of arrival of an EM wave using a phased array. Super-resolution ability means the ability to resolve the directions of arrival of two arriving signals with resolution beyond that can be achieved by a single antenna whose aperture is as large as the aperture of the entire phased array. We shall discuss this technique in Chap. 9.

Phased array may be classified into (a) passive electronically scanned array (PESA) and (b) active electronically scanned array (AESA). A PESA system has one large TRM (transmitter/receiver module). In a PESA system, the individual antenna element has its own phase shifter but is connected to a single TRM. It can only operate at one frequency at a time. An AESA system has multiple TRMs. In an AESA system, each antenna element has its own TRM with phase shifter and amplifier/attenuator module for beam forming and steering. An AESA system can operate at multiple frequencies simultaneously.

References

C.A. Balanis, Antenna theory: A review. Proc. IEEE **80**(1), 7–23 (1992)

C.A. Balanis, *Modern Antenna Handbook* (Wiley, 2011)

S. Gloeckler, Phased array for millimeter wave frequencies. Int. J. Infrared mm Waves **11**, 101–110 (1990)

A. Hajimiri, H. Hashemi, A. Natarajan, X. Guan, A. Komijani, Integrated phased array systems in silicon. Proc. IEEE **93**(9), 1637–1655 (2005)

R.C. Hansen, *Phased Array Antennas* (Wiley, 2009)

J.S. Herd, M.D. Conway, The evolution to modern phased array architectures. Proc. IEEE **104**(3), 519–529 (2015)

J.R. James, P.S. Hall, C. Wood, *Microstrip Antenna: Theory and Design* (IET, 1986)

M.U. Khan, M. Sharawi, R. Mittra, Microstrip patch antenna miniaturisation techniques: a review. IET Microw. Ant. Prop. **9**(9), 913–922 (2015)

R.J. Mailloux, Phased array theory and technology. Proc. IEEE **70**(3), 246–291 (1982)

R.J. Mailloux, *Phased Array Antenna Handbook* (Artech House, 2017)

T.A. Milligan, *Modern Antenna Design* (Wiley, 2005)

D. Parker, D.C. Zimmermann, Phased arrays-part I: theory and architectures. IEEE Trans. Microw. Theory Tech. **50**(3), 678–687 (2002a)

D. Parker, D.C. Zimmermann, Phased arrays-part II: implementations, applications, and future trends. IEEE Trans. Microw. Theory Tech. **50**(3), 688–698 (2002b)

W.L. Stutzman, G.A. Thiele, *Antenna Theory and Design* (Wiley, 2012)
H.J. Visser, *Array and Phased Array Antenna Basics* (Wiley, 2006)
R. Waterhouse, Small microstrip patch antenna. Electron. Lett. **31**(8), 604–605 (1995)

Chapter 7
Velocity and Distance Estimation

7.1 Introduction

The velocity of an object is the rate of change of the object's displacement in a frame of reference with respect to time; it includes the direction of motion. Speed is the magnitude of velocity without information of direction. An object is said to be traveling with constant velocity if it travels with constant speed without changing direction. An object in circular motion with constant speed does not travel with constant velocity, because its direction of motion changes with time; it accelerates toward the center of the circle, although its speed is constant. Acceleration is the rate of change of velocity; it is not the rate of change of speed. For the purpose of simplifying discussion, velocity and speed are often used interchangeably; however, we should know that they are different.

An object's velocity and its distance from an observer are two very closely related and interlinked quantities. An object's velocity is its distance moved per unit time and so may be estimated from its distance moved over a given time interval. The distance of an object is often estimated from a signal reflected from the object (Christian Andreas Doppler 1842a,b,c; Andrade 1959; Cooley and Tukey 1965; Markel 1971; Duhamel and Hollmann 1984; Swarztrauber 1984; Sorensen et al. 1986; Jou and Abraham 1988; Fraser 1989; Van Nee and Coenen 1991; Gough 1994; Offelli and Petri 1992; Lim 1985).

7.2 Frame of Reference

The measurement of velocity requires a frame of reference. A frame of reference in n-dimension can be defined in terms of $n + 1$ distinct points with no relative motion among them. An inertial frame of reference is a frame of reference in which a body with zero net force acting on it will travel with constant velocity. Corollary,

Y. C. Lim et al., *Signal Processing for Autonomous Systems*,
https://doi.org/10.1007/978-3-032-00675-2_7

an object traveling with constant velocity in an inertial frame of reference, has no net force acting on it. There are advantages in using an inertial frame of reference to serve as the frame of reference for an object's motion. In astronomy, all points at infinity are considered stationary; the fixed stars are very far away and for practical purposes may be regarded as points at infinity. Hence, in astronomy, the fixed stars are commonly used as the frame of reference for the study of celestial objects. Nevertheless, the fixed stars are not stationary; their velocity in the direction of the Earth can be measured by the Doppler shift of known electromagnetic wave spectral lines.

The Earth is not stationary and does not travel with constant velocity with respect to the fixed stars. However, the surface of the Earth in the vicinity of an object is commonly used as the frame of reference in the measurement of the velocity of the object. In the scenario consisting of an observer and an object, sometimes it may be more convenient to use the observer as the frame of reference; in this case, the velocity of the object is the relative velocity of the object with respect to the observer.

Velocity is a vector. For mathematical manipulation and for measurement purposes, sometimes it is more convenient to decompose it into its orthogonal components. Two components are said to be orthogonal if changes in one component have no effect on the other. For example, the velocity $\boldsymbol{v}$ of an object P may be decomposed into its components $\boldsymbol{v}_x$ and $\boldsymbol{v}_y$ as shown in Fig. 7.1. The components $\boldsymbol{v}_x$ and $\boldsymbol{v}_y$ are at right angles to each other; i.e., they are orthogonal. The relationships between $\boldsymbol{v}$, $\boldsymbol{v}_x$, $\boldsymbol{v}_y$, $|\boldsymbol{v}|$, and θ are given by

$$|\boldsymbol{v}|^2 = |\boldsymbol{v}_x|^2 + |\boldsymbol{v}_y|^2, \tag{7.1a}$$

$$\tan\theta = \frac{|\boldsymbol{v}_y|}{|\boldsymbol{v}_x|}, \tag{7.1b}$$

$$|\boldsymbol{v}_x| = |\boldsymbol{v}|\cos(\theta), \tag{7.1c}$$

$$|\boldsymbol{v}_y| = |\boldsymbol{v}|\sin(\theta). \tag{7.1d}$$

If an observer Q measures the velocity of P using Doppler shift, Q is measuring the component of the velocity in the direction along a straight line joining P and Q. Hence, it is more convenient to choose the direction of either $\boldsymbol{v}_x$ or $\boldsymbol{v}_y$ to coincide with the straight line joining P and Q. In Fig. 7.2, $\boldsymbol{v}_x$ is in the same direction as the

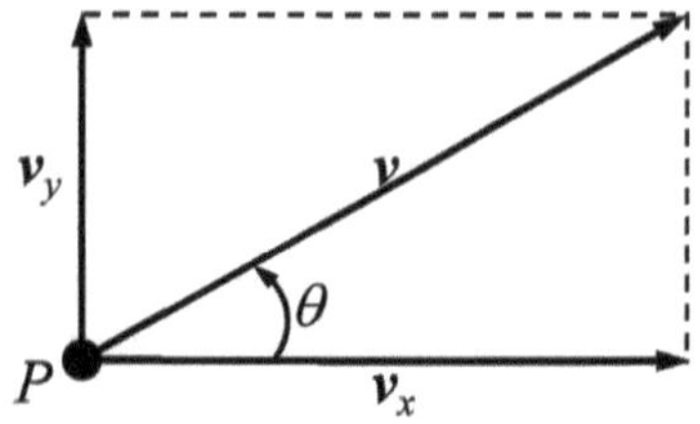

Fig. 7.1 The velocity $\boldsymbol{v}$ is decomposed into $\boldsymbol{v}_x$ and $\boldsymbol{v}_y$

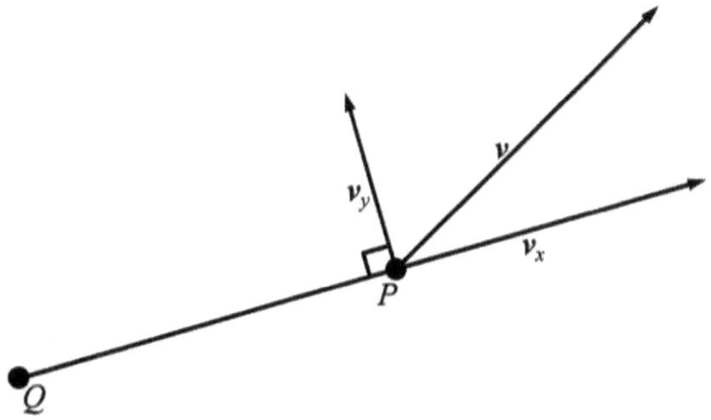

Fig. 7.2 $\boldsymbol{v}_x$ is in the direction of a straight line joining P and Q

straight line joining P and Q. The component $\boldsymbol{v}_y$ produces no Doppler shift and so cannot be measured using Doppler shift method.

7.3 Velocity Estimation

Two of the most commonly used methods for estimating the velocity/speed of a moving object/target are (1) distance moved divided by time and (2) Doppler effect. Both methods have their advantages and disadvantages. The best method will depend on the application.

7.3.1 Distance Moved Divided by Time

The speed of an object relative to an observer may be estimated from first principle by estimating its distances d_1 and d_2 from the observer at two separate time instances. Suppose that the object moves along the straight line joining itself and the observer. The distance d of the object from the observer can be estimated by transmitting a signal such as a lidar pulse (see Chap. 5) from the observer to the object and measure the time τ_d required for the signal to return.

$$d = \frac{c\tau_d}{2}, \tag{7.2}$$

where c is the speed of transmission of the signal and is the speed of light if the signal is a lidar pulse. The distance Δ the object has moved is given by

$$\Delta = |d_1 - d_2|. \tag{7.3}$$

Suppose that τ_1 and τ_2 are the values of τ_d in (7.2) in the measurement of d_1 and d_2, respectively, i.e., $d_1 = \frac{c\tau_1}{2}$ and $d_2 = \frac{c\tau_2}{2}$. If the time lapse between the estimation of d_1 and d_2 is T, the velocity $\boldsymbol{u}$ of the object is given by

$$|\boldsymbol{u}| = \frac{\Delta}{T} = \frac{c(\tau_1 - \tau_2)}{2T} \quad \text{assuming } \tau_1 > \tau_2. \tag{7.4}$$

An application of such a method is speed detection for traffic law enforcement. Speed detection for traffic law enforcement is shifting from the Doppler-based radar to the distance-over-time-based lidar because of the advantages of lidar system such as narrow beamwidth and automatic imaging of registration plate number. In traffic law enforcement lidar, the laser wavelength is about 905 nm, and the power is about 50 mW with a pulse duration of about 30 ns to avoid harming the eyes. The pulse repetition rate is about 1 millisecond. Several hundred pulses are transmitted to improve estimation accuracy.

An object may estimate its own position in a three-dimensional space with the help of a satellite navigation system. Suppose that an object's coordinate in a three-dimensional space changes from (x_1, y_1, z_1) to (x_2, y_2, z_2). The object has moved by a distance Δ given by

$$\Delta = \sqrt{(x_2 - x_1)^2 + (y_2 - y_1)^2 + (z_2 - z_1)^2}. \tag{7.5}$$

Suppose that the move occurred in T unit time. The speed $|\boldsymbol{v}|$ of the object is given by

$$|\boldsymbol{v}| = \frac{\Delta}{T} = \frac{\sqrt{(x_2 - x_1)^2 + (y_2 - y_1)^2 + (z_2 - z_1)^2}}{T}. \tag{7.6}$$

An object may estimate its movement in the same way as a computer optical mouse does so. Currently, there is no vehicle using this method, and so we shall describe the operation of an optical mouse.

The optical mouse of a computer shines a light from an LED to illuminate the surface of the desk. A tiny camera picks up the image of the desk's surface. Figures 7.3a and b show the images captured by the camera sensor at two consecutive instances t_1 and t_2, respectively, with $t_2 > t_1$. Each square in Fig. 7.3 represents an image pixel. In Fig. 7.3, the camera has an (18×18) monochromatic pixel array. For illustrative simplicity, the pixel array in Fig. 7.3 is binary, but the actual optical mouse's image is gray scale. The number of times the images are taken is called the frame rate or refresh rate or polling rate. It is typically between 1500 and 6000 frames per second. Between t_1 and t_2, the camera has moved a distance d, and so the speed is $d/(t_2 - t_1)$. The movement of the mouse can be estimated from the relative displacement between Figs. 7.3a and b. The process of estimating the relative displacement is called motion estimation in digital video processing. To accomplish motion estimation, a block of 8×8 pixels at the centre of Fig. 7.3b is identified. Figure 7.4 shows the 8×8 block identified from Fig. 7.3b. This 8×8 block

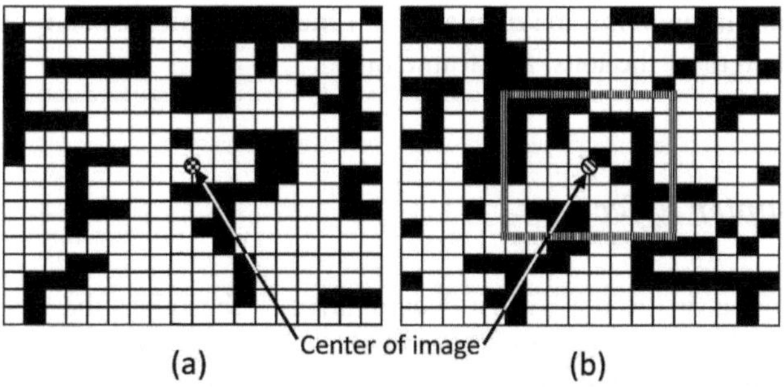

Fig. 7.3 Images recorded by the camera sensor at (**a**) time t_1 and (**b**) time t_2, respectively. $t_2 > t_1$

Fig. 7.4 A block of 8 × 8 pixels for performing motion estimation identified from Fig. 7.3b

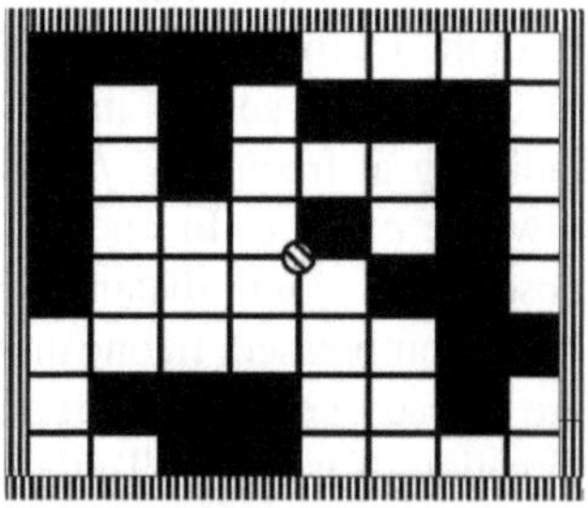

Fig. 7.5 The 8 × 8-pixel block moves by five pixels to the right and four pixels upward

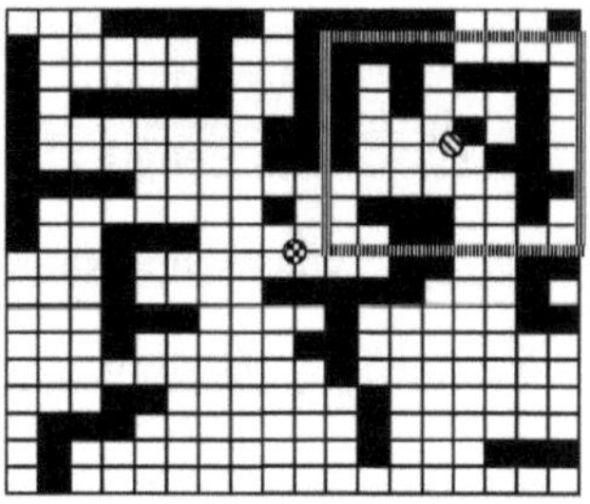

is used to search for a match in Fig. 7.3a; the match is shown in Fig. 7.5. It can be seen from Figs. 7.5 and 7.3b that the mouse moves by five pixels to the right and four pixels upward from t_1 to t_2. In the above example, motion estimation is expressed in terms of an integer number of pixels. In video processing using grayscale images, motion estimation is estimated to 0.25 pixel resolution.

Implementing the above method on a vehicle requires a video camera looking downward at the road surface. Assume that the vehicle moves at 100 km/hour (= 27.778 mm/ms). The separation between the camera's lens and the road surface is called the object distance. If the ratio of the object distance to the focal length of the lens is 30 : 1, the speed of the road surface's image on the video camera sensor will be $= 27.778/30 = 0.9259$ mm/ms $= 925.9$ μm/ms. If the frame rate of the camera is 100 KHz, the road surface's image on the video camera sensor would have moved by 9.259 μm from frame to frame. The displacement of the road surface's image on the video camera sensor is important for deciding the size of the camera sensor. Factors influencing the displacement include vehicle speed, object distance to the focal length of the lens, and frame rate.

7.3.2 Doppler Effect

When the source or the observer moves relative to the other, the frequency of the wave observed by the observer will not be the same as that observed by the source; this phenomenon is called Doppler effect. The difference in frequency is called Doppler shift. The ratio $\frac{\text{Doppler shift}}{\text{Original frequency}}$ is called fractional Doppler shift and is usually very small. The relative motion of the source and observer can be determined from the fractional Doppler shift.

Consider a source moving with velocity v toward a stationary observer and radiating at frequency f_0 cycles per unit time. Suppose that the wave velocity is c where $c \gg v$. In one unit time, the source has moved in the direction of the observer by v unit distance. The wave in the direction of motion of the observer is being compressed. In one unit time, the f_0 cycles of the wave radiated by the source are spread over $c - v$ unit distance. See Fig. 7.6. Therefore, the wavelength in the direction of v is $\frac{c-v}{f_0}$. To the stationary observer at O, the observed frequency f_1 is given by

$$f_1 = \frac{\text{wave velocity}}{\text{wavelength}} = \frac{c}{\frac{c-v}{f_0}} = \frac{f_0}{1 - \frac{v}{c}}, \tag{7.7a}$$

where v = source velocity towards observer.

$$\therefore \quad v = c \times \frac{f_1 - f_0}{f_1} \approx c \times \frac{f_1 - f_0}{f_0} = c \times \text{fractional-Doppler-shift}. \tag{7.7b}$$

Consider another case where the source is stationary but the observer is moving toward the source with velocity u. See Fig. 7.7. The wavelength λ_0 is given by

$$\lambda_0 = \frac{c}{f_0}. \tag{7.8}$$

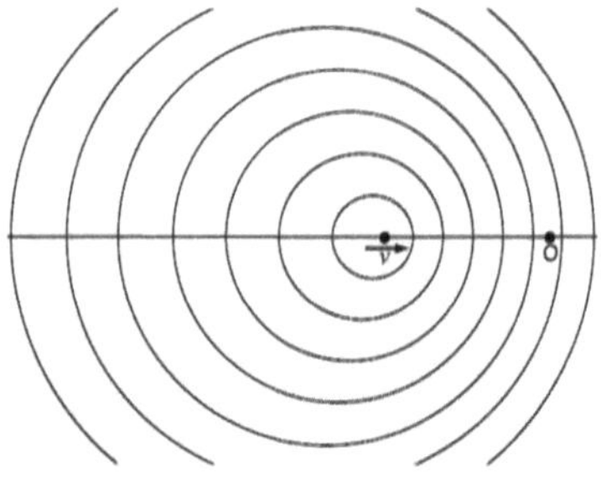

Fig. 7.6 Doppler effect due to source motion

Fig. 7.7 Doppler effect due to observer motion

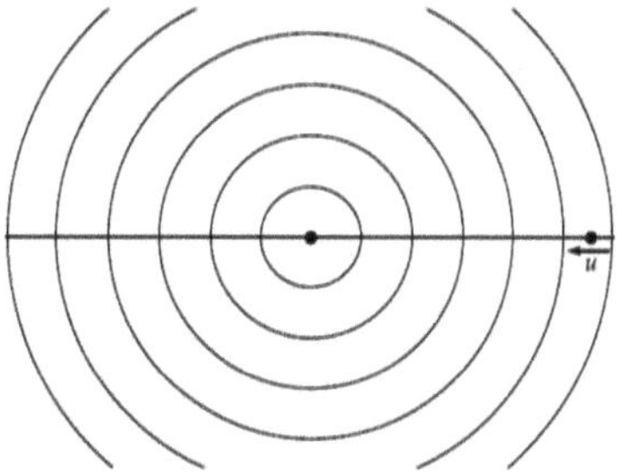

Fig. 7.8 Doppler effect due to both source and observer motions

In one unit time, the observer has moved by u unit distance toward the source. In u unit distance, there are $\frac{u}{\lambda_0}$ cycles of the wave. Hence, in one unit time, the observer has observed $\frac{u}{\lambda_0}$ cycles of the wave in addition to the f_0 cycles if it had stayed stationary. Therefore, the number of cycles f_2 of the wave the observer has observed in one unit time is given by

$$f_2 = f_0 + \frac{u}{\lambda_0} = f_0\left(1 + \frac{u}{\lambda_0 f_0}\right) = \left(1 + \frac{u}{c}\right) f_0, \tag{7.9a}$$

where u = observer velocity towards source.

$$\therefore \quad u = c \times \frac{f_2 - f_0}{f_0} = c \times \text{fractional-Doppler-shift}. \tag{7.9b}$$

Consider the case where the source moves toward the observer with velocity v and at the same time the observer moves toward the source with velocity u. See Fig. 7.8. Following the same arguments as those for source and observer in motion separately, the frequency f_3 observed by the observer is given by

$$f_3 = \frac{1 + \frac{u}{c}}{1 - \frac{v}{c}} f_0. \tag{7.10a}$$

$$\therefore \quad u + v \approx c \times \frac{f_3 - f_0}{f_0} = c \times \text{fractional-Doppler-shift}. \tag{7.10b}$$

$$\text{For } u = v, \quad u = v \approx \frac{c}{2} \times \frac{f_3 - f_0}{f_0} = \frac{c}{2} \times \text{fractional-Doppler-shift}. \tag{7.10c}$$

If the source moves with velocity v toward a stationary reflector which reflects back the wave and the observer moves toward the reflector with velocity u as shown in Fig. 7.9, the frequency f_3 observed by the observer is given by (7.10).

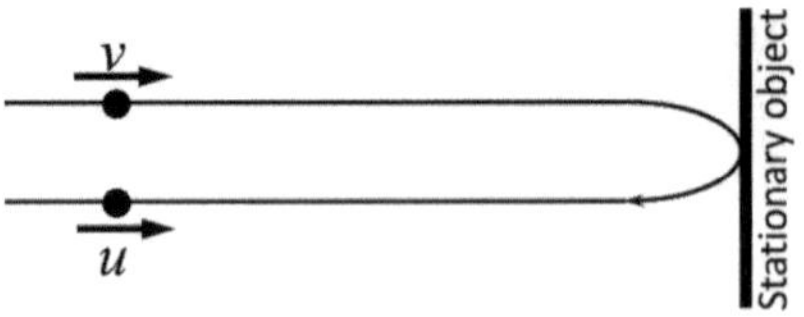

Fig. 7.9 Source and observer are moving. Object is stationary

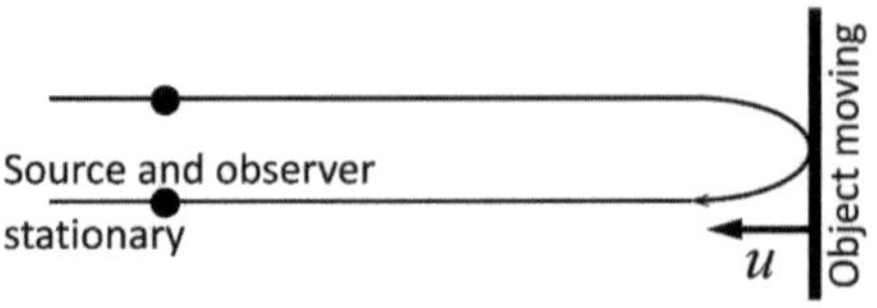

Fig. 7.10 Source and observer are stationary. Object is moving

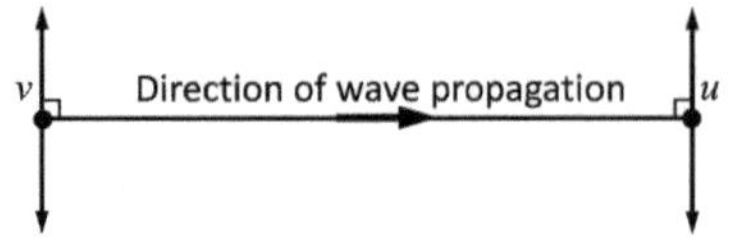

Fig. 7.11 Motion is perpendicular to path joining source and observer

If the source and observer are at the same location and are stationary but the wave is reflected back by a moving object moving with velocity u toward the source and the observer as shown in Fig. 7.10, the frequency f_4 observed by the observer is given by

$$f_4 = \frac{1 + \frac{u}{c}}{1 - \frac{u}{c}} f_0. \tag{7.11a}$$

$$\therefore \quad u \approx \frac{c}{2} \times \frac{f_4 - f_0}{f_0} = \frac{c}{2} \times \text{fractional-Doppler-shift}. \tag{7.11b}$$

If the direction of motion of the source or/and observer is perpendicular to the direct path joining the source and observer as shown in Fig. 7.11, there will be no Doppler effect observed by the observer.

The frequency shifts $f_1 - f_0$, $f_2 - f_0$, $f_3 - f_0$, and $f_4 - f_0$ are called Doppler shifts. The frequency difference $\omega_2 - \omega_1$ of two sinusoids $\cos(\omega_2 t)$ and $\cos(\omega_1 t)$ can be obtained by multiplying $\cos(\omega_2 t)$ and $\cos(\omega_1 t)$ together. Detail approach for doing so is discussed in Sect. 7.4.2.

7.4 Distance Estimation

The distance of a target from an observer may be estimated by transmitting a signal to the target and measure the time lapse for the signal to return. The signal may be a pulse or a linear chirp; a linear chirp is a sinusoid whose frequency varies linearly with time. If it is a pulse, it may be a rectangular pulse or a code, i.e., a finite duration signal pattern. If a linear chirp is transmitted, the time lapse for the transmitted signal to return is proportional to the difference between the frequency of the received signal and the frequency of the currently transmitted signal if the transmitter, receiver, and object are all stationary.

The product of the wave velocity and the duration required by the wave to return is equal to twice the distance of the target. The estimation of the duration required by the wave to return from the target is called TOA (time of arrival) estimation.

7.4.1 Pulse Method

Consider the scenario where a stationary transmitter transmits a finite duration pulse toward a stationary target in the far field as shown in Fig. 7.12. Suppose that the pulse is reflected back to a receiver located close to the transmitter after τ_d unit time. From (7.2), we have $d = c\tau_d/2$. We need to estimate τ_d in order to determine d. The finite duration pulse may be in any form. It may be acoustic or electromagnetic or laser. If it is a laser pulse, the system is usually referred to as a lidar. If the observer and/or the target is/are moving, there will be compression or rarefaction of the received pulse.

A clock is used to measure the time taken by the pulse to return. See Fig. 7.13. The transmitted pulse may either be synchronized to an edge of the clock or may not be synchronized with the clock. Suppose that the transmitted pulse is synchronized to the rising edge of a clock with clock period T unit time. Let N_d denote the number of clock period elapsed until the received pulse is high. Since we are counting clock pulses, N_d is an integer. The duration τ_d satisfies the inequality

$$(N_d - 1)T \leq \tau_d \leq N_d T. \tag{7.12}$$

The uncertainty in the value of τ_d is due to discrete time quantization. The clock frequency may be increased to improve time resolution; nevertheless, N_d is an integer regardless of the frequency of the clock. Let the estimated value of τ_d be

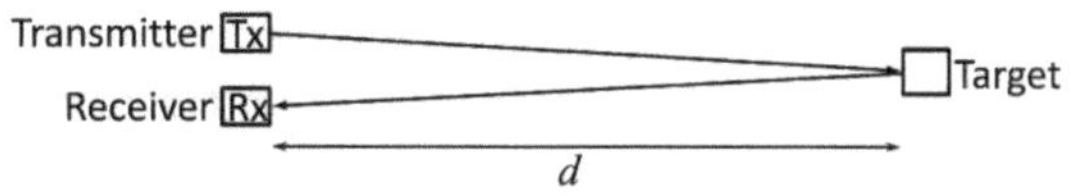

Fig. 7.12 Distance = wave velocity × TOA / 2

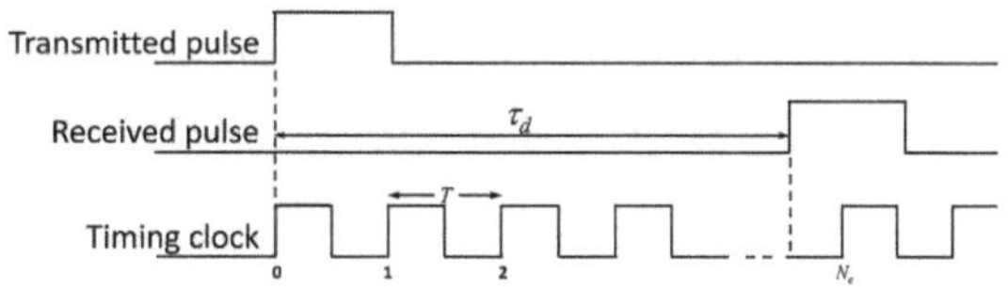

Fig. 7.13 Transmitted pulse is synchronized to rising edge of clock

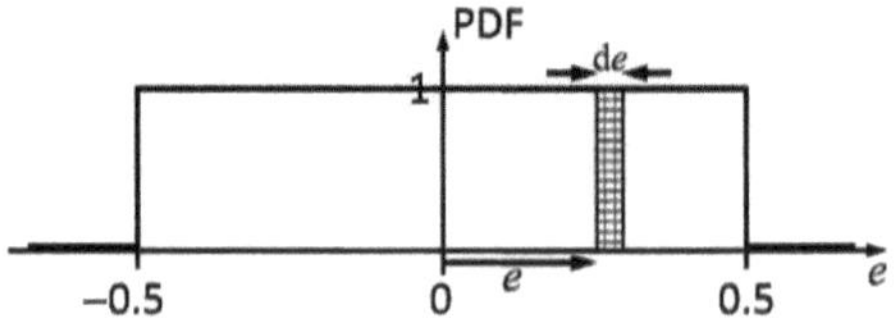

Fig. 7.14 The PDF of e

denoted by $\hat{\tau}_d$. Assume that the value midway between $(N_d - 1)T$ and N_dT is taken to be the estimated τ_d. Under this assumption,

$$\hat{\tau}_d = (N_d - 0.5)T. \tag{7.13}$$

Using (7.13) as the estimated τ_d incurred an estimation error. Let the estimation error be eT. Therefore,

$$\tau_d = \hat{\tau}_d + eT. \tag{7.14}$$

We shall examine the statistical property of eT or specifically the statistical property of e. The PDF (probability density function) of e is shown in Fig. 7.14. It can be easily shown that the mean value of $e = \mathbb{E}\{e\} = 0$. The variance of e , var(e), is given by

$$\begin{aligned} \text{var}(e) &= \mathbb{E}\{(e - \mathbb{E}\{e\})^2\} = \mathbb{E}\{e^2\}, \\ &= \int_{-0.5}^{0.5} e^2 \text{de} = \left[\frac{e^3}{3}\right]_{-0.5}^{0.5} = \frac{1}{12}. \end{aligned} \tag{7.15}$$

In the scheme where the clock starts at the same time as the rising edge of the transmitted pulse, if the measurement is repeated, the same N_d will be recorded if there is no measurement error. Since the same N_d is obtained, it is not possible to improve the accuracy of the estimated τ_d by averaging the values of $\hat{\tau}_d$ obtained from different trials, since every trial gives the same $\hat{\tau}_d$. If there is noise or random measurement error, the value of N_d may change if the error is sufficiently large; a different $\hat{\tau}_d$ may be obtained. However, averaging the values of $\hat{\tau}_d$ may only reduce the error due to noise or measurement error; it cannot reduce the error due to the effect of discrete time.

To improve the accuracy of the estimated τ_d due to the effect of discrete time, the errors from different trials must be uncorrelated. Furthermore, for the estimate to be unbiased, $\mathbb{E}\{\hat{\tau}_d\}$ must be equal to τ_d, the true value of the time elapsed. This can be achieved if the transmitted pulse is not synchronized to the clock. See Fig. 7.15. Suppose that the recorded start time is the first rising edge of the clock after the pulse has been transmitted. The recorded start time is marked as time 0 in Fig. 7.15. The actual start time τ_s is before the rising edge of the clock and satisfies the inequality

$$-T \leq \tau_s \leq 0. \tag{7.16}$$

Fig. 7.15 Transmitted pulse is not synchronized to any clock edge

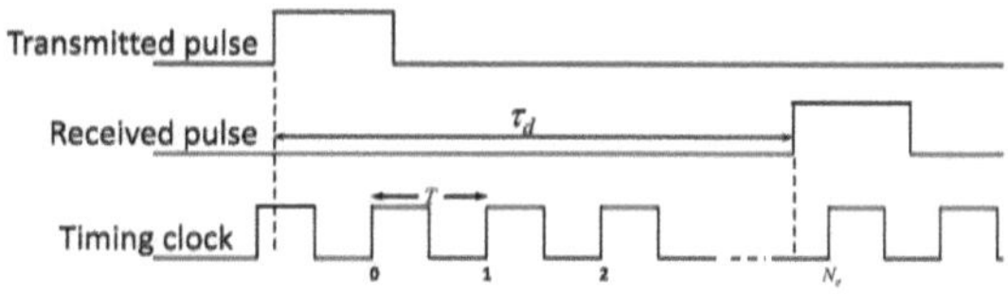

The recorded end time is the first rising edge of the clock after the pulse has returned; it is marked as time N_e in Fig. 7.15. The recorded end time N_e is an integer. The actual end time τ_e satisfies the inequality

$$(N_e - 1)T \le \tau_e \le N_e T. \tag{7.17}$$

Let $\hat{\tau}_s$ and $\hat{\tau}_e$ denote the estimated τ_s and τ_e, respectively. Choose the value midway between $-T$ and 0 to be $\hat{\tau}_s$ and that midway between $(N_e - 1)T$ and $N_e T$ to be $\hat{\tau}_e$. Hence, we have

$$\hat{\tau}_s = -0.5T, \tag{7.18a}$$

$$\mathbb{E}\{\hat{\tau}_s\} = -0.5T, \tag{7.18b}$$

and

$$\hat{\tau}_e = (N_e - 0.5)T. \tag{7.19}$$

As different trial produces different N_e, $\hat{\tau}_e$ changes from trial to trial. The estimated τ_d is given by

$$\hat{\tau}_d = \hat{\tau}_e - \hat{\tau}_s. \tag{7.20}$$

Substitute (7.18a) into (7.20); the estimated τ_d is given by

$$\hat{\tau}_d = \hat{\tau}_e + 0.5T. \tag{7.21}$$

Taking expectation of (7.21),

$$\mathbb{E}\{\hat{\tau}_d\} = \mathbb{E}\{\hat{\tau}_e\} + 0.5T. \tag{7.22}$$

Let the errors for the estimation of τ_s and τ_e be $e_s T$ and $e_e T$, respectively. Hence, we have

$$\tau_s = \hat{\tau}_s + e_s T, \tag{7.23}$$

and

$$\tau_e = \hat{\tau}_e + e_e T. \tag{7.24}$$

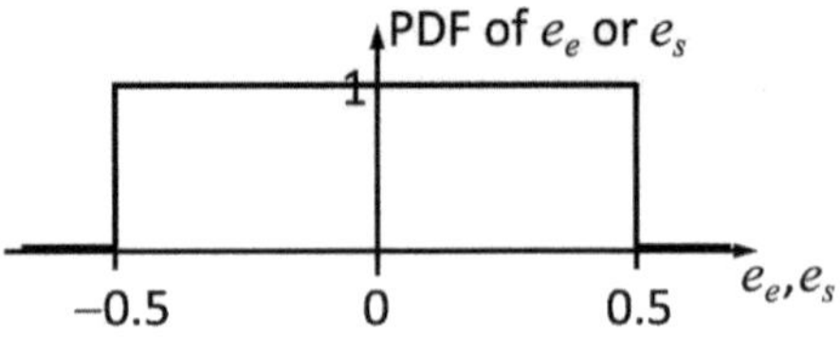

Fig. 7.16 The PDF of e_e and e_s

The true τ_d is given by

$$\tau_d = \tau_e - \tau_s. \tag{7.25}$$

For the estimate to be unbiased, $\mathbb{E}\{\hat{\tau}_d\}$ must be equal to τ_d. For the estimate to improve as the number of trials increases, the variance of $\hat{\tau}_d$ must decrease with the number of trials. Substitute (7.23) and (7.24) into (7.25); we have

$$\tau_d = \hat{\tau}_e - \hat{\tau}_s + (e_e - e_s)T. \tag{7.26}$$

Taking expectation of (7.26),

$$\mathbb{E}\{\tau_d\} = \tau_d \ = \mathbb{E}\{\hat{\tau}_e\} - \mathbb{E}\{\hat{\tau}_s\} + \mathbb{E}\{(e_e - e_s)T\}. \tag{7.27}$$

In (7.27), $\mathbb{E}\{\tau_d\} = \tau_d$ because τ_d is the actual delay and so is a constant. Substitute (7.18b) into (7.27); we have

$$\tau_d = \mathbb{E}\{\hat{\tau}_e\} + 0.5T + \mathbb{E}\{(e_e - e_s)T\}. \tag{7.28}$$

The PDF for e_e and e_s is shown in Fig. 7.16. It can be easily shown that

$$\mathbb{E}\{e_e\} = \mathbb{E}\{e_s\} = 0. \tag{7.29}$$

Furthermore, e_e and e_s are uncorrelated. Substituting (7.29) into (7.28),

$$\tau_d = \mathbb{E}\{\hat{\tau}_e\} + 0.5T. \tag{7.30}$$

Comparing (7.30) and (7.22), we have

$$\mathbb{E}\{\hat{\tau}_d\} = \tau_d. \tag{7.31}$$

Since $\mathbb{E}\{\hat{\tau}_d\} = \tau_d$, the estimate is unbiased.

We shall now examine some statistical properties of e_e and e_s to show that the accuracy of the estimate improves as the number of measurement trials increases. The PDF for $(e_e - e_s)$ is that e_e convolves with that for e_s and is shown in Fig. 7.17. From (7.21), the variance for $\hat{\tau}_d$ is given by

$$\text{var}\{\hat{\tau}_d\} = \text{var}\{\hat{\tau}_e\} + \text{var}\{0.5T\} = \text{var}\{\hat{\tau}_e\}. \tag{7.32}$$

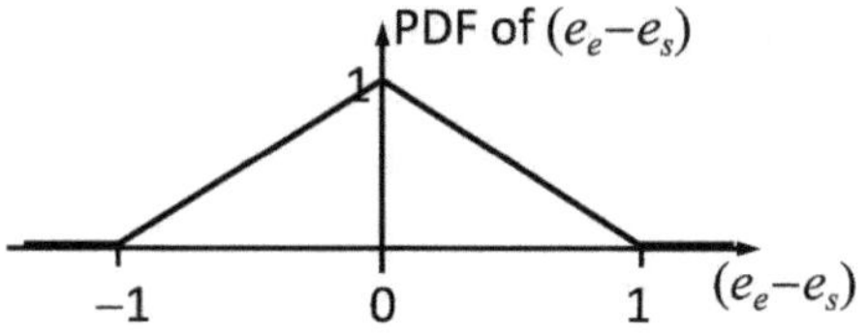

Fig. 7.17 The PDF of $e_e - e_s$

For each trial, the variance of $\hat{\tau}_e$ is $\frac{T^2}{12}$. (See (7.15).) Therefore, for each trial, the variance of $\hat{\tau}_d$ is $\frac{T^2}{12}$, since the variance of $\hat{\tau}_d$ is equal to the variance of $\hat{\tau}_e$ as shown in (7.32). Averaging the results of K trials, $\text{var}\{\hat{\tau}_d\} = \frac{T^2}{12K}$. Thus, the accuracy improves as the number of trials increases, because the variance decreases with increasing K.

7.4.1.1 Matched Filtering

In the above discussion, we assumed that a finite duration pulse is transmitted toward the target. If the pulse is an optical pulse such as a laser pulse in a lidar system, the finite duration pulse is a short rectangular pulse resembling an impulse.

If the pulse is acoustic or electromagnetic, the pulse is usually a wideband signal with considerable duration; this is the type of pulse considered in this section. The returned signal is compared with a template of the transmitted signal in a process called matched filtering to determine the estimated TOA, $\hat{\tau}_d$.

Suppose that the transmitted finite duration signal is $x(t)$ and that a copy of $x(t)$ is saved as a template for comparison with the received reflected signal. Suppose that the received reflected signal is $y(t)$ and that there is no Doppler shift. The signal $y(t)$ is a delayed and scaled version of $x(t)$, i.e.,

$$y(t) = ax(t - \tau_d), \tag{7.33}$$

where a is a real constant. The variable τ_d represents the delay. The cross correlation function $r_{xy}(\tau)$ between $x^*(t)$ and $y(t)$ is computed where $x^*(t)$ is the complex conjugate of $x(t)$.

$$\begin{aligned} r_{xy}(\tau) &= \int_{-\infty}^{\infty} x^*(t-\tau)y(t)\mathrm{d}t, \\ &= a\int_{-\infty}^{\infty} x^*(t-\tau)x(t-\tau_d)\mathrm{d}t. \end{aligned} \tag{7.34}$$

In (7.34), $r_{xy}(\tau)$ reaches its maximum when $\tau = \tau_d$. In digital processing, the integration in (7.34) is replaced by summation and τ is incremented from its minimum possible value to its maximum possible value at discrete step. The estimated value of τ_d is the value of τ that gives the largest $r_{xy}(\tau)$.

7.4.1.2 Effects of Finite Bandwidth and Noise

In the above discussion, we assume that it is possible to transmit a finite duration pulse with infinite bandwidth and that the received signal is not corrupted by noise and/or measurement errors. This scenario is impossible. All practical signals are bandwidth limited. The received signal is always corrupted by noise and/or measurement errors. In this section, we examine the effects of finite bandwidth and noise.

It can be shown that the autocorrelation function $\int_{-\infty}^{\infty} x^*(t-\tau)x(t-\tau_d)\mathrm{d}t$ is the inverse Fourier transform of the power spectrum density of $x(t)$.

Proof Let the Fourier transform of $x(t)$ be $X(f)$. Hence, we have the Fourier transform pair

$$X(f) = \int_{-\infty}^{\infty} x(t)e^{-j2\pi f t}\mathrm{d}t, \tag{7.35a}$$

$$x(t) = \int_{-\infty}^{\infty} X(f)e^{j2\pi f t}\mathrm{d}f. \tag{7.35b}$$

Let the power spectrum density of $x(t)$ be denoted by $S_{xx}(f)$. Hence, we have

$$\begin{aligned} S_{xx}(f) &= X^*(f)X(f), \\ &= \int_{-\infty}^{\infty} x^*(t)e^{j2\pi f t}\mathrm{d}t \int_{-\infty}^{\infty} x(\theta)e^{-j2\pi f \theta}\mathrm{d}\theta, \\ &= \int_{-\infty}^{\infty}\int_{-\infty}^{\infty} x^*(t)x(\theta)e^{j2\pi f(t-\theta)}\mathrm{d}t\mathrm{d}\theta. \end{aligned} \tag{7.36}$$

Let $\tau = \theta - t$. $\therefore \mathrm{d}\tau = -\mathrm{d}t$. Expression (7.36) becomes

$$\begin{aligned} S_{xx}(f) &= -\int_{\infty}^{-\infty}\int_{-\infty}^{\infty} x^*(\theta-\tau)x(\theta)e^{-j2\pi f\tau}\mathrm{d}\tau\mathrm{d}\theta, \\ &= -\int_{\infty}^{-\infty}\int_{-\infty}^{\infty} [x^*(\theta-\tau)x(\theta)\mathrm{d}\theta]e^{-j2\pi f\tau}\mathrm{d}\tau, \\ &= -\int_{\infty}^{-\infty} R_{xx}\{\tau\}e^{-j2\pi f\tau}\mathrm{d}\tau = \int_{-\infty}^{\infty} R_{xx}\{\tau\}e^{-j2\pi f\tau}\mathrm{d}\tau, \end{aligned} \tag{7.37}$$

where

$$R_{xx}\{\tau\} = R_{xx}\{-\tau\} = \int_{-\infty}^{\infty} x^*(\theta-\tau)x(\theta)\mathrm{d}\theta. \tag{7.38}$$

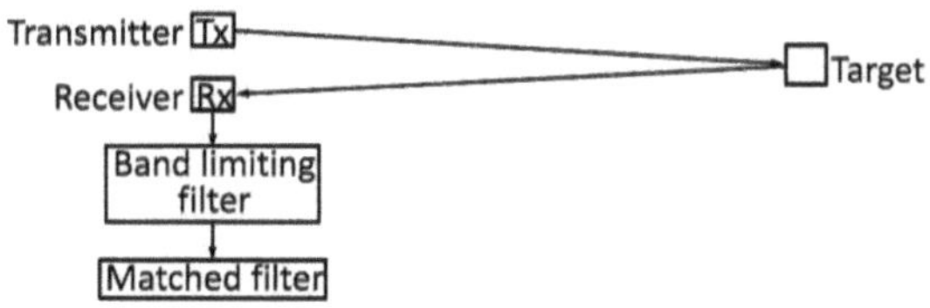

Fig. 7.18 The receiver output is filtered by a bandlimiting filter

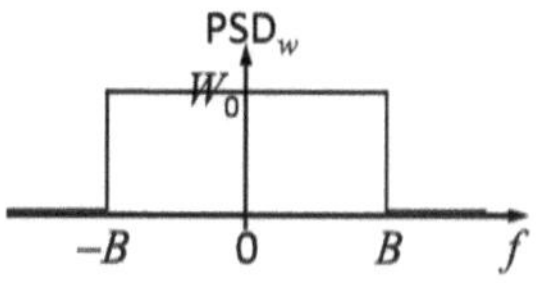

Fig. 7.19 Noise power spectral density after the bandlimiting filter

The expression $\int_{-\infty}^{\infty} R_{xx}\{\tau\}e^{-j2\pi f\tau}\mathrm{d}\tau$ is the Fourier transform of $R_{xx}\{\tau\}$, and, hence, $S_{xx}(f)$ is the Fourier transform of $R_{xx}\{\tau\}$. Since $R_{xx}\{\tau\}$ is real and symmetrical, $R_{xx}\{\tau\}$ is the Fourier transform of $S_{xx}(f)$ (property of Fourier transform).

In due course, we shall show that the accuracy in estimating τ_d increases with increasing bandwidth. An impulse has infinite bandwidth. To maximize bandwidth, it is desirable to choose $S_{xx}(f)$ such that its Fourier transform closely resembles an impulse. In practice, the received signal is corrupted by noise and/or measurement errors. Since the transmitted signal is bandlimited, the received signal is first filtered by a filter whose bandwidth is equal to that of the transmitted signal to remove noise and/or measurement errors with frequency component outside the frequency band of the transmitted signal. The arrangement is shown in Fig. 7.18. Suppose that the bandlimiting filter passes all frequencies from $-B$ Hz to B Hz and rejects all frequencies outside the range $[-B, B]$. Let $w(t)$ be the noise at the output of the bandlimiting filter. The noise $w(t)$ is random, uncorrelated, and bandlimited within the frequency range $[-B, B]$. After the bandlimiting filter, the noise power spectral density (PSD$_w$) is as shown in Fig. 7.19. The noise autocorrelation function $R_{ww}(t)$ after the bandlimiting filter is given by

$$R_{ww}(t) = \int_{-B}^{B} W_0 e^{j2\pi ft}\mathrm{d}f = 2BW_0\frac{\sin(2\pi Bt)}{2\pi Bt}. \tag{7.39}$$

Let the noise power $R_{ww}(0)$ be denoted by σ^2. From (7.39), $R_{ww}(t) = 0$ for $t = n/(2B)$, $n \neq 0$, $n =$ integer, and

$$R_{ww}(0) = 2BW_0 = \sigma^2. \tag{7.40}$$

A plot of $R_{ww}(t)$ is shown in Fig. 7.20. The signal at the output of the bandlimiting filter is $x(t)$ and is given by

$$x(t) = s(t) + w(t). \tag{7.41}$$

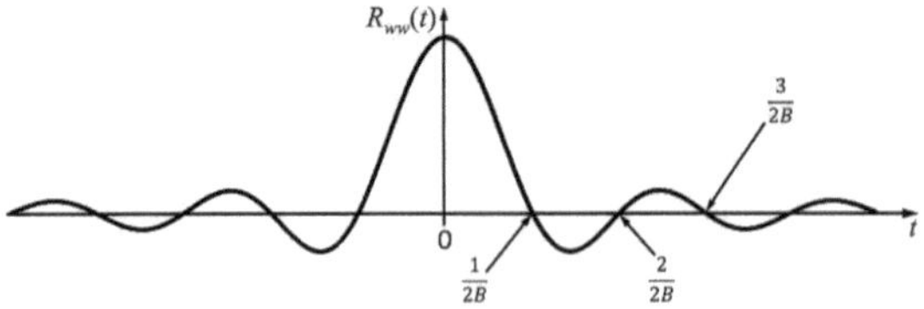

Fig. 7.20 Noise autocorrelation function $R_{ww}(t)$

In (7.41), $s(t)$ is the delayed scaled version of the transmitted signal $s_0(t)$ and is given by

$$s(t) = as_0(t - \tau_d). \tag{7.42}$$

where a is the scale factor and τ_d is the delay. Suppose that the signal after the bandlimiting filter is sampled with sampling period T. Let $x[n]$, $s[n]$, and $w[n]$ be the nth sampled version of $x(t)$, $s(t)$, and $w(t)$ where $x[n] = x(nT)$, $s[n] = s(nT)$, and $w[n] = w(nT)$. We have

$$x[n] = s[n] + w[n] \ , \ \ n = 0, 1, \ldots, N-1. \tag{7.43}$$

Assume that $s[n]$ exists only in the interval $0 \le n \le N-1$, i.e., $s[n] = 0$ for $n < 0$ and $n \ge N$. Both "noise" and "measurement errors" are "errors," and so we shall lump them together as one item and regard it as "noise." Let the "noise power" (variance) of $w[n]$ be σ^2, and let $\hat{\tau}_d$ be the estimated τ_d. Using the standard CRLB formula (See (7.29)) for signal in white Gaussian noise,

$$\text{var}(\hat{\tau}_d) \ge \frac{\sigma^2}{\sum_{n=0}^{N-1} \left(\frac{\partial s[n]}{\partial \hat{\tau}_d} \right)^2}. \tag{7.44}$$

However, $s[n]$ is discrete time, and so $\frac{\partial s[n]}{\partial \hat{\tau}_d}$ does not exist. As partial derivative exists only in the continuous time domain, we need to return to the continuous time domain to determine $\frac{\partial s[n]}{\partial \hat{\tau}_d}$. We have

$$\text{var}(\hat{\tau}_d) \ge \frac{\sigma^2}{\sum_{n=0}^{N-1} \left(\frac{\partial s(t)}{\partial \tau_d} \Big|_{t=nT} \right)^2} = \frac{\sigma^2}{\sum_{n=0}^{N-1} \left(\frac{\partial s(t)}{\partial t} \Big|_{t=nT} \right)^2}. \tag{7.45}$$

In (7.45), note that $\frac{\partial s(t)}{\partial \tau_d} = \frac{\partial a s_0(t-\tau_d)}{\partial \tau_d} = \frac{\partial a s_0(t-\tau_d)}{\partial (t-\tau_d)} \times \frac{\partial (t-\tau_d)}{\partial \tau_d} = -\frac{\partial a s_0(t-\tau_d)}{\partial (t-\tau_d)} = -\frac{\partial s(t)}{\partial t} \times \frac{\partial t}{\partial (t-\tau_d)} = -\frac{\partial s(t)}{\partial t} \times \frac{\partial (t+\tau_d)}{\partial t} = -\frac{\partial s(t)}{\partial t}$. Therefore, $\left(\frac{\partial s(t)}{\partial \tau_d}\right)^2 = \left(\frac{\partial s(t)}{\partial t}\right)^2$. Hence, the bound for $\text{var}(\hat{\tau}_d)$ can be obtained by evaluating (7.45).

Next, we shall proceed to derive a bound on $\text{var}(\hat{\tau}_d)$ in terms of SNR and bandwidth. Suppose that the received signal is sampled at very high rate, i.e.,

T is very small, the summation $\sum_{n=0}^{N-1}\left(\left.\frac{\partial s(t)}{\partial t}\right|_{t=nT-\tau_d}\right)^2$ may be replaced by integration. Expression (7.45) becomes

$$\operatorname{var}(\hat{\tau}_d) \geq \frac{\sigma^2}{\frac{1}{T}\int_0^{(N-1)T}\left(\frac{\partial s(t)}{\partial t}\right)^2 \mathrm{d}t} = \frac{T\sigma^2}{\int_0^{(N-1)T}\left(\frac{\partial s(t)}{\partial t}\right)^2 \mathrm{d}t}. \tag{7.46}$$

Since $s(t)$ is nonzero only within the range $0 \leq t \leq (N-1)T$, (7.46) becomes

$$\operatorname{var}(\hat{\tau}_d) \geq \frac{T\sigma^2}{\int_{-\infty}^{\infty}\left(\frac{\partial s(t)}{\partial t}\right)^2 \mathrm{d}t}. \tag{7.47}$$

Let

$$g(t) = \frac{\partial s(t)}{\partial t}, \tag{7.48}$$

and let the Fourier transform of $g(t)$ be $G(f)$. Parseval's theorem states that

$$\int_{-\infty}^{\infty} |g(t)|^2 \mathrm{d}t = \int_{-\infty}^{\infty} |G(f)|^2 \mathrm{d}f. \tag{7.49}$$

Let the Fourier transform of $s(t)$ be $S(f)$. The Fourier transform of $\frac{\partial s(t)}{\partial t}$, i.e., the Fourier transform of $g(t)$, is given by (Fourier transform property)

$$G(f) = j2\pi f S(f). \tag{7.50}$$

Substituting (7.48) and (7.50) into (7.49),

$$\int_{-\infty}^{\infty}\left|\frac{\partial s(t)}{\partial t}\right|^2 \mathrm{d}t = \int_{-\infty}^{\infty} |j2\pi f S(f)|^2 \mathrm{d}f = \int_{-\infty}^{\infty} |2\pi f S(f)|^2 \mathrm{d}f. \tag{7.51}$$

Substituting (7.51) into (7.47),

$$\operatorname{var}(\hat{\tau}_d) \geq \frac{T\sigma^2}{\int_{-\infty}^{\infty} |2\pi f S(f)|^2 \mathrm{d}f} = \frac{T\sigma^2}{\int_{-\infty}^{\infty} (2\pi f)^2 |S(f)|^2 \mathrm{d}f}. \tag{7.52}$$

Define the RMS BW (RMS bandwidth) B_{rms} as

$$B_{rms} = \sqrt{\frac{\int_{-\infty}^{\infty} (2\pi f)^2 |S(f)|^2 \mathrm{d}f}{\int_{-\infty}^{\infty} |S(f)|^2 \mathrm{d}f}}. \tag{7.53}$$

Substituting (7.53) into (7.52),

$$\mathrm{var}(\hat{\tau}_d) \geq \frac{1}{\frac{\int_{-\infty}^{\infty} |S(f)|^2 \mathrm{d}f}{T\sigma^2} B_{rms}^2}. \tag{7.54}$$

Choose the bandwidth B to be equal to $\frac{1}{T}$, and so the sampling interval T corresponds to the Nyquist frequency range $[-B, B]$. Hence,

$$T = \frac{1}{2B}. \tag{7.55}$$

Therefore,

$$T\sigma^2 = \frac{\sigma^2}{2B}. \tag{7.56}$$

Substituting (7.40) into (7.56),

$$T\sigma^2 = \frac{\sigma^2}{2B} = W_0. \tag{7.57}$$

Substituting (7.57) into (7.54),

$$\mathrm{var}(\hat{\tau}_d) \geq \frac{1}{\frac{\int_{-\infty}^{\infty} |S(f)|^2 \mathrm{d}f}{W_0} B_{rms}^2}. \tag{7.58}$$

$\int_{-\infty}^{\infty} |S(f)|^2 \mathrm{d}f$ is the total signal energy. However, $s(t)$ exists only in the interval $0 \leq t \leq (N-1)T$. Let

$$T_s = (N-1)T. \tag{7.59}$$

The signal power denoted as P_s is given by

$$P_s = \frac{\int_{-\infty}^{\infty} |S(f)|^2 \mathrm{d}f}{T_s}. \tag{7.60}$$

The noise power denoted as P_n is given by

$$P_n = 2BW_0. \tag{7.61}$$

Substituting (7.60) and (7.61) into (7.58),

$$\mathrm{var}(\hat{t}_d) \geq \frac{1}{\frac{T_s P_s}{\frac{P_n}{2B}} B_{rms}^2} = \frac{1}{2BT_s \frac{P_s}{P_n} B_{rms}^2}. \tag{7.62}$$

$\frac{P_s}{P_n}$ is signal-to-noise ratio. Let

$$(SNR) = \frac{P_s}{P_n}. \tag{7.63}$$

Substituting (7.63) into (7.62),

$$\text{var}(\hat{\tau}_d) \geq \frac{1}{2BT_s(SNR)B_{rms}^2}. \tag{7.64}$$

It can be seen from (7.64) that the accuracy in estimating τ_d increases (i.e., variance of $\hat{\tau}_d$ decreases) with increasing sampling rate B (more samples for computation for a given signal duration), longer signal duration T_s (more samples for computation for a given sampling rate), higher SNR, and wider signal bandwidth B_{rms}^2.

7.4.2 Linear Chirp Method: Single Ramp

A chirp is a sinusoid $x(t)$ whose instantaneous frequency $\omega(t)$ at time t changes with t.

$$x(t) = \cos(\omega(t)t + \theta). \tag{7.65}$$

If the frequency $\omega(t)$ is a linear function of t, it is called a linear-frequency chirp or linear chirp. If the frequency increases with time, it is called an up-chirp. If the frequency decreases with time, it is called a down-chirp.

$$\omega(t) = \gamma t + \omega_0, \text{ up-chirp}, \tag{7.66a}$$

$$= -\gamma t + \omega_0, \text{ down-chirp}, \tag{7.66b}$$

$$\gamma = \text{ constant } > 0, \tag{7.66c}$$

where γ and ω_0 are positive constants. The constant γ is called the chirp rate. For linear chirp, the frequency difference $\omega(t_2) - \omega(t_1)$ is constant for any given $t_2 - t_1$.

$$\Delta\omega = \gamma \Delta t, \tag{7.67}$$

where $\Delta\omega = \omega(t_2) - \omega(t_1)$ and $\Delta t = t_2 - t_1$.

If

$$\omega(t) = \omega_0 e^{kt}, \tag{7.68}$$

where k and ω_0 are constants, it is called an exponential chirp. For exponential chirp, the frequency ratio $\frac{\omega(t_2)}{\omega(t_1)}$ is constant for any given $t_2 - t_1$.

If

$$\omega(t) = \frac{\alpha\beta}{1+\beta t} + \eta, \tag{7.69}$$

where α, β, and η are constants, it is called a hyperbolic chirp. Hyperbolic chirp has some advantages in radar applications because of its Doppler tolerance properties.

There are many other types of chirp waveforms. However, in this section, we shall focus only on the application of linear chirp for distance estimation.

Consider the scenario where a stationary transmitter transmits a linear chirp with frequency $\gamma t + \omega_0$ as depicted in (7.66a) toward a stationary target in the far field as shown in Fig. 7.12. If the wave velocity is c, the receiver will receive the reflected chirp after $\frac{2d}{c}$ unit time. Hence, at time t, the received signal was the one transmitted at $t - \frac{2d}{c}$. The frequencies of the received and transmitted signals are shown in Fig. 7.21 for positive γ. The frequency difference $\Delta\omega$ may be obtained by multiplying the transmitted signal $x(t)$ by the received signal $x(t - \Delta t)$.

$$\begin{aligned} x(t) \times x(t-\Delta t) &= \cos\{\omega(t)t + \theta\} \times \cos\{\omega(t-\Delta t)t + \theta\}, \\ &= 0.5\cos\{|\Delta\omega|t\} + 0.5\cos\{[\Delta\omega(t) + \omega(t-\Delta t)]t + 2\theta\}. \end{aligned} \tag{7.70}$$

The signal component $\cos\{[\Delta\omega(t) + \omega(t-\Delta t)]t + 2\theta\}$ has high frequency and so can be eliminated using a low-pass filter as shown in Fig. 7.22. At the output of the low-pass filter is $|\Delta\omega|$ which is the difference between $\omega(t)$ and $\omega(t - \Delta t)$; it does not contain information on whether $\omega(t) > \omega(t - \Delta t)$ or $\omega(t - \Delta t) > \omega(t)$.

From Fig. 7.22, it may seem that Δt can be estimated directly from the edges of the waveforms. However, in practice, the edges of the waveforms are poorly defined leading to poor estimate. The frequency difference $\Delta\omega$ in the interval τ in Figs. 7.22a and b can be estimated accurately. After $\Delta\omega$ has been estimated, Δt is obtained from $\Delta\omega$ via (7.67). Hence, the problem of estimating Δt has become a problem of estimating the frequency of a sinusoid from time domain data samples.

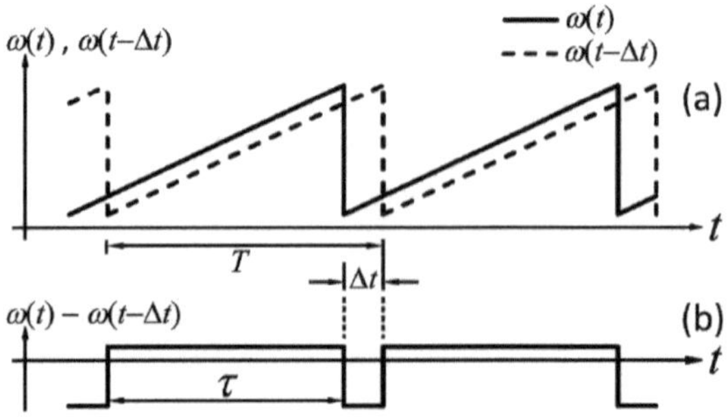

Fig. 7.21 $\omega(t), \omega(t-\Delta t)$, and $\Delta\omega$ plot. $\Delta\omega = |\omega(t) - \omega(t-\Delta t)|$

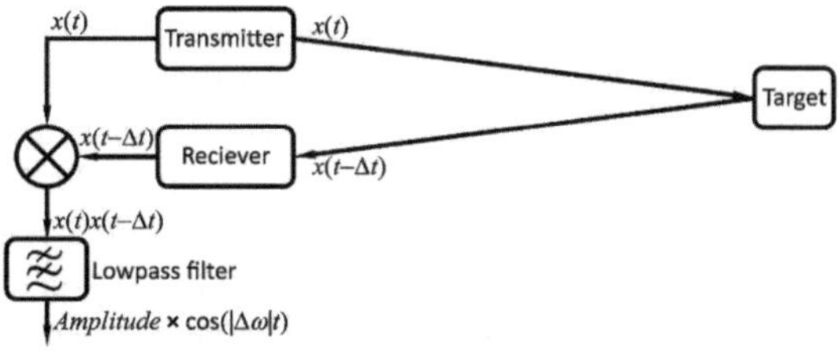

Fig. 7.22 Schematic for detecting $\Delta\omega$

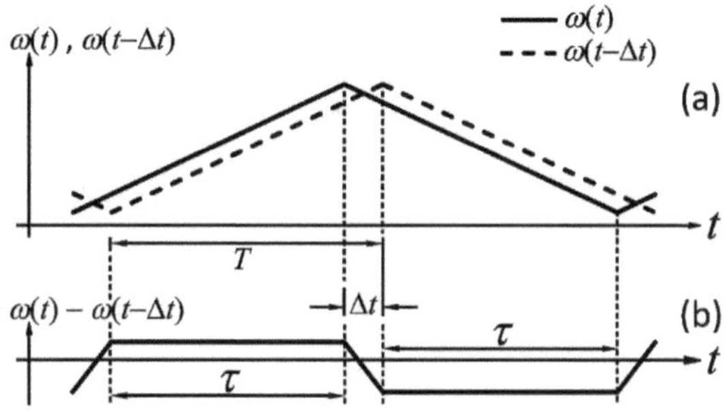

Fig. 7.23 Duo-ramp chirp for velocity and distance estimations of moving object

A simple method for estimating the frequency of a sinusoid is presented in Sect. 7.5. Super resolution methods for frequency estimation are presented in Chap. 8.

7.4.3 Linear Chirp Method: Duo Ramp

The single-ramp chirp method is valid provided that there is no relative motion between the observer and the target. If there is relative motion, Doppler effect will affect the estimate. The duo-ramp chirp method will eliminate the problem of Doppler effect provided that the fractional chirp bandwidth is small and that the target does not move appreciably during a chirp duration.

Figure 7.23 shows the $\omega(t)$ vs t plot for a duo-ramp chirp. It uses both up-chirp and down-chirp as depicted in (7.66a) and (7.66b), respectively. Suppose that the Doppler effect increases the received signal's frequency by ϵ. The Doppler frequency depends on the carrier frequency and so changes as the chirp frequency changes. We shall assume that the chirp bandwidth is small comparing to its center frequency, and so we shall assume that ϵ is constant. Doppler effect is independent of whether the chirp is up-chirp or down-chirp, and so both the up-chirp and down-chirp are affected by the same ϵ. As the object moves, Δt is a function of time. We shall assume that the distance that the object moves in the duration T is negligible, and so Δt is constant. Let the frequency differences be $\Delta\omega_1$ and $\Delta\omega_2$ for the up-chirp and down-chirp, respectively. Hence, we have

$$\Delta\omega_1 = \gamma\Delta t - \epsilon, \tag{7.71a}$$

$$\Delta\omega_2 = -\gamma\Delta t - \epsilon \quad \text{OR} \quad |\Delta\omega_2| = \gamma\Delta t + \epsilon. \tag{7.71b}$$

In (7.71b), whether $\Delta\omega_2 = -\gamma\Delta t - \epsilon$ or $|\Delta\omega_2| = \gamma\Delta t + \epsilon$ should be used depends on the method adopted for the estimation of the frequency of a sinusoid. If the frequency estimation method represents a sinusoid with frequency ω at time n using a complex exponential $e^{j\omega n}$, the method will be able to identify whether a frequency is positive or negative; in this case, $\Delta\omega_2 = -\gamma\Delta t - \epsilon$ should be used. If the frequency estimation method represents a sinusoid using a real trigonometric function $\cos(\omega n)$, there is no difference between positive and negative frequency, and a frequency is always regarded as positive; in this case, $|\Delta\omega_2| = \gamma\Delta t + \epsilon$ should be used. From (7.71), we have

$$\Delta t = \frac{\Delta\omega_1 - \Delta\omega_2}{2\gamma} \quad \text{OR} \quad \Delta t = \frac{\Delta\omega_1 + |\Delta\omega_2|}{2\gamma}, \tag{7.72}$$

and

$$\epsilon = \frac{\Delta\omega_1 + \Delta\omega_2}{2} \quad \text{OR} \quad \epsilon = \frac{\Delta\omega_1 - |\Delta\omega_2|}{2}. \tag{7.73}$$

This method estimates both the delay Δt and Doppler shift ϵ simultaneously. The method for estimating $\Delta\omega_1$ and $\Delta\omega_2$ is the same as that for estimating $\Delta\omega$ described in the single-ramp linear chirp method.

7.5 Estimating the Frequency of a Sinusoid

In the linear chirp method discussed above, it is necessary to estimate the frequency $\Delta\omega_x$ (x is either 1 or 2 or nonexistent) of $\cos\{(\Delta\omega_x)t\}$. The sinusoid $\cos\{(\Delta\omega_x)t\}$ is sampled in the time domain, and the frequency $\Delta\omega_x$ of the sinusoid is estimated from a finite number of samples.

The frequency of a sinusoid can be estimated very accurately using any one of the super resolution methods discussed in Chap. 8. In this section, a simple method is presented.

7.5.1 Frequency Spectrum of a Section of a Sinusoid

Consider the signal $x(t) = \cos(\omega_0 t)$. If $x(t)$ exists for $-\infty \leq t \leq \infty$, i.e., if $x(t)$ is infinite duration, the frequency spectrum of $x(t)$ consists of two impulses, one at $-\omega_0$ and one at ω_0. However, if $x(t)$ exists only over a finite duration, the frequency spectrum of $x(t)$ spreads over a range depending on the duration of $x(t)$.

Suppose that $x(t) = \cos(\omega_0 t) = \cos(2\pi f_0 t), 0 \leq t \leq (N-1)T$ is sampled at a regular time interval T where $\omega_0 = 2\pi f_0$. The sampling rate is $f_s = \frac{1}{T}$ and must be higher than $2f_0$; otherwise, there will be aliasing. When aliasing occurs, the sampled version of $x(t)$ and the original $x(t)$ have different baseband frequency spectrum. There are N samples of $x(t)$, and the samples are $x(0), x(T), x(2T), \ldots, x(nT), \ldots, x((N-1)T)$. To simplify notation, let $x(nT)$ be denoted by x_n. The z-transform of x_n is denoted by $X(z)$ and is given by

$$X(z) = \sum_{n=0}^{N-1} x_n z^{-n}. \tag{7.74}$$

The frequency spectrum of x_n denoted by $X(e^{j\omega T})$ is $X(z)$ evaluated at $z = e^{j\omega T}$.

$$X(e^{j\omega T}) = \sum_{n=0}^{N-1} x_n e^{-jn\omega T}, \tag{7.75a}$$

$$\text{or} \quad X(e^{j\omega}) = \sum_{n=0}^{N-1} x_n e^{-jn\omega}. \tag{7.75b}$$

The difference between (7.75a) and (7.75b) is that ωT in (7.75a) is replaced by ω in (7.75b). Replacing ωT by ω is a usual practice in signal processing. Unless otherwise stated, in this book, ωT is replaced by ω. With ωT replaced by ω, the true frequency f in Hz is given by

$$f = \frac{\omega f_s}{2\pi}. \tag{7.76}$$

The frequency spectrum of $x_n, n = 0, 1, \ldots, N-1$ depends on N. Consider the case where $f_0 = 250$ Hz and $f_s = 1$ KHz. Four data samples are collected per cycle of the sinusoid. Figure 7.24 shows $|X(e^{j\omega})|$ vs. ω plots for $N = 4, 8, 40, 400$, i.e., 1, 2, 10, and 100 cycles of the sinusoid, respectively. The peak of $|X(e^{j\omega})|$ is normalized to unity. The frequency axis in Hz is also shown. Ideally, the frequency spectrum of a sinusoid should be an impulse at the frequency of the sinusoid. It can be seen from Fig. 7.24 that if the data contains only one cycle of the sinusoid, its

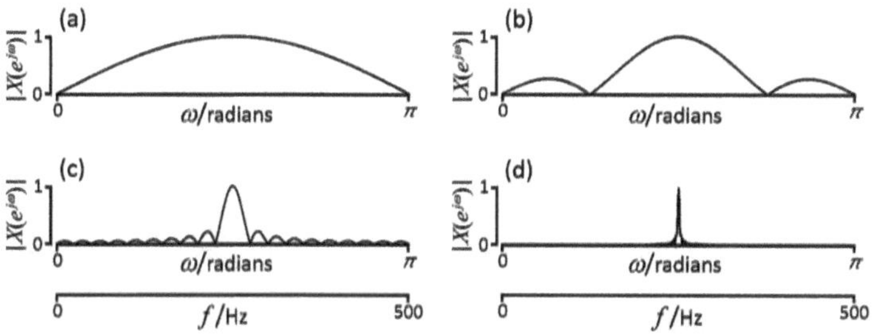

Fig. 7.24 (**a**) $N = 4$ (1 cycle), (**b**) $N = 8$ (2 cycles), (**c**) $N = 40$ (10 cycles), (**d**) $N = 400$ (100 cycles). Maximum of $|X(e^{j\omega})|$ scaled to unity

spectrum spreads out wide. The spreading decreases as the data length increases. It is close to an impulse for $N = 400$.

7.5.2 DFT

The efficiency in computing $|X(e^{j\omega})|$ can be significantly improved by using a digital signal processing technique called FFT (fast Fourier transform). Before embarking on the discussion of FFT, we shall first introduce the concept of DFT (discrete Fourier transform) (Cooley and Tukey 1965; Markel 1971; Duhamel and Hollmann 1984; Swarztrauber 1984; Sorensen et al. 1986; Jou and Abraham 1988; Fraser 1989; Van Nee and Coenen 1991; Gough 1994; Offelli and Petri 1992; Lim 1985).

Let X_k be the kth DFT coefficient of x_n. X_k and x_n are related by the following DFT relationship.

$$X_k = \sum_{n=0}^{N-1} x_n e^{-j2\pi nk/N}, \tag{7.77a}$$

$$x_n = \frac{1}{N}\sum_{k=0}^{N-1} X_k e^{j2\pi nk/N}. \tag{7.77b}$$

The DFT coefficients X_k gives the frequency spectrum values at N evenly spaced discrete points with X_k corresponds to $X(e^{j\omega})$ at $\omega = \frac{2\pi k}{N}$ or $f = \frac{kf_s}{N}$ Hz. Specifically, $k = 0, 1, 2, \ldots, N-1$ corresponds to $\omega = 0, \frac{2\pi}{N}, 2\pi\frac{2}{N}, \ldots, 2\pi\frac{(N-1)}{N}$ or $f = 0, \frac{f_s}{N}, f_s\frac{2}{N}, \ldots, f_s\frac{(N-1)}{N}$ Hz. For real x_n, $X_{N-k} = X_k^*$. Hence, we are only interested in X_k for $0 \le k < N/2$. For small N, the number of available values of X_k may be too small. For example, for $N = 4$, there are only two values of X_k, namely, X_0 and X_1; this is obviously too few for identifying the peak of the spectrum. This problem can be overcome by appending x_n with zero-value elements to any arbitrary desired length N_{desired}. Let

$$y_n = \begin{cases} x_n \text{ for } 0 \le n \le (N-1), \\ 0 \text{ for } N \le n \le N_{\text{desired}} - 1. \end{cases} \tag{7.78}$$

The frequency spectrum of y_n is the same as that of x_n, since y_n and x_n are the same. Let Y_k be the DFT coefficients of y_n. Y_k gives the frequency spectrum values at N_{desired} evenly spaced discrete points. For a smooth frequency response plot, it is common practice to append x_n such that its appended length is between 8 and 16 times its number of nonzero elements.

7.5.3 FFT (Fast Fourier Transform)

If N is highly composite such as $N = 2^{\text{integer}}$, there are extremely efficient methods for computing X_k. Such methods are called FFT. If $N = 2^{\text{integer}}$, FFT requires $\frac{N}{2}\log_2 N$ multiplications which are very much less than N^2 multiplications required by direct computation using the DFT definition of (7.77a). We had pointed out in Sect. 7.5.2 that x_n may be appended with zero-value elements to any arbitrary desired length N_{desired}. Thus, we shall assume that x_n has been appended with zero-value elements such that $N = 2^{\text{integer}}$. There are many FFT algorithms. Presented below is an FFT algorithm called "decimation-in-frequency" algorithm.

To simplify notation, we shall rewrite (7.77a) as

$$X_k = \sum_{n=0}^{N-1} x_n W_N^{nk}, \tag{7.79a}$$

$$\text{where } W_N = e^{-j2\pi/N}. \tag{7.79b}$$

It can be shown that W_N^{nk} is periodic in both n and k with period N. From (7.79b),

$$W_N^k = e^{-j\frac{2\pi}{N}k} = e^{-j\frac{2\pi}{N/k}} = W_{N/k}, \tag{7.80a}$$

$$W_N^{N\times\text{integer}} = e^{-j\frac{2\pi}{N}N\times\text{integer}} = 1, \tag{7.80b}$$

$$W_N^{N/2} = e^{-j\frac{2\pi}{N}N/2} = -1. \tag{7.80c}$$

The summation of (7.79a) may be split into two.

$$\begin{aligned} X_k = \sum_{n=0}^{\frac{N}{2}-1} x_n W_N^{nk} + \sum_{n=\frac{N}{2}}^{N-1} x_n W_N^{nk} &= \sum_{n=0}^{\frac{N}{2}-1} x_n W_N^{nk} + \sum_{n=0}^{\frac{N}{2}-1} x_{n+N/2} W_N^{Nk/2} W_N^{nk}, \\ &= \sum_{n=0}^{\frac{N}{2}-1} (x_n + x_{n+N/2} W_N^{Nk/2}) W_N^{nk}. \end{aligned} \tag{7.81}$$

Replace k by $2k$ in (7.81), and take note of (7.80); we have

$$X_{2k} = \sum_{n=0}^{\frac{N}{2}-1} (x_n + x_{n+N/2} W_N^{Nk}) W_N^{2nk} = \sum_{n=0}^{\frac{N}{2}-1} (x_n + x_{n+N/2}) W_{N/2}^{nk}. \tag{7.82}$$

Comparing (7.77a) and (7.82), it can be seen that X_{2k} are the $\frac{N}{2}$-point DFT coefficients of $x_n + x_{n+N/2}$. Replace k by $2k+1$ in (7.81), and take note of (7.80); we have

$$
\begin{aligned}
X_{2k+1} &= \sum_{n=0}^{\frac{N}{2}-1} \{(x_n + x_{n+N/2} W_N^{Nk} W_N^{N/2}) W_N^n\} W_N^{2nk}, \\
&= \sum_{n=0}^{\frac{N}{2}-1} \{(x_n - x_{n+N/2}) W_N^n\} W_{N/2}^{nk}. \qquad (7.83)
\end{aligned}
$$

Comparing (7.77a) and (7.83), it is clear that X_{2k+1} are the $\frac{N}{2}$-point DFT coefficients of $x_n - x_{n+N/2}$. The computation procedure of (7.82) and (7.83) is summarized in Fig. 7.25a for $N = 8$ using the butterfly representation shown in Fig. 7.25b. The same procedure as in (7.82) and (7.83) can be repeated to each of the four-point DFT in Fig. 7.25a; the result is shown in Fig. 7.25c. The computation flow diagram of the two-point DFT of Fig. 7.25c is drawn in the form of an FFT butterfly, and the final computation flow diagram is shown in Fig. 7.25d. Note from Fig. 7.25d that the final result for X_k is X_0, X_4, X_2, X_6, X_1, X_5, X_3, X_7 is not in natural order. The sequence 0, 4, 2, 6, 1, 5, 3, 7 is in bit-reversed order. See Table 7.1.

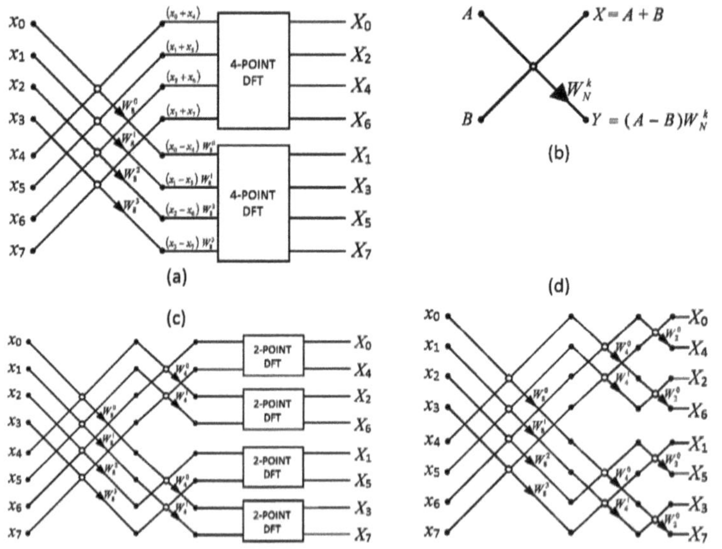

Fig. 7.25 Decimation-in-frequency algorithm for $N = 8$

Table 7.1 Bits-reversed ordering for $N = 8$

Number	Bit pattern	Bit reversal	Bit-reversed number
0	000	000	0
1	001	100	4
2	010	010	2
3	011	110	6
4	100	001	1
5	101	101	5
6	110	011	3
7	111	111	7

Presented below is another FFT algorithm called "decimation-in-time" algorithm. The summation of (7.79a) may be split into two summations with one for N odd and the other one for N even as follows:

$$X_k = \sum_{n=0}^{\frac{N}{2}-1} x_{2n} W_N^{2nk} + \sum_{n=0}^{\frac{N}{2}-1} x_{2n+1} W_N^{(2n+1)k} = \sum_{n=0}^{\frac{N}{2}-1} x_{2n} W_{N/2}^{nk} + \sum_{n=0}^{\frac{N}{2}-1} x_{2n+1} W_N^k W_{N/2}^{nk},$$
$$= \sum_{n=0}^{\frac{N}{2}-1} (x_{2n} + x_{2n+1} W_N^k) W_{N/2}^{nk}. \tag{7.84}$$

Comparing (7.77a) and (7.84), it is obvious that X_{2k} are the $\frac{N}{2}$-point DFT coefficients of $x_n + x_{n+\frac{N}{2}}$. Replace k in (7.84) by $k + \frac{N}{2}$, and take note of (7.80); we have

$$X_{k+\frac{N}{2}} = \sum_{n=0}^{\frac{N}{2}-1} (x_{2n} + x_{2n+1} W_N^{N/2} W_N^k) W_{N/2}^{nN/2} W_{N/2}^{nk},$$
$$= \sum_{n=0}^{\frac{N}{2}-1} (x_{2n} - x_{2n+1} W_N^k) W_{N/2}^{nk}. \tag{7.85}$$

Comparing (7.77a) and (7.85), we note that $X_{k+\frac{N}{2}}$ are the $\frac{N}{2}$-point DFT coefficients of $x_{2n} - x_{2n+1} W_N^k$. The computation procedure of (7.84) and (7.85) is summarized in Fig. 7.26a for $N = 8$ using the butterfly representation shown in Fig. 7.25b.

The same procedure as in (7.84) and (7.85) can be repeated to each of the four-point DFT in Fig. 7.26a; the result is shown in Fig. 7.26c. The computation flow diagram of the two-point DFT of Fig. 7.26c is drawn in the form of an FFT butterfly, and the final computation flow diagram is shown in Fig. 7.26d.

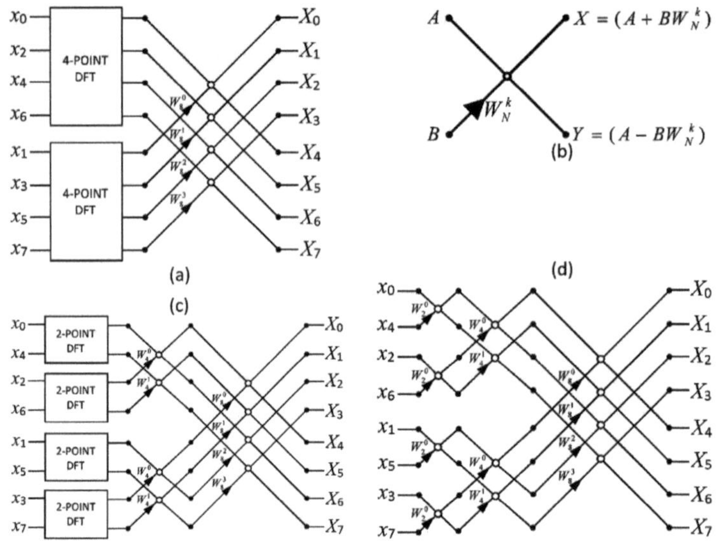

Fig. 7.26 Decimation-in-time algorithm for $N = 8$

7.5.4 *Computation Complexity*

The arithmetic complexity of performing “multiplication” operation is very much higher than that of the “addition” operation. As a consequence, in quantifying arithmetic complexity, we count only the number of multiplication operations. A complex multiplication refers to the multiplication process where both operands are complex. A real multiplication refers to the multiplication process where both operands are real. If one of the operands is complex and the other is real, it is referred to as a mixed multiplication.

Let $A, B, C,$ and D be real numbers. $A + jB$ and $C + jD$ are complex numbers and $(A + jB) \times (C + jD) = AC - BD + j(BC + AD)$. Hence, four real multiplications are performed when performing one complex multiplication. Thus, the arithmetic complexity of performing a complex multiplication is four times that of performing a real multiplication. It is straightforward to show that the arithmetic complexity of performing a mixed multiplication is two times that of performing a real multiplication. In this section, by multiplication, we refer to complex multiplication. Hence, a real multiplication is counted as one quarter of a multiplication, and a mixed multiplication is counted as half a multiplication.

For complex x_n, the computation of X_k using the definition depicted in (7.77a) requires N multiplications to compute each X_k. To compute all the N values of X_k will require N^2 multiplications. If x_n is real, the multiplications are mixed multiplications, since $e^{-j2\pi nk/N}$ is complex; the arithmetic complexity is $\frac{N^2}{2}$ multiplications.

For the FFT, as can be seen from Figs. 7.25 and 7.26, there are $\log_2 N$ columns of butterflies, and each column has $\frac{N}{2}$ butterflies. Hence, in general, there are $\frac{N}{2}\log_2 N$ complex multiplications. Since $N^2 > \frac{N}{2}\log_2 N$, the arithmetic complexity of FFT is lower than that of the definition of (7.77a).

Computing X_k using the definition of (7.77a) has certain advantage if x_n is sparse, because multiplication by zero can be ignored. We shall illustrate this by an example. Suppose that, for a specific example, the number of nonzero x_n is 7 and if only $\frac{1}{16}$ of x_n is nonzero, the total number of x_n will be $7 \times 16 = 112$. The number of multiplication operations is $\frac{7\times 112}{2} = 392$ using the definition of (7.77a). When FFT is used, the FFT will be a 128-point FFT; the number of multiplication operations is $\frac{N}{2}\log_2 N = 448$ and so is larger than 392.

In the computation of X_k using (7.77a), the coefficients $e^{-j2\pi nk/N}$ are precomputed and stored. The coefficients $e^{-j2\pi m/N}$ are periodic with period N where $m = nk =$ integer. Hence, there are at most N values of $e^{-j2\pi m/N}$. Since $e^{-j2\pi(m+N/2)/N} = -e^{-j2\pi m/N}$, only $\frac{N}{2}$ values of $e^{-j2\pi m/N}$, $m = 0, 1, 2, \ldots, \frac{N}{2} - 1$ must be stored.

For the computation of X_k using FFT, $\frac{N}{2}$ values of W_N^m, $m = 0, 1, 2, \ldots, \frac{N}{2} - 1$ must be stored; this storage requirement is the same as that for computation using (7.77a).

7.5.5 FFT Pruning

A multiplication operation where one of the operands is unity or $\pm j$ is trivial. Hence, the arithmetic complexity of the FFT can be reduced by rewriting the FFT computer program to remove all multiplications involving ± 1 and $\pm j$. For the decimation-in-frequency FFT (see, e.g., Fig. 7.25), the butterfly's coefficient values of the last two columns are either $W_2^0 = 1$ or $W_4^0 = 1$ or $W_4^1 = -j$. As a consequence, if trivial multiplication operations are not counted, the arithmetic complicity is reduced to $\frac{N}{2}(\log_2 N - 2)$ multiplications. For the third last column, the coefficient values are W_8^0, W_8^1, W_8^2, and W_8^3. Since $W_8^0 = 1$ and $W_8^2 = -j$, only multiplications involving W_8^1 or W_8^3 are nontrivial, and so 50% of the multiplications in the third last column is trivial. It can be shown that only $2^{2-m} \times 100\%$ multiplications of the mth column $m \geq 3$ counting from the last are nontrivial. For large N, the arithmetic complicity is reduced to $\frac{N}{2}(\log_2 N - 3)$ multiplications. Of course, FFT pruning requires extra efforts in coding the computer program.

Similarly, for the decimation-in-time FFT, the coefficient values for the first two columns are either $W_2^0 = 1$ or $W_4^0 = 1$ or $W_4^1 = -j$. Only $2^{2-m} \times 100\%$ multiplications of the mth column $m \geq 3$ are nontrivial. For large N, the arithmetic complicity is reduced to $\frac{N}{2}(\log_2 N - 3)$ multiplications.

If x_n is sparse, the FFT can be further pruned. We shall illustrate this using a 32-point decimation-in-frequency FFT and a 32-point decimation-in-time FFT as shown in Fig. 7.27 as examples.

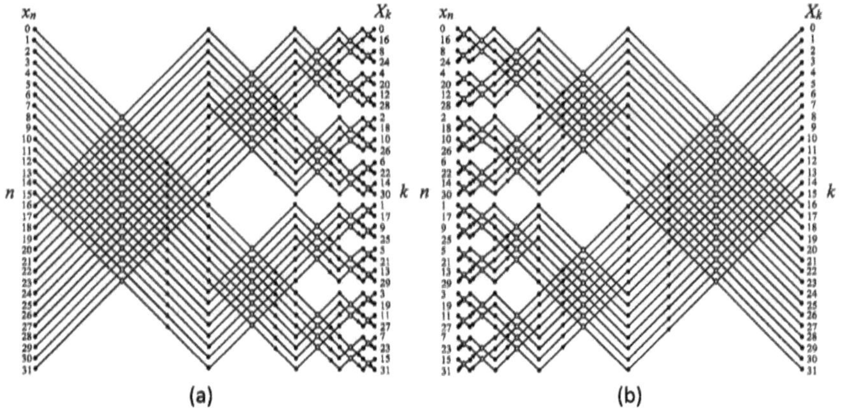

Fig. 7.27 (**a**) 32-point decimation-in-frequency FFT. (**b**) 32-point decimation-in-time FFT

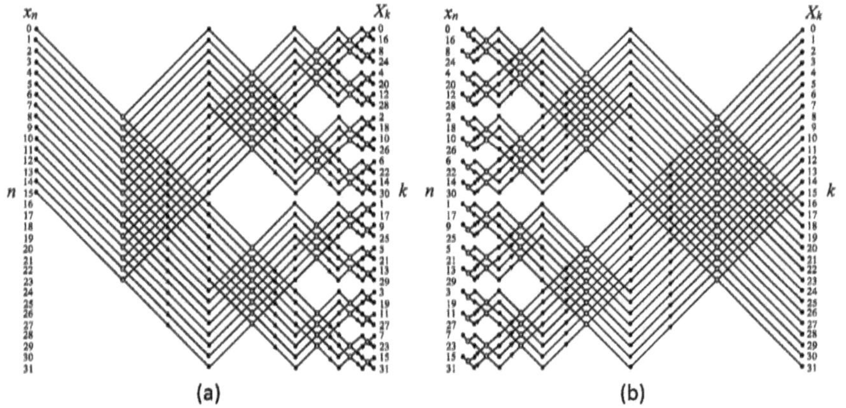

Fig. 7.28 x_n for $n = 16, 17, \ldots, 31$ are deleted. (**a**) Decimation-in-frequency. (**b**) Decimation-in-time

Suppose that $x_n = 0$ for $n = 16, 17, \ldots, 31$ as shown in Fig. 7.28 where x_n for $n = 16, 17, \ldots, 31$ are deleted. As can be seen from Fig. 7.28a, there is no saving in the number of multiplications for the decimation-in-frequency FFT, because no multiplier has been deleted. For the decimation-in-time FFT shown in Fig. 7.28b, all the multiplications by W_2^0 are deleted. However, $W_2^0 = 1$, and so the deleted multiplications are trivial anyway; there is essentially no saving in the number of multiplications.

Consider a second example where $x_n = 0$ for $n = 8, 9, \ldots, 31$. Figure 7.29 shows the scenario where x_n for $n = 8, 9, \ldots, 31$ and all subsequent zero-value nodes are deleted. Comparing Figs. 7.27 and 7.29, it can be seen that for the decimation-in-frequency FFT, W_{32}^m, $m = 8, 9, \ldots, 15$ are deleted. There is a saving of eight multiplications. For the decimation-in-time FFT, all the multiplications in the first two columns are deleted; but the multiplication operations in the first two

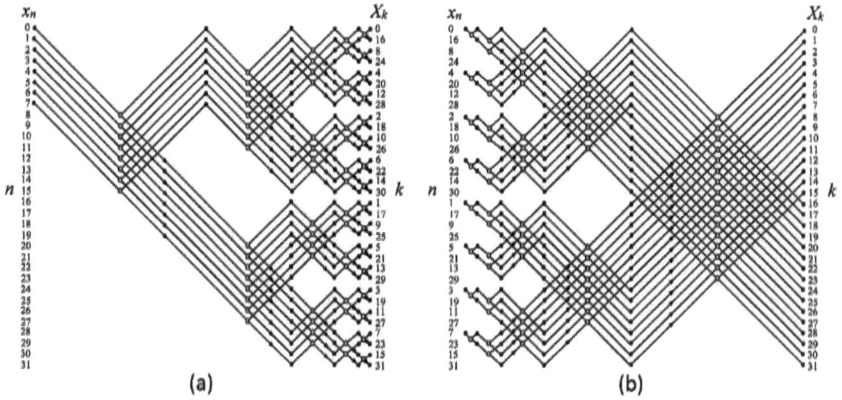

Fig. 7.29 x_n for $n = 8, 9, \ldots, 31$ and all subsequent zero-value nodes are deleted. (**a**) Decimation-in-frequency. (**b**) Decimation-in-time

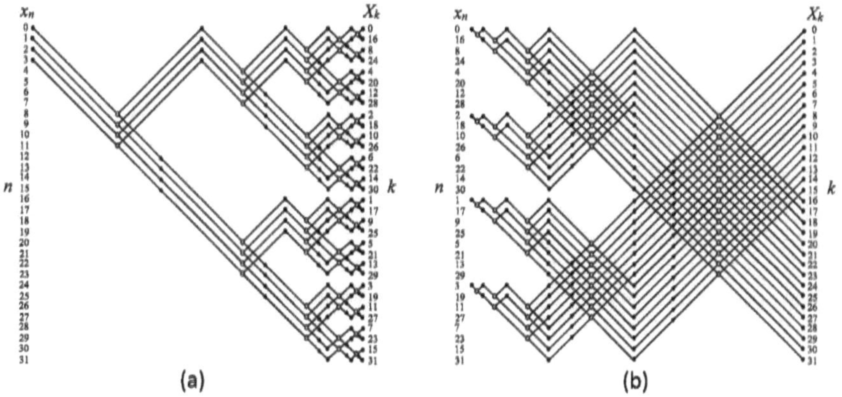

Fig. 7.30 x_n for $n = 4, 5, \ldots, 31$ and all subsequent zero-value nodes are deleted. (**a**) Decimation-in-frequency. (**b**) Decimation-in-time

columns are trivial anyway, and so there is essentially no saving in the number of multiplications.

Consider a third example where $x_n = 0$ for $n = 4, 5, \ldots, 31$. Figure 7.30 shows the scenario where x_n for $n = 4, 5, \ldots, 31$ and all subsequent zero-value nodes are deleted. Comparing Figs. 7.27 and 7.30, it can be seen that for the decimation-in-frequency FFT, W_{32}^m, $m = 4, 5, \ldots, 15$ and W_{16}^m, $m = 4, 5, 6, 7$ are deleted. There is a saving of $12 + 4 \times 2 = 20$ multiplications. For the decimation-in-time FFT, all the multiplications in the first three columns are deleted. However, the multiplication operations in the first two columns and 50% of the multiplication operations in the third column are trivial anyway. Hence, the saving in the number of multiplications is eight.

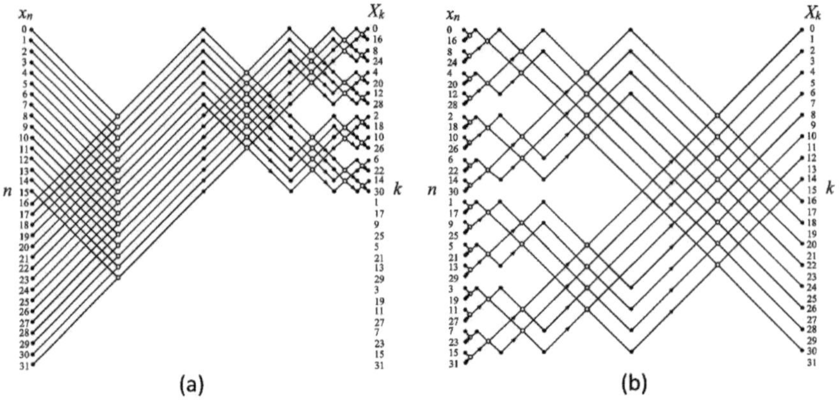

Fig. 7.31 Computation of X_k for k odd has been deleted. (**a**) Decimation-in-frequency. (**b**) Decimation-in-time

In some applications, we may be interested in first computing X_k at a lower density of k. If the density is not sufficiently high, we would like to add the necessary computations to arrive at the higher density result. Ideally, the computing load for computing first at a lower density and then move on to a higher density should be the same as computing at the higher density without going through the lower density solution. We shall examine such possibility for the decimation-in-frequency and decimation-in-time FFT.

Consider the case where only X_k for k even is computed. All the computations leading toward the computation of X_k for k odd are deleted as shown in Fig. 7.31. Comparing Figs. 7.27 and 7.31, it can be seen that, for both the decimation-in-frequency and decimation-in-time algorithms, the computations for X_k where k is even form part of the computations of X_k for all k.

Figure 7.27 shows the computation for $X_k, k = 0, 4, 8, 12, 16, 20, 24, 28$. All the computations leading toward the computation of X_k for $k \neq 0, 4, 8, 12, 16, 20, 24, 28$ have been deleted. Comparing Figs. 7.27 and 7.32, it can be seen that, for both the decimation-in-frequency and decimation-in-time algorithms, the computations for $X_k, k = 0, 4, 8, 12, 16, 20, 24, 28$ form part of the computations of X_k for all k.

7.5.6 *Frequency-Point Interpolation*

The density of X_k may also be increased by interpolation. Let $X_{k+0.5}$ be the frequency spectrum value midway between X_k and X_{k+1}. $X_{k+0.5}$ can be interpolated from values of $X_{k-M}, \ldots, X_{k-1}, X_k, X_{k+1}, \ldots, X_{k+M+1}$ as follows:

$$X_{k+0.5} = \sum_{n=1}^{M} (X_{k+n} + X_{k+1-n})h(n). \tag{7.86}$$

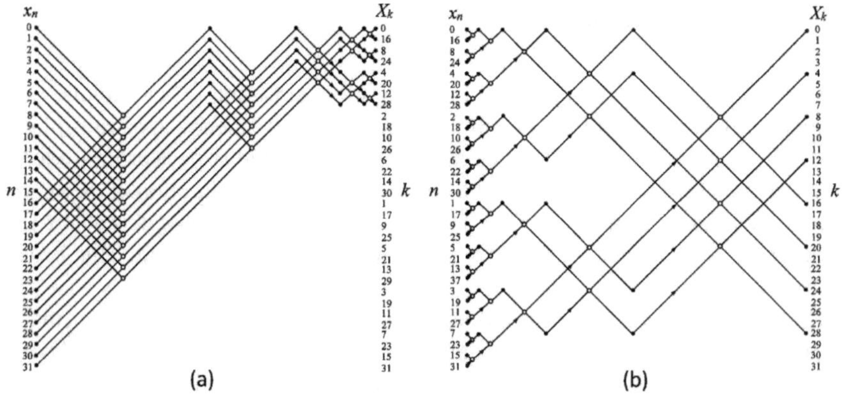

Fig. 7.32 Computation of X_k for $k = 0, 4, 8, 12, 16, 20, 24, 28$. Computations for $k \neq 0, 4, 8, 12, 16, 20, 24, 28$ have been deleted. (**a**) Decimation-in-frequency. (**b**) Decimation-in-time

Table 7.2 X_k interpolator

$x_n = 0$ for $n = d, d+1, d+2, \ldots, N-1$					
$d = \frac{N}{2}$		$d = \frac{N}{4}$		$d = \frac{N}{8}$	
n	$h(n)$	n	$h(n)$	n	$h(n)$
1	0.6187659	1	0.6012887	1	0.5870650
2	−0.1637849	2	−0.1254250	2	−0.0993622
3	0.0609111	3	0.0273063	3	0.0122972
4	−0.0201780	4	−0.003170		
5	0.0049611				
6	−0.0006753				

The interpolator is a low-pass filter with d.c. gain = 2. The normalized bandedge is at 0.25 with zero transition width if x_n has no zero-value element; such a low-pass filter has infinite complexity! If $x_n = 0$ for $n = \frac{N}{2}, \frac{N}{2} + 1, \ldots, N - 1$, the normalized passband edge and stopband edge of the low-pass filter will be at 0.125 and 0.375, respectively. Three interpolators for x_n with different sparsity are tabulated in Table 7.2.

Frequency point interpolation can be applied for any number of times. For example, X_k for $x_n = 0, n = \frac{N}{2}, \frac{N}{2} + 1, \ldots, N - 1$ may be computed using a pruned FFT and then interpolated using the interpolator tabulated in Table 7.2 under $d = \frac{N}{2}$. After that, it can be interpolated again using the interpolator under $d = \frac{N}{4}$ because, after the first interpolation, its sparsity is increased. The result after the second interpolation can be interpolated again using the interpolator under $d = \frac{N}{8}$.

7.5.7 Windowing of Data

The frequency spectrum of an infinite duration sinusoid is an impulse at its frequency. If the sinusoid is not infinite duration, its frequency spectrum plot has a main lobe and many sidelobes. Its main lobe is centered at or near its frequency. Both the main lobe width and sidelobe width decrease as the duration of the sinusoid increases as can be seen in Fig. 7.24. However, the first sidelobe peak is about 13 dB below the main lobe peak regardless of the duration of the sinusoid. This is called Gibbs phenomenon.

Gibbs phenomenon will not cause serious problem if there is only one sinusoid, because the sidelobe peak is significantly smaller than the main lobe peak. However, Gibbs phenomenon will be a serious problem if there are more than one sinusoid. Figure 7.33a shows the spectral plot of a signal containing two sinusoids. Their frequencies are 0.5π and 0.77π, respectively. The magnitude of the 0.77π sinusoid is one-tenth of that of the 0.5π sinusoid. The duration of the signal is ten times the period of the 0.5π sinusoid. Figure 7.33a does not show two distinct peaks, and so it is not obvious from inspecting Fig. 7.33a that there are two sinusoids.

Gibbs phenomenon arises because of the abrupt truncation of the signal. It can be reduced by multiplying the finite duration signal with a window function whose magnitude decreases toward the edges.

Many window functions exist. Among the most popular window functions are the rectangular window given by

$$\begin{aligned} w_n &= 1 \text{ for } 0 \le n \le N-1, \\ &= 0 \text{ elsewhere,} \end{aligned} \tag{7.87}$$

the Hamming window given by

$$\begin{aligned} w_n &= 0.53836 - 0.46164 \cos\left(\frac{2\pi n}{N-1}\right) \text{ for } 0 \le n \le N-1, \\ &= 0 \text{ elsewhere,} \end{aligned} \tag{7.88}$$

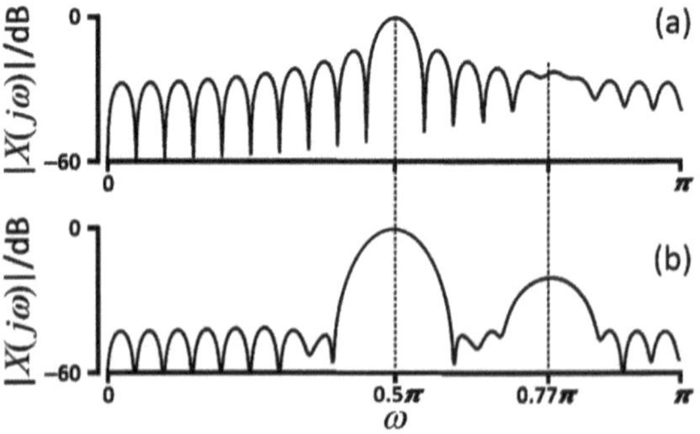

Fig. 7.33 Spectral plot of a signal containing two sinusoids. (**a**) Signal is windowed using rectangular window. (**b**) Signal is windowed using Hamming window

and the Blackman window given by

$$w_n = 0.42 - 0.5\cos\left(\frac{2\pi(n+1)}{N+1}\right) + 0.08\cos\left(\frac{4\pi(n+1)}{N+1}\right) \text{ for } 0 \le n \le N-1, \tag{7.89}$$

$$= 0 \text{ elsewhere.}$$

The windowed data y_n is given by

$$y_n = w_n \times x_n. \tag{7.90}$$

The first column of Fig. 7.34 shows examples of the rectangular, Hamming, and Blackman window functions. The second column shows the windowed time domain signals, and the third column shows $|Y(e^{j\omega})|$, the magnitude of the frequency spectrum of y_n, in dB. The plots in the first, second, and third rows correspond to those of the rectangular, Hamming, and Blackman windows, respectively. The signals are ten cycles of a sinusoid with frequency $\omega = \pi/2$. The window functions and windowed time domain signals are plotted in linear scale with magnitude scaled to unity. The spectral peaks are scaled to 0 dB. It can be seen from Fig. 7.34 that the sidelobe peak for the Blackman window is the lowest and that for the rectangular window is the highest. However, the main lobe width for the rectangular window is the narrowest and that for the Blackman window is the widest. A wider main lobe width results in a larger uncertainty in the frequency estimate.

Figure 7.33b shows the spectral plot of the signal in Fig. 7.33a windowed using a Hamming window. It can be seen in Fig. 7.33b that the sidelobes are very much

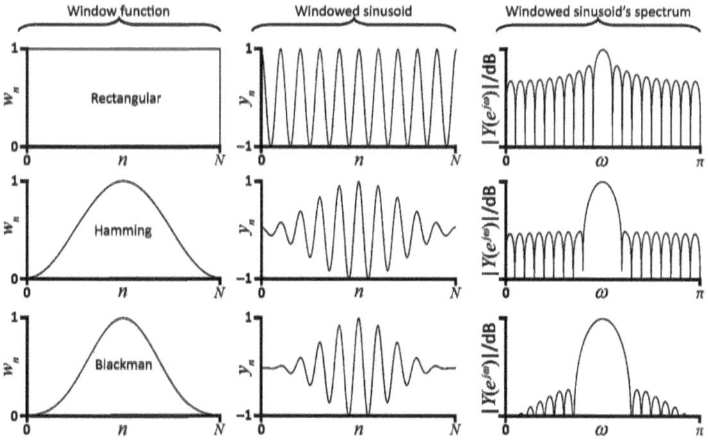

Fig. 7.34 Window functions, windowed time domain signals, and frequency spectra of windowed signals

suppressed and that the 0.77π sinusoid is clearly visible. However, the main lobe widths for both the 0.5π and 0.77π sinusoids are widened.

7.5.8 Spectral Properties of Rectangularly Windowed Sinusoid

Consider a finite duration cosine waveform $x(t)$ which is nonzero only within the time interval $-a \le t \le a$.

$$x(t) = \begin{cases} \cos(\omega_0 t), & -a \le t \le a, \\ 0, \text{otherwise} . \end{cases} \tag{7.91}$$

The Fourier transform of $x(t)$ denoted by $X(j\omega)$ is given by

$$\begin{aligned} X(j\omega) &= \int_{-a}^{a} \cos(\omega_0 t) e^{-j\omega t} \mathrm{d}t, \\ &= \frac{\sin((\omega - \omega_0)a)}{\omega - \omega_0} + \frac{\sin((\omega + \omega_0)a)}{\omega + \omega_0}. \end{aligned} \tag{7.92}$$

Note

$$\max. \left\{ \frac{\sin((\omega - \omega_0)a)}{\omega - \omega_0} \right\} = a \quad , \quad \omega \to \omega_0, \tag{7.93a}$$

$$\max. \left\{ \frac{\sin((\omega + \omega_0)a)}{\omega + \omega_0} \right\} = a \quad , \quad \omega \to -\omega_0, \tag{7.93b}$$

$$X(j\omega) \to a + \frac{\sin(2a\omega_0)}{2\omega_0} \quad , \quad \omega \to \pm\omega_0, \tag{7.93c}$$

$$\to a \quad , \quad \omega \to \pm\omega_0 \text{ and } 2a\omega_0 = \text{integer} \times \pi \tag{7.93d}$$

For $2a\omega_0 \neq$ integer $\times\ \pi$, $X(j\omega)$ peaks at $\omega \approx \pm\omega_0$. For $2a\omega_0 =$ integer $\times\ \pi$, $2a = \frac{\text{integer}\times\pi}{\omega_0} = \frac{\text{integer}\times\pi}{2\pi f_0} = \text{integer} \times \frac{1}{2} \times \frac{1}{f_0}$. The nonzero duration and the period of the sinusoid are $2a$ and $\frac{1}{f_0}$, respectively. Hence, if the nonzero duration of the sinusoid is an integer multiple of half of the period of the sinusoid, $X(j\omega)$ peaks at $\omega = \pm\omega_0$ and $X(j\omega) \to a$ as $\omega \to \pm\omega_0$; the magnitude of the spectral peak is proportional to the nonzero duration of the sinusoid.

In the following analysis, we shall assume that the nonzero duration of the sinusoid is an integer multiple of half of the period of the sinusoid; otherwise, the following analysis is only approximately correct. When $(\omega - \omega_0)a = \pm\pi$, $\frac{\sin((\omega-\omega_0)a)}{\omega-\omega_0} = 0$ and $(\omega + \omega_0)a = (\text{integer} \pm 1) \times \pi$ and so $\frac{\sin((\omega+\omega_0)a)}{\omega+\omega_0} = 0$. Hence, the first set of zeros of $X(j\omega)$ next to $\omega = \omega_0$ is $(\omega - \omega_0)a = \pm\pi$, i.e., one of the zeroes is at $\omega = \omega_0 - \frac{\pi}{a}$ and the other is at $\omega = \omega_0 + \frac{\pi}{a}$. Let B_0 be the

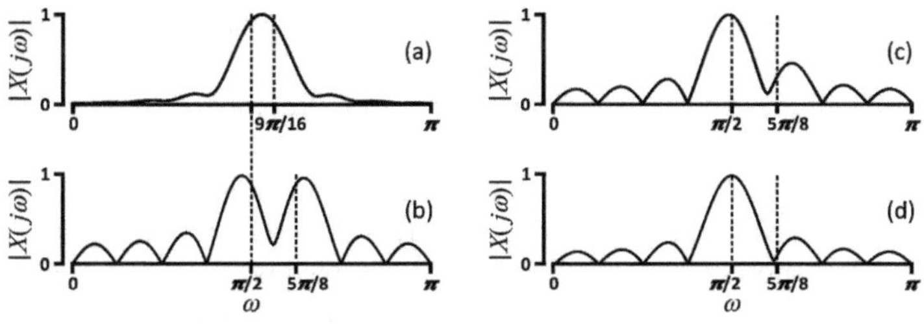

Fig. 7.35 Frequency spectrum of a signal consisting of 16 samples of 2 sinusoids. The signals are scaled such that the spectral peaks are unity. One of the frequencies is $\frac{\pi}{2}$. For (**a**), the other frequency is $\frac{9\pi}{16}$. For (**b**), (**c**), and (**d**), the other frequency is $\frac{5\pi}{8}$. For (**a**) and (**b**), the two sinusoids have the same magnitude. For (**c**) and (**d**), the $\frac{5\pi}{8}$ sinusoid is 10 dB and 20 dB, respectively, below that of (**b**)

frequency separation of the first set of zeros of $X(j\omega)$ next to $\omega = \omega_0$; it is also the spectral main lobe width at $\omega = \omega_0$. We have

$$B_0 = \left(\omega_0 + \frac{\pi}{a}\right) - \left(\omega_0 - \frac{\pi}{a}\right) = \frac{2\pi}{a}. \tag{7.94}$$

Similarly, it can be shown that the spectral main lobe width at $\omega = -\omega_0$ is also $\frac{2\pi}{a}$. Hence, the spectral main lobe width is inversely proportional to the nonzero duration of the sinusoid.

For a signal consisting of two sinusoids with frequency separation less than $\frac{B_0}{2}$, the frequency of one of the sinusoids will be within the main lobe of the frequency spectrum of the other sinusoid. In this case, the spectral plot may not show two frequency peaks.

Figure 7.35a shows the spectral plot of a signal consisting of two sinusoids with equal magnitudes and with frequencies $\frac{\pi}{2}$ and $\frac{9\pi}{16}$, respectively; the frequencies are separated by $\frac{9\pi}{16} - \frac{\pi}{2} = \frac{\pi}{16}$. The duration of the signal is four times the period of the $\frac{\pi}{2}$ sinusoid. The value of B_0 obtained using (7.94) is $\frac{\pi}{4}$ and so $\frac{B_0}{2} = \frac{\pi}{8}$. Hence, the separation of the two frequencies is less $\frac{B_0}{2}$; $\frac{9\pi}{16}$ is within the main lobe of the frequency spectrum of the $\frac{\pi}{2}$ sinusoid. Figure 7.35a does not show two distinct spectral peaks; it is not obvious from Fig. 7.35a that there are two frequencies.

Figure 7.35b shows the case where the sinusoid with frequency $\omega = \frac{9\pi}{16}$ is replaced by a sinusoid with frequency $\omega = \frac{5\pi}{8}$; the frequencies are separated by $\frac{5\pi}{8} - \frac{\pi}{2} = \frac{\pi}{8} = \frac{B_0}{2}$. The frequency $\frac{5\pi}{8}$ is at the spectral null of the frequency spectrum of the $\frac{\pi}{2}$ sinusoid. The two sinusoids have the same magnitude. It can be seen from Fig. 7.35b that there are two distinct spectral peaks indicating that there are two sinusoids. However, the peaks do not accurately represent the frequencies of the sinusoids. In general, if the two sinusoids have comparable magnitudes and if the frequency of one sinusoid is at the spectra null of the frequency spectrum of the other sinusoid, the two frequencies are resolvable on the spectrum plot. If the two frequencies are close together, the spectral peaks may not accurately represent the frequencies.

If the magnitudes of the two frequencies are not comparable, identifying them from a spectral plot may be difficult. Figure 7.35c shows the case where the $\frac{5\pi}{8}$ sinusoid is 10 dB below that of the $\frac{\pi}{2}$ sinusoid. It can be seen that the second peak of the frequency spectrum is not too much larger than the sidelobe peaks.

Figure 7.35d shows the case where the $\frac{5\pi}{8}$ sinusoid is 20 dB below that of the $\frac{\pi}{2}$ sinusoid. The second peak of the frequency spectrum is not visible.

We shall assume that, if a signal contains two sinusoids with comparable magnitudes and with frequencies ω_0 and ω_1, respectively, the two frequencies are resolvable if $|\omega_0 - \omega_1| \geq \frac{B_0}{2}$. Suppose that there are n_0 and n_1 cycles, respectively, of the sinusoids with frequencies ω_0 and ω_1 for a duration of $2a$.

$$n_0 = \frac{a\omega_0}{\pi}, \tag{7.95a}$$

$$n_1 = \frac{a\omega_1}{\pi}. \tag{7.95b}$$

For the two frequencies to be resolvable, the condition that either $\omega_1 \geq \omega_0 + \frac{\pi}{a}$ or $\omega_1 \leq \omega_0 - \frac{\pi}{a}$ must be satisfied. Hence, from (7.95),

$$n_1 \geq \frac{a \times (\omega_0 + \frac{\pi}{a})}{\pi} = \frac{a\omega_0}{\pi} + 1 = n_0 + 1 \ , \quad \text{or} \tag{7.96a}$$

$$n_1 \leq \frac{a \times (\omega_0 - \frac{\pi}{a})}{\pi} = \frac{a\omega_0}{\pi} - 1 = n_0 - 1 \ . \tag{7.96b}$$

From (7.96),

$$|n_1 - n_0| \geq 1. \tag{7.97}$$

Expression (7.97) states that, for a signal containing two sinusoids of comparable magnitudes, the frequencies of the two sinusoids are resolvable if the number of cycles of the two sinusoids differ by more than one within the nonzero duration of the signal.

7.5.9 *Chirp Bandwidth*

Suppose that frequencies ω_1 and ω_2 are the extreme frequencies of a linear chirp. The chirp bandwidth B_c is given by

$$B_c = |\omega_2 - \omega_1|. \tag{7.98}$$

Let the chirp duration be T. See Figs. 7.21 and 7.23. The chirp rate γ is given by

$$\gamma = \frac{B_c}{T}. \tag{7.99}$$

Let $\Delta\omega_s$ denote the frequency difference in the instantaneous frequency between the transmitted and received sinusoids for the single-ramp chirp; $\Delta\omega_s = |\Delta\omega|$ of (7.67). Hence, from (7.67) and (7.99),

$$\Delta\omega_s = \frac{B_c}{T}\Delta t. \tag{7.100}$$

The period of the frequency $\Delta\omega_s$ is $\frac{2\pi}{\Delta\omega_s}$. The signal with frequency $\Delta\omega_s$ will be collected for estimating the value of $\Delta\omega_s$. The number of cycles of the sinusoid used for estimating the frequency of the sinusoid is important, because it determines the spectral main lobe width which determines the frequency resolution. The larger is the number of cycles, the narrower will be the spectral main lobe width and so the higher will be the spectral resolution.

Let n_s be the number of cycles of the frequency $\Delta\omega_s$ within the duration τ.

$$n_s = \tau \times \frac{\Delta\omega_s}{2\pi}. \tag{7.101}$$

From (7.100) and (7.101),

$$n_s = \frac{\tau B_c \Delta t}{2\pi T} = \frac{(T - \Delta t) B_c \Delta t}{2\pi T} = \frac{B_c \Delta t}{2\pi} - \frac{B_c (\Delta t)^2}{2\pi T}, \tag{7.102a}$$

$$\approx \frac{B_c \Delta t}{2\pi} \quad \text{for} \quad \Delta t \ll T. \tag{7.102b}$$

From (7.102b), we note that for $\Delta t \ll T$, the number of cycles collected is independent of T, the duration for collecting the sinusoid.

Suppose that there are two objects at distances d_1 and d_2 from the observer. The chirp signal will be reflected back from the objects after time Δt_1 and Δt_2 and produces two sinusoids with frequencies $\Delta\omega_{s,1}$ and $\Delta\omega_{s,2}$, respectively. The two sinusoids are sampled over a duration of τ for spectral analysis. Suppose that, within the duration τ, there are n_1 and n_2 cycles of the sinusoids corresponding to $\Delta\omega_{s,1}$ and $\Delta\omega_{s,2}$, respectively. The distance resolution is the difference in distances d_1 and d_2 that produces $\Delta\omega_{s,1}$ and $\Delta\omega_{s,2}$ that is resolvable. We wish to determine the distance resolution $\Delta d = d_1 - d_2$. We shall assume (7.102b), i.e., $n_s = \frac{B_c \Delta t}{2\pi}$.

$$n_{s,1} = \frac{B_c \Delta t_1}{2\pi}, \tag{7.103a}$$

$$n_{s,2} = \frac{B_c \Delta t_2}{2\pi}. \tag{7.103b}$$

From (7.103a) and (7.103b),

$$|n_{s,1} - n_{s,2}| = \frac{B_c|\Delta t_1 - \Delta t_2|}{2\pi}. \tag{7.104}$$

For $\Delta\omega_{s,1}$ and $\Delta\omega_{s,2}$ to be resolvable, $|n_{s,1} - n_{s,2}| \geq 1$. Hence,

$$B_c \geq \frac{2\pi}{|\Delta t_1 - \Delta t_2)|}. \tag{7.105}$$

We have

$$|\Delta t_1 - \Delta t_2| = \frac{d_1 - d_2}{c} = \frac{\Delta d}{c}, \tag{7.106}$$

where c is the speed of light. From (7.105) and (7.106),

$$B_c \times \Delta d \geq 2\pi c, \tag{7.107a}$$

$$\text{or} \quad \Delta f \times \Delta d \geq c \quad \text{where} \quad B_c = 2\pi \Delta f. \tag{7.107b}$$

Expression (7.107) shows the relationship between distance resolution and chirp bandwidth. It is interesting to note that the distance resolution is independent of the chirp rate and the duration T as long as $\Delta t \ll T$. Higher distance resolution corresponds to smaller value of Δd. Hence, distance resolution improves with increasing chirp bandwidth.

For the duo-ramp chirp, let $\Delta\omega_d$ denote the frequency difference in the instantaneous frequency between the transmitted and received sinusoids and $\Delta\omega_d = |\Delta\omega_1|$ for the up-chirp of (7.71a) and $\Delta\omega_d = |\Delta\omega_2|$ for the down-chirp of (7.71b). Hence, from (7.71) and (7.99),

$$\Delta\omega_d = \frac{B_c}{T}\Delta t \pm \epsilon. \tag{7.108}$$

The period of the frequency $\Delta\omega_d$ is $\frac{2\pi}{\Delta\omega_d}$. Let n_d be the number of cycles of the frequency $\Delta\omega_d$ within the duration τ.

$$n_d = \tau \times \frac{\Delta\omega_d}{2\pi}. \tag{7.109}$$

From (7.108) and (7.109),

$$n_d = \frac{\tau B_c \Delta t}{2\pi T} \pm \frac{\tau\epsilon}{2\pi} = \frac{(T - \Delta t)B_c\Delta t}{2\pi T} \pm \frac{\tau\epsilon}{2\pi} = \frac{B_c\Delta t}{2\pi} - \frac{B_c(\Delta t)^2}{2\pi T} \pm \frac{\tau\epsilon}{2\pi}, \tag{7.110a}$$

$$\approx \frac{B_c\Delta t}{2\pi} \pm \frac{\tau\epsilon}{2\pi} \quad \text{for} \quad \Delta t \ll T. \tag{7.110b}$$

If there are two objects at distances d_1 and d_2 from the observer, the chirp signal will be reflected back from the objects after time Δt_1 and Δt_2, respectively, producing $n_d = n_{d,1}$ and $n_d = n_{d,2}$ cycles of the sinusoids. Suppose that the Doppler shift due to these two objects are ε_1 and ε_2, respectively. For Δt_1 and Δt_2 very much smaller than T, we have

$$n_{d,1} - n_{d,2} = \frac{B_c(\Delta t_1 - \Delta t_2)}{2\pi} \pm \frac{\tau(\epsilon_1 - \epsilon_2)}{2\pi}. \tag{7.111}$$

If $\left|\frac{\tau(\epsilon_1-\epsilon_2)}{2\pi}\right| \ll \left|\frac{B_c(\Delta t_1-\Delta t_2)}{2\pi}\right|$,

$$|n_{d,1} - n_{d,2}| = \frac{B_c|\Delta t_1 - \Delta t_2|}{2\pi}. \tag{7.112}$$

Expression (7.112) is the same as (7.104) with $n_{s,1}$ and $n_{s,2}$ replaced by $n_{d,1}$ and $n_{d,2}$, respectively. Hence, the relationship between the chirp bandwidth for the duo-ramp chirp is the same as that for the single-ramp chirp shown in (7.107).

In (7.110b), $\frac{\epsilon}{2\pi}$ is the number of cycles per unit time of the Doppler frequency, and so $\frac{\tau\epsilon}{2\pi}$ is the number of cycles due to Doppler shift.

7.5.10 Velocity Resolution

Whenever there is a relative velocity v between an object and an observer, there will be a Doppler shift ϵ. Depending on the scenario as shown in (7.7), (7.9), (7.10), and (7.11), ϵ may be $\frac{v\omega_0}{c}$ or $\frac{2v\omega_0}{c}$ where ω_0 is the frequency of transmission. Let

$$\epsilon = \frac{kv\omega_0}{c} \quad \text{where} \quad k = 1, 2. \tag{7.113}$$

Suppose that there are n_n cycles of the Doppler frequency in the duration τ.

$$n_n = \frac{\tau\epsilon}{2\pi}. \tag{7.114}$$

From (7.113) and (7.114),

$$n_n = \frac{\tau k v\omega_0}{2\pi c}. \tag{7.115}$$

Suppose that there are two objects traveling at velocities v_1 and v_2, respectively. Let n_1 and n_2 be the values of n_n corresponding to v_1 and v_2, respectively. From (7.115),

$$n_1 = \frac{\tau k v_1 \omega_0}{2\pi c}, \tag{7.116a}$$

$$n_2 = \frac{\tau k v_2 \omega_0}{2\pi c}. \tag{7.116b}$$

From (7.116a) and (7.116b),

$$|n_1 - n_2| = \frac{\tau k |v_1 - v_2| \omega_0}{2\pi c}. \tag{7.117}$$

Let $\Delta v = |v_1 - v_2|$; Δv is the velocity resolution. For the two Doppler frequencies to be resolvable, $|n_1 - n_2| \geq 1$ and so

$$k\tau\omega_0\Delta v \geq 2\pi c, \tag{7.118a}$$

$$\text{or} \quad k\tau f_0 \Delta v \geq c \quad \text{where} \quad \omega_0 = 2\pi f_0. \tag{7.118b}$$

It can be seen from (7.118) that the velocity resolution improves with increasing frequency and/or duration τ.

7.5.11 Effect of Noise

So far, in the above discussions on the estimation of the frequency, we assume that there is no noise embedded in the data we used to estimate the frequency. We ignored the effect of noise. In actual practice, noise is unavoidable. In the following discussions, we shall assume that $2a\omega_0 = \text{integer} \times \pi$; otherwise, the result is only approximate.

Over a duration of $2a$, from (7.93) and from the above discussions, the signal spectral peak is proportional to a. Suppose that the noise is white Gaussian noise with flat power spectrum; it is not a function of a. Hence, the signal-spectral-peak to noise-spectral-peak ratio is proportional to a and increases by $10\log(2) = 3$ dB for every doubling of nonzero signal duration.

Figure 7.36 shows three frequency spectra plots of a sinusoid corrupted by white Gaussian noise. The frequency of the sinusoid is $\frac{\pi}{2}$. The signal-to-noise ratio is 0 dB. The plots are scaled such that $X(j\omega)| = 0$ dB at $\omega = 0.5\pi$. The plots for Figs. 7.36a, b, and c correspond to data length of 4 cycles, 10 cycles, and 100 cycles of the sinusoid, respectively. It can be seen that the signal-spectral-peak to noise-spectral-peak ratio improves as the nonzero signal duration increases.

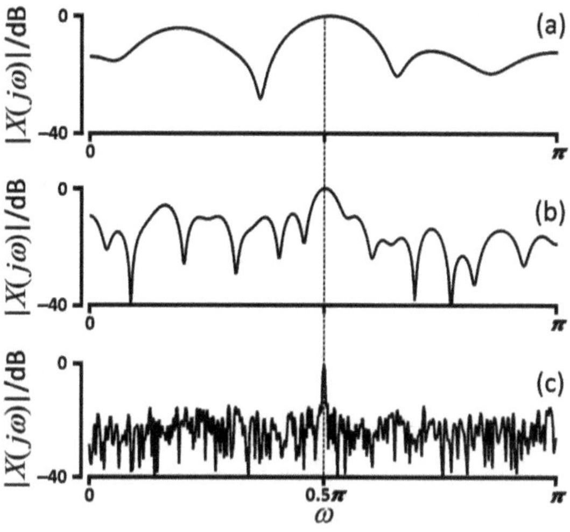

Fig. 7.36 Frequency spectra of a sinusoid corrupted by noise. The sinusoid's frequency is 0.5π, and the SNR is 0 dB. $|X(j\omega)|$ are scaled to 0 dB at $\omega = 0.5\pi$. The nonzero data durations are 4, 10, and 100 cycles of the sinusoid for (**a**), (**b**), and (**c**), respectively

7.5.12 Practical Considerations

There is no international agreed frequency bands for autonomous vehicle radars. Currently, the most popular frequency bands are the 24 GHz ISM band, 75 GHz, 77 GHz, and 79 GHz bands. Regulatory permission differs from country to country. In this section, we shall compute a few parameters derived in the previous sections for chirp frequencies at 24 GHz and 77 GHz. We shall assume that the fractional chirp bandwidth $\frac{B_c}{\omega_0}$ is 1%; i.e., the chirp bandwidths for 24 GHz and 77 GHz bands are 240 MHz and 770 MHz, respectively.

Using (7.107), the distance resolutions are 1.2 m and 0.374 m, respectively, for 240 MHz and 770 MHz chirp bandwidth. The wavelengths of the 24 GHz and 77 GHz carriers are 12.5 mm and 3.896 mm, respectively; the form factors of the antennae are determined by the wavelengths. The chirp signal sent by the observer will travel twice the distance between the observer and the object and so $\Delta t = 2 \times \frac{\text{object distance}}{\text{speed of light}}$; $\Delta t = 6.667$ ns, 0.6667 μs, 6.667 μs, and 66.67 μs, respective, for objects at 1 m, 100 m, 1 km, and 10 km away from the observer.

Consider a stationary observer monitoring the velocity and distance of an object moving at 100 km/hr ($\approx 27.78\ \text{ms}^{-1}$). The Doppler shift will be $2 \times \frac{27.78}{3\times 10^8} \times 24 \times 10^9 = 4445$ Hz and $2 \times \frac{27.78}{3\times 10^8} \times 77 \times 10^9 = 14{,}260$ Hz, respectively, for 24 GHz and 77 GHz carrier frequencies. At 100 km/hr, the object moves by 1 m in 36.00 ms. As we have discussed earlier, as the object moves, Δt changes with time, and so the data for spectral analysis should be collected over a short period such that the change in Δt is small. Suppose that we wish to collect the data over a duration where the object moves by less than 0.1 m; thus, the data should be collected in less than 3.6 ms; T should be less than $\frac{3.6}{2}$ ms $= 1800$ μs for the dual-ramp chirp. Hence, $\Delta t \ll T$ is true even for a distance of 10 km.

When using chirp to estimate distances, signals reflected from objects far beyond the range of interest may corrupt signals reflected from objects within the range of interest if the second chirp is sent, while the reflections from far away objects are still on their way back to the observer. For example, the range of interest may be 100 m corresponding to $\Delta t = 0.6667\ \mu\text{s}$. However, chirp signal reflected from an object 1 km away will arrive after 6.667 μs. This should be taken into consideration in the determination of the chirp repetition rate.

7.6 Other Methods for Velocity and Distance Estimation

Methods for velocity and distance estimation described above belong to the class of methods called "signal processing methods." While signal processing methods are gaining increasing popularity and essential in many cases, many other methods exist and may even be more desirable in specific applications. For example, the speedometer is certainly more appropriate as a speed indicator for a land-based vehicle with wheels. However, an autonomous vehicle need not be confined to be land based; it may be on water surface, or under water, or in air. The pitometer logs and pitot tubes are the instrument currently used in such applications. In this section, we shall discuss some of them.

7.6.1 Vehicle Speed Measurement Using Speedometer

The speedometer in a vehicle is probably one of the most common speed indicator devices encountered in everyday life. Figure 7.37 shows a simplified construction of an analogue speedometer. It uses a rotating permanent magnet driven by a flexible cable connected to the vehicle's transmission. The speed of rotation of the magnet is proportional to the speed of rotation of the vehicle's wheels. As the permanent magnet rotates, it induces eddy current in the aluminium speed cup. The magnetic field of the eddy current interacts with the magnetic field of the permanent magnet producing torque that tends to drive the speed cup to rotate in the same direction as that of the permanent magnet. The torque is proportional to the difference in the

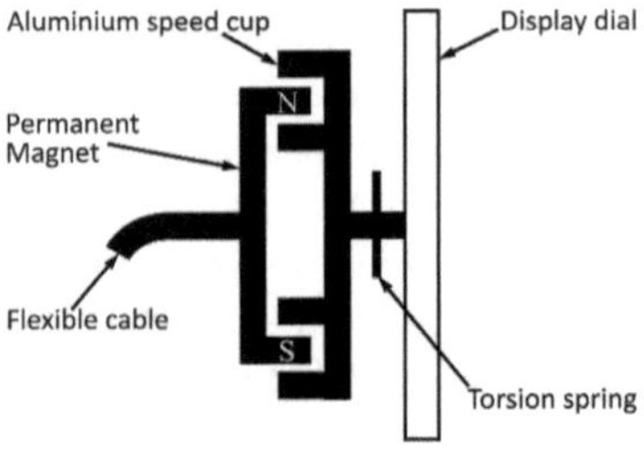

Fig. 7.37 Construction of an analogue speedometer

speed of rotation of the permanent magnet and that of the speed cup. The movement of the speed cup is constrained by a torsion spring. The speed cup will move an angle until the torque due to the eddy current is balanced by that due to the torsion spring. The speed dial attached to the speed cup is calibrated to read the speed of the vehicle assuming a given diameter of the wheel.

Digital speedometer uses a device that produces pulses whose frequency is proportional to the speed of rotation of the vehicle's wheels. A frequency counter counts the frequency of the pulses and is calibrated to read the vehicle's speed assuming a given wheel diameter.

7.6.2 *Fluid Speed Measurement*

A ship or a submarine uses a pitometer logs (or pit logs) to measure its speed relative to the water. Figure 7.38 shows a simplified structure of a pitometer logs. A pitometer logs has two openings. One of the openings faces the direction of the seawater motion; the other one is perpendicular to the seawater motion. The pressure marked as p_t and p_s in Fig. 7.38 are called total pressure (or stagnation pressure) and static pressure, respectively. The total pressure is greater than the static pressure, and the difference $p_t - p_s$ is indicated by the difference in mercury level in Fig. 7.38. Suppose that the flow speed of the seawater is v and the sea water density is ρ. Bernoulli's equation states that

$$\frac{\rho v^2}{2} = p_t - p_s, \tag{7.119a}$$

and, therefore,

$$v = \sqrt{\frac{2(p_t - p_s)}{\rho}}. \tag{7.119b}$$

The velocity v can be computed from measured value of $p_t - p_s$ and known value of ρ using (7.119b). Bernoulli's equation (7.119) applies only to incompressible fluids such as seawater. Gases under certain conditions can be approximated as incompressible.

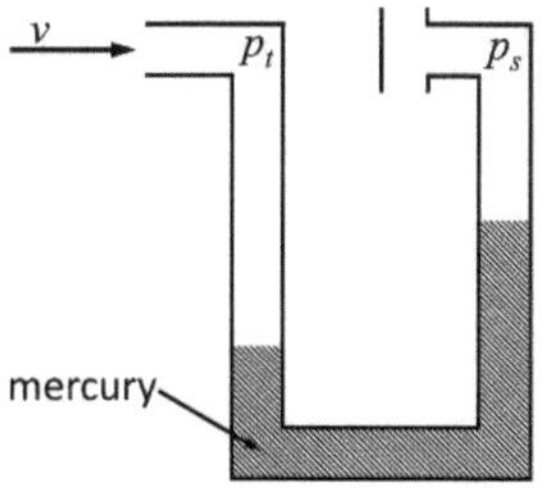

Fig. 7.38 Structure of a pitometer logs. p_t and p_s are the total pressure and static pressure, respectively

The air speed of an aircraft is measured using a device called pitot tube. The operation of a pitot tube is similar to that of a pitometer logs.

7.6.3 Distance Estimation Using Trigonometric Method

The trigonometric method has been used since ancient times for determining the distance of an object. This is the method used in astronomy to estimate the distance of distant stars and galaxies. Consider the system shown in Fig. 7.39. From the trigonometry of the system, we have

$$\theta_3 = \theta_2 - \theta_1, \tag{7.120a}$$

and

$$L = \frac{d\cos(\theta_2)}{\sin(\theta_3)}. \tag{7.120b}$$

The angles θ_1 and θ_2 may be measured using the setup shown in Fig. 7.40 consisting of a lens and a screen. The reference point "O" must first be calibrated before it can be used to measure the angle θ. In Fig. 7.40, the object casts an image on the screen. The angle θ is given by

$$\theta = \tan^{-1}(d/v). \tag{7.121}$$

If $u \gg f$, $v \approx f$ where f is the focal length of the lens. The method depicted in Fig. 7.39 requires two sensor systems shown in Fig. 7.40 spaced d apart. The following "rotating mirror method requires only one sensor system.

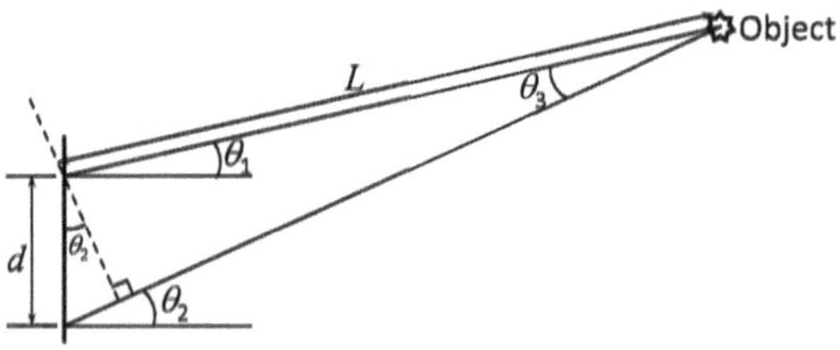

Fig. 7.39 Trigonometric method for object distance estimation

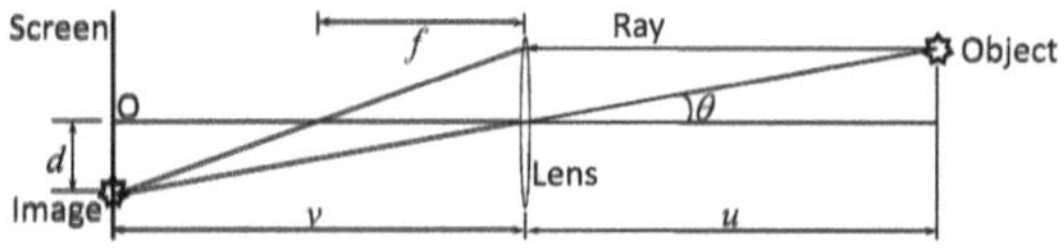

Fig. 7.40 Setup for measuring θ_1 and θ_2 in Fig. 7.39

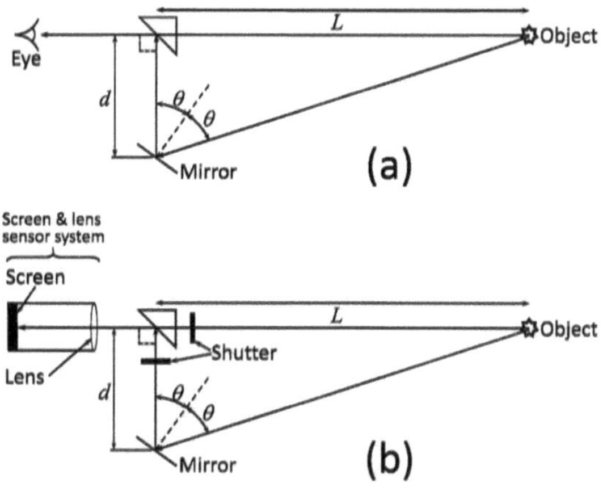

Fig. 7.41 Rotating mirror method

7.6.4 Distance Estimation Using Rotating Mirror Method

Before the era of digital camera, the rotating mirror method was perhaps the most commonly used focusing method in handheld non-SLR cameras. It is essentially the trigonometric method for distance estimation shown in Fig. 7.39 with $\theta_1 = 0$. The basic principle of the rotating mirror method is illustrated in Fig. 7.41. Referring to Fig. 7.41a, the photographer looks at the object through a viewfinder on the camera. There is a prism and a rotating mirror in the viewfinder of the camera. Light from the object travels to the photographer's eye in two paths—one is a direct path through the prism and the other one is via reflection from the mirror and the prism. In general, the eye sees two images, one from each path. The two images can be made to coincide by adjusting the angle θ. If θ is adjusted until the images coincide,

$$L = d\tan(2\theta). \qquad (7.122)$$

In a camera, the rotating mirror is connected to the lens by some mechanism to accurately adjust the image distance v.

Although the rotating mirror method is developed for estimating L for camera focusing application, it can be adapted for other applications by replacing the photographer's eye by electronic sensors as shown in Fig. 7.41b. In Fig. 7.41b, two shutters are added to select between the two images.

References

E.N.d.C. Andrade, Doppler and the Doppler effect. Endeavour **18**(69), 14–19 (1959)

Christian Andreas Doppler, Über das farbige licht der doppelsterne und einiger anderer gestirne des himmels (1842a). [Online]. Available: https://archive.org/details/ueberdasfarbigel00doppuoft/

Christian Andreas Doppler, Über das farbige licht der doppelsterne und einiger anderer gestirne des himmels (1842b). [Online]. Available: https://de.wikisource.org/wiki/%C3%9Cber_das_farbige_Licht_der_Doppelsterne_und_einiger_anderer_Gestirne_des_Himmels
Christian Andreas Doppler, Ueber das farbige licht der doppelsterne und einiger anderer gestirne des himmels. Abhandl d Konigl Bomischen Gesellschaft der Wissenschaften **2**, 465–482 (1842c)
J.W. Cooley, J.W. Tukey, An algorithm for the machine calculation of complex Fourier series. Math. Comput. **19**(90), 297–301 (1965)
P. Duhamel, H. Hollmann, Split-radix FFT algorithm. Electron. Lett. **20**(1), 14–16 (1984)
D. Fraser, Interpolation by the FFT revisited-an experimental investigation. IEEE Trans. ASSP. **37**(5), 665–675 (1989)
P.T. Gough, A fast spectral estimation algorithm based on the FFT. IEEE Trans. ASSP. **42**(6), 1317–1322 (1994)
J.Y. Jou, J.A. Abraham, Fault-tolerant FFT networks. IEEE Trans. Comp. **37**(5), 548–561 (1988)
Y.C. Lim, An interpolation technique for computing the DFT of a sparse sequence. IEEE Trans. Acoust. Speech Signal Process. **33**(6), 1456–1460 (1985)
J. Markel, FFT pruning. IEEE trans. Audio Electro. **19**(4), 305–311 (1971)
C. Offelli, D. Petri, The influence of windowing on the accuracy of multifrequency signal parameter estimation. IEEE Trans. Instrum. Measur. **41**(2), 256–261 (1992)
H.V. Sorensen, M. Heideman, C. Burrus, On computing the split-radix FFT. IEEE Trans. ASSP. **34**(1), 152–156 (1986)
P.N. Swarztrauber, FFT algorithms for vector computers. Parallel Comput. **1**(1), 45–63 (1984)
D.J. Van Nee, A.J. Coenen, New fast GPS code-acquisition technique using FFT. Electron. Lett. **27**(2), 158–160 (1991)

Chapter 8
Frequency Estimation

8.1 Introduction

As discussed in Chap. 7, the estimation of velocity using Doppler effect and the estimation of distance using a chirp is essentially a problem of estimating the frequency of a sinusoid. A simple method for frequency estimation using DFT/FFT is also presented in Chap. 7. Estimating the frequency of a finite duration sinusoid using the DFT/FFT method has the "main lobe width" problem; if the frequency separation between two sinusoids is less than the spectral main lobe width of a rectangular window, the two sinusoids cannot be resolved. See Fig. 7.24 of Chap. 7.

There are many methods that can accurately estimate the frequencies of sinusoids that are separated by less than the spectral main lobe width of a rectangular window (Wu 2012; Gzyl 1995; Ulrych and Bishop 1975; Burg 1972; Andersen 1974; Newman 1977; Ables 1974; Shaman 2010; Brockwell and Dahlhaus 2004; Lee and Wang 1996; Yu et al. 2002; Cybenko 1980; Papaodysseus et al. 1993; Boshnakov and Lambert-Lacroix 2012; Gray and Markel 1973; Lim and Parker 1986, 1984; Lim et al. 2005; Lev-Ari and Kailath 1984; Cho and Lee 1993; Leib et al. 1987; Regalia 1991; Punchalard 2002; Zhu et al. 2015; Reddy et al. 1981; Mojiri and Bakhshai 2004, 2007; Jones and Steinhardt 1985; Myung 2003; Stoica et al. 1989; Abatzoglou 1985; Tufts and Kumaresan 1982; Foias et al. 1988; Umesh and Tufts 1996; Byrne and Fitzgerald 1984; Kay and Nagesha 1994; Turmon and Miller 1994; Abatzoglou 1986). Such methods are called super resolution methods. In this chapter, we shall discuss three of such methods; they are (1) linear prediction (also called autoregressive modeling), (2) Levinson-Durbin algorithm (also called lattice filter or maximum entropy method), and (3) maximum likelihood method.

Y. C. Lim et al., *Signal Processing for Autonomous Systems*,
https://doi.org/10.1007/978-3-032-00675-2_8

8.2 Linear Prediction

Consider an infinite duration discrete time complex signal $x(n)$ at time n, $n = -\infty, \ldots, -1, 0, 1, \ldots, \infty$, consisting of a sum of M complex sinusoids $e^{j\omega_m n}$, $m = 1, 2, \ldots, M$. Let $z_m = e^{j\omega_m}$. Let the z-transform of $x(n)$ be $X(z)$.

$$X(z) = \sum_{n=-\infty}^{\infty} x(n)z^{-n}. \tag{8.1}$$

Since $x(n)$ is a sum of M complex sinusoids, the output will be zero if it is filtered by a filter with z-transform transfer function $H_M^f(z)$ which is a cascade of M notches of the form $(1 - z_m z^{-1})$, $m = 1, 2, \ldots, M$. Hence, we have

$$\begin{aligned} H_M^f(z) &= \prod_{m=1}^{M}(1 - z_m z^{-1}) \ , \quad z_m = e^{j\omega_m}, \\ &= 1 - \sum_{m=1}^{M} h_M^f(m)z^{-m}, \end{aligned} \tag{8.2}$$

and

$$X(z)H_M^f(z) = 0. \tag{8.3}$$

Our aim is to find $H_M^f(z)$ that satisfies (8.3). Once $H_M^f(z)$ is determined, $\omega_m, m = 1, 2, \ldots, M$, can be obtained from the zeroes of $H_M^f(z)$. In general, $h_M^f(m)$ is complex. Let $E_M^f(z) = X(z)H_M^f(z)$, and let $E_M^f(z) = \sum_{n=-\infty}^{\infty} e_M^f(n)z^{-n}$.

$$E_M^f(z) = \sum_{n=-\infty}^{\infty} e_M^f(n)z^{-n} = X(z)H_M^f(z). \tag{8.4}$$

Our aim is to determine $H_M^f(z)$ such that $E_M^f(z) = 0$. From (8.2), (8.3), and (8.4)

$$E_M^f(z) = X(z) - \sum_{m=1}^{M} h_M^f(m)X(z)z^{-m} = 0. \tag{8.5}$$

The inverse z-transform of $X(z)z^{-m}$ is $x(n-m)$. Taking inverse z-transform and equating $e_M^f(n)$ to zero, we have

$$e_M^f(n) = x(n) - \sum_{m=1}^{M} h_M^f(m)x(n-m) = 0. \tag{8.6}$$

From (8.6),

$$x(n) = \sum_{m=1}^{M} h_M^f(m)x(n-m). \tag{8.7}$$

Expression (8.7) states that the value of $x(n)$ can be estimated or "predicted" (in the language of linear prediction) from its M past values $x(n-1), x(n-2), \ldots, x(n-M)$. Since $x(n-1), x(n-2), \ldots, x(n-M)$ are past values of $x(n)$, the prediction is forward in time and is called "forward prediction." $H_M^f(z)$ is the z-transform of the Mth order forward predictor.

The output will also be zero if $x(n)$ is filtered by a filter with z-transform transfer function $H_M^b(z)$ which is a cascade of M notches of the form $(z^{-1} - z_m^{-1}), m = 1, 2, \ldots, M$.

$$\begin{aligned} H_M^b(z) &= \prod_{m=1}^{M}(z^{-1} - z_m^{-1}) \ , \ z_m = e^{j\omega_m}, \\ &= z^{-M} - \sum_{m=1}^{M} h_M^b(m) z^{1-m}, \end{aligned} \tag{8.8}$$

and

$$X(z)H_M^b(z) = 0. \tag{8.9}$$

In general, $h_M^b(m)$ is complex. Let $E_M^b f(z) = X(z)H_M^b(z)$, and let $E_M^b(z) = \sum_{n=-\infty}^{\infty} e_M^b(n)z^{-n}$.

$$E_M^b(z) = \sum_{n=-\infty}^{\infty} e_M^b(n)z^{-1} = X(z)H_M^b(z). \tag{8.10}$$

From (8.8), (8.9), and (8.10),

$$E_M^b(z) = X(z)z^{-M} - \sum_{m=1}^{M} h_M^b(m)X(z)z^{1-m} = 0. \tag{8.11}$$

Taking inverse z-transform and equating $e_M^b(n)$ to zero, we have

$$e_M^b(n) = x(n-M) - \sum_{m=1}^{M} h_M^b(m)x(n-m+1) = 0. \tag{8.12}$$

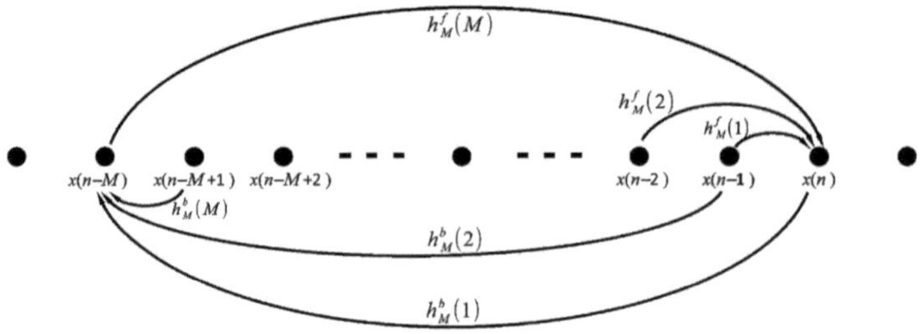

Fig. 8.1 Forward prediction of $x(n)$ and backward prediction of $x(n-M)$

From (8.12),

$$x(n-M) = \sum_{m=1}^{M} h^b_M(m)x(n-m+1). \tag{8.13}$$

Time $n, n-1, \ldots, n-M+1$ are in the future of time $n-M$. Expression (8.13) states that the value of $x(n-M)$ can be "predicted" from its M future values $x(n), x(n-1), \ldots, x(n-M+1)$; the prediction is backward in time and is called "backward prediction." $H^b_M(z)$ is the z-transform of the Mth order backward predictor.

A diagram depicting the forward and backward prediction of $x(n)$ and $x(n-M)$ is shown in Fig. 8.1. Both forward and backward predictions are classified as linear prediction, since $x(n)$ and $x(n-M)$ are predicted using linear combinations of past values of $x(n)$ or future values of $x(n-M)$, respectively.

In general, if $x(n)$ contains noise and/or measurement error, $e^f_M(n)$ and $e^b_M(n)$ will not be zero. Our aim is to find $H^f_M(z)$ or $H^b_M(z)$ to minimize $\sum_{n=-\infty}^{\infty} |e^f_M(n)|^2$ or $\sum_{n=-\infty}^{\infty} |e^b_M(n)|^2$, respectively. The values of ω_m, $m = 1, 2, \ldots, M$ can be determined from either $H^f_M(z)$ or $H^b_M(z)$. We shall represent $[x(n-1), x(n-2), \ldots, x(n-M)]$, $[x(n), x(n-1), \ldots, x(n-M+1)]$, $[h^f_M(1), h^f_M(2), \ldots, h^f_M(M)]$, and $[h^b_M(1), h^b_M(2), \ldots, h^b_M(M)]$ in vector form. Let

$$\boldsymbol{x}_M(n-1) = [x(n-1), x(n-2), \ldots, x(n-M)]^{\mathrm{T}}, \tag{8.14a}$$

$$\boldsymbol{x}_M(n) = [x(n), x(n-1), \ldots, x(n-M+1)]^{\mathrm{T}}, \tag{8.14b}$$

and

$$\boldsymbol{h}^f_M = [h^f_M(1), h^f_M(2), \ldots, h^f_M(M)]^{\mathrm{T}}, \tag{8.15a}$$

$$\boldsymbol{h}^b_M = [h^b_M(1), h^b_M(2), \ldots, h^b_M(M)]^{\mathrm{T}}. \tag{8.15b}$$

The transpose of [•] is denoted by $[\bullet]^{\mathrm{T}}$. Using the definition of (8.14) and (8.15), the forward prediction error $e_M^f(n)$ and backward prediction error $e_M^b(n)$ can be rewritten in vector form as

$$e_M^f(n) = x(n) - \boldsymbol{x}_M^{\mathrm{T}}(n-1)\boldsymbol{h}_M^f = x(n) - [\boldsymbol{h}_M^f]^{\mathrm{T}}\boldsymbol{x}_M(n-1), \tag{8.16a}$$

$$e_M^b(n) = x(n-M) - \boldsymbol{x}_M^{\mathrm{T}}(n)\boldsymbol{h}_M^b = x(n-M) - [\boldsymbol{h}_M^b]^{\mathrm{T}}\boldsymbol{x}_M(n). \tag{8.16b}$$

The signals $x(n), \ldots, x(n-M)$, $e_M^f(n)$, and $e_M^b(n)$ may be collected over a given period, say $n = 1, 2, \ldots, P$. The aim is to find $\boldsymbol{h}_M^f$ or $\boldsymbol{h}_M^b$ that minimizes $\sum_{n=1}^{P} |e_M^f(n)|^2$ or $\sum_{n=1}^{P} |e_M^b(n)|^2$, respectively. Expression (8.16) is in the same form as (8.221), and so $\boldsymbol{h}_M^f$ or $\boldsymbol{h}_M^b$ can be obtained using the least squares method (Appendix A) or the total least squares method (Appendix B). In the presence of noise, a larger P will give a better estimate of the frequencies. There is a minimum number of data points of $x(n)$ that is necessary for determining the frequencies.

If $x(n)$ consists of only one complex sinusoid, $x(n)$ can be estimated using a first-order complex predictor $x(n) = h_M^f(1)x(n-1)$. The coefficient $h_M^f(1)$ can be determined from one set of known values of $x(n)$ and $x(n-1)$. Hence, we need a minimum of two data points, i.e., $x(n)$ and $x(n-1)$ to determine $h_M^f(1)$. If $x(n)$ consists of two complex sinusoids, a second-order complex predictor is needed for estimating $x(n)$, and the expression is $x(n) = h_M^f(1)x(n-1) + h_M^f(2)x(n-2)$. To determine $h_M^f(1)$ and $h_M^f(2)$, we need two equations, since there are two unknowns, namely, $h_M^f(1)$ and $h_M^f(2)$. The second equation is the expression for estimating $x(n-1)$, i.e., the expression $x(n-1) = h_M^f(1)x(n-2) + h_M^f(2)x(n-3)$. Hence, a minimum of four data points, i.e., $x(n)$, $x(n-1)$, $x(n-2)$, and $x(n-3)$ are needed. In general, if there are M complex sinusoids, at least $2M$ data points are needed to determine $\boldsymbol{h}_M^f$. Similarly, it can be shown that at least $2M$ data points are needed to determine $\boldsymbol{h}_M^b$ if there are M complex sinusoids.

If the sinusoids are real, $\boldsymbol{x}_M(n-1)$, $\boldsymbol{h}_M^f$, and $\boldsymbol{h}_M^b$ are also real. For one real sinusoid, $H_M^f(z)$ and $H_M^b(z)$ will be second order with a pair of complex conjugate roots representing the frequency of the sinusoid. At least four data points, namely, $x(n), x(n-1), x(n-2)$, and $x(n-3)$ are needed to determine $\boldsymbol{h}_M^f$, or $\boldsymbol{h}_M^b$. In general, for M real sinusoids, $H_{2M}^f(z)$, and $H_{2M}^b(z)$ will be $2M$th order and $4M$ data points, namely, $x(n)$, $x(n-1)$, …, and $x(n-4M+1)$ are needed to determine $\boldsymbol{h}_{2M}^f$ or $\boldsymbol{h}_{2M}^b$.

The zeroes of $H_M^f(z)$ or $H_M^b(z)$ are the frequencies of $x(n)$. If the zeroes of $H_M^f(z)$ or $H_M^b(z)$ are not clustered together, they may be obtained by evaluating $\left|H_M^f(e^{j\omega})\right|^{-1}$ or $\left|H_M^b(e^{j\omega})\right|^{-1}$ on a dense grid of ω; the peaks are the zeroes and so are the frequencies of $x(n)$. However, if the zeroes are clustered, the peaks of $\left|H_M^f(e^{j\omega})\right|^{-1}$ or $\left|H_M^b(e^{j\omega})\right|^{-1}$ may not accurately indicate the zeroes.

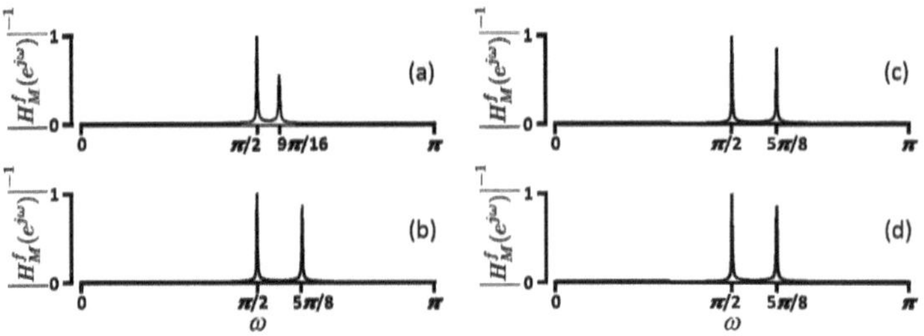

Fig. 8.2 Plots of $\left|H_M^f(e^{j\omega})\right|^{-1}$. The highest peak of $\left|H_M^f(e^{j\omega})\right|^{-1}$ in each plot is scaled to unity. Plots (**a**), (**b**), (**c**), (**d**) correspond to the scenarios in Figs. 7.35a–d, respectively

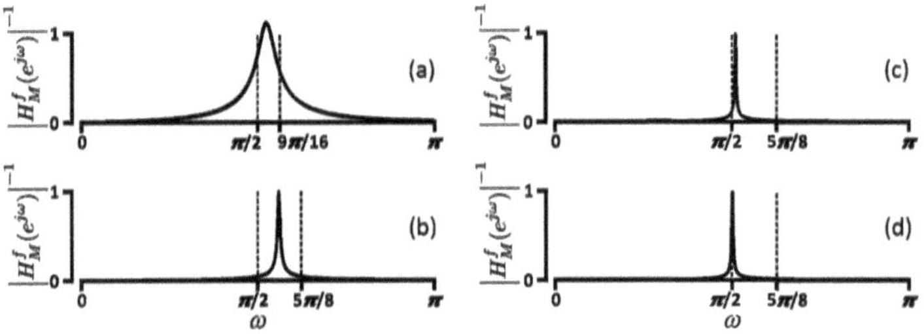

Fig. 8.3 Plots of $\left|H_M^f(e^{j\omega})\right|^{-1}$ for a 2nd-order predictor instead of a 4th-order predictor. The maximum peak of $\left|H_M^f(e^{j\omega})\right|^{-1}$ is scaled to unity in each plot. Plots (**a**), (**b**), (**c**), (**d**) correspond to the sinusoids' scenarios in Figs. 8.2a–d, respectively

The frequencies of the sinusoids in the four scenarios in Fig. 7.35 are estimated using the above linear prediction method for real sinusoids (i.e., real $x(n)$), and the results are presented in Fig. 8.2. The order of the predictor is four. The plots in Figs. 8.2a–d are the plots of $\left|H_M^f(e^{j\omega})\right|^{-1}$ corresponding to the scenarios in Figs. 7.35a–d, respectively. It can be seen that the sinusoids with frequencies $\pi/2$ and $9\pi/16$ are resolved in Fig. 8.2a, whereas they cannot be resolved in Fig. 7.35a. The peaks of $\left|H_M^f(e^{j\omega})\right|^{-1}$ in Fig. 8.2b correctly indicate the frequencies of the sinusoids, whereas the spectral peaks of Fig. 7.35b do not. The peaks of $\left|H_M^f(e^{j\omega})\right|^{-1}$ in Figs. 8.2c and d also correctly indicate the frequencies of the sinusoids; they are not the case in Figs. 7.35c and d.

When using linear prediction, it is necessary to specify the predictor's order. If the number of sinusoid is unknown, the specified order may be wrong. We shall examine the effect of specifying a wrong predictor's order. We shall use the scenarios in Fig. 8.2 except that we use a second-order predictor instead of a fourth-order predictor. The results are shown in Fig. 8.3. The (a), (b), (c), (d) plots in Fig. 8.3 correspond to the (a), (b), (c), (d) scenarios in Fig. 8.2, respectively. Since the order of the predictors is lower than the necessary order, the prediction errors are not zero. The predictors will minimize the energy of the prediction error. If the frequencies of the two sinusoids are close together and if the two sinusoids have the same magnitude such as those in Figs. 8.3a and b, the notch frequency of the predictor is between the frequencies of the two sinusoids. If the magnitudes of the two sinusoids are greatly disparate such as that in 8.3d, the notch frequency of the predictor is very close to the frequency of the larger magnitude sinusoid. If the magnitudes of the two sinusoids are moderately disparate such as that in 8.3c, the notch frequency of the predictor is slightly away from the frequency of the larger magnitude sinusoid and is tilted toward the frequency of the smaller magnitude sinusoid.

If the order of the predictor is higher than necessary and if there is no noise, the matrix $\mathbf{X}^{\mathrm{T}}\mathbf{X}$ in (8.231) is singular and so cannot be inverted.

In the above examples, the effect of noise is not considered. If noise is added to the scenarios of Fig. 7.35 such that the SNR is 40 dB, the spectral plots are shown in Fig. 8.4.

The frequencies in Fig. 8.4 are estimated using a fourth-order linear predictor for real sinusoids, and the results are presented in Fig. 8.5. The plots in Figs. 8.5a–d are the plots of $\left|H_M^f(e^{j\omega})\right|^{-1}$ corresponding to the scenarios in Figs. 8.4a–d,

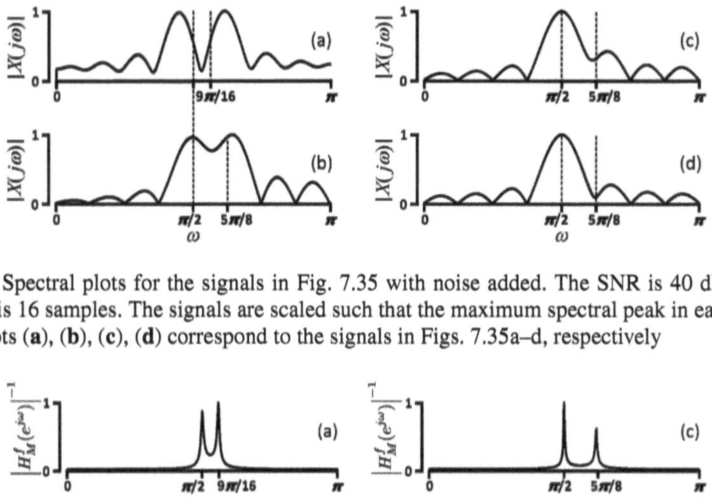

Fig. 8.4 Spectral plots for the signals in Fig. 7.35 with noise added. The SNR is 40 dB. Signal duration is 16 samples. The signals are scaled such that the maximum spectral peak in each plot is unity. Plots (**a**), (**b**), (**c**), (**d**) correspond to the signals in Figs. 7.35a–d, respectively

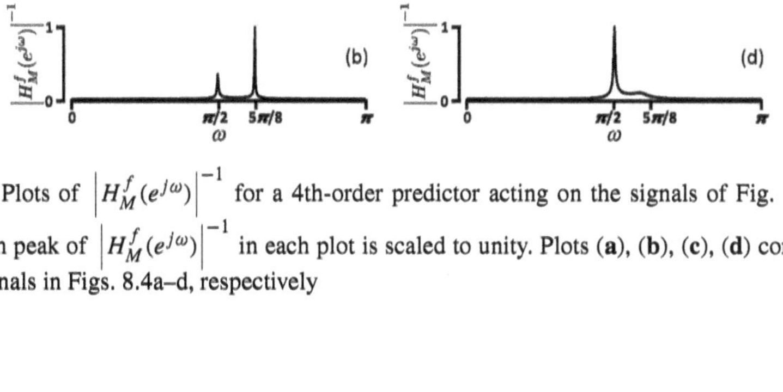

Fig. 8.5 Plots of $\left|H_M^f(e^{j\omega})\right|^{-1}$ for a 4th-order predictor acting on the signals of Fig. 8.4. The maximum peak of $\left|H_M^f(e^{j\omega})\right|^{-1}$ in each plot is scaled to unity. Plots (**a**), (**b**), (**c**), (**d**) correspond to the signals in Figs. 8.4a–d, respectively

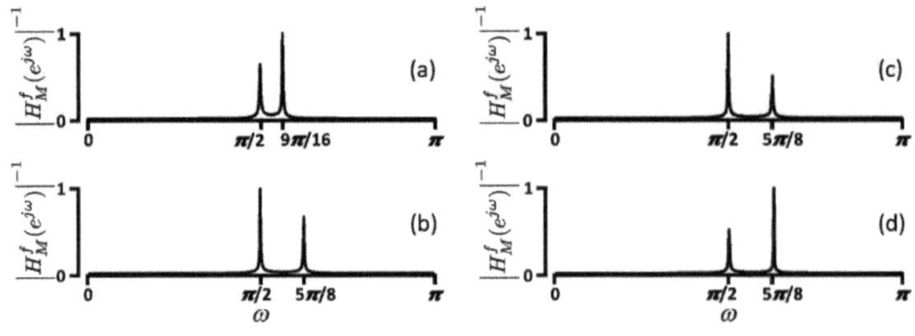

Fig. 8.6 For the same scenarios as in Fig. 8.5 except that the predictor's order is increased to 6. The highest peak of $\left|H_M^f(e^{j\omega})\right|^{-1}$ in each plot is scaled to unity

respectively. The estimated frequencies are the peaks of $\left|H_M^f(e^{j\omega})\right|^{-1}$. It can be seen that the sinusoids in Fig. 8.4a, b, and c are resolved. The sinusoid with frequency $\frac{5\pi}{8}$ in Fig. 8.4d is not resolved.

If the order of the above fourth-order predictor is increased to six, the results are shown in Fig. 8.6. It can be seen in Fig. 8.6 that all the sinusoids are resolved. The estimated frequencies in Figs. 8.5 and 8.6 are listed in Table 8.1. It can be seen from Figs. 8.5, 8.6, and Table 8.1 that, in general, the sixth-order predictor provides better estimates for the frequencies. However, for the 0.5π frequency, the fourth-order predictor performs slightly better. We shall examine the zeroes of $H_M^f(e^{j\omega})$ for Fig. 8.5b and Fig. 8.6b to gain further insight into the issue; their zeroes are shown in Table 8.2. There are three pairs of zeroes for the sixth-order $H_M^f(e^{j\omega})$. The poles of $\left(H_M^f(e^{j\omega})\right)^{-1}$ are the zeroes of $H_M^f(e^{j\omega})$. The spikes in Figs. 8.5 and 8.6 are due mainly to the zeroes of $H_M^f(e^{j\omega}))$ near to the unit circle (i.e., the zeroes with radius close to unity). It can be seen from Table 8.2 that, for Fig. 8.6b, the frequencies of these zeroes are $1.9637 \approx \frac{\pi}{2} \approx 1.9635$ and $1.5705 \approx \frac{5\pi}{8} \approx 1.5708$, respectively. The plot in Fig. 8.6b is the combined effect of all the three pairs of complex conjugate zeroes in Table 8.2. It is the third pair of zeroes at $0.4789e^{\pm j0.5013}$ that cause the peaks of $\left|H_M^f(e^{j\omega})\right|^{-1}$ to shift slightly away from that which would be indicated by the zeroes at $1.0015e^{\pm j1.9637}$ and $1.0015e^{\pm j1.5705}$.

When a higher-order predictor is used, there will be extra poles in $\left|H_M^f(e^{j\omega})\right|^{-1}$. In the above example, the extra poles are far away from the unit circle and so do not produce extra spikes in the $\left|H_M^f(e^{j\omega})\right|^{-1}$ plot. However, in some cases, the extra poles may not be far away from the unit circle. We shall illustrate this using an example. Consider the scenario of Fig. 8.6b. We shall increase the noise such that the SNR is 6 dB. The $\left|H_M^f(e^{j\omega})\right|^{-1}$ plots for fourth-, sixth-, and eighth-order predictors are shown in Fig. 8.7, respectively. The 8th-order predictor is the highest-order predictor that can be used in this example, because there are only 16 data samples.

Table 8.1 Actual frequencies and peak frequency of $\left|H_M^f(e^{j\omega})\right|^{-1}$

Figure	Actual frequency-1	Actual frequency-2	Order of predictor	Peak frequency-1	Peak frequency-2
Fig. 8.5a	$\frac{9\pi}{16}=1.7671$	$\frac{\pi}{2}=1.5708$	4	1.9289	1.4074
Fig. 8.5b	$\frac{5\pi}{8}=1.9635$	$\frac{\pi}{2}=1.5708$	4	2.0232	1.5708
Fig. 8.5c	$\frac{5\pi}{8}=1.9635$	$\frac{\pi}{2}=1.5708$	4	2.0932	1.5708
Fig. 8.5d	$\frac{5\pi}{8}=1.9635$	$\frac{\pi}{2}=1.5708$	4	–	1.5708
Fig. 8.6a	$\frac{9\pi}{16}=1.7671$	$\frac{\pi}{2}=1.5708$	6	1.7705	1.5692
Fig. 8.6b	$\frac{5\pi}{8}=1.9635$	$\frac{\pi}{2}=1.5708$	6	1.9655	1.5692
Fig. 8.6c	$\frac{5\pi}{8}=1.9635$	$\frac{\pi}{2}=1.5708$	6	1.9655	1.5692
Fig. 8.6d	$\frac{5\pi}{8}=1.9635$	$\frac{\pi}{2}=1.5708$	6	1.9718	1.5692

Table 8.2 Estimated zeroes of $H_M^f(e^{j\omega})$ for Figs. 8.5b and 8.6b

Figure	Order of predictor	Estimated Zero-1	Estimated Zero-2	Estimated Zero-3
Fig. 8.5b	4	$0.9987e^{\pm j1.9577}$	$1.0014e^{\pm j1.5704}$	–
Fig. 8.6b	6	$1.0015e^{\pm j1.9637}$	$1.0015e^{\pm j1.5705}$	$0.4789e^{\pm j0.5013}$

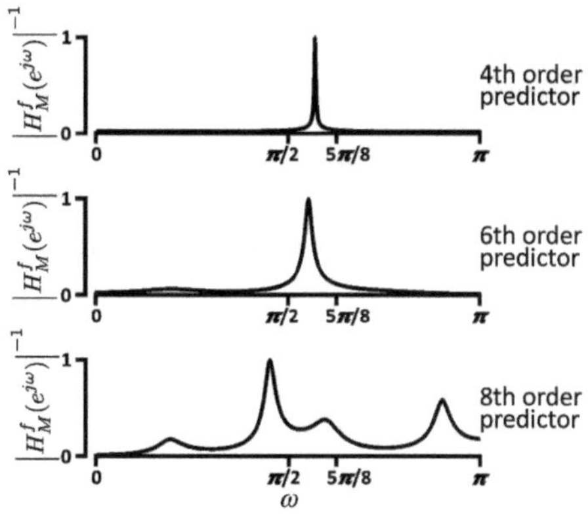

Fig. 8.7 The SNR for the scenario of Fig. 8.6b is changed to 6 dB. The peak of $\left|H_M^f(e^{j\omega})\right|^{-1}$ is scaled to unity for each plot. The order of the predictors is 4, 6, and 8, respectively

Table 8.3 The peak frequencies of $\left|H_M^f(e^{j\omega})\right|^{-1}$

Peak frequency-1	Peak frequency-2	Peak frequency-3	Peak frequency-4
0.6069	1.4246	1.8711	2.8397

It can be seen from Fig. 8.7 that the fourth- and sixth-order predictors produce only one peak. The eighth-order predictor produces four peaks at frequencies 0.6069, 1.4246, 1.8711, and 2.8397, respectively; the frequencies of the peaks are listed in Table 8.3. None of the peaks correspond to the actual frequencies. The four pairs of zeroes of $H_M^f(e^{j\omega})$ are shown in Table 8.4.

Table 8.4 Estimated zeros of $H_M^f(e^{j\omega})$

Estimated zero-1	Estimated zero-2	Estimated zero-3	Estimated zero-4
$0.9099e^{\pm j0.6004}$	$1.0413e^{\pm j1.4192}$	$1.1201e^{\pm j1.8894}$	$0.93922e^{\pm j2.8352}$

8.3 Maximum Entropy Method

The maximum entropy method is also known as the lattice filter method. Consider the minimization of $\mathbb{E}|e_m^f(n)|^2$ where $\mathbb{E}(\bullet)$ denote the expectation operator. From (8.16a),

$$\begin{aligned}|e_m^f(n)|^2 = \left(e_m^f(n)\right)^* e_m^f(n) &= |x(n)|^2 + [\boldsymbol{h}_m^f]^{\mathrm{H}}\boldsymbol{x}_m^*(n-1)\boldsymbol{x}_m^{\mathrm{T}}(n-1)\boldsymbol{h}_m^f \\ &- x^*(n)\boldsymbol{x}_m^{\mathrm{T}}(n-1)\boldsymbol{h}_m^f - [\boldsymbol{h}_m^f]^{\mathrm{H}}\boldsymbol{x}_m^*(n-1)x(n),\end{aligned} \tag{8.17}$$

where $[\bullet]^{\mathrm{H}}$ is the Hermitian transpose or complex conjugate transpose of $[\bullet]$ and $\bullet^*(n)$ is the complex conjugate of $\bullet(n)$. Let the power of the forward prediction error be denoted by P_m^f.

$$P_m^f = \mathbb{E}|e_m^f(n)|^2. \tag{8.18}$$

Taking expectation of (8.17) and substituting the expectation into (8.18),

$$\begin{aligned}P_m^f &= \mathbb{E}\{|x(n)|^2\} + [\boldsymbol{h}_m^f]^{\mathrm{H}}\mathbb{E}\{\boldsymbol{x}_m^*(n-1)\boldsymbol{x}_m^{\mathrm{T}}(n-1)\}\boldsymbol{h}_m^f \\ &- \mathbb{E}\{x^*(n)\boldsymbol{x}_m^{\mathrm{T}}(n-1)\}\boldsymbol{h}_m^f - [\boldsymbol{h}_m^f]^{\mathrm{H}}\mathbb{E}\{\boldsymbol{x}_m^*(n-1)x(n)\}.\end{aligned} \tag{8.19}$$

Let $\mathbb{E}\{x^*(n)x(n-k)\}$ be denoted by $r(k)$, and let the real and imaginary parts of $r(k)$ be $r_r(k)$ and $r_i(k)$, respectively.

$$\mathbb{E}\{x^*(n)x(n-k)\} = r(k) = r_r(k) + jr_i(k). \tag{8.20}$$

It is straightforward to show that

$$r_r(k) = \mathrm{Re}\{x(n)\}\mathrm{Re}\{x(n-k)\} + \mathrm{Im}\{x(n)\}\mathrm{Im}\{x(n-k)\} = r_r(-k), \tag{8.21a}$$

$$r_i(k) = \mathrm{Re}\{x(n)\}\mathrm{Im}\{x(n-k)\} - \mathrm{Im}\{x(n)\}\mathrm{Re}\{x(n-k)\} = -r_i(-k), \tag{8.21b}$$

where $\mathrm{Re}(\bullet)$ and $\mathrm{Im}(\bullet)$ denote the real and imaginary parts of $\bullet$, respectively. It can be easily shown that

$$r(-k) = r^*(k). \tag{8.22}$$

Let $[r(1), r(2), \ldots, r(m)]$ be denoted by $\boldsymbol{r}_m^{\mathrm{T}}$. We have

$$\mathbb{E}\{x(n)^* \boldsymbol{x}_m^{\mathrm{T}}(n-1)\} = [r(1), r(2), \ldots, r(m)] = \boldsymbol{r}_m^{\mathrm{T}} \;, \tag{8.23a}$$

$$\mathbb{E}\{\boldsymbol{x}_m^*(n-1)x(n)\} = [r^*(1), r^*(2), \ldots, r^*(m)]^{\mathrm{T}} = \boldsymbol{r}_m^* \;. \tag{8.23b}$$

Let

$$\begin{aligned}\mathbf{R}_m &= \mathbb{E}\{\boldsymbol{x}_m^*(n-1)\boldsymbol{x}_m^{\mathrm{T}}(n-1)\} \\ &= \begin{bmatrix} r(0) & r(1) & \cdots & r(m-1) \\ r^*(1) & r(0) & \cdots & r(m-2) \\ \vdots & \vdots & \ddots & \vdots \\ r^*(m-1) & r^*(m-2) & \cdots & r(0) \end{bmatrix}.\end{aligned} \tag{8.24}$$

In (8.24), $\mathbf{R}_m$ is Hermitian, i.e., $\mathbf{R}_m^{\mathrm{H}} = \mathbf{R}_m$. The inverse of a Hermitian matrix is also Hermitian, i.e., $[\mathbf{R}_m^{-1}]^{\mathrm{H}} = \mathbf{R}_m^{-1}$. Substitute (8.23) and (8.24) into (8.19),

$$P_m^f = r(0) + [\boldsymbol{h}_m^f]^{\mathrm{H}} \mathbf{R}_m \boldsymbol{h}_m^f - \boldsymbol{r}_m^{\mathrm{T}} \boldsymbol{h}_m^f - [\boldsymbol{h}_m^f]^{\mathrm{H}} \boldsymbol{r}_m^*. \tag{8.25}$$

Note

$$\begin{aligned}&(\boldsymbol{h}_m^f - \mathbf{R}_m^{-1}\boldsymbol{r}_m^*)^{\mathrm{H}} \mathbf{R}_m (\boldsymbol{h}_m^f - \mathbf{R}_m^{-1}\boldsymbol{r}_m^*) \\ &\qquad = [\boldsymbol{h}_m^f]^{\mathrm{H}} \mathbf{R}_m \boldsymbol{h}_m^f + \boldsymbol{r}_m^{\mathrm{T}} \mathbf{R}_m^{-1} \boldsymbol{r}_m^* - [\boldsymbol{h}_m^f]^{\mathrm{H}} \boldsymbol{r}_m^* - \boldsymbol{r}_m^{\mathrm{T}} \boldsymbol{h}_m^f.\end{aligned} \tag{8.26}$$

From (8.25) and (8.26),

$$\begin{aligned}P_m^f &= r(0) + (\boldsymbol{h}_m^f - \mathbf{R}_m^{-1}\boldsymbol{r}_m^*)^{\mathrm{H}} \mathbf{R}_m (\boldsymbol{h}_m^f - \mathbf{R}_m^{-1}\boldsymbol{r}_m^*) - \boldsymbol{r}_m^{\mathrm{T}} \mathbf{R}_m^{-1} \boldsymbol{r}_m^*, \\ &= r(0) + (\boldsymbol{h}_m^f - \mathbf{R}_m^{-1}\boldsymbol{r}_m^*)^{\mathrm{H}} \mathbb{E}\{\boldsymbol{x}_m^*(n-1)\boldsymbol{x}_m^{\mathrm{T}}(n-1)\} (\boldsymbol{h}_m^f - \mathbf{R}_m^{-1}\boldsymbol{r}_m^*) - \boldsymbol{r}_m^{\mathrm{T}} \mathbf{R}_m^{-1} \boldsymbol{r}_m^*, \\ &= r(0) + \mathbb{E}\{|\boldsymbol{x}_m^{\mathrm{T}}(n-1)(\boldsymbol{h}_m^f - \mathbf{R}_m^{-1}\boldsymbol{r}_m^*)|^2\} - \boldsymbol{r}_m^{\mathrm{T}} \mathbf{R}_m^{-1} \boldsymbol{r}_m^*.\end{aligned} \tag{8.27}$$

The aim is to find $\boldsymbol{h}_m^f$ that minimizes P_m^f. In (8.27), the term $\mathbb{E}\{|\boldsymbol{x}_m^{\mathrm{T}}(n-1)(\boldsymbol{h}_m^f - \mathbf{R}_m^{-1}\boldsymbol{r}_m^*)|^2\}$ is the square of magnitude and so $\mathbb{E}\{|\boldsymbol{x}_m^{\mathrm{T}}(n-1)(\boldsymbol{h}_m^f - \mathbf{R}_m^{-1}\boldsymbol{r}_m^*)|^2\} \geq 0$. Therefore, P_m^f reaches its minimum when $\boldsymbol{h}_m^f - \mathbf{R}_m^{-1}\boldsymbol{r}_m^* = 0$ and the minimum value of P_m^f denoted by $P_{m,\mathrm{opt}}^f$ is $r(0) - \boldsymbol{r}_m^{\mathrm{T}} \mathbf{R}_m^{-1} \boldsymbol{r}_m^*$. Let $\boldsymbol{h}_{m,\mathrm{opt}}^f$ denote the value of $\boldsymbol{h}_m^f$ when P_m^f reaches its minimum. Therefore,

$$\boldsymbol{h}_{m,\mathrm{opt}}^f = \mathbf{R}_m^{-1} \boldsymbol{r}_m^* \quad \text{or} \quad \mathbf{R}_m \boldsymbol{h}_{m,\mathrm{opt}}^f = \boldsymbol{r}_m^*. \tag{8.28}$$

Substituting $\boldsymbol{h}^f_{m,\mathrm{opt}} = \mathbf{R}_m^{-1}\boldsymbol{r}_m^*$ into (8.27), we have

$$P^f_{m,\mathrm{opt}} = r(0) - \boldsymbol{r}_m^{\mathrm{T}}\mathbf{R}_m^{-1}\boldsymbol{r}_m^* = r(0) - \boldsymbol{r}_m^{\mathrm{T}}\boldsymbol{h}^f_{m,\mathrm{opt}}. \tag{8.29}$$

Since $\boldsymbol{r}_m^{\mathrm{T}}\boldsymbol{h}^f_{m,\mathrm{opt}}$ is scalar,

$$\boldsymbol{r}_m^{\mathrm{T}}\boldsymbol{h}^f_{m,\mathrm{opt}} = [\boldsymbol{h}^f_{m,\mathrm{opt}}]^{\mathrm{T}}\boldsymbol{r}_m. \tag{8.30}$$

The expressions $\boldsymbol{r}_m^* - \mathbf{R}_m\boldsymbol{h}^f_{m,\mathrm{opt}} = 0$ in (8.28) and $P^f_{m,\mathrm{opt}} = r(0) - \boldsymbol{r}_m^{\mathrm{T}}\boldsymbol{h}^f_{m,\mathrm{opt}}$ in (8.29) can be put in matrix form as

$$\begin{bmatrix} r(0) & \boldsymbol{r}_m^{\mathrm{T}} \\ \boldsymbol{r}_m^* & \mathbf{R}_m \end{bmatrix}\begin{bmatrix} 1 \\ -\boldsymbol{h}^f_{m,\mathrm{opt}} \end{bmatrix} = \begin{bmatrix} P^f_{m,\mathrm{opt}} \\ \mathbf{0} \end{bmatrix}. \tag{8.31}$$

Note

$$\begin{bmatrix} r(0) & \boldsymbol{r}_m^{\mathrm{T}} \\ \boldsymbol{r}_m^* & \mathbf{R}_m \end{bmatrix} = \left[\begin{array}{c|ccc} r(0) & r(1) & \cdots & r(m) \\ \hline r^*(1) & r(0) & \cdots & r(m-1) \\ \vdots & \vdots & \ddots & \vdots \\ r^*(m) & r^*(m-1) & \cdots & r(0) \end{array}\right] = \mathbf{R}_{m+1}. \tag{8.32}$$

From (8.31) and (8.32),

$$\mathbf{R}_{m+1}\begin{bmatrix} 1 \\ -\boldsymbol{h}^f_{m,\mathrm{opt}} \end{bmatrix} = \begin{bmatrix} P^f_{m,\mathrm{opt}} \\ \mathbf{0} \end{bmatrix}. \tag{8.33}$$

Now consider the "backward prediction" problem of minimizing $\mathbb{E}\{|b_m(n)|^2\}$. From (8.16b),

$$\begin{aligned} |b_m(n)|^2 = \left(e^b_m(n)\right)^* e^b_m(n) &= |x(n-m)|^2 + [\boldsymbol{h}^b_m]^{\mathrm{H}}\boldsymbol{x}_m^*(n)\boldsymbol{x}_m^{\mathrm{T}}(n)\boldsymbol{h}^b_m \\ &- x^*(n-m)\boldsymbol{x}_m^{\mathrm{T}}(n)\boldsymbol{h}^b_m - [\boldsymbol{h}^b_m]^{\mathrm{H}}\boldsymbol{x}_m^*(n)x(n-m). \end{aligned} \tag{8.34}$$

Let the power of the backward prediction error be denoted by P^b_m.

$$P^b_m = \mathbb{E}|e^b_m(n)|^2. \tag{8.35}$$

Taking expectation of (8.34) and substitute the expectation into (8.35),

$$P_m^b = \mathbb{E}\{|x(n-m)|^2\} + [\boldsymbol{h}_m^b]^{\mathrm{H}}\mathbb{E}\{\boldsymbol{x}_m^*(n)\boldsymbol{x}_m^{\mathrm{T}}(n)\}\boldsymbol{h}_m^b$$
$$- \mathbb{E}\{x^*(n-m)\boldsymbol{x}_m^{\mathrm{T}}(n)\}\boldsymbol{h}_m^b - [\boldsymbol{h}_m^b]^{\mathrm{H}}\mathbb{E}\{\boldsymbol{x}_m^*(n)x(n-m)\}. \tag{8.36}$$

$\mathbb{E}\{x^*(n-m)\boldsymbol{x}_m^{\mathrm{T}}(n)\}$ and $\mathbb{E}\{\boldsymbol{x}_m^*(n)x(n-m)\}$ are given by

$$\mathbb{E}\{x^*(n-m)\boldsymbol{x}_m^{\mathrm{T}}(n)\} = [r^*(m), \ldots, r^*(2), r^*(1)] = \tilde{\boldsymbol{r}}_m^{\mathrm{H}}, \tag{8.37a}$$

$$\mathbb{E}\{\boldsymbol{x}_m^*(n)x(n-m)\} = [r(m), \ldots, r(2), r(1)]^{\mathrm{T}} = \tilde{\boldsymbol{r}}_m, \tag{8.37b}$$

where $\tilde{\bullet}$ is a vector containing the elements of $\bullet$ placed in reversed order. Substitute (8.24) and (8.37) into (8.36),

$$P_m^b = r(0) + [\boldsymbol{h}_m^b]^{\mathrm{H}}\mathbf{R}_m\boldsymbol{h}_m^b - \tilde{\boldsymbol{r}}_m^{\mathrm{H}}\boldsymbol{h}_m^b - [\boldsymbol{h}_m^b]^{\mathrm{H}}\tilde{\boldsymbol{r}}_m. \tag{8.38}$$

Note

$$(\boldsymbol{h}_m^b - \mathbf{R}_m^{-1}\tilde{\boldsymbol{r}}_m)^{\mathrm{H}}\mathbf{R}_m(\boldsymbol{h}_m^b - \mathbf{R}_m^{-1}\tilde{\boldsymbol{r}}_m)$$
$$= [\boldsymbol{h}_m^b]^{\mathrm{H}}\mathbf{R}_m\boldsymbol{h}_m^b + \tilde{\boldsymbol{r}}_m^{\mathrm{H}}\mathbf{R}_m^{-1}\tilde{\boldsymbol{r}}_m - [\boldsymbol{h}_m^b]^{\mathrm{H}}\tilde{\boldsymbol{r}}_m - \boldsymbol{r}_m^{\mathrm{H}}\boldsymbol{h}_m^b. \tag{8.39}$$

From (8.38) and (8.39),

$$P_m^b = r(0) + (\boldsymbol{h}_m^b - \mathbf{R}_m^{-1}\tilde{\boldsymbol{r}}_m)^{\mathrm{H}}\mathbf{R}_m(\boldsymbol{h}_m^b - \mathbf{R}_m^{-1}\tilde{\boldsymbol{r}}_m) - \tilde{\boldsymbol{r}}_m^{\mathrm{H}}\mathbf{R}_m^{-1}\tilde{\boldsymbol{r}}_m,$$
$$= r(0) + (\boldsymbol{h}_m^b - \mathbf{R}_m^{-1}\tilde{\boldsymbol{r}}_m)^{\mathrm{H}}\mathbb{E}\{\boldsymbol{x}_m^*(n)\boldsymbol{x}_m^{\mathrm{T}}(n)\}(\boldsymbol{h}_m^b - \mathbf{R}_m^{-1}\tilde{\boldsymbol{r}}_m) - \tilde{\boldsymbol{r}}_m^{\mathrm{H}}\mathbf{R}_m^{-1}\tilde{\boldsymbol{r}}_m,$$
$$= r(0) + \mathbb{E}\{|\boldsymbol{x}_m^{\mathrm{T}}(n)(\boldsymbol{h}_m^b - \mathbf{R}_m^{-1}\tilde{\boldsymbol{r}}_m)|^2\} - \tilde{\boldsymbol{r}}_m^{\mathrm{H}}\mathbf{R}_m^{-1}\tilde{\boldsymbol{r}}_m. \tag{8.40}$$

The aim is to find $\boldsymbol{h}_m^b$ that minimizes P_m^b. In (8.40), $\mathbb{E}\{|\boldsymbol{x}_m^{\mathrm{T}}(n)(\boldsymbol{h}_m^b - \mathbf{R}_m^{-1}\tilde{\boldsymbol{r}}_m)|^2\} \geq 0$, since it is a square of magnitude. Therefore, P_m^b reaches its minimum when $\boldsymbol{h}_m^b - \mathbf{R}_m^{-1}\tilde{\boldsymbol{r}}_m = 0$ and the minimum value of P_m^b denoted by $P_{m,\mathrm{opt}}^b$ is $r(0) - \tilde{\boldsymbol{r}}_m^{\mathrm{H}}\mathbf{R}_m^{-1}\tilde{\boldsymbol{r}}_m$. Let $\boldsymbol{h}_{m,\mathrm{opt}}^b$ denote the value of $\boldsymbol{h}_m^b$ when P_m^b reaches its minimum.

$$\boldsymbol{h}_{m,\mathrm{opt}}^b = \mathbf{R}_m^{-1}\tilde{\boldsymbol{r}}_m \quad \text{or} \quad \mathbf{R}_m\boldsymbol{h}_{m,\mathrm{opt}}^b = \tilde{\boldsymbol{r}}_m. \tag{8.41}$$

Note that $\mathbf{R}_m^{-1}$ is Hermitian and so $\mathbf{R}_m^{-1}\tilde{\boldsymbol{r}}_m$ is a vector containing the elements of the complex conjugate of $\mathbf{R}_m^{-1}\boldsymbol{r}_m^*$ placed in reversed order. Hence, from (8.28) and (8.41),

$$\boldsymbol{h}_{m,\mathrm{opt}}^b = [\tilde{\boldsymbol{h}}_{m,\mathrm{opt}}^f]^*. \tag{8.42}$$

Therefore, the optimum backward predictor is the complex conjugate of the optimum forward predictor in reversed order. The minimum P_m^b is given by

$$P_{m,\mathrm{opt}}^b = r(0) - \tilde{\boldsymbol{r}}_m^{\mathrm{H}} \mathbf{R}_m^{-1} \tilde{\boldsymbol{r}}_m = r(0) - \tilde{\boldsymbol{r}}_m^{\mathrm{H}} \boldsymbol{h}_{m,\mathrm{opt}}^b. \tag{8.43}$$

Since $P_{m,\mathrm{opt}}^b$ and $r(0)$ are real, $\tilde{\boldsymbol{r}}_m^{\mathrm{H}} \boldsymbol{h}_{m,\mathrm{opt}}^b$ in (8.43) must also be real. Therefore,

$$\tilde{\boldsymbol{r}}_m^{\mathrm{H}} \boldsymbol{h}_{m,\mathrm{opt}}^b = \left(\tilde{\boldsymbol{r}}_m^{\mathrm{H}} \boldsymbol{h}_{m,\mathrm{opt}}^b\right)^* = \tilde{\boldsymbol{r}}_{m,\mathrm{opt}}^{\mathrm{T}} [\boldsymbol{h}_m^b]^* = \tilde{\boldsymbol{r}}_{m,\mathrm{opt}}^{\mathrm{T}} \tilde{\boldsymbol{h}}_m^f = \boldsymbol{r}_m^{\mathrm{T}} \boldsymbol{h}_{m,\mathrm{opt}}^f. \tag{8.44}$$

In (8.44), we make use of the fact that, for any two vectors $\bullet$ and $\circ$ of the same length, $\bullet^{\mathrm{T}} \circ = \tilde{\bullet}^{\mathrm{T}} \tilde{\circ}$. From (8.43) and (8.44),

$$P_{m,\mathrm{opt}}^b = r(0) - \boldsymbol{r}_m^{\mathrm{T}} \boldsymbol{h}_{m,\mathrm{opt}}^f. \tag{8.45}$$

Comparing (8.29) and (8.45),

$$P_{m,\mathrm{opt}}^b = P_{m,\mathrm{opt}}^f. \tag{8.46}$$

The expressions $\tilde{\boldsymbol{r}}_m - \mathbf{R}_m \boldsymbol{h}_{m,\mathrm{opt}}^b = \mathbf{0}$ and $P_{m,\mathrm{opt}}^b = r(0) - \tilde{\boldsymbol{r}}_m^{\mathrm{H}} \boldsymbol{h}_{m,\mathrm{opt}}^b$ can be put in matrix form as

$$\begin{bmatrix} \mathbf{R}_m & \tilde{\boldsymbol{r}}_m \\ \tilde{\boldsymbol{r}}_m^{\mathrm{H}} & r(0) \end{bmatrix} \begin{bmatrix} -\boldsymbol{h}_{m,\mathrm{opt}}^b \\ 1 \end{bmatrix} = \begin{bmatrix} \mathbf{0} \\ P_{m,\mathrm{opt}}^b \end{bmatrix}. \tag{8.47}$$

Note

$$\begin{bmatrix} \mathbf{R}_m & \tilde{\boldsymbol{r}}_m \\ \tilde{\boldsymbol{r}}_m^{\mathrm{H}} & r(0) \end{bmatrix} = \left[\begin{array}{ccc|c} r(0) & \cdots & r(m-1) & r(m) \\ \vdots & \ddots & \vdots & \vdots \\ r^*(m-1) & \cdots & r(0) & r(1) \\ \hline r^*(m) & \cdots & r(1) & r(0) \end{array}\right] = \mathbf{R}_{m+1}. \tag{8.48}$$

From (8.47) and (8.48),

$$\mathbf{R}_{m+1} \begin{bmatrix} -\boldsymbol{h}_{m,\mathrm{opt}}^b \\ 1 \end{bmatrix} = \begin{bmatrix} \mathbf{0} \\ P_{m,\mathrm{opt}}^b \end{bmatrix}. \tag{8.49}$$

8.3.1 Levinson-Durbin Recursion

The vector $\boldsymbol{h}^f_{m,\text{opt}}$ or $\boldsymbol{h}^b_{m,\text{opt}}$ may be obtained from (8.28) or (8.41), respectively, by computing the $\mathbf{R}^{-1}_{m+1}$ for any arbitrary m, say for $m = M$. However, after obtaining $\boldsymbol{h}^f_{M,\text{opt}}$ or $\boldsymbol{h}^b_{M,\text{opt}}$, if it is desired to obtain $\boldsymbol{h}^f_{M+1,\text{opt}}$ or $\boldsymbol{h}^b_{M+1,\text{opt}}$, the previously obtained $\mathbf{R}^{-1}_M$ is discarded and $\mathbf{R}^{-1}_{M+1}$ will be computed disregarding the efforts already invested in the computation of $\mathbf{R}^{-1}_M$.

In Levinson-Durbin recursion, $\boldsymbol{h}^f_{m,\text{opt}}$ and $\boldsymbol{h}^b_{m,\text{opt}}$ are not computed directly from (8.28) and (8.41), respectively, but are computed by adding a correction term to $\boldsymbol{h}^f_{m-1,\text{opt}}$ and $\boldsymbol{h}^b_{m-1,\text{opt}}$; the computations are not approximations but are the same as that obtained from (8.28) and (8.41). The Levinson-Durbin recursion is developed as follows:

It can be easily verified that

$$\boldsymbol{r}^*_m = \begin{bmatrix} \boldsymbol{r}^*_{m-1} \\ r^*(m) \end{bmatrix}. \tag{8.50}$$

Let

$$\boldsymbol{h}^f_{m,\text{opt}} = \begin{bmatrix} \boldsymbol{h}^f_{m-1,\text{opt}} + \boldsymbol{\varepsilon}^f_{m-1} \\ h^f_{m,\text{opt}}(m) \end{bmatrix}. \tag{8.51}$$

The vector $\boldsymbol{\varepsilon}^f_{m-1}$ in (8.51) is the vector of correction terms. Let the nth elements of $\boldsymbol{\varepsilon}^f_{m-1}$ be $\varepsilon^f_{m-1}(n)$.

$$\boldsymbol{\varepsilon}^f_{m-1} = [\varepsilon^f_{m-1}(1), \varepsilon^f_{m-1}(2), \ldots, \varepsilon^f_{m-1}(m-1)]^{\mathrm{T}}. \tag{8.52}$$

Note that $\boldsymbol{h}^f_{m,\text{opt}} = [h^f_{m,\text{opt}}(1),\ h^f_{m,\text{opt}}(2),\ \ldots,\ h^f_{m,\text{opt}}(m)]^{\mathrm{T}}$. From (8.28), (8.50), and (8.51),

$$\begin{bmatrix} \mathbf{R}_{m-1} & \tilde{\boldsymbol{r}}_{m-1} \\ \tilde{\boldsymbol{r}}^{\mathrm{H}}_{m-1} & r(0) \end{bmatrix} \begin{bmatrix} \boldsymbol{h}^f_{m-1,\text{opt}} + \boldsymbol{\varepsilon}^f_{m-1} \\ h^f_{m,\text{opt}}(m) \end{bmatrix} = \begin{bmatrix} \boldsymbol{r}^*_{m-1} \\ r^*(m) \end{bmatrix}. \tag{8.53}$$

From (8.53), we have

$$\mathbf{R}_{m-1}\boldsymbol{h}^f_{m-1,\text{opt}} + \mathbf{R}_{m-1}\boldsymbol{\varepsilon}^f_{m-1} + h^f_{m,\text{opt}}(m)\tilde{\boldsymbol{r}}_{m-1} = \boldsymbol{r}^*_{m-1}, \tag{8.54}$$

and

$$\tilde{\boldsymbol{r}}^{\mathrm{H}}_{m-1}\boldsymbol{h}^f_{m-1,\text{opt}} + \tilde{\boldsymbol{r}}^{\mathrm{H}}_{m-1}\boldsymbol{\varepsilon}^f_{m-1} + h^f_{m,\text{opt}}(m)r(0) = r^*(m). \tag{8.55}$$

Replace m by $m - 1$ in (8.28), we have $\mathbf{R}_{m-1}\boldsymbol{h}^{f}_{m-1,\text{opt}} = \boldsymbol{r}^{*}_{m-1}$. Replace the $\mathbf{R}_{m-1}\ \boldsymbol{h}^{f}_{m-1,\text{opt}}$ term in (8.54) by $\boldsymbol{r}^{*}_{m-1}$; we have $\boldsymbol{r}^{*}_{m-1} + \mathbf{R}_{m-1}\boldsymbol{\varepsilon}^{f}_{m-1} + h^{f}_{m,\text{opt}}(m)\tilde{\boldsymbol{r}}_{m-1} = \boldsymbol{r}^{*}_{m-1}$. Therefore,

$$\mathbf{R}_{m-1}\boldsymbol{\varepsilon}^{f}_{m-1} = -h^{f}_{m,\text{opt}}(m)\tilde{\boldsymbol{r}}_{m-1}. \tag{8.56}$$

Multiplying both sides of (8.56) by $\mathbf{R}^{-1}_{m-1}$, and taking note of (8.41),

$$\boldsymbol{\varepsilon}^{f}_{m-1} = -h^{f}_{m,\text{opt}}(m)\mathbf{R}^{-1}_{m-1}\tilde{\boldsymbol{r}}_{m-1} = -h^{f}_{m,\text{opt}}(m)\boldsymbol{h}^{b}_{m-1,\text{opt}}. \tag{8.57}$$

The matrix product $\mathbf{R}^{-1}_{m-1}\mathbf{R}_{m-1} = \mathbf{I}$. Multiply the second term of (8.55) by $\mathbf{R}^{-1}_{m-1}\mathbf{R}_{m-1}$ and move its first term to the right-hand side of the equation; we have

$$\tilde{\boldsymbol{r}}^{\text{H}}_{m-1}\mathbf{R}^{-1}_{m-1}\mathbf{R}_{m-1}\boldsymbol{\varepsilon}^{f}_{m-1} + h^{f}_{m,\text{opt}}(m)r(0) = r^{*}(m) - \tilde{\boldsymbol{r}}^{\text{H}}_{m-1}\boldsymbol{h}^{f}_{m-1,\text{opt}}. \tag{8.58}$$

Replacing the $\mathbf{R}_{m-1}\boldsymbol{\varepsilon}^{f}_{m-1}$ factor in (8.58) using the result of (8.56),

$$\begin{aligned} -\tilde{\boldsymbol{r}}^{\text{H}}_{m-1}\mathbf{R}^{-1}_{m-1}h^{f}_{m,\text{opt}}(m)\tilde{\boldsymbol{r}}_{m-1} + h^{f}_{m,\text{opt}}(m)r(0) &= r^{*}(m) - \tilde{\boldsymbol{r}}^{\text{H}}_{m-1}\boldsymbol{h}^{f}_{m-1,\text{opt}}, \\ h^{f}_{m,\text{opt}}(m)[r(0) - \tilde{\boldsymbol{r}}^{\text{H}}_{m-1}\mathbf{R}^{-1}_{m-1}\tilde{\boldsymbol{r}}_{m-1}] &= r^{*}(m) - \tilde{\boldsymbol{r}}^{\text{H}}_{m-1}\boldsymbol{h}^{f}_{m-1,\text{opt}}. \end{aligned} \tag{8.59}$$

Eliminating the $\mathbf{R}^{-1}_{m-1}\tilde{\boldsymbol{r}}_{m-1}$ factor in (8.59) using the result of (8.41),

$$h^{f}_{m,\text{opt}}(m)[r(0) - \tilde{\boldsymbol{r}}^{\text{H}}_{m-1}\boldsymbol{h}^{b}_{m-1,\text{opt}}] = r^{*}(m) - \tilde{\boldsymbol{r}}^{\text{H}}_{m-1}\boldsymbol{h}^{f}_{m-1,\text{opt}}. \tag{8.60}$$

Using the result of (8.43) to eliminate the $[r(0) - \tilde{\boldsymbol{r}}^{\text{H}}_{m-1}\boldsymbol{h}^{b}_{m-1,\text{opt}}]$ factor in (8.60),

$$h^{f}_{m,\text{opt}}(m)P^{f}_{m-1} = r^{*}(m) - \tilde{\boldsymbol{r}}^{\text{H}}_{m-1}\boldsymbol{h}^{f}_{m-1,\text{opt}}. \tag{8.61}$$

Hence, we have

$$h^{f}_{m,\text{opt}}(m) = [P^{f}_{m-1}]^{-1}\left(r^{*}(m) - \tilde{\boldsymbol{r}}^{\text{H}}_{m-1}\boldsymbol{h}^{f}_{m-1,\text{opt}}\right). \tag{8.62}$$

Expression (8.62) provides a method for computing $h^{f}_{m,\text{opt}}(m)$ using previously used information $\tilde{\boldsymbol{r}}^{\text{H}}_{m-1}$, new information $r^{*}(m)$, and previously computed $[P^{f}_{m-1}]^{-1}$ and

$\boldsymbol{h}^f_{m-1,\text{opt}}$. To complete the order update recursion, we need to derive a method for updating P^f_m. From (8.51) and (8.57),

$$\boldsymbol{h}^f_{m,\text{opt}} = \begin{bmatrix} \boldsymbol{h}^f_{m-1,\text{opt}} - h^f_{m,\text{opt}}(m)\boldsymbol{h}^b_{m-1,\text{opt}} \\ h^f_{m,\text{opt}}(m) \end{bmatrix}. \tag{8.63}$$

Substituting $\boldsymbol{h}^f_{m,\text{opt}}$ obtained in (8.63) into (8.45) and taking note of (8.46),

$$\begin{aligned} P^f_m &= r(0) - \begin{bmatrix} \boldsymbol{r}^{\text{T}}_{m-1} & r(m) \end{bmatrix} \begin{bmatrix} \boldsymbol{h}^f_{m-1,\text{opt}} - h^f_{m,\text{opt}}(m)\boldsymbol{h}^b_{m-1,\text{opt}} \\ h^f_{m,\text{opt}}(m) \end{bmatrix}, \\ &= r(0) - \boldsymbol{r}^{\text{T}}_{m-1}\boldsymbol{h}^f_{m-1,\text{opt}} + h^f_{m,\text{opt}}(m)\boldsymbol{r}^{\text{T}}_{m-1}\boldsymbol{h}^b_{m-1,\text{opt}} - r(m)h^f_{m,\text{opt}}(m). \end{aligned} \tag{8.64}$$

Eliminating $r(0) - \boldsymbol{r}^{\text{T}}_{m-1}\boldsymbol{h}^f_{m-1,\text{opt}}$ in (8.64) using the result of (8.45) and (8.46),

$$P^f_m = P^f_{m-1} + h^f_{m,\text{opt}}(m)\boldsymbol{r}^{\text{T}}_{m-1}\boldsymbol{h}^b_{m-1,\text{opt}} - r(m)h^f_{m,\text{opt}}(m). \tag{8.65}$$

Multiplying both sides of (8.61) by $\left(h^f_{m,\text{opt}}(m)\right)^*$,

$$|h^f_{m,\text{opt}}(m)|^2 P^f_{m-1} = \left(h^f_{m,\text{opt}}(m)\right)^* r^*(m) - \left(h^f_{m,\text{opt}}(m)\right)^* \tilde{\boldsymbol{r}}^{\text{H}}_{m-1}\boldsymbol{h}^f_{m-1,\text{opt}}. \tag{8.66}$$

In (8.66), $|h_m(m)|^2 P^f_{m-1}$ is real, and so the conjugate of $|h_m(m)|^2 P^f_{m-1}$ is equal to itself. Taking conjugate of (8.66),

$$h^f_{m,\text{opt}}(m)r(m) = |h^f_{m,\text{opt}}(m)|^2 P^f_{m-1} + h^f_{m,\text{opt}}(m)\tilde{\boldsymbol{r}}^{\text{T}}_{m-1}[\boldsymbol{h}^f_{m-1,\text{opt}}]^*. \tag{8.67}$$

Eliminating the $r(m)h^f_{m,\text{opt}}(m)$ term in (8.65) using (8.67),

$$P^f_m = (1 - |h^f_{m,\text{opt}}(m)|^2)P^f_{m-1} + h^f_{m,\text{opt}}(m)[\boldsymbol{r}^{\text{T}}_{m-1}[\boldsymbol{h}^b_{m-1,\text{opt}}] - \tilde{\boldsymbol{r}}^{\text{T}}_{m-1}[\boldsymbol{h}^f_{m-1,\text{opt}}]^*]. \tag{8.68}$$

Taking note that $[\tilde{\boldsymbol{h}}^f_{m-1,\text{opt}}]^* = \boldsymbol{h}^b_{m-1,\text{opt}}$, (8.68) is simplified to

$$P^f_m = (1 - |h^f_{m,\text{opt}}(m)|^2)P^f_{m-1}. \tag{8.69}$$

Expression (8.69) is the order recursion formula for the prediction error power. The Levinson-Durbin recursion is summarized as follows:

Levinson-Durbin recursion:

1. Initialization: $\boldsymbol{h}^f_{0,\text{opt}} = \boldsymbol{h}^b_{0,\text{opt}} = [\mathbf{0}]_{0\times 1}$, i.e., $\boldsymbol{h}^f_{0,\text{opt}}$ and $\boldsymbol{h}^b_{0,\text{opt}}$ have no element; $P^f_0 = r(0)$; $m = 0$.
2. Compute
   a. increment m.
   b. $h^f_{m,\text{opt}}(m) = \left(P^f_{m-1}\right)^{-1}\left(r^*(m) - \tilde{\boldsymbol{r}}^{\text{H}}_{m-1}\boldsymbol{h}^f_{m-1,\text{opt}}\right)$. See (8.62).
   Note $h^f_{1,\text{opt}}(1) = \frac{r^*(1)}{r(0)}$.
   c. $\boldsymbol{h}^f_{m,\text{opt}} = \begin{bmatrix} \boldsymbol{h}^f_{m-1,\text{opt}} - h^f_{m,\text{opt}}(m)\boldsymbol{h}^b_{m-1,\text{opt}} \\ h^f_{m,\text{opt}}(m) \end{bmatrix}$.
   Note $\boldsymbol{h}^f_{1,\text{opt}} = [h^f_{1,\text{opt}}(1)] = \frac{r^*(1)}{r(0)}$.
   d. $P^f_m = (1 - |h^f_{m,\text{opt}}(m)|^2)P^f_{m-1}$. See (8.69).
3. If m has reached the desired value, stop; otherwise, go to 2.

8.3.2 Lattice Filter

The maximum entropy method is also called the lattice filter method, because it can be put in the form of a lattice filter. From (8.4),

$$E^f_m(z) = X(z)H^f_m(z), \tag{8.70a}$$

$$E^f_{m-1}(z) = X(z)H^f_{m-1}(z). \tag{8.70b}$$

From (8.2),

$$H^f_{m,\text{opt}}(z) = 1 - \sum_{k=1}^{m} h^f_{m,\text{opt}}(k)z^{-k}, \tag{8.71a}$$

$$H^f_{m-1,\text{opt}}(z) = 1 - \sum_{k=1}^{m-1} h^f_{m-1,\text{opt}}(k)z^{-k}. \tag{8.71b}$$

From (8.8),

$$H^b_{m,\text{opt}}(z) = z^{-m} - \sum_{k=1}^{m} h^b_{m,\text{opt}}(k)z^{1-k}, \tag{8.72a}$$

$$H^b_{m-1,\text{opt}}(z) = z^{1-m} - \sum_{k=1}^{m-1} h^b_{m-1,\text{opt}}(k)z^{1-k}. \tag{8.72b}$$

From (8.16),

$$e^f_{m-1,\mathrm{opt}}(n) = x(n) - \boldsymbol{x}^{\mathrm{T}}_{m-1}(n-1)\boldsymbol{h}^f_{m-1,\mathrm{opt}} \,, \tag{8.73a}$$

$$e^b_{m-1,\mathrm{opt}}(n-1) = x(n-m) - \boldsymbol{x}^{\mathrm{T}}_{m-1}(n-1)\boldsymbol{h}^b_{m-1,\mathrm{opt}} \,. \tag{8.73b}$$

From (8.16),

$$e^f_{m,\mathrm{opt}}(n) = x(n) - [\boldsymbol{x}^{\mathrm{T}}_{m-1}(n-1), x(n-m)]\boldsymbol{h}^f_{m,\mathrm{opt}} \,, \tag{8.74a}$$

$$e^b_{m,\mathrm{opt}}(n) = x(n-m) - [x(n), \boldsymbol{x}^{\mathrm{T}}_{m-1}(n-1)]\boldsymbol{h}^b_{m,\mathrm{opt}} \,, \tag{8.74b}$$

where $e^f_{m,\mathrm{opt}}(n)$ and $e^b_{m,\mathrm{opt}}(n)$ are the values of $e^f_m(n)$ and $e^b_m(n)$ when $\boldsymbol{h}^f_m = \boldsymbol{h}^f_{m,\mathrm{opt}}$ and $\boldsymbol{h}^b_m = \boldsymbol{h}^b_{m,\mathrm{opt}}$, respectively. Substituting the result of (8.63) into (8.74a),

$$\begin{aligned} e^f_{m,\mathrm{opt}}(n) = {} & x(n) - \boldsymbol{x}^{\mathrm{T}}_{m-1}(n-1)\boldsymbol{h}^f_{m-1,\mathrm{opt}} \\ & - h^f_{m,\mathrm{opt}}(m)\left(x(n-m) - \boldsymbol{x}^{\mathrm{T}}_{m-1}(n-1)\boldsymbol{h}^b_{m-1}\right). \end{aligned} \tag{8.75}$$

From (8.73) and (8.75),

$$e^f_{m,\mathrm{opt}}(n) = e^f_{m-1,\mathrm{opt}}(n) - h^f_{m,\mathrm{opt}}(m)e^b_{m-1,\mathrm{opt}}(n-1). \tag{8.76}$$

From (8.42) and (8.63),

$$\boldsymbol{h}^b_{m,\mathrm{opt}} = \begin{bmatrix} \left(h^f_{m,\mathrm{opt}}(m)\right)^* \\ \boldsymbol{h}^b_{m-1,\mathrm{opt}} - \left(h^f_{m,\mathrm{opt}}(m)\right)^* \boldsymbol{h}^f_{m-1,\mathrm{opt}} \end{bmatrix}. \tag{8.77}$$

Substituting the result of (8.77) into (8.74b),

$$\begin{aligned} e^b_{m,\mathrm{opt}}(n) = {} & x(n-m) - \boldsymbol{x}^{\mathrm{T}}_{m-1}(n-1)\boldsymbol{h}^b_{m-1,\mathrm{opt}} \\ & - \left(h^f_{m,\mathrm{opt}}(m)\right)^* \left(x(n) - \boldsymbol{x}^{\mathrm{T}}_{m-1}(n-1)\boldsymbol{h}^f_{m-1,\mathrm{opt}}\right). \end{aligned} \tag{8.78}$$

From (8.73) and (8.78),

$$e^b_{m,\mathrm{opt}}(n) = e^b_{m-1,\mathrm{opt}}(n-1) - \left(h^f_{m,\mathrm{opt}}(m)\right)^* e^f_{m-1,\mathrm{opt}}(n). \tag{8.79}$$

We shall simplify the notation and write

$$k_m = h^f_{m,\mathrm{opt}}(m). \tag{8.80}$$

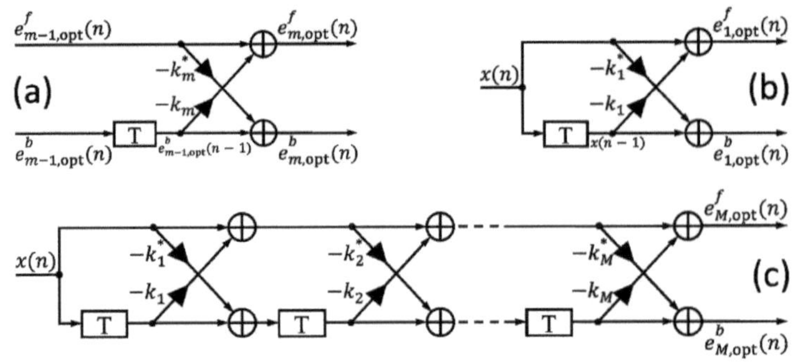

Fig. 8.8 Signal flow diagram for the lattice filter. The box marked "T" represents unit delay

The coefficient k_m is called the reflection coefficient. From (8.76), (8.79), and (8.80),

$$e^f_{m,\text{opt}}(n) = e^f_{m-1,\text{opt}}(n) - k_m e^b_{m-1,\text{opt}}(n-1)\text{ , and} \tag{8.81a}$$

$$e^b_{m,\text{opt}}(n) = e^b_{m-1,\text{opt}}(n-1) - k^*_m e^f_{m-1,\text{opt}}(n). \tag{8.81b}$$

A diagram indicating the signal flow for (8.81) is shown in Fig. 8.8a. From (8.6) and (8.12), and taking note of (8.42) and (8.80),

$$e^f_{1,\text{opt}}(n) = x(n) - k_1 x(n-1)\text{ , and} \tag{8.82a}$$

$$e^b_{1,\text{opt}}(n) = x(n-1) - k^*_1 x(n). \tag{8.82b}$$

A diagram indicating the signal flow for (8.82) is shown in Fig. 8.8b. The complete signal flow diagram is shown in Fig. 8.8c. In Fig. 8.8, the box marked "T" represents unit delay. Let $E^f_{m,\text{opt}}(z)$ and $E^b_{m,\text{opt}}(z)$ be the z-transforms of $e^f_{m,\text{opt}}(n)$ and $e^b_{m,\text{opt}}(n)$, respectively. Taking z-transforms of (8.81) and (8.82),

$$E^f_{m,\text{opt}}(z) = E^f_{m-1,\text{opt}}(z) - k_m E^b_{m-1,\text{opt}}(z) z^{-1} \ , \tag{8.83a}$$

$$E^b_{m,\text{opt}}(z) = E^b_{m-1,\text{opt}}(z) z^{-1} - k^*_m E^f_{m-1,\text{opt}}(z) \ , \tag{8.83b}$$

$$E^f_{1,\text{opt}}(z) = X(z) - k_1 X(z) z^{-1} \ , \tag{8.83c}$$

$$E^b_{1,\text{opt}}(z) = X(z) z^{-1} - k^*_1 X(z) \ . \tag{8.83d}$$

Putting (8.83) in matrix form, we have

$$\begin{bmatrix} E^f_{m,\text{opt}}(z) \\ E^b_{m,\text{opt}}(z) \end{bmatrix} = \begin{bmatrix} 1 & -k_m z^{-1} \\ -k^*_m & z^{-1} \end{bmatrix} \begin{bmatrix} E^f_{m-1,\text{opt}}(z) \\ E^b_{m-1,\text{opt}}(z) \end{bmatrix}, \tag{8.84a}$$

$$\begin{bmatrix} E^f_{1,\text{opt}}(z) \\ E^b_{1,\text{opt}}(z) \end{bmatrix} = \begin{bmatrix} 1 & -k_1 z^{-1} \\ -k_1^* & z^{-1} \end{bmatrix} \begin{bmatrix} X(z) \\ X(z) \end{bmatrix}. \tag{8.84b}$$

From (8.2) and (8.8), let

$$H^f_{M,\text{opt}}(z) = 1 - \sum_{m=1}^{M} h^f_{M,\text{opt}}(m) z^{-m} \text{ , and} \tag{8.85a}$$

$$H^b_{M,\text{opt}}(z) = z^{-M} - \sum_{m=1}^{M} h^b_{M,\text{opt}}(m) z^{1-m}. \tag{8.85b}$$

From (8.84) and (8.85),

$$\begin{aligned} \begin{bmatrix} E^f_{M,\text{opt}}(z) \\ E^b_{M,\text{opt}}(z) \end{bmatrix} &= \begin{bmatrix} 1 & -k_M z^{-1} \\ -k_M^* & z^{-1} \end{bmatrix} \cdots \begin{bmatrix} 1 & -k_2 z^{-1} \\ -k_2^* & z^{-1} \end{bmatrix} \begin{bmatrix} 1 & -k_1 z^{-1} \\ -k_1^* & z^{-1} \end{bmatrix} \begin{bmatrix} X(z) \\ X(z) \end{bmatrix}, \\ &= \begin{bmatrix} H^f_{M,\text{opt}}(z) \\ H^b_{M,\text{opt}}(z) \end{bmatrix} X(z) \; . \end{aligned} \tag{8.86}$$

Hence, $H^f_{M,\text{opt}}(z)$ and $H^b_{M,\text{opt}}(z)$ can be expressed in chain matrix form.

$$\begin{bmatrix} H^f_{M,\text{opt}}(z) \\ H^b_{M,\text{opt}}(z) \end{bmatrix} = \begin{bmatrix} 1 & -k_M z^{-1} \\ -k_M^* & z^{-1} \end{bmatrix} \cdots \begin{bmatrix} 1 & -k_2 z^{-1} \\ -k_2^* & z^{-1} \end{bmatrix} \begin{bmatrix} 1 & -k_1 z^{-1} \\ -k_1^* & z^{-1} \end{bmatrix}. \tag{8.87}$$

The reflection coefficients k_m may also be computed using $e^f_{m-1,\text{opt}}(n)$ and $e^b_{m-1,\text{opt}}(n-1)$. Note that, after obtaining $\boldsymbol{h}^f_{m-1,\text{opt}}$ and $\boldsymbol{h}^b_{m-1,\text{opt}}$, the aim is to determine k_m to minimize $\mathbb{E}|e^f_m(n)|^2$ and $\mathbb{E}|e^b_m(n)|^2$. From (8.81a),

$$\begin{aligned} \mathbb{E}|e^f_m(n)|^2 &= \mathbb{E}\left\{ \left(e^f_m(n)\right)^* e^f_m(n) \right\}, \\ &= \mathbb{E}\Big\{ [e^f_{m-1,\text{opt}}(n) - k_m e^b_{m-1,\text{opt}}(n-1)]^* \\ &\quad \times [e^f_{m-1,\text{opt}}(n) - k_m e^b_{m-1,\text{opt}}(n-1)] \Big\}. \end{aligned} \tag{8.88}$$

Let

$$r_{fb,m-1} = \mathbb{E}\left\{ e^f_{m-1,\text{opt}}(n) \left(e^b_{m-1,\text{opt}}(n)\right)^* \right\}. \tag{8.89}$$

From (8.18), (8.88), and (8.89),

$$\mathbb{E}|e_m^f(n)|^2 = P_{m-1,\text{opt}}^f - k_m^* r_{fb,m-1} - k_m r_{fb,m-1}^* + |k_m|^2 P_{m-1,\text{opt}}^b. \tag{8.90}$$

The expression $|k_m|^2 P_{m-1,\text{opt}}^b - k_m^* r_{fb,m-1} - k_m r_{fb,m-1}^*$ can be put in quadratic form.

$$\begin{aligned}&\left[\left(P_{m-1}^b\right)^{-1} r_{fb,m-1} - k_m\right]^* P_{m-1}^b \left[\left(P_{m-1}^b\right)^{-1} r_{fb,m-1} - k_m\right]\\ &\quad = |k_m|^2 P_{m-1}^b - k_m^* r_{fb,m-1} - k_m r_{fb,m-1}^* + r_{fb,m-1}^* r_{fb,m-1} \left(P_{m-1}^b\right)^{-1}.\end{aligned} \tag{8.91}$$

From (8.90) and (8.91),

$$\begin{aligned}\mathbb{E}|e_m^f(n)|^2 = &\; P_{m-1,\text{opt}}^f + \left[\left(P_{m-1}^b\right)^{-1} r_{fb,m-1} - k_m\right]^*\\ &\; P_{m-1}^b \left[\left(P_{m-1}^b\right)^{-1} r_{fb,m-1} - k_m\right]\\ &\; - r_{fb,m-1}^* r_{fb,m-1} \left(P_{m-1}^b\right)^{-1}.\end{aligned} \tag{8.92}$$

The term $\left[(P_{m-1}^b)^{-1} r_{fb,m-1} - k_m\right]^* P_{m-1}^b \left[(P_{m-1}^b)^{-1} r_{fb,m-1} - k_m\right] \geq 0$ since it is a square. Hence, $\mathbb{E}|e_m^f(n)|^2$ is minimum when $\left(P_{m-1}^b\right)^{-1} r_{fb,m-1} - k_m = 0$.

$$k_m = \left(P_{m-1}^b\right)^{-1} r_{fb,m-1} = \left(P_{m-1}^f\right)^{-1} r_{fb,m-1}. \tag{8.93}$$

Note $P_{m-1}^f = P_{m-1}^b$ (8.46). From (8.81b),

$$\begin{aligned}\mathbb{E}|e_m^b(n)|^2 &= \mathbb{E}\left\{e_m^b(n)\left(e_m^b(n)\right)^*\right\},\\ &= \mathbb{E}\left\{[e_{m-1,\text{opt}}^b(n-1) - k_m^* e_{m-1,\text{opt}}^f(n)]\right.\\ &\qquad \left.[e_{m-1,\text{opt}}^b(n-1) - k_m^* e_{m-1,\text{opt}}^f(n)]^*\right\}.\end{aligned} \tag{8.94}$$

From (8.89) and (8.94),

$$\mathbb{E}|e_m^b(n)|^2 = P_{m-1,\text{opt}}^b - k_m^* r_{fb,m-1} - k_m r_{fb,m-1}^* + |k_m|^2 P_{m-1,\text{opt}}^f. \tag{8.95}$$

The expression $|k_m|^2 P^f_{m-1,\text{opt}} - k^*_m r_{fb,m-1} - k_m r^*_{fb,m-1}$ can be put in quadratic form.

$$\left[\left(P^f_{m-1}\right)^{-1} r_{fb,m-1} - k_m\right]^* P^f_{m-1}\left[\left(P^f_{m-1}\right)^{-1} r_{fb,m-1} - k_m\right]$$
$$= |k_m|^2 P^f_{m-1} - k^*_m r_{fb,m-1} - k_m r^*_{fb,m-1} + r^*_{fb,m-1} r_{fb,m-1}\left(P^f_{m-1}\right)^{-1}. \tag{8.96}$$

From (8.95) and (8.96),

$$\begin{aligned}\mathbb{E}|e^b_m(n)|^2 = \; & P^b_{m-1,\text{opt}} + \left[\left(P^f_{m-1}\right)^{-1} r_{fb,m-1} - k_m\right]^* \\ & P^f_{m-1}\left[\left(P^f_{m-1}\right)^{-1} r_{fb,m-1} - k_m\right] \\ & - r^*_{fb,m-1} r_{fb,m-1}\left(P^f_{m-1}\right)^{-1}.\end{aligned} \tag{8.97}$$

The term $\left[\left(P^f_{m-1}\right)^{-1} r_{fb,m-1} - k_m\right]^* P^f_{m-1}\left[\left(P^f_{m-1}\right)^{-1} r_{fb,m-1} - k_m\right] \geq 0$ since it is a square. Hence, $\mathbb{E}|e^f_m(n)|^2$ is minimum when $\left(P^f_{m-1}\right)^{-1} r_{fb,m-1} - k_m = 0$.

$$k_m = \left(P^f_{m-1}\right)^{-1} r_{fb,m-1} = \left(P^b_{m-1}\right)^{-1} r_{fb,m-1}. \tag{8.98}$$

The result of (8.98) is the same as that of (8.93). Substitute $\left(P^b_{m-1}\right)^{-1} r_{fb,m-1} - k_m = 0$ into (8.92).

$$P^f_{m,\text{opt}} = P^f_{m-1,\text{opt}} - r^*_{fb,m-1} r_{fb,m-1}\left(P^b_{m-1}\right)^{-1} = P^f_{m-1,\text{opt}} - r^*_{fb,m-1} k_m. \tag{8.99}$$

Substituting $r_{fb,m-1} = P^f_{m-1} k_m$ (see (8.93)) into (8.99).

$$P^f_m = P^f_{m-1} - P^f_{m-1} k^*_m k_m = (1 - |k_m|^2) P^f_{m-1} = P^b_m = (1 - |k_m|^2) P^b_{m-1}. \tag{8.100}$$

Expression (8.93) or (8.98) may be written as

$$k_m = \frac{\mathbb{E}\left\{e^f_{m-1,\text{opt}}(n)\left(e^b_{m-1,\text{opt}}(n-1)\right)^*\right\}}{\mathbb{E}\left\{e^b_{m-1,\text{opt}}(n-1)\left(e^b_{m-1,\text{opt}}(n-1)\right)^*\right\}}$$
$$= \frac{\mathbb{E}\left\{e^f_{m-1,\text{opt}}(n)\left(e^b_{m-1,\text{opt}}(n-1)\right)^*\right\}}{\mathbb{E}\left\{e^f_{m-1,\text{opt}}(n)\left(e^f_{m-1,\text{opt}}(n)\right)^*\right\}}. \tag{8.101}$$

The value of k_m may be computed from $e^f_{m-1,\text{opt}}(n)$ and $e^b_{m-1,\text{opt}}(n-1)$ using (8.101). This method of computing k_m has certain finite word-length advantage. As a consequence of finite word-length effect, k_p for $p = 1, 2, \ldots, m-1$ may not be computed accurately or k_p may not be represented accurately. This causes an error in k_p and so k_p is not its optimum value. If k_p is not the optimum value, the effect will be reflected in $e^f_{m-1}(n)$ and $e^b_{m-1}(n-1)$. As k_m is computed using $e^f_{m-1}(n)$ and $e^b_{m-1}(n-1)$, the effect of the error made in k_p, $p = 1, \ldots m-1$ will be partially compensated for by k_m.

The lattice form is particularly useful in analysis-synthesis applications such as linear predictive coding of speech. In such applications, during analysis, $x(n)$ is sectioned into many sections. An Mth order autoregressive model is created for each section of $x(n)$. The M values of k_m, $m = 1, 2, \ldots, M$ are computed for each section of $x(n)$ using the lattice structure. During synthesis, the signal flow path of the lattice structure is reversed. An excitation signal is applied to each set of k_m, $m = 1, \ldots, M$ to produce an output which has similar spectral characteristics as the original $x(n)$. In order to ensure stability during synthesis, the zeros of $H^f_M(z)$ must be inside the unit circle. The zeros of $H^f_M(z)$ are inside the unit circle if and only if $|k_m| < 1$ for $m = 1, 2, \ldots, M$. However, $|k_m|$ obtained from (8.101) may not be always less than unity. This problem can be overcome by computing k_m using Burg's formula in (8.102).

$$k_m = \frac{2\mathbb{E}\left\{e^f_{m-1,\text{opt}}(n)\left(e^b_{m-1,\text{opt}}(n-1)\right)^*\right\}}{\mathbb{E}\left\{e^b_{m-1,\text{opt}}(n-1)\left(e^b_{m-1,\text{opt}}(n-1)\right)^* + e^f_{m-1,\text{opt}}(n)\left(e^f_{m-1,\text{opt}}(n)\right)^*\right\}} \tag{8.102}$$

The lattice filter algorithm is summarized below.

1. Initialization: $e^f_{0,\text{opt}}(n) = e^b_{0,\text{opt}}(n) = x(n)$; $m = 0$.
2. Increment m: $m \leftarrow m + 1$.
3. Compute k_m using either (8.101) or (8.102).
   Update $e^f_{m,\text{opt}}(n)$ and $e^b_{m,\text{opt}}(n)$ using (8.81).
4. If m has reached the desired value, stop; otherwise, go to 2.

We shall now show that $\mathbb{E}\left\{x^*(n)e^f_{m,\text{opt}}(n)\right\} = P^f_{m,\text{opt}}$.

Proof Multiplying both sides of $e^f_{m,\text{opt}}(n) = x(n) - \boldsymbol{x}^{\text{T}}_m(n-1)\boldsymbol{h}^f_{m,\text{opt}}$ (see (8.16a)) by $x^*(n)$, taking expectation, and making use of the result of (8.29),

$$\begin{aligned}\mathbb{E}\left\{x^*(n)e^f_{m,\text{opt}}(n)\right\} &= \mathbb{E}\left\{x^*(n)\left(x(n) - \boldsymbol{x}^{\text{T}}_m(n-1)\boldsymbol{h}^f_{m,\text{opt}}\right)\right\}, \\ &= \mathbb{E}\left\{x^*(n)x(n)\right\} - \mathbb{E}\left\{x^*(n)\boldsymbol{x}^{\text{T}}_m(n-1)\boldsymbol{h}^f_{m,\text{opt}}\right\}, \\ &= r(0) - \boldsymbol{r}^{\text{T}}_m\boldsymbol{h}^f_{m,\text{opt}} = P^f_{m,\text{opt}}.\end{aligned} \tag{8.103}$$

□

8.3.3 Lattice Filter Orthogonality

The optimum mth-order forward prediction error $e^f_{m,\text{opt}}(n)$ is orthogonal to $x^*(n-p)$ for $p = 1, 2, \ldots, m$, i.e., $\mathbb{E}\left\{[x^*_m(n-1), x^*_m(n-2), \ldots, x^*_m(n-m)]^{\text{T}}e^f_{m,\text{opt}}(n)\right\} = \boldsymbol{0}$.

$$\mathbb{E}\left\{\boldsymbol{x}^*_m(n-1)e^f_{m,\text{opt}}(n)\right\} = \boldsymbol{0}. \tag{8.104}$$

Proof Pre-multiply both sides of $e^f_{m,\text{opt}}(n) = x(n) - \boldsymbol{x}^{\text{T}}_m(n-1)\boldsymbol{h}^f_{m,\text{opt}}$ (see (8.16a)) by $\boldsymbol{x}^*_m(n-1)$, we have

$$\boldsymbol{x}^*_m(n-1)e^f_{m,\text{opt}}(n) = \boldsymbol{x}^*_m(n-1)x(n) - \boldsymbol{x}^*_m(n-1)\boldsymbol{x}^{\text{T}}_m(n-1)\boldsymbol{h}^f_{m,\text{opt}}. \tag{8.105}$$

Taking expectation of (8.105),

$$\mathbb{E}\left\{\boldsymbol{x}^*_m(n-1)e^f_{m,\text{opt}}(n)\right\} = \mathbb{E}\left\{\boldsymbol{x}^*_m(n-1)x(n)\right\} - \mathbb{E}\left\{\boldsymbol{x}^*_m(n-1)\boldsymbol{x}^{\text{T}}_m(n-1)\right\}\boldsymbol{h}^f_{m,\text{opt}}. \tag{8.106}$$

Note $\mathbb{E}\left\{\boldsymbol{x}^*_m(n-1)x(n)\right\} = \boldsymbol{r}^*_m$ and $\mathbb{E}\left\{\boldsymbol{x}^*_m(n-1)\boldsymbol{x}^{\text{T}}_m(n-1)\right\} = \mathbf{R}_m$. Hence, (8.106) may be written as

$$\mathbb{E}\left\{\boldsymbol{x}^*_m(n-1)e^f_{m,\text{opt}}(n)\right\} = \boldsymbol{r}^*_m - \mathbf{R}_m\boldsymbol{h}^f_{m,\text{opt}}. \tag{8.107}$$

Making use of the result of (8.28), (8.107) becomes

$$\mathbb{E}\left\{\boldsymbol{x}^*_m(n-1)e^f_{m,\text{opt}}(n)\right\} = \boldsymbol{r}^*_m - \mathbf{R}_m\boldsymbol{h}^f_{m,\text{opt}} = \boldsymbol{0}. \tag{8.108}$$

□

Next, we shall show that $\mathbb{E}\{x^*(n)e^b_{m,\mathrm{opt}}(n)\} = 0$.

Proof Substituting $e^b_{m,\mathrm{opt}}(n)$ given in (8.16b) into $\mathbb{E}\{x^*(n)e^b_{m,\mathrm{opt}}(n)\}$, we have

$$\begin{aligned}\mathbb{E}\left\{x^*(n)e^b_{m,\mathrm{opt}}(n)\right\} &= \mathbb{E}\left\{x^*(n)\left(x(n-m) - \boldsymbol{x}^{\mathrm{T}}_m(n)\boldsymbol{h}^b_{m,\mathrm{opt}}\right)\right\},\\ &= \mathbb{E}\{x^*(n)x(n-m)\}\\ &\quad - \mathbb{E}\left\{x^*(n)x(n-m)\left(x(n), \boldsymbol{x}^{\mathrm{T}}_{m-1}(n-1)\right)\right\},\\ &= r(m) - [r(0), \boldsymbol{r}^{\mathrm{T}}_{m-1}]\boldsymbol{h}^b_{m,\mathrm{opt}}.\end{aligned} \tag{8.109}$$

From (8.23) and (8.24),

$$\mathbf{R}_m = \left[\begin{array}{c:ccc} r(0) & r(1) & \cdots & r(m-1) \\ \hdashline r^*(1) & r(0) & \cdots & r(m-2) \\ \vdots & \vdots & \ddots & \vdots \\ r^*(m-1) & r^*(m-2) & \cdots & r(0) \end{array}\right] = \begin{bmatrix} r(0) & \boldsymbol{r}^{\mathrm{T}}_{m-1} \\ \boldsymbol{r}^*_{m-1} & \mathbf{R}_{m-1} \end{bmatrix}. \tag{8.110}$$

From (8.41) and (8.110),

$$\begin{bmatrix} r(0) & \boldsymbol{r}^{\mathrm{T}}_{m-1} \\ \boldsymbol{r}^*_{m-1} & \mathbf{R}_{m-1} \end{bmatrix} \boldsymbol{h}^b_{m,\mathrm{opt}} = \tilde{\boldsymbol{r}}_m. \tag{8.111}$$

The first row of (8.111) is

$$\begin{bmatrix} r(0) & \boldsymbol{r}^{\mathrm{T}}_{m-1} \end{bmatrix} \boldsymbol{h}^b_{m,\mathrm{opt}} = r(m). \tag{8.112}$$

From (8.109) and (8.112),

$$\mathbb{E}\left\{x^*(n)e^b_{m,\mathrm{opt}}(n)\right\} = r(m) - r(m) = 0. \tag{8.113}$$

□

We shall now show that $\mathbb{E}\left\{\left(e^b_{k,\mathrm{opt}}(n)\right)^* e^b_{m,\mathrm{opt}}(n)\right\} = 0$ for $k = 0, 1, 2, \ldots, m-1$.

Proof Since $e^b_0(n) = x(n)$, it follows from (8.113) that

$$\mathbb{E}\left\{\left(e^b_{0,\mathrm{opt}}(n)\right)^* e^b_{m,\mathrm{opt}}(n)\right\} = 0. \tag{8.114}$$

Eliminate the $e^b_{m-1}(n-1)$ term between (8.81a) and (8.81b), and making use of the result of (8.100), we have

$$\begin{aligned} k_m e^b_{m,\text{opt}}(n) &= e^f_{m-1,\text{opt}}(n) - e^f_{m,\text{opt}}(n) - k_m k^*_m e^f_{m-1,\text{opt}}(n), \\ &= (1-|k_m|^2) e^f_{m-1,\text{opt}}(n) - e^f_{m,\text{opt}}(n), \\ &= \frac{P^f_m}{P^f_{m-1}} e^f_{m-1,\text{opt}}(n) - e^f_{m,\text{opt}}(n). \end{aligned} \tag{8.115}$$

Divide (8.115) by P^f_m, we have

$$\frac{k_m}{P^f_m} e^b_{m,\text{opt}}(n) = \frac{1}{P^f_{m-1}} e^f_{m-1,\text{opt}}(n) - \frac{1}{P^f_m} e^f_{m,\text{opt}}(n). \tag{8.116}$$

Multiplying both sides of (8.116) by $\left(e^b_{k,\text{opt}}(n)\right)^*$,

$$\begin{aligned} \frac{k_m}{P^f_m} \left(e^b_{k,\text{opt}}(n)\right)^* e^b_{m,\text{opt}}(n) =& \frac{1}{P^f_{m-1}} \left(e^b_{k,\text{opt}}(n)\right)^* e^f_{m-1,\text{opt}}(n) \\ &- \frac{1}{P^f_m} \left(e^b_{k,\text{opt}}(n)\right)^* e^f_{m,\text{opt}}(n). \end{aligned} \tag{8.117}$$

Taking complex conjugate of (8.16b), we have

$$\left(e^b_{k,\text{opt}}(n)\right)^* = x^*(n-k) - \boldsymbol{x}^{\mathrm{H}}_k(n) \left(\boldsymbol{h}^b_{k,\text{opt}}\right)^* \quad , \quad k = 1, 2, \ldots, m-1. \tag{8.118}$$

Eliminating $\left(e^b_{k,\text{opt}}(n)\right)^*$ in the right side of (8.117) using (8.118) and taking expectation,

$$\begin{aligned} &\frac{k_m}{P^f_m} \mathbb{E}\left\{\left(e^b_{k,\text{opt}}(n)\right)^* e^b_{m,\text{opt}}(n)\right\} \\ &= \frac{1}{P^f_{m-1}} \left(\mathbb{E}\left\{x^*(n-k) e^f_{m-1,\text{opt}}(n)\right\} - \mathbb{E}\left\{\boldsymbol{x}^{\mathrm{H}}_k(n) e^f_{m-1,\text{opt}}(n)\right\} \left(\boldsymbol{h}^b_{k,\text{opt}}\right)^*\right) \\ &\quad - \frac{1}{P^f_m} \left(\mathbb{E}\left\{x^*(n-k) e^f_{m,\text{opt}}(n)\right\} - \mathbb{E}\left\{\boldsymbol{x}^{\mathrm{H}}_k(n) e^f_{m,\text{opt}}(n)\right\} \left(\boldsymbol{h}^b_{k,\text{opt}}\right)^*\right). \end{aligned} \tag{8.119}$$

We have shown in (8.108) that $\mathbb{E}\left\{\boldsymbol{x}_m^*(n-1)e_{m,\text{opt}}^f(n)\right\} = 0$ and consequently, $\mathbb{E}\{x^*(n-k)e_{m-1,\text{opt}}^f(n)\} = 0$ for $k = 1, 2, \ldots, m-1$ and $\mathbb{E}\{x^*(n-k)e_{m,\text{opt}}^f(n)\} = 0$ for $k = 1, 2, \ldots, m$. Therefore,

$$\begin{aligned}\frac{k_m}{P_m^f}\mathbb{E}\left\{\left(e_{k,\text{opt}}^b(n)\right)^* e_{m,\text{opt}}^b(n)\right\} &= -\frac{1}{P_{m-1}^f}\mathbb{E}\left\{\boldsymbol{x}_k^{\text{H}}(n)e_{m-1,\text{opt}}^f(n)\right\}\left(\boldsymbol{h}_{k,\text{opt}}^b\right)^* \\ &+ \frac{1}{P_m^f}\mathbb{E}\left\{\boldsymbol{x}_k^{\text{H}}(n)e_{m,\text{opt}}^f(n)\right\}\left(\boldsymbol{h}_{k,\text{opt}}^b\right)^*, \quad \text{for } k = 1, 2, \ldots, m-1.\end{aligned} \tag{8.120}$$

In (8.120), $\boldsymbol{x}_k^{\text{H}}(n) = [x^*(n), x^*(n-1), \ldots, x^*(n-k+1)]$. Since $\mathbb{E}\{x^*(n-k)e_{m-1,\text{opt}}^f(n)\} = 0$ for $k = 1, 2, \ldots, m-1$ and $\mathbb{E}\{x^*(n-k)e_{m,\text{opt}}^f(n)\} = 0$ for $k = 1, 2, \ldots, m$ as shown in (8.108), we have

$$\begin{aligned}\mathbb{E}\left\{\boldsymbol{x}_k^{\text{H}}(n)e_{m-1,\text{opt}}^f(n)\right\} &= \left[\mathbb{E}\left\{x^*(n)e_{m-1,\text{opt}}^f(n)\right\}, 0, \ldots, 0\right] \\ &= \left[P_{m-1}^f, 0, \ldots, 0\right]_{1\times k} \quad \text{for } k = 1, 2, \ldots, m,\end{aligned} \tag{8.121}$$

and

$$\begin{aligned}\mathbb{E}\left\{\boldsymbol{x}_k^{\text{H}}(n)e_{m,\text{opt}}^f(n)\right\} &= \left[\mathbb{E}\left\{x^*(n)e_{m,\text{opt}}^f(n)\right\}, 0, \ldots, 0\right] \\ &= \left[P_m^f, 0, \ldots, 0\right]_{1\times k} \quad \text{for } k = 1, 2, \ldots, m+1.\end{aligned} \tag{8.122}$$

From (8.120), (8.121), and (8.122),

$$\begin{aligned}\mathbb{E}\left\{\left(e_{k,\text{opt}}^b(n)\right)^* e_{m,\text{opt}}^b(n)\right\} &= \frac{P_m^f}{k_m}\left(-\frac{1}{P_{m-1}^f}P_{m-1}^f\left(\boldsymbol{h}_{1,\text{opt}}^b\right)^* + \frac{1}{P_m^f}P_m^f\left(\boldsymbol{h}_{1,\text{opt}}^b\right)^*\right), \\ &= 0\,, \text{ for } k = 1, 2, \ldots, m-1.\end{aligned} \tag{8.123}$$

Combining the results of (8.114) and (8.123), we have the very important result that the backward prediction errors are orthogonal.

$$\mathbb{E}\left\{\left(e_{k,\text{opt}}^b(n)\right)^* e_{m,\text{opt}}^b(n)\right\} = 0 \text{ for } k = 0, 1, 2, \ldots, m-1. \tag{8.124}$$

□

8.3.4 Autocorrelation Function of Sinusoids

The maximum entropy method requires the computation of $r(k) = \mathbb{E}\{x^*(n)x(n-k)\}$. (See (8.20).) This can be done using the formula

$$r(k) = \mathbb{E}\{x^*(n)x(n-k)\} = \lim_{K\to\infty} \frac{1}{2K+1} \sum_{n=-K}^{K} x^*(n)x(n-k). \tag{8.125}$$

In (8.125), k is integer, and the increment of n in the summation $\sum_n$ is unity, but we shall see in a moment, n need not be an integer; we shall choose the nature of n to simplify theoretical derivations. The computation of (8.125) requires infinite number of data samples. In practice, the number of data samples available is not infinite. In this section, we shall examine the effect when K is finite.

First, let $x(n)$ at time n be a real sinusoid $\cos(\omega_1 n + \phi)$ with frequency ω_1 and phase angle ϕ. Since $x(n)$ is real, $x^*(n) \times x(n-k) = x(n) \times x(n-k)$ and is given by

$$\cos(\omega_1 n + \phi) \times \cos(\omega_1(n-k) + \phi) = \frac{\cos(\omega_1 k)}{2} + \frac{\cos(2\omega_1 n + 2\phi + \omega_1 k)}{2}. \tag{8.126}$$

The right-hand side of (8.126) has two terms. The first term $\frac{\cos(\omega_1 k)}{2}$ is a constant for given ω_1 and k; its expected value is equal to itself. The second term $\frac{\cos(2\omega_1 n+2\phi+\omega_1 k)}{2}$ is a sinusoid with frequency $2\omega_1$ and phase angle $2\phi + \omega_1 k$; its expected value is zero. Hence, $r(k)$ computed using (8.125) assuming $K \to \infty$ is

$$\mathbb{E}\left\{\cos(\omega_1 n + \phi) \times \cos(\omega_1(n-k) + \phi)\right\} = \frac{\cos(\omega_1 k)}{2}. \tag{8.127}$$

For finite K, in general, $\sum_{n=-K}^{K} \frac{\cos(2\omega_1 n+2\phi+\omega_1 k)}{2}$ is not zero and so $r(k)$ computed using (8.125) for finite K will not be $\frac{\cos(\omega_1 k)}{2}$. We shall investigate the effect of finite K on the computed value of $r(k)$.

The quantity $2K + 1$ is the number of signal samples used in the computation. If the number of data points $2K + 1$ is odd, n and K are integers and $n = -K \;\ldots, -1, 0,\; 1, \ldots, K$. If $2K + 1$ is even, K is (integer $-$ 0.5) and $n = -K \ldots, -1.5, -0.5,\; 0.5, 1.5, \ldots, K$; both n and k are non-integer. For example, if there are four data points, $2K + 1 = 4$, and so $K = 1.5$; in this case, $n = -1.5, -0.5, 0.5, 1.5$. However, the increment of n is unity regardless of whether K is integer. We choose the time reference such that $n = 0$ occurs at the middle of the signal samples to simplify theoretical derivations. It can be shown that the sum of $2K + 1$ samples of a real sinusoid with frequency α and phase angle θ is given by

$$\sum_{n=-K}^{K} \cos(\alpha n + \theta) = \frac{\sin\left((2K+1)\frac{\alpha}{2}\right)}{\sin\left(\frac{\alpha}{2}\right)} \cos(\theta). \tag{8.128}$$

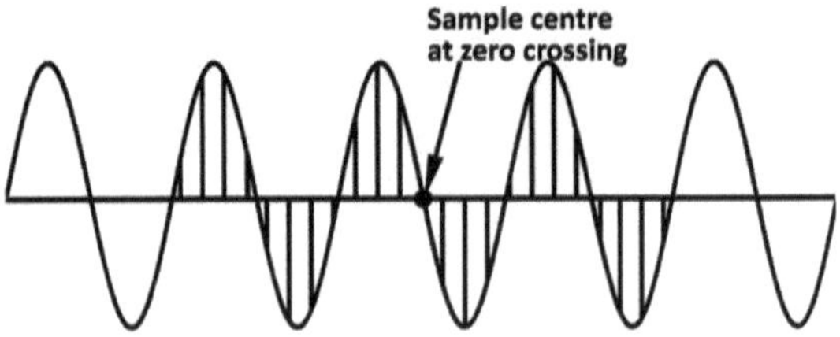

Fig. 8.9 Sampling of sinusoid is centered at a zero crossing

As can be seen from (8.128), the sum is zero if $(2K+1)\frac{\alpha}{2} = \text{integer} \times \pi$ or if $\theta = \frac{\pi}{2}$. For $(2K+1)\frac{\alpha}{2} = \text{integer} \times \pi$, the number of samples $2K+1$ must be equal to integer $\times \frac{2\pi}{\alpha}$ = integer $\times T_\alpha$ where $T_\alpha = \frac{2\pi}{\alpha}$ is the period of the sinusoid with frequency α. Hence, the sum of the samples is zero if an integer number of full cycles is sampled. For $\theta = \frac{\pi}{2}$, $\cos(\alpha n + \theta) = 0$ when $n = 0$, i.e., the sinusoid is sampled centered at a zero crossing of the sinusoid $\cos(\alpha n + \theta)$ as shown in Fig. 8.9.

From (8.126) and (8.128),

$$\begin{aligned}
&\sum_{n=-K}^{K} \frac{\cos(\omega_1 n + \phi) \times \cos(\omega_1 (n-k) + \phi)}{2K+1} \\
&\quad = \frac{\cos(\omega_1 k)}{2} + \sum_{n=-K}^{K} \frac{\cos(2\omega_1 n + 2\phi + \omega_1 k)}{2(2K+1)}, \\
&\quad = \frac{\cos(\omega_1 k)}{2} + \frac{\sin((2K+1)\omega_1)\cos(2\phi + \omega_1 k)}{2(2K+1)\sin(\omega_1)}. \qquad (8.129)
\end{aligned}$$

The $\frac{\sin((2K+1)\omega_1)\cos(2\phi+\omega_1 k)}{2(2K+1)\sin(\omega_1)}$ term in (8.129) is zero if $(2K+1)\omega_1 = \text{integer} \times \pi$ or if $2\phi + \omega_1 k = \frac{\pi}{2}$. For $(2K+1)\omega_1 = \text{integer} \times \pi$, the number of samples $2K+1 = \text{integer} \times \frac{\pi}{\alpha} = \text{integer} \times \frac{T_{\omega_1}}{2}$ where $T_{\omega_1} = \frac{2\pi}{\omega_1}$ is the period of the sinusoid with frequency ω_1. However, ω_1 is unknown until it is estimated. A possibility is to perform the estimation of ω_1 a second time using the first estimated ω_1 to choose a suitable K.

For $2\phi + \omega_1 k = \frac{\pi}{2}$, the center of the sampled data should be as close as possible to a zero crossing of the sinusoid $\cos(2\omega_1 n + 2\phi + \omega_1 k)$.

As a consequence of the $\frac{\sin((2K+1)\omega_1)\cos(2\phi+\omega_1 k)}{2(2K+1)\sin(\omega_1)}$ term, $\mathbf{R}_m$ is no longer Hermitian. (See (8.24).) The Levinson-Durbin recursion assumes that $\mathbf{R}_m$ is Hermitian. If $\mathbf{R}_m$ is not Hermitian, Levinson-Durbin recursion is not applicable. We shall see in a moment how $\mathbf{R}_m$ can be made Hermitian under the condition of finite K.

Next, let $x(n)$ be a complex sinusoid $e^{j(\omega_1 n + \phi)}$ with frequency ω_1 and phase angle ϕ. The product $x^*(n) \times x(n-k)$ is given by

$$e^{-j(\omega_1 n + \phi)} \times e^{j(\omega_1 (n-k) + \phi)} = e^{-j\omega_1 k}. \qquad (8.130)$$

The right-hand side of (8.130) is a constant for given ω_1 and k; its expected value is equal to itself. Thus, for a complex sinusoid, if there is no noise, the result is independent on K. In the presence of noise, larger K provides better noise suppression.

Now, consider the case where $x(n)$ is a sum of two real sinusoids $\cos(\omega_1 n + \phi_1) + \cos(\omega_2 n + \phi_2)$. The product $x(n) \times x(n-k)$ is given by

$$\begin{aligned}
&\{\cos(\omega_1 n + \phi_1) + \cos(\omega_2 n + \phi_2)\} \\
&\quad \times \{\cos(\omega_1 (n-k) + \phi_1) + \cos(\omega_2 (n-k) + \phi_2)\} \\
&\qquad = \sum_{m=1}^{2} \frac{\cos(\omega_m k)}{2} + \sum_{m=1}^{2} \frac{\cos(2\omega_m n + 2\phi_m + \omega_m k)}{2} \\
&\qquad\quad + \sum_{m=1}^{2} \frac{\cos((\omega_1 + \omega_2) n + \phi_1 + \phi_2 - \omega_m k)}{2} \\
&\qquad\quad + \sum_{m=1}^{2} \frac{\cos((\omega_m - \omega_{3-m}) n + \phi_m - \phi_{3-m} - \omega_m k)}{2}.
\end{aligned} \tag{8.131}$$

Following the same reasoning leading to (8.127), we have

$$\begin{aligned}
&\mathbb{E}\{[\cos(\omega_1 n + \phi_1) + \cos(\omega_2 n + \phi_2)] \\
&\qquad \times [\cos(\omega_1 (n-k) + \phi_1) + \cos(\omega_2 (n-k) + \phi_2)]\} \\
&\qquad = \frac{\cos(\omega_1 k)}{2} + \frac{\cos(\omega_2 k)}{2}.
\end{aligned} \tag{8.132}$$

From (8.131),

$$\begin{aligned}
&\sum_{n=-K}^{K} \frac{\begin{array}{c}[\cos(\omega_1 n + \phi_1) + \cos(\omega_2 n + \phi_2)] \\ \times [\cos(\omega_1 (n-k) + \phi_1) + \cos(\omega_2 (n-k) + \phi_2)]\end{array}}{2K+1} \\
&\quad = \sum_{m=1}^{2} \frac{\cos(\omega_m k)}{2} + \sum_{m=1}^{2} \sum_{n=-K}^{K} \frac{\cos(2\omega_m n + 2\phi_m + \omega_m k)}{2(2K+1)} \\
&\qquad + \sum_{m=1}^{2} \sum_{n=-K}^{K} \frac{\cos((\omega_1 + \omega_2) n + \phi_1 + \phi_2 - \omega_m k)}{2(2K+1)} \\
&\qquad + \sum_{m=1}^{2} \sum_{n=-K}^{K} \frac{\cos((\omega_m - \omega_{3-m}) n + \phi_1 - \phi_{3-m} - \omega_m k)}{2(2K+1)}.
\end{aligned} \tag{8.133}$$

The last three terms in the right-hand side of the above expression consist of sinusoids with frequencies $2\omega_m$, $(\omega_1 + \omega_2)$, and $|\omega_1 - \omega_2|$. Let the periods of the sinusoids corresponding to these frequencies be $T_{2\omega_m}$, $T_{\omega_1+\omega_2}$, and $T_{\omega_1-\omega_2}$, respectively. For these last three terms to vanish simultaneously, the number of signal samples $2K + 1$ must be an integer multiple of $T_{2\omega_m}$, $T_{\omega_1+\omega_2}$, and $T_{\omega_1-\omega_2}$. If $2K + 1$ is an integer multiple of T_{ω_1} and T_{ω_2}, it will also be an integer multiple of $T_{2\omega_m}$, $T_{\omega_1+\omega_2}$, and $T_{\omega_1-\omega_2}$.

Now, consider the case where $x(n)$ is a sum of two complex sinusoids $e^{j(\omega_1 n+\phi_1)} + e^{j(\omega_2 n+\phi_2)}$. The product $x^*(n) \times x(n-k)$ is given by

$$\left\{e^{-j(\omega_1 n+\phi_1)} + e^{-j(\omega_2 n+\phi_2)}\right\} \times \left\{e^{j(\omega_1 (n-k)+\phi_1)} + e^{j(\omega_2 (n-k)+\phi_2)}\right\}$$

$$= \sum_{m=1}^{2} e^{j\omega_m k} + \sum_{m=1}^{2} e^{j((\omega_m-\omega_{3-m})n+\phi_m-\phi_{3-m}-\omega_m k)}, \tag{8.134}$$

and $\sum_{n=-K}^{K} \frac{x^*(n)\times x(n-k)}{2K+1}$ is given by

$$\sum_{n=-K}^{K} \frac{\left\{e^{-j(\omega_1 n+\phi_1)} + e^{-j(\omega_2 n+\phi_2)}\right\} \times \left\{e^{j(\omega_1 (n-k)+\phi_1)} + e^{j(\omega_2 (n-k)+\phi_2)}\right\}}{2K+1}$$

$$= \sum_{m=1}^{2} e^{j\omega_m k} + \sum_{m=1}^{2} \sum_{n=-K}^{K} \frac{e^{j((\omega_m-\omega_{3-m})n+\phi_m-\phi_{3-m}-\omega_m k)}}{2K+1}. \tag{8.135}$$

It can be shown that the sum of $2K+1$ samples of a complex sinusoid with frequency α and phase angle θ is given by

$$\sum_{n=-K}^{K} e^{j(\alpha n+\theta)} = \frac{\sin\left((2K+1)\frac{\alpha}{2}\right)}{\sin\left(\frac{\alpha}{2}\right)} e^{j\theta}. \tag{8.136}$$

Following the same reasoning for (8.128), the above sum is zero if the number of samples $2K + 1$ is equal to an integer multiple of T_α where T_α is the period of the complex sinusoid $e^{j(\alpha n+\theta)}$. Hence, the $\sum_{m=1}^{2} \sum_{n=-K}^{K}(\bullet)$ term in (8.135) is zero if $2K + 1$ is an integer multiple of the period of the frequency difference between the two sinusoids.

Let $[\bullet]_{i,j}$ be the element at the ith row and jth column of matrix $[\bullet]$, and let $[\bullet]_i$ be the ith element of the vector $[\bullet]$. Levinson-Durbin recursion assumes that $\mathbf{R}_M$ is Hermitian, i.e., (1) $[\mathbf{R}_M]_{i,i} = [\mathbf{R}_M]_{j,j}$, $i, j = 1, 2, \ldots, M$, (2) $[\mathbf{R}_M]_{i,j} = [\mathbf{R}_M]_{i+1,j+1} = [\mathbf{R}_M]^*_{j,i} = [\mathbf{R}_M]^*_{j+1,i+1}$, $i = 1, 2, \ldots, M-1$, $j = i+1, \ldots, M$. Furthermore, Levinson-Durbin recursion assumes that (3) $[\mathbf{R}_M]_{1,j} = [\boldsymbol{r}_M]^*_{j-1}$, $j = 2, 3, \ldots, M$.

Suppose that L samples of $x(n)$ are available. To simplify notation, we shall label these L samples as $x(L)$, $x(L-1)$, $\ldots$, $x(2)$, $x(1)$. Let

$$\boldsymbol{y} = \begin{bmatrix} x(L) \\ x(L-1) \\ \vdots \\ x(M+2) \\ x(M+1) \end{bmatrix}, \quad \mathbf{X} = \begin{bmatrix} x(L-1) & x(L-2) & \cdots & x(L-M) \\ x(L-2) & x(L-3) & \cdots & x(L-M-1) \\ \vdots & \vdots & \ddots & \vdots \\ x(M+1) & x(M) & \cdots & x(2) \\ x(M) & x(M-1) & \cdots & x(1) \end{bmatrix}. \tag{8.137}$$

It can be seen from (8.137) that

$$[\boldsymbol{y}]_i = x(L+1-i), \tag{8.138}$$

and

$$[\mathbf{X}]_{i,j} = x(L+1-i-j). \tag{8.139}$$

First, we shall determine the condition for $[\mathbf{X}^{\mathrm{H}}\mathbf{X}]$ to be Hermitian. From (8.138) and (8.139),

$$[\mathbf{X}^{\mathrm{H}}\mathbf{X}]_{i,j} = \sum_{p=1}^{L-M} x^*(L+1-i-p) \times x(L+1-j-p), \tag{8.140a}$$

$$[\mathbf{X}^{\mathrm{H}}\boldsymbol{y}]_i = \sum_{p=1}^{L-M} x^*(L+1-i-p) \times x(L+1-p). \tag{8.140b}$$

Consider the $[\mathbf{X}^{\mathrm{H}}\mathbf{X}]_{i,i}$ terms for $i = 1, 2, \ldots, M$.

$$[\mathbf{X}^{\mathrm{H}}\mathbf{X}]_{1,1} = x^*(L-1)x(L-1) + \sum_{p=M}^{L-2} x^*(p)x(p), \tag{8.141a}$$

$$[\mathbf{X}^{\mathrm{H}}\mathbf{X}]_{2,2} = x^*(M-1)x(M-1) + \sum_{p=M}^{L-2} x^*(p)x(p), \tag{8.141b}$$

$$[\mathbf{X}^{\mathrm{H}}\mathbf{X}]_{i,i} = \ldots, \tag{8.141c}$$

$$[\mathbf{X}^{\mathrm{H}}\mathbf{X}]_{M-1,M-1} = x^*(L-M+1)x(L-M+1) + \sum_{p=2}^{L-M} x^*(p)x(p), \tag{8.141d}$$

$$[\mathbf{X}^{\mathrm{H}}\mathbf{X}]_{M,M} = x^*(1)x(1) + \sum_{p=2}^{L-M} x^*(p)x(p). \tag{8.141e}$$

From (8.141a) and (8.141b), for $[\mathbf{X}^{\mathrm{H}}\mathbf{X}]_{1,1} = [\mathbf{X}^{\mathrm{H}}\mathbf{X}]_{2,2}$, in general, $x^*(L-1)x(L-1)$ and $x^*(M-1)x(M-1)$ must be zero, i.e., $x(L-1) = 0$ and $x(M-1) = 0$. From (8.141d) and (8.141e), for $[\mathbf{X}^{\mathrm{H}}\mathbf{X}]_{M-1,M-1} = [\mathbf{X}^{\mathrm{H}}\mathbf{X}]_{M,M}$, in general, $x^*(L-M+1)x(L-M+1)$ and $x^*(1)x(1)$ must be zero, i.e., $x(L-M+1) = 0$ and $x(1) = 0$. Similarly, it can be shown that if $x(L-1) = x(L-2) = \ldots = x(L-M+2) = x(L-M+1) = 0$ and $x(M-1) = x(M-2) = \ldots = x(2) = x(1) = 0$, $[\mathbf{X}^{\mathrm{H}}\mathbf{X}]$ is Hermitian.

Next, we shall determine the condition for $[\mathbf{R}_M]_{1,j} = [\boldsymbol{r}_M]^*_{j-1}, j = 2, 3, \ldots, M$.

$$[\mathbf{X}^{\mathrm{H}}\boldsymbol{y}]_i = \sum_{p=1+M-i}^{L-i} x^*(p)x(i+p). \tag{8.142}$$

From (8.142) and (8.140a),

$$[\mathbf{X}^{\mathrm{H}}\mathbf{X}]_{1,2} = x(M-1)x^*(M) + \sum_{p=M}^{L-2} x(p)x(1+p)^*, \tag{8.143a}$$

$$[\mathbf{X}^{\mathrm{H}}\boldsymbol{y}]^*_1 = x(L-1)x(L)^* + \sum_{p=M}^{L-2} x(p)x^*(1+p). \tag{8.143b}$$

If $x(M-1)x^*(M) = 0$ and $x(L-1)x(L)^* = 0$, then $[\mathbf{X}^{\mathrm{H}}\boldsymbol{y}]^*_1 = [\mathbf{X}^{\mathrm{H}}\mathbf{X}]_{1,2}$. It can be shown that if the condition for $[\mathbf{X}^{\mathrm{H}}\mathbf{X}]$ to be Hermitian is satisfied, the condition for $[\mathbf{R}_M]_{1,j} = [\boldsymbol{r}_M]^*_{j-1}, j = 2, 3, \ldots, M$ will also be satisfied.

Hence, for a finite length $x(n)$, at least $M-1$ samples of zero must be added to precede the first sample and at least M samples of zero must be appended after the last sample in the computation of $\mathbf{R}_M$ and $\boldsymbol{r}_M$. If there are L samples of nonzero $x(n)$, it is convenient to regard $x(n)$ as being $L+2M-1$ samples where the nonzero samples are labeled as $x(M), \ldots, x(L+M-1)$; $x(1) = \ldots = x(M-1) = 0$ and $x(L+M) = \ldots = x(L+2M-1) = 0$ are appended to the original $x(n)$.

Appending the available signal samples with zero-valued samples is equivalent to windowing the original infinite length signal using a rectangular window. This causes spectral spread. If the signal contains several sinusoids with closely spaced frequencies, it will be difficult to resolve the frequencies. If the available signal is sufficiently long, a better frequency resolution can be achieved by extracting a shorter length signal with small edge discontinuity from the original signal by discarding a few signal samples at the two ends of the original signal.

If the available signal consists of many pieces of discontinuous short signal segments, it will not be possible to extract an even shorter length signal. In this case, a good way for computing the autocorrelation factors is to compute them from all available signal samples. For example, $r(k)$ is computed as the sum of all available $x^*(n)x(n-k)$ product terms divided by the total number of available relevant product terms. The accuracy of $r(k)$ thus obtained will depend on the number of available product terms. If the number of signal segments is large, the number of available product terms will also be large, and $r(k)$ can be computed accurately.

If the available signal is a single piece of long signal samples, extracting a shorter length signal with small edge discontinuity from the original signal and appending the extracted signal with the appropriate number of zero-valued samples produces a better result than computing $r(k)$ from all available $x^*(n)x(n-k)$ product terms.

8.3.5 Windowing of Finite Duration Signal

It is common practice in signal processing to apply a window to a finite duration signal sequence to reduce the effect of abrupt discontinuity at the edges of the sequence. (See Sect. 7.5.7 of Chap. 7.) Windowing of signal sequence may work well for estimation method such as the FFT but may produce erroneous results for super resolution methods. The reason is that windowing a signal sequence involves multiplying the signal by a window function; this generates spurious frequencies which, although cannot be resolved by methods such as the FFT, can be resolved by super resolution methods. We shall illustrate this effect by using an example.

Consider windowing an L-sample sinusoid $x_0(n) = 2\cos(\omega_0 n)$, $0 \leq n \leq L-1$ using a Hamming window $w(n) = a_0 + a_1 \cos\left(\frac{2\pi n}{L-1}\right)$ where $a_0 = 0.53836$ and $a_1 = -0.46164$. The windowed signal $x_w(n) = 2\cos(\omega_0 n)\left(a_0 + a_1\cos\left(\frac{2\pi n}{L-1}\right)\right) = 2a_0\cos(\omega_0 n) + 2a_1\cos(\omega_0 n)\cos\left(\frac{2\pi n}{L-1}\right) = 2a_0\cos(\omega_0 n) + a_1\cos\left(\left(\omega_0 + \frac{2\pi}{L-1}\right)n\right) + a_1\cos\left(\left(\omega_0 - \frac{2\pi}{L-1}\right)n\right), 0 \leq n \leq L-1$. The windowed $x_w(n)$ consists of three sinusoids. If a second-order real estimator is used, $\left|H_M^f(e^{j\omega})\right|^{-1}$ will peak at ω_0 as it is the strongest sinusoid. Windowing improves the estimate. If a fourth-order real estimator is used, $\left|H_M^f(e^{j\omega})\right|^{-1}$ will peak at two frequencies—one between ω_0 and $\omega_0 - \frac{2\pi}{L-1}$ and the other one between ω_0 and $\omega_0 + \frac{2\pi}{L-1}$; the estimates are wrong.

In general, for $x_0(n)$ consisting of L samples of S sinusoids,

$$x_0(n) = \sum_{s=1}^{S} b_s \cos(\omega_s n + \theta_s) \ , \quad 0 \leq n \leq L-1 \ , \tag{8.144}$$
$$= 0 \text{ otherwise} \ ,$$

where b_s, ω_s, and θ_s are the amplitude, frequency, and phase, respectively, of the sth sinusoid. Suppose that the window function $w(n)$ is expressible is a sum of C cosine terms shown below.

$$w(n) = a_0 + \sum_{c=1}^{C} a_c \cos(\phi_c n + \varphi_c). \tag{8.145}$$

where a_c, ϕ_c, and φ_c are the amplitude, frequency, and phase, respectively, of the Fourier series terms of the window function. The windowed signal sequence $x_w(n)$ is given by

$$\begin{aligned} x_w(n) &= \left(\sum_{s=1}^{S} b_s \cos(\omega_s n + \theta_s)\right)\left(a_0 + \sum_{c=1}^{C} a_c \cos(\phi_c n + \varphi_c)\right), \\ &= \sum_{s=1}^{S} a_0 b_s \cos(\omega_s n + \theta_s) + \sum_{s=1}^{S}\sum_{c=1}^{C} a_c b_s \cos\left((\omega_s + \phi_c)n + \theta_s + \varphi_c\right) \\ &\quad + \sum_{s=1}^{S}\sum_{c=1}^{C} \cos\left((\omega_s - \phi_c)n + \theta_s - \varphi_c\right). \end{aligned} \tag{8.146}$$

There are $S(2C + 1)$ frequencies. Hence, unless there is only one frequency and a second-order real estimator is used, in general, windowing will lead to erroneous estimate.

If a window is not used, in general, there will be large discontinuities at the edges of the sampled signal. This discontinuity tends to drive the predictor's coefficients toward zero and will cause error in the estimated frequencies; this is particularly problematic if the frequencies of the sinusoids are clustered together. If it is possible to choose the edges of the signal, choosing the edges to occur at smaller signal magnitudes reduces the discontinuities and so produces more accurate results.

Over modeling is an approach to overcome the problem of large discontinuity. For a real signal, if there are S sinusoids clustered together, use a model with order $> 2S$ instead of $2S$. For a complex signal, if there are S sinusoids clustered together, use a model with order $> S$ instead of S.

We shall examine such a problem associated with estimating the frequencies of a signal $x(n)$ consisting of two sinusoids ω_1 and ω_2 which are close together. Let $x(n)$ be

$$\begin{aligned} x(n) &= e^{j(\omega_1 n + \theta_1)} + e^{j(\omega_2 n + \theta_2)}, \\ &= 2e^{j(\frac{\omega_1+\omega_2}{2} n + \phi_a)} \cos\left(\frac{\omega_1 - \omega_2}{2} n + \phi_b\right), \phi_a = \frac{\theta_1 + \theta_1}{2}, \phi_b = \frac{\theta_1 - \theta_1}{2}. \end{aligned} \tag{8.147}$$

The factor $e^{j(\frac{\omega_1+\omega_2}{2} n + \phi_a)}$ in (8.147) is a complex sinusoid with frequency midpoint between ω_1 and ω_2. The factor $\cos\left(\frac{\omega_1-\omega_2}{2} n + \phi_b\right)$ is the envelop of the complex sinusoid $e^{j(\frac{\omega_1+\omega_2}{2} n + \phi_a)}$. If the discontinuities at the signal's edges are large and the two frequencies ω_1 and ω_2 are not well separated, a single frequency $\frac{\omega_1+\omega_2}{2}$ will be identified instead of two separate frequencies ω_1 and ω_2 by a second-order complex Levinson-Durbin algorithm. This problem can be overcome by increasing the order of the model. Similar effect can be derived for $x(n)$ consists of two real sinusoids.

8.3.6 Maximum Entropy Examples

This subsection illustrates some numerical examples of the maximum entropy method.

Example 8.1 Consider estimating the frequency of the sinusoid $x(n) = \cos\left(\frac{9\pi n}{16}\right)$. The period T_{ω_1} is $\frac{32}{9}$ samples and so $\frac{T_{\omega_1}}{2} = \frac{16}{9}$ samples. Integer $\times \frac{T_{\omega_1}}{2} = 16$ for "integer" $= 9$. Hence, the $\frac{\sin((2K+1)\omega_1)\cos(2\phi+\omega_1 k)}{2(2K+1)\sin(\omega_1)}$ term in (8.129) vanishes if $r(k)$ is obtained by averaging 16 consecutive $x(n) \times x(n-k)$ terms. The results for $\frac{r(k)}{r(0)}$ obtained by averaging 14 ∼ 19 and 4001 consecutive terms of $x(n) \times x(n-k)$ are tabulated in Table 8.5. The difference between the estimated frequency and the actual frequency is tabulated as $|\Delta\omega|$. When $L = 16$, $|\Delta\omega| = 0$ as expected. Although 4001 is not an integer multiple of $\frac{T_{\omega_1}}{2}$, the computed values of $\frac{r(1)}{r(0)}$ and $\frac{r(2)}{r(0)}$ approach that of the case for 16 terms because of the large number of terms.

In the above, each $r(k)$ is computed by averaging the same fixed number of $x(n) \times x(n-k)$ terms. For example, for 16 $x(n) \times x(n-k)$ terms, $r(0)$, $r(1)$, and $r(2)$ are obtained as $r(0) = \frac{1}{16}\sum_{m=0}^{15}|x(n-m)|^2$ using 16 signal samples, $r(1) = \frac{1}{16}\sum_{m=0}^{15}x(n-m)x(n-m-1)$ using 17 signal samples, and $r(2) = \frac{1}{16}\sum_{m=0}^{15}x(n-m)x(n-m-2)$ using 18 signal samples. The $x(n)$ samples used are $x(n), x(n-1), \ldots, x(n-17)$, i.e., a total of 18 $x(n)$ samples. The computation requires prior knowledge of the frequency which, in practice, is unknown.

Now consider the case where L signal samples are available and $x(n)$ is appended with the appropriate number of zero samples as described in Sect. 8.3.4 for the computation of $r(k)$. The results are tabulated in Table 8.6.

Table 8.7 shows the results if a fourth-order $H_M^f(e^{j\omega})$ is used. Comparing the results of Tables 8.6 and 8.7, it can be seen that there is an improvement in the values of $|\Delta\omega|$ when a fourth-order $H_M^f(e^{j\omega})$ is used.

Table 8.8 shows the results where $x(n)$ is windowed by a Hamming window and then appended with the appropriate number of zero samples. Comparing the $|\Delta\omega|$

Table 8.5 $r(k)$ obtained by averaging 14, …, 18, 19, and 4001 $x(n) \times x(n-k)$ terms. $\frac{9\pi}{16} = 1.767$

	$\frac{r(k)}{r(0)}$ obtained by averaging L terms of $x(n) \times x(n-k)$						
k	$L = 14$	$L = 15$	$L = 16$	$L = 17$	$L = 18$	$L = 19$	$L = 4001$
0	1	1	1	1	1	1	1
1	− 0.1727	− 0.1466	− 0.1951	− 0.1406	− 0.1993	− 0.1513	− 0.1953
2	− 0.9326	− 0.9428	− 0.9239	− 0.9452	− 0.9222	− 0.9410	− 0.9238
Estimated frequency	1.744	1.718	1.767	1.712	1.771	1.723	1.767
$\lvert\Delta\omega\rvert$	0.023	0.049	0.000	0.055	0.004	0.048	0.000

Table 8.6 $r(k)$ obtained by appending $x(n)$ with the appropriate number of zero samples

k	$\frac{r(k)}{r(0)}$ obtained from L available $x(n)$ samples						
	$L = 15$	$L = 16$	$L = 17$	$L = 18$	$L = 19$	$L = 20$	$L = 4001$
0	1	1	1	1	1	1	1
1	− 0.1942	− 0.1707	− 0.1734	− 0.1943	− 0.1593	− 0.2048	− 0.1950
2	− 0.8034	− 0.8132	− 0.8255	− 0.8177	− 0.8406	− 0.8258	− 0.9234
Estimated frequency	1.767	1.743	1.746	1.767	1.731	1.778	1.767
$\lvert\Delta\omega\rvert$	0.000	0.024	0.021	0.000	0.036	0.011	0.000

Table 8.7 The same as Table 8.6 except that $H_M^f(e^{j\omega})$ is fourth order

	Appropriate number of zero valued samples appended to L samples of $x(n)$						
	$L = 15$	$L = 16$	$L = 17$	$L = 18$	$L = 19$	$L = 20$	$L = 4001$
Estimated frequency	1.770	1.770	1.770	1.770	1.770	1.769	1.767
$\lvert\Delta\omega\rvert$	0.003	0.003	0.003	0.003	0.003	0.002	0.000

Table 8.8 $x(n)$ is windowed using Hamming window and appended with the appropriate number of zero samples

	$\frac{r(k)}{r(0)}$ obtained from L available Hamming windowed $x(n)$ samples						
k	$L = 15$	$L = 16$	$L = 17$	$L = 18$	$L = 19$	$L = 20$	$L = 4003$
0	1	1	1	1	1	1	1
1	− 0.3603	− 0.3628	− 0.3661	− 0.3692	− 0.3703	− 0.3736	− 0.3902
2	− 0.8967	− 0.9079	− 0.9196	− 0.9273	− 0.9353	− 0.9415	− 1.0000
Estimated frequency	1.763	1.763	1.763	1.764	1.764	1.764	1.767
$\lvert\Delta\omega\rvert$	0.004	0.004	0.004	0.003	0.003	0.003	0.000

values in Tables 8.6 and 8.8, it can be seen that the values of $|\Delta\omega|$ for Table 8.8 are smaller; windowing improves accuracy in this case.

Consider the case where 15 samples of $x(n) = \cos\left(\frac{9\pi n}{16}\right)$ is windowed by using a Hamming window. The windowed sequence $x_w(n)$ is $x_w(n) = \cos\left(\frac{9\pi n}{16}\right)\left(0.53836 - 0.46164\cos\left(\frac{\pi n}{7}\right)\right) = 0.53836\cos(\frac{9\pi n}{16}) - 0.23082\cos\left(\left(\frac{9\pi}{16} + \frac{\pi}{7}\right)n\right) - 0.23082\cos\left(\left(\frac{9\pi}{16} - \frac{\pi}{7}\right)n\right)$, $0 \leq n \leq 14$. There are three frequencies, namely, $\frac{9\pi}{16} - \frac{\pi}{7} = 1.318$, $\frac{9\pi}{16} = 1.767$, and $\frac{9\pi}{16} + \frac{\pi}{7} = 2.216$. If a fourth-order real estimator is used to estimate the frequencies, the estimated frequencies from the peaks of $\left|H_M^f(e^{j\omega})\right|^{-1}$ are 1.534 and 1.967, respectively. If a sixth-order real estimator is used, the estimated frequencies from the peaks

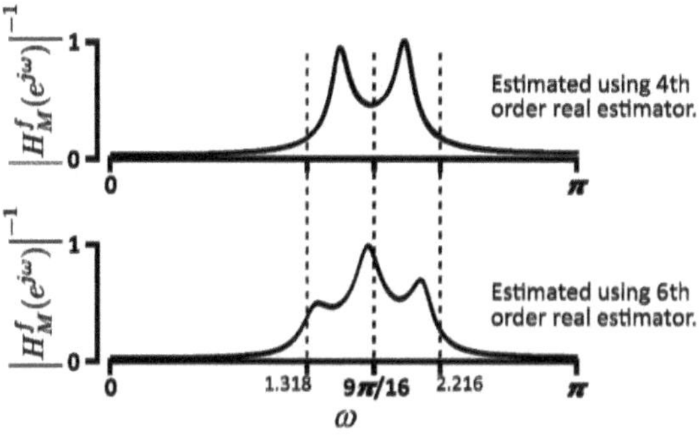

Fig. 8.10 15 samples of sinusoid windowed using Hamming window and the frequency estimated using a 4th- and 6th-order real estimators, respectively

Table 8.9 Comparison of estimated frequencies for three methods. SNR = 20 dB

	A, B : 2nd order; C : 1st order.			A, B : 4th order; C : 2nd order		
	A	B	C	A	B	C
Estimated frequencies	1.763	1.737	1.762	1.779	1.784	1.752
$\|\Delta\omega\|$	0.004	0.030	0.005	0.012	0.017	0.015

A: Linear prediction. 16 real signal samples. 2nd or 4th order
B: Levinson-Durbin. 16 real signal samples. 2nd or 4th order
C: Levinson-Durbin. 8 complex signal samples. 1st or 2nd order

of $\left|H_M^f(e^{j\omega})\right|^{-1}$ are 1.390, 1.725, and 2.079, respectively, and from the zeros of $H_M^f(e^{j\omega})$ are 1.353, 1.723, and 2.097, respectively. Note: The zeroes of $H_M^f(e^{j\omega})$ indicate the frequencies; when the zeroes are too near to each other, the peaks of $\left|H_M^f(e^{j\omega})\right|^{-1}$ no longer coincide with the zeroes. The respective $\left|H_M^f(e^{j\omega})\right|^{-1}$ functions for the fourth-order and sixth-order real estimators are plotted in Fig. 8.10.

Suppose that $x(n) = \cos\left(\frac{9\pi n}{16}\right)$ is corrupted by noise and the SNR is 20 dB. A comparison of the estimates obtained using three different methods, namely, A, B, and C, are tabulated in Table 8.9. For A and B, the estimators are linear prediction and Levinson-Durbin, respectively; the data used are 16 samples of real signal samples. For C, the estimator is Levinson-Durbin; the data used are eight samples of complex signal samples. Second- and fourth-order estimator results are tabulated for A and B. First- and second-order estimator results are tabulated for C. It can be seen from Table 8.9 that increasing the estimator order may not necessarily improve estimation accuracy.

Now consider estimating the frequencies of a signal consisting of two sinusoids with different frequencies. Let $x(n) = \cos\left(\frac{\pi n}{2}\right) + \cos\left(\frac{5\pi n}{8}\right)$. Linear prediction will give the correct result if there is no noise. As discussed in Sect. 8.3.4, if the number of $x(n) \times x(n-k)$ terms used to compute $r(k)$ is an integer multiple of the periods of both the sinusoids, Levinson-Durbin algorithm will also give the correct

Table 8.10 SNR = 40 dB. 2 frequencies: $\frac{\pi}{2} = 1.571$ and $\frac{5\pi}{8} = 1.963$

	A	B	C
Estimated	1.573	1.570	1.569
frequencies	1.962	1.963	1.962
$\lvert\Delta\omega\rvert$	0.002	0.001	0.002
	0.001	0.000	0.002

A: Linear prediction, 4th order, 16 real signal samples
B: Levinson-Durbin, 4th order, real signal, $r(k)$ obtained by averaging 16 terms of $x(n) \times x(n-k)$
C: Levinson-Durbin, 4th order, complex signal, $r(k)$ obtained by averaging 16 terms of $x^*(n) \times x(n-k)$

Table 8.11 Estimated frequencies for small and large discontinuities with different signal data length. SNR = 40 dB. 2nd-order complex Levinson-Durbin

	A	B	C	D	E	F
	16 samples	32 samples	16 samples	144 samples	270 samples	528 samples
Estimated	1.581	1.573	1.766	1.759	1.605	1.572
frequencies	1.950	1.960	1.787	1.781	1.928	1.961
$\lvert\Delta\omega\rvert$	0.010	0.002	0.195	0.188	0.034	0.001
	0.013	0.003	0.173	0.182	0.035	0.002

A and B: Discontinuity at envelop zero crossing
C, D, E, and F: Discontinuity at envelop maxima

results if there is no noise. The integer 16 is the lowest common multiple of $\frac{\pi n}{2}$ and $\frac{5\pi n}{8}$; hence, $r(k)$ is computed by averaging 16 $x(n) \times x(n-k)$ terms. Suppose that the signal is subject to noise and the SNR is 40 dB. The results are tabulated in Table 8.10.

Now consider the case where $x(n)$ is complex with frequencies $\frac{\pi}{2} = 1.571$ and $\frac{5\pi}{8} = 1.963$. The period of the envelop of (8.147) is 32 samples, and so the envelop's consecutive zero crossings and consecutive maxima of opposite signs are separated by 16 samples. Suppose that 16 signal samples are available and $x(n)$ is appended with the appropriate number of zero samples for the computation of $r(k)$. The accuracy of the estimate will depend on the size of the discontinuity. If a discontinuity occurs at a zero crossing of the envelop of (8.147), the discontinuity is small. If it occurs at a maxima, the discontinuity is large. We shall compare the effects of the discontinuity for various signal length for a second-order complex Levinson-Durbin algorithm. The results are tabulated in Table 8.11. The estimated frequencies are computed from the zeroes of $H_M^f(e^{j\omega})$. It can be seen from Table 8.11 that, for small discontinuity, a relatively short signal duration is able to achieve better accuracy than that for large discontinuity at very much longer signal duration.

Table 8.12 Estimated frequencies for small and large discontinuities with different estimator order at the same signal data length. SNR = 40 dB. Complex signal. Levinson-Durbin

	A	*B*	*C*	*D*	*E*
	2nd order	4th order	2nd order	4th order	8th order
Estimated	1.581	1.587	1.766	1.435	1.473
frequencies	1.950	1.951	1.787	2.092	2.061
$\lvert\Delta\omega\rvert$	0.010	0.016	0.195	0.136	0.098
	0.013	0.012	0.176	0.129	0.098

A and *B*: 16 samples. Discontinuity at envelop zero crossing
C, *D*, and *E*: 16 samples. Discontinuity at envelop maxima

Table 8.13 Estimated frequencies for large discontinuities with different estimator order and different signal data length. SNR = 40 dB. Real signal. Levinson-Durbin

	16 samples	16 samples	32 samples	64 samples	64 samples
	4th order	8th order	8th order	16th order	32nd order
Estimated	1.226	1.474	1.525	1.562	1.576
frequencies	1.882	2.057	2.011	1.970	1.957
$\lvert\Delta\omega\rvert$	0.345	0.097	0.046	0.009	0.005
	0.081	0.094	0.048	0.007	0.006

Table 8.12 shows the frequencies estimated using Levinson-Durbin algorithm for different estimator order with signal length kept at 16 samples. It can be seen from Table 8.12 that increasing the estimator order improves estimation accuracy for large discontinuity but not for small discontinuity.

Table 8.13 shows the frequencies estimated using Levinson-Durbin algorithm for different estimator order and signal data length for $x(n)$ consisting of two real sinusoids with the same magnitude. The frequencies are $\frac{\pi}{2} = 1.571$ and $\frac{5\pi}{8} = 1.963$, respectively. The discontinuity occurs at the envelop's maxima. It can be seen from Table 8.13 that increasing the estimator order and signal data length improves estimation accuracy.

Table 8.14 shows the results for a signal consisting of 19 segments of short signals with 13 complex samples per segment. The signal contains two sinusoids with frequencies $\frac{\pi}{2}$ and $\frac{5\pi}{8}$, respectively. The SNR is 40 dB. Two methods for computing the values of $r(k)$ are used. For one of the methods, each 13-sample segment is appended with the appropriate number of zero-value elements for computing $r(k)$, $k = 0, 1, 2$. The final value of $r(k)$ for any particular k is obtained by averaging the values of $r(k)$ for that particular k from all the 19 segments. For the other method, each $r(k)$, $k = 0, 1, 2$ is obtained by averaging all the available $x^*(n)x(n-k)$ product terms. It can be seen from Table 8.14 that the method where each 13-sample segment is appended with the appropriate number of zero-value elements fails. The method where each $r(k)$ is obtained by averaging all the available $x^*(n)x(n-k)$ product terms produces good frequency estimates.

Table 8.14 Data consists of 19 segments of signals. Each segment has 13 complex samples. SNR = 40 dB. 2nd-order complex Levinson-Durbin

	$r(k)$ computed by appending Zeroes to each segment	$r(k)$ computed from all Available product terms
Estimated	1.767	1.563
frequencies	1.767	1.965
$\|\Delta\omega\|$	0.196	0.008
	0.196	0.002

8.4 Maximum Likelihood Method

8.4.1 CRLB for a Sinusoid

We shall estimate the CRLB (Cramér-Rao lower bound) for the scenario with only one sinusoid subject to the presence of white Gaussian noise. (Please refer to Chap. 2, Sect. 2.6, for details on CRLB.) Let the true signal (signal without noise) at time n with frequency ω be denoted by v_n.

$$v_n = v_0 e^{j\omega n}\ , n = -K, 1-K, 2-K \ldots, K-2, K-1, K, \tag{8.148}$$

$$K = \begin{cases} \left(\text{Integer} - \frac{1}{2}\right) & \text{for } 2K+1 = \text{ even,} \\ \text{Integer for } 2K+1 = \text{ odd,} \end{cases}$$

$$2K+1 \text{ is the total number of data points.}$$

As already explained in Sect. 8.3.4, n and K need not be integer. The actual signal received is $v_n + w_n$ where w_n is a zero mean white Gaussian noise with variance σ^2.

$$x_n = v_0 e^{j\omega n} + w_n\ ,\quad w_n \sim N(0, \sigma^2)\ ,\quad n = -K, \ldots, K\ . \tag{8.149}$$

The notation $w_n \sim N(0, \sigma^2)$ means w_n is normal distributed with zero mean and variance σ^2. In general, v_0 is complex. Let

$$v_0 = ae^{jb}, \tag{8.150}$$

where a and b are deterministic real constants, but we do not know their values. Substituting (8.150) into (8.149),

$$x_n = ae^{jb}e^{jn\omega} + w_n. \tag{8.151}$$

Our aim is to estimate ω, but to achieve that using the maximum likelihood method, we also need to estimate a and b. Let

$$\boldsymbol{\phi} = [a, b, \omega]^{\mathrm{T}}. \tag{8.152}$$

Define

$$\boldsymbol{v} = \begin{bmatrix} v_K \\ v_{K-1} \\ \vdots \\ v_{-K} \end{bmatrix}, \quad \boldsymbol{x} = \begin{bmatrix} x_K \\ x_{K-1} \\ \vdots \\ x_{-K} \end{bmatrix}, \quad \boldsymbol{w} = \begin{bmatrix} w_K \\ w_{K-1} \\ \vdots \\ w_{-K} \end{bmatrix}, \quad \boldsymbol{u} = \begin{bmatrix} e^{jK\omega} \\ e^{j(K-1)\omega} \\ \vdots \\ e^{-jK\omega} \end{bmatrix}. \tag{8.153}$$

The expressions (8.148) and (8.151) for $n = -K, \ldots, K$ can be written in vector form as

$$\boldsymbol{v} = ae^{jb}\boldsymbol{u}, \tag{8.154}$$

and

$$\boldsymbol{x} = ae^{jb}\boldsymbol{u} + \boldsymbol{w}. \tag{8.155}$$

The vector $\boldsymbol{x}$ is observed data. It is not a function of $\boldsymbol{\phi}$. It is a "constant" once the data is observed. Hence,

$$\frac{\partial \boldsymbol{x}}{\partial a} = \frac{\partial \boldsymbol{x}}{\partial b} = \frac{\partial \boldsymbol{x}}{\partial \omega} = 0, \tag{8.156}$$

$$\mathbb{E}\{\boldsymbol{x}\} = \mathbb{E}\{\boldsymbol{v}\} = \boldsymbol{v} = ae^{jb}\boldsymbol{u}, \tag{8.157}$$

and

$$\boldsymbol{u}^{\mathrm{H}}\boldsymbol{u} = 2K + 1. \tag{8.158}$$

The vector $\boldsymbol{u}$ is neither a function of a nor b. Therefore,

$$\frac{\partial \boldsymbol{u}}{\partial a} = \frac{\partial \boldsymbol{u}}{\partial b} = 0. \tag{8.159}$$

It can be easily shown by differentiating $\boldsymbol{u}$ with respect to ω that

$$\frac{\partial \boldsymbol{u}}{\partial \omega} = j \begin{bmatrix} Ke^{jK\omega} \\ (K-1)e^{j(K-1)\omega} \\ \vdots \\ -Ke^{-jK\omega} \end{bmatrix}. \tag{8.160}$$

To simplify notation, let

$$\boldsymbol{u}_1 = \frac{\partial \boldsymbol{u}}{\partial \omega}. \tag{8.161}$$

It can be easily verified from (8.153) and (8.160) that

$$\boldsymbol{u}_1^{\mathrm{H}}\boldsymbol{u} = \boldsymbol{u}^{\mathrm{H}}\boldsymbol{u}_1 = -j \sum_{k=-K}^{K} k = 0, \tag{8.162}$$

and that

$$\boldsymbol{u}_1^{\mathrm{H}}\boldsymbol{u}_1 = \sum_{k=-K}^{K} k^2 = \frac{K(K+1)(2K+1)}{3}. \tag{8.163}$$

The likelihood function $p_X(\boldsymbol{x};\boldsymbol{\phi})$ is given by

$$p_X(\boldsymbol{x};\boldsymbol{\phi}) = Ce^{-(\boldsymbol{x}-\boldsymbol{v})^{\mathrm{H}}\mathbf{R}^{-1}(\boldsymbol{x}-\boldsymbol{v})}, \tag{8.164a}$$

$$\text{where} \quad \mathbf{R} = \sigma^2\mathbf{I}. \tag{8.164b}$$

In (8.164), C is a constant, and $\mathbf{I}$ is the identity matrix. The log-likelihood function is given by

$$\ln(p_X(\boldsymbol{x};\boldsymbol{\phi})) = \ln(C) + \frac{ae^{-jb}\mathbf{u}^{\mathrm{H}}\mathbf{x} + ae^{jb}\mathbf{x}^{\mathrm{H}}\mathbf{u} - a^2\mathbf{u}^{\mathrm{H}}\mathbf{u}}{\sigma^2}. \tag{8.165}$$

Let

$$g(\boldsymbol{\phi}) = \ln(C) + \frac{ae^{-jb}\mathbf{u}^{\mathrm{H}}\mathbf{x} + ae^{jb}\mathbf{x}^{\mathrm{H}}\mathbf{u} - a^2\mathbf{u}^{\mathrm{H}}\mathbf{u}}{\sigma^2}. \tag{8.166}$$

The Fisher information matrix, $\mathbf{I}(\boldsymbol{\phi})$, is given by

$$\mathbf{I}(\boldsymbol{\phi}) = -\mathbb{E}\begin{bmatrix} \frac{\partial^2 g}{\partial a^2} & \frac{\partial^2 g}{\partial a \partial b} & \frac{\partial^2 g}{\partial a \partial \omega} \\ \frac{\partial^2 g}{\partial b \partial a} & \frac{\partial^2 g}{\partial b^2} & \frac{\partial^2 g}{\partial b \partial \omega} \\ \frac{\partial^2 g}{\partial \omega \partial a} & \frac{\partial^2 g}{\partial \omega \partial b} & \frac{\partial^2 g}{\partial \omega^2} \end{bmatrix}. \tag{8.167}$$

(Please refer to Chap. 2 for details on Fisher nformation matrix.) We shall obtain all the terms in the Fisher information matrix. Differentiating (8.166) with respect to a, we have

$$\frac{\partial g}{\partial a} = \frac{e^{-jb}\mathbf{u}^{\mathrm{H}}\mathbf{x} + e^{jb}\mathbf{x}^{\mathrm{H}}\mathbf{u} - 2a\mathbf{u}^{\mathrm{H}}\mathbf{u}}{\sigma^2}. \tag{8.168}$$

Differentiating (8.168) with respect to a, we have

$$\frac{\partial^2 g}{\partial a^2} = \frac{-2\boldsymbol{u}^{\mathrm{H}}\boldsymbol{u}}{\sigma^2}. \tag{8.169}$$

From (8.169) and (8.158),

$$\mathbb{E}\left[\frac{\partial^2 g}{\partial a^2}\right] = \frac{-2\boldsymbol{u}^{\mathrm{H}}\boldsymbol{u}}{\sigma^2} = \frac{-2(2K+1)}{\sigma^2}. \tag{8.170}$$

Differentiating (8.168) with respect to b, we have

$$\frac{\partial^2 g}{\partial a \partial b} = \frac{-je^{-jb}\boldsymbol{u}^{\mathrm{H}}\boldsymbol{x} + je^{jb}\boldsymbol{x}^{\mathrm{H}}\boldsymbol{u}}{\sigma^2}. \tag{8.171}$$

From (8.155) and (8.171),

$$\mathbb{E}\left[\frac{\partial^2 g}{\partial a \partial b}\right] = \frac{-je^{-jb}\boldsymbol{u}^{\mathrm{H}}ae^{jb}\boldsymbol{u} + je^{jb}ae^{-jb}\boldsymbol{u}^{\mathrm{H}}\boldsymbol{u}}{\sigma^2} = 0. \tag{8.172}$$

Differentiating (8.168) with respect to ω, we have

$$\frac{\partial^2 g}{\partial a \partial \omega} = \frac{e^{-jb}\boldsymbol{u}^{\mathrm{H}}\boldsymbol{x} + e^{jb}\boldsymbol{x}^{\mathrm{H}}\boldsymbol{u}_1 - 2a\boldsymbol{u}_1^{\mathrm{H}}\boldsymbol{u} - 2a\boldsymbol{u}^{\mathrm{H}}\boldsymbol{u}_1}{\sigma^2}. \tag{8.173}$$

From (8.155) and (8.173),

$$\mathbb{E}\left[\frac{\partial^2 g}{\partial a \partial \omega}\right] = \frac{e^{-jb}\boldsymbol{u}_1^{\mathrm{H}}ae^{jb}\boldsymbol{u} + e^{jb}ae^{-jb}\boldsymbol{u}^{\mathrm{H}}\boldsymbol{u}_1 - 2a\boldsymbol{u}_1^{\mathrm{H}}\boldsymbol{u} - 2a\boldsymbol{u}^{\mathrm{H}}\boldsymbol{u}_1}{\sigma^2} = 0. \tag{8.174}$$

Differentiating (8.166) with respect to ω, we have

$$\frac{\partial g}{\partial \omega} = \frac{ae^{-jb}\boldsymbol{u}_1^{\mathrm{H}}\boldsymbol{x} + ae^{jb}\boldsymbol{x}^{\mathrm{H}}\boldsymbol{u}_1 - a^2\boldsymbol{u}_1^{\mathrm{H}}\boldsymbol{u} - a^2\boldsymbol{u}^{\mathrm{H}}\boldsymbol{u}_1}{\sigma^2}. \tag{8.175}$$

Differentiating (8.175) with respect to ω, we have

$$\frac{\partial^2 g}{\partial \omega^2} = \frac{ae^{-jb}\boldsymbol{u}_2^{\mathrm{H}}\boldsymbol{x} + ae^{jb}\boldsymbol{x}^{\mathrm{H}}\boldsymbol{u}_2 - a^2\boldsymbol{u}_2^{\mathrm{H}}\boldsymbol{u} - 2a^2\boldsymbol{u}_1^{\mathrm{H}}\boldsymbol{u}_1 - a^2\boldsymbol{u}^{\mathrm{H}}\boldsymbol{u}_2}{\sigma^2}. \tag{8.176}$$

From (8.155) and (8.176),

$$\mathbb{E}\left[\frac{\partial^2 g}{\partial \omega^2}\right] = \frac{ae^{-jb}\boldsymbol{u}_2^{\mathrm{H}}ae^{jb}\boldsymbol{u} + ae^{jb}ae^{-jb}\boldsymbol{u}^{\mathrm{H}}\boldsymbol{u}_2 - a^2\boldsymbol{u}_2^{\mathrm{H}}\boldsymbol{u} - 2a^2\boldsymbol{u}_1^{\mathrm{H}}\boldsymbol{u}_1 - a^2\boldsymbol{u}^{\mathrm{H}}\boldsymbol{u}_2}{\sigma^2},$$

$$= \frac{-2a^2\boldsymbol{u}_1^{\mathrm{H}}\boldsymbol{u}_1}{\sigma^2}. \tag{8.177}$$

Substitute the results of (8.163) into (8.177),

$$\mathbb{E}\left[\frac{\partial^2 g}{\partial \omega^2}\right] = \frac{-2a^2K(K+1)(2K+1)}{3\sigma^2}. \tag{8.178}$$

Differentiating (8.175) with respect to b, we have

$$\frac{\partial^2 g}{\partial \omega \partial b} = \frac{-jae^{-jb}\boldsymbol{u}_1^{\mathrm{H}}\boldsymbol{x} + jae^{jb}\boldsymbol{x}^{\mathrm{H}}\boldsymbol{u}_1}{\sigma^2}. \tag{8.179}$$

From (8.155) and (8.179),

$$\mathbb{E}\left[\frac{\partial^2 g}{\partial \omega \partial b}\right] = \frac{-jae^{-jb}\boldsymbol{u}_1^{\mathrm{H}}e^{jb}\boldsymbol{u} + jae^{jb}e^{-jb}\boldsymbol{u}^{\mathrm{H}}\boldsymbol{u}_1}{\sigma^2} = 0. \tag{8.180}$$

Differentiating (8.166) with respect to b, we have

$$\frac{\partial g}{\partial b} = \frac{-jae^{-jb}\boldsymbol{u}^{\mathrm{H}}\boldsymbol{x} + jae^{jb}\boldsymbol{x}^{\mathrm{H}}\boldsymbol{u}}{\sigma^2}. \tag{8.181}$$

Differentiating (8.181) with respect to b, we have

$$\frac{\partial^2 g}{\partial b^2} = \frac{-ae^{-jb}\boldsymbol{u}^{\mathrm{H}}\boldsymbol{x} + -ae^{jb}\boldsymbol{x}^{\mathrm{H}}\boldsymbol{u}}{\sigma^2}. \tag{8.182}$$

From (8.155) and (8.182),

$$\mathbb{E}\left[\frac{\partial^2 g}{\partial b^2}\right] = \frac{-ae^{-jb}\boldsymbol{u}^{\mathrm{H}}ae^{jb}\boldsymbol{u} - ae^{jb}ae^{-jb}\boldsymbol{u}^{\mathrm{H}}\boldsymbol{u}}{\sigma^2} = \frac{-2a^2\boldsymbol{u}^{\mathrm{H}}\boldsymbol{u}}{\sigma^2},$$

$$= \frac{-2a^2(2K+1)}{\sigma^2}. \tag{8.183}$$

From the above results, the Fisher information matrix is

$$\mathbf{I}(\boldsymbol{\phi}) = -\mathbb{E}\begin{bmatrix} \frac{2(2K+1)}{\sigma^2} & 0 & 0 \\ 0 & \frac{2a^2(2K+1)}{\sigma^2} & 0 \\ 0 & 0 & \frac{2a^2K(K+1)(2K+1)}{3\sigma^2} \end{bmatrix}. \tag{8.184}$$

$\mathbf{I}(\boldsymbol{\phi})$ is diagonal. Its inverse is also diagonal. The diagonal elements of its inverse are the reciprocal of its diagonal elements. CRLB states that the covariance matrix for $\bar{\phi}$ denoted as $[\mathbf{C}_{\bar{\phi}}]_{n,n}$ and defined as $[\mathbf{C}_{\bar{\phi}}]_{n,n} = \mathbb{E}\{[\bar{\phi} - \phi][\bar{\phi} - \phi]^{\mathrm{T}}\}$ is larger than or equal to $[\mathbf{I}^{-1}(\boldsymbol{\phi})]_{n,n}$ where $\bar{\phi}$ is the estimated $\boldsymbol{\phi}$. We are interested in $\mathbf{I}^{-1}(\boldsymbol{\phi})$ corresponding to the $\mathbb{E}\left[\frac{\partial^2 g}{\partial \omega^2}\right]$ term as it gives the CRLB bound for the variance of $\bar{\omega}$, the unbiased estimate of ω. We are not interested in those terms corresponding to $\mathbb{E}\left[\frac{\partial^2 g}{\partial a^2}\right]$ and $\mathbb{E}\left[\frac{\partial^2 g}{\partial b^2}\right]$ as we are not estimating ae^{jb}. Hence, we have

$$\begin{aligned} \operatorname{var}(\bar{\omega}) &\geq \left(\frac{2a^2K(K+1)(2K+1)}{3\sigma^2}\right)^{-1}, \\ &\geq \frac{3\sigma^2}{2a^2K(K+1)(2K+1)}. \end{aligned} \tag{8.185}$$

As can be seen from (8.185), $\operatorname{var}(\bar{\omega})$ decreases, i.e., better estimate, as

1. SNR $\frac{a^2}{\sigma^2}$ increases.
2. Number of data points $2K+1$ increases.

We have spent many lines of mathematics to arrive at the seemingly obvious result; namely, increasing the SNR and increasing the number of data points will result in a better estimate, but we have not yet derived a method to estimate the frequency.

8.4.2 Estimating the Frequency of a Single Sinusoid

We shall now develop a method for estimating the frequency of a single sinusoid in the presence of white Gaussian noise. Let the PDF for w_n be

$$p(w) = \frac{1}{\pi\sigma^2} e^{-w^* w/\sigma^2}. \tag{8.186}$$

See Appendix C for further details on the PDF of complex random variables. The likelihood function $p_X(\boldsymbol{x}; \boldsymbol{\phi})$ is given by

$$p_X(\boldsymbol{x}; \boldsymbol{\phi}) = \frac{1}{\pi^{2K+1}\sigma^2(2K+1)} e^{-(x-v)^{\mathrm{H}}\mathbf{R}^{-1}(x-v)}, \tag{8.187}$$

where

$$\mathbf{R} = \sigma^2 \mathbf{I}. \tag{8.188}$$

In (8.188), $\mathbf{I}$ is a $(2K+1) \times (2K+1)$ identity matrix. We wish to find $\boldsymbol{\phi}$ that maximizes the likelihood function $p_X(\boldsymbol{x}; \boldsymbol{\phi})$. Alternatively, we wish to find $\boldsymbol{\phi}$ that minimizes $(x - v)^{\mathrm{H}}\mathbf{R}^{-1}(x - v)$. Since $\mathbf{R} = \sigma^2\mathbf{I}$, the result remains unchanged if we find $\boldsymbol{\phi}$ that minimizes $(x - v)^{\mathrm{H}}(x - v)$. We shall cast this minimization problem into a least squares problem. From (8.148) and (8.150), we have

$$\begin{aligned} v_n &= ae^{j(b+n\omega)}, \\ &= a\cos(b)\cos(n\omega) + a\sin(b)\sin(n\omega) \\ &\quad + j[a\sin(b)\cos(n\omega) - a\cos(b)\sin(n\omega)]. \end{aligned} \tag{8.189}$$

Let

$$\alpha_1 = a\cos(b), \tag{8.190a}$$

$$\alpha_2 = a\sin(b), \tag{8.190b}$$

$$c_n(\omega) = \cos(n\omega), \tag{8.190c}$$

$$s_n(\omega) = \sin(n\omega). \tag{8.190d}$$

Using the notations of (8.190), (8.189) may be written as

$$v_n = \alpha_1 c_n(\omega) + \alpha_2 s_n(\omega) + j[\alpha_2 c_n(\omega) - \alpha_1 s_n(\omega)]. \tag{8.191}$$

Let the real and imaginary parts of x_n be $x_{r,n}$ and $x_{i,n}$, respectively, and those of v_n be $v_{r,n}$ and $v_{i,n}$, respectively. We have

$$x_n = x_{r,n} + jx_{i,n}, \tag{8.192a}$$

$$v_n = v_{r,n} + jv_{i,n}. \tag{8.192b}$$

From (8.191) and (8.192),

$$v_{r,n} = \alpha_1 c_n(\omega) + \alpha_2 s_n(\omega), \tag{8.193a}$$

$$v_{i,n} = \alpha_2 c_n(\omega) - \alpha_1 s_n(\omega). \tag{8.193b}$$

Let

$$\epsilon_n = x_n - v_n, \tag{8.194a}$$

$$= x_{r,n} - v_{r,n} + j[x_{i,n} - v_{i,n}]. \tag{8.194b}$$

Define

$$\boldsymbol{\epsilon} = \begin{bmatrix} \epsilon_K \\ \epsilon_{K-1} \\ \vdots \\ \epsilon_{-K} \end{bmatrix}, \quad \boldsymbol{x}_r = \begin{bmatrix} x_{r,K} \\ x_{r,K-1} \\ \vdots \\ x_{r,-K} \end{bmatrix}, \quad \boldsymbol{x}_i = \begin{bmatrix} x_{i,K} \\ x_{i,K-1} \\ \vdots \\ x_{i,-K} \end{bmatrix},$$

$$\boldsymbol{v}_r = \begin{bmatrix} v_{r,K} \\ v_{r,K-1} \\ \vdots \\ v_{r,-K} \end{bmatrix}, \quad \boldsymbol{u} = \begin{bmatrix} v_{i,K} \\ v_{i,K-1} \\ \vdots \\ v_{i,-K} \end{bmatrix}. \tag{8.195}$$

Using the definitions of (8.153), (8.192) and (8.193), we have

$$(\boldsymbol{x} - \boldsymbol{v})^{\mathrm{H}}(\boldsymbol{x} - \boldsymbol{v}) = \boldsymbol{\epsilon}^{\mathrm{H}}\boldsymbol{\epsilon}, \tag{8.196a}$$

$$= (\boldsymbol{x}_r - \boldsymbol{v}_r)^{\mathrm{T}}(\boldsymbol{x}_r - \boldsymbol{v}_r) + (\boldsymbol{x}_i - \boldsymbol{v}_i)^{\mathrm{T}}(\boldsymbol{x}_i - \boldsymbol{v}_i). \tag{8.196b}$$

From (8.196), the minimization of $(\boldsymbol{x} - \boldsymbol{v})^{\mathrm{H}}(\boldsymbol{x} - \boldsymbol{v})$ can be achieved by minimizing $(\boldsymbol{x}_r - \boldsymbol{v}_r)^{\mathrm{T}}(\boldsymbol{x}_r - \boldsymbol{v}_r) + (\boldsymbol{x}_i - \boldsymbol{v}_i)^{\mathrm{T}}(\boldsymbol{x}_i - \boldsymbol{v}_i)$. Define

$$\mathbf{A}_r = \begin{bmatrix} c_K(\omega) & , & s_K(\omega) \\ c_{K-1}(\omega) & , & s_{K-1}(\omega) \\ \vdots & , & \vdots \\ c_{-K}(\omega) & , & s_{-K}(\omega) \end{bmatrix}, \quad \mathbf{A}_i = \begin{bmatrix} -s_K(\omega) & , & c_K(\omega) \\ -s_{K-1}(\omega) & , & c_{K-1}(\omega) \\ \vdots & , & \vdots \\ -s_{-K}(\omega) & , & c_{-K}(\omega) \end{bmatrix},$$

$$\mathbf{A} = \begin{bmatrix} \mathbf{A}_r \\ \mathbf{A}_i \end{bmatrix}, \quad \boldsymbol{\alpha} = \begin{bmatrix} \alpha_1 \\ \alpha_2 \end{bmatrix}, \quad \boldsymbol{y} = \begin{bmatrix} \boldsymbol{x}_r \\ \boldsymbol{x}_i \end{bmatrix}. \tag{8.197}$$

We wish to minimize

$$J(\alpha_1, \alpha_2, \omega) = (\boldsymbol{x} - \boldsymbol{v})^{\mathrm{H}}(\boldsymbol{x} - \boldsymbol{v}) = (\boldsymbol{x}_r - \boldsymbol{v}_r)^{\mathrm{T}}(\boldsymbol{x}_r - \boldsymbol{v}_r) + (\boldsymbol{x}_i - \boldsymbol{v}_i)^{\mathrm{T}}(\boldsymbol{x}_i - \boldsymbol{v}_i),$$

$$= (\boldsymbol{y} - \mathbf{A}\boldsymbol{\alpha})^{\mathrm{T}}(\boldsymbol{y} - \mathbf{A}\boldsymbol{\alpha}). \tag{8.198}$$

The least squares optimum solution in the minimization of $J(\alpha_1, \alpha_2, \omega)$ is

$$\hat{\boldsymbol{\alpha}} = (\mathbf{A}^{\mathrm{T}}\mathbf{A})^{-1}\mathbf{A}^{\mathrm{T}}\boldsymbol{y}. \tag{8.199a}$$

Let

$$\hat{\boldsymbol{\alpha}} = [\hat{\alpha}_1, \hat{\alpha}_2]^{\mathrm{T}}. \tag{8.199b}$$

The vector y is observed data, and so we know y. From (8.197), $\mathbf{A}$ is a function of ω, and so we need to estimate $\mathbf{A}$ first. Substituting (8.199) into (8.198), we have

$$\begin{aligned} J(\hat{\alpha}_1, \hat{\alpha}_2, \omega) &= (y - \mathbf{A}\hat{\boldsymbol{\alpha}})^{\mathrm{T}}(y - \mathbf{A}\hat{\boldsymbol{\alpha}}), \\ &= (y - \mathbf{A}(\mathbf{A}^{\mathrm{T}}\mathbf{A})^{-1}\mathbf{A}^{\mathrm{T}}y)^{\mathrm{T}}(y - \mathbf{A}(\mathbf{A}^{\mathrm{T}}\mathbf{A})^{-1}\mathbf{A}^{\mathrm{T}}y), \\ &= y^{\mathrm{T}}(\mathbf{I} - \mathbf{A}(\mathbf{A}^{\mathrm{T}}\mathbf{A})^{-1}\mathbf{A}^{\mathrm{T}})^{\mathrm{T}}(\mathbf{I} - \mathbf{A}(\mathbf{A}^{\mathrm{T}}\mathbf{A})^{-1}\mathbf{A}^{\mathrm{T}})y. \end{aligned} \tag{8.200}$$

A matrix $\mathbf{P}$ is idempotent if $\mathbf{P}^2 = \mathbf{P}$. It can be shown that $(\mathbf{I} - \mathbf{A}(\mathbf{A}^{\mathrm{T}}\mathbf{A})^{-1}\mathbf{A}^{\mathrm{T}})$ is symmetric and idempotent. Hence,

$$\begin{aligned} J(\hat{\alpha}_1, \hat{\alpha}_2, \omega) &= y^{\mathrm{T}}(\mathbf{I} - \mathbf{A}(\mathbf{A}^{\mathrm{T}}\mathbf{A})^{-1}\mathbf{A}^{\mathrm{T}})y, \\ &= y^{\mathrm{T}}y - y^{\mathrm{T}}\mathbf{A}(\mathbf{A}^{\mathrm{T}}\mathbf{A})^{-1}\mathbf{A}^{\mathrm{T}}y. \end{aligned} \tag{8.201}$$

Let

$$J_0(\omega) = y^{\mathrm{T}}\mathbf{A}(\mathbf{A}^{\mathrm{T}}\mathbf{A})^{-1}\mathbf{A}^{\mathrm{T}}y. \tag{8.202}$$

Substituting (8.202) into (8.201), we have

$$J(\hat{\alpha}_1, \hat{\alpha}_2, \omega) = y^{\mathrm{T}}y - J_0(\omega). \tag{8.203}$$

From (8.203), minimizing $J(\hat{\alpha}_1, \hat{\alpha}_2, \omega)$ can be achieved by maximizing $J_0(\omega)$. One method for finding the value of ω that maximizes $J_0(\omega)$ is by evaluating (8.202) on a dense grid of ω to find the value of ω that maximizes $J_0(\omega)$. This approach is computationally intensive. We shall examine another approach. Let

$$\boldsymbol{c} = \begin{bmatrix} c_K(\omega) \\ c_{K-1}(\omega) \\ \vdots \\ c_{-K}(\omega) \end{bmatrix}, \quad \boldsymbol{s} = \begin{bmatrix} s_K(\omega) \\ s_{K-1}(\omega) \\ \vdots \\ s_{-K}(\omega) \end{bmatrix}. \tag{8.204}$$

Substituting (8.204) into (8.197), we have

$$\mathbf{A} = \begin{bmatrix} \boldsymbol{c} & , \boldsymbol{s} \\ -\boldsymbol{s} & , \boldsymbol{c} \end{bmatrix}. \tag{8.205}$$

From (8.205),

$$\begin{aligned} \mathbf{A}^{\mathrm{T}}\mathbf{A} = \begin{bmatrix} \boldsymbol{c}^{\mathrm{T}} & , -\boldsymbol{s}^{\mathrm{T}} \\ \boldsymbol{s}^{\mathrm{T}} & , \boldsymbol{c}^{\mathrm{T}} \end{bmatrix} \begin{bmatrix} \boldsymbol{c} & , \boldsymbol{s} \\ -\boldsymbol{s} & , \boldsymbol{c} \end{bmatrix} &= \begin{bmatrix} \boldsymbol{c}^{\mathrm{T}}\boldsymbol{c} + \boldsymbol{s}^{\mathrm{T}}\boldsymbol{s} & , \boldsymbol{c}^{\mathrm{T}}\boldsymbol{s} - \boldsymbol{s}^{\mathrm{T}}\boldsymbol{c} \\ \boldsymbol{s}^{\mathrm{T}}\boldsymbol{c} - \boldsymbol{c}^{\mathrm{T}}\boldsymbol{s} & , \boldsymbol{c}^{\mathrm{T}}\boldsymbol{c} + \boldsymbol{s}^{\mathrm{T}}\boldsymbol{s} \end{bmatrix}, \\ &= \begin{bmatrix} \boldsymbol{c}^{\mathrm{T}}\boldsymbol{c} + \boldsymbol{s}^{\mathrm{T}}\boldsymbol{s} & , \quad 0 \\ 0 & , \boldsymbol{c}^{\mathrm{T}}\boldsymbol{c} + \boldsymbol{s}^{\mathrm{T}}\boldsymbol{s} \end{bmatrix}. \end{aligned} \tag{8.206}$$

From (8.190) and (8.204),

$$\begin{aligned} c^{\mathrm{T}}c + s^{\mathrm{T}}s &= \sum_{n=-K}^{K} \{\cos^2(n\omega) + \sin^2(n\omega)\}, \\ &= 2K + 1. \end{aligned} \tag{8.207}$$

Substituting (8.207) into (8.206), we have

$$\mathbf{A}^{\mathrm{T}}\mathbf{A} = \begin{bmatrix} 2K+1\,, & 0 \\ 0 & , 2K+1 \end{bmatrix}. \tag{8.208}$$

Substituting (8.208) into (8.202), we have

$$\begin{aligned} J_O(\omega) &= \frac{1}{2K+1}[y^{\mathrm{T}}\mathbf{A}][\mathbf{A}^{\mathrm{T}}y], \\ &= \frac{1}{2K+1}\left[x_r^{\mathrm{T}}\,,\, x_i^{\mathrm{T}}\right]\begin{bmatrix} c & , s \\ -s & , c \end{bmatrix}\begin{bmatrix} c^{\mathrm{T}} & , -s^{\mathrm{T}} \\ s^{\mathrm{T}} & , c^{\mathrm{T}} \end{bmatrix}\begin{bmatrix} x_r \\ x_i \end{bmatrix}, \\ &= \frac{1}{2K+1}\left[\left(x_r^{\mathrm{T}}c\right)^2 + \left(x_i^{\mathrm{T}}s\right)^2 + \left(x_r^{\mathrm{T}}s\right)^2 + \left(x_i^{\mathrm{T}}c\right)^2\right], \\ &= \frac{1}{2K+1}\left[\left(\sum_{n=-K}^{K} x_{r,n}c_n\right)^2 + \left(\sum_{n=-K}^{K} x_{r,n}s_n\right)^2 \right. \\ &\quad \left. + \left(\sum_{n=-K}^{K} x_{i,n}c_n\right)^2 + \left(\sum_{n=-K}^{K} x_{i,n}s_n\right)^2\right]. \end{aligned} \tag{8.209}$$

In (8.209), we have summation of the form $(\sum x_{r,n}\cos(n\omega))^2 + (\sum x_{r,n}\sin(n\omega))^2$. It can be seen from Fig. 8.11 that $x_{r,n}\cos(n\omega)$ is the projection of $x_{r,n}e^{jn\omega}$ onto the horizontal axis and that $x_{r,n}\sin(n\omega)$ is the projection of $x_{r,n}e^{jn\omega}$ onto the vertical axis. Hence, $(\sum x_{r,n}\cos(n\omega))^2 = \alpha^2$ and that $(\sum x_{r,n}\sin(n\omega))^2 = \beta^2$. Therefore, $(\sum x_{r,n}\cos(n\omega))^2 + (\sum x_{r,n}\sin(n\omega))^2 = r^2 = |\sum x_{r,n}e^{jn\omega}|^2$. Similarly, $(\sum x_{i,n}\cos(n\omega))^2 + (\sum x_{i,n}\sin(n\omega))^2 = |\sum x_{i,n}e^{jn\omega}|^2$. Hence,

$$\left(\sum_{n=-K}^{K} x_{r,n}c_n\right)^2 + \left(\sum_{n=-K}^{K} x_{r,n}s_n\right)^2 = \left|\sum_{n=-K}^{K} x_{r,n}e^{jn\omega}\right|^2, \tag{8.210a}$$

$$\left(\sum_{n=-K}^{K} x_{i,n}c_n\right)^2 + \left(\sum_{n=-K}^{K} x_{i,n}s_n\right)^2 = \left|\sum_{n=-K}^{K} x_{i,n}e^{jn\omega}\right|^2. \tag{8.210b}$$

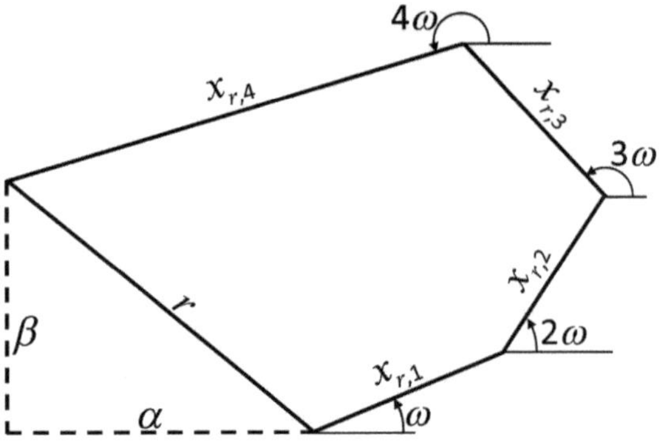

Fig. 8.11 $(\sum x_{r,n}\cos(n\omega))^2 + (\sum x_{r,n}\sin(n\omega))^2$

Substituting (8.210) into (8.209), an alternate form for (8.209) is

$$J_0(\omega) = \frac{1}{2K+1}\left|\sum_{n=-K}^{K} x_{r,n}e^{jn\omega}\right|^2 + \frac{1}{2K+1}\left|\sum_{n=-K}^{K} x_{i,n}e^{jn\omega}\right|^2. \tag{8.211}$$

The expression (8.211) may be evaluated on a dense grid of ω for $0 \le \omega \le \pi$. The value of ω that maximizes $J_0(\omega)$, denoted as $\hat{\omega}$, is the desired value of ω. Note that $J_0(\omega) = J_0(-\omega)$. If the search is in the range $-\pi \le \omega \le \pi$, there will be two peaks at $\pm\hat{\omega}$. The next step is to resolve between $\hat{\omega}$ and $-\hat{\omega}$.

From (8.148) and (8.150), the sinusoid is $ae^{jb}e^{jn\hat{\omega}}$; its amplitude is a. The contribution of a particular frequency is determined by its amplitude a. From (8.199) and (8.208), the estimated $\boldsymbol{\alpha}$, denoted by $\hat{\boldsymbol{\alpha}}$, is given by

$$[\hat{\alpha}_1, \hat{\alpha}_2]^{\mathrm{T}} = \hat{\boldsymbol{\alpha}} = \frac{1}{2K+1}\mathbf{A}^{\mathrm{T}}\mathbf{y}. \tag{8.212}$$

From (8.190a), (8.190b), and (8.212), the estimated a, denoted by $\hat{a}$, is given by

$$\hat{a} = \sqrt{\hat{\alpha}_1^2 + \hat{\alpha}_2^2}. \tag{8.213}$$

When the values $\hat{\omega}$ and $-\hat{\omega}$ are substituted into (8.212), respectively, one of them will return $\hat{a} \approx 0$ signifying negligible contribution; the other will return a finite $\hat{a}$ indicating effective contribution. The frequency that returns a finite $\hat{a}$ is the estimated frequency. We shall illustrate this by using an example.

Consider the case where $x(n)$ is complex with frequency $\frac{5\pi}{8}$ and 40 dB SNR. The simulation result of $J_0(\omega)$ for $0 \le \omega \le \pi$ is shown in Fig. 8.12. $J_0(\omega)$ peaks at $\omega = \pm 1.9635$ with $\omega = -1.9635$ produces $a \approx 0$ and $\omega = 1.9635$ produces a finite a. Hence, the estimated frequency is 1.9635.

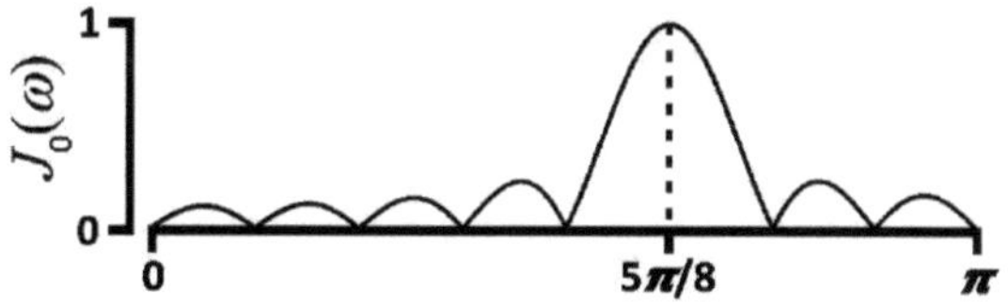

Fig. 8.12 Plot of $J_0(\omega)$ for $0 \leq \omega \leq \pi$

Appendix A: Complex Least Squares Optimization

Mathematical Background

Before embarking on deriving the least squares method, we shall revisit some important results on matrix differentiation. Let $\boldsymbol{a}$ and $\boldsymbol{x}$ be $(M \times 1)$ column vectors.

$$\boldsymbol{a} = [a_1, a_2, \ldots, a_M]^{\mathrm{T}}, \qquad \boldsymbol{x} = [x_1, x_2, \ldots, x_M]^{\mathrm{T}}. \tag{8.214}$$

Let scalar p be given by

$$p = \boldsymbol{a}^{\mathrm{T}}\boldsymbol{x} = \boldsymbol{x}^{\mathrm{T}}\boldsymbol{a} = a_1x_1 + a_2x_2 + \ldots + a_Mx_M. \tag{8.215}$$

The derivative $\frac{\mathrm{d}p}{\mathrm{d}\boldsymbol{x}}$ is given by

$$\frac{\mathrm{d}p}{\mathrm{d}\boldsymbol{x}} = \left[\frac{\mathrm{d}p}{\mathrm{d}x_1}, \frac{\mathrm{d}p}{\mathrm{d}x_2}, \ldots, \frac{\mathrm{d}p}{\mathrm{d}x_M}\right]^{\mathrm{T}} = [a_1, a_2, \ldots, a_M]^{\mathrm{T}} = \boldsymbol{a}. \tag{8.216}$$

From (8.215) and (8.216),

$$\frac{\mathrm{d}\boldsymbol{a}^{\mathrm{T}}\boldsymbol{x}}{\mathrm{d}\boldsymbol{x}} = \frac{\mathrm{d}\boldsymbol{x}^{\mathrm{T}}\boldsymbol{a}}{\mathrm{d}\boldsymbol{x}} = \boldsymbol{a}. \tag{8.217}$$

Let $\mathbf{A}$ be an $(M \times M)$ symmetric square matrix.

$$\mathbf{A} = \begin{bmatrix} a_{1,1} & a_{1,2} & \cdots & a_{1,M} \\ a_{2,1} & a_{2,2} & \cdots & a_{2,M} \\ \vdots & \vdots & \ddots & \vdots \\ a_{M,1} & a_{M,2} & \cdots & a_{M,M} \end{bmatrix}, \quad a_{m,n} = a_{n,m}. \tag{8.218}$$

Let scalar q be given by

$$q = \boldsymbol{x}^{\mathrm{T}}\mathbf{A}\boldsymbol{x} = \sum_{n=1}^{M}\sum_{m=1}^{M} a_{m,n}x_nx_m. \tag{8.219}$$

It can be shown that the derivative $\frac{d\boldsymbol{x}^{\mathrm{T}}\mathbf{A}\boldsymbol{x}}{d\boldsymbol{x}}$ is given by

$$\frac{d\boldsymbol{x}^{\mathrm{T}}\mathbf{A}\boldsymbol{x}}{d\boldsymbol{x}} = \frac{dq}{d\boldsymbol{x}} = \left[\frac{\partial q}{\partial x_1}, \frac{\partial q}{\partial x_2}, \ldots, \frac{\partial q}{\partial x_M}\right]^{\mathrm{T}} = 2\mathbf{A}\boldsymbol{x}. \tag{8.220}$$

The Least Squares Problem Let $\boldsymbol{x}_M(n)$ and $\boldsymbol{h}_M$ be $(M \times 1)$ complex column vectors. Suppose that we wish to find $\boldsymbol{h}_M$ such that $\boldsymbol{h}_M^{\mathrm{T}}\boldsymbol{x}_M(n) = \boldsymbol{x}_M^{\mathrm{T}}(n)\boldsymbol{h}_M$ is the optimum least squares approximation to a complex scalar $y(n)$ for $n = 1, 2, \ldots, K$ where $K > M$. Let the approximation error be $e(n)$ where $e(n)$ is another complex scalar. Hence, we have

$$e(n) = y(n) - \boldsymbol{x}_M^{\mathrm{T}}(n)\boldsymbol{h}_M = y(n) - \boldsymbol{h}_M^{\mathrm{T}}\boldsymbol{x}_M(n). \tag{8.221}$$

Let $e_R(n)$, $y_R(n)$, $\boldsymbol{x}_{R,M}(n)$, and $\boldsymbol{h}_{R,M}$ be the real parts of $e(n)$, $y(n)$, $\boldsymbol{x}_M(n)$, and $\boldsymbol{h}_M$, respectively, and let $e_I(n)$, $y_I(n)$, $\boldsymbol{x}_{I,M}(n)$, and $\boldsymbol{h}_{I,M}$ be the imaginary parts of $e(n)$, $y(n)$, $\boldsymbol{x}_M(n)$, and $\boldsymbol{h}_M$, respectively.

$$\begin{aligned} e(n) &= e_R(n) + je_I(n), \quad y(n) = y_R(n) + jy_I(n), \\ \boldsymbol{x}_M(n) &= \boldsymbol{x}_{R,M}(n) + j\boldsymbol{x}_{I,M}(n), \quad \boldsymbol{h}_M = \boldsymbol{h}_{R,M} + j\boldsymbol{h}_{I,M}. \end{aligned} \tag{8.222}$$

Note: $e_R(n)$, $e_I(n)$, $y_R(n)$, $y_I(n)$, $\boldsymbol{x}_{R,M}(n)$, $\boldsymbol{x}_{I,M}(n)$, $\boldsymbol{h}_{R,M}$, and $\boldsymbol{h}_{I,M}$ are real. Using the definitions of (8.222), (8.221) becomes

$$e_R(n) = y_R(n) - \left(\boldsymbol{x}_{R,M}^{\mathrm{T}}(n)\boldsymbol{h}_{R,M} - \boldsymbol{x}_{I,M}^{\mathrm{T}}(n)\boldsymbol{h}_{I,M}\right), \tag{8.223a}$$

$$e_I(n) = y_I(n) - \left(\boldsymbol{x}_{I,M}^{\mathrm{T}}(n)\boldsymbol{h}_{R,M} + \boldsymbol{x}_{R,M}^{\mathrm{T}}(n)\boldsymbol{h}_{I,M}\right). \tag{8.223b}$$

Let

$$\begin{aligned} \boldsymbol{e}_R &= [e_R(1), e_R(2), \ldots, e_R(K)]^{\mathrm{T}}, \quad \boldsymbol{e}_I = [e_I(1), e_I(2), \ldots, e_I(K)]^{\mathrm{T}}, \\ \boldsymbol{y}_R &= [y_R(1), y_R(2), \ldots, y_R(K)]^{\mathrm{T}}, \quad \boldsymbol{y}_I = [y_I(1), y_I(2), \ldots, y_I(K)]^{\mathrm{T}}, \\ \boldsymbol{h}_{R,M} &= \left[h_{R,M}(1), \ldots, h_{R,M}(M)\right]^{\mathrm{T}}, \quad \boldsymbol{h}_{I,M} = \left[h_{I,M}(1), \ldots, h_{I,M}(M)\right]^{\mathrm{T}}, \\ \mathbf{X}_{R,M} &= \left[\boldsymbol{x}_{R,M}(1), \ldots, \boldsymbol{x}_{R,M}(K)\right]^{\mathrm{T}}, \quad \mathbf{X}_{I,M} = \left[\boldsymbol{x}_{I,M}(1), \ldots, \boldsymbol{x}_{I,M}(K)\right]^{\mathrm{T}}. \end{aligned} \tag{8.224}$$

Note: $\boldsymbol{e}_R$, $\boldsymbol{e}_I$, $\boldsymbol{y}_R$, and $\boldsymbol{y}_I$ are $(K \times 1)$ real column vectors, $\boldsymbol{h}_{R,M}$, and $\boldsymbol{h}_{I,M}$ are $(M \times 1)$ real column vectors, and $\mathbf{X}_{R,M}$ and $\mathbf{X}_{I,M}$ are $(K \times M)$ real matrix. Expression (8.223) can now be written in matrix form as

$$\begin{bmatrix} \boldsymbol{e}_R \\ \boldsymbol{e}_I \end{bmatrix} = \begin{bmatrix} \boldsymbol{y}_R \\ \boldsymbol{y}_I \end{bmatrix} - \begin{bmatrix} \mathbf{X}_{R,M} & -\mathbf{X}_{I,M} \\ \mathbf{X}_{I,M} & \mathbf{X}_{R,M} \end{bmatrix} \begin{bmatrix} \boldsymbol{h}_{R,M} \\ \boldsymbol{h}_{I,M} \end{bmatrix}. \tag{8.225}$$

Define

$$e = [e_R^T \ , \ e_I^T]^T, \quad y = [y_R^T \ , \ y_I^T], \quad h = [h_{R,M}^T \ , \ h_{I,M}^T]^T,$$
$$\mathbf{X} = \begin{bmatrix} \mathbf{X}_{R,M} & , & -\mathbf{X}_{I,M} \\ \mathbf{X}_{I,M} & , & \mathbf{X}_{R,M} \end{bmatrix}. \tag{8.226}$$

Note: e and y are $(2K \times 1)$ real column vector, h is a $(2M \times 1)$ real column vector, and $\mathbf{X}$ is a $(2K \times 2M)$ real matrix. From (8.225) and (8.226), we have

$$e = y - \mathbf{X}h. \tag{8.227}$$

The vector inner product $e^T e$ is the sum of all the squares of the $2K$ elements of e, i.e., $e^T e = \sum_{n=1}^{K} \{e_R^2(n) + e_I^2(n)\}$.

$$e^T e = (y^T - h^T \mathbf{X}^T)(y - \mathbf{X}h) = y^T y - y^T \mathbf{X}h - h^T \mathbf{X}^T y + h^T \mathbf{X}^T \mathbf{X}h. \tag{8.228}$$

Differentiate $e^T e$ with respect to h.

$$\frac{\mathrm{d}e^T e}{\mathrm{d}h} = \frac{\mathrm{d}y^T y}{\mathrm{d}h} - \frac{\mathrm{d}y^T \mathbf{X}h}{\mathrm{d}h} - \frac{\mathrm{d}h^T \mathbf{X}^T y}{\mathrm{d}h} + \frac{\mathrm{d}h^T \mathbf{X}^T \mathbf{X}h}{\mathrm{d}h}. \tag{8.229}$$

Since y is constant and so is independent of h; hence, $\frac{\mathrm{d}y^T y}{\mathrm{d}h} = 0$. From (8.217), $\frac{\mathrm{d}y^T \mathbf{X}h}{\mathrm{d}h} = \frac{\mathrm{d}h^T \mathbf{X}^T y}{\mathrm{d}h} = \mathbf{X}^T y$. From (8.220), $\frac{\mathrm{d}h^T \mathbf{X}^T \mathbf{X}h}{\mathrm{d}h} = 2\mathbf{X}^T \mathbf{X}h$. Hence,

$$\frac{\mathrm{d}e^T e}{\mathrm{d}h} = 2\mathbf{X}^T \mathbf{X}h - 2\mathbf{X}^T y. \tag{8.230}$$

Let the optimum h for minimizing $e^T e$ be h_{opt}. Equate the derivative $\frac{\mathrm{d}e^T e}{\mathrm{d}h}$ to zero to determining h_{opt}. From (8.230), we have

$$\mathbf{X}^T \mathbf{X} h_{\mathrm{opt}} = \mathbf{X}^T y. \tag{8.231a}$$

$$\therefore \quad h_{\mathrm{opt}} = \left[\mathbf{X}^T \mathbf{X}\right]^{-1} \mathbf{X}^T y. \tag{8.231b}$$

If $x_M(n)$ and h_M are real, $\mathbf{X}_{I,M} = 0$ and $y_I = 0$. In this case, $\mathbf{X} = \mathbf{X}_{R,M}$ and $y = y_R$. The optimum solution is given by $h_{\mathrm{opt}} = \left[\mathbf{X}_{R,M}^T \mathbf{X}_{R,M}\right]^{-1} \mathbf{X}_{R,M}^T y_R$. If h_M is real, but $x_M(n)$ is complex, (8.227) becomes $e = y - \begin{bmatrix} \mathbf{X}_{R,M} & , & -\mathbf{X}_{I,M} \\ \mathbf{X}_{I,M} & , & \mathbf{X}_{R,M} \end{bmatrix}$

$\begin{bmatrix} \boldsymbol{h}_{R,M} \\ \mathbf{0} \end{bmatrix} = \boldsymbol{y} - \begin{bmatrix} \mathbf{X}_{R,M} \\ \mathbf{X}_{I,M} \end{bmatrix} \boldsymbol{h}_{R,M}$. In this case, $\mathbf{X} = \begin{bmatrix} \mathbf{X}_{R,M} \\ \mathbf{X}_{I,M} \end{bmatrix}$. For the general complex case, let

$$\mathbf{P} = \mathbf{X}_{R,M}^{\mathrm{T}} \mathbf{X}_{R,M} + \mathbf{X}_{I,M}^{\mathrm{T}} \mathbf{X}_{I,M}, \tag{8.232a}$$

$$\mathbf{Q} = \mathbf{X}_{R,M}^{\mathrm{T}} \mathbf{X}_{I,M} - \mathbf{X}_{I,M}^{\mathrm{T}} \mathbf{X}_{R,M}. \tag{8.232b}$$

From (8.226) and (8.232), we have

$$\mathbf{X}^{\mathrm{T}}\mathbf{X} = \begin{bmatrix} \mathbf{P} & -\mathbf{Q} \\ \mathbf{Q} & \mathbf{P} \end{bmatrix}. \tag{8.233}$$

From (8.226) and (8.233), (8.231) may be written as

$$\boldsymbol{h}_{\mathrm{opt}} = \begin{bmatrix} \mathbf{P} & -\mathbf{Q} \\ \mathbf{Q} & \mathbf{P} \end{bmatrix}^{-1} \begin{bmatrix} \mathbf{X}_{R,M}^{\mathrm{T}} & \mathbf{X}_{I,M}^{\mathrm{T}} \\ -\mathbf{X}_{I,M}^{\mathrm{T}} & \mathbf{X}_{R,M}^{\mathrm{T}} \end{bmatrix} \boldsymbol{y}. \tag{8.234}$$

The matrix $\begin{bmatrix} \mathbf{P} & -\mathbf{Q} \\ \mathbf{Q} & \mathbf{P} \end{bmatrix}^{-1}$ may be computed either directly or in a more efficient way using the following formula.

$$\begin{bmatrix} \mathbf{P} & -\mathbf{Q} \\ \mathbf{Q} & \mathbf{P} \end{bmatrix}^{-1} = \begin{bmatrix} \mathbf{R} & \mathbf{S} \\ -\mathbf{S} & \mathbf{R} \end{bmatrix}, \tag{8.235a}$$

$$\text{where } \mathbf{R} = [\mathbf{P} + \mathbf{Q}\mathbf{P}^{-1}\mathbf{Q}]^{-1} \text{ and } \mathbf{S} = \mathbf{P}^{-1}\mathbf{Q}\mathbf{R}. \tag{8.235b}$$

$\mathbf{P}$ and $\mathbf{R}$ are $(M \times M)$ symmetric real matrices. $\mathbf{Q}$ and $\mathbf{S}$ are $(M \times M)$ skew-symmetric or antisymmetric real matrices. It can also be shown that

$$\begin{bmatrix} \mathbf{P} & -\mathbf{Q} \\ \mathbf{Q} & \mathbf{P} \end{bmatrix}^{-1} \begin{bmatrix} \mathbf{X}_{R,M}^{\mathrm{T}} & \mathbf{X}_{I,M}^{\mathrm{T}} \\ -\mathbf{X}_{I,M}^{\mathrm{T}} & \mathbf{X}_{R,M}^{\mathrm{T}} \end{bmatrix} = \begin{bmatrix} \mathbf{T} & \mathbf{U} \\ -\mathbf{U} & \mathbf{T} \end{bmatrix}, \tag{8.236a}$$

$$\text{where } \mathbf{T} = \mathbf{R}\mathbf{X}_{R,M}^{\mathrm{T}} - \mathbf{S}\mathbf{X}_{I,M}^{\mathrm{T}} \text{ and } \mathbf{U} = \mathbf{R}\mathbf{X}_{I,M}^{\mathrm{T}} + \mathbf{S}\mathbf{X}_{R,M}^{\mathrm{T}}. \tag{8.236b}$$

Appendix B: Total Least Squares Optimization

Mathematical Background

We shall revisit an important mathematical tool called SVD (singular value decomposition). The SVD of an $(M \times N)$ matrix $\mathbf{X}$ is the product of three matrices.

$$\mathbf{X} = \mathbf{U}\boldsymbol{\Sigma}\mathbf{V}^{\mathrm{T}}. \tag{8.237}$$

where $\mathbf{U}$, $\boldsymbol{\Sigma}$, and $\mathbf{V}$ are $(M \times M)$, $(M \times N)$, and $(N \times N)$ matrices, respectively. $\mathbf{U}$, $\boldsymbol{\Sigma}$, and $\mathbf{V}$ are given by

$$\mathbf{U} = \begin{bmatrix} \boldsymbol{u}_1 \ \boldsymbol{u}_2 \ \ldots \ \boldsymbol{u}_M \end{bmatrix} \ , \ \boldsymbol{\Sigma} = \begin{bmatrix} \sigma_1 & 0 & \cdots & 0 \\ 0 & \sigma_2 & \cdots & 0 \\ \vdots & \vdots & \ddots & \vdots \\ 0 & 0 & \cdots & \sigma_N \\ & & \mathbf{0}_{(M-N)\times N} & \end{bmatrix} \ , \ \mathbf{V} = \begin{bmatrix} \boldsymbol{v}_1^{\mathrm{T}} \\ \boldsymbol{v}_2^{\mathrm{T}} \\ \vdots \\ \boldsymbol{v}_N^{\mathrm{T}} \end{bmatrix}, \tag{8.238}$$

where $\mathbf{0}_{(M-N)\times N}$ represents an $((M-N) \times N)$ null matrix. $\mathbf{U}$ and $\mathbf{V}$ are orthonormal matrices, i.e.,

$$\mathbf{U}^{\mathrm{T}}\mathbf{U} = \mathbf{I}, \tag{8.239}$$

and

$$\mathbf{V}^{\mathrm{T}}\mathbf{V} = \mathbf{I}. \tag{8.240}$$

$\boldsymbol{\Sigma}$ is diagonal with its elements arranged such that

$$\sigma_1 \geq \sigma_2 \geq \ldots \geq \sigma_N \geq 0. \tag{8.241}$$

Since the last $M - N$ rows of $\boldsymbol{\Sigma}$ are zeroes, (8.237) may also be written as

$$\mathbf{X} = \begin{bmatrix} \boldsymbol{u}_1 \ \boldsymbol{u}_2 \ \ldots \ \boldsymbol{u}_N \end{bmatrix}_{M\times N} \begin{bmatrix} \sigma_1 & 0 & \cdots & 0 \\ 0 & \sigma_2 & \cdots & 0 \\ \vdots & \vdots & \ddots & \vdots \\ 0 & 0 & \cdots & \sigma_N \end{bmatrix}_{N\times N} \begin{bmatrix} \boldsymbol{v}_1^{\mathrm{T}} \\ \boldsymbol{v}_2^{\mathrm{T}} \\ \vdots \\ \boldsymbol{v}_N^{\mathrm{T}} \end{bmatrix}_{N\times N} . \tag{8.242}$$

The SVD of a matrix can also be used to find the least squares solution as illustrated in (8.243).

$$\begin{aligned} &\boldsymbol{y} = \mathbf{X}\boldsymbol{h} = \mathbf{U}\boldsymbol{\Sigma}\mathbf{V}^{\mathrm{T}}\boldsymbol{h} = [\mathbf{V}\boldsymbol{\Sigma}^{-1}\mathbf{U}^{\mathrm{T}}]^{-1}[\mathbf{V}\boldsymbol{\Sigma}^{-1}\mathbf{U}^{\mathrm{T}}]\mathbf{U}\boldsymbol{\Sigma}\mathbf{V}^{\mathrm{T}}\boldsymbol{h} = [\mathbf{V}\boldsymbol{\Sigma}^{-1}\mathbf{U}^{\mathrm{T}}]^{-1}\boldsymbol{h}. \\ &\therefore \quad \boldsymbol{h} = \mathbf{V}\boldsymbol{\Sigma}^{-1}\mathbf{U}^{\mathrm{T}}\boldsymbol{y}. \end{aligned} \tag{8.243}$$

The Total Least Squares Problem
In least squares, we minimize $\boldsymbol{e}^{\mathrm{T}}\boldsymbol{e}$ where $\boldsymbol{e} = \boldsymbol{y} - \mathbf{X}\boldsymbol{h}$. (See (8.227).) The least squares solutions in (8.231) and (8.234) are based on the assumption that $\boldsymbol{y}$ contains error (e.g., measurement noise), but there is no error in $\mathbf{X}$. The solution is the same if we assume that there is error in $\mathbf{X}$ but no error in $\boldsymbol{y}$. In the following discussions, we shall derive the optimum $\boldsymbol{h}$ assuming both $\boldsymbol{y}$ and $\mathbf{X}$ contain errors. Let the errors

in $\mathbf{X}$ and $\boldsymbol{y}$ be $\tilde{\mathbf{X}}$ and $\tilde{\boldsymbol{y}}$, respectively. Hence, we have

$$\boldsymbol{y} + \tilde{\boldsymbol{y}} = [\mathbf{X} + \tilde{\mathbf{X}}]\boldsymbol{h}, \tag{8.244a}$$

$$\Longrightarrow [\mathbf{X} + \tilde{\mathbf{X}} \quad \boldsymbol{y} + \tilde{\boldsymbol{y}}]\begin{bmatrix} \boldsymbol{h} \\ -1 \end{bmatrix} = \mathbf{0}. \tag{8.244b}$$

The vectors $\boldsymbol{y}$ and $\tilde{\boldsymbol{y}}$ are $(M \times 1)$ vectors. The vector $\boldsymbol{h}$ is an $(N \times 1)$ vector. Matrices $\mathbf{X}$ and $\tilde{\mathbf{X}}$ are $(M \times N)$ matrices. Assume that the vector $\boldsymbol{y}$ does not lie entirely in the subspace spanned by $\mathbf{X}$. The matrix $[\mathbf{X} \quad \boldsymbol{y}]$ has $N + 1$ columns and has rank $N + 1$. For the solution $\boldsymbol{h}$ to be unique, $[\mathbf{X} + \tilde{\mathbf{X}} \quad \boldsymbol{y} + \tilde{\boldsymbol{y}}]$ must have exactly N linearly independent columns; i.e., the rank of $[\mathbf{X}+\tilde{\mathbf{X}} \ , \ \boldsymbol{y}+\tilde{\boldsymbol{y}}]$ must be N. However, $[\mathbf{X}+\tilde{\mathbf{X}} \quad \boldsymbol{y}+\tilde{\boldsymbol{y}}]$ has $N+1$ columns. Our aim is to find the "smallest" matrix $[\tilde{\mathbf{X}} \quad \tilde{\boldsymbol{y}}]$ such that, when added to $[\mathbf{X} \quad \boldsymbol{y}]$, reduces the rank of $[\mathbf{X} \quad \boldsymbol{y}]$ from $N + 1$ to that of $[\mathbf{X} + \tilde{\mathbf{X}} \quad \boldsymbol{y} + \tilde{\boldsymbol{y}}]$ with rank N. SVD of $[\mathbf{X} \quad \boldsymbol{y}]$ is

$$\left[\mathbf{X} \ \boldsymbol{y}\right] = \left[\mathbf{U}_N \ \boldsymbol{u}_{N+1}\right]\begin{bmatrix} \boldsymbol{\Sigma}_N & \mathbf{0} \\ \mathbf{0} & \sigma_{N+1} \end{bmatrix}\begin{bmatrix} V_N & \boldsymbol{v}_{N+1} \\ \hat{\boldsymbol{v}}_{N+1} & v_{N+1} \end{bmatrix}^{\mathrm{T}}. \tag{8.245}$$

Matrix $\mathbf{U}_N$ is $(M \times N)$; vector $\boldsymbol{u}_{N+1}$ is $(M \times 1)$; matrix $\boldsymbol{\Sigma}_N$ is $(N \times N)$; scalar σ_{N+1} is (1×1); matrix $\mathbf{V}_N$ is $(N \times N)$; vector $\boldsymbol{v}_{N+1}$ is $(N \times 1)$; and vector $\hat{\boldsymbol{v}}_{N+1}$ is $(1 \times N)$. Post multiply (8.245) by $\begin{bmatrix} V_N & \boldsymbol{v}_{N+1} \\ \hat{\boldsymbol{v}}_{N+1} & v_{N+1} \end{bmatrix}$ and taking note that $\begin{bmatrix} V_N & \boldsymbol{v}_{N+1} \\ \hat{\boldsymbol{v}}_{N+1} & v_{N+1} \end{bmatrix}^{\mathrm{T}}$ $\begin{bmatrix} V_N & \boldsymbol{v}_{N+1} \\ \hat{\boldsymbol{v}}_{N+1} & v_{N+1} \end{bmatrix} = \mathbf{I}$, where $\mathbf{I}$ is an identity matrix, we have

$$\left[\mathbf{X} \ \boldsymbol{y}\right]\begin{bmatrix} V_N & \boldsymbol{v}_{N+1} \\ \hat{\boldsymbol{v}}_{N+1} & v_{N+1} \end{bmatrix} = \left[\mathbf{U}_N \ \boldsymbol{u}_{N+1}\right]\begin{bmatrix} \boldsymbol{\Sigma}_N & \mathbf{0} \\ \mathbf{0} & \sigma_{N+1} \end{bmatrix}. \tag{8.246}$$

Equating the last column of (8.246), we have

$$\left[\mathbf{X} \ \boldsymbol{y}\right]\begin{bmatrix} \boldsymbol{v}_{N+1} \\ v_{N+1} \end{bmatrix} = \boldsymbol{u}_{N+1}\sigma_{N+1}. \tag{8.247}$$

Let $[\mathbf{X}+\tilde{\mathbf{X}} \quad \boldsymbol{y}+\tilde{\boldsymbol{y}}]$ be the closest rank N approximation to $[\mathbf{X} \quad \boldsymbol{y}]$ in the Frobenius norm; i.e., $[\mathbf{X}+\tilde{\mathbf{X}} \quad \boldsymbol{y}+\tilde{\boldsymbol{y}}]$ has the same singular vectors as $[\mathbf{X} \quad \boldsymbol{y}]$, but only the N positive singular values contained in $\boldsymbol{\Sigma}_N$.

$$\left[\mathbf{X} + \tilde{\mathbf{X}} \ \boldsymbol{y} + \tilde{\boldsymbol{y}}\right] = \left[\mathbf{U}_N \ \boldsymbol{u}_{N+1}\right]\begin{bmatrix} \boldsymbol{\Sigma}_N & \mathbf{0} \\ \mathbf{0} & 0 \end{bmatrix}\begin{bmatrix} V_N & \boldsymbol{v}_{N+1} \\ \hat{\boldsymbol{v}}_{N+1} & v_{N+1} \end{bmatrix}^{\mathrm{T}}. \tag{8.248}$$

$\left[\tilde{\mathbf{X}}\ \tilde{y}\right]$ can be written as

$$\left[\tilde{\mathbf{X}}\ \tilde{y}\right] = \left[\mathbf{X}+\tilde{\mathbf{X}}\ y+\tilde{y}\right] - \left[\mathbf{X}\ y\right]. \tag{8.249}$$

Substitute (8.245) and (8.248) into (8.249).

$$\left[\tilde{\mathbf{X}}\ \tilde{y}\right] = \left[\mathbf{U}_N\ \boldsymbol{u}_{N+1}\right]\begin{bmatrix}\boldsymbol{\Sigma}_N - \boldsymbol{\Sigma}_N & \mathbf{0}\\ \mathbf{0} & 0-\sigma_{N+1}\end{bmatrix}\begin{bmatrix}V_N & \boldsymbol{v}_{N+1}\\ \hat{\boldsymbol{v}}_{N+1} & v_{N+1}\end{bmatrix}^{\mathrm{T}}, \tag{8.250a}$$

$$= \left[\mathbf{0}\ -\boldsymbol{u}_{N+1}\sigma_{N+1}\right]\begin{bmatrix}V_N & \boldsymbol{v}_{N+1}\\ \hat{\boldsymbol{v}}_{N+1} & v_{N+1}\end{bmatrix}^{\mathrm{T}}, \tag{8.250b}$$

$$= -\boldsymbol{u}_{N+1}\sigma_{N+1}\left[\boldsymbol{v}_{N+1}\ v_{N+1}\right]. \tag{8.250c}$$

Substitute for $\boldsymbol{u}_{N+1}\sigma_{N+1}$ in (8.250c) using (8.247).

$$\left[\tilde{\mathbf{X}}\ \tilde{y}\right] = -\left[\mathbf{X}\ y\right]\begin{bmatrix}\boldsymbol{v}_{N+1}\\ v_{N+1}\end{bmatrix}\begin{bmatrix}\boldsymbol{v}_{N+1}\\ v_{N+1}\end{bmatrix}^{\mathrm{T}}. \tag{8.251}$$

Substitute (8.251) into (8.249).

$$\left[\mathbf{X}+\tilde{\mathbf{X}}\ y+\tilde{y}\right] = \left[\mathbf{X}\ y\right] - \left[\mathbf{X}\ y\right]\begin{bmatrix}\boldsymbol{v}_{N+1}\\ v_{N+1}\end{bmatrix}\begin{bmatrix}\boldsymbol{v}_{N+1}\\ v_{N+1}\end{bmatrix}^{\mathrm{T}}. \tag{8.252}$$

Post multiply (8.252) by $[\boldsymbol{v}_{N+1}\ \ v_{N+1}]^{\mathrm{T}}$.

$$\left[\mathbf{X}+\tilde{\mathbf{X}}\ y+\tilde{y}\right]\begin{bmatrix}\boldsymbol{v}_{N+1}\\ v_{N+1}\end{bmatrix} = \left[\mathbf{X}\ y\right]\begin{bmatrix}\boldsymbol{v}_{N+1}\\ v_{N+1}\end{bmatrix} - \left[\mathbf{X}\ y\right]\begin{bmatrix}\boldsymbol{v}_{N+1}\\ v_{N+1}\end{bmatrix}\begin{bmatrix}\boldsymbol{v}_{N+1}\\ v_{N+1}\end{bmatrix}^{\mathrm{T}}\begin{bmatrix}\boldsymbol{v}_{N+1}\\ v_{N+1}\end{bmatrix}. \tag{8.253}$$

Note that $\mathbf{V}\mathbf{V}^{\mathrm{T}} = \mathbf{V}^{\mathrm{T}}\mathbf{V} = \mathbf{I}$. Hence,

$$\begin{bmatrix}\boldsymbol{v}_{N+1}\\ v_{N+1}\end{bmatrix}^{\mathrm{T}}\begin{bmatrix}\boldsymbol{v}_{N+1}\\ v_{N+1}\end{bmatrix} = 1. \tag{8.254}$$

Substituting (8.254) into (8.253), we have

$$\left[\mathbf{X}+\tilde{\mathbf{X}}\ y+\tilde{y}\right]\begin{bmatrix}\boldsymbol{v}_{N+1}\\ v_{N+1}\end{bmatrix} = \left[\mathbf{X}\ y\right]\begin{bmatrix}\boldsymbol{v}_{N+1}\\ v_{N+1}\end{bmatrix} - \left[\mathbf{X}\ y\right]\begin{bmatrix}\boldsymbol{v}_{N+1}\\ v_{N+1}\end{bmatrix} = \mathbf{0}. \tag{8.255}$$

Dividing (8.255) by v_{N+1}, we have

$$\left[\mathbf{X}+\tilde{\mathbf{X}}\ \mathbf{y}+\tilde{\mathbf{y}}\right]\begin{bmatrix}-\frac{\boldsymbol{v}_{N+1}}{v_{N+1}}\\-1\end{bmatrix}=\mathbf{0}. \tag{8.256}$$

From (8.244b) and (8.256), we have

$$\begin{bmatrix}\boldsymbol{h}\\-1\end{bmatrix}=\begin{bmatrix}-\frac{\boldsymbol{v}_{N+1}}{v_{N+1}}\\-1\end{bmatrix}. \tag{8.257}$$

Denote the total least squares optimum solution as $\boldsymbol{h}_{\mathrm{TLS}}$. From (8.257),

$$\boldsymbol{h}_{\mathrm{TLS}}=-\frac{\boldsymbol{v}_{N+1}}{v_{N+1}}. \tag{8.258}$$

Appendix C: The PDF of Complex Random Variable

A complex random variable z has both real and imaginary parts. Let its real and imaginary parts be x and y, respectively.

$$z=x+jy, \tag{8.259}$$

where x and y are real random variables. The PDF of z depends on the PDFs of x and y. Thus, there are many different expressions for the PDF of "Gaussian distributed" z depending on the nature of x and y. If x and y are independent normal (also called Gaussian) distributed random variables, with mean μ_x and μ_y, respectively, and variances σ_x^2 and σ_y^2, respectively, the PDF of z is the product of the pdfs of x and y. In this case,

$$p(z)=\frac{1}{2\pi\sigma_x\sigma_y}e^{-(x-\mu_x)^2/(2\sigma_x^2)}e^{-(y-\mu_y)^2/(2\sigma_y^2)}. \tag{8.260}$$

For $\mu_x=\mu_y=\mu$ and $2\sigma_x^2=2\sigma_y^2=\sigma^2$,

$$p(z)=\frac{1}{\pi\sigma^2}e^{-(x-\mu)^2/\sigma^2}e^{-(y-\mu)^2/\sigma^2}. \tag{8.261}$$

For $\mu_x=\mu_y=0$ and $2\sigma_x^2=2\sigma_y^2=\sigma^2$,

$$p(z)=\frac{1}{\pi\sigma^2}e^{-(x^2+y^2)/\sigma^2}=\frac{1}{\pi\sigma^2}e^{-|z|^2/\sigma^2}=\frac{1}{\pi\sigma^2}e^{-z^*z/\sigma^2}. \tag{8.262}$$

where z^* is the complex conjugate of z. Let

$$z = [z_1, z_2, \ldots, z_N]^{\mathrm{T}}. \tag{8.263}$$

If $z_1, z_2, \ldots, z_N$ are uncorrelated noise, $\mathrm{E}\{z_i z_i^*\} = \sigma_i^2$ and $\mathrm{E}\{z_i z_j^*\} = 0$ for $i \neq j$; the vector form for (8.262) is

$$p(z) = \frac{1}{\pi^N \prod_{n=1}^{N} \sigma_n^2} e^{-z^{\mathrm{H}}\mathbf{R}^{-1}z}, \tag{8.264a}$$

$$\mathbb{E}\{z\} = \mathbf{0}, \tag{8.264b}$$

$$\mathrm{var}\{z\} = [\sigma_1^2, \sigma_2^2, \ldots, \sigma_N^2]^{\mathrm{T}}, \tag{8.264c}$$

$$\mathbf{R} = \mathbb{E}\{zz^{\mathrm{H}}\}, \tag{8.264d}$$

where z^{H} is the Hermitian conjugate transpose of z. In (8.264b), the null vector $\mathbf{0}$ is an $N \times 1$ column vector, and in (8.264d), $\mathbf{R}$ is an $N \times N$ matrix. In (8.264a), for the purpose of determining CRLB (Cramér-Rao lower bound) and the frequency of the sinusoid, the $\frac{1}{\pi^N \prod_{n=1}^{N} \sigma_n^2}$ factor is unimportant. In order to simplify notation, it is usually replaced by a constant. The expression for (8.264a) may be written as

$$p(z) = constant \times e^{-z^{\mathrm{H}}\mathbf{R}^{-1}z}. \tag{8.265}$$

References

T. Abatzoglou, A fast maximum likelihood algorithm for frequency estimation of a sinusoid based on Newton's method. IEEE Trans. Acoust. Speech Sig. Process. **33**(1), 77–89 (1985)

T.J. Abatzoglou, Fast maximum likelihood joint estimation of frequency and frequency rate. IEEE Trans. Aero. Electron. Syst. **22**(6), 708–715 (1986)

J.G. Ables, Maximum entropy spectral analysis. Astro. Astrophys. Suppl. **15**, 383 (1974)

N. Andersen, On the calculation of filter coefficients for maximum entropy spectral analysis. Geophysics **39**(1), 69–72 (1974)

G.N. Boshnakov, S. Lambert-Lacroix, A periodic Levinson–Durbin algorithm for entropy maximization. Comp. Stat. Data Anal. **56**(1), 15–24 (2012)

P.J. Brockwell, R. Dahlhaus, Generalized Levinson–Durbin and Burg algorithms. J. Econom. **118**(1–2), 129–149 (2004)

J.P. Burg, The relationship between maximum entropy spectra and maximum likelihood spectra. Geophysics **37**(2), 375–376 (1972)

C.L. Byrne, R.M. Fitzgerald, Spectral estimators that extend the maximum entropy and maximum likelihood methods. SIAM J. Appl. Math. **44**(2), 425–442 (1984)

N.I. Cho, S.U. Lee, On the adaptive lattice notch filter for the detection of sinusoids. IEEE Trans. Circuit Syst. II: Ana. Digit. Sig. Process. **40**(7), 405–416 (1993)

G. Cybenko, The numerical stability of the Levinson-Durbin algorithm for Toeplitz systems of equations. SIAM J. Sci. Stat. Comp. **1**(3), 303–310 (1980)

C. Foias, A.E. Frazho, P.J. Sherman, A geometric approach to the maximum likelihood spectral estimator for sinusoids in noise. IEEE Trans. Inform. Theo. **34**(5), 1066–1970 (1988)

A. Gray, J. Markel, Digital lattice and ladder filter synthesis. IEEE Trans. Audio Electro. **21**(6), 491–500 (1973)

H. Gzyl, *The Method of Maximum Entropy* (World Scientific, 1995)

W. Jones, A. Steinhardt, Finding the poles of the lattice filter. IEEE Trans. Acoust. Speech Sig. Process. **33**(5), 1328–1331 (1985)

S.M. Kay, V. Nagesha, Maximum likelihood estimation of signals in autoregressive noise. IEEE Trans. Sig. Process. **42**(1), 88–101 (1994)

L.M. Lee, H.C. Wang, An extended Levinson-Durbin algorithm for the analysis of noisy autoregressive process. IEEE Sig. Process. Lett. **3**(1), 13–15 (1996)

H. Leib, M. Eizenman, S. Pasupathy, J. Krolik, Adaptive lattice filter for multiple sinusoids in white noise. IEEE Trans. Acoust. Speech Sig. Process. **35**(7), 1015–1023 (1987)

H. Lev-Ari, T. Kailath, Lattice filter parameterization and modeling of nonstationary processes. IEEE Trans. Inform. Theo. **30**(1), 2–16 (1984)

Y.C. Lim, On the synthesis of iir digital filters derived from single channel AR lattice network. IEEE Trans. Acoust. Speech Sig. Process. **32**(4), 741–749 (1984)

Y.C. Lim, S.R. Parker, On the synthesis of lattice parameter digital filters. IEEE Trans. Circuit Syst. **31**(7), 593–601 (1984)

Y.C. Lim, S.R. Parker, On the identification of systems from data measurements using ARMA lattice models. IEEE Trans. Acoust. Speech Sig. Process. **34**(4), 824–828 (1986)

Y.C. Lim, Y.X. Zou, N. Zheng, A piloted adaptive notch filter. IEEE Trans. Sig. Process. **53**(4), 1310–1323 (2005)

M. Mojiri, A.R. Bakhshai, An adaptive notch filter for frequency estimation of a periodic signal. IEEE Trans. Auto. Contr. **49**(2), 314–318 (2004)

M. Mojiri, A.R. Bakhshai, Estimation of n frequencies using adaptive notch filter. IEEE Trans. Circuit Syst. II: Exp. Brief. **54**(4), 338–342 (2007)

I.J. Myung, Tutorial on maximum likelihood estimation. J. Math. Psychol. **47**(1), 90–100 (2003)

W. Newman, Extension to the maximum entropy method. IEEE Trans. Inform. Theo. **23**(1)), 89–93 (1977)

C.N. Papaodysseus, E.B. Koukoutsis, C.N. Triantafyllou, Error sources and error propagation in the Levinson-Durbin algorithm. IEEE Trans. Sig. Process. **41**(4), 1635–1651 (1993)

R. Punchalard, On adaptive IIR lattice notch filter using a robust variable step-size for the detection of sinusoid, in *Proc 2002 Int Conf. Comm. Syst.*, vol. 2, pp. 800–804 (Nov 2002)

V. Reddy, B. Egardt, T. Kailath, Optimized lattice-form adaptive line enhancer for a sinusoidal signal in broad-band noise. IEEE Trans. Circuit Syst. **28**(6), 542–550 (1981)

P.A. Regalia, An improved lattice-based adaptive IIR notch filter. IEEE Trans. Sig. Process. **39**(9), 2124–2128 (1991)

P. Shaman, Generalized Levinson–Durbin sequences, binomial coefficients and autoregressive estimation. J. Multivariate Anal. **101**(5), 1263–1273 (2010)

P. Stoica, R.L. Moses, B. Friedlander, T. Soderstrom, Maximum likelihood estimation of the parameters of multiple sinusoids from noisy measurements. IEEE Trans. Acoust. Speech Sig. Process. **37**(3), 378–392 (1989)

D.W. Tufts, R. Kumaresan, Estimation of frequencies of multiple sinusoids: Making linear prediction perform like maximum likelihood. Proc. IEEE **70**(9), 975–989 (1982)

M.J. Turmon, M.I. Miller, Maximum-likelihood estimation of complex sinusoids and Toeplitz covariances. IEEE Trans. Sig. Process. **42**(5), 1074–1086 (1994)

T.J. Ulrych, T.N. Bishop, Maximum entropy spectral analysis and autoregressive decomposition. Rev. Geophys. **13**(1), 183–200 (1975)

S. Umesh, D.W. Tufts, Estimation of parameters of exponentially damped sinusoids using fast maximum likelihood estimation with application to NMR spectroscopy data. IEEE Trans. Sig. Process. **44**(9), 2245–2259 (1996)

N. Wu, *The Maximum Entropy Method* (Springer, 2012)

R. Yu, X. Lin, C. Ko, A multi-stage Levinson-Durbin algorithm, in *36th Asilomar Conf. Sig. Syst. Comp.*, vol. 1, pp. 218–221 (Nov 2002)

R. Zhu, F. Yang, J. Yang, An RLS-based lattice-form complex adaptive notch filter. IEEE Sig. Process. Lett. **23**(2), 217–221 (2015)

Chapter 9
Array Beamforming

9.1 Introduction

In this chapter, array and beamformer are used interchangeably. When two or more signals are added, they may be added constructively if they are in phase or destructively if they are out of phase. The term array in this chapter refers to a transducer array. A transducer array is a set of transducers where each transducer element is connected to a phase shifter and a gain scaler in such a way that when the signal impinging on (or radiating from) the array is added together, the signals from (or radiating to) a particular direction is added constructively but those in other directions are added destructively to achieve spatial selectivity. The process of configuring the phase shifters and gain scalers of the sensors to form a spatially selective array is called beamforming. If the transducers are transmitters, it is called a transmitting array or radiating array. If the transducers are receivers, it is called a receiver array (Van Veen and Buckley 1988; Chen et al. 2002; Mucci 1984; Steyskal 1988; Mudumbai et al. 2009; Xu et al. 2017; Kutty and Sen 2015; Litva and Lo 1996; Dolph 1946; Kaiser 1974; Saramäki 1989; Deczky 2001; Bergen and Antoniou 2004, 2005; Taber et al. 1985; Lim et al. 2021a,b, 2022a,b).

A point radiator or point receiver that has no directional preference is called an isotropic radiator or receiver. It radiates or receives uniformly to or from all directions. An omnidirectional radiator or receiver is one that radiates or receives without azimuth angle preference but varies with elevation angle. Please see Sect. 6.3 for definitions of azimuth angle and elevation angle.

Y. C. Lim et al., *Signal Processing for Autonomous Systems*,
https://doi.org/10.1007/978-3-032-00675-2_9

9.2 Beamforming

9.2.1 Transmission-End Beamforming

Consider a signal $x_1(t) = Ae^{j\omega t}$ at position x_1 traveling losslessly in the direction from x_1 to x_2 as shown in Fig. 9.1. Suppose that it takes Δ unit time for the signal to travel from x_1 to x_2. The signal at x_2 is given by

$$x_2(t) = Ae^{j\omega(t-\Delta)} = e^{-j\omega\Delta}x_1(t). \tag{9.1}$$

Suppose that the distance between x_1 and x_2 is d and the wave velocity is c. Hence,

$$\Delta = d/c. \tag{9.2}$$

From (9.1) and (9.2),

$$x_2(t) = e^{-j\omega d/c}x_1(t). \tag{9.3}$$

From the law of physics, the relationship between wavelength (λ), frequency (ω), and velocity (c) is given by

$$c = \frac{\lambda\omega}{2\pi}. \tag{9.4}$$

Thus,

$$x_2(t) = e^{-j\omega d/c}x_1(t) = e^{-j2\pi d/\lambda}x_1(t). \tag{9.5}$$

Consider the scenario with two radiating elements located at x_a and x_b as shown in Fig. 9.2. For a point C very far away from x_a and x_b, the angles θ_a and θ_b will be approximately equal. For practical purposes, we shall assume that $\theta_a = \theta_b = \theta$. In this case,

$$r_a - r_b = d\sin(\theta). \tag{9.6}$$

Consider a plane wave radiating from an array consisting of N evenly spaced identical isotropic elements at an angle of θ as shown in Fig. 9.3. Suppose that the separation between two neighboring elements is d. Let the signal radiating from x_n

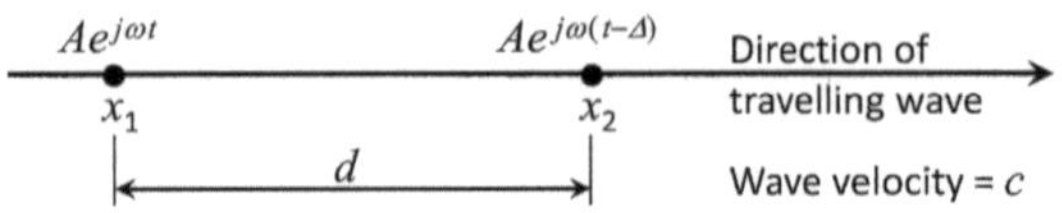

Fig. 9.1 Signal traveling from x_1 to x_2 losslessly

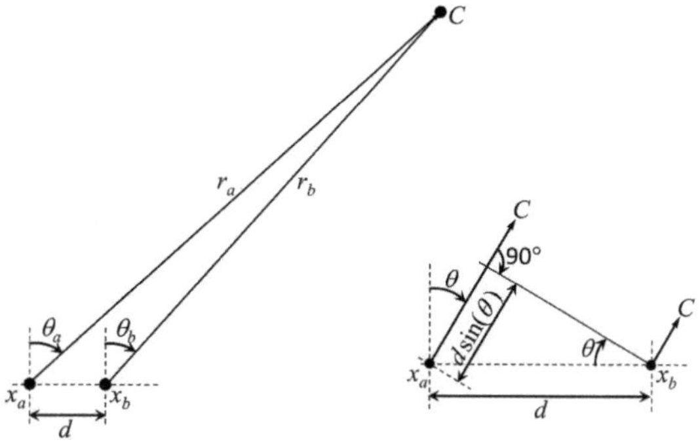

Fig. 9.2 Signals radiating from x_a and x_b

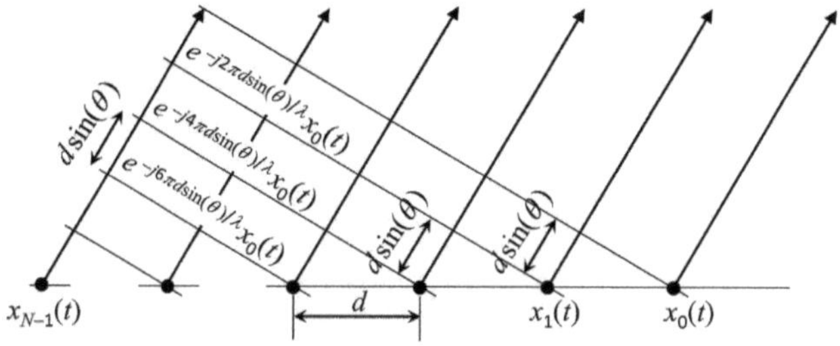

Fig. 9.3 Signals radiating from an N-element array

be $x_n(t)$, $n = 0, 1, 2, \ldots, N-1$, and the effect of $x_0(t)$ received at a point C at the far field is $\rho x_0(t)$ where ρ represents the signal attenuation from x_0 to C. Let $x_c(t)$ denote the sum of the effect of the signals $x_n(t)$, $n = 0, 1, \ldots, N-1$, at C. Using the results of (9.5) and (9.6), it can be shown that

$$x_c(t) = \rho\{x_0(t) + e^{-j2\pi d\sin(\theta)/\lambda}x_1(t) + \ldots + e^{-j2(N-1)\pi d\sin(\theta)/\lambda}x_{(N-1)}(t)\}. \tag{9.7}$$

Suppose that $x_n(t) = h_n x(t)$ as shown in Fig. 9.4, (9.7) may be written as

$$x_c(t) = \rho x(t)\sum_{n=0}^{N-1}\{e^{-j2n\pi d\sin(\theta)/\lambda}h_n\}. \tag{9.8}$$

The factor $\sum_{n=0}^{N-1}\{e^{-j2n\pi d\sin(\theta)/\lambda}h_n\}$ is called the array factor (AF). Let $H_\theta(\theta)$ denote the array factor as a function of θ:

$$H_\theta(\theta) = \sum_{n=0}^{N-1}\{e^{-j2n\pi d\sin(\theta)/\lambda}h_n\} \tag{9.9}$$

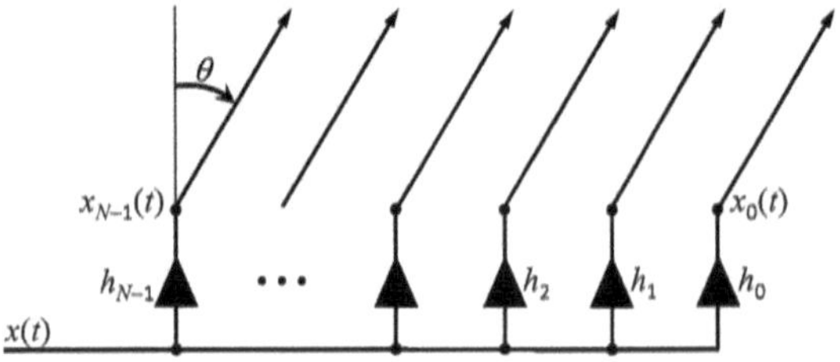

Fig. 9.4 $x_n(t) = h_n x(t)$

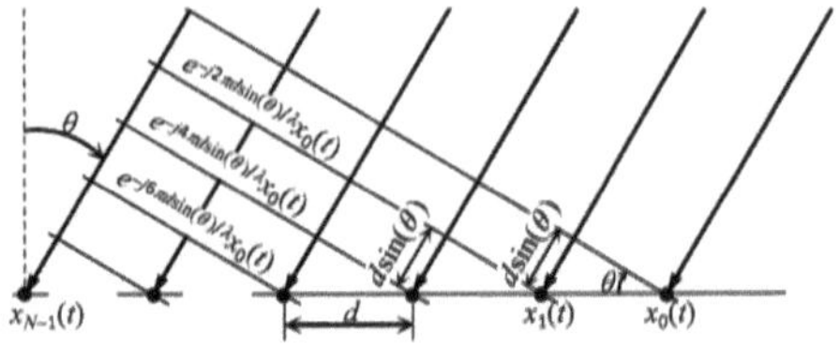

Fig. 9.5 An N-element receiver array

Let

$$\phi = 2\pi d \sin(\theta)/\lambda. \tag{9.10}$$

Let $H_\phi(\phi)$ denote the array factor expressed as a function of ϕ:

$$H_\phi(\phi) = \sum_{n=0}^{N-1} e^{-jn\phi} h_n. \tag{9.11}$$

Since $e^{-jn\phi} = e^{-jn(\phi+2\pi\times\text{integer})}$, $H_\phi(\phi)$ is periodic in ϕ with period 2π. $H_\phi(\phi)$ is a warped version of $H_\theta(\theta)$. Depending on the ratio d/λ, $H_\phi(\phi)$ and $H_\theta(\theta)$ may look very different.

9.2.2 Reception-End Beamforming

Consider an array consisting of N identical isotropic elements. Suppose that a plane wave strikes the N-element array at an angle of θ as shown in Fig. 9.5. Suppose that the array elements are evenly spaced and the separation between two neighboring elements is d. For the wavefront, the distance between two neighboring elements is $d\sin(\theta)$. From Fig. 9.5, we have

$$x_m(t) = e^{-j2m\pi d \sin(\theta)/\lambda} x_0(t) \ , \quad m = 1, 2, \ldots, N-1. \tag{9.12}$$

Suppose that $h_0x_0(t)$, $h_1x_1(t)$, …and $h_{N-1}x_{N-1}(t)$ are added to form $y(t)$ as shown in Fig. 9.6:

$$y(t) = \sum_{n=0}^{N-1} \{e^{-j2n\pi d \sin(\theta)/\lambda} h_n\} x_0(t) = H_\theta(\theta) x_0(t) \tag{9.13}$$

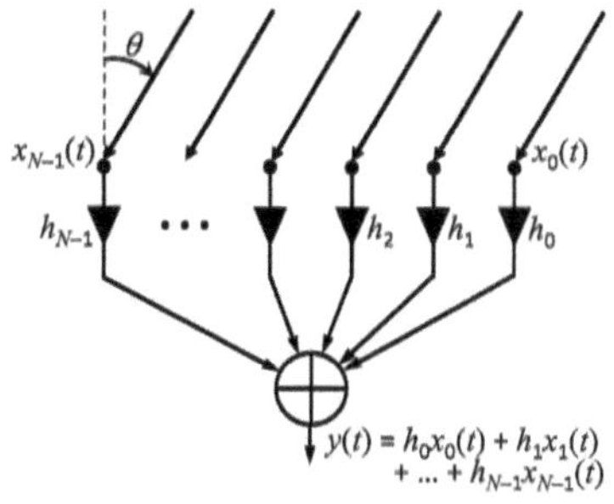

Fig. 9.6 Receiver array beamformer

Comparing (9.9) and (9.13), it can be seen that the array factor expression for receiving array is the same as that for radiating array. Thus, an array factor is independent of whether the array is a radiating array or a receiver array.

9.3 Symmetrical Arrays

Consider an array where all the coefficient values are real and symmetrical, i.e., $h_n = h_{N-1-n}$. There are two possibilities of symmetry depending on whether N is odd or even. For N even, the array factor $H_\phi(\phi)$ is given by

$$\begin{aligned} H_\phi(\phi) &= e^{-j(N-1)\phi/2} \\ &\quad \times \{h_0(e^{j(N-1)\phi/2} + e^{-j(N-1)\phi/2}) + \ldots + h_{N/2-1}(e^{j\phi/2} + e^{-j\phi/2})\}, \\ &= 2e^{-j(N-1)\phi/2}\{h_0\cos((N-1)\phi/2) + \ldots + h_{N/2-1}\cos(\phi/2)\}. \end{aligned} \tag{9.14}$$

The $e^{-j(N-1)\phi/2}$ factor in $H_\phi(\phi)$ is unimportant since $|e^{-j(N-1)\phi/2}| = 1$ and so has no spatial selectivity. We shall drop the $e^{-j(N-1)\phi/2}$ factor from $H_\phi(\phi)$ and write

$$H_\phi(\phi) = 2h_0\cos((N-1)\phi/2) + 2h_1\cos((N-3)\phi/2) + \ldots + 2h_{N/2-1}\cos(\phi/2). \tag{9.15}$$

Let $\alpha_n = 2h_{N/2-n}$, $n = 1, 2, \ldots, N/2$. $H_\phi(\phi)$ in terms of α_n is given by

$$H_\phi(\phi) = \sum_{n=1}^{N/2} \alpha_n \cos((2n-1)\phi/2). \tag{9.16}$$

For N odd, the array factor $H_\phi(\phi)$ is given by

$$\begin{aligned} H_\phi(\phi) &= e^{-j(N-1)\phi/2}\{h_0(e^{j(N-1)\phi/2} + e^{-j(N-1)\phi/2}) + \dots \\ &\quad + h_{(N-3)/2}(e^{j\phi} + e^{-j\phi}) + h_{(N-1)/2}\}, \\ &= e^{-j(N-1)\phi/2}\{2h_0 \cos((N-1)\phi/2) + \dots \\ &\quad + 2h_{(N-3)/2} \cos(\phi) + h_{(N-1)/2}\}. \end{aligned} \tag{9.17}$$

We shall drop the $e^{-j(N-1)\phi/2}$ factor from $H_\phi(\phi)$ for the same reason as that for N even. Let $\alpha_n = 2h_{(N-1)/2-n}$, $n = 1, 2, \dots, (N-1)/2$ and let $\alpha_0 = h_{(N-1)/2}$. $H_\phi(\phi)$ in terms of α_n is given by

$$H_\phi(\phi) = \alpha_0 + \sum_{n=1}^{(N-1)/2} \alpha_n \cos(n\phi). \tag{9.18}$$

Note that the definitions of α_n for N odd and N even are different. We shall illustrate a special case of $H_\phi(\phi)$ plot where $N = 2$, $h_0 = h_1 = 1$. For this special case, $H_\phi(\phi) = 2\cos(\phi/2)$. The $H_\phi(\phi)$ plot is shown in Fig. 9.7. If the array element separation d is $\lambda/2$, substituting $d = \lambda/2$ into (9.10), we have $\phi = \pi \sin(\theta)$, and, hence, $\cos(\phi/2) = \cos(\pi \sin(\theta)/2)$. Cartesian and polar plots for $H_\theta(\theta)$ for $-\pi/2 \le \theta \le \pi/2$ and for $-\pi \le \theta \le \pi$ are shown in Fig. 9.8. Note that $H_\phi(\phi)$ is not a function of the array element separation d, whereas $H_\theta(\theta)$ is. The $H_\phi(\phi)$ plot does not show directly the angular selectivity, whereas the $H_\theta(\theta)$ plot is a plot showing the angular selectivity. The simplest form of an array is the uniform array where h_n

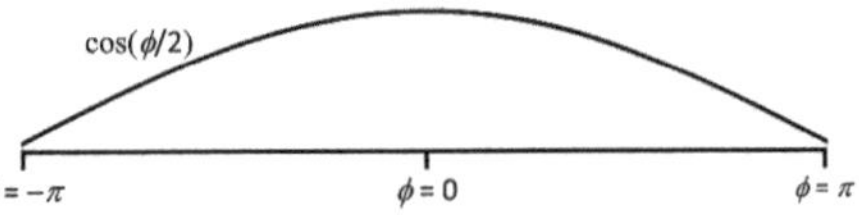

Fig. 9.7 $H_\phi(\phi) = 2\cos(\phi/2)$ plot

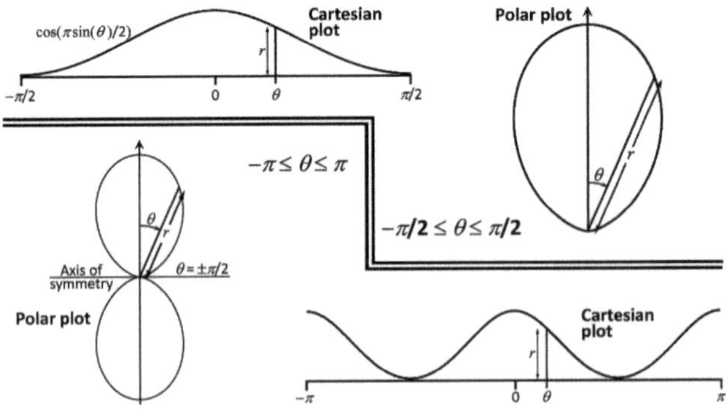

Fig. 9.8 $H_\theta(\theta)$ plot for a two-element array. $\lambda = 2d$

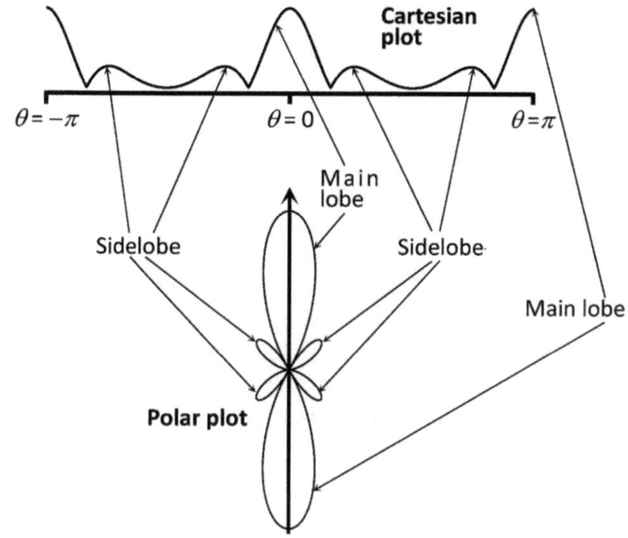

Fig. 9.9 $H_\theta(\theta)$ plot for a four-element uniform array. $\lambda = 2d$

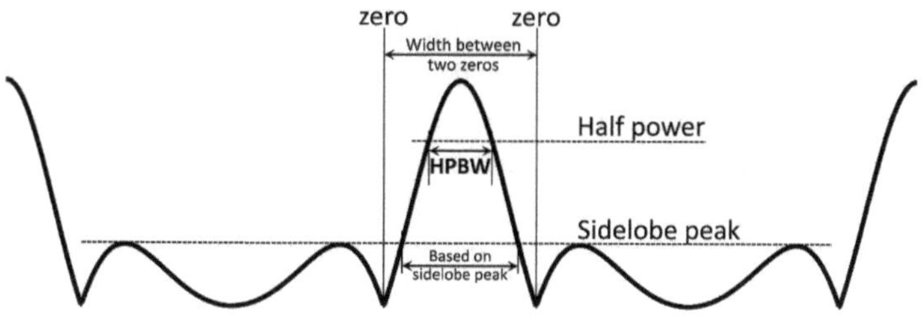

Fig. 9.10 Beamwidth of an array

$= 1$ for $n = 0, 1, 2, \ldots, N - 1$. Figure 9.9 shows a $H_\theta(\theta)$ plot of a four-element uniform array for $\lambda = 2d$. The response lobe with the largest magnitude (maximum gain) is called the main lobe. Response lobes other than the main lobe are called the sidelobes. The axis of maximum gain is called the boresight of the array. There are many definitions for array beamwidth. Different definition best suits different application. Some of the commonly used definitions are as follows:

1. Separation between two -3 dB points on the main lobe. This is called half-power beamwidth (HPBM).
2. Separation between two zeros immediately adjacent to the main lobe.
3. Separation between two points on the main lobe that have the same gain as the peak sidelobe gain.

These three definitions are illustrated in Fig. 9.10.

9.4 Array Element Separation

The array factor $H_\phi(\phi)$ is used in the design of beamformer, whereas the angle θ is the actual direction of propagation. Since $\phi = 2\pi d \sin(\theta)/\lambda$, the ratio d/λ is important in the mapping between ϕ and θ. We shall illustrate this by a few examples based on a Dolph-Chebyshev beamformer with 16 elements.

Since $\phi = 2\pi d \sin(\theta)/\lambda$, for small θ, $\phi = 2\pi \theta d/\lambda$. Furthermore, as $-1 \leq \sin(\theta) \leq 1$, ϕ is bounded by $-2\pi d/\lambda \leq \phi \leq 2\pi d/\lambda$. The range of interest for the values of θ is $-\pi \leq \theta \leq \pi$.

Consider the case where $\lambda = 2d$. For this case, $\phi = \pi \sin(\theta)$. For small θ, $\sin(\theta) \approx \theta$, and so $\phi \approx \pi\theta$. Consequently, the main lobe width of $H_\phi(\phi)$ is π times that of $H_\theta(\theta)$. As θ increases from 0 to $\pi/2$, ϕ increases from 0 to π. Hence, the plot of $H_\theta(\theta)$ for $0 \leq \theta \leq \pi/2$ is the warped version of $H_\phi(\phi)$ for $0 \leq \phi \leq \pi$. As θ increases from $\pi/2$ to π, ϕ decreases from π to 0. Hence, the plot of $H_\theta(\theta)$ for $\pi/2 \leq \theta \leq \pi$ is the warped version of $H_\phi(\phi)$ for $0 \leq \phi \leq \pi$ in reversed direction. This is shown in Fig. 9.11.

Consider the case where $\lambda = d$. For this case, $\phi = 2\pi \sin(\theta)$. For small θ, $\phi \approx 2\pi\theta$. Consequently, the main lobe width of $H_\phi(\phi)$ is 2π times that of $H_\theta(\theta)$. As θ increases from 0 to $\pi/2$, ϕ increases from 0 to 2π. Hence, the plot of $H_\theta(\theta)$ for $0 \leq \theta \leq \pi/2$ is the warped version of $H_\phi(\phi)$ for $0 \leq \phi \leq 2\pi$. As θ increases from $\pi/2$ to π, ϕ decreases from 2π to 0. Hence, the plot of $H_\theta(\theta)$ for $\pi/2 \leq \theta \leq \pi$ is the warped version of $H_\phi(\phi)$ for $0 \leq \phi \leq 2\pi$ in reversed direction. This is shown in Fig. 9.12. It can be seen in Fig. 9.12 that the large magnitude of $H_\phi(\phi)$ near $\phi = 2\pi$ is mapped into $H_\theta(\theta)$ near $\theta = \pi/2$. This undesirable mapping of a large magnitude portion of $H_\phi(\phi)$ into $H_\theta(\theta)$ is called aliasing. However, aliasing is not observed in the case of $\lambda = 2d$ of Fig. 9.11. From Figs. 9.11 and 9.12, we observed that the main lobe width decreases (i.e., the spatial selectivity of the array increases) as the ratio

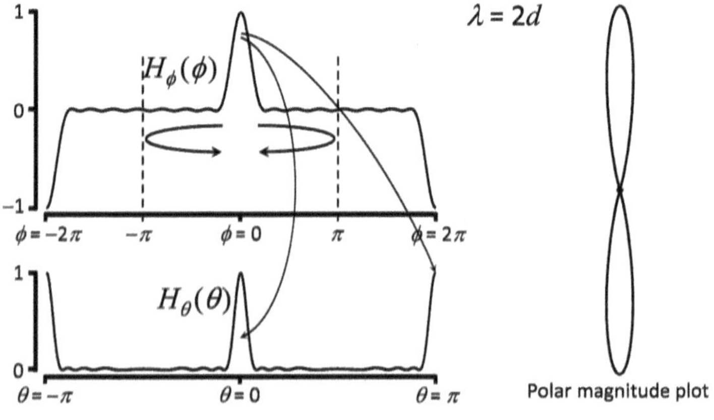

Fig. 9.11 Plots of $H_\phi(\phi)$ and $H_\theta(\theta)$ for $\lambda = 2d$

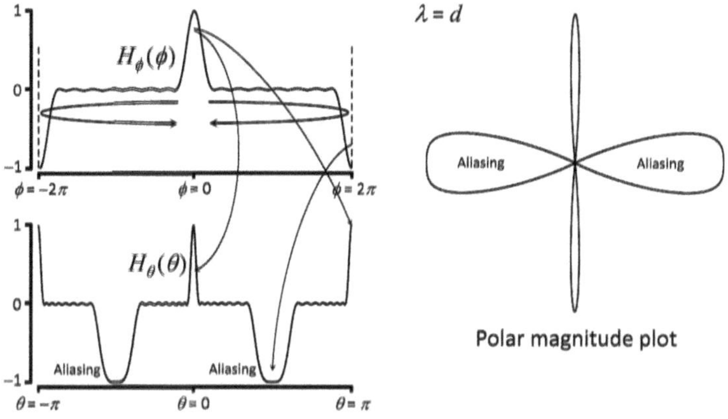

Fig. 9.12 Plots of $H_\phi(\phi)$ and $H_\theta(\theta)$ for $\lambda = d$

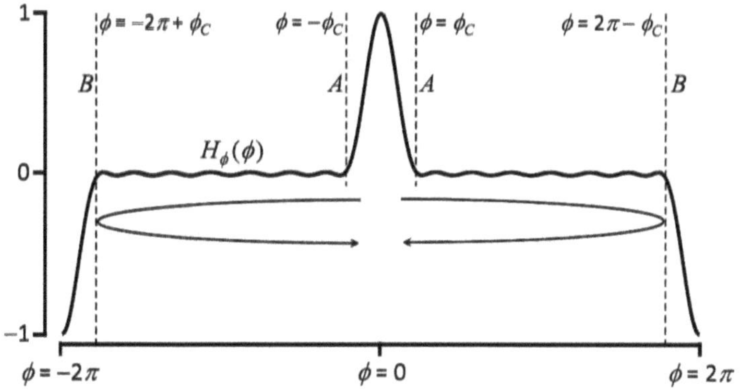

Fig. 9.13 Aliasing free maximum d/λ ratio without steering

d/λ increases. However, if the ratio d/λ is too large, it may cause aliasing. We wish to determine the maximum d/λ ratio that does not cause aliasing.

Consider Fig. 9.13. Suppose that the main lobe cutoff value for ϕ is ϕ_C (see dotted lines marked A on the figure). To avoid aliasing, ϕ must be bounded by $-2\pi + \phi_C \le \phi \le 2\pi - \phi_C$ (see dotted lines marked B on the figure). Since $\phi = 2\pi d \sin(\theta)/\lambda$, the maximum value of ϕ is $2\pi d/\lambda$. Hence, for no aliasing,

$$2\pi d/\lambda \le 2\pi - \phi_C \Longrightarrow \frac{d}{\lambda} \le r_{\max}, \tag{9.19a}$$

$$\text{where } r_{\max} = \frac{2\pi - \phi_C}{2\pi}. \tag{9.19b}$$

The maximum ratio of d/λ in (9.19) is obtained without considering beam steering. The ratio will be reduced if beam steering is considered.

The boresight of the array factor $H_\phi(\phi)$ can be steered from $\phi = 0$ to $\phi = \phi_0$ by replacing ϕ in (9.11) by $\phi - \phi_0$,

$$H_\phi(\phi - \phi_0) = \sum_{n=0}^{N-1} e^{-jn(\phi-\phi_0)} h_n = \sum_{n=0}^{N-1} e^{-jn\phi}(e^{jn\phi_0} h_n), \tag{9.20a}$$

$$= \sum_{n=0}^{N-1} e^{-jn\phi} \hat{h}_n, \tag{9.20b}$$

where

$$\hat{h}_n = e^{jn\phi_0} h_n. \tag{9.21}$$

$\hat{h}_n$ is the nth steered coefficient value of the array. If the 16-element array of Fig. 9.11 with $\lambda = 2d$ is steered by an angle of $\pi/6$ (i.e., $\phi_0 = 2\pi d \sin(\pi/6)/\lambda = 0.5\pi$), the result is shown in Fig. 9.14. Figure 9.15 shows the result of steering

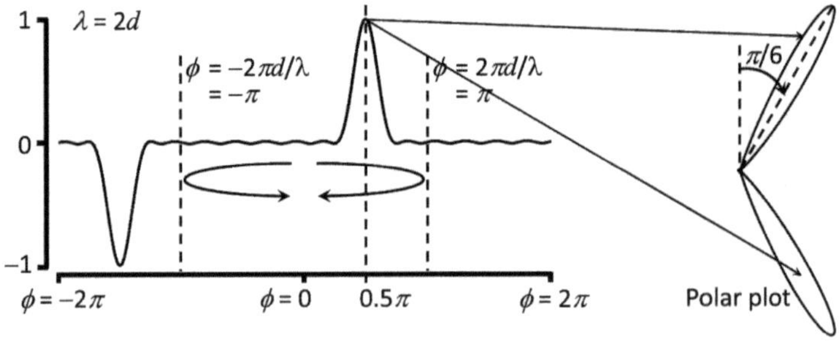

Fig. 9.14 Steering the array of Fig. 9.11 by an angle of $\theta = \pi/6$

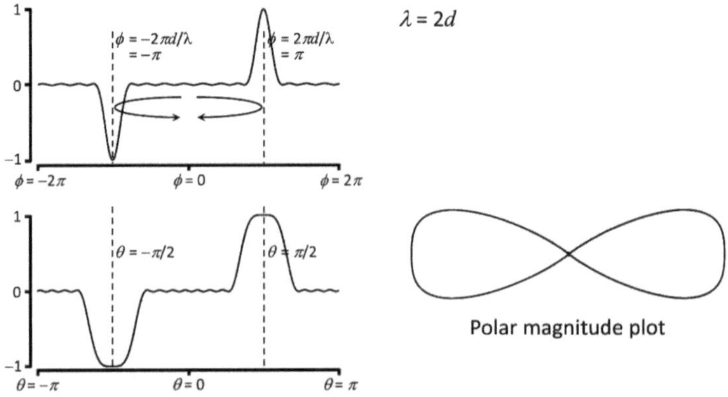

Fig. 9.15 Array of Fig. 9.11 is steered by $\theta = \pi/2$ making it an end-fire array

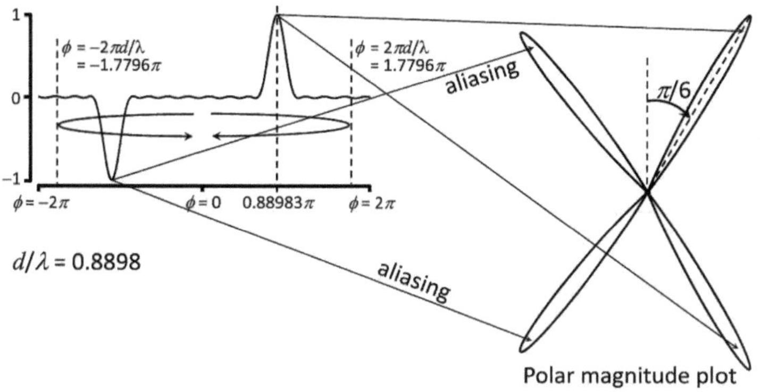

Fig. 9.16 Steering a 16-element array with $d/\lambda = 0.88983$

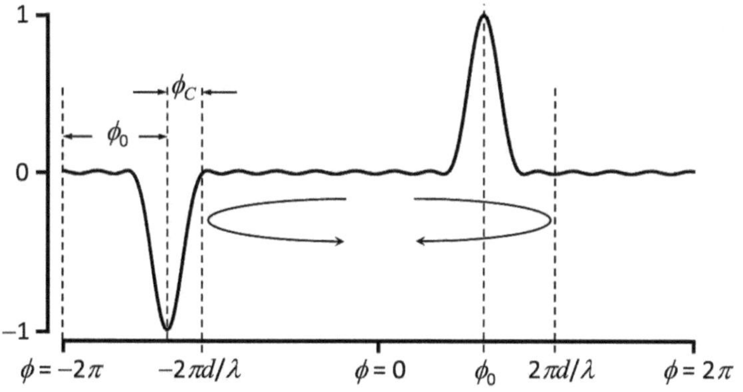

Fig. 9.17 Aliasing free maximum d/λ ratio with steering

the array by an angle of $\pi/2$; such an array where the boresight is along the array is called an end-fire array. An array whose boresight is perpendicular to the array such as that shown in Fig. 9.11 is called a broadside array.

The 16-element array of Fig. 9.11 has $\phi_C = 0.69219$. Substituting $\phi_C = 0.69219$ into (9.19), the maximum d/λ ratio without aliasing is 0.88983. Under this maximum ratio, there is no aliasing provided that the array is not steered. Figure 9.16 shows the result if the array with $d/\lambda = 0.88983$ is steered by an angle of $\pi/6$; aliasing is obvious. If the array is steered, (9.19) is no longer valid for determining the maximum d/λ ratio. Suppose that it is necessary to steer an array by an angle of $\theta = \theta_0$, and we wish to find the maximum d/λ ratio without aliasing. Suppose that $\phi = \phi_0$ when $\theta = \theta_0$. Let ϕ_C be the cutoff angle of the array. From Fig. 9.17, it can be seen that $-2\pi + \phi_0 + \phi_C \leq -2\pi d/\lambda$ for no aliasing. Hence, for aliasing free steering,

$$2\pi d/\lambda \leq 2\pi - \phi_0 - \phi_C \Longrightarrow \frac{d}{\lambda} \leq \frac{r_{\max}}{1 + \sin(\theta_0)}. \tag{9.22}$$

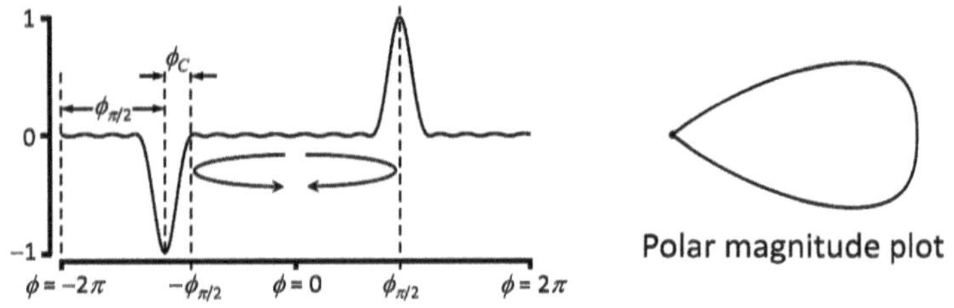

Fig. 9.18 Unidirectional end-fire array

In (9.22), $r_{max} = \frac{2\pi - \phi_C}{2\pi}$ as defined in (9.19b).

While a one-dimensional broadside array with isotropic array elements cannot be made unidirectional since its array factor on one side of the array is the mirror image of that on the other side, it is possible to design end-fire array with isotropic array elements unidirectional, i.e., radiating or receiving in only one direction, by a suitable choice of d/λ ratio. Let ϕ_C be the cutoff value of ϕ when the array is in a broadside mode with steering angle $\theta = 0$. Let $\phi_{\pi/2}$ be the value of ϕ when the array is steered to an end-fire mode, i.e., when the steering angle $\theta_0 = \pi/2$. Hence,

$$\phi_{\pi/2} = 2\pi d \sin(\pi/2)/\lambda = 2\pi d/\lambda. \tag{9.23}$$

From Fig. 9.18, it can be seen that $-2\pi + \phi_{\pi/2} + \phi_C \le -\phi_{\pi/2}$ for no aliasing at $\theta = \pi/2$. Therefore,

$$\phi_{\pi/2} = 2\pi d/\lambda \le \pi - \phi_{\pi/2} \Longrightarrow \frac{d}{\lambda} \le \frac{r_{max}}{2}. \tag{9.24}$$

In (9.24), $r_{max} = \frac{2\pi - \phi_C}{2\pi}$ as defined in (9.19b). Expression (9.24) may also be obtained by substituting $\theta_0 = \pi/2$ into (9.22). Consider the example where $\phi_C = 0.69219$, $r_{max} = \frac{2\pi - 0.69219}{2\pi} = 0.88983$. Substituting $r_{max} = 0.88983$ into (9.24), $d/\lambda \le \frac{0.88983}{2} = 0.44492$. Any choice of $d/\lambda \le 0.44492$ will produce unidirectional beam.

9.5 Beamwidth of Broadside and End-Fire Arrays

Suppose that the broadside beamwidth of the array is $2\theta_b$ and that $\phi = \phi_b$ when $\theta = \theta_b$. Therefore,

$$\phi_b = 2\pi d \sin(\theta_b)/\lambda. \tag{9.25}$$

Suppose that $\phi = \phi_0$ when $\theta = \pi/2$. Therefore, $\phi_0 = 2\pi d/\lambda$. The array beamwidth is a function of the beam direction. Suppose that when the array is steered to become an end-fire array, its beamwidth is $2\theta_e$. Therefore,

$$\phi_0 - \phi_b = 2\pi d \sin(\pi/2 + \theta_e)/\lambda = 2\pi d \cos(\theta_e)/\lambda. \tag{9.26}$$

Substituting $\phi_0 = 2\pi d/\lambda$ and $\phi_b = 2\pi d \sin(\theta_b)/\lambda$ into (9.26), we have

$$\begin{aligned} &2\pi d/\lambda - 2\pi d \sin(\theta_b)/\lambda = 2\pi d \cos(\theta_e)/\lambda, \\ &\qquad \Longrightarrow 1 - \sin(\theta_b) = \cos(\theta_e) \quad [\text{independent of } \ d/\lambda\ ]. \end{aligned} \tag{9.27}$$

Suppose that both θ_b and θ_e are small, i.e., $\sin(\theta_b) \approx \theta_b$ and $\cos(\theta_e) \approx 1 - \theta_e^2/2$. Therefore,

$$1 - \theta_b \approx 1 - \theta_e^2/2 \Longrightarrow \theta_b \approx \theta_e^2/2 \Longrightarrow \theta_e \approx \sqrt{2\theta_b}\ . \tag{9.28}$$

For $\theta_b < 2$, $\theta_e > \theta_b$, i.e., end-fire beamwidth is larger than broadside beamwidth, for example, for $\theta_b = 0.18$, $\theta_e \approx 0.6$. Figure 9.19 shows the responses of a broadside array and an end-fire array derived from the same 16-element array factor. We have shown in (9.25) that the relationship between broadside beamwidth $2\theta_b$ and array factor beamwidth $2\phi_b$ is $\phi_b = 2\pi d \sin(\theta_b)/\lambda$. The relationship between end-fire beamwidth and array factor beamwidth may be obtained by substituting (9.27) into (9.25). Hence, we have

$$\phi_b = 2\pi d(1 - \cos(\theta_e))/\lambda. \tag{9.29}$$

Figure 9.20 shows steering the beam of a 16-element Dolph-Chebyshev beamformer with $d/\lambda = 0.44492$, $\phi_C = 0.69219$, and 40 dB attenuation by five different angles,

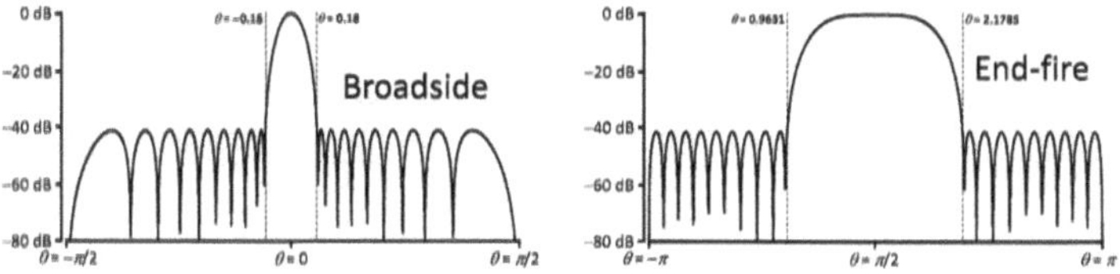

Fig. 9.19 Responses of a broadside array and an end-fire array derived from the same 16-element array factor

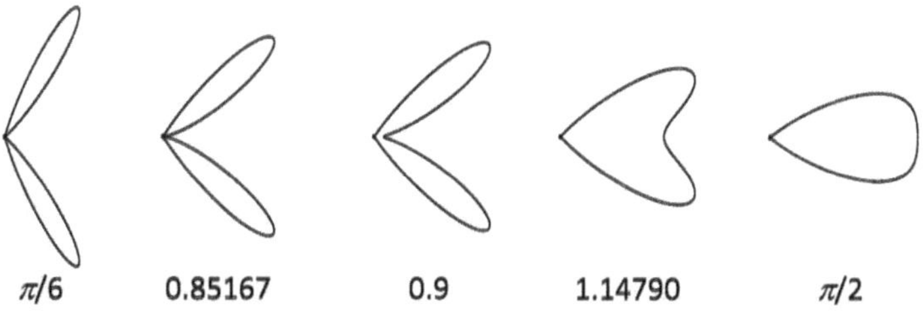

Fig. 9.20 Steering the beam of an array by $\theta = \pi/6$, 0.85167, 0.9, 1.14790, and $\pi/2$, respectively

namely, $\theta = \pi/6$, 0.85167, 0.9, 1.14790, and $\pi/2$, respectively. It can be seen from the plots that, for steering angles of 0.9 and 1.1479, the two branches of the beam partially overlap. The two branches completely overlap when the steering angle is $\pi/2$.

Suppose that $\theta_{\max}$ is the maximum that the beamformer can be steered with no overlap. Suppose that $\phi = \phi_{\max}$ when $\theta = \theta_{\max}$, i.e., $\phi_{\max} = 2\pi d \sin(\theta_{\max})/\lambda$. Let ϕ_C be the beamwidth of the array factor and that $\phi_C = 2\pi d \sin(\theta_C)/\lambda$ where θ_C is the broadside critical beamwidth of the array. For no overlap of the branches, the following condition must be satisfied:

$$\phi_{\max} + \phi_C \leq 2\pi d \sin(\pi/2)/\lambda = 2\pi d\lambda \tag{9.30}$$

Substituting $\phi_{\max} = 2\pi d \sin(\theta_{\max})/\lambda$ and $\phi_C = 2\pi d \sin(\theta_C)/\lambda$ into (9.30), we have

$$\begin{aligned} &2\pi d \sin(\theta_{\max})/\lambda + 2\pi d \sin(\theta_C)/\lambda \leq 2\pi d\lambda, \\ &\Longrightarrow \sin(\theta_{\max}) \leq 1 - \sin(\theta_C). \end{aligned} \tag{9.31}$$

Thus, the maximum steer without overlap of the beam branches is $1 - \sin(\theta_C)$. We wish to find the condition of maximum steer, i.e., $\sin(\theta_{\max}) = 1 - \sin(\theta_C)$ and at the same time without aliasing. We have shown in (9.22) that for steering up to an angle of $\theta_{\max}(= \theta_0)$ without aliasing $\frac{d}{\lambda} \leq \frac{r_{\max}}{1+\sin(\theta_{\max})}$. Substituting (9.31) into (9.22), we have

$$\begin{aligned} &\frac{d}{\lambda} \leq \frac{r_{\max}}{1 + \sin(\theta_{\max})} = \frac{r_{\max}}{2 - \frac{\phi_C}{2\pi d/\lambda}}, \\ \Longrightarrow &2\frac{d}{\lambda} - \frac{\phi_C}{2\pi} \leq r_{\max} \Longrightarrow \frac{d}{\lambda} \leq \frac{2\pi r_{\max} + \phi_C}{4\pi} = \frac{1}{2}. \end{aligned} \tag{9.32}$$

Thus, the condition for maximum steer without overlap and without aliasing is $\lambda \geq 2d$. We shall illustrate this by using an example. Consider an example where $\phi_C = 0.69219$. Substituting $\phi_C = 0.69219$ into (9.10), we have $\sin(\theta_C) = \frac{\phi_C}{2\pi d/\lambda} = \frac{0.69219}{2\pi d/\lambda}$. From (9.31), the maximum $\sin(\theta_{\max}) = 1 - \sin(\theta_C) = 1 - \frac{0.69219}{2\pi d/\lambda}$. To maximize $\sin(\theta_{\max})$, choose d/λ to be as large as possible, i.e., choose $d/\lambda = 0.5$ since $\lambda \geq 2d$. Hence, $\sin(\theta_{\max}) = 1 - \frac{0.69219}{\pi} = 0.77967$. Therefore, $\theta_{\max} = 0.8941$ rad.

9.6 Two-Dimensional Rectangular Array

In the foregoing sections, we discuss a class of arrays whose elements lie on a straight line. This class of arrays is called linear array. If the elements of an array form a two-dimensional rectangular plane, the array is called a two-dimensional rectangular array. Figure 9.21 shows an $(N_x \times N_y)$-element two-dimensional rectangular array. The signal at the (n_x, n_y)th element is $x_{n_x,n_y}(t)$. Assume that

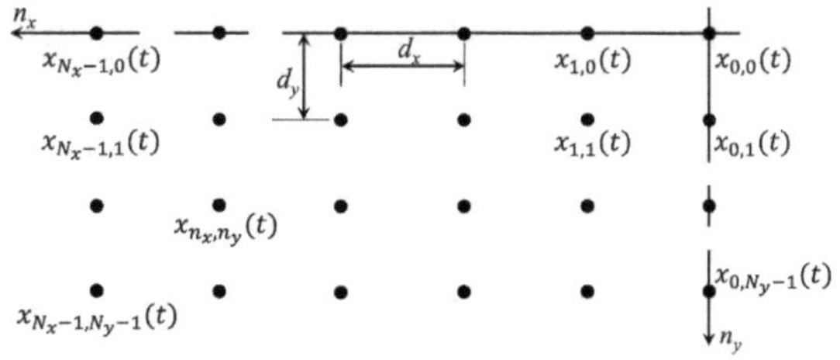

Fig. 9.21 A two-dimensional rectangular array

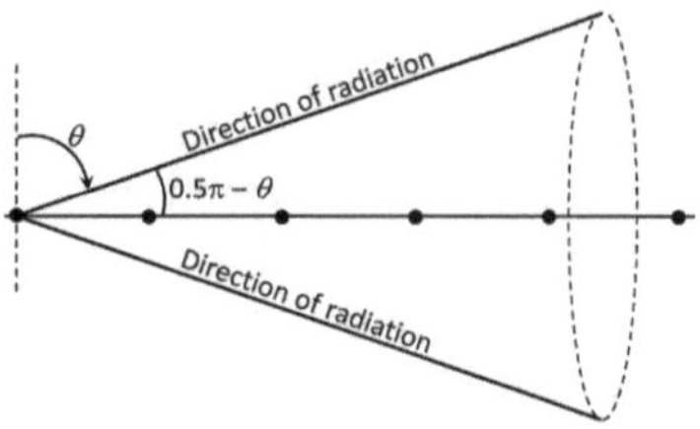

Fig. 9.22 Radiation pattern of a linear array

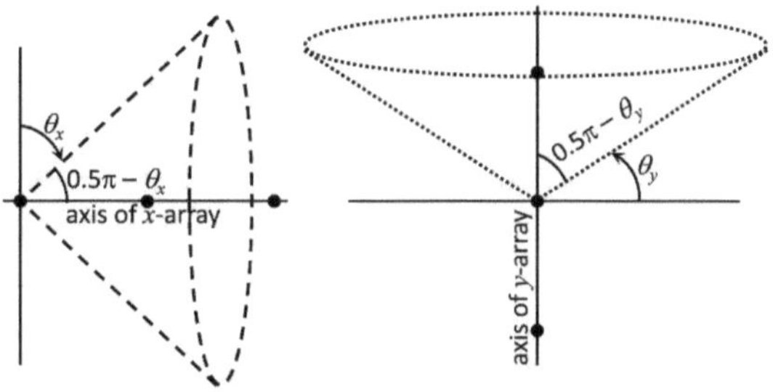

Fig. 9.23 The cones representing possible θ_x and θ_y angles of arrival

the n_x-axis is orthogonal to the n_y-axis. Let the element spacing on the n_x and n_y directions be d_x and d_y, respectively.

It is easy to visualize the angle θ of a plane wave radiating from or impinging on a linear array. It is straightforward to represent the angle θ of a linear array on a two-dimensional diagram. See, for example, Figs. 9.2, 9.3, or 9.5. In a three-dimensional space, the radiation pattern of a linear array is the surface of a cone. An example is shown in Fig. 9.22.

For a two-dimensional array, the angle θ of the linear array is replaced by two angles θ_x and θ_y as explained below. The array shown in Fig. 9.21 may be regarded as a collection of N_y one-dimensional x-arrays each of which has N_x elements. Each x-array consists of one row of the array elements shown in Fig. 9.21. For a far-field signal arriving from infinity, the signal will impinge on all the N_y x-arrays at the same angle θ_x (known as θ in the one-dimensional case); see Figs. 9.5, 9.6, 9.22, and 9.23. The array shown in Fig. 9.21 may also be regarded as a collection of N_x one-dimensional y-arrays each of which has N_y elements. Each y-array consists of one column of the array elements shown in Fig. 9.21. For a far-field signal arriving from infinity, the signal will impinge on all the N_x y-arrays at the same angle θ_y; see Fig. 9.23.

9.7 Relations Between θ_x, θ_y, Azimuth, and Elevation Angles

The angles θ_x and θ_y are neither elevation angle nor azimuth angle. In this section, we shall derive relationships relating θ_x, θ_y, elevation, and azimuth angles. Superimpose the two cones in Fig. 9.23 to form Fig. 9.24 assuming that both θ_x and θ_y are positive. In Fig. 9.24, the x-axis and y-axis are coplanar and are parallel to the broadside of the x-array and y-array, respectively (broadside: the direction perpendicular to the array elements). Let the plane span by the x-axis and y-axis be the (x, y)-plane. Let the z-axis be perpendicular to the (x, y)-plane and in the direction coming out of the (x, y)-plane. There are two lines of intersection for the two cones; one is behind the (x, y)-plane with negative z, and the other one is in front of the (x, y)-plane with positive z. We shall ignore the one with negative z. The two cones will intersect provided that $\theta_x + \theta_y \leq \frac{\pi}{2}$. If $\theta_x + \theta_y = \frac{\pi}{2}$, the two cones touch each other at the (x, y)-plane. The scenarios for negative θ_x and/or θ_y are shown in Fig. 9.25. From Figs. 9.24 and 9.25, for the two cones to intersect,

$$|\theta_x| + |\theta_y| \leq \frac{\pi}{2}. \tag{9.33}$$

The region of support for (9.33) is the region inside the rhombus shown in Fig. 9.26. Consider a point with (x, y, z) coordinates given by (x_0, y_0, z_0) on a line of intersection between the two cones. A three-dimensional plot of the line of intersection of the two cones and the point (x_0, y_0, z_0) is shown in Fig. 9.27. In Fig. 9.27, θ_A (defined over the range $-\pi < \theta_A \leq \pi$) is the azimuth angle, and

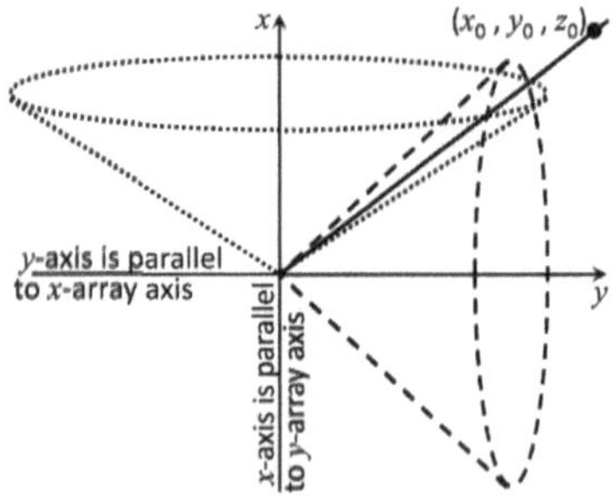

Fig. 9.24 The intersection of the two cones represents the direction of arrival of the impinging signal

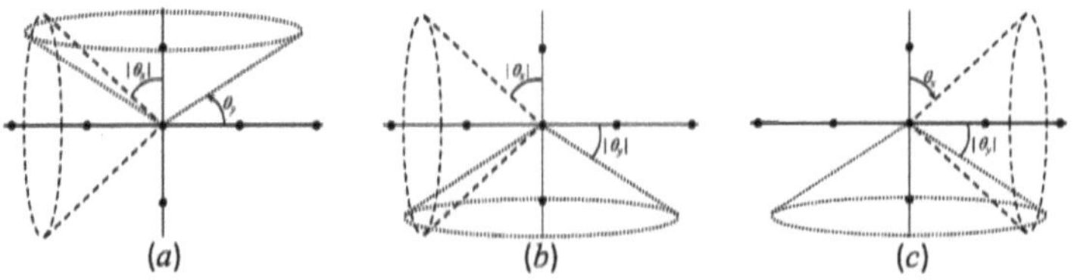

Fig. 9.25 (**a**) θ_x negative and θ_y positive. (**b**) θ_x and θ_y both negative. (**c**) θ_x positive and θ_y negative

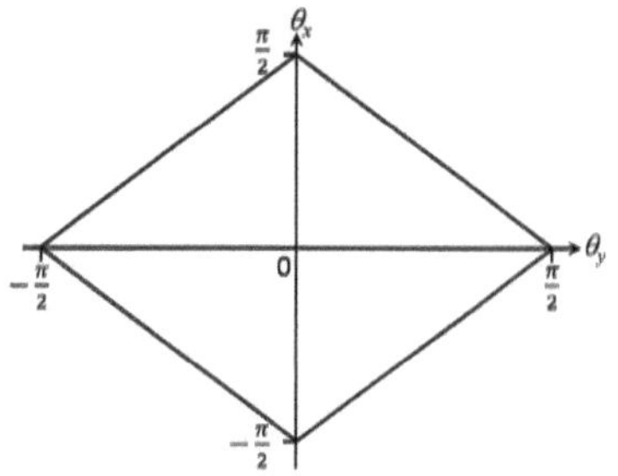

Fig. 9.26 Region of support for (9.33) is inside the rhombus

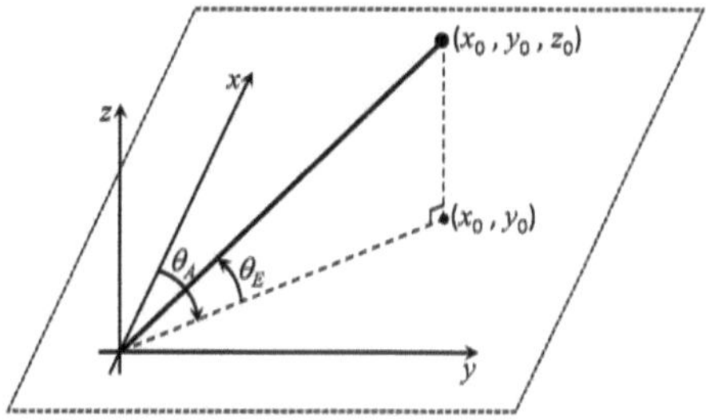

Fig. 9.27 Three-dimensional plot of the line of intersection of the two cones and the point (x_0, y_0, z_0). $-\pi < \theta_A \leq \pi$ and $0 \leq \theta_E \leq \frac{\pi}{2}$

θ_E (defined over the range $0 \leq \theta_E \leq \frac{\pi}{2}$) is the elevation angle. The perpendicular projection of the point (x_0, y_0, z_0) on the (x, y)-plane is (x_0, y_0).

From Figs. 9.24 and 9.27, for the x-array, we have

$$x_0^2 + z_0^2 = y_0^2 \tan^2\left(\frac{\pi}{2} - \theta_x\right) = y_0^2 \cot^2(\theta_x). \tag{9.34}$$

For the y-array, we have

$$y_0^2 + z_0^2 = x_0^2 \tan^2\left(\frac{\pi}{2} - \theta_y\right) = x_0^2 \cot^2(\theta_y). \tag{9.35}$$

Subtracting (9.34) from (9.35),

$$y_0^2 - x_0^2 = x_0^2 \cot^2(\theta_y) - y_0^2 \cot^2(\theta_x).$$

$$\therefore \quad \frac{y_0^2}{x_0^2} = \tan^2(\theta_A) = \frac{1 + \cot^2(\theta_y)}{1 + \cot^2(\theta_x)}. \tag{9.36a}$$

$$\therefore \quad \theta_A = \pm \arctan\left(\sqrt{\frac{1 + \cot^2(\theta_y)}{1 + \cot^2(\theta_x)}}\right) = \pm \arctan\left(\frac{\sin(\theta_x)}{\sin(\theta_y)}\right). \tag{9.36b}$$

Taking note that the principal angle of arctangent is between $-\frac{\pi}{2}$ and $\frac{\pi}{2}$, the azimuth angle is given by

$$\begin{aligned}\theta_A &= \arctan\left(\frac{\sin(\theta_x)}{\sin(\theta_y)}\right) \quad \text{for } \theta_y \text{ positive ,}\\ &= \arctan\left(\frac{\sin(\theta_x)}{\sin(\theta_y)}\right) + \pi \quad \text{for } \theta_x \text{ positive and } \theta_y \text{ negative,}\\ &= \arctan\left(\frac{\sin(\theta_x)}{\sin(\theta_y)}\right) - \pi \quad \text{for } \theta_x \text{ and } \theta_y \text{ negative.}\end{aligned} \tag{9.37}$$

From (9.34) and (9.35), eliminating y_0^2, we have

$$x_0^2 = \frac{z_0^2\{1 + \cot^2(\theta_x)\}}{\cot^2(\theta_x)\cot^2(\theta_y) - 1}. \tag{9.38}$$

From (9.34) and (9.35), eliminating x_0^2, we have

$$y_0^2 = \frac{z_0^2\{1 + \cot^2(\theta_y)\}}{\cot^2(\theta_x)\cot^2(\theta_y) - 1}. \tag{9.39}$$

From (9.38) and (9.39), we have

$$\frac{z_0^2}{x_0^2 + y_0^2} = \tan^2(\theta_E) = \frac{\cot^2(\theta_x)\cot^2(\theta_y) - 1}{2 + \cot^2(\theta_x) + \cot^2(\theta_y)}. \tag{9.40a}$$

$$\begin{aligned}\therefore \quad \theta_E &= \arctan\left(\sqrt{\frac{\cot^2(\theta_x)\cot^2(\theta_y) - 1}{2 + \cot^2(\theta_x) + \cot^2(\theta_y)}}\right),\\ &= \arctan\left(\sqrt{\frac{1 - \sin^2(\theta_x) - \sin^2(\theta_y)}{\sin^2(\theta_x) + \sin^2(\theta_y)}}\right).\end{aligned} \tag{9.40b}$$

Next, we shall derive expressions for θ_x and θ_y in terms of θ_E and θ_A. From (9.36a) and (9.40a), eliminating $\cot^2(\theta_y)$, we have

$$\tan^2(\theta_E) = \frac{\cot^2(\theta_x)\tan^2(\theta_A) - 1}{1 + \tan^2(\theta_A)}. \tag{9.41a}$$

$$\begin{aligned}\therefore \quad \theta_x &= \operatorname{arccot}\left(\sqrt{\tan^2(\theta_E) + \cot^2(\theta_A)\{1 + \tan^2(\theta_E)\}}\right) \quad \text{for } 0 \leq \theta_A \leq \pi,\\ &= -\operatorname{arccot}\left(\sqrt{\tan^2(\theta_E) + \cot^2(\theta_A)\{1 + \tan^2(\theta_E)\}}\right) \quad \text{for } -\pi < \theta_A < 0.\end{aligned}$$

$$\sin(\theta_x) = \sin(\theta_A)\cos(\theta_E). \tag{9.41b}$$

From (9.36a) and (9.41a), eliminating $\cot^2(\theta_x)$, we have

$$\cot^2(\theta_y) = \tan^2(\theta_E) + \tan^2(\theta_A)\{1 + \tan^2(\theta_E)\}. \tag{9.42a}$$

$$\therefore\ \theta_y = \text{arccot}\left(\sqrt{\tan^2(\theta_E) + \tan^2(\theta_A)\{1 + \tan^2(\theta_E)\}}\right) \text{ for } \frac{-\pi}{2} \le \theta_A \le \frac{\pi}{2},$$

$$= -\text{arccot}\left(\sqrt{\tan^2(\theta_E) + \tan^2(\theta_A)\{1 + \tan^2(\theta_E)\}}\right) \text{ otherwise.}$$

$$\sin(\theta_y) = \cos(\theta_A)\cos(\theta_E). \tag{9.42b}$$

Expressions (9.37), (9.40), (9.41), and (9.42) provide the relationships for conversion between (θ_x, θ_y) and (θ_E, θ_A). Let δ_x, δ_y, δ_E, and δ_A be small changes in θ_x, θ_y, θ_E, and θ_A, respectively. Differentiating (9.41b) and (9.42b), we have

$$\delta_x = \frac{\cos(\theta_A)\cos(\theta_E)}{\cos(\theta_x)}\delta_A - \frac{\sin(\theta_A)\sin(\theta_E)}{\cos(\theta_x)}\delta_E. \tag{9.43a}$$

$$\delta_y = -\frac{\sin(\theta_A)\cos(\theta_E)}{\cos(\theta_y)}\delta_A - \frac{\cos(\theta_A)\sin(\theta_E)}{\cos(\theta_y)}\delta_E. \tag{9.43b}$$

$$\delta_A = \sin(\theta_A)\cos(\theta_A)\{\cot(\theta_x)\delta_x - \cot(\theta_y)\delta_y\}. \tag{9.43c}$$

$$\delta_E = -\frac{\sin(\theta_x)\cos(\theta_x)}{\sin(\theta_E)\cos(\theta_E)}\delta_x - \frac{\sin(\theta_y)\cos(\theta_y)}{\sin(\theta_E)\cos(\theta_E)}\delta_y, \tag{9.43d}$$

$$= -\frac{\sin(2\theta_x)}{\sin(2\theta_E)}\delta_x - \frac{\sin(2\theta_y)}{\sin(2\theta_E)}\delta_y. \tag{9.43e}$$

Equation (9.43) provides the sensitivity measures for changes in θ_x, θ_y, θ_E, and θ_A with respect to the respective changes in θ_x, θ_y, θ_E, and θ_A. In (9.43a), $\cos(\theta_x)$ appears in the denominator. However, it does not mean that the sensitivity approaches infinity when $\theta_x \to \frac{\pi}{2}$. This is because when $\theta_x \to \frac{\pi}{2}$ both the denominator and the numerator approach zero. We shall prove this as follows:

First, we shall examine the term $\frac{\cos(\theta_A)\cos(\theta_E)}{\cos(\theta_x)}$. Let $\theta_x = \frac{\pi}{2} - \varepsilon$ where ε is positive. Hence, $\cos(\theta_x) = \cos(\frac{\pi}{2} - \varepsilon) = \sin(\varepsilon)$. From (9.33), $\theta_x + \theta_y \le \frac{\pi}{2}$. Therefore, $\frac{\pi}{2} - \varepsilon + \theta_y \le \frac{\pi}{2}$. Hence, $\theta_y \le \varepsilon$. From (9.42b) and $\theta_y \le \varepsilon$, we have $\cos(\theta_A)\cos(\theta_E) = \sin(\theta_y) \le \sin(\varepsilon)$. Thus, $\frac{\cos(\theta_A)\cos(\theta_E)}{\cos(\theta_x)} \le \frac{\sin(\varepsilon)}{\sin(\varepsilon)} = 1$.

Next, we shall examine the term $\frac{\sin(\theta_A)\sin(\theta_E)}{\cos(\theta_x)}$ for $\cos(\theta_x) \to 0$. Let $\theta_x = \frac{\pi}{2} - \varepsilon$ where ε is positive and let $\varepsilon \to 0$. Since $\theta_x = \frac{\pi}{2} - \varepsilon$, $\sin(\theta_x) = \sin(\frac{\pi}{2} - \varepsilon) = \cos(\varepsilon)$ and $\to (1 - \frac{\varepsilon^2}{2})$ as $\varepsilon \to 0$. From (9.41b), $\sin(\theta_A)\cos(\theta_E) = \sin(\theta_x)$. Hence, $\sin(\theta_A)\cos(\theta_E) \to 1 - \frac{\varepsilon^2}{2}$ as $\varepsilon \to 0$, i.e., $\cos(\theta_E) \to \frac{1-\frac{\varepsilon^2}{2}}{\sin(\theta_A)}$. Hence, $\sin(\theta_E) \to \frac{\varepsilon}{\sin(\theta_A)}$ as $\to 0$. Thus, $\frac{\sin(\theta_A)\sin(\theta_E)}{\cos(\theta_x)} \to 1$ as $\varepsilon \to 0$.

From the above, it can be seen that the sensitivity of θ_x with respect to changes in θ_A and θ_E does not approach infinity when $\cos(\theta_x) \to 0$. Similarly, it can be shown for (9.43b), (9.43c), and (9.43d) that the sensitivities do not approach infinity when the respective denominators approach zero.

9.8 Two-Dimensional Array Factor

Consider the n_yth row of the array elements in Fig. 9.21 with array element signals $x_{0,n_y}(t), x_{1,n_y}(t), \ldots, x_{n_x,n_y}(t), \ldots, x_{N_x-1,n_y}(t)$. The signal $x_{n_x,n_y}(t)$ may be expressed in terms of $x_{0,n_y}(t)$ using (9.12) as

$$x_{n_x,n_y}(t) = e^{-j2n_x\pi d_x \sin(\theta_x)/\lambda} x_{0,n_y}(t). \tag{9.44}$$

Consider the n_xth column of the array elements in Fig. 9.21 with array element signals $x_{n_x,0}(t), x_{n_x,1}(t), \ldots, x_{n_x,n_y}(t), \ldots, x_{n_x,N_y}(t)$. The signal $x_{n_x,n_y}(t)$ may be expressed in terms of $x_{n_x,0}(t)$ using (9.12) as

$$x_{n_x,n_y}(t) = e^{-j2n_y\pi d_y \sin(\theta_y)/\lambda} x_{n_x,0}(t). \tag{9.45}$$

Either letting $n_y = 0$ in (9.44) and substituting into (9.45) or letting $n_x = 0$ in (9.45) and substituting into (9.44), we have

$$x_{n_x,n_y}(t) = e^{-j2\pi n_x d_x \sin(\theta_x)/\lambda} e^{-j2\pi n_y d_y \sin(\theta_y)/\lambda} x_{0,0}(t). \tag{9.46}$$

Suppose that $x_{n_x,n_y}(t)$ is weighted by coefficient h_{n_x,n_y} and summed over all n_x and n_y to form $x_0(t)$. We have

$$x_0(t) = x_{0,0}(t) \sum_{n_x=0}^{N_x-1} \sum_{n_y=0}^{N_y-1} e^{-j2\pi n_x d_x \sin(\theta_x)/\lambda} e^{-j2\pi n_y d_y \sin(\theta_y)/\lambda} h_{n_x,n_y}. \tag{9.47}$$

The factor $\sum_{n_x=0}^{N_x-1} \sum_{n_y=0}^{N_y-1} e^{-j2\pi n_x d_x \sin(\theta_x)/\lambda} e^{-j2\pi n_y d_y \sin(\theta_y)/\lambda} h_{n_x,n_y}$ is the array factor for the two-dimensional rectangular array. Let $H_\theta(\theta_x, \theta_y)$ and $H_{A,E}(\theta_A, \theta_E)$ denote the array factor as functions of (θ_x, θ_y) and (θ_A, θ_E), respectively:

$$H_\theta(\theta_x, \theta_y) = \sum_{n_x=0}^{N_x-1} \sum_{n_y=0}^{N_y-1} e^{-j2\pi n_x d_x \sin(\theta_x)/\lambda} e^{-j2\pi n_y d_y \sin(\theta_y)/\lambda} h_{n_x,n_y}. \tag{9.48a}$$

$$H_{A,E}(\theta_A, \theta_E) = \sum_{n_x=0}^{N_x-1} \sum_{n_y=0}^{N_y-1} e^{-j2\pi n_x d_x \sin(\theta_A)\cos(\theta_E)/\lambda} e^{-j2\pi n_y d_y \cos(\theta_A)\cos(\theta_E)/\lambda} h_{n_x,n_y}. \tag{9.48b}$$

Let

$$\phi_x = 2\pi d_x \sin(\theta_x)/\lambda, \tag{9.49a}$$

$$\text{and} \quad \phi_y = 2\pi d_y \sin(\theta_y)/\lambda. \tag{9.49b}$$

Let $H_\phi(\phi_x, \phi_y)$ denote the array factor expressed as a function of (ϕ_x, ϕ_y):

$$H_\phi(\phi_x, \phi_y) = \sum_{n_x=0}^{N_x-1} \sum_{n_y=0}^{N_y-1} e^{-jn_x\phi_x} e^{-jn_y\phi_y} h_{n_x,n_y} \tag{9.50}$$

As we have discussed, the array shown in Fig. 9.21 may be regarded as a collection of N_y one-dimensional x-arrays each of which has N_x elements. Beamforming may be performed on each of the one-dimensional x-array. Suppose that the beamformer coefficients for all the N_y x-arrays are the same and the beamformer output for the n_yth x-array $\{x_{0,n_y}(t), x_{1,n_y}(t), \ldots, x_{N_x-1,n_y}(t)\}$ is $y_{n_y}(t)$. We have

$$y_{n_y}(t) = \sum_{n_x=0}^{N_x-1} x_{n_x,n_y}(t) h_{x,n_x}, \tag{9.51a}$$

$$= \sum_{n_x=0}^{N_x-1} e^{-j2n_x\pi d_x \sin(\theta_x)/\lambda} x_{0,n_y}(t) h_{x,n_x}, \tag{9.51b}$$

where h_{x,n_x} is the x-array beamformer coefficient values.

The N_y x-array outputs $\{y_0(t), y_1(t), \ldots, y_{N_y-1}(t)\}$ may be beamformed using coefficient values $\{h_{y,0}, h_{y,1}, \ldots, h_{y,N_y-1}\}$ to form $x_0(t)$. We have

$$x_0(t) = \sum_{n_y=0}^{N_y-1} y_{n_y}(t) h_{y,n_y}, \tag{9.52a}$$

$$= \sum_{n_y=0}^{N_y-1} \sum_{n_x=0}^{N_x-1} e^{-j2n_x\pi d_x \sin(\theta_x)/\lambda} x_{0,n_y}(t) h_{x,n_x} h_{y,n_y}. \tag{9.52b}$$

Expression (9.52b) is obtained by substituting $y_{n_y}(t)$ in (9.51b) into (9.52a). Since $x_{0,n_y}(t) = e^{-j2n_y\pi d_y \sin(\theta_y)/\lambda} x_{0,0}(t)$, (9.52b) becomes

$$x_0(t) = \sum_{n_y=0}^{N_y-1} \sum_{n_x=0}^{N_x-1} e^{-j2n_x\pi d_x \sin(\theta_x)/\lambda} e^{-j2n_y\pi d_y \sin(\theta_y)/\lambda} x_{0,0}(t) h_{x,n_x} h_{y,n_y}, \tag{9.53a}$$

$$= \sum_{n_y=0}^{N_y-1} \sum_{n_x=0}^{N_x-1} e^{-j2n_x\pi d_x \sin(\theta_x)/\lambda} e^{-j2n_y\pi d_y \sin(\theta_y)/\lambda} x_{0,0}(t) h_{n_x,n_y}, \tag{9.53b}$$

$$\text{where} \quad h_{n_x,n_y} = h_{x,n_x} h_{y,n_y}. \tag{9.53c}$$

Array factor with $h_{n_x,n_y} = h_{x,n_x} h_{y,n_y}$ is said to be separable. For separable array factor, (9.48) may be written as

$$H_\theta(\theta_x, \theta_y) = H_{\theta_x}(\theta_x) H_{\theta_y}(\theta_y), \tag{9.54a}$$

$$\text{where} \quad H_{\theta_x}(\theta_x) = \sum_{n_x=0}^{N_x-1} e^{-j2\pi n_x d_x \sin(\theta_x)/\lambda} h_{x,n_x}, \tag{9.54b}$$

$$H_{\theta_y}(\theta_y) = \sum_{n_y=0}^{N_y-1} e^{-j2\pi n_y d_y \sin(\theta_y)/\lambda} h_{y,n_y}. \tag{9.54c}$$

and

$$H_\phi(\phi_x, \phi_y) = H_{\phi_x}(\phi_x) H_{\phi_y}(\phi_y), \tag{9.55a}$$

$$\text{where} \quad H_{\phi_x}(\phi_x) = \sum_{n_x=0}^{N_x-1} e^{-jn_x\phi_x} h_{x,n_x}, \tag{9.55b}$$

$$H_{\phi_y}(\phi_y) = \sum_{n_y=0}^{N_y-1} e^{-jn_y\phi_y} h_{y,n_y}. \tag{9.55c}$$

Since, for separable array factor, $H_\theta(\theta_x, \theta_y) = H_{\theta_x}(\theta_x) H_{\theta_y}(\theta_y)$ or $H_\phi(\phi_x, \phi_y) = H_{\phi_x}(\phi_x) H_{\phi_y}(\phi_y)$, the separable two-dimensional array factor is equal to the product of two linear arrays, one in the n_x-direction and one in the n_y-direction. Assume that $H_{\phi_x}(\phi_x)$ and $H_{\phi_y}(\phi_y)$ are not steered, i.e., $|H_{\phi_x}(\phi_x)|$ and $|H_{\phi_y}(\phi_y)|$ reach their maximum when $\phi_x = 0$ and $\phi_y = 0$, respectively. Assume also that the coefficient values of $H_{\phi_x}(\phi_x)$ and $H_{\phi_y}(\phi_y)$ are normalized such that $|H_{\phi_x}(\phi_x)| = 1$ when $\phi_x = 0$ and $|H_{\phi_y}(\phi_y)| = 1$ when $\phi_y = 0$. $|H_\phi(\phi_x, \phi_y)|$ is very small when both $|H_{\phi_x}(\phi_x)|$ and $|H_{\phi_y}(\phi_y)|$ are small; this occurs when both ϕ_x and ϕ_y are not within the respective main lobes of $H_{\phi_x}(\phi_x)$ and $H_{\phi_y}(\phi_y)$. When $\phi_x = 0$, $H_{\phi_x}(\phi_x) = 1$ and $H_\phi(\phi_x, \phi_y) = H_{\phi_y}(\phi_y)$. When $\phi_y = 0$, $H_{\phi_y}(\phi_y) = 1$ and $H_\phi(\phi_x, \phi_y) = H_{\phi_x}(\phi_x)$. As a consequence, the sidelobe magnitudes along the n_x and n_y axes are significantly larger than those away from the n_x and n_y axes.

We shall illustrate this by using an example with $N_x = N_y = 10$ and $h_{n_x,n_y} = h_{x,n_x} = h_{y,n_y} = 1$ for $0 \le n_x \le 9$ and $0 \le n_y \le 9$. Ignoring the linear phase term $4e^{-j9(\phi_x+\phi_y)/2}$, the array factor $H_\phi(\phi_x, \phi_y)$ is given by

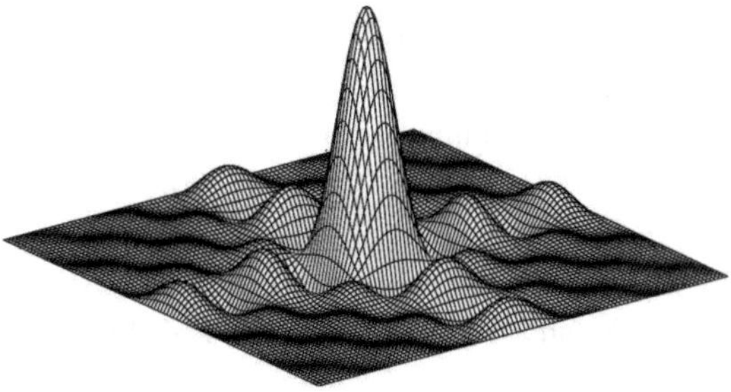

Fig. 9.28 $H_\phi(\phi_x, \phi_y)$ plot for $h_{n_x,n_y} = 1$, $N_x = N_y = 10$

$H_\phi(\phi_x, \phi_y) = \sum_{n_x=0}^{4} \sum_{n_y=0}^{4} \cos((n_x + 0.5)\phi_x)\cos((n_y + 0.5)\phi_y)$. $H_\phi(\phi_x, \phi_y)$ is plotted in Fig. 9.28. It can be seen from Fig. 9.28 that the sidelobe magnitudes along the n_x and n_y axes are significantly larger than those away from the n_x and n_y axes.

For linear array, the array factor expression for radiating array is the same as that for receiving array. It is also the case for two-dimensional array; the two-dimensional array factor expression for radiating array is the same as that for receiving array.

The boresight of the two-dimensional array factor $H_\phi(\phi_x, \phi_y)$ can be steered from $\phi_x = 0$ and $\phi_y = 0$ to $\phi_x = \phi_{x0}$ and $\phi_y = \phi_{y0}$ by replacing ϕ_x and ϕ_y in (9.50) by $\phi_x - \phi_{x0}$ and $\phi_y - \phi_{y0}$, respectively. Hence, the two-dimensional version of (9.20) is given by

$$H_\phi(\phi_x - \phi_{x0}, \phi_y - \phi_{y0}) = \sum_{n_x=0}^{N_x-1} \sum_{n_y=0}^{N_y-1} e^{-jn_x(\phi_x - \phi_{x0})} e^{-jn_y(\phi_y - \phi_{y0})} h_{n_x,n_y}, \tag{9.56a}$$

$$= \sum_{n_x=0}^{N_x-1} \sum_{n_y=0}^{N_y-1} e^{-jn_x\phi_x} e^{-jn_y\phi_y} \{e^{jn_x\phi_{x0}} e^{jn_y\phi_{y0}} h_{n_x,n_y}\},$$

$$= \sum_{n_x=0}^{N_x-1} \sum_{n_y=0}^{N_y-1} e^{-jn_x\phi_x} e^{-jn_y\phi_y} \hat{h}_{n_x,n_y}, \tag{9.56b}$$

$$\text{where} \quad \hat{h}_{n_x,n_y} = e^{jn_x\phi_{x0}} e^{jn_y\phi_{y0}} h_{n_x,n_y}. \tag{9.56c}$$

Since $e^{-jn_x\phi_x} e^{-jn_y\phi_y} = e^{-jn_x(\phi_x + 2\pi \times \text{integer-1})} e^{-jn_y(\phi_y + 2\pi \times \text{integer-2})}$, where integer-1 and integer-2 are two arbitrary integers, the array factor $H_\phi(\phi_x, \phi_y)$ is periodic in both ϕ_x and ϕ_y with period 2π.

9.9 Symmetrical Two-Dimensional Arrays

Consider a two-dimensional array where the coefficient values are real and symmetrical, i.e., $h_{n_x,n_y} = h_{N_x-1-n_x,n_y} = h_{n_x,N_y-1-n_y} = h_{N_x-1-n_x,N_y-1-n_y}$. Ignoring the $e^{-j\frac{N_x-1}{2}\phi_x}e^{-j\frac{N_y-1}{2}\phi_y}$ factor, the two-dimensional array factor $H_\phi(\phi_x, \phi_y)$ shown in (9.50) may be written in a trigonometric form as follows:

For N_x and N_y both even, $H_\phi(\phi_x, \phi_y)$ is given by

$$H_\phi(\phi_x, \phi_y) = \sum_{n_x=1}^{\frac{N_x}{2}} \sum_{n_y=1}^{\frac{N_y}{2}} \alpha_{n_x,n_y} \cos\left(\frac{2n_x - 1}{2}\phi_x\right) \cos\left(\frac{2n_y - 1}{2}\phi_y\right), \tag{9.57}$$

where $\alpha_{n_x,n_y} = 4h_{\frac{N_x}{2}-n_x,\frac{N_y}{2}-n_y}$.

For N_x and N_y both odd, $H_\phi(\phi_x, \phi_y)$ is given by

$$H_\phi(\phi_x, \phi_y) = \sum_{n_x=0}^{\frac{N_x-1}{2}} \sum_{n_y=0}^{\frac{N_y-1}{2}} \alpha_{n_x,n_y} \cos(n_x\phi_x) \cos(n_y\phi_y), \tag{9.58}$$

where $\alpha_{0,0} = h_{\frac{N_x-1}{2},\frac{N_y-1}{2}}$; $\alpha_{n_x,0} = 2h_{\frac{N_x-1}{2}-n_x,\frac{N_y-1}{2}}$, $n_x \neq 0$; $\alpha_{0,n_y} = 2h_{\frac{N_x-1}{2},\frac{N_y-1}{2}-n_y}$, $n_y \neq 0$; and $\alpha_{n_x,n_y} = 4h_{\frac{N_x-1}{2}-n_x,\frac{N_y-1}{2}-n_y}$, $n_x \neq 0, n_y \neq 0$.

For N_x even and N_y odd, $H_\phi(\phi_x, \phi_y)$ is given by

$$H_\phi(\phi_x, \phi_y) = \sum_{n_x=1}^{\frac{N_x}{2}} \sum_{n_y=0}^{\frac{N_y-1}{2}} \alpha_{n_x,n_y} \cos\left(\frac{2n_x - 1}{2}\phi_x\right) \cos(n_y\phi_y), \tag{9.59}$$

where $\alpha_{n_x,0} = 2h_{\frac{N_x}{2}-n_x,\frac{N_y}{2}}$, and $\alpha_{n_x,n_y} = 4h_{\frac{N_x}{2}-n_x,\frac{N_y}{2}-n_y}$, $n_y \neq 0$.

For N_x odd and N_y even, $H_\phi(\phi_x, \phi_y)$ is given by

$$H_\phi(\phi_x, \phi_y) = \sum_{n_x=0}^{\frac{N_x-1}{2}} \sum_{n_y=1}^{\frac{N_y}{2}} \alpha_{n_x,n_y} \cos(n_x\phi_x) \cos\left(\frac{n_y - 1}{2}\phi_y\right), \tag{9.60}$$

where $\alpha_{0,n_y} = 2h_{\frac{N_x}{2},\frac{N_y}{2}-n_y}$, and $\alpha_{n_x,n_y} = 4h_{\frac{N_x}{2}-n_x,\frac{N_y}{2}-n_y}$, $n_x \neq 0$.

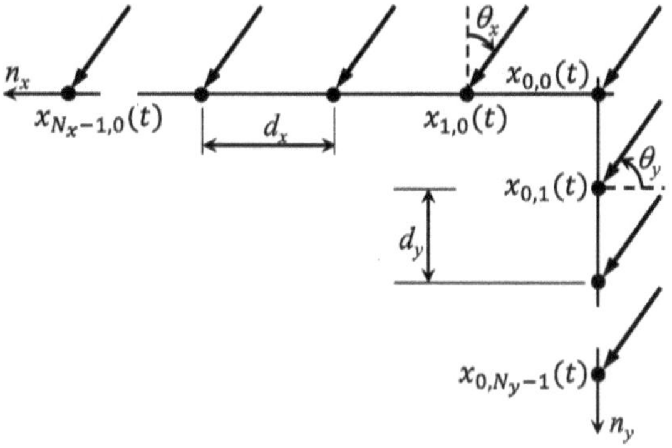

Fig. 9.29 L-shaped array

9.10 L-Shaped Arrays

Figure 9.29 shows an L-shaped array. The total number of array elements for an L-shaped array is $N_x + N_y - 1$ where N_x and N_y are the number of array elements for the x-array and y-array, respectively. The equivalents of (9.48) and (9.50) for the L-shaped array are given by

$$H_\Theta(\theta_x, \theta_y) = h_{0,0} + \sum_{n_x=1}^{N_x-1} e^{-j2\pi n_x d_x \sin(\theta_x)/\lambda} h_{n_x,0} + \sum_{n_y=1}^{N_y-1} e^{-j2\pi n_y d_y \sin(\theta_y)/\lambda} h_{0,n_y}. \tag{9.61a}$$

$$H_{A,E}(\theta_A, \theta_E) = h_{0,0} + \sum_{n_x=1}^{N_x-1} e^{-j2\pi n_x d_x \sin(\theta_A)\cos(\theta_E)/\lambda} h_{n_x,0} + \sum_{n_y=1}^{N_y-1} e^{-j2\pi n_y d_y \cos(\theta_A)\cos(\theta_E)/\lambda} h_{0,n_y}. \tag{9.61b}$$

$$H_\Phi(\phi_x, \phi_y) = h_{0,0} + \sum_{n_x=1}^{N_x-1} e^{-jn_x\phi_x} h_{n_x,0} + \sum_{n_y=1}^{N_y-1} e^{-jn_y\phi_y} h_{0,n_y}. \tag{9.61c}$$

9.11 Classical Beamformer

Beamforming involves the derivation of the coefficient values of $H_\phi(\phi)$ to combine signals from the array elements in such a way that those in a particular direction are added constructively and those in other directions are added destructively to achieve a given spatial selectivity specification. The coefficients of a classical beamformer is the same as the coefficients of the window function of a classical window in signal processing for sectioning an infinite duration signal into a finite duration signal. Many beamforming techniques have been developed over the past decades. We shall present some of the most common linear beamformers in this section. In the following, we shall use the notation $H_{\phi,\text{name}}(\phi)$ to denote the array factor of a particular beamformer called "name," for example, the array factor of a uniform beamformer is denoted as $H_{\phi,\text{uniform}}(\phi)$.

9.11.1 Uniform Array

For the uniform array, $h_n = 1 \ \forall \ n$ within the support of the array. Its array factor $H_{\phi,\text{uniform}}(\phi)$ is given by

$$H_{\phi,\text{uniform}}(\phi) = \frac{\sin(N\phi/2)}{\sin(\phi/2)}. \tag{9.62}$$

As can be seen from (9.62), $H_{\phi,\text{uniform}}(\phi) \longrightarrow N$ as $\phi \longrightarrow 0$. As $|\phi|$ increases from 0, $H_{\phi,\text{uniform}}(\phi)$ decreases from N and $H_{\phi,\text{uniform}}(\phi) = 0$ when $\phi = \pm 2\pi/N$. Thus, the first zeros of $H_{\phi,\text{uniform}}(\phi)$ occur at $\phi = \pm 2\pi/N$. When $\phi = \pm 2\pi/N$, $\theta = \pm \sin^{-1}(\lambda/Nd)$. If N is large, $\sin^{-1}(\lambda/(Nd) \approx \lambda/(Nd) \propto 1/N$. The beamwidth, defined as the separation between the two zeros immediately adjacent to the main lobe, is $2\sin^{-1}(\lambda/Nd) \approx 2\lambda/(Nd) \propto 1/N$ for large N; it is inversely proportional to N.

To determine the half-power beamwidth, we note that the maximum magnitude of $H_{\phi,\text{uniform}}(\phi)$ is N. At half-power point, its magnitude is $N/\sqrt{2}$. Let $\phi = \phi_{\text{half}}$ and $\theta = \theta_{\text{half}}$ at half-power point. Hence, we have

$$\begin{aligned} H_{\phi,\text{uniform}}(\phi_{\text{half}}) &= \frac{\sin(N\phi_{\text{half}}/2)}{\sin(\phi_{\text{half}}/2)} = \frac{N}{\sqrt{2}}, \\ \Longrightarrow \sin(N\phi_{\text{half}}/2) &= \frac{N\sin(\phi_{\text{half}}/2)}{\sqrt{2}}. \end{aligned} \tag{9.63}$$

The half-power point occurs between the positive first zero and negative first zero. Hence, $|\phi_{\text{half}}| < 2\pi/N$. As a consequence, for large N, $|\phi_{\text{half}}|$ is small and $\sin(|\phi_{\text{half}}|/2) \approx |\phi_{\text{half}}|/2$. From (9.63), for large N,

$$\begin{aligned} &\sin(N\phi_{\text{half}}/2) \approx \frac{N\phi_{\text{half}}}{2\sqrt{2}}, \\ &\Longrightarrow \phi_{\text{half}} \approx \pm 2.7831/N \propto 1/N, \\ &\Longrightarrow \theta_{\text{half}} \approx \pm 1.3916\lambda/(\pi d N) \propto 1/N. \end{aligned} \tag{9.64}$$

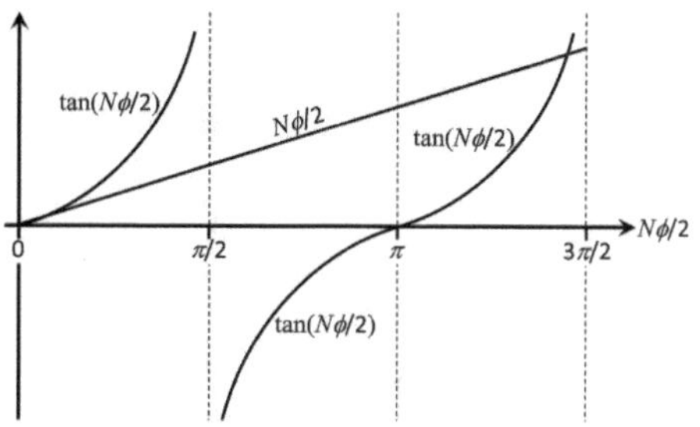

Fig. 9.30 $N\phi/2$ and $\tan(N\phi/2)$ plots

The peak of the first sidelobe of $H_{\phi,\text{uniform}}(\phi)$ occurs near $\tan(N\phi/2) = 3\pi/2$. To determine the peak, differentiate $H_{\phi,\text{uniform}}(\phi)$ with respect to ϕ, and equate the derivative to 0.

$$\frac{\mathrm{d}H_{\phi,\text{uniform}}(\phi)}{\mathrm{d}\phi} = \frac{\sin(\phi/2)\cos(N\phi/2)N/2 - \sin(N\phi/2)\cos(\phi/2)/2}{\sin^2(\phi/2)},$$
$$= 0 \Longrightarrow N\tan(\phi/2) = \tan(N\phi/2). \tag{9.65}$$

The second zero of $H_{\phi,\text{uniform}}(\phi)$ occurs at $\phi = 4\pi/N$. The first sidelobe peak occurs before the second zero. Hence, for large N, $|\phi|$ is small. From (9.65), for large N, $\tan(\phi/2) \approx \phi/2$. Substituting $\tan(\phi/2) \approx \phi/2$ into (9.65), we have $N\phi_{\text{1st-peak}}/2 \approx \tan(N\phi_{\text{1st-peak}}/2)$ where $\phi_{\text{1st-peak}}$ is the value of ϕ at the first sidelobe peak. To determine $\phi_{\text{1st-peak}}$, we need to solve for $N\phi/2 = \tan(N\phi/2)$. Figure 9.30 shows plots for $N\phi/2$ and $\tan(N\phi/2)$. There is a solution at $N\phi/2 = 0$; this corresponds to the main lobe peak. The solution corresponding to the first sidelobe peak is the one in the third quadrant, i.e., $\pi \le N\phi/2 \le 3\pi/2$. The solution is $N\phi/2 \approx \pi + \tan^{-1}(N\phi/2)$ or $\phi \approx 8.9868/N$ or $\theta \approx 4.4934\lambda/(\pi dN)$. The gain at the first sidelobe peak is $H_{\phi,\text{uniform}}(8.9869/N) \approx -0.2172N$. The normalized gain at the first sidelobe peak (for large N) $\approx \left|\frac{H_{\phi,\text{uniform}}(8.9869/N)}{H_{\phi,\text{uniform}}(0)}\right| \approx 0.2172 \approx -13.26$ dB. Since $H_{\phi,\text{uniform}}(\phi) = \frac{\sin(N\phi/2)}{\sin(\phi/2)}$, sidelobes reach their maximum when $|\sin(N\phi_{\text{side-peak}}/2)| \approx 1$ where $\phi_{\text{side-peak}}$ is the value of ϕ at sidelobe peak. For small $|\phi/2|$, $\sin(\phi/2) \approx \phi/2$. Hence, for small $|\phi_{\text{side-peak}}/2|$, $H_{\phi,\text{uniform}}(\phi_{\text{side-peak}}) \approx 2/\phi_{\text{side-peak}}$. Hence, sidelobe peaks of $H_{\phi,\text{uniform}}(\phi)$ roll off at 6 dB per octave.

The array factor plots for $N = 20$ and $N = 4$ of uniform beamformers are shown in Fig. 9.31. It can be seen from the plots that when N is increased from 4 to 20, the reduction in the main lobe width is significant, whereas the reduction in the first sidelobe peak magnitude is small. The peak magnitude of the first sidelobe is about -13 dB for large N. The half-power angle, first zero, and first sidelobe peak for the uniform array for N ranging from 2 to 12 are shown in Table 9.1. In Table 9.1, "Simulated" values are values obtained from computer simulations, whereas "Predicted" values are values obtained from the

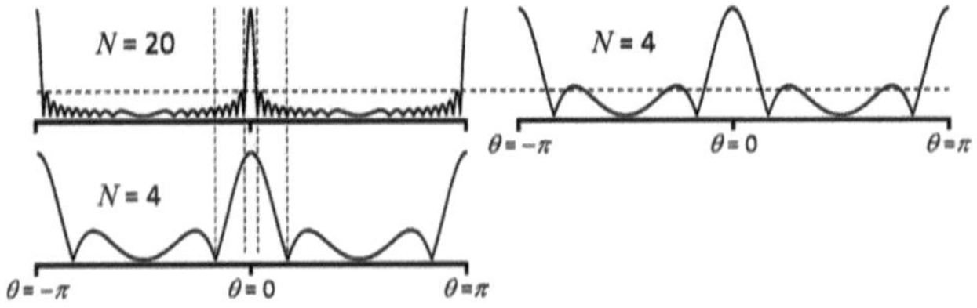

Fig. 9.31 Array factor plots for $N = 20$ and $N = 4$ of uniform beamformers

Table 9.1 Uniform array half-power angle, first zero, and first sidelobe for $\lambda = 2d$. Half power: $\theta \approx \sin^{-1}(1.3916\lambda/(\pi Nd))$ for $N \gg 1$; $\theta = \sin^{-1}(\lambda/4d)$ for $N = 2$. First sidelobe peak: $\theta \approx \sin^{-1}(4.4934\lambda/(\pi Nd))$ for $N \gg 1$

N	Half-power angle (radian)		First zero angle/rad	First sidelobe peak angle/rad		First sidelobe peak gain/dB	
	Simulated	Predicted	$\sin^{-1}(\lambda/Nd)$	Simulated	Predicted	Simulated	Predicted
2	0.5236	0.5236	$\pi/2$				
3	0.3157	0.300	0.72973	$\pi/2$ (1.5708)	1.265	−9.54	−13.26
4	0.2297	0.223	0.52360	0.8217	0.797	−11.30	−13.26
5	0.1813	0.1781	0.41152	0.6193	0.6091	−12.04	−13.26
6	0.1500	0.1482	0.33984	0.5023	0.4970	−12.43	−13.26
7	0.1280	0.1266	0.28975	0.4241	0.4210	−12.65	−13.26
8	0.1117	0.1110	0.25268	0.3677	0.3657	−12.80	−13.26
9	0.0991	0.0986	0.22409	0.3249	0.3235	−12.90	−13.26
10	0.0891	0.0897	0.20136	0.2911	0.2901	−12.96	−13.26
11	0.0809	0.0806	0.18284	0.2638	0.2631	−13.02	−13.26
12	0.0741	0.0739	0.16745	0.2413	0.2407	−13.06	−13.26

expressions "half-power $\theta \approx \sin^{-1}(1.3916\lambda/(\pi Nd))$" and "first sidelobe peak $\theta \approx \sin^{-1}(4.4934\lambda/(\pi Nd))$."

9.11.2 Binomial Array

It can be seen from Figs. 9.7 and 9.8 that the array factor $H_\phi(\phi) = 1 + e^{-j\phi}$ has no sidelobe. Since $1 + e^{-j\phi}$ has no sidelobe, the binomial array factor

$$H_{\phi,\mathrm{binomial}}(\phi) = (1 + e^{-j\phi})^M = \sum_{r=0}^{M} \{e^{-jr\phi} h_r\} \tag{9.66}$$

will have no sidelobe. In (9.66), $M = N - 1$ and

$$h_r = C_r^M = \frac{M!}{r!(M-r)!} \ . \tag{9.67}$$

Define the factorial of zero $0! = 1! = 1$. Note that $h_0 = h_{N-1} = C_0^{N-1} = \frac{(N-1)!}{0!(N-1)!}$ $= 1$ and $h_1 = h_{N-2} = C_1^{N-1} = \frac{(N-1)!}{1!(N-2)!} = N - 1$. Furthermore, we have

$$|H_{\phi,\text{binomial}}(\phi)| = |1 + e^{-j\phi}|^M = |\cos(\phi/2)|^M = |\cos(\phi/2)|^{N-1}. \tag{9.68}$$

Let θ_{half} be the value of θ at half-power point, and let the corresponding value of ϕ at half-power point be ϕ_{half}. Substituting $\phi = \phi_{\text{half}}$ into (9.68), $|H_{\phi,\text{binomial}}(\phi_{\text{half}})|$ $= |\cos(\phi_{\text{half}}/2)|^{N-1}$. At half-power point, $|H_{\phi,\text{binomial}}(\phi_{\text{half}})| = 1/\sqrt{2}$. Therefore, $\cos(\phi_{\text{half}}/2) = \frac{1}{2(N-1)\sqrt{2}}$. Hence, $\phi_{\text{half}} = 2\cos^{-1}\left(\frac{1}{2(N-1)\sqrt{2}}\right)$. Consequently,

$$\theta_{\text{half}} = \sin^{-1}\left(\frac{\lambda\cos^{-1}\left(\frac{1}{2(N-1)\sqrt{2}}\right)}{\pi d}\right) = \sin^{-1}\left(\frac{\lambda\cos^{-1}\left((\sqrt{2})^{-1/(N-1)}\right)}{\pi d}\right). \tag{9.69}$$

HPBW (half-power beamwidth) is equal to twice θ_{half}. A comparison between the half-power angle of uniform array and binomial array for $\lambda = 2d$ is shown in Table 9.2. It can be seen from Table 9.2 that, for a given N, the half-power angle of the uniform array is narrower than that of the binomial array. The binomial coefficient values are shown in the Pascal triangle in Fig. 9.32. (Pascal's triangle

Table 9.2 Comparison between HPBW angles of uniform array and binomial array

$\lambda = 2d$		
	Half-power angle (unit = radian)	
N	Uniform array	Binomial array
2	0.5236	0.5236
3	0.3157	0.3726
4	0.2297	0.3049
5	0.1813	0.2643
6	0.1500	0.2365
7	0.1280	0.2169
8	0.1117	0.2000
9	0.0991	0.1871
10	0.0891	0.1765
11	0.0809	0.1674
12	0.0741	0.1596

Fig. 9.32 Pascal triangle of binomial coefficient values

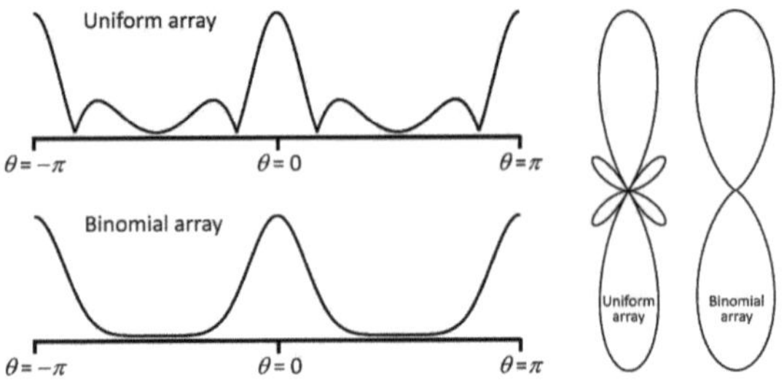

Fig. 9.33 Comparison between the spatial selectivity of uniform array and binomial array. Four array elements

was known well before Pascal's time. It is known as Yang Hui's triangle in China, Halayudha or Mahaviracharya or Bhattotpala triangle in India, Al-Karaji or Khayyam triangle in Iran, Tartaglia's triangle in Italy, etc.) The sum of the Nth row coefficient values of the Pascal triangle is 2^{N-1}, for example, for $N = 5$, the coefficients are 1, 4, 6, 4, and 1; the sum is $1 + 4 + 6 + 4 + 1 = 16 = 2^{5-1}$. Note that the sum is contributed mostly by the coefficients at the center. This means that the magnitude disparity of the coefficient values is very large for large N. The large magnitude disparity causes serious implementation problem. A comparison between the spatial selectivity of the uniform array and the binomial array for a four-element array is shown in Fig. 9.33. It can be seen from Fig. 9.33 that the binomial array has no sidelobe but its beamwidth is wider than that of the uniform array with the same N. For N larger than 10 and $\lambda = 2d$, the beamwidth of the binomial array is more than twice that of the uniform array. For a given N, there are many other types of beamformer with beamwidth narrower than that of the binomial beamformer and with peak sidelobe magnitude smaller than that of the uniform beamformer.

9.11.3 Triangular Array

Let $H_{\phi,\text{uniform},N}(\phi)$ denote the array factor of a uniform array with N elements. We have

$$H_{\phi,\text{uniform},N}(\phi) = \sum_{n=0}^{N-1} e^{-j\phi n}. \tag{9.70}$$

The triangular array factor $H_{\phi,\text{triangular},2N-1}(\phi)$ for a $(2N - 1)$-element triangular array is given by

$$\begin{aligned} H_{\phi,\text{triangular},2N-1}(\phi) &= \left(H_{\phi,\text{uniform},N}(\phi)\right)^2 , \\ &= \sum_{n=0}^{N-1}(n+1)e^{-j\phi n} + \sum_{n=N}^{2N-2}(2N-1-n)e^{-j\phi n} . \end{aligned} \tag{9.71}$$

It has triangular coefficient values, for example, $H_{\phi,\text{triangular},7}(\phi) = 1 + 2e^{-j\phi} + 3e^{-2j\phi} + 4e^{-3j\phi} + 3e^{-4j\phi} + 2e^{-5j\phi} + e^{-6j\phi}$. The coefficient values are 1, 2, 3, 4, 3, 2, and 1; it is a triangle. The sum of the coefficient values of $H_{\phi,\text{triangular},2N-1}(\phi)$ is N^2, for example, for $N = 4$, the sum of the coefficients is $1 + 2 + 3 + 4 + 3 + 2 + 1 = 16 = 4^2$. The magnitude disparity of the coefficient values for the triangular beamformer is very much smaller than that for the binomial beamformer. Since $H_{\phi,\text{triangular},2N-1}(\phi) = \left(H_{\phi,\text{uniform},N}(\phi)\right)^2$, for large N, the first sidelobe peak of $H_{\phi,\text{triangular},2N-1}(\phi)$ is $2 \times (-13.26)\text{dB} = -26.52\text{dB}$. The zeros of $H_{\phi,\text{triangular},2N-1}(\phi)$ and those of $H_{\phi,\text{uniform},N}(\phi)$ are at the same locations. Similarly, the sidelobe peaks of $H_{\phi,\text{triangular},2N-1}(\phi)$ and those of $H_{\phi,\text{uniform},N}(\phi)$ are also at the same locations.

9.11.4 Hann Array

The coefficient values h_r of the Hann (or Hanning or raised cosine) beamformer are given by

$$h_r = 0.5 - 0.5\cos\left(\frac{2\pi r}{N-1}\right) \quad , \quad r = 0, 1, \ldots, N-1 \ . \tag{9.72}$$

It can be easily verified that $h_0 = h_{N-1} = 0$. Thus, there are only $N - 2$ nonzero coefficients. In order to have N nonzero coefficients, we shall design an $(N+2)$-element array and discard the two zero-value coefficients. Hence, (9.72) for the Hanning beamformer is redefined as

$$h_r = 0.5 - 0.5\cos\left(\frac{2\pi(r+1)}{N+1}\right) \quad , \quad r = 0, 1, \ldots, N-1 \ . \tag{9.73}$$

For large N, the sidelobe peak roll-off of the Hanning beamformer is 18 dB per octave; the first sidelobe peak is − 31.5 dB. Its main lobe width is approximately twice that of the uniform beamformer.

9.11.5 Hamming Array

The expression shown in (9.72) may be generalized to

$$h_r = (1-\alpha) - \alpha\cos\left(\frac{2\pi r}{N-1}\right) \quad , \quad r = 0, 1, \ldots, N-1 \ . \tag{9.74}$$

The value of α may be selected to minimize sidelobe. The best solution that minimizes peak sidelobe magnitude is $\alpha = 0.46164$. This is called Hamming

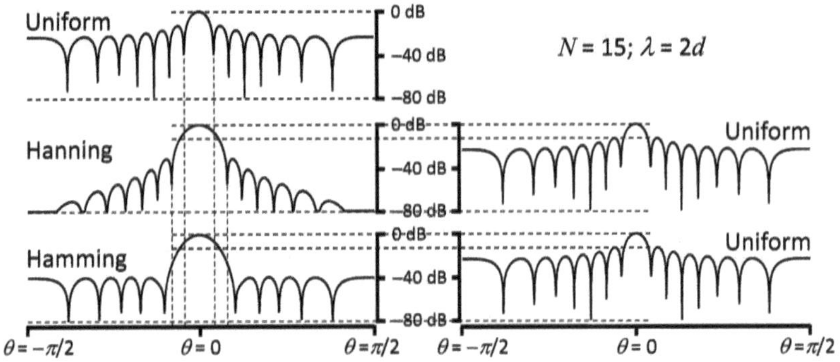

Fig. 9.34 Spatial selectivity of a 15-element uniform, Hanning, and Hamming beamformer

beamformer or Hamming window. Hence, the coefficient value for Hamming beamformer is

$$h_r = 0.53836 - 0.46164 \cos\left(\frac{2\pi r}{N-1}\right) \quad , \quad r = 0, 1, \ldots, N-1 \ . \tag{9.75}$$

For large N, the first sidelobe peak is $-$ 41 dB. Figure 9.34 compares the spatial selectivity of a 15-element uniform, Hanning, and Hamming beamformer with $\lambda = 2d$.

9.11.6 Blackman and Blackman-Harris Array

The expression shown in (9.74) may be further generalized by adding another cosine term:

$$h_r = (1 - \alpha_1 - \alpha_2) - \alpha_1 \cos\left(\frac{2\pi r}{N-1}\right) + \alpha_2 \cos\left(\frac{4\pi r}{N-1}\right) \quad , \quad r = 0, 1, \ldots, N-1 \tag{9.76}$$

In Blackman array, the values of α_1 and α_2 are selected to optimize sidelobe peak and sidelobe peak roll-off rate. For large N, the first sidelobe peak is $-$ 58 dB, and the sidelobe roll-off is 18 dB per octave. The coefficient values for Blackman array are

$$h_r = 0.42 - 0.5 \cos\left(\frac{2\pi (r+1)}{N+1}\right) + 0.08 \cos\left(\frac{4\pi (r+1)}{N+1}\right) \quad , \quad r = 0, 1, \ldots, N-1. \tag{9.77}$$

In the three-term Blackman-Harris beamformer, the values of α_1 and α_2 are selected to optimize only the sidelobe peak. For large N, the sidelobe peak is about -72 dB. The coefficient values are given by

$$h_r = 0.4243801 - 0.4973406 \cos\left(\frac{2\pi r}{N-1}\right) + 0.0782793 \cos\left(\frac{4\pi r}{N-1}\right) \quad , \quad r = 0, 1, \ldots, N-1. \tag{9.78}$$

The coefficient values for the four-term Blackman-Harris beamformer are given by

$$h_r = 0.35875 - 0.48829 \cos\left(\frac{2\pi r}{N-1}\right) + 0.14128 \cos\left(\frac{4\pi r}{N-1}\right) - 0.01168 \cos\left(\frac{6\pi r}{N-1}\right) \quad , \quad r = 0, 1, \ldots, N-1. \tag{9.79}$$

9.11.7 Exponential Array

In uniform, binomial, triangular, Hanning, Hamming, Blackman, and Blackman-Harris beamformers, different beamformer offers different sidelobe rejection characteristics. There is no parameter for trading off beamwidth and sidelobe rejection. The selection of the type of beamformer is determined by the desired sidelobe characteristics. After the beamformer type is selected, the number of array elements N is selected to meet the desired beamwidth. Some beamformers provide a parameter for trading off sidelobe rejection for beamwidth. Such a parameter exists in the exponential, cosh, and Kaiser beamformers. The performance of the exponential, cosh, AND Kaiser beamformers is similar.

The coefficient values of the exponential beamformer are given by

$$h_r = \frac{e^{\left(\beta\sqrt{1-\left(\frac{2r}{N-1}-1\right)^2}\right)}}{e^{\beta}} \quad , \quad r = 0, 1, \ldots, N-1 \ . \tag{9.80}$$

It has two parameters, N and β. The value of β depends on the sidelobe rejection A dB and is given by

$$\beta = \begin{cases} 0 & , \quad A < 13.26 \ , \\ -0.001513A^2 + 0.2809A - 3.398 & , \quad 13.26 < A < 50 \ , \\ -0.0001085A^2 + 0.1506A - 0.304 & , \quad 50 > A \ . \end{cases} \tag{9.81}$$

9.11.8 Cosh Array

The coefficient values of the cosh beamformer are given by

$$h_r = \frac{\cosh\left(\beta\sqrt{1-\left(\frac{2r}{N-1}-1\right)^2}\right)}{\cosh(\beta)} \quad , \quad r = 0, 1, \ldots, N-1 \ . \tag{9.82}$$

It has two parameters, N and β. The value of β depends on the sidelobe rejection A dB and is given by

$$\beta = \begin{cases} 0 & , \quad A < 13.26 \ , \\ 0.4611(A-13.26)^{0.4} + 0.1165(A-13.26) & , \quad 13.26 < A < 50 \ , \\ -0.00008763A^2 + 0.1469A - 0.1461 & , \quad 50 > A \ . \end{cases} \tag{9.83}$$

9.11.9 Kaiser Array

The coefficient values of the Kaiser beamformer are given by

$$h_r = \frac{I_0\left(\beta\sqrt{1-\left(\frac{2r}{N-1}-1\right)^2}\right)}{I_0(\beta)} \quad , \quad r = 0, 1, \ldots, N-1 \ . \tag{9.84}$$

$I_0(x)$ is the modified zeroth-order Bessel function of the first kind. The value of β depends on the sidelobe rejection A dB and is given by

$$\beta = \begin{cases} 0 & , \quad A < 21 \ , \\ 0.5842(A-21)^{0.4} + 0.07886(A-21) & , \quad 21 < A < 50 \ , \\ 0.1102(A-8.7) & , \quad 50 > A \ . \end{cases} \tag{9.85}$$

The modified zeroth-order Bessel function of the first kind may be evaluated using the following formula:

$$I_0(x) = 1 + \sum_{k=1}^{\infty} \frac{\left(\frac{x^2}{4}\right)^k}{(k!)^2} \tag{9.86}$$

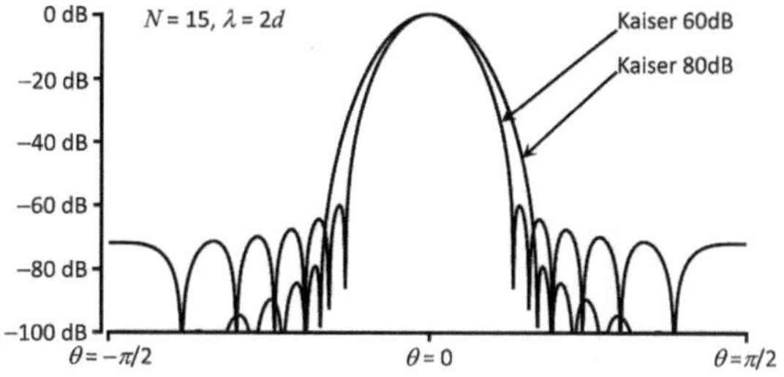

Fig. 9.35 Responses of Kaiser beamformer

The responses of two Kaiser beamformers are shown in Fig. 9.35; one of them has peak sidelobe magnitude of -60 dB and the other one has peak sidelobe magnitude of -80 dB.

9.11.10 Dolph-Chebyshev Array

In exponential, cosh, and Kaiser beamformers, there is a parameter for controlling the sidelobe rejection. However, the control is only approximate. It is not possible to precisely specify the sidelobe rejection, although this may not be necessary. In the Dolph-Chebyshev beamformer, it is possible to precisely specify the sidelobe rejection. The sidelobe of the Dolph-Chebyshev beamformer is equiripple, i.e., all sidelobes have the same peak magnitude. The Dolph-Chebyshev beamformer is optimum in the sense that, for a given peak sidelobe magnitude, it has the minimum beamwidth; the beamwidth is defined as the width of the beam with gain larger than the peak sidelobe gain.

The Dolph-Chebyshev array is developed based on the Chebyshev polynomial $T_M(x)$ of the first kind defined as

$$T_M(x) = \begin{cases} (-1)^M \cosh(M \cosh^{-1}(-x)), & x < -1 \ , \\ \cos(M \cos^{-1}(x)) \ , & -1 \le x \le 1 \ , \\ \cosh(M \cosh^{-1}(x)) \ , & x > 1 \ , \end{cases} \tag{9.87a}$$

$$M = N - 1 \ . \tag{9.87b}$$

The function $T_M(x)$ may be expressed in the form

$$T_M(x) = \begin{cases} \sum_{m=0}^{M} a_m x^m \ , \\ m = 0, 2, 4, \ldots \text{ for } M \text{ even}, \\ m = 1, 3, 5, \ldots \text{ for } M \text{ odd}, \end{cases} \tag{9.88}$$

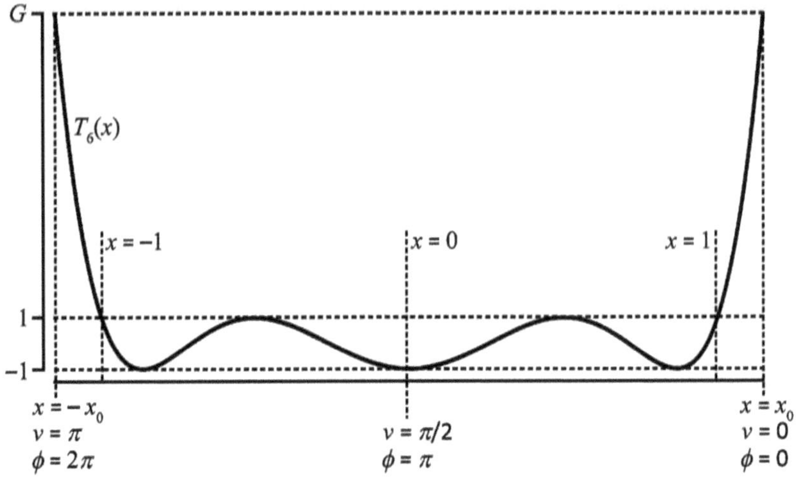

Fig. 9.36 Plot of $T_6(x)$

and can be computed recursively using (9.89),

$$T_m(x) = 2xT_{m-1} - T_{m-2}(x), m = 2, 3, \ldots, M, \tag{9.89}$$

with $T_0(x) = 1$ and $T_1(x) = x$. A plot of $T_6(x)$ is shown in Fig. 9.36. $T_M(x)$ oscillates between 1 and -1 for $-1 \leq x \leq 1$. $|T_M(x)|$ increases as $|x|$ increases beyond unity. Let $x = x_0 \cos(v)$. The array factor of the Dolph-Chebyshev array is $T_M(x_0 \cos(v))$ where $\phi = 2v$; its maximum gain, denoted by G, occurs at $v = 0$ and is equal to $T_M(x_0)$. The value of G is determined by the desired sidelobe rejection. Since the peak sidelobe gain is unity, the sidelobe rejection $= G$.

The first step in the design of Dolph-Chebyshev array is to make a decision on the value of N. Note that $M = N - 1$. After M has been selected, $T_M(x) = \cosh(M \cosh^{-1}(x))$ is evaluated on a dense grid of x for $x > 1$ to determine x_0 which gives $T_M(x_0) = G$. After x_0 has been obtained, $T_M(x_0 \cos(v))$ is evaluated on N evenly spaced values of v between $v = 0$ and $v = \pi$ including $v = 0$ but excluding $v = \pi$, i.e., $T_M(x_0 \cos(v))$ is evaluated on $v = 0, \frac{\pi}{N}, \ldots, \frac{k\pi}{N}, \ldots, \frac{(N-1)\pi}{N}$. Let

$$Y_k = T_M\left(x_0 \cos\left(\frac{k\pi}{N}\right)\right) \quad , \quad k = 0, 1, \ldots, \frac{(N-1)\pi}{N} \quad . \tag{9.90}$$

After obtaining Y_k, compute the IDCT (inverse discrete cosine transform) coefficients y_n. For N odd,

$$y_n = \frac{1}{N}\left(Y_0 + 2\sum_{k=1}^{\frac{N-1}{2}} Y_k \cos\left(\frac{2\pi}{N}nk\right)\right) \quad , n = 0, 1, \ldots, \frac{N-1}{2} \quad . \tag{9.91a}$$

$$h_{\frac{N-1}{2}+n} = h_{\frac{N-1}{2}-n} = y_n \ , \quad n = 0, 1, \ldots, \frac{N-1}{2} \ . \tag{9.91b}$$

For N even,

$$y_n = \frac{1}{N}\left(Y_0 + 2\sum_{k=1}^{N/2-1} Y_k \cos\left(\frac{2\pi}{N}(n+0.5)k\right)\right) \ , \quad n = 0, 1, \ldots, \frac{N}{2} - 1 \ . \tag{9.92a}$$

$$h_{\frac{N}{2}-1-n} = h_{\frac{N}{2}+n} = y_n \ , \quad n = 0, 1, \ldots, \frac{N}{2} - 1 \ . \tag{9.92b}$$

Tables 9.3 and 9.4 show the result for $N = 7$ and $N = 6$, respectively, and $x_0 = 1.2$ for both of them. The values of $T_6(x_0 \cos(v))$ for $v = k\pi/7, k = 0, 1, \ldots, 6$, the values of Y_k for $k = 0, 1, \ldots, 6$, and the values of y_n for $n = 0, 1, \ldots, 6$ are listed in Table 9.3. However, only the values of $T_6(x_0 \cos(v))$ for $v = k\pi/7, k = 0, 1, 2, 3$, the values of Y_k for $k = 0, 1, 2, 3$, and the values of y_n for $n = 0, 1, 2, 3$ are needed by (9.91). Listed in Table 9.4 are the values of $T_5(x_0 \cos(v))$ for $v = k\pi/6, k = 0, 1, \ldots, 5$, the values of Y_k for $k = 0, 1, \ldots, 5$, and the values of y_n for $n = 0, 1, \ldots, 5$. However, only the values of $T_5(x_0 \cos(v))$ for $v = k\pi/6, k = 0, 1, 2$, the values of Y_k for $k = 0, 1, 2$, and the values of y_n for $n = 0, 1, 2$ are needed by (9.92). The other values are listed for the purpose of showing the symmetry of $T_M(x_0 \cos(v))$ and Y_k.

Table 9.3 $x_0 = 1.2$ and $N = 7$. Only $Y_k, k = 0, 1, 2, 3$ are used in (9.91)

v	0	$\pi/7$	$2\pi/7$	$3\pi/7$	$4\pi/7$	$5\pi/7$	$6\pi/7$
$T_6(\bullet)$	20.9387	5.5642	− 0.3519	0.0510	0.0510	− 0.3519	5.5642
k	0	1	2	3	4	5	6
Y_k	20.9387	5.5642	− 0.3519	0.0510	0.0510	− 0.3519	5.5642
n	0	1	2	3	4	5	6
y_n	4.4950	3.9917	2.7372	1.4930	1.4930	2.7372	3.9917
h_n	1.4930	2.7372	3.9917	4.4950	3.9917	2.7372	1.4930

Table 9.4 $x_0 = 1.2$ and $N = 6$. Only $Y_k, k = 0, 1, 2$ are used in (9.92)

v	0	$\pi/6$	$2\pi/6$	$3\pi/6$	$4\pi/6$	$5\pi/6$
$T_5(\bullet)$	11.2531	2.1433	− 0.0758	0.0000	0.0758	− 2.1433
n	0	1	2	3	4	5
y_n	2.4816	1.9008	1.2442	1.2442	1.9008	2.4816
h_n	1.2442	1.9008	2.4816	2.4816	1.9008	1.2442

9.12 MUSIC Beamformer

MUSIC and its derivatives are perhaps currently the most commonly used super resolution method for DOA estimation. MUSIC is the abbreviation for MUltiple SIgnal Classifier. For an N-element array, MUSIC classifies the N-dimensional vector space into signal subspace and noise subspace. It assumes that the number of array elements N is more than the number of impinging plane waves M. The N-dimensional vector space has M eigenvalues corresponding to the signal subspace and $N - M$ smallest repeated eigenvalues corresponding to the noise subspace. The number of array elements N must be larger than the number of impinging plane waves M, i.e., $N \geq M + 1$.

9.12.1 *Original MUSIC*

Suppose that there are M plane waves each with a different frequency ω_m impinging on the array and each plane wave impinging the array at a different angle θ_m where the subscript m denote the mth plane wave the variable or vector is associated with. Assume that $\omega_i \neq \omega_j$ for $i \neq j$. From Fig. 9.5, it can be easily shown that

$$\begin{bmatrix} x_0(t) \\ x_1(t) \\ \vdots \\ x_{N-1}(t) \end{bmatrix} = \begin{bmatrix} x_0(t) \\ x_0(t)e^{-j\omega_m d \sin(\theta_m)/c} \\ \vdots \\ x_0(t)e^{-j(N-1)\omega_m d \sin(\theta_m)/c} \end{bmatrix} = \boldsymbol{a}_m x_0(t) \quad \text{for the } m\text{th plane wave} \tag{9.93a}$$

$$\text{where } \boldsymbol{a}_m = \begin{bmatrix} 1 \\ e^{-j\omega_m d \sin(\theta_m)/c} \\ \vdots \\ e^{-j(N-1)\omega_m d \sin(\theta_m)/c} \end{bmatrix}. \tag{9.93b}$$

In (9.93b), $\boldsymbol{a}_m$ is an $N \times 1$ steering vector. Suppose that $s_m(t)$ is the component of the mth plane wave in $x_0(t)$. Thus, if there is no noise, we have

$$x_0(t) = s_0(t) + s_1(t) + \ldots + s_{M-1}(t). \tag{9.94}$$

For all the M plane waves, the collective effect on the array elements is given by

$$\begin{bmatrix} x_0(t) \\ x_1(t) \\ \vdots \\ x_{N-1}(t) \end{bmatrix} = \boldsymbol{a}_0 s_0(t) + \cdots + \boldsymbol{a}_{M-1} s_{M-1}(t) = \begin{bmatrix} \boldsymbol{a}_0 \ \boldsymbol{a}_1 \ \cdots \ \boldsymbol{a}_{M-1} \end{bmatrix} \begin{bmatrix} s_0(t) \\ s_1(t) \\ \vdots \\ s_{M-1}(t) \end{bmatrix}. \tag{9.95}$$

Define

$$\boldsymbol{x}(t) = \begin{bmatrix} x_0(t) \\ x_1(t) \\ \vdots \\ x_{N-1}(t) \end{bmatrix}, \quad \mathbf{A} = \begin{bmatrix} \boldsymbol{a}_0 \ \boldsymbol{a}_1 \ \cdots \ \boldsymbol{a}_{M-1} \end{bmatrix}, \quad \text{and} \quad \boldsymbol{s}(t) = \begin{bmatrix} s_0(t) \\ s_1(t) \\ \vdots \\ s_{M-1}(t) \end{bmatrix}. \tag{9.96}$$

From (9.94), (9.95), and (9.96), we have

$$\boldsymbol{x}(t) = \mathbf{A}\boldsymbol{s}(t). \tag{9.97}$$

In (9.97), $\boldsymbol{x}(t)$ is an $N \times 1$ vector, $\mathbf{A}$ is an $N \times M$ matrix, and $\boldsymbol{s}(t)$ is an $M \times 1$ vector. Let the noise associated with $x_n(t)$ be $n_n(t)$ where $\mathbb{E}\{n_i(t)n_j(t)\} = 0$ for $i \neq j$, i.e., $n_i(t)$ and $n_j(t)$ are uncorrelated. Define the noise vector $\boldsymbol{n}(t)$

$$\boldsymbol{n}(t) = \begin{bmatrix} n_0(t) \\ n_1(t) \\ \vdots \\ n_{N-1}(t) \end{bmatrix}. \tag{9.98}$$

In (9.98), $\boldsymbol{n}(t)$ is an $N \times 1$ vector. Adding $\boldsymbol{n}(t)$ to $\boldsymbol{x}(t)$, we have

$$\boldsymbol{x}(t) = \mathbf{A}\boldsymbol{s}(t) + \boldsymbol{n}(t). \tag{9.99}$$

The conjugate transpose of $\boldsymbol{x}(t)$ is given by

$$\boldsymbol{x}^{\mathrm{H}}(t) = \boldsymbol{s}^{\mathrm{H}}(t)\mathbf{A}^{\mathrm{H}} + \boldsymbol{n}^{\mathrm{H}}(t). \tag{9.100}$$

From (9.99) and (9.100),

$$\begin{aligned} \boldsymbol{x}(t)\boldsymbol{x}^{\mathrm{H}}(t) &= [\mathbf{A}\boldsymbol{s}(t) + \boldsymbol{n}(t)][\boldsymbol{s}^{\mathrm{H}}(t)\mathbf{A}^{\mathrm{H}} + \boldsymbol{n}^{\mathrm{H}}(t)] \\ &= \mathbf{A}\boldsymbol{s}(t)\boldsymbol{s}^{\mathrm{H}}(t)\mathbf{A}^{\mathrm{H}} + \mathbf{A}\boldsymbol{s}(t)\boldsymbol{n}^{\mathrm{H}}(t) + \boldsymbol{n}(t)\boldsymbol{s}^{\mathrm{H}}(t)\mathbf{A}^{\mathrm{H}} + \boldsymbol{n}(t)\boldsymbol{n}^{\mathrm{H}}(t). \end{aligned} \tag{9.101}$$

Taking expectation of (9.101),

$$\begin{aligned} \mathbb{E}\{\boldsymbol{x}(t)\boldsymbol{x}^{\mathrm{H}}(t)\} =& \mathbf{A}\mathbb{E}\{\boldsymbol{s}(t)\boldsymbol{s}^{\mathrm{H}}(t)\}\mathbf{A}^{\mathrm{H}} + \mathbf{A}\mathbb{E}\{\boldsymbol{s}(t)\boldsymbol{n}^{\mathrm{H}}(t)\} \\ &+ \mathbb{E}\{\boldsymbol{n}(t)\boldsymbol{s}^{\mathrm{H}}(t)\}\mathbf{A}^{\mathrm{H}} + \mathbb{E}\{\boldsymbol{n}(t)\boldsymbol{n}^{\mathrm{H}}(t)\}. \end{aligned} \tag{9.102}$$

Assume that signal and noise are uncorrelated and have zero mean. Thus,

$$\mathbb{E}\{\boldsymbol{s}(t)\boldsymbol{n}^{\mathrm{H}}(t)\} = \mathbb{E}\{\boldsymbol{n}(t)\boldsymbol{s}^{\mathrm{H}}(t)\} = \mathbf{0} \tag{9.103}$$

where $\mathbf{0}$ denotes a null matrix of appropriate dimension. From (9.102) and (9.103),

$$\mathbb{E}\{\boldsymbol{x}(t)\boldsymbol{x}^{\mathrm{H}}(t)\} = \mathbf{A}\mathbb{E}\{\boldsymbol{s}(t)\boldsymbol{s}^{\mathrm{H}}(t)\}\mathbf{A}^{\mathrm{H}} + \mathbb{E}\{\boldsymbol{n}(t)\boldsymbol{n}^{\mathrm{H}}(t)\}. \tag{9.104}$$

We have assumed that $\mathbb{E}\{n_i(t)n_j(t)\} = 0$ for $i \neq j$. Thus, $\mathbb{E}\{\boldsymbol{n}(t)\boldsymbol{n}^{\mathrm{H}}(t)\}$ is diagonal. We shall also assume that $\mathbb{E}\{n_i(t)n_i(t)\} = \eta^2$ for all i. Hence,

$$\mathbb{E}\{\boldsymbol{n}(t)\boldsymbol{n}^{\mathrm{H}}(t)\} = \eta^2\mathbf{I} \tag{9.105}$$

where $\mathbf{I}$ is an identity matrix of appropriate dimension. From (9.104) and (9.105),

$$\mathbb{E}\{\boldsymbol{x}(t)\boldsymbol{x}^{\mathrm{H}}(t)\} = \mathbf{A}\mathbb{E}\{\boldsymbol{s}(t)\boldsymbol{s}^{\mathrm{H}}(t)\}\mathbf{A}^{\mathrm{H}} + \eta^2\mathbf{I}. \tag{9.106}$$

For $\boldsymbol{s}(t)$ as defined in (9.96), $\mathbb{E}\{\boldsymbol{s}(t)\boldsymbol{s}^{\mathrm{H}}(t)\}$ is given by

$$\mathbb{E}\{\boldsymbol{s}(t)\boldsymbol{s}^{\mathrm{H}}(t)\} = \begin{bmatrix} \mathbb{E}\{s_0(t)s_0^*(t)\} & \mathbb{E}\{s_0(t)s_1^*(t)\} & \cdots & \mathbb{E}\{s_0(t)s_{M-1}^*(t)\} \\ \mathbb{E}\{s_1(t)s_0^*(t)\} & \mathbb{E}\{s_1(t)s_1^*(t)\} & \cdots & \mathbb{E}\{s_1(t)s_{M-1}^*(t)\} \\ \vdots & \vdots & \ddots & \vdots \\ \mathbb{E}\{s_{M-1}(t)s_0^*(t)\} & \mathbb{E}\{s_{M-1}(t)s_1^*(t)\} & \cdots & \mathbb{E}\{s_{M-1}(t)s_{M-1}^*(t)\} \end{bmatrix}. \tag{9.107}$$

We have already assume that the frequency of $s_m(t)$ is ω_m and that $\omega_i \neq \omega_j$ for $i \neq j$. Hence, $\mathbb{E}\{s_i(t)s_j^*(t)\} = 0$ for $i \neq j$. Note: $\mathbb{E}\{s_i(t)\} = 0$ (i.e., $s_i(t)$ has zero mean) since $s_i(t)$ is a complex sinusoid with frequency ω_i. Let $\mathbb{E}\{s_m(t)s_m^*(t)\} = \sigma_m^2$. Hence,

$$\mathbb{E}\{s_i(t)s_j^*(t)\} = 0 \quad \text{for} \quad i \neq j, \tag{9.108a}$$

$$\mathbb{E}\{s_m(t)s_m^*(t)\} = \sigma_m^2, \tag{9.108b}$$

and

$$\mathbb{E}\{\boldsymbol{s}(t)\boldsymbol{s}^{\mathrm{H}}(t)\} = \begin{bmatrix} \sigma_0^2 & 0 & \cdots & 0 \\ 0 & \sigma_1^2 & \cdots & 0 \\ \vdots & \vdots & \ddots & \vdots \\ 0 & 0 & \cdots & \sigma_{M-1}^2 \end{bmatrix} = \mathbf{R}_{ss}. \tag{9.109}$$

In (9.109), $\mathbf{R}_{ss}$ is diagonal. From (9.106) and (9.109),

$$\mathbb{E}\{\boldsymbol{x}(t)\boldsymbol{x}^{\mathrm{H}}(t)\} = \mathbf{A}\mathbb{E}\{\boldsymbol{s}(t)\boldsymbol{s}^{\mathrm{H}}(t)\}\mathbf{A}^{\mathrm{H}} + \eta^2\mathbf{I} = \mathbf{A}\mathbf{R}_{ss}\mathbf{A}^{\mathrm{H}} + \eta^2\mathbf{I}. \tag{9.110}$$

Let $\mathbb{E}\{\boldsymbol{x}(t)\boldsymbol{x}^{\mathrm{H}}(t)\}$ be denoted by $\mathbf{R}_{xx}$.

$$\mathbf{R}_{xx} = \mathbb{E}\{\boldsymbol{x}(t)\boldsymbol{x}^{\mathrm{H}}(t)\}. \tag{9.111}$$

From (9.110) and (9.111),

$$\mathbf{R}_{xx} = \mathbf{A}\mathbf{R}_{ss}\mathbf{A}^{\mathrm{H}} + \eta^2\mathbf{I}. \tag{9.112}$$

Rearranging (9.112),

$$\mathbf{A}\mathbf{R}_{ss}\mathbf{A}^{\mathrm{H}} = \mathbf{R}_{xx} - \eta^2\mathbf{I}. \tag{9.113}$$

Since there are M independent plane waves, the rank of $\mathbf{R}_{ss}$ is M. Hence, the rank of $\mathbf{A}\mathbf{R}_{ss}\mathbf{A}^{\mathrm{H}}$ is M. Assume that the number of array elements N is more than the number of impinging plane waves M, i.e., $N > M$. The dimension of $\mathbf{A}\mathbf{R}_{ss}\mathbf{A}^{\mathrm{H}}$ is $N \times N$. Hence, for $N > M$, the square matrix $\mathbf{A}\mathbf{R}_{ss}\mathbf{A}^{\mathrm{H}}$ is singular. Therefore,

$$\det[\mathbf{A}\mathbf{R}_{ss}\mathbf{A}^{\mathrm{H}}] = 0. \tag{9.114}$$

In (9.114), $\det[\bullet]$ denotes the determinant of $\bullet$. Since $\mathbf{A}\mathbf{R}_{ss}\mathbf{A}^{\mathrm{H}} = \mathbf{R}_{xx} - \eta^2\mathbf{I}$ and $\det[\mathbf{A}\mathbf{R}_{ss}\mathbf{A}^{\mathrm{H}}] = 0$,

$$\det[\mathbf{R}_{xx} - \eta^2\mathbf{I}] = 0. \tag{9.115}$$

Since $\det[\mathbf{R}_{xx} - \eta^2\mathbf{I}] = 0$, η^2 is an eigenvalue of $\mathbf{R}_{xx}$. The dimension of the null space of $\mathbf{A}\mathbf{R}_{ss}\mathbf{A}^{\mathrm{H}}$ is $N - M$. Therefore, $\mathbf{R}_{xx}$ has $N - M$ eigenvalues that make $\det[\mathbf{R}_{xx} - \eta^2\mathbf{I}] = 0$, i.e., the eigenvalue η^2 is repeated $N - M$ times. In practice, they are not equal and become "more equal" when more data points are used in the computation. The smallest eigenvalue of $\mathbf{R}_{xx}$ is chosen as η^2. Let λ_p and $\boldsymbol{v}_p$ be the pth eigenvalue and pth eigenvector of $\mathbf{R}_{xx}$. Therefore,

$$\mathbf{R}_{xx}\boldsymbol{v}_p = \lambda_p\boldsymbol{v}_p. \tag{9.116}$$

From (9.112) and (9.116),

$$\mathbf{R}_{xx}\boldsymbol{v}_p = [\mathbf{A}\mathbf{R}_{ss}\mathbf{A}^{\mathrm{H}} + \eta^2\mathbf{I}]\boldsymbol{v}_p = \lambda_p\boldsymbol{v}_p. \tag{9.117}$$

Since $\mathbf{R}_{xx}$ is positive definite, $\lambda_p > 0$. Rearranging terms in (9.117),

$$\mathbf{A}\mathbf{R}_{ss}\mathbf{A}^{\mathrm{H}}\boldsymbol{v}_p = (\lambda_p - \eta^2)\boldsymbol{v}_p. \tag{9.118}$$

In (9.118), $\lambda_p - \eta^2 \geq 0$ since η^2 is the smallest eigenvalue. Classify the N-dimensional vector space into signal subspace and noise subspace. The signal

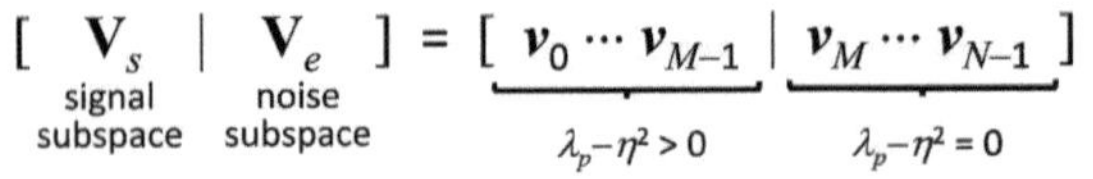

Fig. 9.37 The N-dimensional vector space is classified into signal subspace and noise subspace

subspace comprises of the M vectors where $\lambda_p - \eta^2 > 0$ and the noise subspace corresponds to the $N - M$ vectors where $\lambda_p - \eta^2 = 0$:

$$\mathbf{A}\mathbf{R}_{ss}\mathbf{A}^{\mathrm{H}}\boldsymbol{v}_p = (\lambda_p - \eta^2)\boldsymbol{v}_p \ , \quad \lambda_p - \eta^2 > 0, \ \ p = 0, \ldots, M-1, \tag{9.119a}$$

$$= 0 \ , \quad p = M, M+1 \ldots, N-1 \tag{9.119b}$$

Let

$$\mathbf{V}_e = [\boldsymbol{v}_M, \ldots, \boldsymbol{v}_{N-1}] \quad \text{and} \tag{9.120a}$$

$$\mathbf{V}_s = [\boldsymbol{v}_0, \ldots, \boldsymbol{v}_{M-1}]. \tag{9.120b}$$

See Fig. 9.37. From (9.119) and (9.120),

$$\mathbf{A}\mathbf{R}_{ss}\mathbf{A}^{\mathrm{H}}\mathbf{V}_e = \mathbf{0}, \tag{9.121a}$$

$$\mathbf{V}_e^{\mathrm{H}}\mathbf{A}\mathbf{R}_{ss}\mathbf{A}^{\mathrm{H}}\mathbf{V}_e = [\mathbf{A}^{\mathrm{H}}\mathbf{V}_e]^{\mathrm{H}}\mathbf{R}_{ss}[\mathbf{A}^{\mathrm{H}}\mathbf{V}_e] = \mathbf{0}. \tag{9.121b}$$

Since $\mathbf{R}_{ss}$ is positive definite, this means that

$$\mathbf{A}^{\mathrm{H}}\mathbf{V}_e = \mathbf{0}. \tag{9.122}$$

Note from (9.96) that the columns of $\mathbf{A}$ are $\boldsymbol{a}_m$. Since $\mathbf{A}^{\mathrm{H}}\mathbf{V}_e = \mathbf{0}$, the $N - M$ noise eigenvectors of $\mathbf{R}_{xx}$ (corresponding to the $N - M$ smallest repeated eigenvalues of $\mathbf{R}_{xx}$) are orthogonal to all the M steering vectors $\boldsymbol{a}_m$. Define

$$\boldsymbol{a}(\theta) = \begin{bmatrix} 1 \\ e^{-j\omega d \sin(\theta)/c} \\ \vdots \\ e^{-j(N-1)\omega d \sin(\theta)/c} \end{bmatrix}. \tag{9.123}$$

The value of θ that makes $|\boldsymbol{a}^{\mathrm{H}}(\theta)\mathbf{V}_e| = 0$ is a DOA. This is the basis for MUSIC. MUSIC computes $P(\theta)$ given by

$$P(\theta) = \frac{1}{\boldsymbol{a}^{\mathrm{H}}(\theta)\mathbf{V}_e\mathbf{V}_e^{\mathrm{H}}\boldsymbol{a}(\theta)}. \tag{9.124}$$

Peaks of $P(\theta)$ are the DOA.

MUSIC algorithm:

1. Find eigenvalues and eigenvectors of $\mathbf{R}_{xx}$.
2. Eigenvectors corresponding to the $N - M$ smallest repeated eigenvalues of $\mathbf{R}_{xx}$ are the noise eigenvectors. Let $\boldsymbol{v}_q, q = M, \ldots, N-1$ denote these $N - M$ noise eigenvectors.
3. Define $\mathbf{V}_e = [\boldsymbol{v}_M, \ldots, \boldsymbol{v}_{N-1}]$.
4. Define $N \times 1$ steering vector $\boldsymbol{a}(\theta) = \begin{bmatrix} 1 \\ e^{-j\omega d \sin(\theta)/c} \\ \vdots \\ e^{-j(N-1)\omega d \sin(\theta)/c} \end{bmatrix}$.
5. Compute $P(\theta) = \frac{1}{\boldsymbol{a}^{\mathrm{H}}(\theta)\mathbf{V}_e\mathbf{V}_e^{\mathrm{H}}\boldsymbol{a}(\theta)}$ on a dense grid of θ.
6. Peaks of $P(\theta)$ are spatial angles of all incident waves.

After obtaining the directions of arrival of all the plane waves, $\mathbf{A}$ is known. $\mathbf{R}_{ss}$ can be computed as follows. We have shown in (9.111) that $\mathbf{A}\mathbf{R}_{ss}\mathbf{A}^{\mathrm{H}} = \mathbf{R}_{xx} - \eta^2\mathbf{I}$. Therefore,

$$\mathbf{R}_{ss} = (\mathbf{A}^{\mathrm{H}}\mathbf{A})^{-1}\mathbf{A}^{\mathrm{H}}(\mathbf{R}_{xx} - \eta^2\mathbf{I})\mathbf{A}(\mathbf{A}^{\mathrm{H}}\mathbf{A})^{-1}. \tag{9.125}$$

Since $\mathbf{A}\mathbf{R}_{ss}\mathbf{A}^{\mathrm{H}}\boldsymbol{v}_p = (\lambda_p - \eta^2)\boldsymbol{v}_p$ (see (9.119a)), $\boldsymbol{v}_p$ (eigenvector of $\mathbf{R}_{xx}$) is also eigenvector of $\mathbf{A}\mathbf{R}_{ss}\mathbf{A}^{\mathrm{H}}$ with eigenvalue $(\lambda_p - \eta^2)$.

9.12.2 Root MUSIC

The original MUSIC finds the DOA (direction of arrival) by evaluating $\boldsymbol{a}^{\mathrm{H}}(\theta)\mathbf{V}_e\mathbf{V}_e^{\mathrm{H}}\boldsymbol{a}(\theta)$ on a dense grid of θ where $\boldsymbol{a}(\theta) = [1, e^{-j\omega d \sin(\theta)/c}, \ldots, e^{-j(N-1)\omega d \sin(\theta)/c}]^{\mathrm{T}}$. The computation load is very high. Root MUSIC follows the same basic principle as the original MUSIC but θ is derived from the root of a polynomial. Define

$$\mathbf{V} = \mathbf{V}_e\mathbf{V}_e^{\mathrm{H}} \tag{9.126}$$

and

$$z = e^{-j\omega d \sin(\theta)/c}. \tag{9.127}$$

Let $\boldsymbol{z}$ denote the $N \times 1$ column vector consisting of powers of z defined in (9.127):

$$\boldsymbol{z} = [1, z, z^2, \ldots, z^{N-1}]^{\mathrm{T}} \tag{9.128}$$

The vector $\boldsymbol{a}(\theta)$ is the vector $\boldsymbol{z}$ evaluated on $z = e^{-j\omega d \sin(\theta)/c}$. In root MUSIC, DOA is obtained from the roots of

$$\boldsymbol{z}^{\mathrm{T}}\mathbf{V}\boldsymbol{z} = 0. \tag{9.129}$$

Let the elements of $\mathbf{V}$ be $v_{i,j}$. Hence,

$$\mathbf{V} = \begin{bmatrix} v_{1,1} & v_{1,2} & \cdots & v_{1,N} \\ v_{2,1} & v_{2,2} & \cdots & v_{2,N} \\ \vdots & \vdots & \ddots & \vdots \\ v_{N,1} & v_{N,2} & \cdots & v_{N,N} \end{bmatrix}. \tag{9.130}$$

It can be easily verified that

$$\boldsymbol{z}^{\mathrm{T}}\mathbf{V}\boldsymbol{z} = \sum_{i=1}^{N}\sum_{j=1}^{N} z^{j-i} v_{i,j}. \tag{9.131}$$

Let $l = j - i$. Equation (9.131) can be rewritten as

$$\boldsymbol{z}^{\mathrm{T}}\mathbf{V}\boldsymbol{z} = \sum_{l=1-N}^{N-1}\left(\sum_{j-i=l} v_{i,j}\right) z^{l}, \tag{9.132a}$$

$$\text{where} \quad l = j - i. \tag{9.132b}$$

If l is positive, $\left(\sum_{j-i=l} v_{i,j}\right)$ is the sum of the diagonal elements of $\mathbf{V}$ containing $v_{1,l+1}$. If l is negative, $\left(\sum_{j-i=l} v_{i,j}\right)$ is the sum of the diagonal elements of $\mathbf{V}$ containing $v_{1-l,1}$. See Fig. 9.38. The expression $\boldsymbol{z}^{\mathrm{T}}\mathbf{V}\boldsymbol{z} = \sum_{l=1-N}^{N-1}\left(\sum_{j-i=l} v_{i,j}\right) z^{l}$ is a $(2N-2)$-degree polynomial and so $\boldsymbol{z}^{\mathrm{T}}\mathbf{V}\boldsymbol{z}$ has $(2N-2)$ zeros. However, the zeros are not all independent. If z_0 is a zero, $1/z_0^*$ will also be a zero where z_0^* is the complex conjugate of z_0; the zeros appear in pairs. The zeros z_0 and $1/z_0^*$ have the same phase and reciprocal magnitude. Let

$$z_0 = r_0 e^{j\phi_0}. \tag{9.133}$$

In root MUSIC, only ϕ_0 is relevant. The magnitude r_0 is discarded. DOA θ_0 is obtained by equating $e^{j\phi_0}$ to $e^{-j\omega d \sin(\theta_0)/c}$:

$$\phi_0 = -\omega d \sin(\theta_0)/c, \tag{9.134a}$$

$$\theta_0 = -\sin^{-1}\{\phi_0 c/(\omega d)\} \tag{9.134b}$$

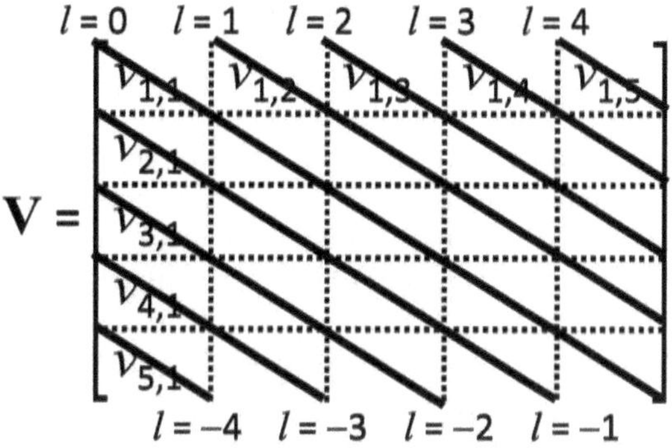

Fig. 9.38 Sum of the diagonal elements of **V** for l positive/negative

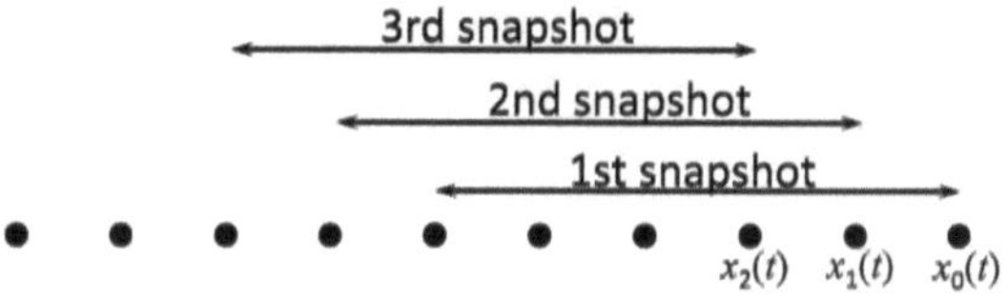

Fig. 9.39 Simulated array movement

Root MUSIC algorithm:

1. Find eigenvalues and eigenvectors of $\mathbf{R}_{xx}$.
2. Eigenvectors corresponding to the $N - M$ smallest repeated eigenvalues of $\mathbf{R}_{xx}$ are the noise eigenvectors. Let $\boldsymbol{v}_q, q = M, \ldots, N-1$ denote these $N - M$ noise eigenvectors.
3. Define $\mathbf{V}_e = [\boldsymbol{v}_M, \ldots, \boldsymbol{v}_{N-1}]$.
4. Define $\mathbf{V} = \mathbf{V}_e\mathbf{V}_e^{\mathrm{H}}$, $z = e^{-j\omega d\sin(\theta)/c}$, and $\boldsymbol{z} = [1, z, z^2, \ldots, z^{N-1}]^{\mathrm{T}}$.
5. Find the zeros of $\boldsymbol{z}^{\mathrm{T}}\mathbf{V}\boldsymbol{z} = \sum_{l=1-N}^{N-1}\left(\sum_{j-i=l} v_{i,j}\right)z^l$.
6. Let the M zeros inside and nearest the unit circle be $r_l e^{j\phi_l}, i = 1, 2, \ldots, M$.
7. DOA $\theta_i = -\sin^{-1}\{\phi_i c/(\omega d)\}$, $i = 1, 2, \ldots, M$.

9.12.3 *Smooth MUSIC*

The MUSIC algorithm assumes that the signals are uncorrelated. This may not be the case in a multipath scenario, and the receiver and transmitter are stationary; there is no Doppler shift. One possibility for overcoming the signal correlation problem is to “simulate the array motion” “without physically moving” the array. This can be done by taking snapshots of signals from different parts of the array at different time as shown in Fig. 9.39 for computing $\mathbf{R}_{xx}$. The array “appears” to be moving. See Appendix A. Obviously, more array elements are needed.

9.13 Simulation Results

Some simulation results are shown in this section to help understand some of the important properties of the MUSIC algorithm. In the derivation of the MUSIC algorithm, it is assumed that there is noise and the signals are uncorrelated. The number of snapshots for computing R_{xx} for an N-element array should be at least $N-1$. As many snapshots as possible should be taken for computing R_{xx} to improve SNR as resources permit. The snapshots should also be spread out over a duration comparable to $\frac{1}{\Delta}$ where Δ is the frequency difference between the signals impinging on the array. See Appendix A. While for Levinson-Durbin algorithm it is desired to achieve a Toeplitz matrix, achieving a Toeplitz matrix is not necessary for the MUSIC algorithm. If the frequencies of the incident waves are not evenly spaced, Δ is the highest common factor of the frequency differences. The array sampling frequency (i.e., the rate at which the snapshots of the array data are taken) f_s is given by $f_s \lesssim P\Delta$ where P is the number of snapshots to be taken. Figure 9.40 shows simulation results for three incident waves at angles $-\pi/4$, $\pi/4$, and $\pi/3$, respectively. In Fig. 9.40, the frequencies of the three waves are 1251 MHz, 1252 MHz, and 1253 MHz, respectively. The frequency difference is 1 MHz. Since there are four array elements, at least three snapshots must be taken. Although as many snapshots as possible should be used to compute R_{xx} to improve SNR, in this and subsequent examples, we use the minimum number of snapshots only for illustrative purposes. Suppose that three snapshots are taken. Therefore, the sampling frequency is 3×1 MHz = 3 MHz. The three snapshots are taken over a duration of 1 microsecond. The eigenvalues of R_{xx} are 0, 0.35, 4.26, and 7.38, respectively. There are three large and one small eigenvalues. If only one snapshot is taken, the eigenvalues of R_{xx} would be 0, 0, 0, and 17.5, respectively. There would be one large and three small eigenvalues.

If the sampling frequency is increased to 30 MHz and three array snapshots are taken, the three snapshots would spread over 0.1 microsecond, i.e., 10% of $\frac{1}{\text{frequency difference}}$. The eigenvalues are 0, 0.00024, 0.517, and 17.8, respectively. The simulation result is shown in Fig. 9.41. Comparing Figs. 9.40 and 9.41, it can be seen that the result of Fig. 9.41 is seriously degraded.

If the sampling frequency is 6 MHz instead of 30 MHz and three array snapshots are taken, the three snapshots would spread over 0.5 microsecond, i.e., 50% of $\frac{1}{\text{frequency difference}}$. The eigenvalues are 0, 0.024, 4.0, and 12.9, respectively. The

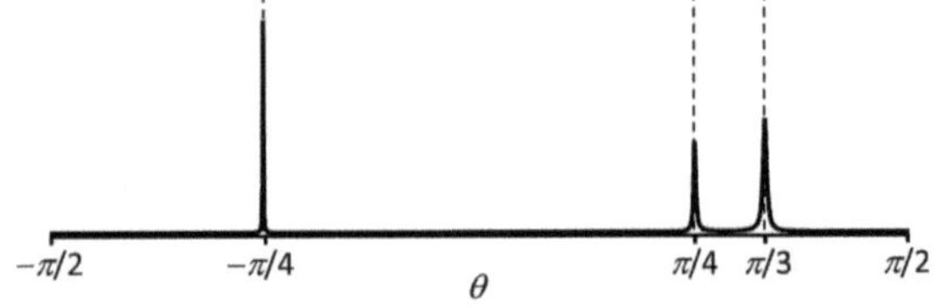

Fig. 9.40 Four array elements and 50 dB SNR. Array spacing = half wavelength

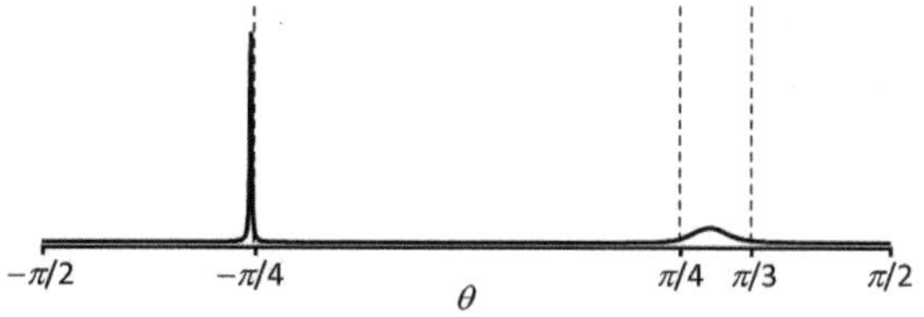

Fig. 9.41 Four array elements and 50 dB SNR; array snapshots taken over 10% of $\frac{1}{\text{frequency difference}}$. Array spacing = half wavelength

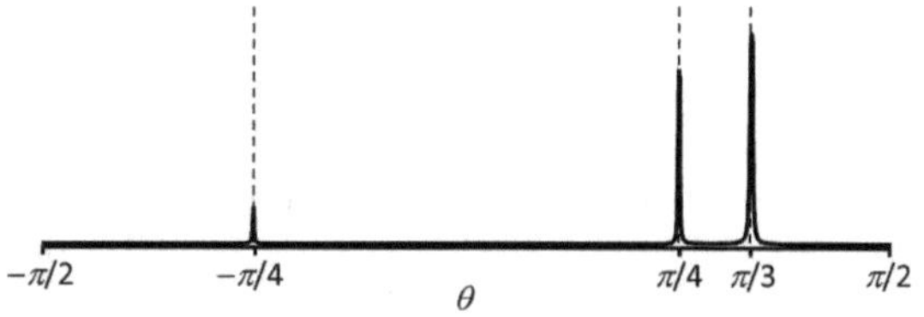

Fig. 9.42 Four array elements and 50 dB SNR; array snapshots taken over 50% of $\frac{1}{\text{frequency difference}}$. Array spacing = half wavelength

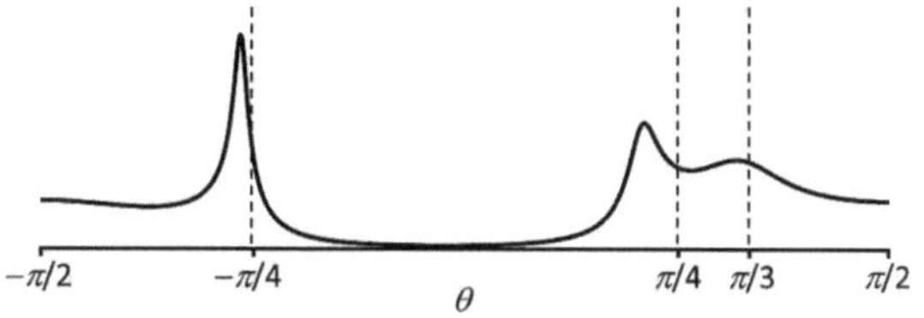

Fig. 9.43 Four array elements and 20 dB SNR. Array spacing = half wavelength

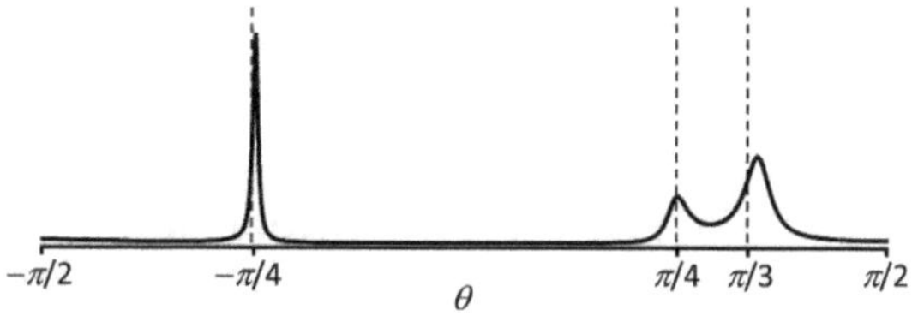

Fig. 9.44 Six array elements and 20 dB SNR. Array spacing = half wavelength

simulation result is shown in Fig. 9.42. Comparing Figs. 9.40, 9.41, and 9.42, the importance of the duration over which the snapshots are taken can be easily seen.

Suppose that the noise level for Fig. 9.40 is increased from 50 dB SNR to 20 dB SNR with sampling frequency set at 3 MHz. The eigenvalues are 0, 0.39, 4.3, and 6.9, respectively. The simulation result is shown in Fig. 9.43. Comparing Figs. 9.40 and 9.43, the effect of noise can be easily seen.

Figure 9.44 shows the result for six array elements at 20 dB SNR. The sampling rate remains at 3 MHz. The eigenvalues are 0, 0.0092, 0.043, 1.8, 5.9, and 10.9, respectively. There are three large eigenvalues and three small eigenvalues as expected. Comparing Figs. 9.43 and 9.44, it can be seen that increasing the number of array elements improves the result.

Increasing the number of array elements to 10 while keeping the SNR at 20 dB and sampling rate at 3 MHz, the results improve further to that shown in Fig. 9.45. The eigenvalues are 0, 0.0029, 0.0058, 0.012, 0.024, 0.027, 0.058, 7.5, 10.1, and 12.4, respectively. As expected, three of the eigenvalues are large and the remaining seven are small. Comparing Figs. 9.44 and 9.45, it can be seen that increasing the number of array elements from six to ten further improves the result.

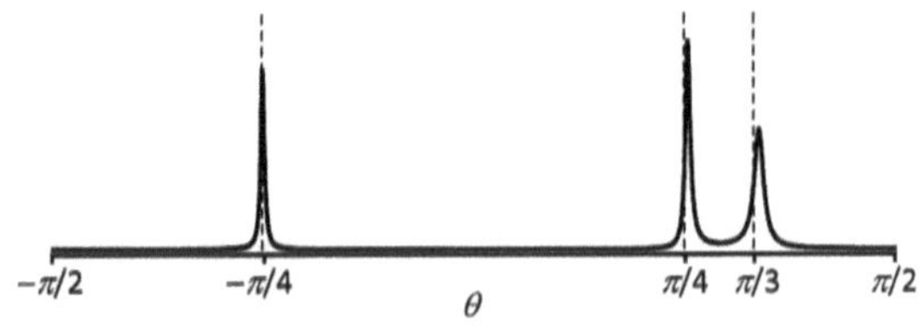

Fig. 9.45 Ten array elements and 20 dB SNR. Array spacing = half wavelength

When a radar signal shines on a moving target, as a result of Doppler effect, the frequency of the returned signal will be shifted. The frequency shift is proportional to the carrier frequency of the radar and the speed of the target. In vacuum, the frequency shift is 1.85314 Hz for every GHz of radar carrier frequency and every km/hour of the target speed. For a carrier frequency of 30 GHz and a speed difference of 1 km/hour, the frequency difference is 55.59 Hz.

At 20°C and 1 atmospheric pressure, sound travels at a speed of about 343 meters per second in air. The speed of sound depends strongly on temperature but varies slightly with pressure. It also varies by several meters per second depending on the type of gas it is traveling in. This causes a frequency difference Δ of 1.620 Hz for every kHz of carrier frequency and every km/hour speed difference for the targets. For a carrier frequency of 40 kHz and a speed difference of 1 km/hour, the frequency difference is 64.8 Hz.

9.14 MUSIC for Two-Dimensional Rectangular Arrays

MUSIC algorithm for the one-dimensional array discussed in the foregoing sections may be extended into two dimensions in a straightforward manner. With reference to Fig. 9.21, the two-dimensional equivalent of (9.93) for the mth plane wave is

$$\begin{bmatrix} x_{0,0}(t) \\ x_{1,0}(t) \\ \vdots \\ x_{N_x-1,0}(t) \\ x_{0,1}(t) \\ x_{1,1}(t) \\ \vdots \\ x_{N_x-1,1}(t) \\ \vdots \\ x_{N_x-1,N_y-1}(t) \end{bmatrix} = \begin{bmatrix} x_{0,0}(t) \\ x_{0,0}(t)e^{-j\omega_m d_x \sin(\theta_{x,m})/c} \\ \vdots \\ x_{0,0}(t)e^{-j(N_x-1)\omega_m d_x \sin(\theta_{x,m})/c} \\ x_{0,0}(t)e^{-j\omega_m d_y \sin(\theta_{y,m})/c} \\ x_{0,0}(t)e^{-j\omega_m d_x \sin(\theta_{x,m})/c}e^{-j\omega_m d_y \sin(\theta_{y,m})/c} \\ \vdots \\ x_{0,0}(t)e^{-j(N_x-1)\omega_m d_x \sin(\theta_{x,m})/c}e^{-j\omega_m d_y \sin(\theta_{y,m})/c} \\ \vdots \\ x_{0,0}(t)e^{-j(N_x-1)\omega_m d_x \sin(\theta_{x,m})/c}e^{-j(N_y-1)\omega_m d_y \sin(\theta_{y,m})/c} \end{bmatrix}$$

$$= \boldsymbol{a}_m x_{0,0}(t) \tag{9.135a}$$

$$\text{where } \boldsymbol{a}_m = \begin{bmatrix} 1 \\ e^{-j\omega_m d_x \sin(\theta_{x,m})/c} \\ \vdots \\ e^{-j(N_x-1)\omega_m d_x \sin(\theta_{x,m})/c} \\ e^{-j\omega_m d_y \sin(\theta_{y,m})/c} \\ e^{-j\omega_m d_x \sin(\theta_{x,m})/c} e^{-j\omega_m d_y \sin(\theta_{y,m})/c} \\ \vdots \\ e^{-j(N_x-1)\omega_m d_x \sin(\theta_{x,m})/c} e^{-j\omega_m d_y \sin(\theta_{y,m})/c} \\ \vdots \\ e^{-j(N_x-1)\omega_m d_x \sin(\theta_{x,m})/c} e^{-j(N_y-1)\omega_m d_y \sin(\theta_{y,m})/c} \end{bmatrix}. \tag{9.135b}$$

In (9.135), $\theta_{x,m}$ and $\theta_{y,m}$ are, respectively, the corresponding values of θ_x and θ_y for the mth plane wave. See Sect. 9.6 for θ_x and θ_y and Sect. 9.7 for the relationship between θ_x, θ_y, azimuth, and elevation angles. The $\sin(\theta_{x,m})$ and $\sin(\theta_{y,m})$ terms in (9.135b) may be replaced by $\sin(\theta_{A,m})\cos(\theta_{E,m})$ and $\cos(\theta_{A,m})\cos(\theta_{E,m})$, respectively, where $\theta_{A,m}$ and $\theta_{E,m}$ are θ_A and θ_E for the mth plane wave as in (9.136) to obtain the optimum solution directly in the (θ_E, θ_A) domain:

$$\begin{aligned} & e^{-jn_x\omega_m d_x \sin(\theta_{x,m})/c} \; e^{-jn_y\omega_m d_y \sin(\theta_{y,m})/c} \\ &= e^{-jn_x\omega_m d_x \sin(\theta_{A,m})\cos(\theta_{E,m})/c} \; e^{-jn_y\omega_m d_y \cos(\theta_{A,m})\cos(\theta_{E,m})/c}. \end{aligned} \tag{9.136}$$

The associated derivations following (9.93) remain the same with $\boldsymbol{a}_m$ in (9.93) replaced by $\boldsymbol{a}_m$ in (9.135). $P(\theta)$ of (9.124) is evaluated and the spikes indicate the directions of arrivals (DOA).

We shall show some simulation results. Consider a scenario with four array elements and two arriving sinusoidal signals with equal magnitudes. The four array elements were positioned at the four corners of a square. The frequencies of the two arriving signals were 1.000,000 and 1.000,001 arbitrary unit, respectively. There were two DOAs, viz., DOA-1 and DOA-2. The elevation and azimuth angles (θ_E, θ_A) for DOA-1 and DOA-2 were (1.000, 0.500) and (0.600, −2.000), respectively. The unit for angular measurement is radian. The array element separations to signal wave-length ratio d_x and d_y were the same and were each 0.5. Ten snapshots of the array signals were taken over a duration of $\frac{1}{0.000{,}001} =$ 1,000,000 unit time for computing $\mathbf{R}_{xx}$. Results for 40dB and 10dB SNR are shown in Figs. 9.46 and 9.47, respectively. The SNR values were the SNR of each array element. For the 40dB SNR case of Fig. 9.46, the detected (θ_E, θ_A) for DOA-1 and DOA-2 were (1.000, 0.496) and (0.600, −2.001), respectively. The corresponding angular errors were 0.002 rad and 0.001 rad, respectively. For the 10dB SNR case of Fig. 9.47, the detected (θ_E, θ_A) for DOA-1 and DOA-2 were (1.008, 0.365) and (0.576, −2.026), respectively. The corresponding angular errors were 0.073 rad and

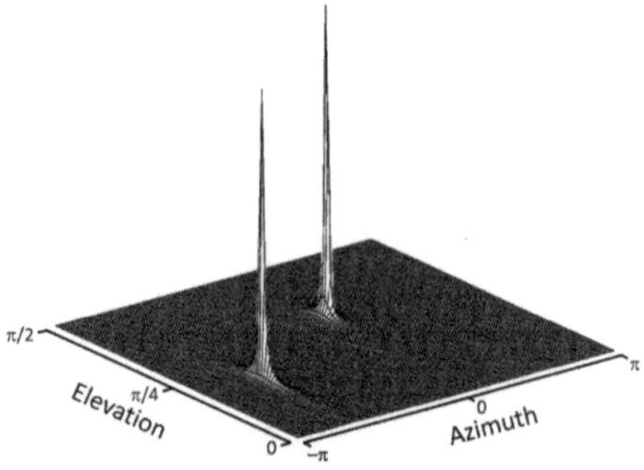

Fig. 9.46 (2x2)-element 2D rectangular array. 40dB SNR

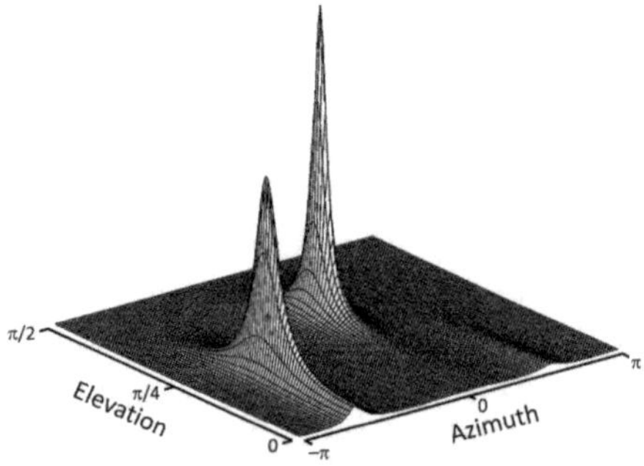

Fig. 9.47 (2x2)-element 2D rectangular array. 10dB SNR

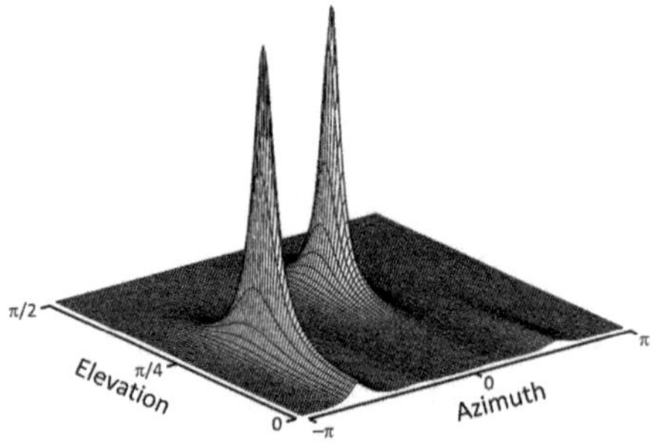

Fig. 9.48 (3x3)-element 2D rectangular array. 10dB SNR

0.032 rad, respectively. As expected, the accuracy of the estimate decreases as the SNR decreases.

Figure 9.48 shows the result where the SNR value of each array element remained at 10dB but the array was replaced by a three-by-three array. The detected (θ_E, θ_A) for DOA-1 and DOA-2 were (1.014, 0.545) and (0.611, −2.022), respectively. The corresponding angular errors were 0.028 rad and 0.021 rad, respectively. Comparing to the results shown in Fig. 9.47 with angular errors 0.073 rad and 0.032 rad, the errors of Fig. 9.48 were smaller. The results showed that, for a given SNR, increasing the number of array elements improved the accuracy of the estimate.

9.15 MUSIC for L-Shaped Arrays

With reference to Fig. 9.29, the L-shape equivalent of (9.93) for the mth plane wave is

$$\begin{bmatrix} x_{0,0}(t) \\ x_{1,0}(t) \\ \vdots \\ x_{N_x-1,0}(t) \\ x_{0,1}(t) \\ x_{0,2}(t) \\ \vdots \\ x_{0,N_y-1}(t) \end{bmatrix} = \begin{bmatrix} x_{0,0}(t) \\ x_{0,0}(t)e^{-j\omega_m d_x \sin(\theta_{x,m})/c} \\ \vdots \\ x_{0,0}(t)e^{-j(N_x-1)\omega_m d_x \sin(\theta_{x,m})/c} \\ x_{0,0}(t)e^{-j\omega_m d_y \sin(\theta_{y,m})/c} \\ x_{0,0}(t)e^{-2j\omega_m d_y \sin(\theta_{y,m})/c} \\ \vdots \\ x_{0,0}(t)e^{-j(N_y-1)\omega_m d_y \sin(\theta_{y,m})/c} \end{bmatrix} = \boldsymbol{a}_m x_{0,0}(t), \quad (9.137a)$$

$$\text{where } \boldsymbol{a}_m = \begin{bmatrix} 1 \\ e^{-j\omega_m d_x \sin(\theta_{x,m})/c} \\ \vdots \\ e^{-j(N_x-1)\omega_m d_x \sin(\theta_{x,m})/c} \\ e^{-j\omega_m d_y \sin(\theta_{y,m})/c} \\ e^{-2j\omega_m d_y \sin(\theta_{y,m})/c} \\ \vdots \\ e^{-j(N_y-1)\omega_m d_y \sin(\theta_{y,m})/c} \end{bmatrix}. \quad (9.137b)$$

The $e^{-jn_x\omega_m d_y \sin(\theta_{x,m})/c}$ and $e^{-jn_y\omega_m d_y \sin(\theta_{y,m})/c}$ terms in (9.137) may be replaced by the $e^{-jn_x\omega_m d_y \sin(\theta_{A,m})\cos(\theta_{E,m})/c}$ and $e^{-jn_y\omega_m d_y \cos(\theta_{A,m})\cos(\theta_{E,m})/c}$ terms, respectively, to optimize directly in the (θ_A, θ_E) domain:

$$e^{-jn_x\omega_m d_y \sin(\theta_{x,m})/c} = e^{-jn_x\omega_m d_y \sin(\theta_{A,m})\cos(\theta_{E,m})/c}. \quad (9.138a)$$

$$e^{-jn_y\omega_m d_y \sin(\theta_{y,m})/c} = e^{-jn_y\omega_m d_y \cos(\theta_{A,m})\cos(\theta_{E,m})/c} \quad (9.138b)$$

The associated derivations following (9.93) remain the same with $\boldsymbol{a}_m$ in (9.93) replaced by $\boldsymbol{a}_m$ in (9.137). $P(\theta)$ of (9.124) is evaluated and the spikes indicate the directions of arrivals.

We shall show some simulation results. Consider the scenario of Fig. 9.46 with the two-by-two rectangular array replaced by an L-shaped array with $N_x = N_y = 2$. The result is shown in Fig. 9.49. The detected (θ_E, θ_A) for DOA-1 and DOA-2 were $(0.997, 0.500)$ and $(0.589, -2.003)$, respectively. The corresponding angular errors were 0.003 rad and 0.011 rad, respectively; they are larger than the angular errors of 0.002 rad and 0.001 rad in Fig. 9.46. For a given N_x and N_y, the L-shaped array requires less array element at the expense of an increase in the angular error of the estimated DOA.

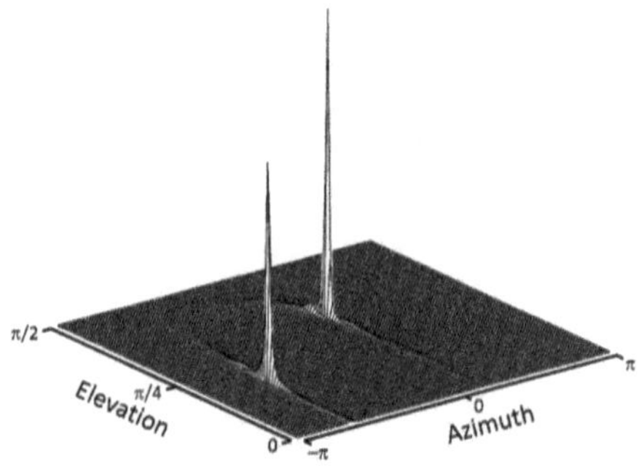

Fig. 9.49 L-shaped array. $N_x = N_y = 2$. 40dB SNR

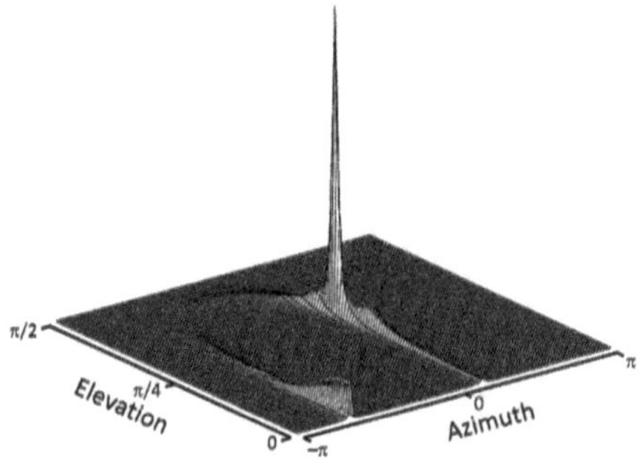

Fig. 9.50 L-shaped array. $N_x = N_y = 2$. 10dB SNR

The result when the SNR is decreased to 10dB is shown in Fig. 9.50. The detected (θ_E, θ_A) for DOA-1 was (0.899, 0.520). The corresponding angular error was 0.102 rad; this was larger than the angular error of 0.073 rad of Fig. 9.47 as expected. Detection of DOA-2 on the (θ_E, θ_A) plane failed. Detection of DOA-2 on the (θ_x, θ_y) plane produces $\theta_x = -1.2557$ and $\theta_y = -0.5951$. Since $|\theta_x| + |\theta_y| = |-1.2557| + |-0.5951| = 1.8508 > \frac{\pi}{2}$, solution for (θ_E, θ_A) does not exist; see (9.33).

DOA-2 at 10dB SNR can be detected if the L-shaped array is replaced by another L-shaped array with $N_x = N_y = 3$. The result is shown in Fig. 9.51. The detected (θ_E, θ_A) for DOA-1 and DOA-2 were (1.027, 0.473) and (0.650, −1.968), respectively. The corresponding angular errors were 0.031 rad and 0.056 rad, respectively; the results were better than the angular errors of 0.073 rad and 0.032 rad for the two-by-two rectangular array of Fig. 9.47 but not as good as the angular errors of 0.028 rad and 0.021 rad for the three-by-three rectangular array of Fig. 9.48.

9.16 Separate θ_x and θ_y Detection

An $N_x \times N_y$ array may also be regarded as a collection of N_y x-arrays. Each x-array will produce a correlation matrix. The N_y correlation matrices of the x-arrays may be averaged for estimating θ_x for all the arriving signals using a one-dimensional MUSIC algorithm if the number of arriving signals is less than N_x.

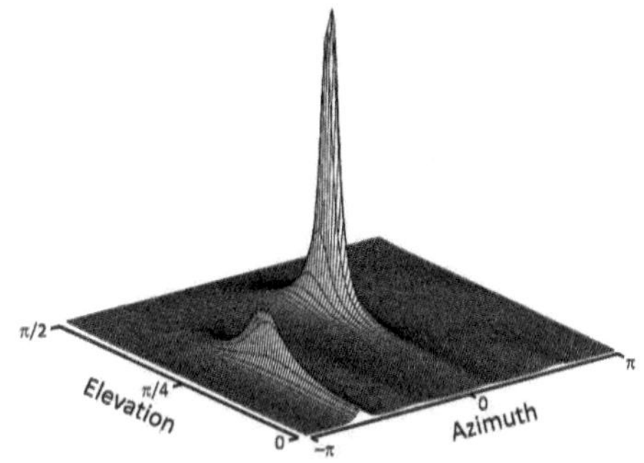

Fig. 9.51 L-shaped array. $N_x = N_y = 3$. 10dB SNR

After θ_x for all the arriving signals have been estimated, a particular θ_x, say, θ_{x0}, may be selected. After that, each x-array may be beamformed to peak in the θ_{x0} direction and reject signals from all other directions. Suppose that the nth x-array with array signals $x_{0,n}(t), x_{1,n}(t), \ldots, x_{N_x-1,n}(t)$ is beamformed using beamformer coefficients $c_0, c_1, \ldots, c_{n_x}, \ldots, c_{N_x-1}$ to form $\tilde{y}_n(t)$ as shown in (9.139):

$$\tilde{y}_n(t) = \sum_{i=0}^{N_x-1} c_i x_{i,n}(t) \tag{9.139}$$

Suppose that all the N_y x-arrays use the same beamformer coefficient values. Since there are N_y x-arrays, there are N_y values of $\tilde{y}_n(t)$, $n = 0, 1, \ldots, N_y - 1$, forming a y-array. The signals $\tilde{y}_0(t), \tilde{y}_1(t), \ldots, \tilde{y}_n(t), \ldots, \tilde{y}_{N_y-1}(t)$ may be put in a vector form as shown in (9.140):

$$\mathbf{y}(t) = \begin{bmatrix} \tilde{y}_0(t) \\ \vdots \\ \tilde{y}_n(t) \\ \vdots \\ \tilde{y}_{N_y-1}(t) \end{bmatrix} = \begin{bmatrix} \sum_{i=0}^{N_x-1} c_i x_{i,0} \\ \vdots \\ \sum_{i=0}^{N_x-1} c_i x_{i,n} \\ \vdots \\ \sum_{i=0}^{N_x-1} c_i x_{i,N_y-1} \end{bmatrix} \tag{9.140}$$

Let $\mathbf{y}(t)\mathbf{y}^{\mathrm{H}}(t)$ be denoted by $\mathbf{Y}(t)$, and let the nth row mth column of $\mathbf{Y}(t)$ be $y_{n,m}(t)$ where $\mathbf{y}^{\mathrm{H}}(t)$ denote the Hermitian transpose of $\mathbf{y}(t)$, i.e., $\mathbf{y}^{\mathrm{H}}(t) = [\mathbf{y}^*(t)]^{\mathrm{T}}$:

$$\mathbf{Y}(t) = \mathbf{y}(t)\mathbf{y}^{\mathrm{H}}(t) \equiv \begin{bmatrix} y_{0,0}(t) & \cdots & y_{0,m}(t) & \cdots & y_{0,N_y-1}(t) \\ \vdots & \ddots & \vdots & \ddots & \vdots \\ y_{n,0}(t) & \cdots & y_{n,m}(t) & \cdots & y_{n,N_y-1}(t) \\ \vdots & \ddots & \vdots & \ddots & \vdots \\ y_{N_y-1,0}(t) & \cdots & y_{N_y-1,m}(t) & \cdots & y_{N_y-1,N_y-1}(t) \end{bmatrix} \tag{9.141}$$

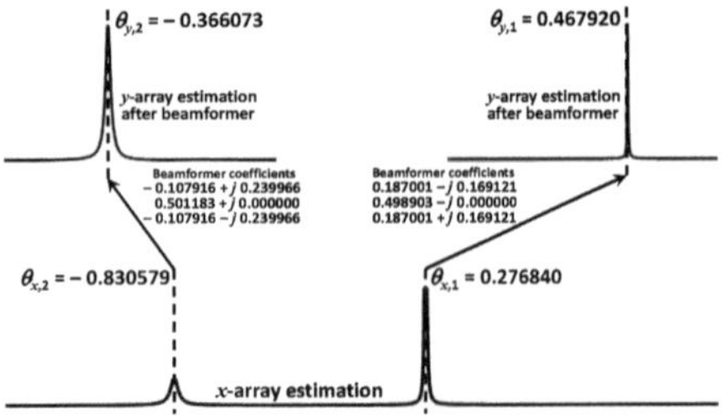

Fig. 9.52 (3x3)-element 2D array. 10 dB SNR. Separate estimation of θ_x and θ_y

and

$$y_{n,m}(t) = \sum_{i=0}^{N_x-1} \sum_{j=0}^{N_x-1} \hat{c}_{i,j} x_{i,n}(t) x_{j,m}^*(t) \tag{9.142a}$$

$$\text{where} \quad \hat{c}_{i,j} = c_i c_j^* \tag{9.142b}$$

Expression (9.142) shows that the elements of the correlation matrix $\mathbf{Y}(t)$ are the weighted average of the elements of the correlation matrix of the array signals; the weight factors are $\hat{c}_{i,j}$.

The correlation matrix $\mathbf{Y}(t)$ will be used to determine the particular θ_y, say, θ_{y0} associated with θ_{x0} using the one-dimensional MUSIC. The process will be repeated for all values of detected θ_x.

We shall illustrate this by using the three-by-three rectangular array with 10dB SNR. Two directions of arrival at $\theta_{x,1} = 0.276840$ and $\theta_{x,2} = -0.830579$ were detected. See Fig. 9.52.

Each x-array was beamformed to reject signal with $\theta_{x,2} = -0.830579$ using the coefficients $(0.18700117 - j0.16912052)$, $(0.49890330 + j0)$, and $(0.18700117 + j0.16912052)$. The outputs of the three x-arrays after beamforming constitute a y-array; application of one-dimensional MUSIC obtained $\theta_{y,1} = 0.467920$. The corresponding θ_E and θ_A were 1.015 and 0.545, respectively. The corresponding angular error was 0.028 rad and was similar to the angular error of 0.028 for the three-by-three rectangular array of Fig. 9.48.

Next, each x-array was beamformed to reject signal with $\theta_{x,1} = 0.276840$ using the coefficients $(-0.10791620 + j0.23996626)$, $(0.50118320 + j0)$, and $(-0.10791620 - j0.23996626)$. The outputs of the three x-arrays after beamforming constitute a y-array; the application of one-dimensional MUSIC obtained $\theta_{y,2} = -0.366073$. The corresponding θ_E and θ_A were 0.608 and $-$ 2.022, respectively. The corresponding angular error was 0.020 rad and was similar to the angular error of 0.021 for the three-by-three rectangular array of Fig. 9.48.

9.17 ESPRIT

ESPRIT is the abbreviation for Estimation of Signal Parameters via Rotational Invariance Technique. ESPRIT partitions a linear array into two sub-arrays displaced from each other by a fixed amount. For the same number of plane waves, the number of array elements required by ESPRIT is at least twice as many as that required by MUSIC, i.e., $N \geq 2(M+1)$.

In ESPRIT, the steering vector $\boldsymbol{a}_m$ defined in (9.93b) is partitioned into two vectors (whether overlapping or nonoverlapping) $\boldsymbol{a}_{0,m}$ and $\boldsymbol{a}_{1,m}$. The vector $\boldsymbol{a}_{0,m}$ is the first $N-1$ elements of $\boldsymbol{a}_m$. The vector $\boldsymbol{a}_{1,m}$ is the last $N-1$ elements of $\boldsymbol{a}_m$:

$$\boldsymbol{a}_m = \begin{bmatrix} 1 \\ e^{-j\omega_m d \sin(\theta_m)/c} \\ \vdots \\ e^{-j(N-1)\omega_m d \sin(\theta_m)/c} \end{bmatrix} = \begin{bmatrix} \boldsymbol{a}_{0,m} \\ e^{-j(N-1)\omega_m d \sin(\theta_m)/c} \end{bmatrix} \tag{9.143a}$$

$$= \begin{bmatrix} 1 \\ e^{-j\omega_m d \sin(\theta_m)/c} \\ \vdots \\ e^{-j(N-1)\omega_m d \sin(\theta_m)/c} \end{bmatrix} = \begin{bmatrix} 1 \\ \boldsymbol{a}_{1,m} \end{bmatrix} \tag{9.143b}$$

In (9.143), $\boldsymbol{a}_{0,m} = \begin{bmatrix} 1 \\ e^{-j\omega_m d \sin(\theta_m)/c} \\ \vdots \\ e^{-j(N-2)\omega_m d \sin(\theta_m)/c} \end{bmatrix}$ and $\boldsymbol{a}_{1,m} = \begin{bmatrix} e^{-j\omega_m d \sin(\theta_m)/c} \\ \vdots \\ e^{-j(N-1)\omega_m d \sin(\theta_m)/c} \end{bmatrix}$.

Two matrices $\mathbf{A}_0$ and $\mathbf{A}_1$ are formed as follows:

$$\mathbf{A}_0 = \begin{bmatrix} \boldsymbol{a}_{0,0} \; \boldsymbol{a}_{0,1} \; \cdots \; \boldsymbol{a}_{0,M-1} \end{bmatrix}, \tag{9.144a}$$

$$\mathbf{A}_1 = \begin{bmatrix} \boldsymbol{a}_{1,0} \; \boldsymbol{a}_{1,1} \; \cdots \; \boldsymbol{a}_{1,M-1} \end{bmatrix} \tag{9.144b}$$

Define the $M \times M$ matrix Φ given by

$$\Phi = \begin{bmatrix} z_0 & 0 & \cdots & 0 \\ 0 & z_1 & \cdots & 0 \\ \vdots & \vdots & \ddots & \vdots \\ 0 & 0 & \cdots & z_{M-1} \end{bmatrix}, \tag{9.145a}$$

$$\text{where} \quad z_m = e^{-j\omega_m d \sin(\theta_m)/c}. \tag{9.145b}$$

From (9.143), (9.144), and (9.145), it can be easily verified that

$$\mathbf{A}_1 = \mathbf{A}_0 \Phi. \tag{9.146}$$

Since $\mathbf{V}_s$ (see (9.120b)) and $\mathbf{A}$ (see (9.96)) span the same subspace, each column vector of $\mathbf{V}_s$ may be expressed as a linear combination of the columns of $\mathbf{A}$. Hence,

$$\mathbf{V}_s = \mathbf{AC} \tag{9.147}$$

where $\mathbf{C}$ is an invertible $M \times M$ square matrix. Define $\mathbf{V}_{0,s}$ and $\mathbf{V}_{1,s}$ in a similar way as $\mathbf{A}_0$ and $\mathbf{A}_1$:

$$\mathbf{V}_s = \begin{bmatrix} \mathbf{V}_{0,s} \\ \text{Last row of } \mathbf{V}_s \end{bmatrix}, \tag{9.148a}$$

$$= \begin{bmatrix} \text{First row of } \mathbf{V}_s \\ \mathbf{V}_{1,s} \end{bmatrix} \tag{9.148b}$$

$\mathbf{V}_{0,s}$ and $\mathbf{V}_{1,s}$ are $(N-1) \times M$ matrices. From (9.148a), removing the last row of $\mathbf{V}_s$ and taking note of (9.143), (9.144), and (9.147), we have

$$\mathbf{V}_{0,s} = \mathbf{A}_0 \mathbf{C}. \tag{9.149}$$

From (9.148b), removing the first row of $\mathbf{V}_s$ and taking note of (9.143), (9.144), and (9.147), we have

$$\mathbf{V}_{1,s} = \mathbf{A}_1 \mathbf{C}. \tag{9.150}$$

From (9.146) and (9.150), we have

$$\mathbf{V}_{1,s} = \mathbf{A}_0 \Phi \mathbf{C}. \tag{9.151}$$

Multiplying both sides of (9.151) by $\mathbf{C}^{-1}\Phi^{-1}\mathbf{C}$,

$$\mathbf{V}_{1,s}\mathbf{C}^{-1}\Phi^{-1}\mathbf{C} = \mathbf{A}_0 \Phi \mathbf{C}\mathbf{C}^{-1}\Phi^{-1}\mathbf{C} = \mathbf{A}_0 \mathbf{C}. \tag{9.152}$$

From (9.149) and (9.152),

$$\mathbf{V}_{1,s}\mathbf{C}^{-1}\Phi^{-1}\mathbf{C} = \mathbf{V}_{0,s}. \tag{9.153}$$

Let

$$\Psi^{-1} = \mathbf{C}^{-1}\Phi^{-1}\mathbf{C}. \tag{9.154a}$$

$$\therefore \quad \Psi = \mathbf{C}^{-1}\Phi\mathbf{C}. \tag{9.154b}$$

Since $\Psi = \mathbf{C}^{-1}\Phi\mathbf{C}$ (see (9.154b)) and since Φ is diagonal (see (9.145a)), the diagonal elements of Φ are the eigenvalues of Ψ, i.e., Φ is a diagonal matrix of eigenvalues of Ψ. Substitute (9.154a) into (9.152) and (9.152) into (9.149)

$$\mathbf{V}_{1,s}\Psi^{-1} = \mathbf{V}_{0,s}. \tag{9.155a}$$

$$\therefore \quad \mathbf{V}_{1,s} = \mathbf{V}_{0,s}\Psi. \tag{9.155b}$$

$\mathbf{V}_{0,s}$ and $\mathbf{V}_{1,s}$ are $(N-1) \times M$ matrices and Ψ is a $M \times M$ matrix and $N > M$. The elements of Ψ may be estimated from the values of $\mathbf{V}_{0,s}$ and $\mathbf{V}_{1,s}$ using least squares method or more appropriately using total least squares since both $\mathbf{V}_{0,s}$ and $\mathbf{V}_{1,s}$ are estimated values containing errors.

ESPRIT algorithm:

1. Find eigenvalues and eigenvectors of $\mathbf{R}_{xx}$.
2. Eigenvectors corresponding to the largest M eigenvalues of $\mathbf{R}_{xx}$ are the signal eigenvectors. Let $\boldsymbol{v}_p$, $p = 0, \ldots, M-1$ denote these M signal eigenvectors.
3. Define $\mathbf{V}_s = [\boldsymbol{v}_0, \boldsymbol{v}_1, \ldots, \boldsymbol{v}_{M-1}]$.
4. Estimate Ψ using the expression $\mathbf{V}_{1,s} = \mathbf{V}_{0,s}\Psi$. Use least squares or total least squares to estimate Ψ.
5. Find the eigenvalues of Ψ. The eigenvalues of Ψ are the diagonal elements of Φ.
6. Obtain DOA using $\Phi = \begin{bmatrix} z_0 & 0 & \cdots & 0 \\ 0 & z_1 & \cdots & 0 \\ \vdots & \vdots & \ddots & \vdots \\ 0 & 0 & \cdots & z_{M-1} \end{bmatrix}$ where $z_m = e^{-j\omega_m d \sin(\theta_m)/c}$.

In (9.143), we partition a linear array into two sub-arrays displaced by one array element, for example, for a linear array with ten elements, we partitioned it into two sub-arrays each with nine elements as shown in Fig. 9.53. The linear array can also be partitioned into two sub-arrays displaced by any number of array elements, for example, it can be partitioned into two eight-element sub-arrays displaced by two array elements as shown in Fig. 9.54. The linear array can also be partitioned into two sub-arrays without overlap, for example, a ten-element array may be partitioned into two sub-arrays each with five elements as shown in Fig. 9.55. The derivation for the algorithm remains largely unchanged with the exception that d in (9.145) should be replaced by $(number\_of\_array\_element\_separation\_between\_two\_sub - arrays) \times d$.

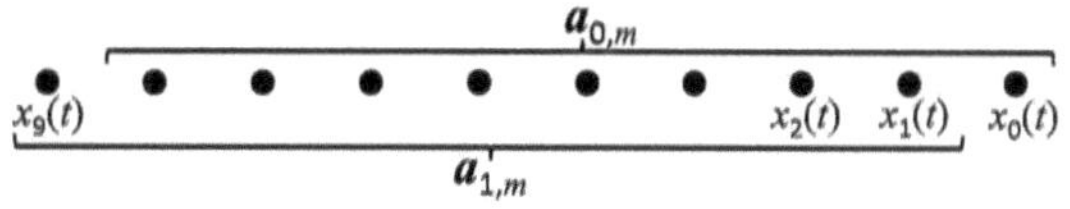

Fig. 9.53 Partition into two sub-arrays displaced by one array element

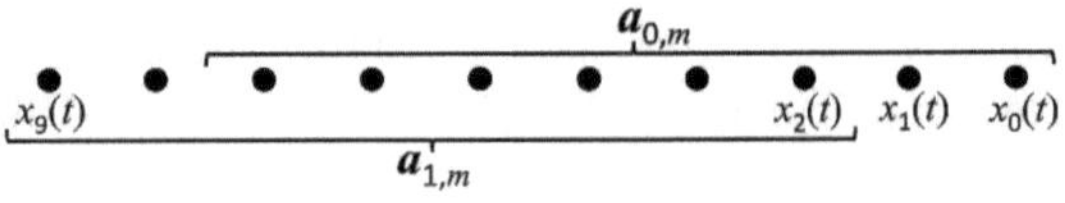

Fig. 9.54 Partition into two sub-arrays displaced by two array element

Fig. 9.55 Partition into two sub-arrays without overlap

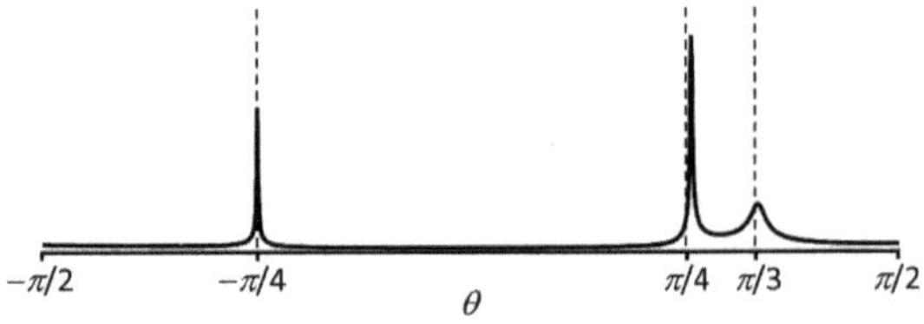

Fig. 9.56 Estimating DOA of three incident waves with 50 dB SNR using a third-order complex linear predictor

9.18 DOA Estimation as Frequency Estimation

From (9.10) and (9.12), we have

$$x_m(t) = e^{-jm\phi}x_0(t) \ , \quad m = 1, 2, \ldots, N - 1. \tag{9.156}$$

The value of $x_m(t), m = 1, 2, \ldots, N - 1$ for any particular value of t is a sinusoid with frequency ϕ. The value of ϕ can be estimated by estimating the frequency of $x_m(t), m = 1, 2, \ldots, N - 1$ at any given t. The frequencies of $x_m(t)$ can be estimated using any one of the methods discussed in Sect. 7.5 and Chap. 8.

First, we shall show an example where the frequency estimation is done using linear prediction. We shall assume that $x_m(t)$ is complex. As discussed in Sect. 8.2, a minimum of $2M$ data points are needed for estimating the frequencies of a signal containing M sinusoids. Hence, for M directions of arrival, the array must have at least $2M$ elements.

Consider the same signal scenario as that in Fig. 9.40 with the exception that the number of array elements is six instead of four, i.e., the three incident waves are at angles $-\pi/4$, $\pi/4$, and $\pi/3$, respectively, and the SNR is 50 dB. Figure 9.56 shows a $\left|\frac{1}{D(e^{j\omega})}\right|$ plot for a third-order predictor estimating the DOAs of the three incident waves. As can be seen from Fig. 9.56, the positions of the peaks are shifted. The three zeros of $D(e^{j\omega})$ must attenuate the three incident waves while at the same time attenuate the noise. Hence, the peaks are more seriously affected. It requires more array elements, and the result is less accurate than MUSIC. The only advantage it has is that the carrier frequencies of the incident waves may be the same.

9.19 Rotman Lens

A microwave lens is a type of beamforming network that can focus EM wave from an omnidirectional/isotropic source into a directional beam. A typical example is shown in Fig 9.57a. In Fig. 9.57a, the beam contour is the contour where the omnidirectional radiator is placed. It may be an array of omnidirectional radiators but with only one of them radiating. Three omnidirectional sources G_1, G_0, and G_{-1} are show in Fig. 9.57a. The radiator will radiate EM wave to an array of receivers $R_i(x_i, y_i)$ placed on the inner receiver contour. Each receiver $R_i(x_i, y_i)$ will be connected to a linear array element P_i through a coaxial cable of length l_i. The coaxial cables have different lengths. Figures 9.57b and c show the scenario when G_0 and G_1 are radiating, respectively. The angle Ψ may or may not be the same as the angle α.

We have shown in (9.3) that for a wave traveling from x_1 to x_2 at a distance d away from x_1, $x_2(t) = e^{-j\omega d/c}x_1(t)$ where $x_1(t)$ and $x_2(t)$ are the signals at x_1 and x_2, respectively, and c is the speed of light in vacuum. If the relative permittivity of the medium is ϵ, (9.3) becomes (9.157),

$$x_2(t) = e^{-j\omega\sqrt{\epsilon}d/c}x_1(t) = e^{-j2\pi\sqrt{\epsilon}d\lambda_0}x_1(t), \tag{9.157}$$

where λ_0 is the wavelength of the sinusoid in free space. The signal at x_2 is that at x_1 with a phase shift of $-\omega\sqrt{\epsilon}d/c$ or $-2\pi\sqrt{\epsilon}d\lambda_0$. Let ϵ_r, ϵ_e, and ϵ_p be the relative permittivity of the medium between the beam contour and inner receiver contour, the medium of coaxial cables linking $R_i(x_i, y_i)$ and P_i, and the medium

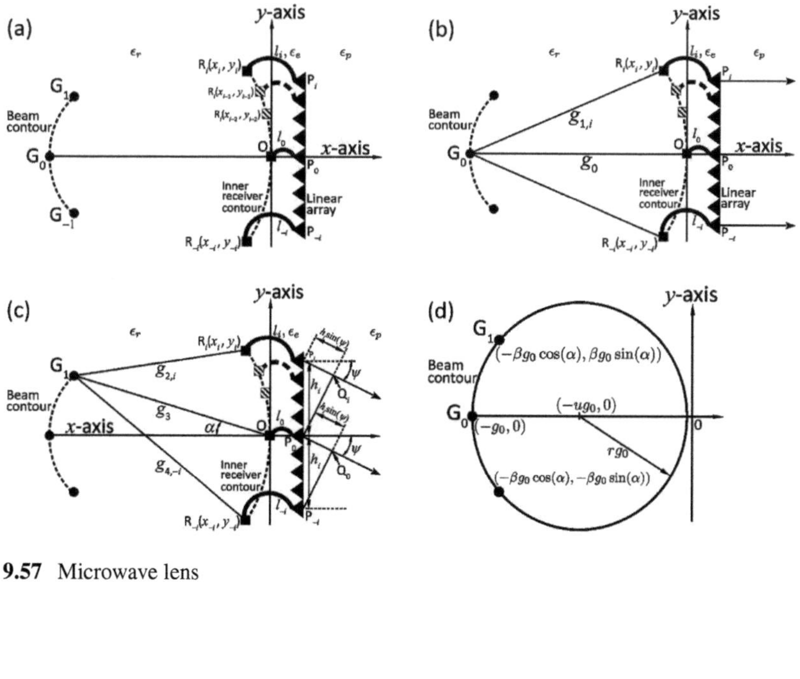

Fig. 9.57 Microwave lens

the linear array radiates into, respectively. If the linear array is radiating into free space, $\epsilon_p = 1$. Some examples of microwave lenses are Rotman lens, Ruze lens, Gent Bootlace lens, and Luneburg lens.

In this section, we shall discuss only the Rotman lens. The Rotman lens is a trifocal lens; there are three focal points on the beam contour, i.e., three points where the phase of the signal radiated from a radiator at the beam contour is exactly matched at the linear array. The focal points G_{-1} and G_1 are symmetrical about the x-axis. Also, $x_{-i} = x_i$ and $y_{-i} = -y_i$. The aim is to determine the inner contour (x_i, y_i) and l_i for all i given the coordinates of G_1, G_0, and G_{-1}.

9.19.1 *Inner Receiver Contour*

Referring to Fig. 9.57b, the phase difference between the signal at G_0 and that at P_0 is $2\pi\sqrt{\epsilon_r}g_0\lambda_0 + 2\pi\sqrt{\epsilon_e}l_0\lambda_0$. The phase difference between the signal at G_0 and that at P_i is $2\pi\sqrt{\epsilon_r}g_{1,i}\lambda_0 + 2\pi\sqrt{\epsilon_e}l_i\lambda_0$. For the focused beam to be radiating in parallel to the x-axis, these phase differences must be equal. Hence,

$$2\pi\sqrt{\epsilon_r}g_0\lambda_0 + 2\pi\sqrt{\epsilon_e}l_0\lambda_0 = 2\pi\sqrt{\epsilon_r}g_{1,i}\lambda_0 + 2\pi\sqrt{\epsilon_e}l_i\lambda_0. \tag{9.158}$$

Referring to Fig. 9.57c, the phase difference between the signal at G_1 and that at P_0 is $2\pi\sqrt{\epsilon_r}g_3\lambda_0 + 2\pi\sqrt{\epsilon_e}l_0\lambda_0$. The phase difference between the signal at G_1 and that at Q_i is $2\pi\sqrt{\epsilon_r}g_{2,i}\lambda_0 + 2\pi\sqrt{\epsilon_e}l_i\lambda_0 + 2\pi\sqrt{\epsilon_p}h_i\sin(\Psi)\lambda_0$. For the focused beam to be radiating in the direction shown in Fig. 9.57c, these phase differences must be equal. Hence,

$$2\pi\sqrt{\epsilon_r}g_3\lambda_0 + 2\pi\sqrt{\epsilon_e}l_0\lambda_0 = 2\pi\sqrt{\epsilon_r}g_{2,i}\lambda_0 + 2\pi\sqrt{\epsilon_e}l_i\lambda_0 + 2\pi\sqrt{\epsilon_p}h_i\sin(\Psi)\lambda_0. \tag{9.159}$$

The phase difference between the signal at G_1 and that at P_{-i} is $2\pi\sqrt{\epsilon_r}g_{4,-i}\lambda_0 + 2\pi\sqrt{\epsilon_e}l_{-i}\lambda_0$. The phase difference between the signal at G_1 and that at Q_0 is $2\pi\sqrt{\epsilon_r}g_3\lambda_0 + 2\pi\sqrt{\epsilon_e}l_0\lambda_0 + 2\pi\sqrt{\epsilon_p}h_i\sin(\Psi)\lambda_0$. For the focused beam to be radiating in the direction shown in Fig. 9.57c, these phase differences must be equal. Hence,

$$2\pi\sqrt{\epsilon_r}g_{4,-i}\lambda_0 + 2\pi\sqrt{\epsilon_e}l_{-i}\lambda_0 = 2\pi\sqrt{\epsilon_r}g_3\lambda_0 + 2\pi\sqrt{\epsilon_e}l_0\lambda_0 + 2\pi\sqrt{\epsilon_p}h_i\sin(\Psi)\lambda_0. \tag{9.160}$$

Eliminating the $2\pi\lambda_0$ factor in (9.158), (9.159), and (9.160), we have

$$\sqrt{\epsilon_r}g_0 + \sqrt{\epsilon_e}l_0 = \sqrt{\epsilon_r}g_{1,i} + \sqrt{\epsilon_e}l_i, \tag{9.161a}$$

$$\sqrt{\epsilon_r}g_3 + \sqrt{\epsilon_e}l_0 = \sqrt{\epsilon_r}g_{2,i} + \sqrt{\epsilon_e}l_i + \sqrt{\epsilon_p}h_i\sin(\Psi), \tag{9.161b}$$

$$\sqrt{\epsilon_r}g_{4,-i} + \sqrt{\epsilon_e}l_{-i} = \sqrt{\epsilon_r}g_3 + \sqrt{\epsilon_e}l_0 + \sqrt{\epsilon_p}h_i\sin(\Psi). \tag{9.161c}$$

Dividing (9.161) by $\sqrt{\epsilon_r} g_0$ and rearranging terms, we have

$$\frac{g_{1,i}}{g_0} = 1 - \frac{l_i - l_0}{g_0}\frac{\sqrt{\epsilon_e}}{\sqrt{\epsilon_r}}, \tag{9.162a}$$

$$\frac{g_{2,i}}{g_0} = \frac{g_3}{g_0} - \frac{l_i - l_0}{g_0}\frac{\sqrt{\epsilon_e}}{\sqrt{\epsilon_r}} - \frac{h_i \sin(\Psi)}{g_0}\frac{\sqrt{\epsilon_p}}{\sqrt{\epsilon_r}}, \tag{9.162b}$$

$$\frac{g_{4,-i}}{g_0} = \frac{g_3}{g_0} - \frac{l_i - l_0}{g_0}\frac{\sqrt{\epsilon_e}}{\sqrt{\epsilon_r}} + \frac{h_i \sin(\Psi)}{g_0}\frac{\sqrt{\epsilon_p}}{\sqrt{\epsilon_r}}. \tag{9.162c}$$

Let $\beta = \frac{g_3}{g_0}$ and $w_i = \frac{l_i - l_0}{g_0}$; (9.162) may be written as

$$\frac{g_{1,i}}{g_0} = 1 - w_i\frac{\sqrt{\epsilon_e}}{\sqrt{\epsilon_r}}, \tag{9.163a}$$

$$\frac{g_{2,i}}{g_0} = \beta - w_i\frac{\sqrt{\epsilon_e}}{\sqrt{\epsilon_r}} - \frac{h_i \sin(\Psi)}{g_0}\frac{\sqrt{\epsilon_p}}{\sqrt{\epsilon_r}}, \tag{9.163b}$$

$$\frac{g_{4,-i}}{g_0} = \beta - w_i\frac{\sqrt{\epsilon_e}}{\sqrt{\epsilon_r}} + \frac{h_i \sin(\Psi)}{g_0}\frac{\sqrt{\epsilon_p}}{\sqrt{\epsilon_r}},. \tag{9.163c}$$

$$\beta = \frac{g_3}{g_0} \quad \text{and} \quad w_i = \frac{l_i - l_0}{g_0}. \tag{9.163d}$$

From Fig. 9.57 and noting that for $\mathrm{R}_i(x_i, y_i)$ x_i is negative, y_i is positive, and $y_{-i} = -y_i$ we have

$$g_{1,i}^2 = (g_0 + x_i)^2 + y_i^2, \tag{9.164a}$$

$$\begin{aligned} g_{2,i}^2 &= (g_3\cos(\alpha) + x_i)^2 + (y_i - g_3\sin(\alpha))^2 \\ &= g_3^2 + x_i^2 + y_i^2 + 2g_3x_i\cos(\alpha) - 2g_3y_i\sin(\alpha), \end{aligned} \tag{9.164b}$$

$$\begin{aligned} g_{4,-i}^2 &= (g_3\cos(\alpha) + x_i)^2 + (g_3\sin(\alpha) + y_i)^2 \\ &= g_3^2 + x_i^2 + y_i^2 + 2g_3x_i\cos(\alpha) + 2g_3y_i\sin(\alpha). \end{aligned} \tag{9.164c}$$

Dividing (9.164) by g_0^2 and letting $X_i = x_i/g_0$ and $Y_i = y_i/g_0$, we have

$$\frac{g_{1,i}^2}{g_0^2} = (1 + X_i)^2 + Y_i^2, \tag{9.165a}$$

$$\frac{g_{2,i}^2}{g_0^2} = \beta^2 + X_i^2 + Y_i^2 + 2\beta X_i\cos(\alpha) - 2\beta Y_i\sin(\alpha), \tag{9.165b}$$

$$\frac{g_{4,-i}^2}{g_0^2} = \beta^2 + X_i^2 + Y_i^2 + 2\beta X_i \cos(\alpha) + 2\beta Y_i \sin(\alpha), \tag{9.165c}$$

$$X_i = \frac{x_i}{g_0} \quad \text{and} \quad Y_i = \frac{y_i}{g_0}. \tag{9.165d}$$

Squaring (9.163) and equating the result to (9.165), we have

$$(1 + X_i)^2 + Y_i^2 = \left(1 - w_i \frac{\sqrt{\epsilon_e}}{\sqrt{\epsilon_r}}\right)^2, \tag{9.166a}$$

$$\begin{aligned}
&\beta^2 + X_i^2 + Y_i^2 + 2\beta X_i \cos(\alpha) - 2\beta Y_i \sin(\alpha) \\
&\quad = \left(\beta - w_i \frac{\sqrt{\epsilon_e}}{\sqrt{\epsilon_r}} - \frac{h_i \sin(\Psi)}{g_0} \frac{\sqrt{\epsilon_p}}{\sqrt{\epsilon_r}}\right)^2 \\
&\quad = \beta^2 + w_i^2 \frac{\epsilon_e}{\epsilon_r} + \left(\frac{h_i \sin(\Psi)}{g_0}\right)^2 \frac{\epsilon_p}{\epsilon_r} - 2\beta w_i \frac{\sqrt{\epsilon_e}}{\sqrt{\epsilon_r}} - 2\beta \frac{h_i \sin(\Psi)}{g_0} \frac{\sqrt{\epsilon_p}}{\sqrt{\epsilon_r}} \\
&\qquad + 2 \frac{w_i h_i \sin(\Psi)}{g_0} \frac{\sqrt{\epsilon_e \epsilon_p}}{\epsilon_r},
\end{aligned} \tag{9.166b}$$

$$\begin{aligned}
&\beta^2 + X_i^2 + Y_i^2 + 2\beta X_i \cos(\alpha) + 2\beta Y_i \sin(\alpha) \\
&\quad = \left(\beta - w_i \frac{\sqrt{\epsilon_e}}{\sqrt{\epsilon_r}} + \frac{h_i \sin(\Psi)}{g_0} \frac{\sqrt{\epsilon_p}}{\sqrt{\epsilon_r}}\right)^2 \\
&\quad = \beta^2 + w_i^2 \frac{\epsilon_e}{\epsilon_r} + \left(\frac{h_i \sin(\Psi)}{g_0}\right)^2 \frac{\epsilon_p}{\epsilon_r} - 2\beta w_i \frac{\sqrt{\epsilon_e}}{\sqrt{\epsilon_r}} + 2\beta \frac{h_i \sin(\Psi)}{g_0} \frac{\sqrt{\epsilon_p}}{\sqrt{\epsilon_r}} \\
&\qquad - 2 \frac{w_i h_i \sin(\Psi)}{g_0} \frac{\sqrt{\epsilon_e \epsilon_p}}{\epsilon_r}.
\end{aligned} \tag{9.166c}$$

The above expressions can be simplified by summing (9.166b) and (9.166c) to produce (9.167b) and subtracting (9.166b) from (9.166c) to produce (9.167c):

$$2X_i + X_i^2 + Y_i^2 = w_i^2 \frac{\epsilon_e}{\epsilon_r} - 2w_i \frac{\sqrt{\epsilon_e}}{\sqrt{\epsilon_r}}, \tag{9.167a}$$

$$X_i^2 + Y_i^2 + 2\beta X_i \cos(\alpha) = w_i^2 \frac{\epsilon_e}{\epsilon_r} + \left(\frac{h_i \sin(\Psi)}{g_0}\right)^2 \frac{\epsilon_p}{\epsilon_r} - 2\beta w_i \frac{\sqrt{\epsilon_e}}{\sqrt{\epsilon_r}}, \tag{9.167b}$$

$$\beta Y_i \sin(\alpha) = \beta \frac{h_i \sin(\Psi)}{g_0} \frac{\sqrt{\epsilon_p}}{\sqrt{\epsilon_r}} - \frac{w_i h_i \sin(\Psi)}{g_0} \frac{\sqrt{\epsilon_e \epsilon_p}}{\epsilon_r} \tag{9.167c}$$

Dividing (9.167c) by $\beta \sin(\alpha)$ to obtain Y_i and subtracting (9.167a) from (9.167b), we have

$$
\begin{aligned}
Y_i &= \frac{h_i \sin(\Psi)}{g_0 \sin(\alpha)} \frac{\sqrt{\epsilon_p}}{\sqrt{\epsilon_r}} - \frac{w_i h_i \sin(\Psi)}{\beta g_0 \sin(\alpha)} \frac{\sqrt{\epsilon_e \epsilon_p}}{\epsilon_r} \\
&= \frac{h_i \sin(\Psi)}{g_0 \sin(\alpha)} \frac{\sqrt{\epsilon_p}}{\sqrt{\epsilon_r}} \left(1 - \frac{w_i \sqrt{\epsilon_e}}{\beta \sqrt{\epsilon_r}}\right) = \zeta_i \frac{\sqrt{\epsilon_p}}{\sqrt{\epsilon_r}} \left(1 - \frac{w_i \sqrt{\epsilon_e}}{\beta \sqrt{\epsilon_r}}\right), \qquad (9.168a) \\
\zeta_i &= \frac{h_i \sin(\Psi)}{g_0 \sin(\alpha)}, \qquad (9.168b) \\
X_i &= \frac{\epsilon_p \zeta_i^2 S^2}{2\epsilon_r (\beta C - 1)} + \frac{(1-\beta) w_i}{\beta C - 1} \frac{\sqrt{\epsilon_e}}{\sqrt{\epsilon_r}}, \qquad (9.168c) \\
S &= \sin(\alpha) \quad \text{and} \quad C = \cos(\alpha). \qquad (9.168d)
\end{aligned}
$$

Let

$$
E_p = \frac{\epsilon_p}{\epsilon_r} \quad \text{and} \quad E_e = \frac{\epsilon_e}{\epsilon_r}. \qquad (9.169)
$$

Let

$$
\begin{aligned}
a &= 1 - \frac{(1-\beta)^2}{(\beta C - 1)^2} - \frac{\zeta^2 E_p}{\beta^2}, \qquad (9.170a) \\
b &= -2 + \frac{2\zeta^2 E_p}{\beta} + \frac{2(1-\beta)}{1 - \beta C} - \frac{\zeta^2 S^2 (1-\beta) E_p}{(1-\beta C)^2}, \qquad (9.170b) \\
c &= \left(-\zeta^2 + \frac{\zeta^2 S^2}{1 - \beta C} - \frac{\zeta^4 S^4}{4(1-\beta C)^2}\right) E_p. \qquad (9.170c)
\end{aligned}
$$

Substituting (9.168a) and (9.168c) into (9.167a) and making use of the definitions in (9.169) and (9.170), we have

$$
a E_e w_i^2 + b \sqrt{E_e} w_i + c = 0 \qquad (9.171a)
$$

$$
\text{or} \quad a \left(\sqrt{E_e} w_i\right)^2 + b \left(\sqrt{E_e} w_i\right) + c = 0. \qquad (9.171b)
$$

The solutions are

$$
\sqrt{E_e} w_i = \frac{-b \pm \sqrt{b^2 - 4ac}}{2a} \qquad (9.172a)
$$

$$
\text{or} \quad w_i = \frac{\sqrt{\epsilon_r}}{\sqrt{\epsilon_e}} \times \frac{-b \pm \sqrt{b^2 - 4ac}}{2a}. \qquad (9.172b)
$$

9.19.2 Beam Contour

In the original Rotman lens, the beam contour is the arc of a circle. It has also been noticed that if it is an arc of an ellipse, some optimization may be possible by manipulating the eccentricity of the ellipse. In this section, we shall assume that the beam contour is the arc of a circle as shown in Fig. 9.57d. The coordinates of G_0 and G_1 are $(-g_0, 0)$ and $(-\beta g_0 \cos(\alpha), \beta g_0 \sin(\alpha))$, respectively, where $\beta g_0 = g_3$ as defined in (9.163d). Let the center and radius of the circle be $(-ug_0, 0)$ and rg_0, respectively. Let (x, y) be the coordinates of a point on the circumference of the circle. We have

$$(x + ug_0)^2 + y^2 = r^2 g_0. \tag{9.173}$$

G_0 and G_1 are on the circumference. Substituting the coordinates of G_0 and G_1 into (9.173), we have

$$(-g_0 + ug_0)^2 + 0 = r^2 g_0^2$$

$$\text{or} \quad u^2 - 2u + 1 = r^2, \tag{9.174a}$$

$$(-\beta g_0 \cos(\alpha) + ug_0)^2 + \beta^2 g_0^2 \sin^2(\alpha) = r^2$$

$$\text{or} \quad u^2 - 2u\beta \cos(\alpha) + \beta^2 = r^2. \tag{9.174b}$$

From (9.174a) and (9.174b), we have

$$u = \frac{1 - \beta^2}{2(1 - \beta \cos(\alpha))}. \tag{9.175}$$

Appendix A: Multipath and Simulated Array Movement

Signal processing algorithm such as the MUSIC algorithm requires the signals to be uncorrelated. In some cases, for example, if the objects are moving with different velocities, even if the signals at the objects have the same frequency, as a consequence of Doppler shift, the signals observed by the array will have slightly different frequencies and so are uncorrelated. However, the signals remain correlated if they are multipath signals from a stationary object.

Figure 9.58 shows an example where the signal from a source travels to an observer via two paths. One of the paths is a direct path from the source to the observer. The second path is a reflected path where the signal from the object is reflected by an obstacle before reaching the observer. In general, the signal from a source may travel to an observer via several paths. Such a phenomenon is called multipath effect. If the signal from the source is a sinusoid, the signals

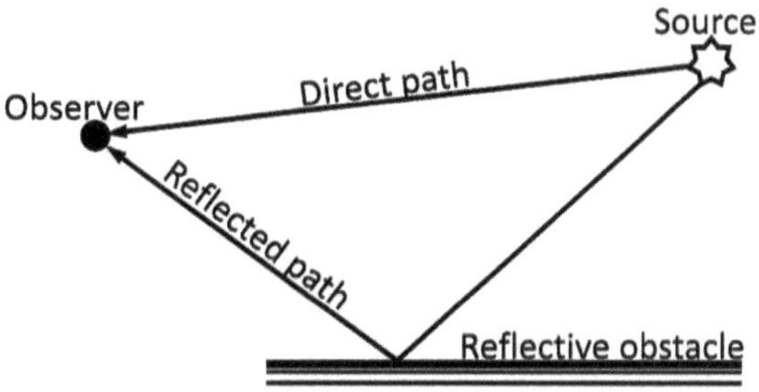

Fig. 9.58 Signal from a source travels to the observer via (1) a direct path and (2) a reflected path

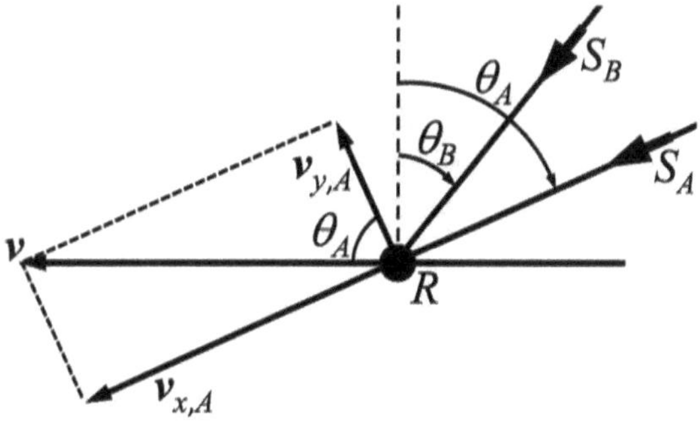

Fig. 9.59 The angles between the boresight of the array and the directions of propagation of S_A and S_B are θ_A and θ_B, respectively

received by the observer will be sinusoids from different directions with the same frequency, assuming the observer, the object, and the obstacle reflecting the signal are stationary; in this case, the observer's received signals from different directions are correlated.

The signals can be made uncorrelated by Doppler shift if the receiver moves. Consider the scenario as shown in Fig. 9.59 consisting of a receiver R and two signals S_A and S_B, respectively. The receiver R moves with velocity $\boldsymbol{v}$ along the axis of the array. The angles of arrival of S_A and S_B are θ_A and θ_B, respectively, as indicated on Fig. 9.58. The angles θ_A and θ_B are not equal. The velocity $\boldsymbol{v}$ may be decomposed into two orthogonal components $\boldsymbol{v}_{x,A}$ and $\boldsymbol{v}_{y,A}$, respectively. The component $\boldsymbol{v}_{x,A}$ is in the same direction of propagation of S_A, and $\boldsymbol{v}_{y,A}$ is orthogonal to the direction of propagation of S_A. The magnitude $|\boldsymbol{v}_{x,A}|$ is equal to $|\boldsymbol{v}\sin(\theta_A)|$. Suppose that S_A and S_B have the same frequency. Let the frequency of S_A and S_B be f_0. As a result of Doppler shift, according to (7.9), the frequency of S_A observed by R denoted by f_A is given by

$$f_A = \left(1 - \frac{|\boldsymbol{v}|\sin(\theta_A)}{c}\right) f_0, \tag{9.176}$$

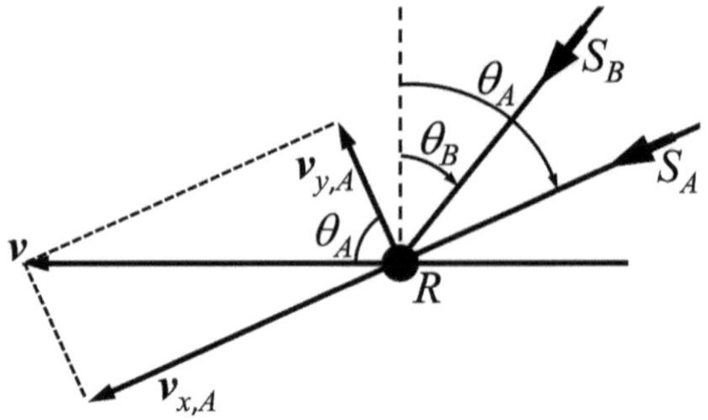

Fig. 9.60 Simulated array movement

where c is the speed of propagation of S_A. The component $\boldsymbol{v}_{y,A}$ produces no Doppler shift since it is perpendicular to the direction of propagation of the wave. Similarly, it can be shown that the frequency of S_B observed by R denoted by f_B is given by

$$f_B = \left(1 - \frac{|\boldsymbol{v}|\sin(\theta_B)}{c}\right) f_0. \tag{9.177}$$

Since $f_A \neq f_B$, the two signals S_A and S_B observed by the receiver will have different frequencies and so are not correlated. The difference in frequency $f_A - f_B$ is given by

$$f_A - f_B = \frac{|\boldsymbol{v}|}{c} \times (\sin(\theta_B) - \sin(\theta_A)) \times f_0. \tag{9.178}$$

The receiver may not be physically moving. The motion can be simulated. Consider the array shown in Fig. 9.60. In Fig. 9.60, the array has $N + KL$ elements with element separation d. However, it is not used as an $(N + KL)$-element array. Instead, it is used as an N-element array, for example, at time t_0, the signals $x_0(t), x_1(t), \ldots, x_{N-1}(t)$ are captured as array signals. At time t_1, the signals $x_L(t), x_{L+1}(t), \ldots, x_{N-1+L}(t)$ are captured as array signals. At time t_n, the signals $x_{nL}(t), x_{nL+1}(t), \ldots, x_{N-1+nL}(t)$ are captured as array signals. Although the array does not move physically, the signal captured is the same as an N-element array that has moved. Let the sampling frequency of the array signal snapshots be f_s, i.e., f_s snapshots of the array signal are sampled per second. Hence, the interval between snapshots is $\frac{1}{f_s}$. The spatial displacement of the array between consecutive snapshots is L array elements or Ld unit distance. Thus, $\boldsymbol{v}$, the effective velocity of the array motion, is given by

$$\boldsymbol{v} = Ldf_s. \tag{9.179}$$

Assume the multipath scenario shown in Fig. 9.58 and the array motion as shown in Figure 9.59. Let Δ denote the frequency difference $f_A - f_B$. Substituting (9.179) into (9.178), we have

$$\Delta = \frac{Ldf_s f_0}{c} \times (\sin(\theta_B) - \sin(\theta_A)) \tag{9.180a}$$

$$= \frac{Ldf_s}{\lambda_0} \times (\sin(\theta_B) - \sin(\theta_A)). \tag{9.180b}$$

The snapshots should be collected over a duration at least comparable to $\frac{1}{|\Delta|}$. Since the array is traveling at a velocity of Ldf_s (see (9.179)), the distance D the array travels in duration $\frac{1}{|\Delta|}$ is given by

$$D = \frac{Ldf_s}{|\Delta|} = \frac{\lambda_0}{|\sin(\theta_B) - \sin(\theta_A)|}. \tag{9.181}$$

Since the separation between consecutive array elements is d, the equivalent number of array elements the array has traveled is $\frac{D}{d}$ given by

$$\frac{D}{d} = \frac{\lambda_0}{d} \times \frac{1}{|\sin(\theta_B) - \sin(\theta_A)|}. \tag{9.182}$$

The total number of array elements $N + KL$ is given by

$$N + KL = N + \left\lceil \frac{D}{d} \right\rceil = N + \left\lceil \frac{\lambda_0}{d} \times \frac{1}{|\sin(\theta_B) - \sin(\theta_A)|} \right\rceil \tag{9.183}$$

where $\lceil x \rceil$ is the smallest integer larger than or equal to x. It is interesting to note that KL, the additional array elements needed to resolve multipath effect, is not a function of f_s, i.e., it is independent on the array snapshot sampling rate. The array snapshot sampling rate should be as high as possible to improve SNR.

References

S.W.A. Bergen, A. Antoniou, Design of ultraspherical window functions with prescribed spectral characteristics. EURASIP J. Appl. Signal Process. **2004**(13), 2053–2065 (2004)

S.W.A. Bergen, A. Antoniou, Design of nonrecursive digital filters using the ultraspherical window function. EURASIP J. Appl. Signal Process. **2005**(12), 1910–1922 (2005)

J.C. Chen, K. Yao, R.E. Hudson, Source localization and beamforming. IEEE Sig. Process. Mag. **19**(2), 30–39 (2002)

A.G. Deczky, Unispherical windows, in *Proc. IEEE Int. Symp. Circuits Syst.*, Sydney, Australia, pp. 85–88 (May 2001)

C.L. Dolph, A current distribution for broadside arrays which optimizes the relationship between beam width and side-lobe level. Proc. IRE **34**, 335–348 (1946)

J.F. Kaiser, Nonrecursive digital filter design using the I_0-sinh window function, in *Proc. IEEE Int. Symp. Circuits Syst.*, San Francisco, CA, USA, pp. 20–23 (April 1974)
S. Kutty, D. Sen, Beamforming for millimeter wave communications: An inclusive survey. IEEE Comm. Surv. Tut. **18**(2), 949–973 (2015)
Y.C. Lim, T. Saramäki, P.S.R. Diniz, Q. Liu, A method for scaling window sidelobe magnitude. IEEE Signal Process. Lett. **28**, 1056–1069 (2021a)
Y.C. Lim, T. Saramäki, P.S.R. Diniz, Q. Liu, Fast convergence method for scaling window sidelobe magnitude. IEEE Signal Process. Lett. **28**, 2078–2081 (2021b)
Y.C. Lim, Q. Liu, P.S.R. Diniz, T. Saramäki, Efficient design of scaled rectangular (Saramäki) window. IEEE Signal Process. Lett. **29**, 489–492 (2022a)
Y.C. Lim, Q. Liu, P.S.R. Diniz, T. Saramäki, Efficient scaling of window function expressed as sum of exponentials. IEEE Signal Process. Lett. **29**, 1814–1817 (2022b)
J. Litva, T.K. Lo, *Digital Beamforming in Wireless Communications* (Artech House, Inc, 1996)
R. Mucci, A comparison of efficient beamforming algorithms. IEEE Trans. Acoust. Speech Sig. Process. **32**(3), 548–558 (1984)
R. Mudumbai, D.R. Iii, U. Madhow, H.V. Poor, Distributed transmit beamforming: challenges and recent progress. IEEE Comm. Mag. **47**(2), 102–110 (2009)
T. Saramäki, A class of window functions with nearly minimum sidelobe energy for designing FIR filters, in *Proc. IEEE Int. Symp. Circuits Syst.*, Portland, OR, USA, pp. 359–365 (1989)
H. Steyskal, Digital beamforming, in *Proc. 18th Euro. μW Conf.*, pp. 49–57 (Sep 1988)
R.C. Taber, H. Void, D.L. Brown, G.T. Rocklin, Exponential window FIR burst random excitation, in *Proc. Third International Modal Analysis Conference*, Orlando, FL, USA, pp. 989–995 (1985)
B.D. Van Veen, K.M. Buckley, Beamforming: A versatile approach to spatial filtering. IEEE Acoust. Speech Sig. Process. Mag. **5**(2), 4–24 (1988)
Q. Xu, C. Jiang, Y. Han, B. Wang, K.R. Liu, Waveforming: An overview with beamforming. IEEE Comm. Surv. Tut. **20**(1), 132–149 (2017)

Chapter 10
Array Signal Preprocessing

10.1 Introduction

In the foregoing discussions, many algorithms are developed based on the assumption that the signals are complex. However, the signals received in the antennae in terms of voltage or current are real. A real sinusoid $\cos(\omega t)$ may be decomposed into two complex sinusoids $e^{j\omega t}$ and $e^{-j\omega t}$ using the Euler's formula $\cos(\omega t) = \left(\frac{e^{j\omega t}+e^{-j\omega t}}{2}\right)$. When the signal is assumed to be complex, a sinusoid with frequency ω is represented as $e^{j\omega t}$. Hence, the input real sinusoid $\cos(\omega t)$ must be preprocessed removing the $e^{-j\omega t}$ term. Such preprocessing will be discussed in Sect. 10.2.

The frequency of the signal at the antenna is often higher than the limit a practical ADC (analogue-to-digital converter) can convert. Therefore, the antenna signal must be brought down to a lower frequency for analogue-to-digital conversion. The effect of frequency translation will be discussed in Sect. 10.3.

If signal processing is done digitally, the analogue signal must be converted into digital form through an ADC. In the conversion, an analogue value is represented by its nearest digital value. The difference between the true analogue value and the digital value representing the analogue value is the quantization error (Diniz et al. 2010; Mitra 2001; Rabiner and Gold 1975; Proakis 2007; Crystal and Ehrman 1968; Pei and Shyu 1993; Chen and Parks 1987; Lim and Constantinides 1979; Lim et al. 1982; Lim and Parker 1983a,b; Lim 1983, 1986, 1990; Gray 1990; Claasen and A. Jongepier 1981; Widrow et al. 1996). The effect of quantization error is discussed in Sect. 10.4.

In a narrowband array, the signal bandwidth is always very small, and so the signal may be down sampled. The effects of down sampling on SNR and accuracy of signal correlation matrix as well as the effect of the filter's impulse response duration are discussed in Sects. 10.5 and 10.6, respectively. A design is illustrated in Sect. 10.6.

Y. C. Lim et al., *Signal Processing for Autonomous Systems*,
https://doi.org/10.1007/978-3-032-00675-2_10

10.2 Extracting Complex Sinusoid from Real Sinusoid

Let $x(n)$ be the sampled version of a real signal at time n in digital form, and let $X(e^{j\omega})$ be its frequency spectrum at frequency ω. Assume that the sampling rate is one sample per unit time. For any real signal, the frequency spectrum at negative frequency $-\omega$ is always the complex conjugate of its frequency spectrum at positive frequency ω. Since $X(e^{-j\omega}) = X^*(e^{j\omega})$, $X(e^{-j\omega})$ carries no additional information once $X(e^{j\omega})$ for $\omega \geq 0$ is known. As a consequence, $X(e^{-j\omega})$ may be discarded without loss of information.

Figure 10.1a shows the spectrum of two sinusoids with frequencies ω_1 and ω_2 immersed in broadband noise assuming that $\omega_2 > \omega_1 > 0$. The spectrum of the broadband noise is triangular centered at $\omega = 0$. Figure 10.1b shows the spectrum of Fig. 10.1a with the negative frequency components removed. The time domain signal with frequency spectrum shown in Fig. 10.1b is complex. Removing the negative frequency components of $X(e^{j\omega})$ does not alter the SNR because both the signal power and noise power are halved.

10.2.1 Real Signal Filtered by Complex Filter

There are two commonly used methods for removing the negative frequency components of a signal. One of the methods is by using a complex filter that passes only positive frequencies. An example of such a filter's frequency response is $H_P(e^{j\omega})$ shown in Fig. 10.2b. Let $H_P(z)$ be the z-transform transfer function of the filter whose frequency response is $H_P(e^{j\omega})$. $H_P(z)$ may be obtained from $H(z)$, the z-transform transfer function of a lowpass filter with real coefficients whose frequency response is shown in Fig. 10.2a, by replacing z in $H(z)$ by $e^{-j\pi/2}z$, i.e., $H_P(z) = H(e^{-j\pi/2}z)$. $H(z)$ is usually a half-band filter. A half-band filter is a symmetrical impulse response filter with an odd number of coefficients; with the exception of the center coefficient, every other coefficient value is zero.

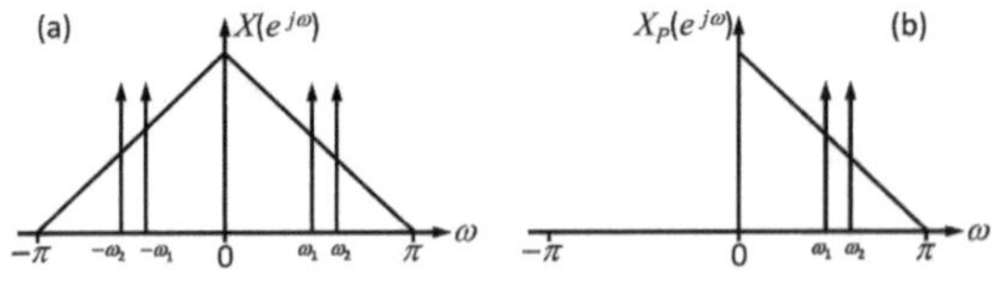

Fig. 10.1 (**a**) Spectrum of two sinusoids immersed in broadband noise: $\omega_2 > \omega_1 > 0$. (**b**) Spectrum of (**a**) with negative frequency components removed

Fig. 10.2
$H_P(e^{j\omega}) = H(e^{j\omega-\pi/2})$

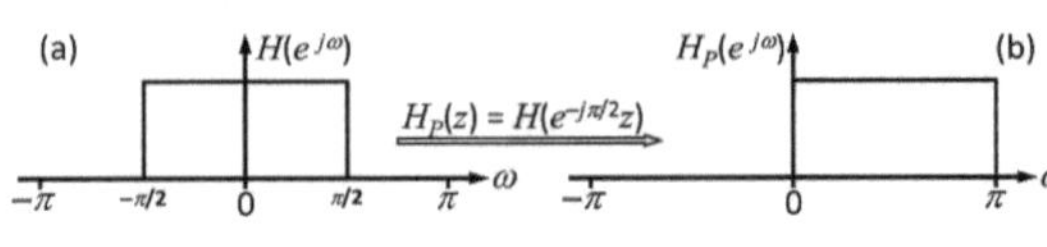

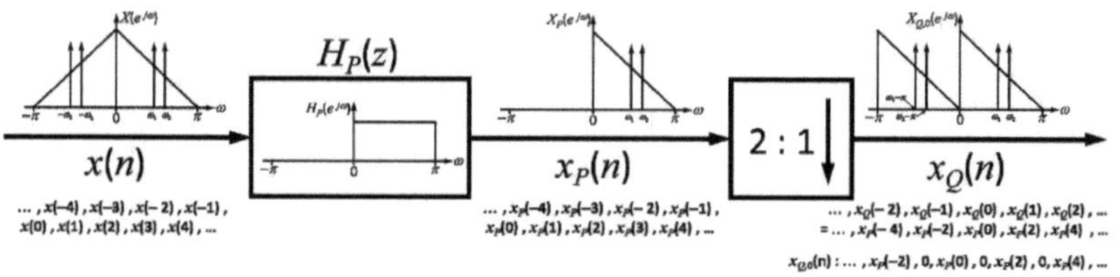

Fig. 10.3 Real $x(n)$ is filtered and down sampled to produce complex $x_Q(n)$

Let the time domain signal of $X_P(e^{j\omega})$ be $x_P(n)$. Since the bandwidth of $x_P(n)$ is only half of that of $x(n)$, $x_P(n)$ is usually down sampled by a factor of two to form $x_Q(n)$ as shown in Fig. 10.3. The signals $x_P(n)$ and $x_Q(n)$ are complex. To facilitate theoretical analysis, we shall define an intermediate signal $x_{Q,0}(n)$ given by

$$\begin{aligned} x_{Q,0}(n) &= x_P(n) \quad , \quad n = \text{ even ;} \\ &= 0 \quad , \quad n = \text{ odd .} \end{aligned} \tag{10.1}$$

Let the z-transforms of $x_P(n)$ and $x_{Q,0}(n)$ be $X_P(z)$ and $X_{Q,0}(z)$, respectively. $X_{Q,0}(z)$ may be expressed in terms of $X_P(z)$ as

$$X_{Q,0}(z) = \frac{X_P(z) + X_P(-z)}{2}. \tag{10.2}$$

Since $X_P(-z) = X_P(e^{\pm j\pi} z)$, $X_P(-e^{j\omega}) = X_P(e^{j(\pm\pi+\omega)})$.

The signal $x_Q(n)$ in Fig. 10.3 is the compacted version of $x_{Q,0}(n)$ after discarding those samples corresponding to n odd; the discarded samples are zero. The sampling rate of $x_Q(n)$ is half that of $x_{Q,0}(n)$. As the system involves more than one sampling rate, we should be very careful in interpreting the frequency of the signal. The frequency variable ω of a digital system is a normalized quantity assuming that the sampling rate is one sample per unit time. The actual physical frequency is equal to normalized frequency times sampling rate, for example, if $\omega = 0.4\pi$ radians per second and the sampling rate is 1000 samples per second, the actual physical frequency will be $0.4\pi \times 1000 = 400\pi$ radians per second. Fig. 10.4a and b show the frequency spectrum plot for $x_Q(n)$ with frequencies normalized with respect to sampling rate before 2:1 $\downarrow$ down sampling and after $2:1 \downarrow$ down sampling, respectively. For the purpose of illustration, let the normalized frequency $\omega_1 = 0.4\pi$ and assume that the sampling rate before $2:1 \downarrow$ down sampling is 1000 samples per second. The actual physical frequency will be 400π radians per second. After $2:1 \downarrow$ down sampling the sampling rate will become 500 samples per second. Normalizing 400π radians per second against 500 samples per second gives $\omega_1 = 0.8\pi$.

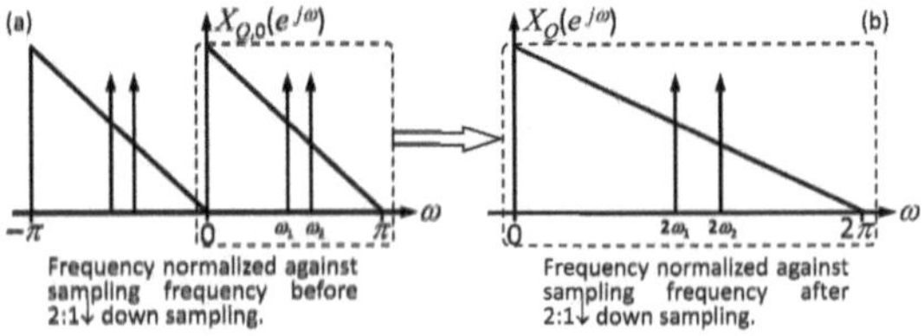

Fig. 10.4 Frequency spectrum for $x_Q(n)$ normalized with respect to sampling rate (**a**) before down sampling and (**b**) after down sampling

Consider a symmetrical impulse response filter with z-transform transfer function $H(z)$ and filter length $2L + 1$ as shown in (10.3):

$$H(z) = z^{-L}\left\{h_0 + \sum_{n=1}^{L} h_n(z^n + z^{-n})\right\};$$

$$= z^{-2N}\left\{h_0 + \sum_{n=1}^{N} h_{2n}\{(z^2)^n + (z^2)^{-n}\} + z^{-1}\sum_{n=1}^{N} h_{2n-1}\{(z^2)^n + (z^2)^{1-n}\}\right\}$$

$$\text{for} \quad L = 2N; \tag{10.3a}$$

$$= z^{-2N-1}\left\{h_0 + \sum_{n=1}^{N} h_{2n}\{(z^2)^n + (z^2)^{-n}\} + z^{-1}\sum_{n=1}^{N+1} h_{2n-1}\{(z^2)^n + (z^2)^{1-n}\}\right\}$$

$$\text{for} \quad L = 2N + 1 \tag{10.3b}$$

Let

$$H_E(z^2) = (z^2)^{-N}\left\{h_0 + \sum_{n=1}^{N} h_{2n}\{(z^2)^n + (z^2)^{-n}\}\right\} \tag{10.4}$$

regardless of whether $L = 2N$ or $L = 2N + 1$. Let

$$H_O(z^2) = (z^2)^{-N}\left\{\sum_{n=1}^{N} h_{2n-1}\{(z^2)^n + (z^2)^{1-n}\}\right\} \quad \text{for} \quad L = 2N; \tag{10.5a}$$

$$= (z^2)^{-(N+1)}\left\{\sum_{n=1}^{N+1} h_{2n-1}\{(z^2)^n + (z^2)^{1-n}\}\right\} \quad \text{for} \quad L = 2N + 1. \tag{10.5b}$$

Substituting (10.4) and (10.5) into (10.3), we have

$$H(z) = H_E(z^2) + z^{-1}H_O(z^2) \quad \text{for} \quad L = 2N; \tag{10.6a}$$

$$= z^{-1}H_E(z^2) + H_O(z^2) \quad \text{for} \quad L = 2N + 1. \tag{10.6b}$$

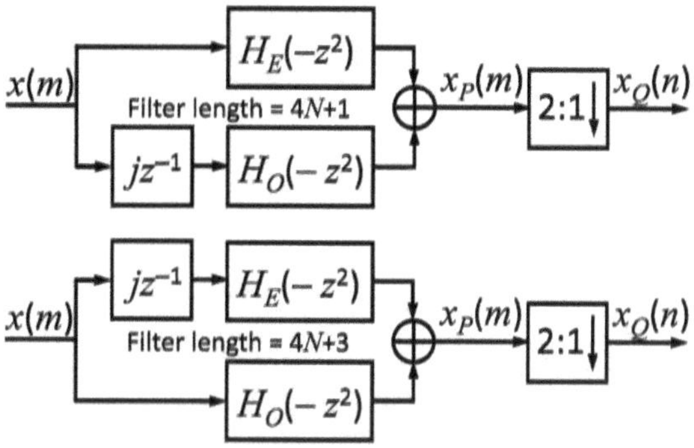

Fig. 10.5 Implementation structure for $H_P(z)$ based on (10.7)

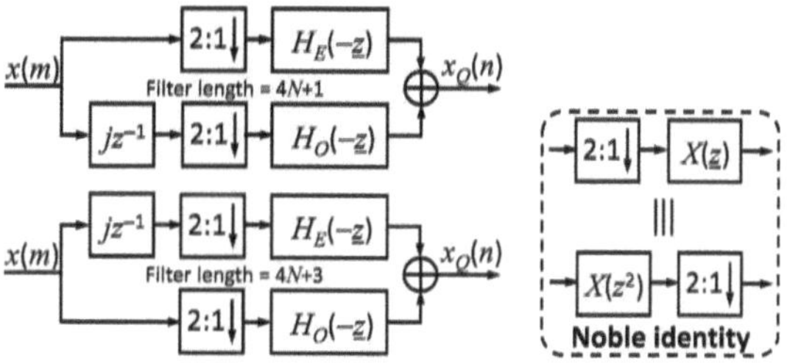

Fig. 10.6 Implementation structure for $H_P(z)$ after moving the 2 : 1 ↓ down sampling process to precede $H_P(z)$. The variable $\underline{z}$ is the z-transform variable at the lower sampling rate

Replacing z by $e^{-j\pi/2}z$ in $H(z)$, we have

$$H_P(z) = H(e^{-j\pi/2}z) = H_E(-z^2) + jz^{-1}H_O(-z^2) \quad \text{for} \quad L = 2N; \tag{10.7a}$$

$$= jz^{-1}H_E(-z^2) + H_O(-z^2) \quad \text{for} \quad L = 2N + 1. \tag{10.7b}$$

Based on (10.7), an implementation structure for $H_P(z)$ is shown in Fig. 10.5. Note that the coefficients of $H_E(-z^2)$ and $H_O(-z^2)$ are real. Since $x(n)$ is real, the output of $H_E(-z^2)$ is real, and the output of $H_O(-z^2)$ is imaginary if the filter length is $4N + 1$; the output of $H_E(-z^2)$ is imaginary, and the output of $H_O(-z^2)$ is real if the filter length is $4N + 3$. The adder simply amalgamates a real number and an imaginary number together to form a complex number; no actual add operation is performed.

Noble identity states that a 2 : 1 ↓ down sampling following a transfer function is identically equal to a 2 : 1 ↓ down sampling preceding the transfer function with every z^2 factor of the transfer function replaced by z. Applying noble identity, the block diagram of Fig. 10.5 is converted into that shown in Fig. 10.6. Two sampling rates are involved in Fig. 10.6. The variables z and $\underline{z}$ are used to represent the z-transform variables at the higher and lower sampling rates, respectively.

If $H(z)$ is a half-band filter, $h_{2n} = 0$ for $n \neq 0$. Hence, $H_E(z^2) = (z^2)^{-N}h_0$. Since $h_{2n} = 0$ for $n \neq 0$, the filter length is $4N + 3$; the first and last coefficients of the filter with length $4N + 1$ will be zero. Figure 10.6 is simplified to Fig. 10.7.

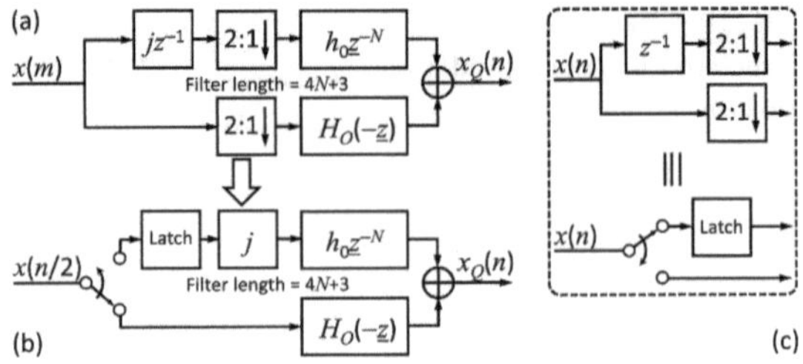

Fig. 10.7 Implementation structure for $H_P(z)$ if $H(z)$ is half-band

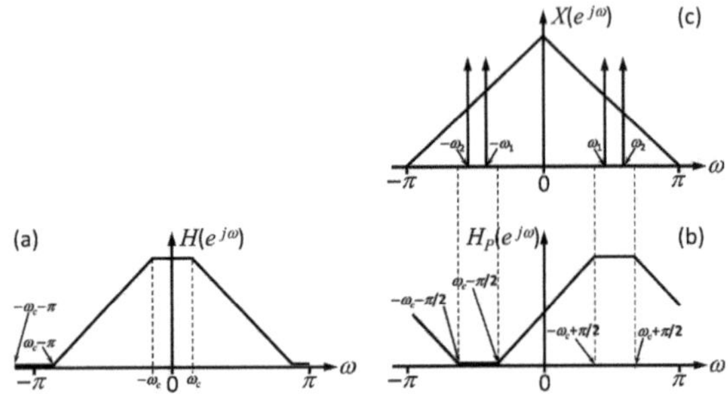

Fig. 10.8 Diagram illustrating the relationship between ω_1, ω_2, and the band edges of $H(z)$ and those of $H_P(z)$

Let the passband cutoff frequency of the half-band filter, $H(z)$, be $\pm\,\omega_c$. Its stopband cutoff frequency will be $\pm\,(\pi \pm \omega_c)$ as shown in Fig. 10.8a. Figure 10.8b shows the frequency response of $H_P(z)$ (where $H_P(z)$ is the shifted version of $H(z)$), and Fig. 10.8c shows the frequency spectrum $X(e^{j\omega})$ of the input signal. $H_P(e^{j\omega})$ must pass both ω_1 and ω_2 but must reject both $-\,\omega_1$ and $-\,\omega_2$. Hence, both (10.8a) and (10.8b) must be satisfied simultaneously:

$$\omega_c \geq \omega_2 - \frac{\pi}{2}; \tag{10.8a}$$

$$\omega_c \geq \frac{\pi}{2} - \omega_1 \tag{10.8b}$$

The bandwidth of $x_Q(n)$ is narrow and so $x_Q(n)$ can be further down sampled. However, $x_Q(n)$ must be filtered to remove the noise from aliasing into the signal if further down sampling is desired. It should be noted that once the noise is aliased into the same frequency band as the signal, the noise cannot be removed by subsequent filtering.

10.2.2 Complex Signal Filtered by Real Filter

In the forgoing section, we discussed a method for obtaining $X_P(e^{j\omega})$ by filtering $X(e^{j\omega})$ using a complex filter with frequency response $H_P(e^{j\omega})$. The z-transform transfer function of the complex filter is obtained from that of a lowpass filter by replacing z in $H(z)$ by $e^{-j\pi/2}z$.

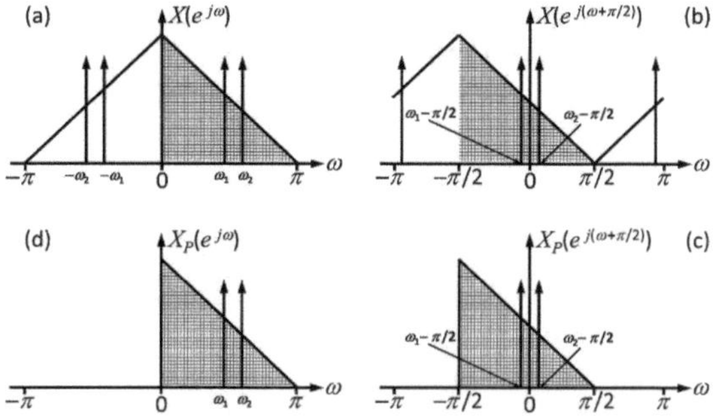

Fig. 10.9 (**a**) $X(e^{j\omega})$, (**b**) $X(e^{j(\omega+\pi/2)})$, (**c**) $X_P(e^{j(\omega+\pi/2)})$, (**d**) $X_P(e^{j\omega})$

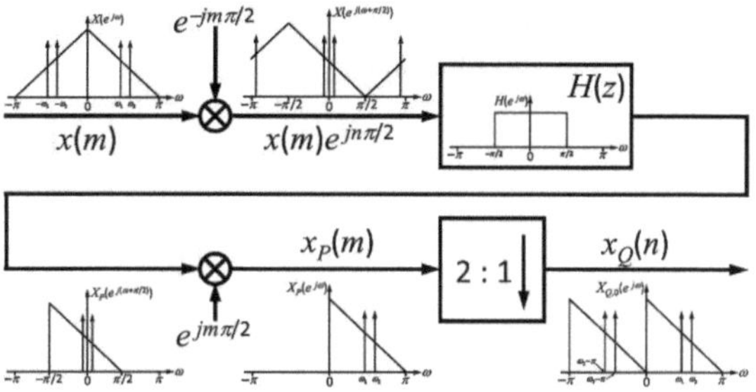

Fig. 10.10 The spectrum of $x(m)$ is rotated, filtered, rotated, and down sampled to produce complex $x_Q(n)$

In this section, we shall describe another method. In this method, $x(n)$ with spectrum $X(e^{j\omega})$ shown in Fig. 10.9a is modulated by $e^{-jn\pi/2}$. The spectrum of $x(n)e^{-jn\pi/2}$ is $X(e^{j(\omega+\pi/2)})$ and is shown in Fig. 10.9b. The modulated signal $x(n)e^{-jn\pi/2}$ is filtered by $H(z)$ with frequency response $H(e^{j\omega})$ shown in Fig. 10.2a. The frequency spectrum of the filtered output is denoted by $X_P(e^{j(\omega+\pi/2)})$ and is shown in Fig. 10.9c. The filtered output is then modulated by $e^{jn\pi/2}$ to produce $x_P(n)$ whose frequency spectrum $X_P(e^{j\omega})$ is shown in Fig. 10.9d; as we shall see in due course, modulation by $e^{jn\pi/2}$ at the output is not necessary for DOA estimation, i.e., the signal whose spectrum is $X_P(e^{j(\omega+\pi/2)})$ can be used in DOA estimation. Finally, $x_P(n)$ is down sampled by two to produce $x_Q(n)$. The entire process is shown in Fig. 10.10.

For $H_E(z^2)$, $H_O(z^2)$, and $H(z)$ as defined in (10.4), (10.5), and (10.6), the implementation structure for Fig. 10.10 is shown in Fig. 10.11. The $2 : 1 \downarrow$ down sampler may be moved forward as shown in Fig. 10.12. The values of $e^{-jn\pi/2}$, $e^{\pm jn\pi}$, and $(-1)^n$ are tabulated in Table 10.1. Taking note of Table 10.1 and Figs. 10.7c, and 10.12 can be redrawn as Fig. 10.13. If $H(z)$ is a half-band filter, its length will be $4N + 3$, and Fig. 10.13 simplifies to that shown in Fig. 10.14. In Fig. 10.7, the sign of alternate coefficient values of $H(z)$ is flipped. In Fig. 10.14, the sign of alternate samples of the down-sampled input signal is flipped and filtered, and the sign of the alternate output samples flipped again.

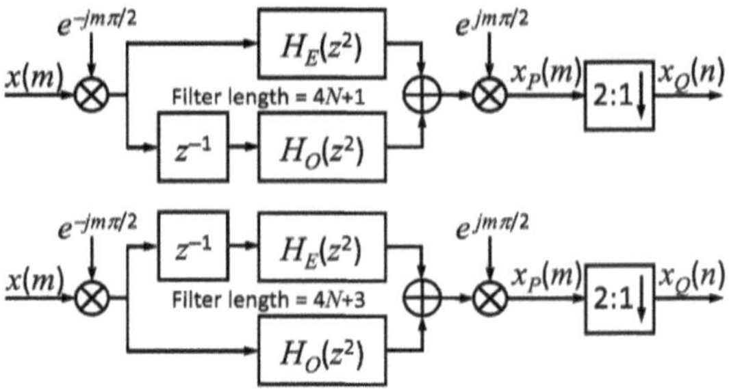

Fig. 10.11 Implementation structure for Fig. 10.10

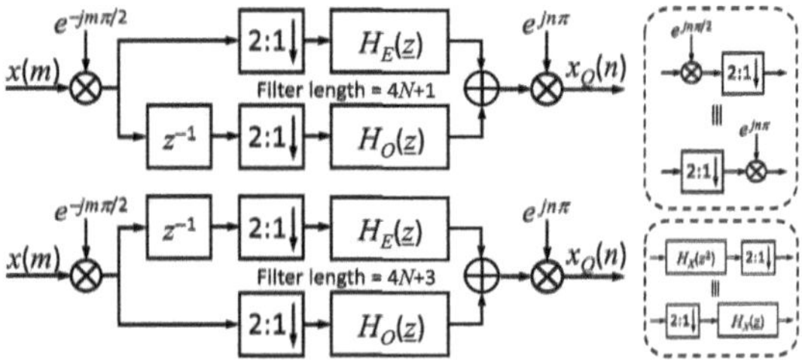

Fig. 10.12 Moving the down sampler of Fig. 10.11 forward

Table 10.1 Values of $e^{-jn\pi/2}$, $e^{\pm jn\pi}$, and $(-1)^n$ for various n

n	...	-1	0	1	2	3	4	5	6	7	...
$e^{-jn\pi/2}$	...	j	1	$-j$	-1	j	1	$-j$	-1	j	...
$e^{\pm jn\pi}$	...	-1	1	-1	1	-1	1	-1	1	-1	...
$(-1)^n$	...	-1	1	-1	1	-1	1	-1	1	-1	...

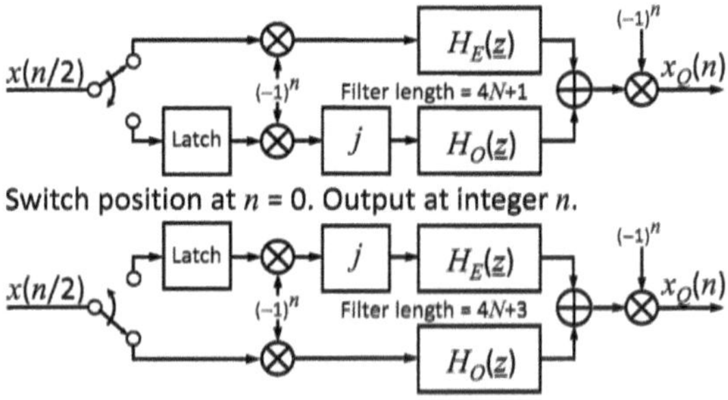

Fig. 10.13 The latch stores data of $x(-n/2)$, $x(n/2)$, ..., etc

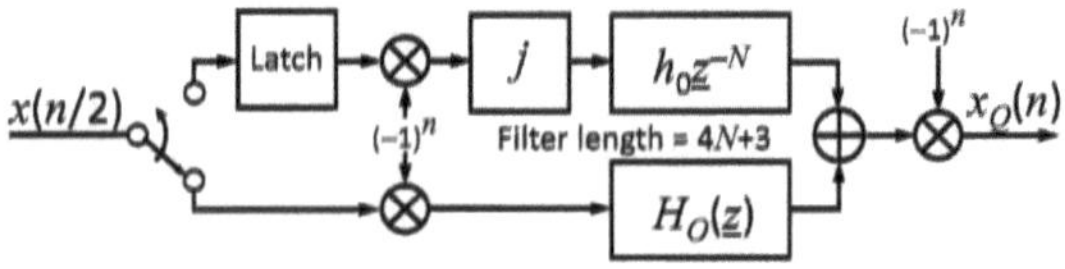

Fig. 10.14 Structure for half-band $H(z)$

10.3 Frequency Translation

The frequencies of the signals at the array elements are often higher than the operation limit of an ADC. Thus, it is often necessary to modulate the array element signals down to a much lower frequency to facilitate analogue-to-digital conversion. As we shall see in due course, shifting the array signals down to a lower frequency will not affect DOA estimation provided that the signals of all the array elements are shifted by the same frequency and phase.

Consider the signals $x_0(t)$ and $x_1(t)$ received by two array elements as shown in Fig. 10.15. The separation between the two array elements is d and the DOA is θ. Let $x_0(t) = e^{j(\omega_0 t+\alpha)}$. The phase difference between $x_0(t)$ and $x_1(t)$ is $2\pi d\sin(\theta)/\lambda_0$ where λ_0 is the wavelength of the sinusoid with frequency ω_0. Hence, $x_1(t) = e^{j(\omega_0 t+\alpha-2\pi d\sin(\theta)/\lambda_0)}$. The cross correlation of $x_0(t)$ and $x_1(t)$ is given by $\mathbb{E}\{x_0(t)x_1^*(t)\} = \mathbb{E}\{e^{j(\omega_0 t+\alpha)}e^{-j(\omega_0 t+\alpha-2\pi d\sin(\theta)/\lambda_0)}\} = \mathbb{E}\{e^{j2\pi d\sin(\theta)/\lambda_0}\}$.

Suppose that the frequency of $x_0(t)$ and $x_1(t)$ is shifted downward by modulating each of them by $e^{-j(\omega_L t+\beta)}$ to become $\hat{x}_0(t)$ and $\hat{x}_1(t)$, respectively. Hence, $\hat{x}_0(t) = e^{-j(\omega_L t+\beta)}x_0(t) = e^{j((\omega_0-\omega_L)t+\alpha-\beta)}$ and $\hat{x}_1(t) = e^{-j(\omega_L t+\beta)}x_1(t) = e^{j((\omega_0-\omega_L)t+\alpha-\beta)} \times e^{-j2\pi d\sin(\theta)/\lambda_0}$. The cross correlation of $\hat{x}_0(t)$ and $\hat{x}_1(t)$ is given by $\mathbb{E}\{\hat{x}_0(t)\hat{x}_1^*(t)\} = \mathbb{E}\{e^{j2\pi d\sin(\theta)/\lambda_0}\}$. Thus, $\mathbb{E}\{x_0(t)x_1^*(t)\} = \mathbb{E}\{\hat{x}_0(t)\hat{x}_1^*(t)\}$. The cross correlation and, hence, the DOA estimation are not affected by frequency translation. In Fig. 10.11, 10.12, 10.13, and 10.14, the modulation after filtering by $H_E(z)$ and $H_O(z)$ performs frequency translation. Since DOA estimation is not affected by frequency translation, the modulation may be omitted.

It should be noted that the signal of each and every array element must be translated by the same frequency and phase. Any mismatch will cause an error in the cross correlation and, hence, the DOA estimation. As processing before ADC is done in the analogue domain, the analogue circuitry processing the signals must be very carefully matched.

Figure 10.16 shows the sequence in which signals from the antenna are processed. A narrowband of signal from the antenna is selected by the tuned circuit and mixed with the local oscillator output. The mixer output consists of (a) a signal whose frequency is the "difference" between that of the local oscillator and that of the input, (b) a signal whose frequency is the sum of that of the local oscillator and that of the input, as well as (c) harmonics of various signals. The bandpass filter

Fig. 10.15 Signal impinges on two array elements

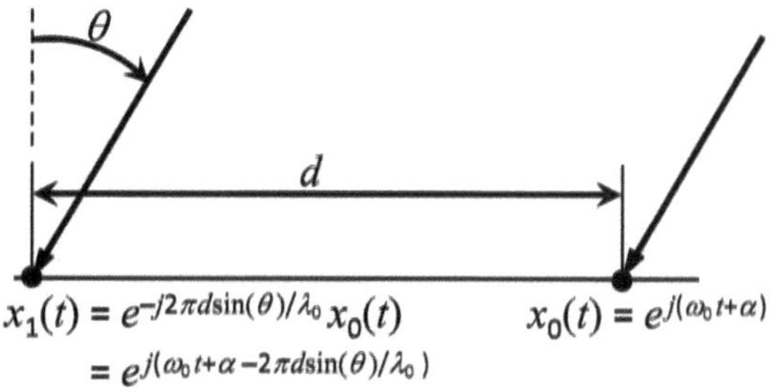

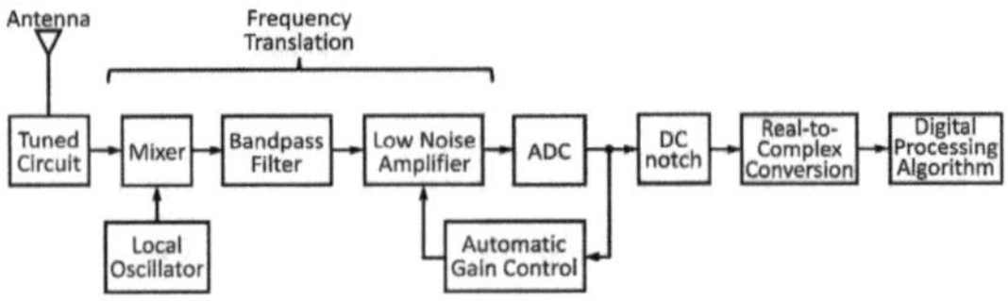

Fig. 10.16 Block diagram showing signal processing sequence starting from antenna

following the mixer passes the "difference" component and rejects all the rest. This "difference" component is the IF (intermediate frequency) of a superheterodyne receiver. The "difference" is at a much lower frequency compared to the original from the antenna. The output of the bandpass filter is very weak, and a low-noise amplifier amplifies the signal to maximize the usage of the dynamic range available with the ADC. The amplification factor of the low-noise amplifier is automatically adjusted based on the output of the ADC in an automatic-gain-control (AGC) loop to avoid saturating the ADC and to achieve high utilization of the dynamic range of the ADC. The digital output of the ADC usually contains a dc component which drifts with temperature and component aging. It can be removed by a dc notch filter. The digital output of the ADC is real and is converted to complex using any one of the methods discussed above.

10.4 ADC Quantization Noise

The ADC output is digital. A digital signal has a finite number of quantized levels, and so it cannot precisely represent a continuous analogue signal. The difference between a digital signal and its original analogue form is called the quantization error e where

$$e = \text{Original\_analogue\_value} - \text{Quantized\_digital\_value}. \tag{10.9}$$

Quantization error is noise-like and so it is classified as quantization noise. The quantization noise power depends on its quantization step size, Q. We assume that the original analogue signal is rounded to its nearest digital level. Hence, its quantization error can be any value from $-\frac{Q}{2}$ to $\frac{Q}{2}$. We shall assume that all values of quantization error between $-\frac{Q}{2}$ and $\frac{Q}{2}$ are equally probable; under this assumption, the probability density function of the quantization error e is shown in Fig. 10.17. The d.c. component η of the quantization noise is the mean of e given by

$$\eta = \mathbb{E}\{e\} = \frac{1}{Q}\int_{-Q/2}^{Q/2} e de = 0. \tag{10.10}$$

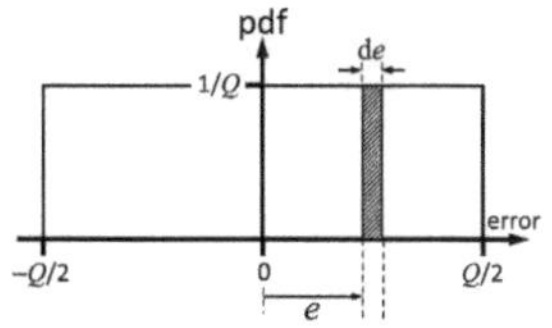

Fig. 10.17 Probability density function (pdf) of quantization error, e

The quantization noise power denoted by σ^2 is the variance of e and is given by

$$\sigma^2 = \mathbb{E}\{e^2\} = \frac{1}{Q}\int_{-Q/2}^{Q/2} e^2 de = \frac{Q^2}{12}. \tag{10.11}$$

The output of a B-bit ADC lies approximately within the range $\pm 2^{B-1}Q$. The magnitude of the largest magnitude real sinusoid with zero mean that can be represented is approximately $2^{B-1}Q$; its power is $\frac{(2^{B-1})^2 Q}{2} = 2^{2B-3}Q^2$. Hence, the maximum SNR denoted by $\mathrm{SNR}_{\mathrm{max,sin}}$ of a B-bit ADC when converting a sinusoid is given by

$$\mathrm{SNR}_{\mathrm{max,sin}} = 3 \times 2^{2B-1} = (4.771 + 3.010(2B-1))\mathrm{dB}. \tag{10.12}$$

Consider a signal with frequency spectrum $X_{W,A}(j\Omega)$ and bandwidth W radians per second sampled at f_s samples per second. Its normalized frequency spectrum $X_{W,D}(e^{j\omega})$ is shown in Fig. 10.18. Its normalized bandwidth is W/f_s. The quantization noise power $\frac{Q^2}{12}$ is spread uniformly over the entire normalized frequency bandwidth of 2π. While the quantization noise outside the signal band can be removed by filtering, the noise spectrum occupying the same frequency band as the signal, called the "in-band noise," cannot be removed without also removing the signal. The in-band noise power is the minimum noise power that can be achieved by linear bandpass filtering. For a normalized signal bandwidth of W/f_s, the minimum quantization noise power is $\frac{Q^2 W}{24\pi f_s}$. (Note: W is in radians per second and f_s is in sample per second or hertz.) Let the signal power be P_W. The maximum quantization SNR, denoted by $\mathrm{SNR}_{\mathrm{max},P_W}$, that can be achieved by linear bandpass

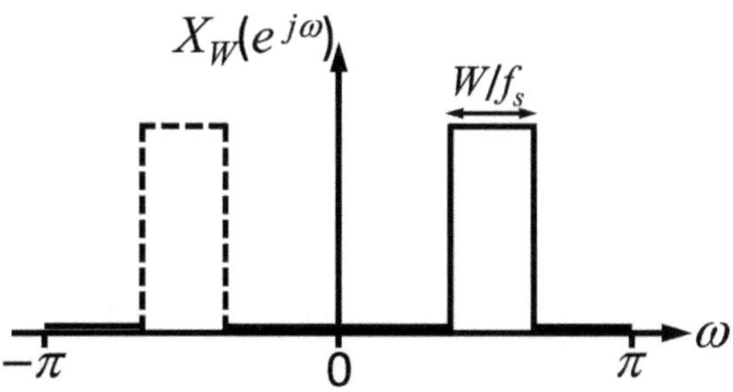

Fig. 10.18 Normalized bandwidth of a signal with bandwidth W radians per second sampled at f_s samples per second

filtering is given by

$$\mathrm{SNR}_{\max,P_W} = \frac{24\pi f_s P_W}{Q^2 W} = \left(18.77 + 10\log_{10}\left(\frac{f_s}{W}\right) + 10\log_{10}\left(\frac{P_W}{Q^2}\right)\right)\ \mathrm{dB}. \tag{10.13}$$

As can be seen from (10.13), for a given W and P_W, $\mathrm{SNR}_{\max,P_W}$ increases with increasing sampling rate f_s and decreasing quantization step size Q.

10.5 Sampling Rate Reduction

Depending on application, the narrowband phased array signal bandwidth may be very narrow. In the extreme case, the signal spectrum is only one spectral line, i.e., only one single frequency. The SNR can be improved by filtering using a narrowband filter that passes only the frequency of interest and rejects all other frequencies. Since the bandwidth of the signal of interest is narrow, the array signal is often down sampled to reduce computation load in subsequent processing.

Consider the signal $x(n)$ with z-transform $X(z)$ given by (10.14):

$$X(z) = \sum_{n=-\infty}^{\infty} x(n) z^{-n} \tag{10.14}$$

Define the signal $x_{Q,0}(n)$ given by

$$\begin{aligned} x_{Q,0}(n) &= x(n) \quad , \quad n = \text{ integer multiple of } Q\ ; \\ &= 0 \qquad\ \ , \quad \text{otherwise}\ . \end{aligned} \tag{10.15}$$

The sampling rates of $x_{Q,0}(n)$ and that of $x(n)$ are the same. Let the z-transform of $x_{Q,0}(n)$ be $X_{Q,0}(z)$. It can be shown that

$$X_{Q,0}(z) = \frac{1}{Q}\sum_{k=0}^{Q-1} X(e^{j2\pi k/Q} z) \tag{10.16a}$$

$$= \frac{1}{Q}\left(X(z) + X(e^{j2\pi/Q} z) + X(e^{j4\pi/Q} z) + \ldots\right). \tag{10.16b}$$

An example of the frequency spectrum for $Q = 2$ is shown in Fig. 10.19. Figure 10.19a shows an example of $X(e^{j\omega})$ consisting of a narrowband signal centered at ω_0 immersed in a broadband noise whose spectrum is triangular. If the sampling rate is f_s, the normalized frequency ω_0 corresponds to physical frequency $\omega_0 f_s$. Figure 10.19b shows a $\frac{1}{2}\left(X(e^{j\omega}) + X(e^{j\omega-\pi})\right)$ plot. The spectrum

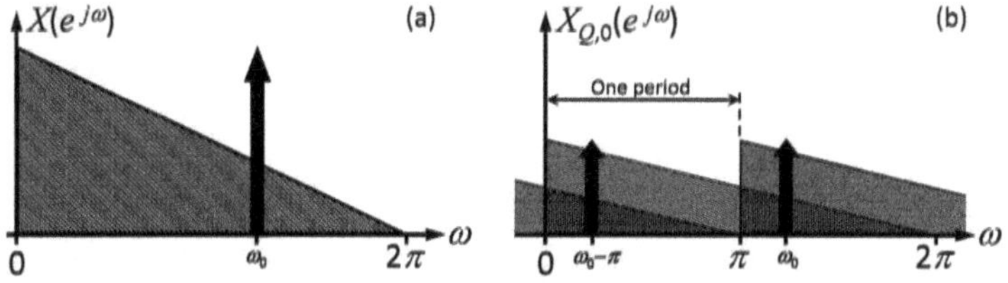

Fig. 10.19 (**a**) $X(e^{j\omega})$. (**b**) $X_{Q,0}(e^{j\omega})$ for $Q = 2$

of $X_{2,0}(e^{j\omega})$ is periodic with period π. When $x_{2,0}(n)$ is compacted by discarding those values where n is odd, the spectrum of the compacted signal is that of Fig. 10.19b from 0 to π with the frequency axis relabeled from 0 to 2π, and the normalized frequency of the signal becomes centered at $2\omega_0 - 2\pi$. The normalized frequency $2\omega_0 - 2\pi$ corresponds to physical frequency $(\omega_0/\pi - 1)f_s$ hertz and is different from $\omega_0 f_s/2\pi$ hertz. The changed in frequency is unimportant for DOA estimation as already discussed in Sect. 10.3. However, it should be noted that the down-sampling process must not cause two distinct frequencies before down sampling to be aliased so that they become indistinguishable. Once they become indistinguishable, it is impossible to make them distinct again. This means that the sampling rate must be higher than the original complex signal's bandwidth.

In Fig. 10.19b, the signal magnitudes centered at ω_0 and $\omega_0 - \pi$ are half that in Fig. 10.19a, and so each of its power is one-quarter that of Fig. 10.19a. Since there are two of them, one at ω_0 and the other one at $\omega_0 - \pi$, the total signal power is half $(= 2 \times \frac{1}{4})$ that of Fig. 10.19a. This is expected since alternate samples of $x_2(n)$ is set to zero. The noise in Fig. 10.19b consists of two components. One is the original noise component of Fig. 10.19a with magnitude scaled down by a factor of two; for this component, since its magnitude is half that in Fig. 10.19a, its power is one-quarter that of Fig. 10.19a. The other noise component is the aliased version of the noise component of that in Fig. 10.19a scaled down by a factor of two, and so its power is scaled down by a factor of four. The total noise power is also half of that in Fig. 10.19a. Hence, the SNR remains unchanged since both signal power and noise power are scaled down by the same factor.

It should be noted that, after down sampling, the in-band noise component at $\omega_0 - \pi$ comprises of the original noise component at $\omega_0 - \pi$ and the aliased version of the original noise component at ω_0. The in-band noise power has increased. Once aliased, the in-band noise cannot be removed by subsequent filtering.

Although down sampling does not affect SNR, down sampling affects the accuracy of the signal correlation matrix used in the estimation of DOA. Suppose that the signal plus noise of the ith and kth array elements are $Ae^{j(\omega n+\theta_i)} + \eta_i(n)$ and $Ae^{j(\omega n+\theta_k)} + \eta_k(n)$, respectively, where $\eta_i(n)$ and $\eta_k(n)$ are, respectively, the noise at the ith and kth array element. Assume that $\eta_i(n)$ and $\eta_k(n)$ are zero mean, mutually uncorrelated and uncorrelated to $Ae^{j(\omega n+\theta_i)}$ and

$Ae^{j(\omega n+\theta_k)}$. The correlation between $Ae^{j(\omega n+\theta_i)}+\eta_i(n)$ and $Ae^{j(\omega n+\theta_k)}+\eta_k(n)$ is $\mathbb{E}\left\{\left(Ae^{j(\omega n+\theta_i)}+\eta_i(n)\right)\left(Ae^{-j(\omega n+\theta_k)}+\eta_k^*(n)\right)\right\}$ and is, in general, given by

$$\mathbb{E}\left\{\left(Ae^{j(\omega n+\theta_i)}+\eta_i(n)\right)\left(Ae^{-j(\omega n+\theta_k)}+\eta_k^*(n)\right)\right\}=A^2e^{j(\theta_i-\theta_k)}$$
$$+\mathbb{E}\left\{Ae^{j(\omega n+\theta_i)}\eta_k^*(n)\right\}+\mathbb{E}\left\{Ae^{-j(\omega n+\theta_k)}\eta_i(n)\right\}+\mathbb{E}\left\{\eta_i(n)\eta_k^*(n)\right\}. \tag{10.17}$$

If there is no noise, $\mathbb{E}\left\{\left(Ae^{j(\omega n+\theta_i)}+\eta_i(n)\right)\left(Ae^{-j(\omega n+\theta_k)}+\eta_k^*(n)\right)\right\} = A^2e^{j(\theta_i-\theta_k)}$; the diagonal elements of the correlation matrix are A^2; the off-diagonal elements are $A^2e^{j(\theta_i-\theta_k)}$.

If an infinite number of sample values are available for computing the expectation values, $\mathbb{E}\left\{Ae^{j(\omega n+\theta_i)}\eta_k^*(n)\right\} = \mathbb{E}\left\{Ae^{-j(\omega n+\theta_k)}\eta_i(n)\right\} = 0$. $\mathbb{E}\left\{\eta_i(n)\eta_k^*(n)\right\} = 0$ for $i \neq k$. For $i = k$, $\mathbb{E}\left\{\eta_k(n)\eta_k^*(n)\right\}$ is not zero if the noise is not zero. Let $\mathbb{E}\left\{\eta_i(n)\eta_i^*(n)\right\} = \sigma^2$ for $i = k$. The quantity σ^2 is called the noise power. Hence, in the presence of noise and given an infinite number of sample values for computing the signal correlation matrix, the diagonal terms of the signal correlation matrix are $A^2+\sigma^2$ and are different from the noiseless case; the off-diagonal terms are $A^2e^{j(\theta_i-\theta_k)}$ and are the same as the noiseless case.

In a practical situation, only a finite number of sample values are available for computing the expectation values of the signal correlation matrix; in this case, the expectation value is replaced by the average. We shall examine the effect of replacing expectation by averaging. Suppose that we are computing the expected values by using the average of M samples. We have

$$\mathbb{E}\left\{Ae^{j(\omega n+\theta_i)}\eta_k^*(n)\right\}=\frac{1}{M}\sum_{n=1}^{M}Ae^{j(\omega n+\theta_i)}\eta_k^*(n) \tag{10.18a}$$

$$\mathbb{E}\left\{Ae^{-j(\omega n+\theta_k)}\eta_i(n)\right\}=\frac{1}{M}\sum_{n=1}^{M}Ae^{-j(\omega n+\theta_k)}\eta_i(n) \tag{10.18b}$$

$$\mathbb{E}\left\{\eta_i(n)\eta_k^*(n)\right\}=\frac{1}{M}\sum_{n=1}^{M}\eta_i(n)\eta_k^*(n). \tag{10.18c}$$

The averages in (10.18) will not be zero if M is finite. The spread of the computed result of (10.18) away from zero is indicated by the variance of the computed result. The variances of $\sum_{n=1}^{M}\frac{Ae^{j(\omega n+\theta_i)}\eta_k^*(n)}{M}$, $\sum_{n=1}^{M}\frac{Ae^{-j(\omega n+\theta_k)}\eta_i(n)}{M}$, and $\sum_{n=1}^{M}\frac{\eta_i(n)\eta_k^*(n)}{M}$ denoted by $\operatorname{var}\left\{\sum_{n=1}^{M}\frac{Ae^{j(\omega n+\theta_i)}\eta_k^*(n)}{M}\right\}$, $\operatorname{var}\left\{\sum_{n=1}^{M}\frac{Ae^{-j(\omega n+\theta_k)}\eta_i(n)}{M}\right\}$, and

$\text{var}\left\{\sum_{n=1}^{M}\frac{\eta_i(n)\eta_k^*(n)}{M}\right\}$, respectively, are given by

$$\text{var}\left\{\sum_{n=1}^{M}\frac{Ae^{j(\omega n+\theta_i)}\eta_k^*(n)}{M}\right\} = \frac{1}{M^2}\sum_{n=1}^{M}\mathbb{E}\left\{Ae^{j(\omega n+\theta_i)}\eta_k^*(n)\left(Ae^{j(\omega n+\theta_i)}\eta_k^*(n)\right)^*\right\}$$

$$= \frac{A^2}{M^2}\sum_{n=1}^{M}\mathbb{E}\left\{e^{j(\omega n+\theta_i)}\eta_k^*(n)e^{-j(\omega n+\theta_i)}\eta_k(n)\right\}$$

$$= \frac{A^2}{M^2}\sum_{n=1}^{M}\mathbb{E}\left\{\eta_k^*(n)\eta_k(n)\right\} = \frac{A^2\sigma^2}{M}. \tag{10.19a}$$

$$\text{var}\left\{\sum_{n=1}^{M}\frac{Ae^{-j(\omega n+\theta_k)}\eta_i(n)}{M}\right\} = \frac{A^2\sigma^2}{M}. \tag{10.19b}$$

$$\text{var}\left\{\sum_{n=1}^{M}\frac{\eta_i(n)\eta_k^*(n)}{M}\right\} = \frac{\sum_{n=1}^{M}\mathbb{E}\left\{[\eta_i(n)\eta_k^*(n)]^2\right\}}{M^2}$$

$$= \frac{\mathbb{E}\left\{[\eta_i(n)\eta_k^*(n)]^2\right\}}{M}, \quad i \neq k. \tag{10.19c}$$

$$\text{var}\left\{\sum_{n=1}^{M}\frac{\eta_i(n)\eta_i^*(n)}{M}\right\} = \frac{\sum_{n=1}^{M}\mathbb{E}\left\{[\eta_i(n)\eta_i^*(n) - \sigma^2]^2\right\}}{M^2}$$

$$= \frac{\mathbb{E}\left\{[\eta_i(n)\eta_i^*(n)]^2\right\} - \sigma^4}{M}. \tag{10.19d}$$

Note that in (10.19d) $\mathbb{E}\left\{[\eta_i(n)\eta_i^*(n)]^2\right\} = \mathbb{E}\left\{|\eta_i(n)|^4\right\}$ and $\sigma^4 = |\mathbb{E}\left\{|\eta_i(n)|^2\right\}|^2$. In can be seen from (10.19) that the variances decrease with increasing number of samples M used in the computation. Smaller variance implies lower uncertainty or higher accuracy. Hence, as many samples as computational resources permit should be used in the computation of the correlation matrix even though down sampling does not deteriorate SNR.

10.6 Filter Bandwidth and Impulse Response Duration

SNR can be improved by keeping only the desired signal and rejecting as much unwanted noise as possible. This implies designing a filter with a passband to pass only the desired signal and with as wide a stop band as possible. Since the passband is fixed, increasing the width of the stop band will result in decreasing the width of the transition band. The transition band is the gap between the passband and the stop band. Decreasing the width of the transition band will result in increasing

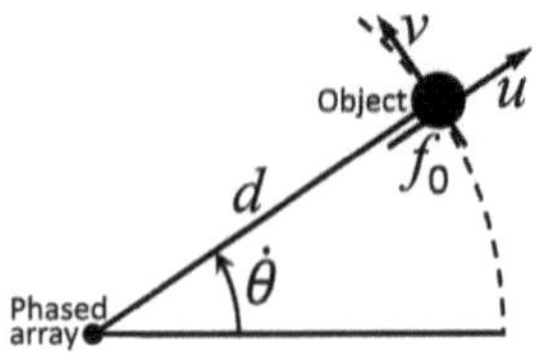

Fig. 10.20 Object with tangential speed v, radial speed u, and distance d from phase array

the impulse response length of the filter. If the impulse response of the filter is too long, outdated information on DOA will remain in the filter for too long. We shall illustrate this by using an example consisting of an object moving with tangential speed v, radial speed u, and distance d from a phased array as shown in Fig. 10.20. Suppose that the object radiates a wave with f_0 nominal frequency.

Supposed that $|v| < 100$ km/hr, $|u| < 0.05$ speed of sound, $d = 100$ m, and nominal $f_0 = 40$ kHz. The angular speed $\dot{\theta}$ observed by the phased array given by $\frac{\text{tangential speed}}{\text{radius}}$ is less than $\frac{100{,}000}{60\times60\times100} = 0.27778$ rad/s. Suppose that the object is tracked by a classical beamformer with 0.1 rad beam width. At an angular speed of 0.27778 rad/s, the object takes 360 ms to travel 0.1 rad. Signal from the antenna will be processed by a series of analogue stages and digital filters in cascade before it is used for computing the signal correlation matrix. The effective impulse response duration of the entire processing path (including both analogue and digital stages) must be significantly less than the time the object takes to travel one beam width, i.e., 360 ms in this case. In this example, we shall assume that the effective impulse response duration of the digital filter must be less than 20 ms.

The fractional Doppler shift is $\frac{|u|}{\text{speed of sound}}$ and so is 0.05. After taking Doppler effect into consideration, the frequency of the wave impinges on the phased array is between $(1 - 0.05) \times 40$ kHz and $(1 + 0.05) \times 40$ kHz. Suppose that the sampling frequency is 160 kHz. The normalized frequency will be between $(1 - 0.05)\pi/2$ and $(1 + 0.05)\pi/2$. The frequency spectrum $X(e^{j\omega})$ is shown in Fig. 10.21a. In Fig. 10.21a, $\alpha = 0.05$. Since the signal is a narrowband signal centered at $\frac{\pi}{2}$ normalized frequency, a filter with z-transform transfer function $0.5(1 - z^{-2})$ may be used as the dc notch to remove the dc component from the ADC; this dc notch has 0.027 dB attenuation at $0.95\frac{\pi}{2}$ and $1.05\frac{\pi}{2}$. This attenuation is immaterial since signals from all the antennae elements are attenuated by the same amount. After removing the dc component, the ADC output is filtered by a filter with z-transform transfer function $H_P(z)$ to remove the negative frequency component of the signal. The frequency response $H_P(e^{j\omega})$ of the filter is shown in Fig. 10.21b. Suppose that, in Fig. 10.21b, we let $\beta = \alpha$. Hence, $\beta = 0.05\pi/2$. Suppose that $H_P(e^{j\omega})$ is derived from a half-band filter with z-transform transfer function $H(z)$ and frequency response $H(e^{j\omega})$ shown in Fig. 10.21c.

Figure 10.22a shows the frequency response plot for a seven-tap half-band filter with $\beta = 0.05\pi/2$. The ripple magnitude is approximately -121 dB. Its coefficient values are shown beside its frequency response plot.

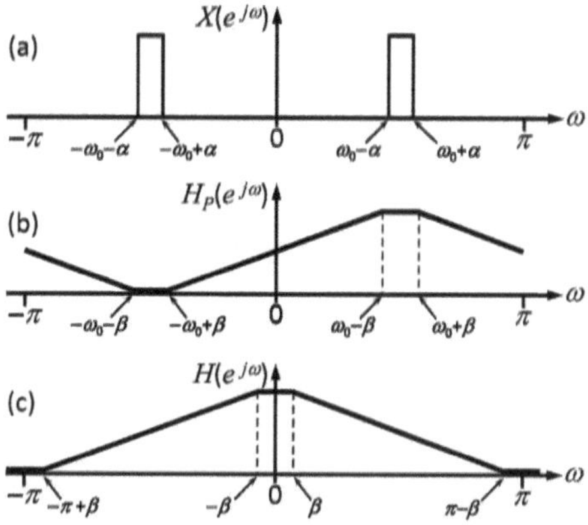

Fig. 10.21 (**a**) Original real frequency spectrum of antenna signal. (**b**) Frequency response of filter to remove the negative frequency component of signal. (**c**) Frequency response of prototype half-band filter

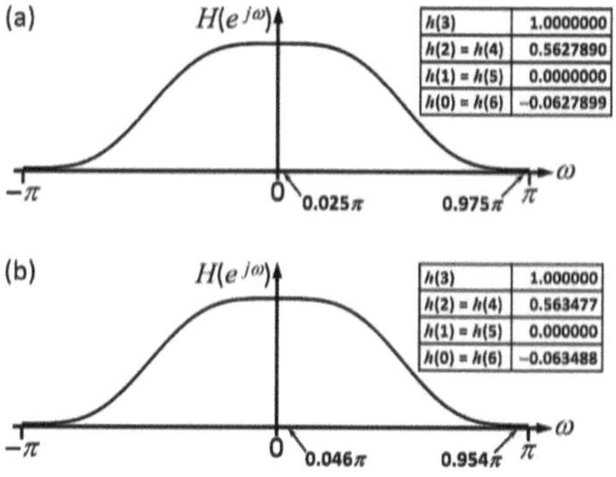

Fig. 10.22 Seven-tap half-band low-pass filter. (**a**) Bandedges at 0.025π and 0.975π. (**b**) Bandedges at 0.046π and 0.954π

The impulse response duration for the dc notch with z-transform transfer function $0.5(1 - z^{-2})$ is $\frac{1}{80}$ ms. The impulse response for the seven-tap filter is $\frac{3}{80}$ ms. The impulse response duration for both of them combined is $\frac{1}{20}$ ms.

If $\beta = \alpha$, when the output of $H_P(z)$ is down sampled, the aliased frequency spectrum of the noise becomes immediately next to the signal's frequency band. This may render difficulty to subsequent removal of noise. If $\beta > \alpha$, when the output of $H_P(z)$ is down sampled, there is a gap between the signal band and the aliased version of the noise. This will relieve some of the problems encountered in the design of subsequent filters. Figure 10.22b shows the frequency response plot for a seven-tap half-band filter with $\beta = 0.046\pi$. The ripple magnitude is approximately − 100 dB. Its coefficient values are shown beside its frequency response plot.

In the above example, the value of β was relaxed. Alternatively, the frequency response $H(e^{j\omega})$ may be jointly optimized with the frequency responses of subsequent filters to achieve an optimum design.

After removing the signal's negative frequency component and down sampling by a factor of two, the frequency spectrum $X_Q(e^{j\omega})$ is as shown in Fig. 10.23a; its sampling frequency becomes 80 kHz.

The signal whose frequency spectrum is $X_Q(e^{j\omega})$ shown in Fig. 10.23a may be further filtered by a bandpass filter to remove noise. The bandpass filter should

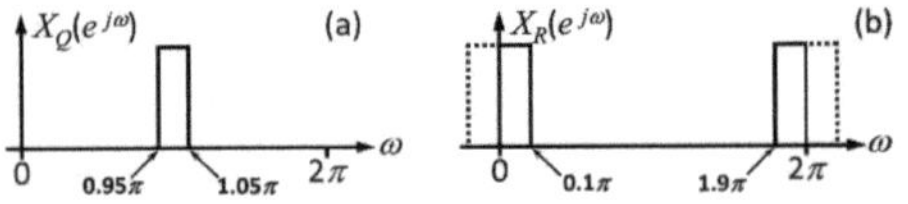

Fig. 10.23 (**a**) Frequency spectrum after down sampling by 2. Frequency spectrum after down sampling by 4

pass the normalized frequency band from 0.95π to 1.05π and remove all other frequencies as much as possible. Its impulse response duration should be less than $20 - \frac{1}{20} = 19.95$ ms.

If the signal is further down sampled by another factor of two, its frequency spectrum will be as shown in Fig. 10.23b. The signal is now centered at dc.

References

X. Chen, T. Parks, Design of FIR filters in the complex domain. IEEE Trans. Acoust. Speech Sig. Process. **35**(2), 144–153 (1987)

T.A. Claasen, A. Jongepier, Model for the power spectral density of quantization noise. IEEE Trans. Acoust. Speech Sig. Process. **29**(4), 914–917 (1981)

T. Crystal, L. Ehrman, The design and applications of digital filters with complex coefficients. IEEE Trans. Audio Electroac. **16**(3), 315–320 (1968)

P.S.R. Diniz, E.A.B. da Silva, S.L. Netto, *Digital Signal Processing* (Cambridge University, Cambridge, 2010)

R.M. Gray, Quantization noise spectra. IEEE Trans. Inform. Theo. **36**(6), 1220–1244 (1990)

Y.C. Lim, Efficient special purpose linear programming for FIR filter design. IEEE Trans. Acoust. Speech Sig. Process. **31**(4), 963–968 (1983)

Y.C. Lim, Frequency response masking approach for the synthesis of sharp linear phase digital filters. IEEE Trans. Circ. Syst. **33**(4), 357–364 (1986)

Y.C. Lim, Design of discrete-coefficient-value linear phase FIR filters with optimum normalized peak ripple magnitude. IEEE Trans. Circ. Syst. **37**(12), 1480–1486 (1990)

Y.C. Lim, A.G. Constantinides, New integer programming scheme for nonrecursive digital filter design. Electron. Lett. **15**(25), 812–813 (1979)

Y.C. Lim, S.R. Parker, FIR filter design over a discrete powers-of-two coefficient space. IEEE Trans. Acoust. Speech Sig. Process. **31**(3), 583–591 (1983a)

Y.C. Lim, S.R. Parker, Discrete coefficient FIR digital filter design based upon an LMS criteria. IEEE Trans. Circ. Syst. **30**(10), 723–739 (1983b)

Y.C. Lim, S.R. Parker, A.G. Constantinides, Finite wordlength FIR filter design using integer programming over a discrete coefficient space. IEEE Trans. Acoust. Speech Sig. Process. **30**(4), 661–664 (1982)

S.K. Mitra, *Digital Signal Processing: A Computer-Based Approach* (McGraw-Hill Higher Education, New York, 2001)

S.C. Pei, J.J. Shyu, Complex eigenfilter design of arbitrary complex coefficient FIR digital filters. IEEE Trans. Circ. Syst. II: Anal. Dig. Sig. Process. **40**(1), 32–40 (1993)

J.G. Proakis, *Digital Signal Processing: Principles, Algorithms, and Applications* (Pearson Education India, Noida, 2007)

L.R. Rabiner, B. Gold, *Theory and Application of Digital Signal Processing* (Englewood Cliffs, Prentice-Hall, 1975)

B. Widrow, I. Kollar, M.C. Liu, Statistical theory of quantization. IEEE Trans. Instrument. Measure. **45**(2), 353–361 (1996)

Part III
Introduction to Wireless Communications: MIMO Systems and Radar

Chapter 11
Communication Principles

11.1 Introduction

The communication systems differ in their configuration according to the environment they are meant to operate, for example, wired connections consist of private or shared physical infrastructure such as coaxial cables, optical fibers, twisted-pair wires, etc. On the other hand, wireless connections require distinct concepts for reusing radio resources and careful interference mitigation since many types of services and users should share the channel. In the wireless case, the communication occurs primarily through the air and underwater.

In this part of the text, we will cover a few basic principles of communications which are somewhat relevant to understanding some developments related to autonomous vehicle technology. Among these concepts, we can list:

1. Multiple access
2. Cellular systems
3. Wireless channels
4. CDMA systems
5. Orthogonal frequency division multiplex (OFDM)
6. MIMO systems
7. Radar signal processing

11.2 Communication Systems

Figure 11.1 shows the general setup of a digital communication system, where each building block represents a set of tasks to be performed to allow efficient information transfer. The system starts by acquiring the information source to be transmitted, and depending on the type of source, the appropriate encoder is chosen.

Y. C. Lim et al., *Signal Processing for Autonomous Systems*,
https://doi.org/10.1007/978-3-032-00675-2_11

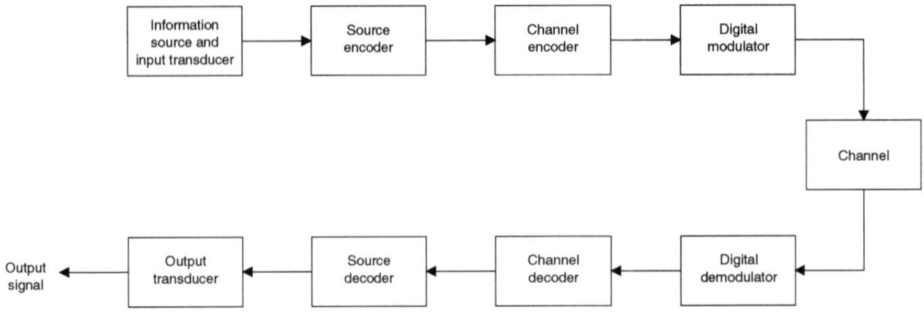

Fig. 11.1 Digital communication systems

For each family of sources, a different type of source encoder applies. Examples of sources are speech, audio, image, video, bank account data, etc. Each type of source requires its own performance in terms of the quality of the received data. The task of the source encoder is to make the information to be transmitted more compact so that the communication resources are utilized efficiently. The next building block is the channel encoder, which includes some redundancy in the data originating from the source encoder. The channel encoding allows for error detection and correction of the data at the receiver end. Then, the baseband signal is modulated to traverse the channel in a prescribed frequency range either established by the regulations or to benefit from the best channel response. At this point, the signal is prepared to illuminate the channel and reach the receiver. At the receiver end, the channel impairments affect the transmitted signal. Therefore, the signal is first demodulated and decoded with possible error correction. After the channel decoding, the source decoder brings the information back to its original format to feed an output transducer possibly.

The channel is formed by propagating electromagnetic energy through the air in wireless communications. The electromagnetic energy radiates to the propagation medium via one or multiple antennas (Proakis and Salehi 1994). A few issues to consider to warrant proper communication are as follows:

- The antenna must be longer than $1/10$ the carrier wavelength for efficient radiation.
- There are three forms of propagation of electromagnetic waves in the open space and atmosphere: (i) line-of-sight (LOS) propagation, (ii) sky wave propagation, and (iii) ground wave propagation.

For each range of frequency, there are several considerations for its use, such as the permission from radio regulation inherent to the law of the nations, propagation loss, environmental and human-made noise interference, multipath reflections, signal scattering, etc. An important aspect is that while the propagation loss increases with the frequency, the size of the antennas as well as their separation in an array configuration tends to decrease. Therefore, there are many issues to take into consideration before deploying a reliable communication system.

11.3 Symbol Constellations

In digital communication systems, the transmitted symbols denoted by $s(k)$ belong to a discrete set of complex numbers, $s(k) \in \mathbb{C}$. These numbers originate from information one wishes to transmit and belong to a set of codewords or symbols, which are often called constellations.

The simplest case is the pulse amplitude modulation (PAM), where M-PAM constellation points are represented by $s_i = \tilde{a}_i$, with $\tilde{a}_i = \pm\tilde{d}, \pm 3\tilde{d}, \ldots, \pm(M-1)\tilde{d}$. The parameter $\tilde{d}$ represents half of the distance between two points in the constellation. Figure 11.2 shows a PAM constellation with M symbols.

The code words might have a complex representation where the information is a four quadrature amplitude modulation (4-QAM) constellation as depicted in Fig. 11.3a. The representation can be extended to M-QAM constellation points as $s_i = \tilde{a}_i + \jmath\tilde{b}_i$, with $\tilde{a}_i = \pm\tilde{d}, \pm 3\tilde{d}, \ldots, \pm(\sqrt{M}-1)\tilde{d}$, and $\tilde{b}_i = \pm\tilde{d}, \pm 3\tilde{d}, \ldots, \pm(\sqrt{M}-1)\tilde{d}$. The parameter $\tilde{d}$ represents half of the distance between two points in the constellation. Figure 11.3b illustrates the 16-QAM where each symbol in the constellation represents four bits. Typically, neighboring symbols in any constellation represent symbols differing from each other in only one bit, and this representation is called Gray code.

Another form of complex representation is the phase shift keying (PSK), whose constellations in the binary case and octal phase cases are depicted in Figs. 11.4a and b, respectively. The key feature of the PSK is that all symbols require the same power.

Fig. 11.2 PAM

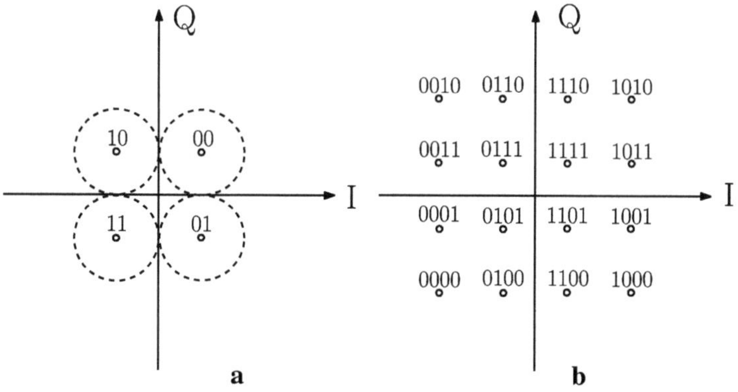

Fig. 11.3 **(a)** 4-QAM, **(b)** 16-QAM

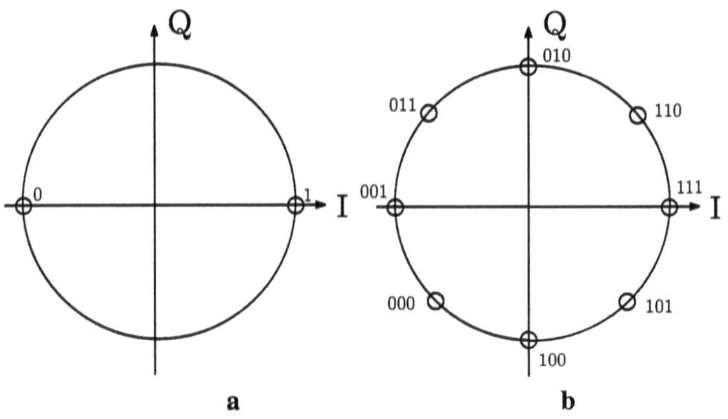

Fig. 11.4 (**a**) BPSK, (**b**) 8-PSK

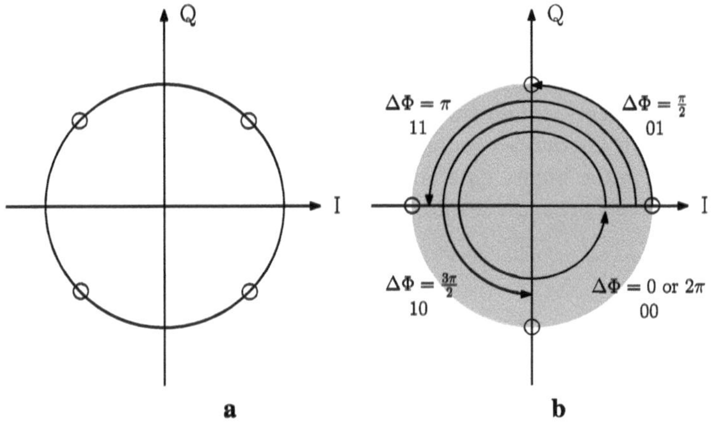

Fig. 11.5 (**a**) QPSK, (**b**) DPSK

A representation related to the PSK is the quadriphase-shift keying (QPSK), as illustrated in Fig. 11.5a. Through a differential encoding of the input binary signals, it is possible by shifting the carrier according to the current bit pair, as seen in Fig. 11.5b. This strategy relaxes the synchronization between transmitter and receiver.

The detailed description of the code words is beyond the scope of this text. For each constellation, it is possible to derive an expression for the probability of error when detecting the symbol after traversing a channel corrupted by additive white Gaussian noise (AWGN). The interested reader might consult textbooks such as Gallager (2008), Madhow (2008), Nguyen and Shwedyk (2009), Proakis and Salehi (1994), Haykin (2001), Barry et al. (2003), Proakis (2001), and Anderson (1999).

In many implementations, the transmitted data is differentially encoded, that is,

$$d(k) = b(k)d(k-1) \tag{11.1}$$

where the transmitted information is the data transition from instant $k-1$ to k determined by the current bit $b(k)$. Note that $d(k)$ and $d(k-1)$ represent the current and the last symbol, respectively. Initializing the differential encoder with $d(-1)$ (assuming transmission starts at $k=0$), the decoding is

$$b(k) = d(k)d(k-1). \tag{11.2}$$

This technique eases demodulation under phase uncertainty, but there is some loss in performance (Haykin 2001).

11.4 Remarks

The overall concept of digital communication permeates the actual implementation of automatic vehicles, which involve multiple devices and systems. This brief introduction paves the way to enlighten the concepts taking part in the practical solutions.

References

J.B. Anderson, *Digital Transmission Engineering* (IEEE Press, New York, NY, 1999)

J.B. Barry, E.A. Lee, S. Messerschmitt, *Digital Communications*, 3rd edn. (Kluwer Academic Publishers, Norwell, MA, 2003)

P.S.R. Diniz, W.A. Martins, M.V.S. Lima, *Block Transceivers: OFDM and Beyond* (Springer, New York, NY, 2012)

R.G. Gallager, *Principles of Digital Communication* (Cambridge University Press, Cambridge, UK, 2008)

S. Haykin, *Communication Systems*, 4th ed. (Wiley, New York, NY, 2001)

U. Madhow, *Fundamentals of Digital Communication* (Cambridge University Press, Cambridge, UK, 2008)

H. Nguyen, E. Shwedyk, *A first Course in Digital Communications* (Cambridge University Press, Cambridge, UK, 2009)

J.G. Proakis, *Digital Communications*, 4th edn. (McGraw Hill, New York, NY, 2001)

J.G. Proakis, M. Salehi, *Communication Systems Engineering* (Prentice Hall, Upper Saddle River, NJ, 1994)

Chapter 12
Multiple-Access Methods

12.1 Introduction

Multiple-access methods are the enablers to allow access to services to several users under the same and limited network environment. The appropriate choice of the access method is critical to achieving high-data-rate transmissions, which will increase the capacity to provide multiple services to users at different locations. In particular, in wireless cellular systems, the information faces very diverse channel conditions.

This chapter briefly describes the main multiple-access methods under use, namely:

- Time Division Multiple Access (TDMA)
- Frequency Division Multiple Access (FDMA)
- Hybrids of TDMA and FDMA
- Code Division Multiple Access (CDMA)

  Two main cases:

  Direct Sequence CDMA (DS-CDMA)

  Frequency-Hopping CDMA (FH-CDMA)

  The CDMA system is also known as the Spread-Spectrum communication system.
- Orthogonal Frequency Division Multiple Access (OFDMA)

The multiple-access method must share the available bandwidth for services in a given network in such a way that every user can have its share and no user utilizes all the resources available.

Y. C. Lim et al., *Signal Processing for Autonomous Systems*,
https://doi.org/10.1007/978-3-032-00675-2_12

12.2 TDMA

In TDMA, the transmission time is shared among the users. Each user has a time slot to transmit its information within a larger frame. Every other user is silent whenever a given user is communicating. A user communicates only in his/her assigned time slot, which requires synchronism among the users (Fig. 12.1).

The data in a TDMA system is organized in frames, where each slot has its own type of information. Figure 12.2 illustrates a typical data frame organization.

Figure 12.3 describes how the symbols are possibly shuffled in subsequent times. This procedure warrants more robustness in information transport.

The data of each user in a TDMA system, as well as in other systems, is divided into slots, for instance, as seen in Fig. 12.4.

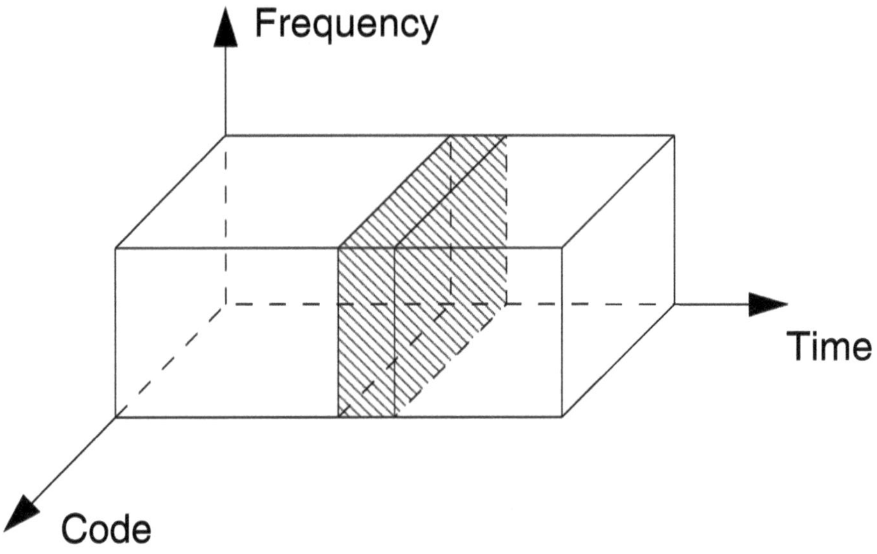

Fig. 12.1 Channel sharing in TDMA

Header	Data slots	Error detection bits

Fig. 12.2 A frame of the TDMA system

s_1	s_2	s_3	s_4	s_5	s_6	s_7	s_8	s_9	s_{10}

s_9	s_3	s_7	s_{10}	s_5	s_1	s_4	s_8	s_6	s_2

Fig. 12.3 Sub-partition of the data TDMA frame at a given time, and in a moment later

s_i

Trail bits	Sync. bits	DATA	Guard period

Fig. 12.4 Slot division for each user in TDMA

12.3 FDMA

FDMA consists of dividing the available frequency spectrum among the users. That is, the users utilize separate frequency channels where each channel occupies part of the available bandwidth, and the users have access to services without significant interference from each other. In FDMA, each user has its own frequency band for communicating. This way, orthogonality is straightforward to maintain. FDMA was the multiple-access method of choice for first-generation cellular phones (AMPS in the USA, NMT in Finland). Figure 12.5 depicts the channel division using FDMA.

12.3.1 Hybrid TDMA-FDMA

The 2G GSM mobile communication system utilized subbands of 200 kHz of channel bandwidth, initially utilized in Europe. These subbands were, in turn, shared among eight users using TDMA. In the IS-54 system (in the USA), each 30-kHz channel bandwidth was shared among three TDMA users.

Many other aspects should be considered when choosing a multiple-access system. For instance, in any scenario, the Duplex method has to be defined, that is, TDD (time-division duplex) and FDD (frequency-division duplex) in order to define how the uplink and downlink connections are made. Alternatively, OFDM can be applied with multiple-access methods leading to, for example, CDMA-based and TDMA-based OFDM systems which will be discussed later.

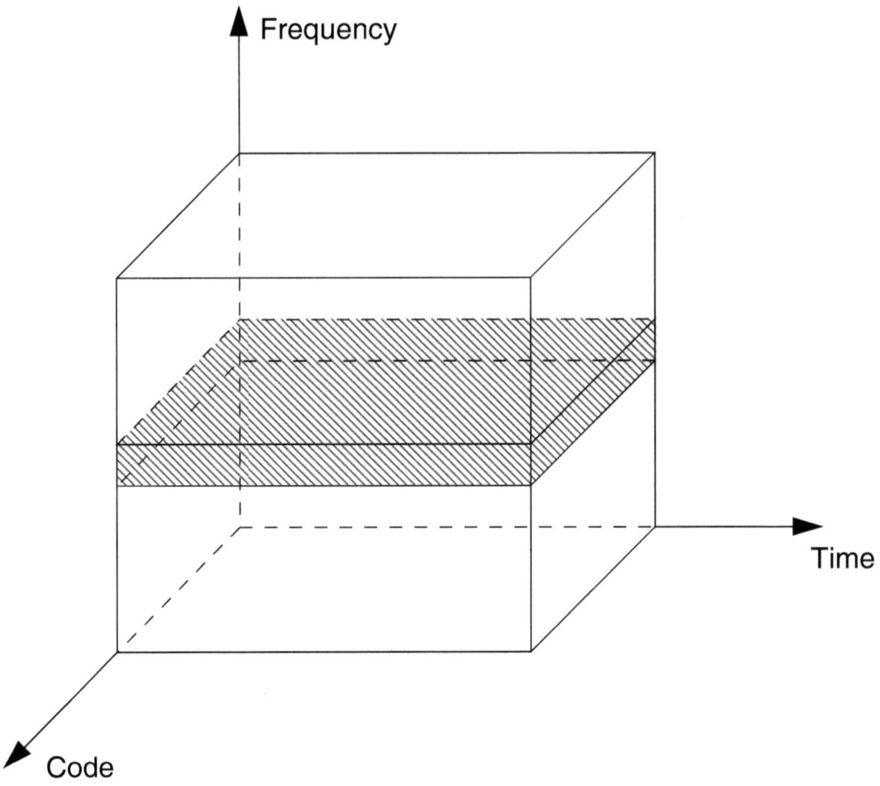

Fig. 12.5 Channel sharing in FDMA

As the number and types of services available to the users increase, the fixed channel assignment inherent to FDMA and TDMA is not efficient, and also more spectrum would be required. Even if some flexible channel assignment is incorporated in the current FDMA and TDMA, there is always a fixed upper bound in the number of users.

12.4 CDMA

The CDMA system spreads the power spectral density (PDS) of the baseband signal as illustrated in Fig. 12.6. By assuming a spread spectrum interference to the reference CDMA signal, as per Fig. 12.7, we can show in the following the benefits of CDMA in this situation. The in-band interference is reduced by using the spread spectrum, as illustrated in Fig. 12.8.

In CDMA, the users share all the available bandwidth simultaneously. The way to separate them is by using a code for each user.

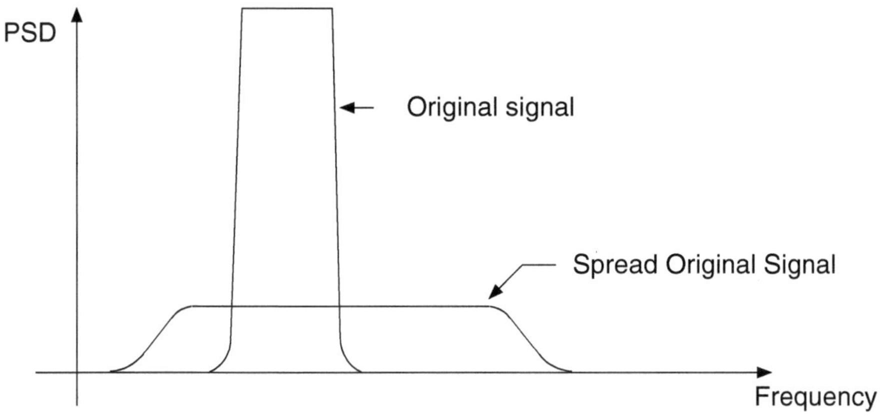

Fig. 12.6 Spreading the spectrum in CDMA

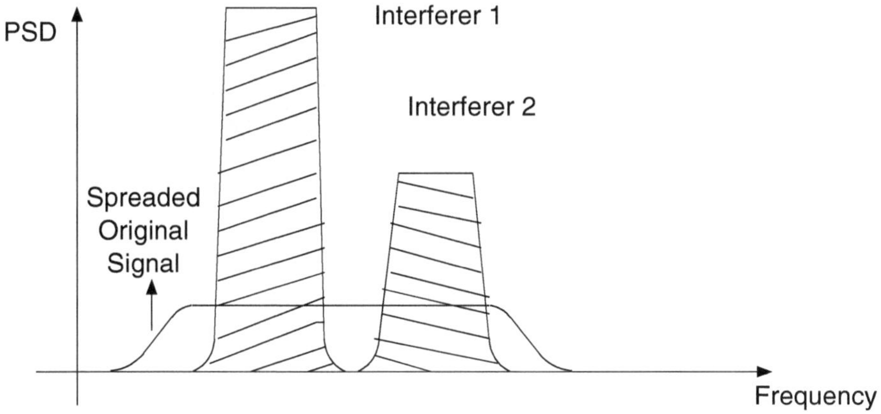

Fig. 12.7 Interference in CDMA

12.4.1 Direct-Sequence CDMA

Each user is assigned a unique code sequence that will be multiplied by the bitstream before transmission. These code sequences have a simple format consisting of plus and minus 1 as elements and have a much higher rate than the bit rate. Implementing the Direct-Sequence CDMA is quite simple, as illustrated in Fig. 12.9.

Some second-generation cellular systems, such as the IS-95 wireless system, utilized a hybrid of FDMA and DS-CDMA.

Multiple access in the CDMA systems is provided by assigning a spreading code to each user, as illustrated in Fig. 12.10.

It is possible to come up with a myriad of configurations to provide multiple access. Some of these alternatives are discussed in the CDMA chapter.

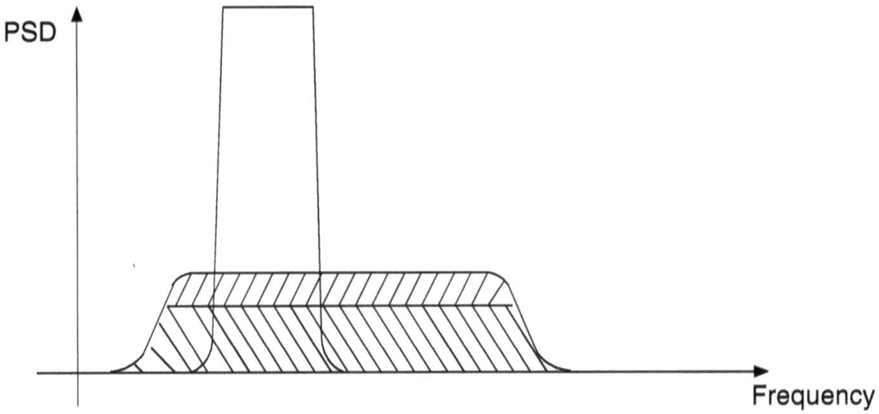

Fig. 12.8 Despreading in CDMA

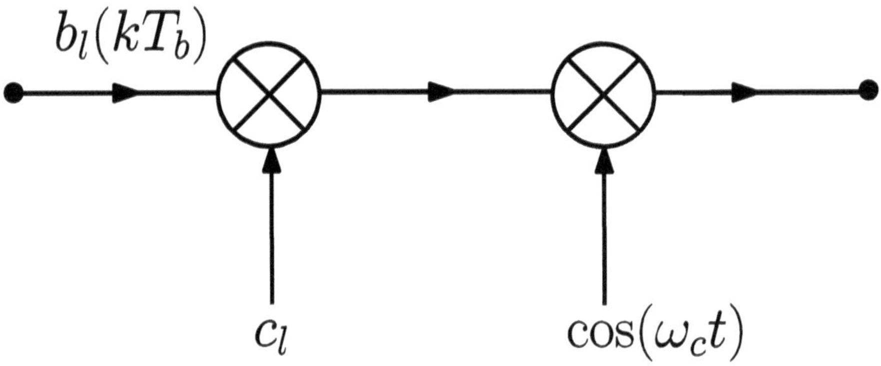

Fig. 12.9 Direct-sequence CDMA

The main advantages of the CDMA, when applied to cellular systems, are:

- CDMA is wideband compared to FDMA: the use of several paths in frequency-selective channels can be considered as frequency diversity.
- The symbol time in CDMA is longer than in TDMA, so CDMA is less affected by systems with long impulse responses and requires simpler equalizers.
- TDMA and FDMA have a fixed number of simultaneous users. CDMA does not have fixed numbers. If the number of users increases, the service gracefully degrades.
- In cellular systems, CDMA allows soft handover between neighboring cells. Close to the frontier where the signal is weaker, but since the user communicates with two base stations, the diversity helps compensate for it.
- In TDMA and FDMA, neighboring cells must use different frequencies because they control interference based on spatial attenuation. All the cells might use

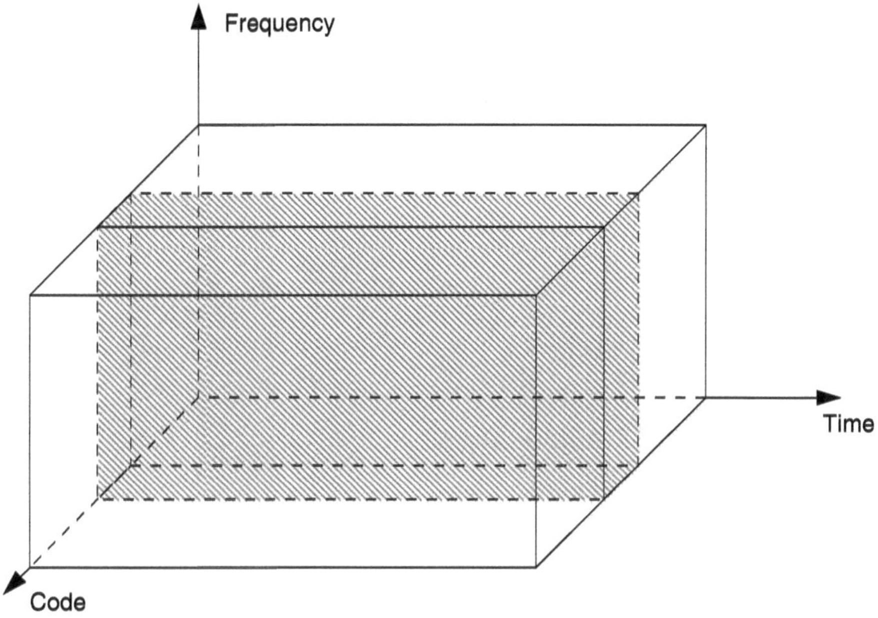

Fig. 12.10 Channel sharing in CDMA

the same frequency range in CDMA, eliminating the need for frequency use planning, whereas TDMA and FDMA may use adaptive frequency reallocation.
- CDMA requires power control and base station synchronization. However, intercell interference can be suppressed at the receiver.

12.4.2 Frequency-Hopping CDMA

Each user employs a different frequency band at a given time frame. During transmission, the user frequency band hops to different bands according to a prescribed hopping pattern (code). The receiver hops synchronously with the transmitter with the knowledge of the code. Figure 12.11 illustrates the basic configuration of the frequency-hopping CDMA.

The hopping can be slow, where hopping occurs at the bit rate. Otherwise, it can be fast, where more than one hop occurs during a bit duration. The latter case is challenging to implement.

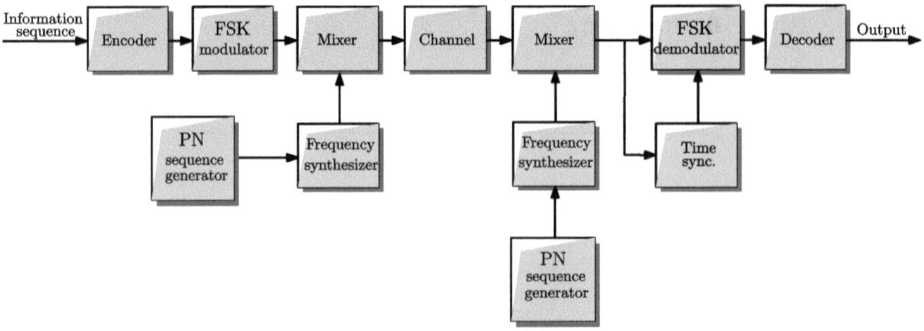

Fig. 12.11 Frequency-hopping CDMA

12.4.3 Wideband Multiple Access

There are CDMA systems operating on part of the available spectrum and are called narrowband CDMA. However, when the entire band is utilized to spread the information of each user, the system is called wideband CDMA. The use of wideband multiple access brings about some benefits, as outlined below. Wideband MA systems can operate in the frequency range of existing narrowband channels, allowing flexibility in the number of users and services provided to each user. Furthermore, they allow improved interference rejection, such as multiuser, multipath, and narrowband interferences. All the benefits come at the cost of requiring more advanced technology for implementation.

12.5 OFDMA

The orthogonal frequency-division multiplex (OFDM) is the most widely used multicarrier transceiver. The multicarrier transceivers split the channel into very narrowband subchannels so that each subchannel appears to be flat, and depending on its signal-to-noise ratio (SNR), the Shannon capacity might be nearly accomplished. These issues will be discussed later. In some standards, the OFDM subchannels can be assigned to distinct users, leading to a multiple-access system (OFDMA—orthogonal frequency-division multiple access). In the simplest case, multiple access can be implemented in a TDMA format where, at a given time slot, a given user is allowed to employ all the subchannels to transmit its information, as illustrated in Fig. 12.12.

It is also possible to assign distinct frequency ranges for the users such that the users utilize the subchannels belonging to its assigned range and in FDMA. In this multiple-access case, the subchannels are distributed as depicted in Fig. 12.13.

The most efficient way to assign the subchannels to multiple users is random assignment so that all users might enjoy the same quality of service. In this case,

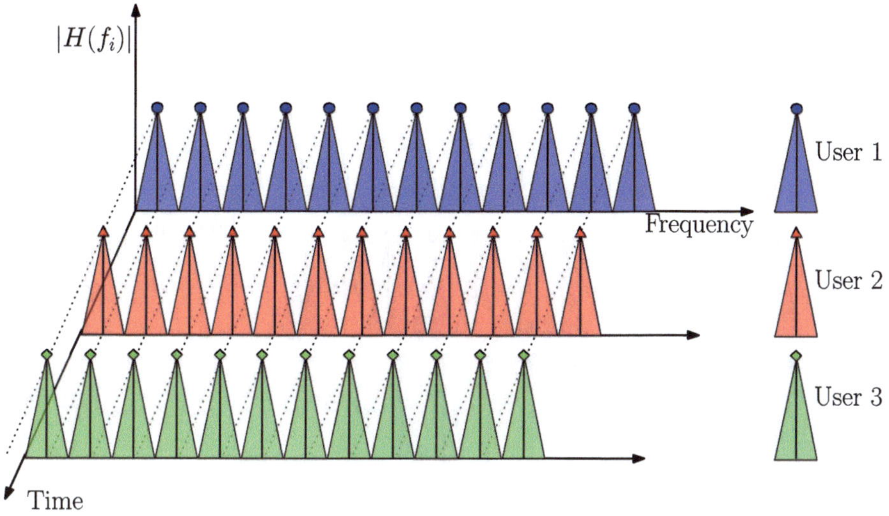

Fig. 12.12 OFDMA system, TDMA case

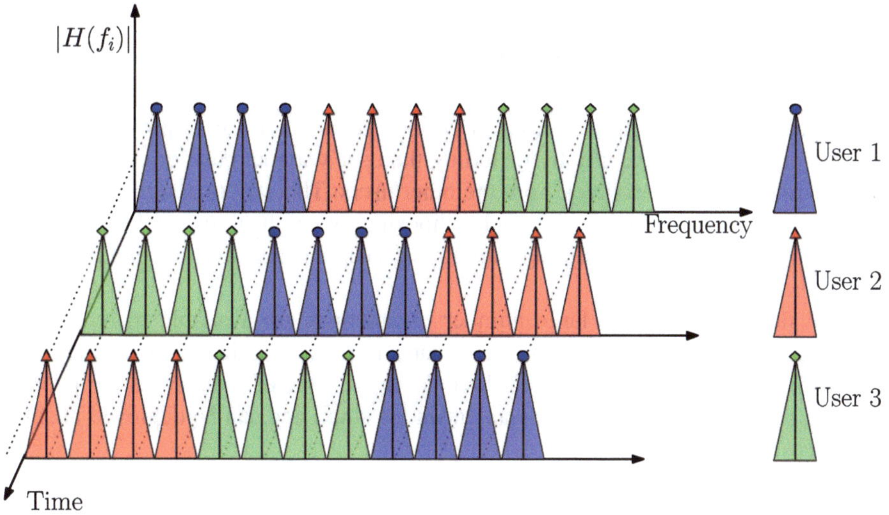

Fig. 12.13 OFDMA system, FDMA case

each user's data has increased in diversity in the form in which it is transmitted. Figure 12.14 illustrates the random assignment of subchannels.

An alternative way to represent the assignment of radio resources in OFDMA utilizes a time-frequency plane as illustrated in Fig. 12.15. The visualization, borrowed from Diniz et al. (2012), is widely used and quite handy in some practical discussions.

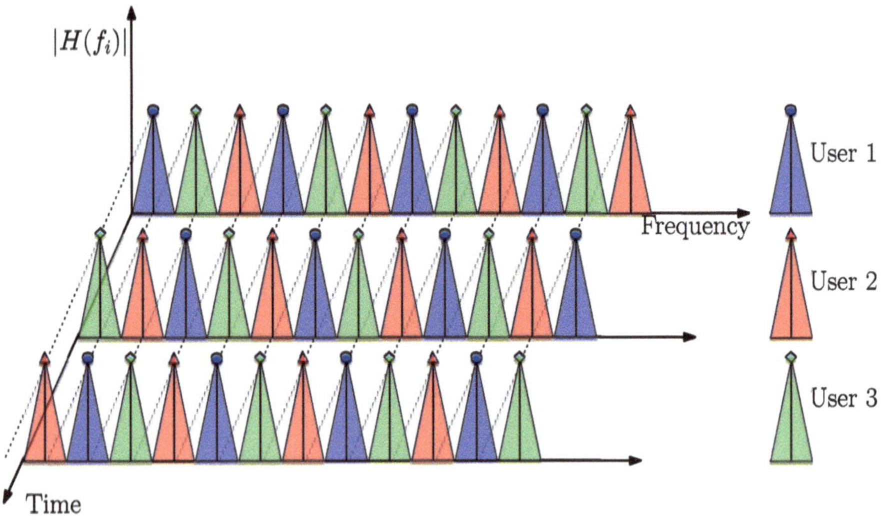

Fig. 12.14 OFDMA system, random assignment

12.6 Space-Division Multiple Access

The space-division multiple access (SDMA) is yet another way to provide multiple access or multiplex/diversity by including the user's spatial position as an element to separate its radio resources from the other users. The way to assign resources to the user can be viewed in Fig. 12.16, by choosing two of the three available multiple access tools. The possible solutions are time division, space division, and frequency division.

For situations where a large number of antennas are available for communications, particularly in the cases where carriers in the gigahertz (GHz) range are utilized, the spatial beams can be directed at the user of interest, as depicted in Fig. 12.17. In such a situation, the antennas are small and can be deployed at small distances.

12.7 Duplex Systems

A requirement to define how two terminals communicate relates to how information travels back and forth, called the duplex method. Typically, the most widely used forms are the frequency and time duplexes, as explained below.

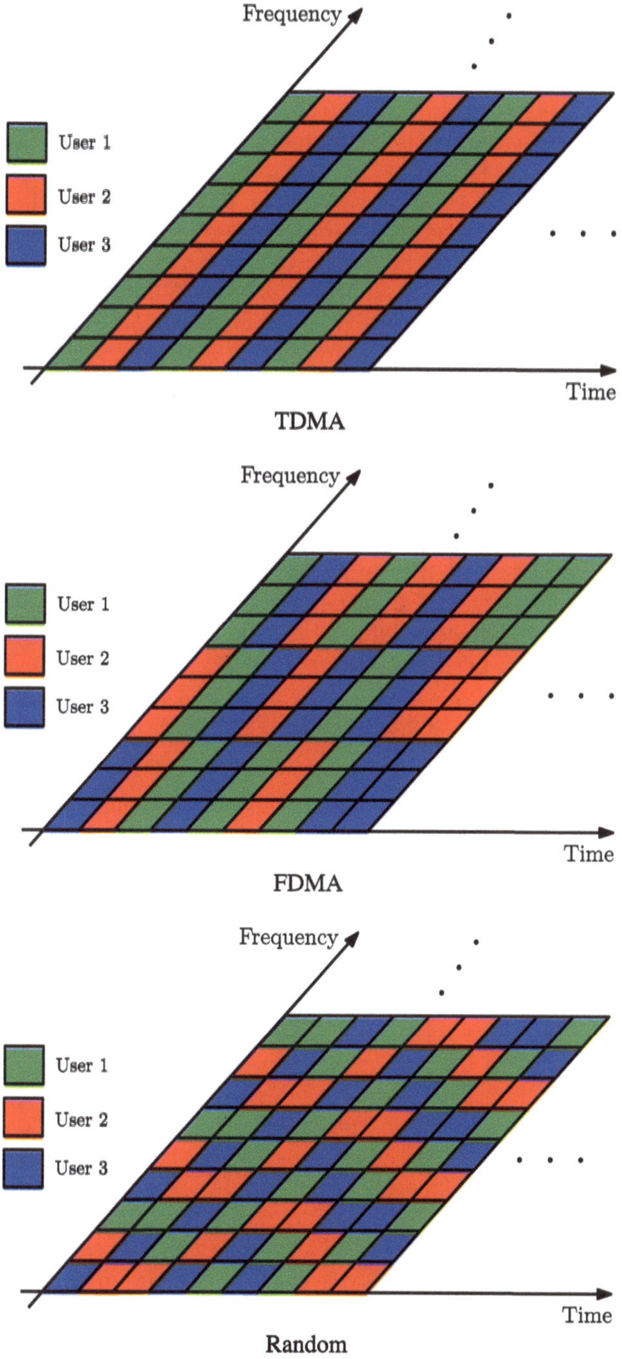

Fig. 12.15 Time-frequency plane representation of OFDMA systems

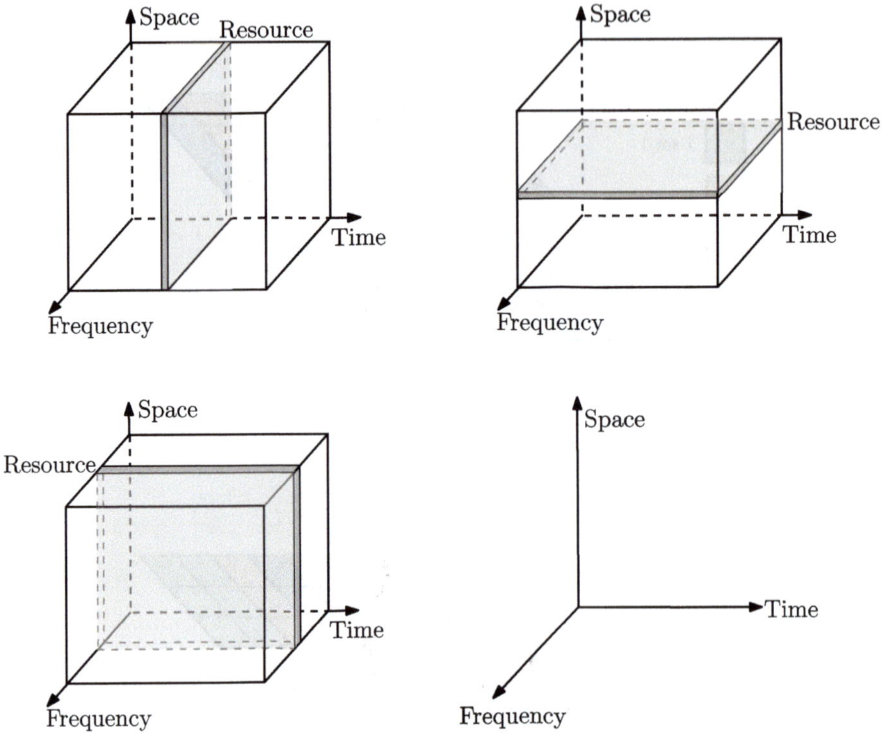

Fig. 12.16 Time-frequency-space plane representation of multiple-access systems

Fig. 12.17 Massive multiple-antenna configuration

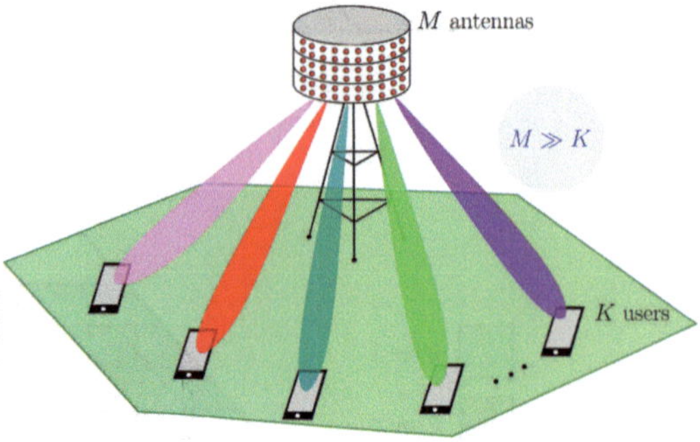

12.7.1 FDD

Frequency-division duplex (FDD) is one of the techniques to separate the information sent by the mobile to the base station (uplink), from the information sent by the base station to the mobile station (downlink) (Fig. 12.18).

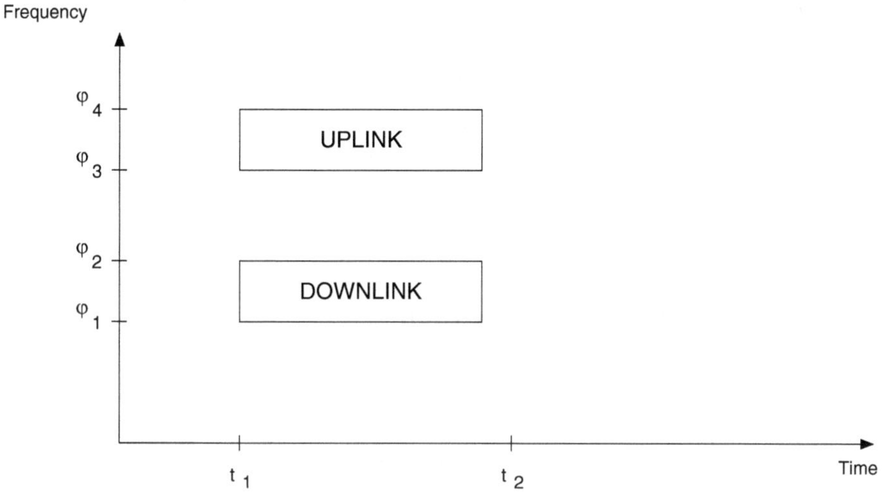

Fig. 12.18 Frequency-division duplex (FDD)

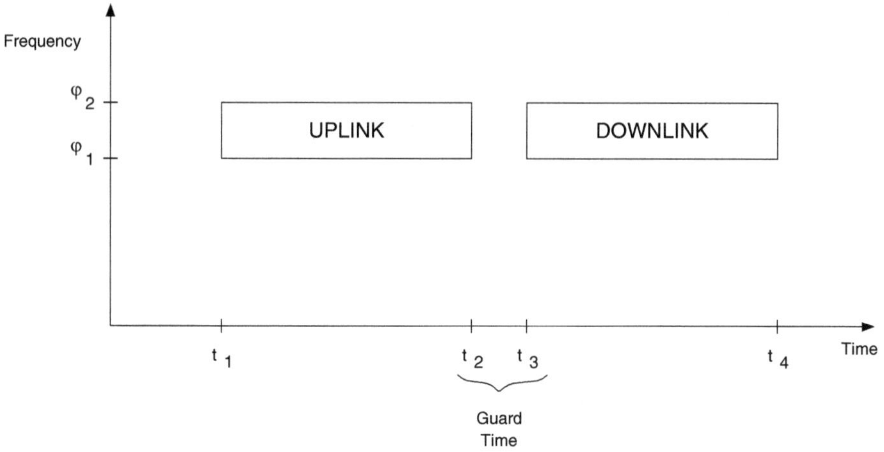

Fig. 12.19 Time-division duplex (TDD)

12.7.2 TDD

Time-division duplex (TDD) is the other strategy to separate the information sent by the mobile to the base station, i.e., uplink, from the information sent by the base station to the mobile, i.e., downlink (Fig. 12.19).

12.7.3 *Full Duplex*

Another method is the full duplex, where the terminal communicates with each other simultaneously and utilizes the same frequency range. An essential requirement in this case is the availability of sophisticated signal processing tools to mitigate the interference. The autonomous vehicle ecosystem includes several sensing devices and transceivers that would benefit from full-duplex communications.

12.8 Full-Duplex Communication

The growing importance of full-duplex connections warrants some discussion, given their potential to allow better use of the available spectrum. The autonomous vehicle ecosystem includes several sensing devices and transceivers that would benefit from full-duplex communications.

12.8.1 *Motivation*

In the era of big data, the number of sensing devices has increased, giving rise to a broader range of possible solutions for highly integrated heterogeneous analog-digital systems. In addition, some devices sit on the analog domain, such as sensors, antennas, medical implants, and other transducers. Accordingly, their neighboring building blocks might rely on analog designs consisting of filters, amplifiers, oscillators, automatic gain control, sample-and-hold circuits, anti-aliasing filters, and mixers.

This subsection deals with analog and digital signal processing principles and their coexistence in real applications. The presentation tries to emphasize current solutions on the analog and digital parts, composing many mixed-mode applications. It is also essential to determine where the analog and digital implementations meet each other and what defines the frontier between them. The solution lies in practical issues related to the processed signals frequency range, required speed, power consumption constraints, and precision of analog-to-digital (A/D) and digital-to-analog (D/A) conversions, restrictions on the physical size, etc. The main inherent drawbacks of keeping all the signal processing tasks in the analog domain consist of their limited flexibility, reliability, reproducibility, and low accuracy due to the signal-to-noise ratio (SNR). The latter is equivalent to around 13 fixed-point bits in a well-designed circuit due to the random deviations in physical parameters, such as resistance, capacitance, and transconductance, caused by fabrication process variations. With the correct data acquisition performed in the analog domain, it is possible to benefit from the many tools currently available in the digital world. However, if low power, high frequency, and small dimensions are desirable, challenging the engineering solution, several constraints remain.

This section discusses the available design solutions for the analog and digital signal processing building blocks. In addition, it provides some insights into how the analog-digital frontier is drawn in some practical systems. Hopefully, the discussion will pave the way to exciting and innovative analog and digital signal processing techniques that can be incorporated in autonomous systems.

12.8.2 Duplex Methods Revisited

One of the most relevant technological problems of the current times is to address the spectrum scarcity related to the increasing demand for wireless communications. In many wireless communications applications, the channel capacity is approaching the limit measured by the number of bits per second per Hertz, which are reliably transmitted through the physical channel. However, most currently available wireless systems utilize half-duplex transmissions where the time or frequency allocations for communication between two terminals are distinct. Indeed, these allocation approaches are called duplex methods, where the main ones are the TDD and the FDD. As previously mentioned, the TDD schemes assign different time intervals to the transmit and receive connections, where in both cases, the communication devices use the whole channel bandwidth during their transmissions. In the FDD method, different frequency bands are assigned for transmitting and receiving. Both TDD and FDD systems require a *guard time* and a *frequency guard*, respectively, to avoid interference between these two connections that might be caused by propagation effects such as multipath. Figure 12.20 illustrates the communication between two wireless devices where each device has its time or frequency slot to perform the data transfer.

These duplex schemes lead to spectrum sub-utilization, making full-duplex communications highly attractive, but their actual implementation in the so-called in-band setup brings about technical challenges that can only be addressed by

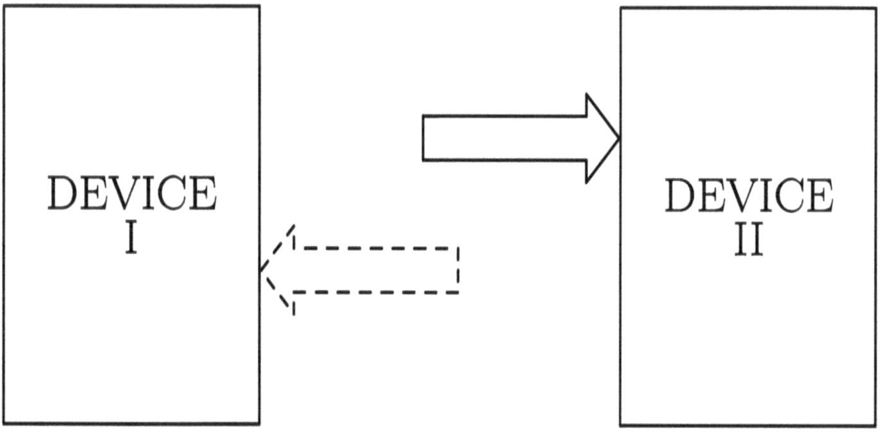

Fig. 12.20 Half-duplex communication

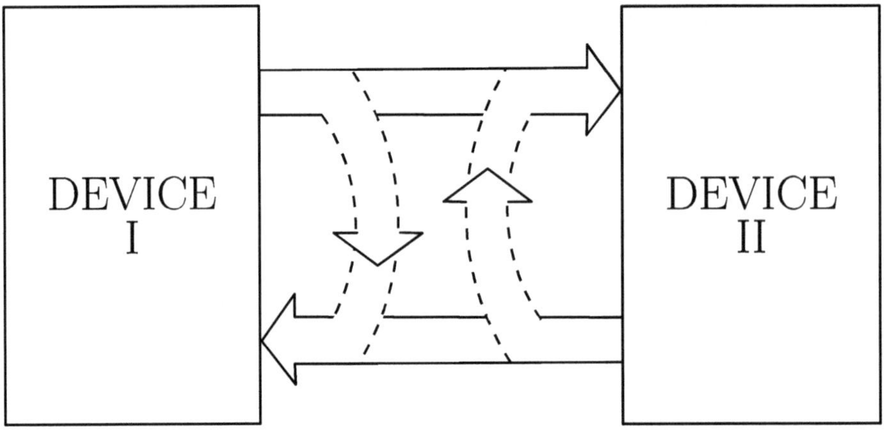

Fig. 12.21 Full-duplex communication

state-of-the-art mixed-mode designs. In the mixed-mode solution, part of the self-interference cancellation should be performed in the continuous-time domain and part in the discrete-time domain. In both cases, the solution includes several stages of cancellation in both domains. As seen in Fig. 12.21, the full-duplex configuration allows communication between the devices utilizing the complete radio resources available. This desirable feature comes with the challenge of coping with the self-interference boosted by the higher power of the transmitted signal compared to the received signal. The latter reaches the wireless device after facing the inherent fading of the physical channel.

It is worth mentioning that the same concept of full-duplex communication can be applied to relay devices, which recover and retransmit information from a wireless device to another one when their direct communication is not feasible or faces drastic fading (Riihonen et al. 2011; Sabharwal et al. 2014). Yet another practical situation appears in cellular systems, as the base station operates in full duplex, it has to include an interference cancellation method, even if the mobile devices do not operate in full duplex. In both the relay and cellular systems, mobile devices may interfere with each other to some extent; such a case requires proper modeling.

12.8.3 Full-Duplex Transceivers

As previously discussed, full-duplex transmission has the potential to enhance the spectral usage by allowing the communication between wireless devices utilizing the entire in-band spectrum. The implementation of the full-duplex solution is challenging, and it is currently a subject of intense research, highlighting the cases

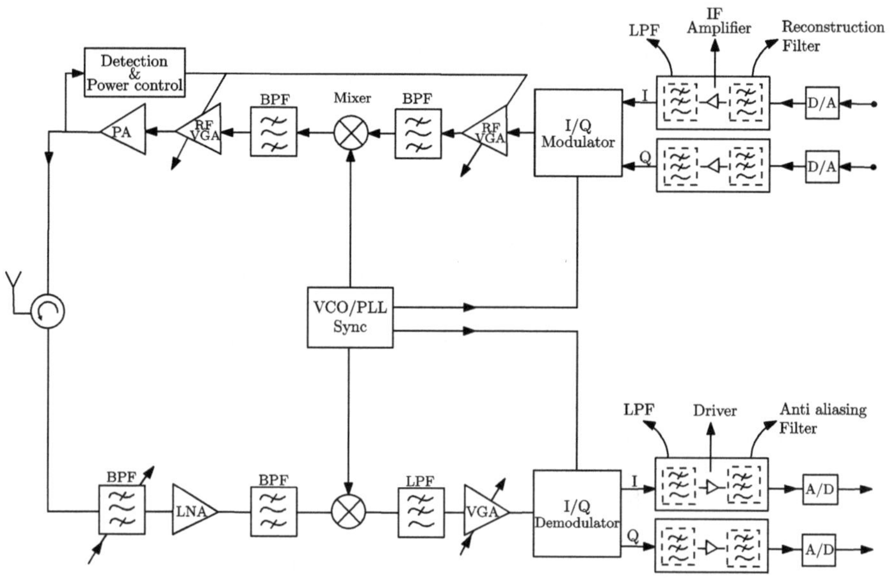

Fig. 12.22 Simplified full-duplex communication transceiver

where we have to mitigate the in-band self-interference in the physical layer of the wireless systems (Smida et al. 2024).

Figure 12.22 shows a full-duplex scheme with a single antenna including several main components. The antenna connects directly with a circulator, which in turn provides a passive self-interference cancellation (S-IC).[1] The circulator tries to reduce the transmitted signal leakage to the income signal, a difficult task given that the former signal has high power. The circulators are usually implemented in passive form, employing magnetic materials with nonlinear propagation features and achieving mild cancellation levels. It is also possible to implement S-IC using hybrid transformers with similar features as those used in plain twisted-pair telephone systems. Depending on the carrier frequency and bandwidth, typically, the passive interference cancellation is in the range of 20–40 dB, according to Everett et al. (2014).

The low-noise amplifier (LNA) increases the RF signal received at the antenna with expected low power. Therefore, the LNA should enhance the desired signal without introducing substantial component noise. Typically, the noise figure (NF) of an LNA ranges from less than 1 dB at the frequencies below 1 GHz and a few dBs for higher frequencies (Patyuchenko 2021a,b).

The several variable gain amplifiers (VGAs) implement automatic gain control to cope with changes in the amplifier gain in the presence of variations in temperature and other disturbances. In addition, some VGAs are controlled in the analog domain,

[1] Not to be confused with the serial interference cancellation, often called SIC in the open literature.

particularly when there is no digital control or its granularity is unacceptable. The power amplifier (PA) at the output of the transmitter must be efficient in the sense of providing the required power with low dissipation of heat. The input-output power curve of the PA tends to saturate according to the linear behavior of the physical amplifiers. One can define the power level at which the linear and nonlinear curves differ by 1 dB as p_1, and this differs from the saturation level at p_2. This nonlinearity gives rise to undesired harmonics at the transmitted signal. The transceiver also includes driver types and intermediate frequency (IF) amplifiers.

Several bandpass filters (BPF) perform frequency selection, a couple with the central frequency at the carrier and two others in the baseband. Additional low-pass filters (LPF) play the roles of anti-aliasing, reconstruction, and plain low-pass filters.

In Fig. 12.22, the voltage-controlled oscillator (VCO), along with the phase-locked loop (PLL) designed to provide stability to the VCO, implements the mixing feature to shift the baseband signal to the carrier frequency range. The in-phase and quadrature (I/Q) modulator/demodulator building blocks integrate the data coming/feeding the digital-to-analog (D/A) and the analog-to-digital (A/D) converters, respectively Patyuchenko (2021b).

For multiple-antenna configurations, Fig. 12.23 illustrates two possible locations where the S-IC can be performed, one in the baseband in the digital domain and the other in-band (at the carrier band) in the analog domain. It should be mentioned that some full-duplex configurations implement the upconversion from the baseband to the central carrier frequency through several stages of intermediate frequency layers, where each layer includes an S-IC. The digital domain, in the right-hand side of Fig. 12.23, includes an S-IC in the baseband followed by all-digital signal processing and digital communications tools required by the full-duplex transceiver.

The question to address is how to implement full-duplex transceivers meeting desired specifications with the currently available technology. The possible answer starts by discussing the effect of all types of interferences corrupting the desired

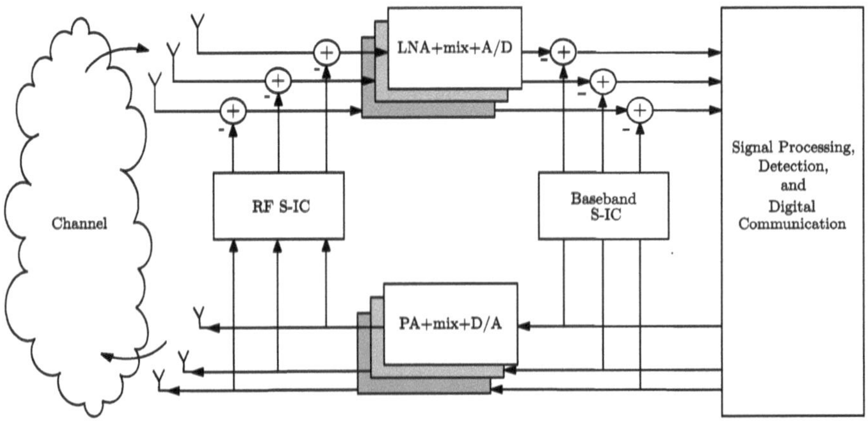

Fig. 12.23 MIMO transceiver including interference cancellation at the baseband and in-band

received signal. Some solutions come from established interference cancellation methods.

In classical multiuser detection receivers, the SIC receiver detects the stronger user with a conventional matched filter detector and then subtracts the signal due to that user from the received waveform. By repeating this process to the resulting signal, it is possible to remove the second strongest user, and so on. Note that the interference of the previous stronger user should be substantially reduced if it was demodulated correctly. The method requires accurate user relative power estimation; otherwise, the cancellation might fail. The parallel interference canceller (PIC) at the receiver attempts to detect all users in parallel by mitigating the interference of all remaining users while detecting the desired user symbols. The knowledge originating from multiuser detection provides insights into the issues discussed here; see Verdu (1988).

12.8.4 Power-Capacity Issues

A crucial discussion in future communications systems is the trade-off between spectral and energy efficiencies. For this particular issue, there is a growing discussion in the literature for attractive solutions, particularly in MIMO systems, entailing the use of low-resolution A/D converters and addressing the nonlinear effects that originated in the power amplifiers; see Stoica et a. (2022).

Usually, the receivers require high-speed A/D converters, and if they have high resolution, the power consumption is also high. At the transmitter, the power amplifiers feed the antennas, producing some heating that ultimately consumes energy that is not really useful.

Assuming real-valued signals at a transmitter and receiver, the channel capacity of a band-limited and power-limited Gaussian channel is given by

$$
\begin{aligned}
C &= \frac{1}{2}\log_2\left(1+\frac{P}{\sigma_{\mathrm{n}}^2}\right)\\
&= \frac{1}{2}\log_2\left(1+\frac{P}{BWI_0}\right)
\end{aligned}
\tag{12.1}
$$

in bits per real dimension, where BW is the channel bandwidth in Hertz. The noise variance relates to the channel bandwidth by $\sigma_{\mathrm{n}}^2 = BWI_0$, where $\frac{I_0}{2}$ is the noise spectral density of the real additive white Gaussian noise (AWGN) channel.[2] The quantity $\frac{P}{2BW}$ represents the power constraint per real symbol; see Tse and Viswanath (2005).

[2] Usually measured at room temperature of $290^\circ K$, where K denotes degrees in Kelvin.

In the case of complex data, the capacity is

$$C = \log_2\left(1 + \frac{P}{\sigma_n^2}\right) = \log_2\left(1 + \frac{P}{BWI_0}\right) \tag{12.2}$$

in bits per complex dimension. In case we transmit BW complex samples per second, the capacity in bits per second is

$$C = BW\log_2\left(1 + \frac{P}{\sigma_n^2}\right) = BW\log_2\left(1 + \frac{P}{BWI_0}\right) \tag{12.3}$$

in bits per second.

The signal-to-noise ratio per complex data is $\mathrm{SNR} = \frac{P}{BWI_0}$, so that the capacity of a standard Gaussian channel can be expressed as

$$C = \log_2(1 + \mathrm{SNR})\mathrm{bits/s/Hz} \tag{12.4}$$

in bits per second per Hertz, or bits per complex dimension for AWGN channels (Tse and Viswanath 2005). This expression is a good measure of spectrum efficiency, whereas a good measure of the power efficiency is

$$P_{\mathrm{eff}} = \frac{C}{\mathrm{SNR}} = \frac{\log_2(1 + \mathrm{SNR})}{\mathrm{SNR}} \tag{12.5}$$

The idea is to maximize the product:

$$CP_{\mathrm{eff}} = \frac{(\log_2(1 + \mathrm{SNR}))^2}{\mathrm{SNR}} \tag{12.6}$$

where the maximum value of the cost function with respect to the SNR leads to a trade-off between the capacity and the power consumption in a single-antenna transmission; see Nossek (2021); Bai and Nossek (2015). The maximum SNR in dBs is around 5.94 corresponding to 3.92 in the linear scale; see Fig. 12.24. This is equivalent to a capacity of $C = 2.3$ bits/s/Hz for complex digital modulations such as PSK and QAM.

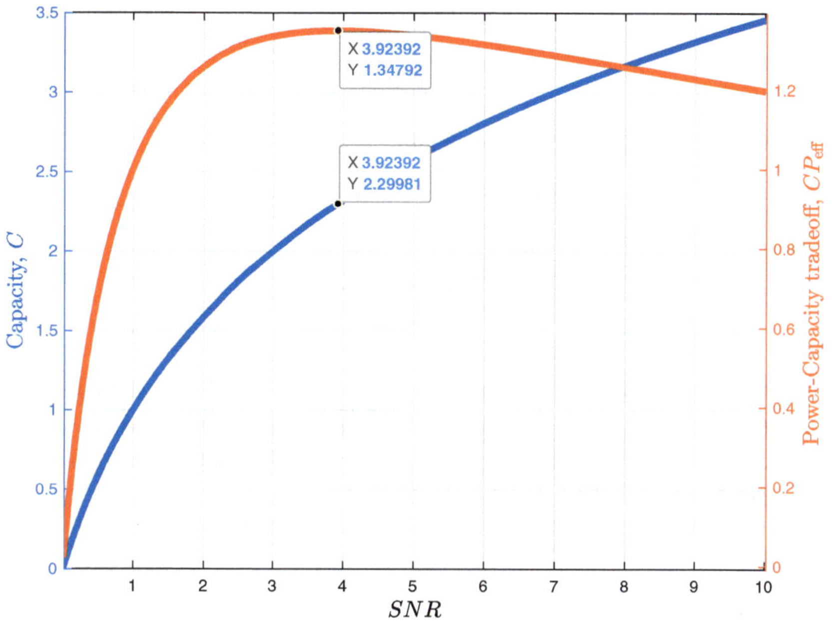

Fig. 12.24 Power-capacity trade-off in single-antenna systems

By taking into consideration the available bandwidth

$$C = BW \log_2\left(1 + \frac{G_c P}{I_0 BW}\right) \text{ bits/s}$$
$$= BW \log_2\left(1 + \frac{G_c E_b C}{I_0 BW}\right) \tag{12.7}$$

where G_c is the channel gain, $E_b = \frac{P}{C}$ is the energy per bit, and C is the data rate. From the expression above, we can deduce that

$$\frac{E_b}{I_0} = (2^{\frac{C}{BW}} - 1)\frac{1}{G_c \frac{C}{BW}}$$
$$= (2^a - 1)\frac{1}{G_c a} \tag{12.8}$$

where $a = \frac{C}{BW}$ is just an auxiliary variable. By calculating the derivative of $\frac{E_b}{I_0}$ with respect to a, we obtain:

$$\frac{\partial E_b/I_0}{\partial a} = -\frac{(2^a - 1)}{G_c a^2} + \frac{2^a \ln 2}{G_c a} \tag{12.9}$$

Then, by equating the result to zero, it follows that

$$2^a - 1 - 2^a a \ln 2 = 0 \tag{12.10}$$

By applying this result in (12.8), we obtain:

$$\begin{aligned}\frac{E_\text{b}}{I_0} &= (2^a - 1)\frac{1}{G_c a} \\ &= \frac{a 2^a \ln 2}{G_c a} = \frac{2^a \ln 2}{G_c}\end{aligned} \tag{12.11}$$

For $BW \rightarrow \infty$, the $a \rightarrow 0$ so that $\left[\frac{E_\text{b}}{I_0}\right]_\text{min} = \frac{\ln 2}{G_c}$. This discussion allows us to infer that the minimum energy per bit is reached at infinite bandwidth. This theoretical limit neglects the A/D converter precision and the electronics power consumption; see Mezghani and Nossek (2011).

Assuming that $G_c = 1$ without loss of generality, for a 1-bit converter, it can be shown that (Mezghani 2015)

$$\begin{aligned}\left[\frac{E_\text{b}}{I_0}\right]_\text{min, for 1-bit} &= \frac{\pi}{2}\left[\frac{E_\text{b}}{I_0}\right]_\text{min} \\ &= \frac{\pi}{2}\ln 2 \\ &= \lim_{BW \rightarrow \infty}\left[\frac{E_\text{b}}{I_0}\right]_\text{min, for 1-bit}\end{aligned} \tag{12.12}$$

This result reflects that the ratio of the mutual information between quantized and unquantized, i.i.d. Gaussian transmissions converge to $\frac{2}{\pi}$ for wideband or low SNR conditions; see (Mezghani 2015) for details. The result indicates a mild capacity loss under the discussed conditions. For a thorough discussion on these issues, refer to Stein et al. (2018); Rao et al. (2021); Van Vleck and Middleton (1966).

It is possible to formulate an optimization to reduce the power consumption in the analog domain by considering the interplay among the components of the transceivers, such as the power amplifier, the low-noise amplifier, the signal decoding, and, in particular, the bit resolution of the A/D converter, as, for instance, discussed in Mezghani and Nossek (2011).

If power consumption is the primary concern, one can consider using multiple antennas at the transmitter and receiver. In agreement with what happens with the one antenna transmissions, for multiple antennas, the mutual information for the case of 1-bit quantized A/D converter reduces by $\frac{2}{\pi}$ in comparison with the unquantized case for wideband transmissions. Therefore, in the low SNR regime, the use of a quantized channel representation leads to a penalty in the power efficiency of less than 2 dB in comparison with the unquantized channel representation. The ratio $\frac{2}{\pi}$ also appears when comparing, asymptotically and in the low SNR case, the ergodic capacity of a 1-bit transmission with the unquantized case using QPSK code; see Mezghani (2015); Nossek (2021).

The discussion in this subsection illustrates a current trend to search for solutions requiring low power. The outcome provides attractive solutions to broaden the use of wireless devices in future generations of IoT systems. However, there is a complementary challenge of exploiting the available spectrum optimally. Therefore, we switch gears to focus on the challenges of designing full-duplex solutions, relying on the previously discussed channel capacity concepts. In the following, the main discussion is how to reach an efficient configuration to implement full-duplex communication systems.

12.8.5 Signal-to-Noise Ratio Issues

We return to the technological demands to allow a full-duplex transceiver design. The critical ingredient to developing a successful design is considering the interferences affecting the desired received signal. The received signal including the typical interferences in full-duplex devices can be modeled as

$$r(t) = d(t) + s(t) + n(t) + f(t) \tag{12.13}$$

where $r(t)$ is the baseband received signal, including the desired signal $d(t)$, the self-interference $s(t)$, the environment noise $n(t)$, and the electronics-related induced and quantization noise $f(t)$. The variances of the signals corrupting the desired signal play a crucial role in the success of the full-duplex system design. Let's denote these variances, as σ_r^2, σ_d^2, σ_s^2, σ_n^2, and σ_f^2, respectively. It should be noted that this is a simplified but useful model for the present discussion.

The dynamic range of the A/D converter is $6.02(b_{\text{eff}} - 2)$ dB, where b_{eff} is the effective number of bits of the A/D converter (Gu 2005). The A/D converter using a reasonable number of bits follows a design objective of generating a quantization error of 6.02 dB[3] corresponding to one bit below the noise power for a signal with peak-to-average-ratio $PAPR$ of 5.0 dB, as demonstrated in Sabharwal et al. (2014). The $PAPR$ takes part in the derivation, and the given value of 5.0 dB falls within the range measured in wireless communication commercial systems.

The A/D converter generates a quantization noise of

$$\sigma_q^2 = \kappa_x \sigma_x^2 10^{-\frac{6.02(b_{\text{eff}}-2)}{10}} \tag{12.14}$$

where κ_x represents a gain related to the $PAPR$ of the input signal x to the A/D converter, such that $\sqrt{\kappa_x}\sigma_x$ quantifies the maximum value of the voltage of x. According to Sabharwal et al. (2014), κ_x controls the saturation level (clipping)

[3] This relates to $20 \log_{10} \frac{1}{2} = -6.02$ dB, since it measures the value ratio.

of the A/D converters. It is worth mentioning that $\sigma_f^2 = \sigma_q^2 + \sigma_e^2$, where σ_e^2 represents solely the electronics generated noise that is assumed uncorrelated to the quantization noise.

Following the approach of Sabharwal et al. (2014), if the self-interference is totally removed from the (12.13), we have:

$$y(t) = r(t) - s(t) = d(t) + n(t) + f(t) \tag{12.15}$$

Assuming the quantization is such that

$$\sigma_q^2 = \frac{\sigma_n^2}{2^{2b_{noise}}} \tag{12.16}$$

where b_{noise} determines how much lower the quantization noise should be compared with the noise from the environment. For $b_{noise} = 1$, the quantization noise variance is 6.02 dB below σ_n^2. In the following discussion, if there is an electronic noise figure, we will use:

$$\sigma_q^2 = \frac{\sigma_n^2 + \sigma_e^2}{2^{2b_{noise}}} \tag{12.17}$$

by considering that the electronics-generated noise adds to the thermal noise.

The overall signal-to-interference-noise ratio (SINR) becomes:

$$\begin{aligned}\text{SINR} &= \frac{\sigma_d^2}{\sigma_n^2 + \sigma_q^2 + \sigma_e^2 + \sigma_s^2} \\ &= \frac{\sigma_d^2}{(2^{2b_{noise}} + 1)\sigma_q^2 + \sigma_s^2} \\ &= \frac{\sigma_d^2}{(2^{2b_{noise}} + 1)\kappa_x \sigma_x^2 10^{-\frac{6.02(b_{\text{eff}}-2)}{10}} + \sigma_s^2}\end{aligned} \tag{12.18}$$

where the variance of the input signal to the A/D converter is

$$\begin{aligned}\sigma_x^2 &= \sigma_d^2 + \sigma_q^2 + \sigma_e^2 + \sigma_s^2 + \sigma_n^2 \\ &= \sigma_d^2 + (2^{2b_{noise}} + 1)\sigma_q^2 + \sigma_s^2\end{aligned} \tag{12.19}$$

with $d(t)$, $q(t)$, $e(t)$, $s(t)$, and $n(t)$ assumed uncorrelated. Note that we did not consider the self-interference canceled in this expression.

Figure 12.25 describes how these noise variances interplay to clarify how to address a design objective.

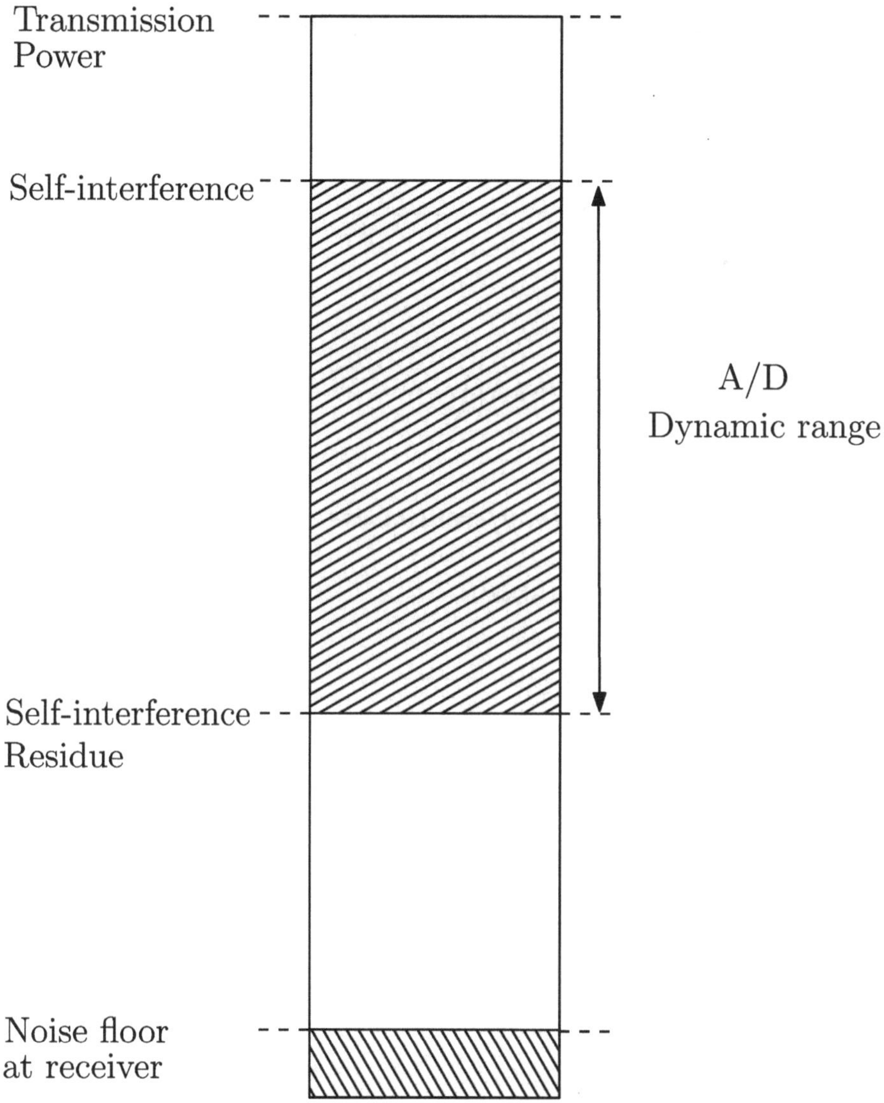

Fig. 12.25 Self-interference evaluation. Given the transmission power in dBm, it causes self-interference in a certain level. Self-interference varies according to the transmitter class, such as those in wireless communication cells. The cancellation of self-interference in the digital domain is limited by the A/D dynamic range $6.02(b_{\text{eff}} - 2)$ in dB, leaving a gap between the self-interference residue and the noise floor that would be witnessed in the standard half-duplex implementation. A way to remove the gap is to reduce the self-interference prior to the A/D conversion in the in-band domain. Typically, the self-interference can be 100 dB higher than the desired signal

12.8.6 An Upper Bound for Strong Self-Interference

For $\sigma_s^2 \gg \sigma_d^2 + (2^{2b_{noise}} + 1)\sigma_q^2 + \sigma_e^2$, the case of strong self-interference, a tight upper bound to the SINR is

$$\text{SINR} = \frac{\sigma_d^2}{\kappa_x \sigma_s^2 10^{-\frac{6.02(b_{\text{eff}}-2)}{10}} + \sigma_s^2} \tag{12.20}$$

In dB, we get:

$$\text{SINR}_{\text{dB}} = \left.\frac{\sigma_d^2}{\sigma_s^2}\right|_{\text{dB}} - 10\log_{10}\left[\left(\kappa_x 10^{-\frac{6.02(b_{\text{eff}}-2)}{10}} + 1\right)\right]_{\text{dB}} \tag{12.21}$$

12.8.7 The Small Self-Interference Case

For small self-interference, the SINR is

$$\begin{aligned} \text{SINR} &= \frac{\sigma_d^2}{\sigma_n^2 + \sigma_q^2 + \sigma_e^2} \\ &= \frac{\sigma_d^2}{(2^{2b_{noise}} + 1)\sigma_q^2} \\ &= \frac{\sigma_d^2}{(2^{2b_{noise}} + 1)\kappa_x \sigma_x^2 10^{-\frac{6.02(b_{\text{eff}}-2)}{10}}} \end{aligned} \tag{12.22}$$

where for this case

$$\begin{aligned} \sigma_x^2 &= \sigma_d^2 + \sigma_q^2 + \sigma_e^2 + \sigma_n^2 \\ &= \sigma_d^2 + (2^{2b_{noise}} + 1)\sigma_q^2 \end{aligned} \tag{12.23}$$

The result in dB is given by

$$\text{SINR}_{\text{dB}} = \left.\frac{\sigma_d^2}{\sigma_x^2}\right|_{\text{dB}} - 10\log_{10}\left[(2^{2b_{noise}} + 1)\kappa_x 10^{-\frac{6.02(b_{\text{eff}}-2)}{10}}\right]_{\text{dB}} \tag{12.24}$$

12.8.8 Other Issues

The power in the A/D converter can be modeled as

$$P_{\mathrm{A/D}} = 2c_{\mathrm{imp}} I_0 2^{2b} BW \tag{12.25}$$

where c_{imp} is a constant that depends on the technology implementation, I_0 is the noise spectral density, b is the number of bits of the A/D converter and BW is the bandwidth (Lee and Sodini 2008). This number of bits should consider b_{noise} whenever it is chosen according to the noise floor level.

A simplified model is

$$P_{\mathrm{A/D}} = 2c_{\mathrm{imp2}} 2^{2b_{\mathrm{eff}}} BW \tag{12.26}$$

where c_{imp2} is also a constant that depends on the technology implementation, b_{eff} is again the effective number of bits of the A/D converter, and BW is the bandwidth, Lee and Sodini (2008).

According to Korbi et al. (2014), the power of the quantization noise is

$$P_{\mathrm{q}} = P_{\mathrm{in}} - (6.02b - 4.76 - PAPR) \tag{12.27}$$

where $PAPR$ is the peak-to-average power in the A/D converter, b is the number of bits, and P_{q} is the power of the quantization noise power quantifying the resolution in bits with respect to the received power. In this context, P_{in} represented the maximum average power of the A/D converter input before facing saturation, i.e., clipping.

The work of Mezghani and Nossek (2011) argues that sometimes the A/D converter resolution is determined such that its quantization noise lies below 10 dB of the remaining noise sources. However, the authors point out that this solution has power consumption issues since it would require a high $PAPR$.

12.8.9 Typical Solution

By examining some current state-of-the-art solutions, we can provide a toy example to access the interplay of the technological requirements to implement a full-duplex communication system. There are some practical contributions where the authors offer numerical and experimental results related to the application of full-duplex solutions in diverse environments (Korbi et al. 2014,a; Barneto et al. 2019; Ahmed et al. 2013).

Let us assume an environment that faces the following situation:

- Transmit power: 125 mW
- Path gain: 10^{-2}

Table 12.1 Typical parameters

N_f	σ_n^2	σ_q^2	σ_s^2	b	b_{eff}	κ_x
50	$-90.97dBm$	$\frac{\sigma_n^2+\sigma_e^2}{2^{2b_{noise}}}$	1	12	10	1

- Circulator interference cancellation: 20 dB
- Temperature: 290 degrees Kelvin
- $PAPR$: 5 dB
- $BW = 20$ MHz
- $N_f = 50$ dB and variable
- $b_{noise} = 2$
- $\kappa = 1.3810^{-23}$ is the Boltzmann constant in Joules/Kelvin; see some insights related to capacity addressed in Sect. 15.6.4

With the fixed parameters given in Table 12.1, it is possible to compute the SINR using (12.18) using the following parameters:

- $\sigma_n^2 = \kappa TBW = 8.004010^{-13}$ is the noise variance
- $\sigma_e^2 = 8.003810^{-08}$ represents the electronics noise
- $\sigma_q^2 = \frac{\sigma_e^2+\sigma_n^2}{2^{2b_{noise}}} = \frac{8.003810^{-08}+8.004010^{-13}}{8} = 1.000510^{-8}$
- $\sigma_d^2 = 10^{-4} \times 125\ 10^{-3}$ represents the power gain squared times the transmit power
- $\sigma_s^2 = 1$ and the attenuation by the circulator is $10^{-\frac{10}{20}} = 0.312$. In the (12.18), we should use $\sigma_s^2 = 0.312$

The SINR with these values is SINR $= 4.005110^{-05} = -13.9738$dBm or $-$ 43.9738 dB. In case the self-interference is eliminated, the result would be SINR $= 0.1248 = 20.9635$dBm or $-$ 9.0365 dB.

The σ_s^2 value reduces if successful RF cancellers and digital domain cancellers are designed and applied. The work (Barneto et al. 2019) describes achievements of 50 dB by the RF canceller and reaching levels superior to 70 dB by using both RF and digital cancellers. Furthermore, adaptive filtering (Diniz et al. 2010) allows higher cancellation levels, also considering the power amplifiers' nonlinear features.

To wrap up this discussion, we show some plots related to the self-interference to the desired-signal ratio (SI-DS) aiming at providing some insights into the design requirements inherent to full-duplex systems.

By employing (12.18) for an SI-DS ratio close to one, Fig. 12.26 shows the SINR as a function of the effective number of bits and the standard deviation of the desired signal. As can be observed, the SINR is relatively flat for a wide range of values of the effective number of bits, where in this particular plot $\sigma_s = 1.01\sigma_d$.

Figure 12.27 shows a similar plot for a high SI-DS ratio, according to (12.22). As can be observed, the self-interference plays an important role in determining the SINR level. We can prescribe the SI cancellers' specifications once we know the required sensitivity to detect the received signal. In this example, we chose $\sigma_s = 10.1\sigma_d$. Furthermore, it is possible to verify that the SINR is low and relatively flat for a wide range of values of the effective number of bits.

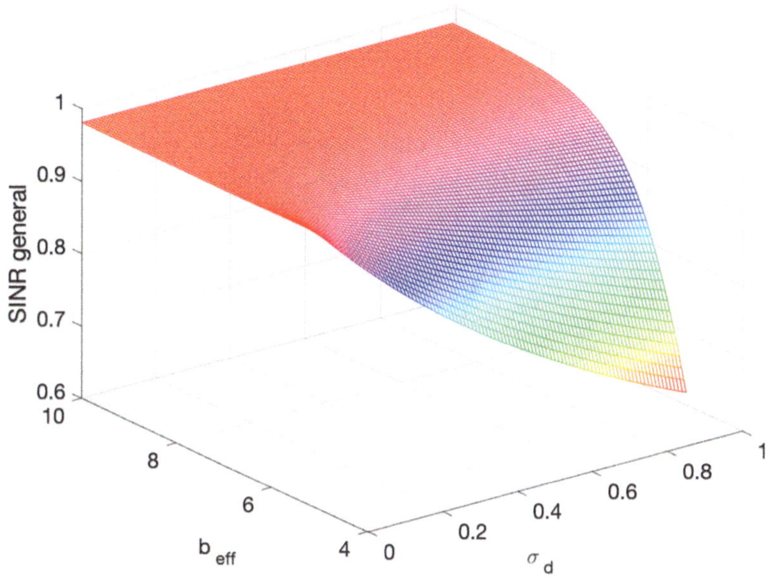

Fig. 12.26 SINR trade-off for general SI-DS

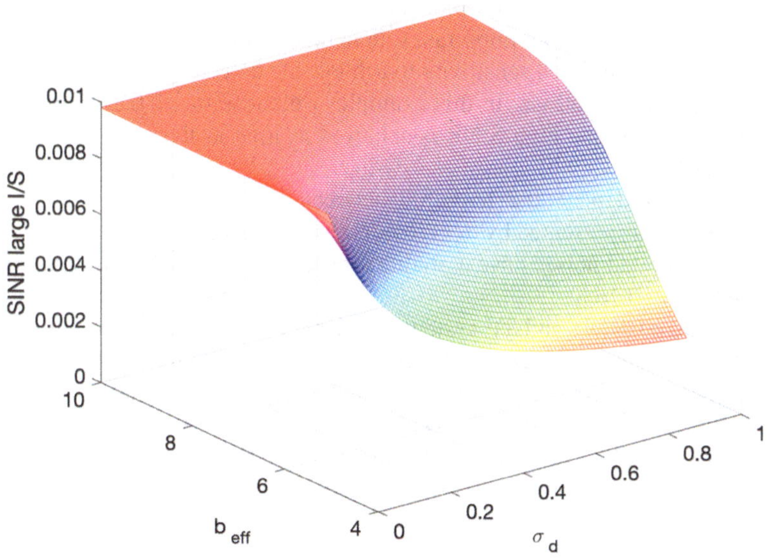

Fig. 12.27 SINR trade-off for high SI-DS

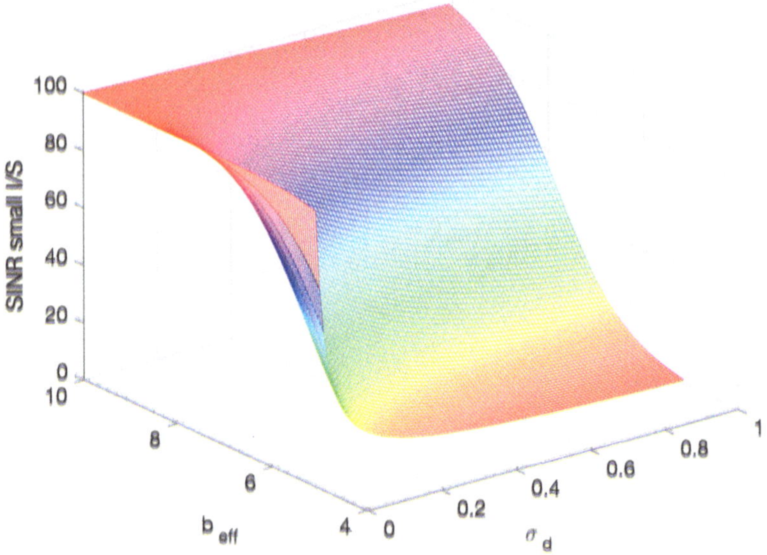

Fig. 12.28 SINR trade-off for low SI-DS

Figure 12.28 shows a similar plot for a low SI-DS ratio as per (12.20). As can be observed, the SI plays an important role in determining the SINR level. As such, once we know the required sensitivity to achieve the detection, we can prescribe the SI cancellers' specifications. In this example, we chose $\sigma_s = 0.1\sigma_d$. Indeed, it is possible to observe that high SINR is achieved as long as there are many effective bits.

Figure 12.29 depicts the trade-off between the noise figure related to the electronics noise represented by σ_q. Provided we have a number of effective bits, the internally generated noises have a predictable effect on the SINR. In this example, we chose $\sigma_s = 10.1\sigma_d$.

As an illustration, in an LTE system requiring a sensitivity level of 18.60 dB for a 64-QAM modulation scheme and a code rate of $4/5$ (see Korbi et al. (2014) and the references therein) with $\sigma_d = 0.401$ and $b_{\text{eff}} = 10$, according to Fig. 12.28, the SINR is 9.78×10^{-3} or -20.09 dB. That means the SI interference should be reduced by $-20.09 - 18.60 = -38.69$ dBs to reach the sensitivity level. However, if an additional 5dBs above the sensitivity level is required, the SI reduction should be -43.69 dBs. For the case of Fig. 12.26, the SINR is -0.088 dBs requiring an extra SI cancellation of $-0.088 - 18.60 - 5.0 = -23.688$ dBs. The sensitivity is higher for higher modulations, and for illustration, we chose a relatively high modulation. It is worth mentioning that a higher SINR reduces the required sensitivity.

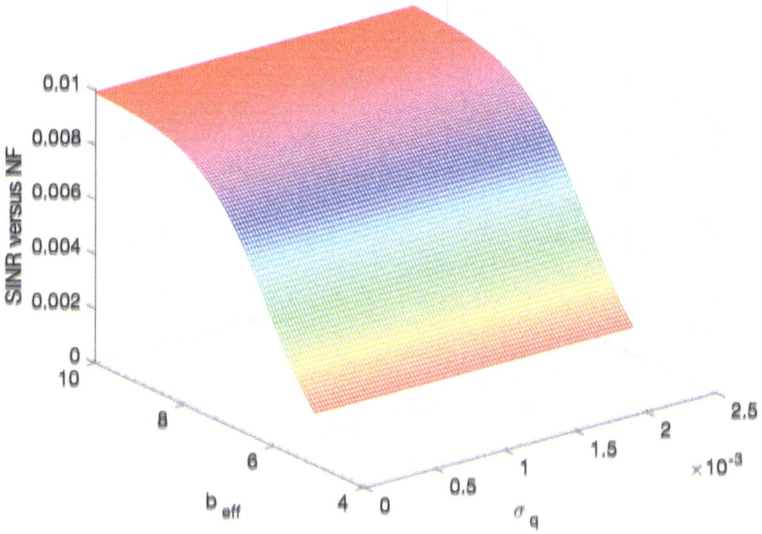

Fig. 12.29 SINR trade-off for variable noise figure (NF)

12.8.10 Constraints and Challenges

The major constraints inherent to the widespread use of full-duplex systems relate to implementation costs, which vary over time, power consumption, and technological limitations. It is certain that well-informed decisions play the leading role in balancing when to exploit the benefit of operating in full-duplex, in practical deployments.

The challenge lies in optimal solutions originating from industry and academy, in a joint effort to design faster devices along with improving their noise figure performances. Consult, for instance, the work of Theuwissen and Meynants (2021) related to the design of A/D converters for CMOS image sensors, where the authors discuss some issues related to design requirements for the particular application.

As an illustration related to a current commercial product, a high-speed A/D converter with 800 mega samples per second (MSPS) of 12 bits with a single input channel has a power consumption of around 830 mW, and analog input full power bandwidth of 6 GHz,[4] as per the time of this writing; see ADC12SJ800-Q1 (2022). This A/D converter is suitable to automotive applications.

The discussion in this section enlightens the challenges inherent to efficient adoption of a full-duplex solution in power-limited communication and sensing nodes.

---

[4] The quantity is defined as the input frequency where the reconstructed output of the A/D converter drops 3 dB below the power of a full-scale input signal, at low frequencies at the input.

12.9 Wireless Networks

The availability of wireless services everywhere, with the best quality of service, requires organized planning and standards to be followed. The standards play a pivotal role in setting the requirements to be met by service and equipment providers. In addition, all deployments depend on the available technology to implement the standards. Figure 12.30 provides an illustration of how some systems integrate the overall network.

12.9.1 Wireless Standards

A set of complementary wireless standards are available in order to ensure their interoperability. Their main classes are:

- Personal Area Networks (PANs)
- Local Area Networks (LAN)
- Metropolitan Area Networks (MAN)
- Wide Area Networks (WAN)

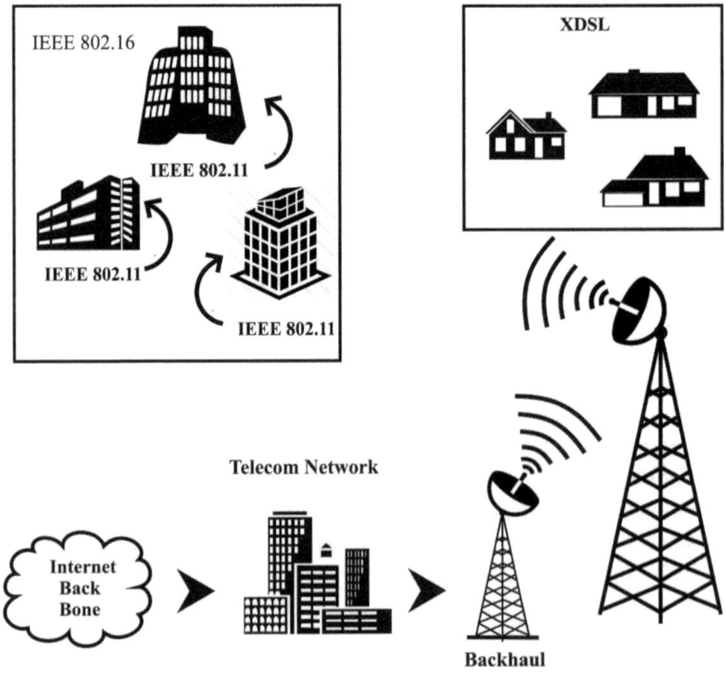

Fig. 12.30 Wireless network integration

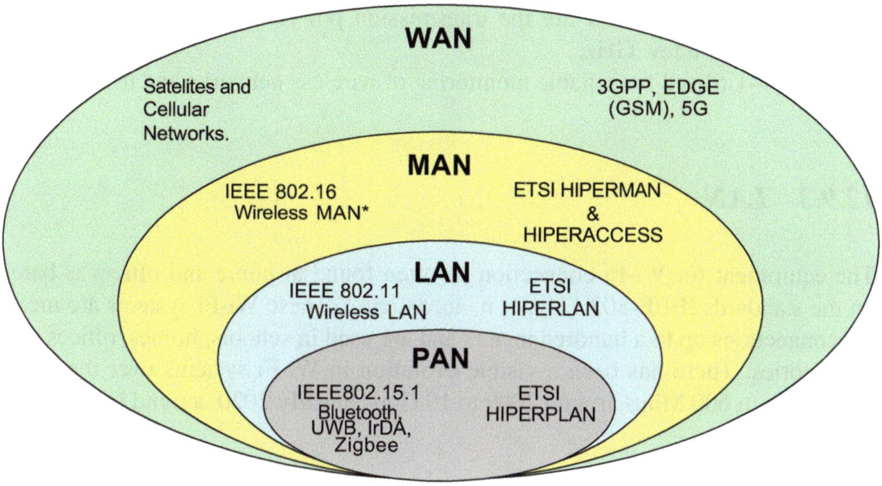

Fig. 12.31 Wireless network systems

Many of the wireless network standards use OFDM in the air interface.

The Institute of Electrical and Electronics Engineering (IEEE) and the International Telecommunications Union (ITU) developed the main standards employed in wireless communications. The International Mobile Communications (IMT) describes the requirements and specifications for high-speed mobile broadband communications. The IMTs are issued by the ITU radiocommunication sector (ITU-R). The IMT-2000, IMT-Advanced, IMT-2020, and IMT-2030 determine the framework and aim for the development of the *nicknamed* 3G, 4G, 5G, and the forthcoming 6G wireless network systems, respectively. The IEEE 802.11 and 802.16 families of standards aimed at LAN and MAN, respectively. The ETSI family is the European counterpart of the equivalent class of standards.

Figure 12.31 illustrates the assumed coverage areas and the specific standards to serve them. These systems might take part in autonomous systems networks.

The question that remains is: what are the primary purposes of these several wireless systems? Here are some short answers.

12.9.2 PAN

Several standards exist for PAN; some of them are described below.

- Bluetooth—Used in a wireless mouse, keyboards, headphones, etc.,
- IrDA—Used for point-to-point communication between devices for data transfer and synchronization,
- UWB—Used for wideband connection for multimedia.

Some features of UWB are the transmission power in Watt per MHz over a bandwidth of a few GHz.

- Zigbee—Utilized for reliable monitoring of wireless networks and their control.

12.9.3 LAN

The equipment for Wi-Fi connection so often found at home and offices is based on the standards IEEE-802.11(g, b, n, ac, ax, etc.). These Wi-Fi systems are meant for connections up to a hundred meters and are used in schools, homes, offices, and laboratories. There has been a visible evolution in Wi-Fi systems over the years, ranging from 600 Mbps around 2009 to 10 Gbps in early 2020, a trend for the years to come.

12.9.4 MAN

The WiMAX (Worldwide Interoperability for Microwave Access) standard initially aimed to provide last-mile wireless broadband access to compete with wired XDSL (including several types of digital subscriber line systems) or wired broadband. The goal was to provide up to 40 Mbps for low mobility access and up to 15 Mbps for mobile access. The original idea of WiMAX was to provide high-speed wideband connections from point to multipoint to feed urban networks at distances up to 50 km. WiMAX is based on the standard IEEE-802.16.

WiMAX employs an adaptive modulation scheme as shown in Fig. 12.32 as well as a multiple-access approach based on OFDM (OFDMA).

The wireless MAN standards (IEEE 802.16 and HiperMAN) were guided by an association called WiMAX forum, aiming for similar success of WLAN (IEEE 802.11) and DSL sparked by the Wi-Fi and DSL forums. The aim was to provide fixed wireless broadband service in an area of 50 kilometers (about 30 miles) of radius. The data transfer could reach up to 70 megabits per second, equivalent to 45 T1 (1.544 megabits per second) services.

The main targets included:

- Broadband on Demand: MAN can accelerate the deployment of LAN hotspots, especially at locations not served by DSL or cable, and the phone companies cannot provide broadband service at short notice.
- Cellular Operator Backhaul: These operators usually use wired and leased connections from third-party service providers. These could be replaced by MAN.
- Residential Broadband: In many areas, it is challenging to provide wired broadband service.
- Provide services in rural and scarcely populated areas.

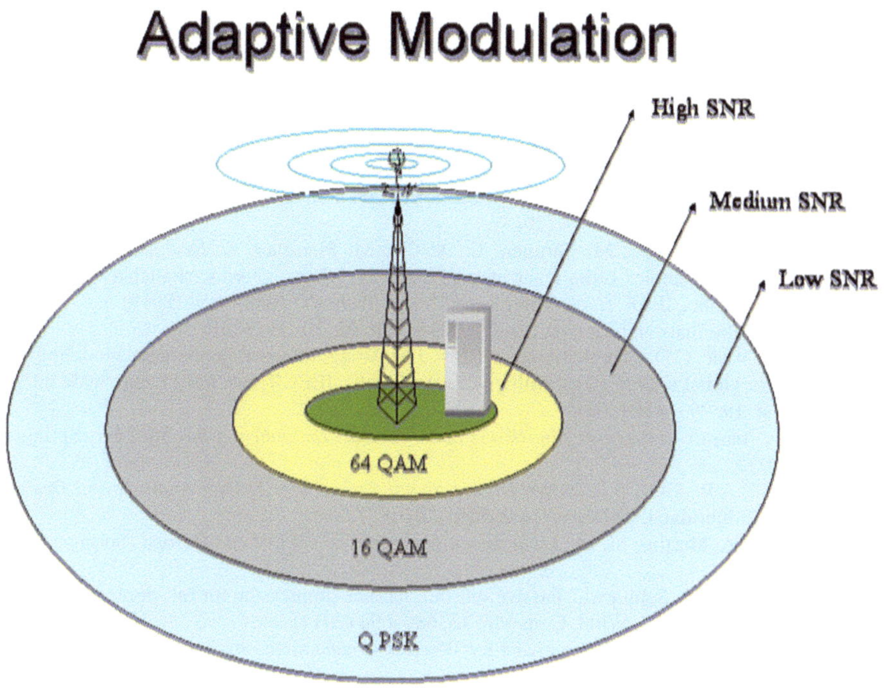

Fig. 12.32 Adaptive modulation in WiMAX

- Wireless Everywhere: As the number of hotspots increases, the demand for wireless service in areas not covered by LAN also grows.

12.9.5 WAN

The WAN services are meant to be provided by the cellular and satellite networks. WAN is the solution for distances over 50 km between the unit elements and the networks. Internet providers usually control this type of service.

12.10 Remarks

This chapter describes some ways to allow several agents to share the radio frequency spectrum while considering interference issues. In addition, it briefly describes a few licensed wireless networks that will play a key role in autonomous vehicle solutions.

References

ADC12SJ800-Q1, Automotive single-channel, 12-bit, 800-msps adc with jesd204c interface, Texas Instruments, Tech. Rep. (2022). https://www.ti.com/product/ADC12SJ800-Q1

E. Ahmed, A.M. Eltawil, A. Sabharwal, Rate gain region and design tradeoffs for full-duplex wireless communications. IEEE Trans. Wirel. Commun. **12**, 3556–3565 (2013)

Q. Bai, J.A. Nossek, Energy efficiency maximization for 5G multi-antenna receivers. Trans. Emerg. Telecommun. Technol. **26**, 3–14 (2015)

C.B. Barneto, T. Riihonen, M. Turunen, L. Anttila, M. Fleischer, K. Stadius, J. Ryynänen, M. Valkama, Full-duplex ofdm radar with LTE and 5G NR waveforms: challenges, solutions, and measurements. IEEE Trans. Microwave Theory Tech. **67**, 4042–4054 (2019)

J. Brand, The radio that can hear over itself. IEEE Spectr. **32**, 30–35 (2021)

S.Z. Denic, B. Vasic, C.D. Charalambous, J. Chen, J.M. Wang, Information theoretic modeling and analysis for global interconnects with process variations. IEEE Trans. Very Large Scale Integr. VLSI Syst. **19**, 397–410 (2011)

P.S.R. Diniz, *Adaptive Filtering: Algorithms and Practical Implementation*, 5th edn. (Springer, Berlin, 2020)

P.S.R. Diniz, E.A.B. Silva, S.L. Netto, *Digital Signal Processing: System Analysis and Design*, 2nd edn. (Cambridge University, Cambridge, 2010)

P.S.R. Diniz, W.A. Martins, M.V.S. Lima, *Block Transceivers: OFDM and Beyond* (Springer, New York, 2012)

E. Everett, A. Sahai, A. Sabharwal, Passive self-interference suppression for full-duplex infrastructure nodes. IEEE Trans. Wirel. Commun. **13**, 680–694 (2014)

Q. Gu, *RF System Design of Transceivers for Wireless Communications*, 1st edn. (Springer, New York, 2005)

R. Hagde, N.R. Shanbhag, Toward achieving energy efficiency in the presence of deep submicron noise. IEEE Trans. Very Large Scale Integr. VLSI Syst. **8**, 389–391 (2000)

D. Korbi, T. Riihonen, V. Syrjälä, L. Anttila, M. Valkama, R. Wichman, Full-duplex transceiver system calculations: analysis of ADC and linearity challenges. IEEE Trans. Wirel. Commun. **13**, 3821–3836 (2014)

D. Korbi, L. Anttila, V. Syrjälä, L. Anttila, M. Valkama, Widely linear digital self-interference cancellation in direct-conversion full-duplex transceiver. IEEE J. Sel. Areas Commun. **32**, 1674–1687 (2014)

H.-S. Lee, C.C. Sodini, Analog-to-digital converters: digitizing the analog world. Proc. IEEE **96**, 323–334 (2008)

A. Mezghani, *Information-Theoretic Analysis and Signal Processing Techniques for Quantized MIMO Communications*, Ph.D. dissertation, Technical University of Munich, TUM (2015)

A. Mezghani, J. Nossek, Power efficiency in communication systems from a circuit perspective, in *ISCAS 2011—IEEE International Symposium on Circuits and Systems* (2011), pp. 1896–1899

A.F. Molisch, *Wireless Communications*, 1st edn. (Wiley, Hoboken, 2005)

C. Motz, T. Paireder, M. Huemer, Low-complex digital cancellation of transmitter harmonics in LTE-a/5g transceivers. IEEE Open J. Commun. Soc. **2**, 949–963 (2021)

J. Nossek, How to achieve energy efficiency in multi-antenna systems? in *XXXIX Brazilian Telecommunications and Signal Processing Symposium* (2021), p. Plenary talk

A. Patyuchenko, RF signal chain discourse: properties and performance metrics, in *Analog Devices*, Tech. Rep. (2021). https://www.analog.com/en/analog-dialogue/articles/rf-signal-chain-discourse.html

A. Patyuchenko, RF signal chain discourse-part 2: essential building blocks, in *Analog Devices*, Tech. Rep. (2021), analog Dialogue. https://www.analog.com/en/analog-dialogue/articles/rf-signal-chain-discourse-part-2-essential-building-blocks.html

O. Ploder, C. Motz, T. Paireder, C. Auer, H. Pretl, M. Huemer, Digitally intensive mixed-signal approach for self-interference cancellation in LTE-a/5g-transceivers, in *IEEE International Symposium on Circuits and Systems ISCAS-2021* (2021)

S. Rao, G. Seco-Granados, H. Pirzadeh, J.A. Nossek, A. Lee Swindlehurst, Massive MIMO channel estimation with low-resolution spatial sigma-delta ADCS. IEEE Access **9**, 109320–109334 (2021)

T. Riihonen, S. Werner, R. Wichman, Mitigation of loopback self-interference in full-duplex mimo relays. IEEE Trans. Signal Process. **59**, 5983–5993 (2011)

A. Sabharwal, P. Schniter, D. Guo, D.W. Bliss, S. Rangarajan, R. Wichman, In-band full-duplex wireless: challenges and opportunities. IEEE J. Sel. Areas Commun. **32**, 1637–1652 (2014)

C. Shannon, A mathematical theory of communications. Bell Syst. Tech. J. **27**, 379–423 (1948)

B. Smida, R. Wichman, K.E. Kolodziej, H.A. Suraweera, T. Riihonen, A. Sabharwal, In-band full-duplex: the physical layer. Proc. IEEE **112**, 433–462 (2024)

S.R. Sridhara, N.R. Shanbhag, Coding for system-on-chip networks: a unified framework. IEEE Trans. Very Large Scale Integr. VLSI Syst. **13**, 655–667 (2005)

M.S. Stein, S. Bar, J.A. Nossek, J. Tabrikian, Performance analysis for channel estimation with 1-bit ADC and unknown quantization threshold. IEEE Trans. Signal Process. **66**, 2557–2571 (2018)

P. Stoica, X. Shang, Y. Cheng, The cramér-rao bound for signal parameter estimation from quantized data. IEEE Signal Process. Mag. **39**, 118–125 (2022)

A. Theuwissen, G. Meynants, Welcome to the world of single-slope column-level analog-to-digital converters for cmos image sensors. Found. Trends® Integr. Circuits Syst. **1**(1), 1–71 (2021)

D. Tse, P. Viswanath, *Fundamentals of Wireless Communication*, 2nd edn. (Cambridge University, Cambridge, 2005)

A.H.M. van Roermund, Shifting the frontiers of analog and mixed-signal electronics. Adv. Electro. **2014**, 1–16 (2014)

J.H. Van Vleck, D. Middleton, The spectrum of clipped noise. Proc. IEEE **54**, 2–19 (1966)

S. Verdu, *Multiuser Detection*, 1st edn. (Cambridge University, Cambridge, 1988)

Y. Wang, R. Wu, J. Pang, D. You, A.V. Fadila, R. Saengchan, X. Fu, D. Matsumoto, T. Nakamura, R. Kubozoe, M. Kawabuchi, B. Liu, H. Zhang, J. Qiu, H. Liu, N. Oshima, K. Motoi, S. Hori, K. Kunihiro, T. Kaneko, A. Shirane, K. Okada, A 39-Ghz 64-element phased-array transceiver with built-in phase and amplitude calibrations for large-array 5g nr in 65-nm CMOS. IEEE J. Solid State Circuits **55**, 1249–1269 (2020)

Chapter 13
Cellular Systems

13.1 Introduction

The chapter describes a few basic concepts of some generations of public wireless networks utilizing licensed frequency bands. No detailed description of the most current system is provided.

In autonomous vehicle applications, terrestrial and non-terrestrial wireless networks play a key role. Combined with radar, lidar, and navigation satellite technologies, the many wireless communication services are essential for precise and reliable positioning.

13.2 Cellular Evolution

Here, we briefly discuss the evolution of cellular systems, highlighting the features of each generation.

13.2.1 First Generation: Early 1980s

The earlier licensed cellular systems utilized an analog FM technology. The main systems that divided the licensed frequency band into subbands were the Nordic system employing a 25 kHz of bandwidth and the advanced mobile system (AMPS) using 30 kHz of bandwidth.

The AMPS building block is shown in Fig. 13.1. The AMPS high-pass filter has six dBs/dec of attenuation from 300 to 3000 Hz, whereas the limiter is responsible for removing frequencies above 10 or 12 kHz.

Y. C. Lim et al., *Signal Processing for Autonomous Systems*,
https://doi.org/10.1007/978-3-032-00675-2_13

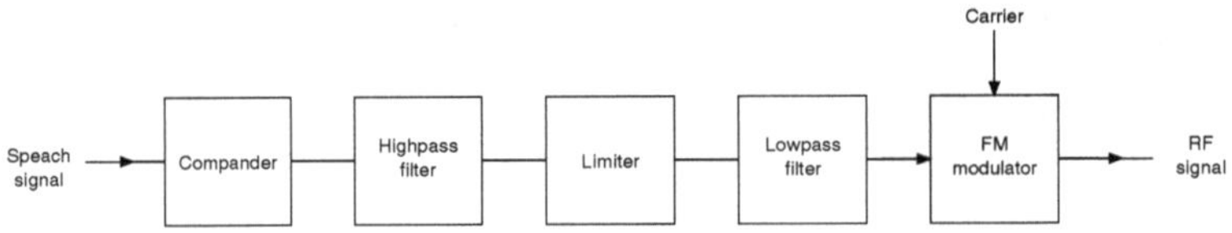

Fig. 13.1 Analog AMPS

13.2.2 Second Generation: Early 1990s

The second generation (2G) brought about many improvements, a consequence of developments in channel coding with interleaving (to reduce Rayleigh fading), low-rate voice coders, and efficient modulation using constant envelope.

The main 2G systems are briefly described to help us understand the forthcoming developments in this area.

13.2.2.1 European System

The European Conference of Post and Telecommunications allocated 900 MHz for the second-generation cellular system in Europe. TDMA was the choice because it allows sharing of the transceiver at the base station; has better spectrum efficiency; and combats slow Rayleigh fading through interleaving and frequency hopping (FH—which also reduces intercell interference). This system is digital and allows data transmission and security in voice services. In addition, the soft handover (or handoff) was possible, leading to a reduction in the network signaling tasks, allowing very small cells. The early versions failed due to multipath propagation that generated delay spread, making reception virtually impossible. By employing an adaptive equalizer, this problem could be solved and the TDMA system could be implemented and deployed.

In 1987, Europe chose a narrowband TDMA using 200 kHz for channel spacing. The 200-kHz band was meant to fit 8 voice channels at a full rate of 16 Kbps (kilobit per second) and 16 half-rate voice calls. The modulation was GMSK, with convolutional channel coding and the frequency bands were 900, 1800, and 1900 MHz. This system called GSM (global system for mobility) was the most widely used in the world, especially in Europe and Asia. Each 32ms of data is coded using Reed-Solomon code (12,8) is divided into 192-Kbps slots assembled via TDMA interleaver, and each slot is then frequency hopped for transmission (Mouly and Pautet 1992; Halonen et al. 2003).

13.2.2.2 US and Japan Systems

A similar US system, called D-AMPS, was used but with 30-kHz bandwidth to follow the previous AMPS system, where each 30 kHz provided three voice channels, each with 8 Kbps. The modulation was QPSK, with convolutional channel coding, and the frequency bands were 800 and 1900 MHz (Harte et al. 1998). Later, the IS-95 narrowband CDMA system was deployed in the USA, which was the system provided in Brazil, for instance (Yang 2004; Lee and Miller 1998).

In Japan, a system using 25 kHz of bandwidth was used. The modulation was QPSK, with convolutional channel coding, and the frequency bands were 700 and 1500 MHz.

13.2.3 Third Generation: Early 2000s

It aims provision of bit rates ranging from 0.384 to 2 Mbps, including image, audio, voice, data, and Internet services. These are typically high-bit-rate applications with high bursts of data in short periods. Thus, the natural choice is packet transmission.

13.2.3.1 European System

The International Telecommunication Union (ITU—project—IMT2000) and the European Standardization Union (ETSI, project UMTS, Universal Mobile Telecommunication System) merged the standards discussions. For the UMTS/IMT2000 system, a part of the 2 GHz band (carrier) had been allocated.

GSM was enhanced to provide a high data rate. This system aimed to provide high-bit-rate services in a wide coverage area, which will go up to 115 Kbps through packed radio service using high-speed circuit-switched data and packet-switched data. Using an improved modulation in the standard TDMA system, the bit rate can be increased up to 384Kbps (mainly in highly populated areas). The enhanced GSM coexisted with the new UMTS standard, which provided even higher data service but in limited areas and low mobility users (up to 2Mbps).

The typical architecture of the GSM/GPRS system is shown in Fig. 13.2. As observed, the upper part of the figure shows the original GSM architecture, which was later expanded to provide data services; see the lower part of the same figure.

The enhanced GSM coexisted with the new UMTS standard, which provided even higher data service but in limited areas and low mobility users (up to 2 Mbps) (Holma and Toskala 2007, 2006). GSM handsets were meant to get lower bit rates, but the coverage was wider, whereas UMTS handsets were designed to provide higher bit rates but lower coverage. Dual-mode handsets would have both.

Europe and Asia allocated a part of the 2-GHz carrier frequency band for third-generation systems (1920–1980 MHz paired with 2110–2170 MHz). For UMTS, ETSI chose wideband CDMA (WCDMA) employing 5-MHz radio carriers.

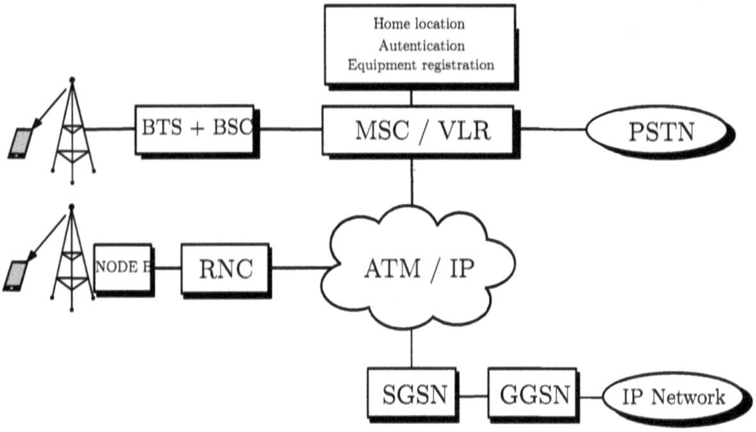

Fig. 13.2 GSM/GPRS system. It includes a public switched telephone network (PSTN), gateway GPRS support node (GGSN), serving GPRS support node (SGSN), mobile switching center (MSC), base transceiver station (BTS), base station controller (BSC), internet protocol (IP), asynchronous transfer System (ATM), radio network controller (RNC), visitor location register (VLR), Node B, generalized packet radio system (GPRS)

WCDMA has proved to be optimum for high-speed packet-data services, including video, text, etc. It is also more efficient in exploiting multipath propagation. It has a wider coherence bandwidth and allows fast power control and coherent demodulation in the uplink and downlink connections. The WCDMA has twice the capacity of narrowband CDMA.

Antenna arrays, handover between different radio frequencies, and layered cell system can also improve performance. UMTS/IMTS2000 employs an asynchronous transfer system (ATM) mini-cell transmission protocol between radio-base station and mobile switching centers. This feature allows for more efficient transmissions and lower delay.

13.2.3.2 US System

The standard IS-136, which promised to provide up to 0.384 Mps (Harte et al. 1998), is an enhanced D-AMPS. Up to the third generation of wireless systems, Fig. 13.3 illustrates the big picture of this where several 3G and 2G systems coexisted, complementing each other. For UMTS, ETSI chose wideband CDMA (WCDMA) employing 5-MHz radio carriers. At the time of this writing, we are witnessing the discontinuation of the 3G network in some countries, so that we can comment that in some well-deployed networks, the expected 2 Mbps was beyond reach, whereas 0.4 Mbps was typical. At the same time, some 5G networks can provide 5 Gps, demonstrating the impressive evolution of wireless networks.

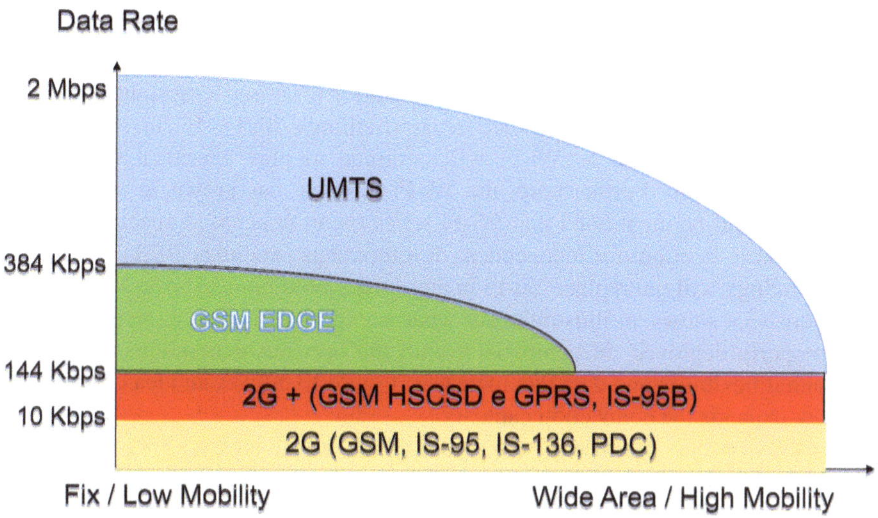

Fig. 13.3 General evolution picture

13.2.4 4G Standard

Just to give an idea of what 4G wireless communications promise, we list the performance aimed at the long-term evolution (LTE) standards (Hara and Prasad 2003; Dahlman et al. 2011). For the LTE system, the performance requirements are:

- Downlink peak spectrum efficiency 30 bits/s/Hz (max eight antennas)
- Uplink peak spectrum efficiency
- 15 bits/s/Hz (max four transmit antennas) Uplink cell edge user spectral efficiency 0.04–0.07 bits/s/Hz
- Downlink cell edge spectral efficiency 0.07–0.12 bits/s/Hz
- User plane latency 10 ms

A comparison among the several generations of cellular networks shows that roughly the first two generations aimed at voice communications. In contrast, the third and fourth generations had data communications as their primary goal. Each new generation brings about data-rate increases and improved quality of service. Therefore, the next generations aim to widen machine-to-machine communication and remote control.

13.2.5 5G Standard

The fifth generation (5G) of wireless communication promised a substantial increase in data rates concentrated in limited areas (Stallings 2021). 5G in combination with local systems such as Wi-Fi will continue to play a critical role in our wireless connection. Furthermore, the Wi-Fi industry has grown in the last few years. It should be mentioned that Wi-Fi relies on an unlicensed spectrum, which provides more freedom for independent developments (Stallings 2021). Therefore, 5G technology will not replace Wi-Fi in any foreseeable future, if ever.

Figure 13.4 shows an illustration of what is expected from a 5G network when it is properly deployed. In comparison with the previous generations, 5G allows substantial flexibility and data rate increases. Figure 13.5 gives an idea of how a 5G network expands the connection capacity.

The 5G networks promise to surpass the 1 Gb/s download mark for mobile users and allow low-latency connections to allow some delay-sensitive applications to become a reality. For instance, it would be possible to perform remote surgery if the network connection is reliable and attains low delay. The more encompassing connection among machines allows the implementation of the so-called Industry 4.0 with automated production lines, along with precision agriculture, massive IoT, and more sophisticated autonomous vehicles. In practical terms, 5G enables downloads a hundred times faster than the 4G network, as well as 70-fold faster communication between vehicles (Dahlman et al. 2024).

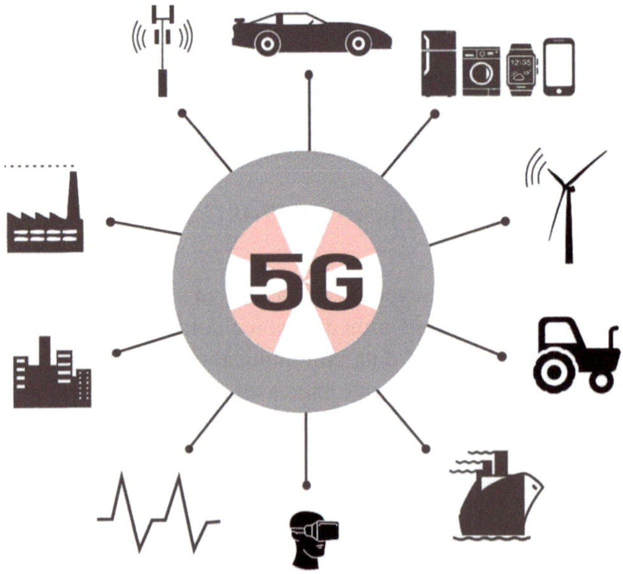

Fig. 13.4 5G objectives

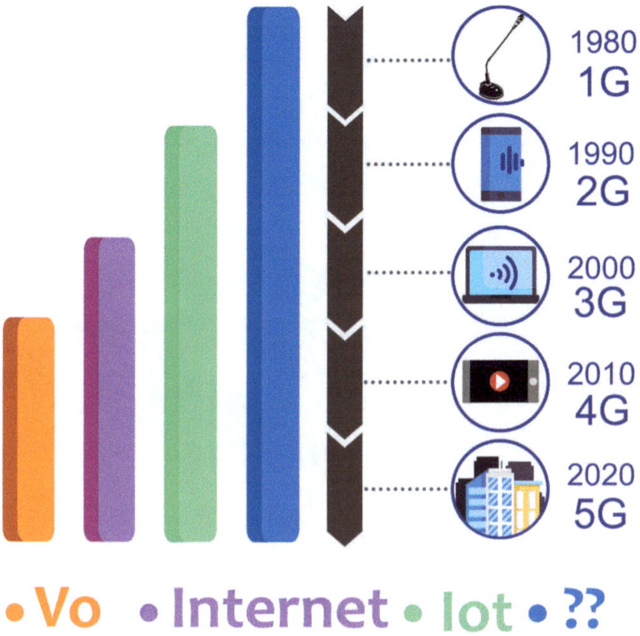

Fig. 13.5 5G capability

As illustrated in Fig. 13.6, the 5G network is changing the current way of life substantially by allowing communication among many devices. The 5G networks require the antennas in more proximity than the 4G systems (Dahlman et al. 2021). The new paradigm inherent to 5G gives a head start for many new business opportunities that will find room in the forthcoming 6G networks.

The predicted demand for wireless services in 5G is expected to grow from around 80 Mbps in 2019 to about 600 Mbps by 2030. This expectation is at least tenfold higher than what 4G and other wireless systems can provide. To meet the growing demand for enhanced mobile broadband (eMBB), ultra-reliable and low latency (URLLC), along with massive machine type of communications (mMTC), more sophisticated control and user planes are required, as depicted in Fig. 13.7. Figure 13.7 highlights the service-based representation where the various services and functionalities inherent to the network are mentioned. The details of these control plane building blocks fall beyond the scope of this book; see (Stallings 2021; Dahlman et al. 2011) for further information.

The user plane allows the connection of the user equipment (UE) with the radio access network (RAN) and a direct connection between the UE and the control network through the access and mobility management function (AMF), which in turn congregates the section and connection originating from the UE. The AMF handles connection, mobility management, reachability, and registration tasks. All information concerning the session management is directed to the SMF.

Fig. 13.6 5G urban environment

The session management function (SMF) handles packet-data unit (PDU) session establishment, modification, and release between a UE and a data network. A PDU session is an association between the UE and the Internet that provides a PDU connectivity service to exchange PDUs between a UE and a data network (Internet in Fig. 13.7). The user plane function (UPF) handles the user plane path of PDU sessions. The RAN allows access to a 5G core network, including a 5G RAN and other wireless and wired access networks. As a result, the UE might be logically connected to the internet (data network) and other services by a session.

The SCP enables multiple network functions to communicate with each other and with user plane entities in a highly distributed multi-access edge compute cloud environment. This provides features such as routing control, security, scalability, resiliency, and observability to the core network.

The interface between the UE and the network is the radio interface. Figure 13.8 shows the radio access architecture of the 5G, whose base stations are called next-generation radio access network (NG-RAN) nodes. The gNB (generalized NodeB in new radio (NR)) nodes provide a 5G user plane and control plane protocol terminations toward the UE. The so-called ng-eNB nodes allow 4G (E-UTRA) user

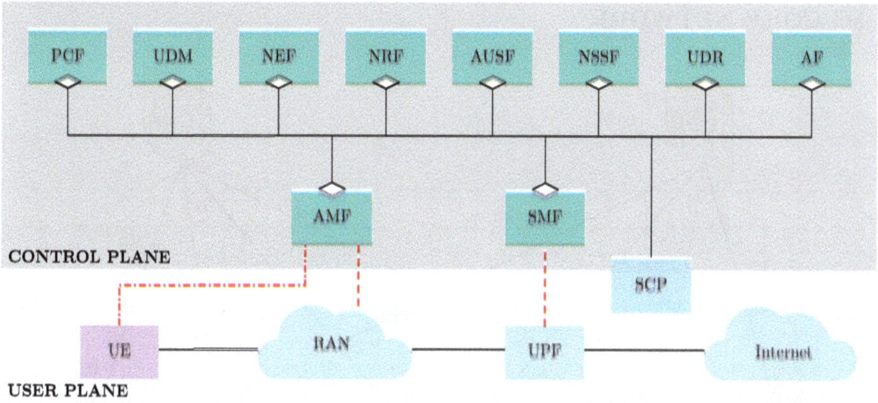

Fig. 13.7 5G system architecture, non-roaming case. The control plane includes the authentication server function (AUSF), application function (AF), access and mobility management function (AMF), network exposure function (NEF), network repository function (NRF), network slice selection function (NSSF), policy control function (PCF), unified data management (UDM), unified data repository (UDR), user plane function (UPF), service communication proxy (SCP). The diamond-shaped symbol denotes service-based interfaces that other network functions can access

plane and control plane protocol terminations toward the UE and connect via the NG interface to the 5G core, allowing 5G networks to support UE using the 4G air interface.

The somewhat simplified RAN architecture of the 5G network of Fig. 13.8 is accompanied by a more sophisticated 5G system architecture of Fig. 13.7 required to meet the widespread service demand over the network.

13.2.6 6G Standard

The ITU has released a head start on the IMT-2030, which will ultimately lead to the sixth generation (6G) of wireless communications systems (M.2160-0 2023). The main directives of the recommendation include:

- Inclusivity
  This relates to providing connectivity to everyone.
- Ubiquitous connectivity
  This means connecting the unconnected with minimum broadband service.
- Sustainability
  This relates to the overall minimization of greenhouse gas emissions and other environmental issues. It also includes energy efficiency related to the amount of information bits transported per unit of energy consumption.

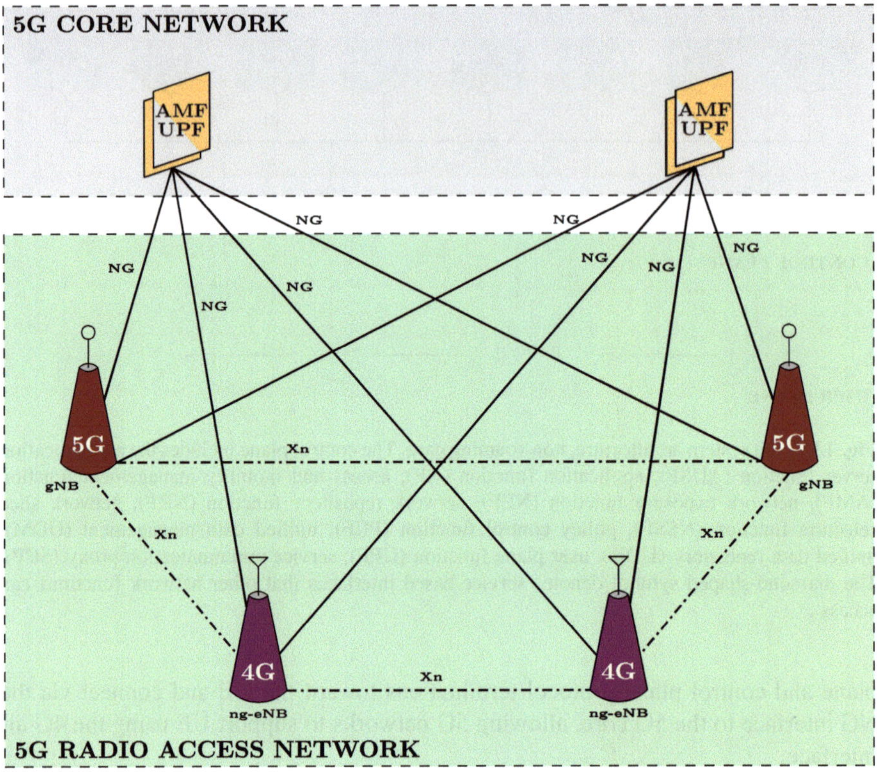

Fig. 13.8 5G radio access network, non-roaming case. The nodes interfaces are depicted

- Innovation
  This means inducing innovation with technologies that facilitate connectivity and productivity, among other things.
- Enhanced security, privacy, and resilience
  This means achieving security and privacy-preserving by design, with continuous operation when facing a disruptive event.
- Standardization and interoperability
  This means transparent and member-inclusive standardized and interoperable interfaces so that different parts of the network, regardless of the vendors, lead to a fully functional system.
- Interworking
  This means supporting service continuity and providing users with flexibility through close interworking with non-terrestrial network implementations, existing IMT systems, and other non-IMT access systems.

The 6G concept envisions connecting all citizens on Earth; therefore, on top of integrating all possible terrestrial wireless infrastructure (IMT or non-IMT), it foresees interworking with non-terrestrial network implementations.

The technical specifications for the demands inherent to the 5G systems are related to eMBB, URLLC, and mMTC. In the 6G, these demands are expanded to *immersive communications*, *hyper-reliable low-latency communications*, and *massive communications*, respectively. New features shall be incorporated, such as *integrated sensing and communications*, *integrated AI and communications*, and *massive communications*. Therefore, it is possible to predict that the control plane of the 6G system architecture will be more sophisticated than the current 5G case.

The 6G systems will take part and play a key role in the development of autonomous vehicles, particularly due to their integration of wireless networks and their predicted key performance indicators.

13.3 Cellular Division

The primary tool to provide services to a large number of subscribers to licensed spectrum is to allow for frequency reuse, where a coverage region is divided into cell clusters within which each cell is assigned a different channel group. This strategy benefits from the natural radio signal attenuation inherent to a wireless connection. Assuming we have N cells per cluster

$$N_t = kN \tag{13.1}$$

where N_t is the total number of duplex channels available for distribution and k represents the number of duplex channels per cell. Then, for M clusters, the total capacity for the given coverage area is

$$C = MN_t = MkN \tag{13.2}$$

13.3.1 Handover Strategy

An essential issue in cellular systems is the radio resource, or channel assignment. It can be *fixed*, but this case is inefficient because if a cell is overloaded, the new calls will be blocked. A solution is to borrow channels from neighboring cells (borrowing strategy). An MSC (mobile switching center)[1] is required to supervise the borrowing.

The assignment can be *dynamic* so that no cell has fixed channels, and when a call comes, the base station requests a channel to the MSC. It increases channel usage, but the MSC has several tasks: monitoring received signal strength indicator (RSSI), access channel occupancy, and traffic distribution. The MSC also ensures

[1] Also known as MTSO: mobile telephone switching office. These terms belong to 2G systems.

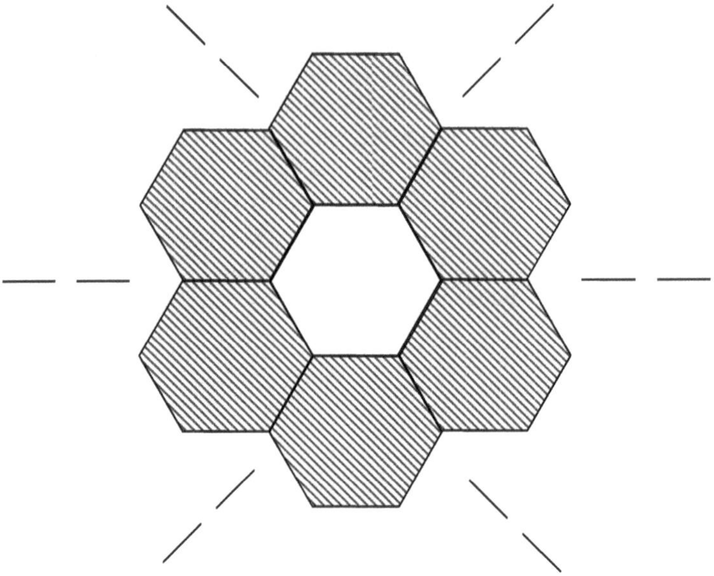

Fig. 13.9 Cellular division

that given or neighboring frequencies are not distributed to the same or adjacent cells. The described action relates to radio resource management.

Figure 13.9 shows how neighboring hexagonal cells are placed. These neighboring cells should not use the same radio resources to mitigate intercell interference.

By properly configuring a set of cells that utilize different radio resources, it is possible to establish a coverage region with some interference control. Figure 13.10 illustrates the coverage area of a cellular system.

In many situations mobile users are moving, requiring a technical strategy to keep them connected whenever their communication with a base station becomes weak. A handover strategy implements power tracking as exemplified in Fig. 13.11.

When the mobile moves toward the frontier of two cells, the signal connection with the original cell becomes weak. Before it reaches the minimum level of acceptance in quality (for instance, around $-95\,\text{dBm} \pm 5\text{dBm}$ for voice service), a handover to the neighboring cell should be made without interruption.

The margin to decide to perform and undertake a handover cannot be large; otherwise, frequent handovers might occur, since the MSC gets overloaded. On the other hand, the margin cannot be small. Otherwise, handover time might be short and the signal quality reaches a low level before the handover is made. Traditional MSC should measure average signal strength so that handover is not made due to short-time fading. The measurement is optimized in the sense that handover is made before quality or connection is lost, while unnecessary handovers are avoided.

Irrespective of the cellular system generation, Fig. 13.12 represents well a typical infrastructure environment.

Fig. 13.10 Coverage region

Fig. 13.11 Handover strategy

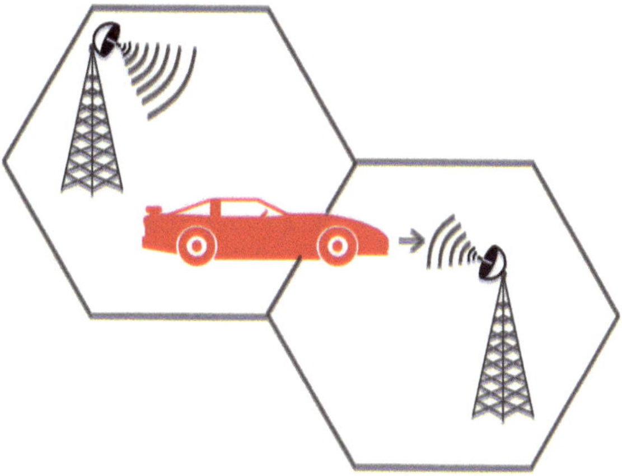

A recurrent concept in wireless communication is the *Dwell Time*. It is the time over which a call may be maintained on a given cell without handover. This time depends on interference, propagation, the distance between the mobile and base station, etc. For example, in highways, the interference is a random variable with a small variance, whereas it has a large variance and a small average in cities.

In analog systems (first generation), there was an extra locator receiver at base stations controlled by the MSC to monitor neighboring cell users' signal strength. Based on this information from each base station, the MSC decides the handover. In the second generation (using TDMA), the mobile measures the power received from the base stations around it and reports to its serving base station (*mobile-assisted handover*). It releases the MSC and allows faster handover.

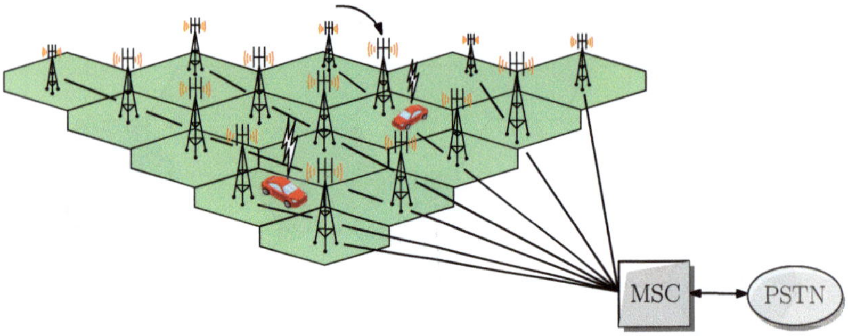

Fig. 13.12 Cellular infrastructure showing the mobile switching center and the public switched telephone network

Handovers must be prioritized in order to avoid forced call termination. For that, we have either a *guard channel concept* or queuing of handover requests. The first case leads to inefficient channel utilization, which can be relieved when using dynamic channel allocation. However, the second case does not guarantee that forced termination is avoided. Another concept is the so-called intersystem handover, which consists of the action taken when a mobile moves into a zone where a different system is used. In 5G Advanced, as well as the previous 4G and 5G systems, the handover procedures aim at reducing the signaling overhead and are standardized by 3GPP.

Yet another concept, the *Umbrella Cell*, is a larger cell embedding several cells to deal with high-speed traffic while providing small area coverage to users traveling at low speeds. It is usually implemented with a taller antenna.

The *Handover Time* varies from generation to generation and becomes tighter at each new deployment. For instance, it would last approximately 10 seconds in the first generation for a 6–12-dB handover margin. The 2G systems prescribed around 2 seconds for a handover margin of 0–6 dB. The soft handover was employed in 3G WCDMA systems, as explained in Fig. 13.13. It is possible to observe that as time progresses, the mobile faces a reducing power connection, and before it reaches a handover level at t_2, an approaching base station provides another connection starting at t_1. The mobile benefits from both connections up to the moment when the original connection power prescribes a handover, completed at t_3.

The handover time in 5G requires a lot of signaling, which could entail tens of milliseconds.

13.4 Cellular Communication

Many additional issues should be considered when deploying a cellular communication system. In this section, we discuss a few problems.

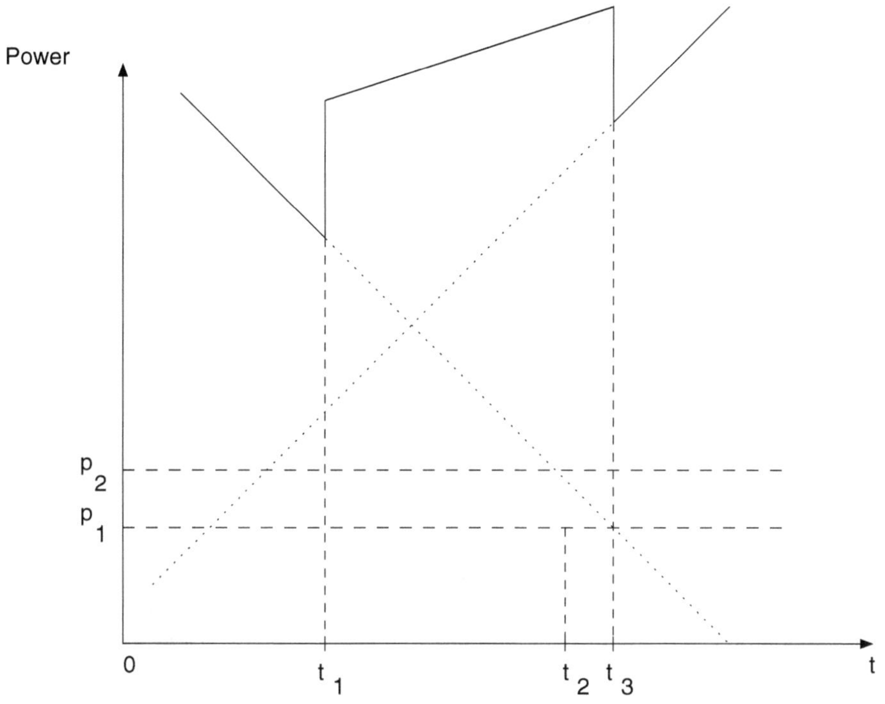

Fig. 13.13 Soft handover

13.4.1 Co-Channel Interference

The co-channel interference originates from the reuse of the radio resources in not-far-away cells. Let's define the co-channel reuse ratio for hexagonal cell geometry as

$$Q = \frac{D}{R} = \sqrt{3N} \tag{13.3}$$

where D is the distance of the closest co-channel cells, R is the cell radius, and N is the cluster size. For hexagonal cells, a commonly used and important quantity is the signal-to-interference ratio given by

$$\frac{S}{I} = \frac{\sqrt{3N}^{\gamma}}{N_i} \tag{13.4}$$

where N_i is the number of neighboring interferers and γ is the path loss exponent.

Define the receiver power at a distance d_0 from transmitting antenna as

$$P_r = P_0\left(\frac{d_0}{d_{\text{ref}}}\right)^{-\gamma} \tag{13.5}$$

where P_0 is the power measured at a close distance of reference d_{ref}. Considering all neighboring interferers as equidistant from the user of interest

$$P_r = P_0\left(\frac{R^{-\gamma}}{N_i D_i^{-\gamma}}\right) = P_0\frac{\sqrt{3N}^{\gamma}}{N_i} \tag{13.6}$$

where D_i represents the distance to N_i.

13.4.2 Adjacent Channel Interference

The adjacent channel interference comes from the spectral leakage of neighboring spectrum radio transmissions. Despite coming from different frequency channels, the separation filters might not be efficient enough (*near-far problem*). A possible strategy to solve this problem is to separate the frequency channels of neighboring cells as much as possible.

The adjacent channel interference can be quantified as

$$\frac{S}{I} \approx (cl)^{-\gamma} \tag{13.7}$$

where (cl) is a measure of how closely a mobile is to a base station compared to another one. This quantity measures the desired signal to the interference from another mobile inside the band of interest of the desired mobile.

In all cellular systems, power control plays a crucial role in keeping interferences within bearable levels.

13.4.3 Other Issues

In modern-day data-intensive wireless communications, many other issues might play a role in the perception of the quality of services (QoS). For instance, if we have a backbone utilizing some sort of trunking to allow a large number of users to share a relatively small number of channels, the grade of service in the backbone affects the chances of a user accessing a trucked system during the busiest hour.

In the air interface, the network might improve the QoS by applying cell splitting in busy regions, and by sectoring with directional antennas.

13.5 Remarks

The ongoing development in communications and sensing predicts the fusion of information originating from multiple sources in an unprecedented way. The networks will inherently include terrestrial and non-terrestrial stations, and information will also be acquired from sensing systems deployed everywhere. Autonomous vehicle solutions, for certain, will require access to areas' networks that are within reach and benefit from all information originating from sensors. It is worth mentioning that the duplex methods are important in the task of exploiting the available spectrum as much as possible, particularly in Wi-Fi routers and cellular networks.

References

E. Dahlman, S. Parkwall, J. Sköld, *4G LTE/LTE-Advanced for Mobile Broadband* (Academic Press, Oxford, 2011)

E. Dahlman, S. Parkwall, J. Sköld, *5G NR The Next Generation Wireless Access Technology* (Academic Press, London, 2021)

E. Dahlman, S. Parkwall, J. Sköld, *5G/5G-Advanced The New Generation Wireless Access Technology* (Academic Press, London, 2024)

T. Halonen, J. Romero, J. Melero, *GSM, GPRS, and EDGE Performance, Evolution Towards 3G/UMTS*, 2nd edn. (Wiley, Chischester, UK, 2003)

S. Hara, R. Prasad, *Multicarrier Techniques for 4G Mobile Communications* (Artech House Publishers, Boston, 2003)

L.J. Harte, A.D. Smith, C.A. Jacobs, *IS-136 TDMA Technology Economics and Services* (Artech House Publishers, Boston, 1998)

H. Holma, A. Toskala, *HSDPA/HSUPA for UMTS: High Speed Radio Access for Mobile Communications* (Wiley, New York, 2006)

H. Holma, A. Toskala, *WCDMA for UMTS*, 4th edn. (Wiley, New York, 2007)

J.S. Lee, L.E. Miller, *CDMA Systems Engineering Handbook* (Artech House Publishers, Boston, 1998)

R. I.-R. M.2160-0, Framework and overall objectives of the future development of IMT for 2030 and beyond, ITU, Tech. Rep. (2023). https://www.itu.int/rec/R-REC-M.2160-0-202311-I/en

G. Madyam, J. Lai, *Third Generation CDMA for Enhanced Data Services* (Academic Press, New York, 2002)

M. Mouly, M.-B. Pautet, *The GSM System for Mobile Communications* (Cell & System, New York, 1992)

T. Ojanperä, R. Prasad, *WCDMA: Towards IP Mobility and Mobile Internet* (Artech House Publishers, Boston, 2001)

A. Richardson, *WCDMA Handbook* (Cambridge University, Cambridge, 2005)

W. Stallings, *5G Wireless: A Comprehensive Introduction* (Addison Wesley, Boston, 2021)

V. Vanghi, A. Damnjanovic, B. Vojcic, *The CDMA 2000 System for Mobile Communications* (Prentice Hall, Upper Saddle River, NJ, 2004)

S.C. Yang, *CDMA RF System Engineering* (Artech House Publishers, Boston, 1998)

S.C. Yang, *3G CDMA 2000 Wireless System Engineering* (Artech House Publishers, Boston, 2004)

Chapter 14
Wireless Channels

14.1 Introduction

Autonomous vehicles rely on precise positioning, navigation, and timing (PNT) information. Although traditional commercial wired and wireless communication systems were not designed for this role, they can be incorporated into existing systems using lidar, global navigation satellites, and camera and computer vision systems. The wireless and, eventually, the wired systems provide signals of opportunity that allow for improvements and more accurate PNT solutions. Like any information traversing a physical layer, the transmitted signals face multipath channels with random behavior that should be modeled appropriately statistically.

This chapter covers the basic concepts of wireless channel models, which are used in simulation tools to plan network deployment. Many books address wireless channel modeling with varying levels of detail (Lee, 1998; Pätzold, 2012). In the following, we briefly describe some issues related to wireless channel modeling. The channel characteristics relate to the environment where the wireless system is deployed, and an illustrative example is depicted in Fig. 14.1.

14.2 Types of Fading

This section describes the main types of fading witnessed by a signal transmitted through a physical channel. The discussion in this section provides an overview of the big picture. The following sections provide more details.

Y. C. Lim et al., *Signal Processing for Autonomous Systems*,
https://doi.org/10.1007/978-3-032-00675-2_14

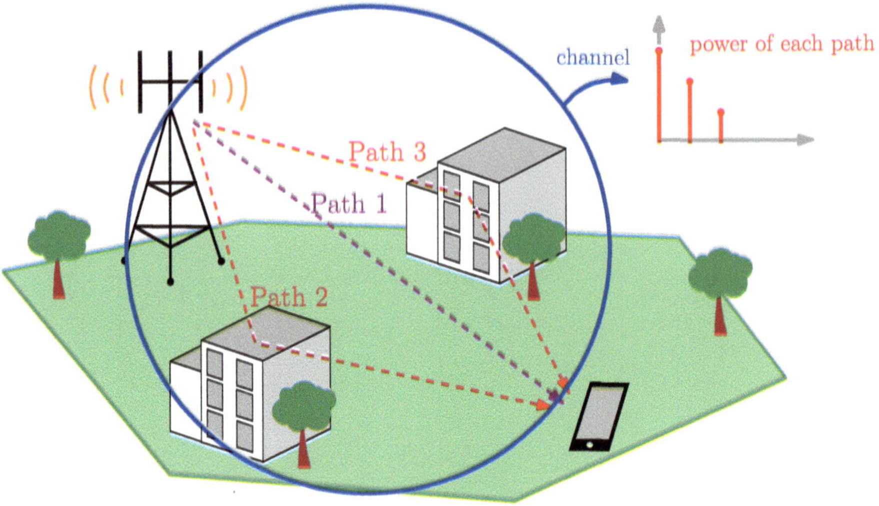

Fig. 14.1 Wireless channel

14.2.1 Large-Scale Fading

It is a measure of variation in the mean value of the signal power level for any arbitrary distance between the transmitter and receiver. This mean value represents the average signal power measured over a range of a few λ to around 50λ where λ is the carrier wavelength. Large-scale fading is caused by shadowing and the inherent path loss as the signal travels through the air. Therefore, the path loss and the shadowing to be discussed are propagation models falling into the large-scale fading category.

The large-scale fading usually leads to a frequency-flat fading, meaning that the channel model is not frequency-selective (Tse and Viswanath, 2005). In the following, we describe the typical model in this case.

Assume the amplitude of the transmitted symbol is modeled as

$$G_l = \pm\sqrt{2E_l} \tag{14.1}$$

where E_l represents the energy of the symbol, for the l-th user of the system. The received signal amplitude has a random behavior:

$$R_l(kT_b) = g_l r(kT_b) \tag{14.2}$$

where g_l is a constant representing the large-scale fading characteristics. The symbol $r(kT_b)$ represents the small-scale fading (Marzetta et al., 2016) being a random variable whose probability density function (PDF) usually follows one of the following discussed formats.

Typically, the PDFs are log-normal distribution when there is shadowing by obstacles in direct line of sight; Rayleigh when the phases of the users' signals are uniformly distributed, or the users' signals are faced with multipath with low relative delay; Rice when there is a strong direct line of sight component; and Nakagami matches the PDF well in urban areas when compared to the Rayleigh distribution. This chapter provides further details regarding these distributions.

14.2.2 Small-Scale Fading

Small-scale fading is caused by the rapid phase changes that each multipath component undergoes, affecting the exponential terms of the channel model whenever the carrier frequency f_c times the m-th path delay τ_m becomes significant, that is $f_c\tau_m \gg 1$. This model will be discussed at the end of this chapter.

Small-scale fading is a measure of rapid variations in the power value of the received signal over very short distances (comparable to the wavelength) and/or short time frames (Tse and Viswanath, 2005). It also includes the random modulation caused by the different Doppler shifts observed in the signals arriving from distinct paths, and the multipath delays that cause time dispersion. The analysis of small-scale fading obviates the consideration of large-scale fading, as the former encompasses the latter.

The related propagation models are called small-scale or fading models (Rappaport, 2002). Small-scale fading can be classified into two categories: flat fading and frequency-selective fading. In addition, small-scale fading models can be classified as *fast* or *slow* fading if the Doppler spread is *comparable* to the signal bandwidth of the baseband or *not*.

The fast and slow fadings do not indicate if the channel is flat or frequency-selective. The slow-fading channel can also be called a time-flat channel, whereas a fast-fading channel is time-selective. For example, the multipath components vary faster than the symbol duration in a fast-fading, frequency-selective channel. Therefore, fast fading usually occurs in very low data rates whenever the symbol duration is relatively long. The fast fading is always related to (originates from) rapid changes of the channel in time, and frequency spread such as the time-varying fading caused by the Doppler effect. It is worth mentioning that the concepts of fast and slow fading have no firm agreement in the literature, according to Tse and Viswanath (2005), who considers a channel fast fading when the coherence time (the time span during which the channel model changes significantly) is shorter than the application delay requirement. For instance, in remote surgeries the delay requirements is less than 100 ms so that the data rate should be high enough to avoid fast-fading.

14.2.2.1 Frequency-Selective Fading

In rough environments, the received signal undergoes fading which is frequency-selective (it has memory).

An accepted model for the frequency-selective fading is a linear time-varying model for the channel impulse response:

$$r_l(t) = \int_0^t s_l(t)h_l(t, \tau)d\tau \tag{14.3}$$

where $h_l(t, \tau)$ is the lth user channel impulse response. In mobile systems, time-varying channels are a common case.

Frequency-selective fading occurs when the baseband signal bandwidth is greater than the bandwidth of the (multipath) channel. The received signal will be distorted in frequency, but on the other hand, the signal strength will not fade much over a small area. In this case, a coherent combination of the several path signals guarantees the signal strength.

14.3 Models

The degree of sophistication of the channel model depends on factors such as whether the channel is stationary or not. Does the channel consist of free space? Is the mobile in motion? Are there signal reflections? Does the radio signal face scattering? Does scattering occur locally? Remotely? Or both?

Mobile signals are reflected in several types of obstacles, such as local buildings or structures around a few hundred wavelengths from the mobile. There are also dominant reflectors, such as hills and buildings. Together, these structures generate multipath propagation.

Time-varying multipath is due to the moving mobile and/or moving reflectors, generating a *Doppler effect* leading to a rate of variation in the received signal. Each path has a different delay, generating a *delay spread*. In addition, if the receiver has an antenna array, the direction of arrival can be explored due to *angle spread*.

In any case, for proper channel model simulation, we have to take into consideration the large-scale fading features such as the path loss that depends only on the distance between the mobile and the base station and shadowing fading caused by the obstruction of the line of sight between the mobile and the base station. Both types of fading belong to the class of large-scale path loss, and in the latter case, the signal variation follows a log-normal distribution. The large-scale fading models are crucial to network planning in terms of cell sizes, sites, and splitting.

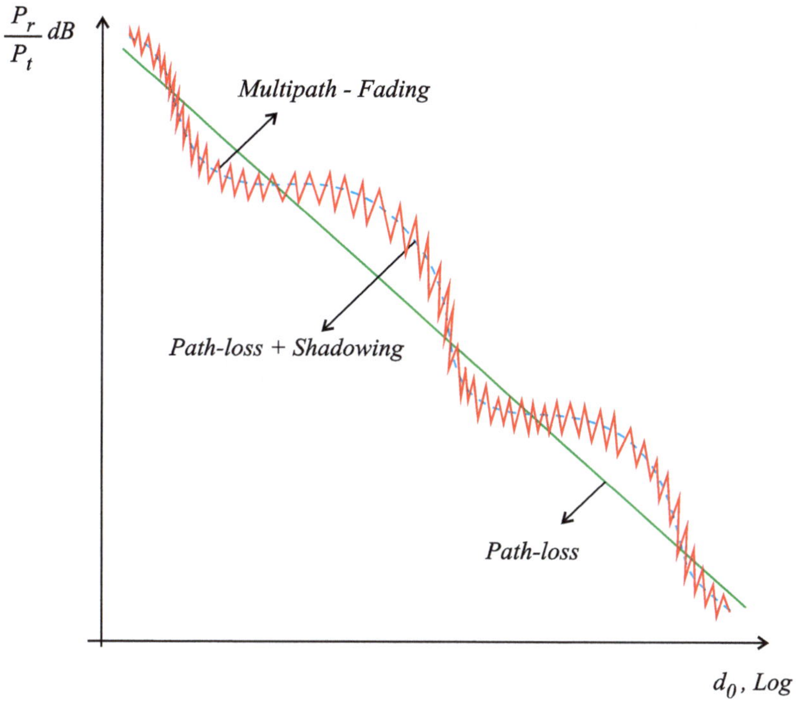

Fig. 14.2 Power loss in a wireless environment, where P_r and P_t represent the received and transmitted powers, respectively

The small-scale fading deals with the multipath fading that causes signal amplitude variations over distances proportional to signal wavelengths. The main issues related to small-scale fading are characterized by:

- Flat fading or amplitude-varying channel.
- Doppler spread.
- Time-delay spread.
- Angle spread.

An illustrative behavior of path loss, shadowing, and small-scale fading is depicted in Fig. 14.2. The mentioned quantities are depicted as a function of distance.

In theory, the propagation of electromagnetic waves requires the solution of coupled differential equations, known as Maxwell equations, to model the reflections and refractions in the medium. In order to avoid the use of the Maxwell equations, a possible simplification employs the ray-tracing technique, where the possible paths between the transmitter and receiver are taken into consideration through environment-dependent multiple reflections. However, this method might not apply to many environments where some empirical methods must be considered.

14.4 Large-Scale Path Loss

The large-scale path loss characterizes the mean signal strength for any given distance between the transmitter and receiver. This term "large" originates from the characterization of large distances ranging from hundreds to a thousand meters.

14.4.1 Path Loss

Caused by the transport of the signal between the transmitter and receiver via the line of sight, and indirectly from reflections on the ground and surface wave. This model is called the two-ray ray-tracing model. This model is more accurate when the receiver is far from the closest scatterer (several wavelengths of distance). The scatterers should be considered large and smooth (Goldsmith, 2005). The ray-tracing concept can be generalized for many reflections, scatterers, and diffraction, but it is not very accurate to determine the delay spread in multipath channels (Fig. 14.3).

If the signals' delays in the direct and the reflected paths are much smaller than the inverse of the signal bandwidth, the received signal power is

$$P_r = g\left(\frac{\lambda}{2\pi d_0}\right)^2 |1 + Re^{j\Delta\phi}| \tag{14.4}$$

where g is the product of the antenna gains with the transmitted signal power, R is the ground reflection coefficient, λ is the wavelength and $\Delta\phi$ is the phase difference between the line of sight and reflected signals, with

$$\Delta\phi = \frac{2\pi(r_o + r_1 - d_1)}{\lambda} \tag{14.5}$$

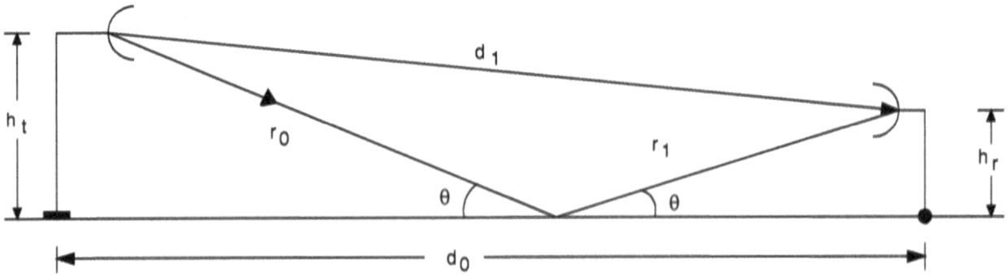

Fig. 14.3 Pass-loss environment

The ground reflection coefficient is given by

$$R = \frac{\sin\theta - Z}{\sin\theta + Z} \tag{14.6}$$

where

$$Z = \frac{\sqrt{\epsilon_r - \cos^2\theta}}{\epsilon_r} \tag{14.7}$$

for vertical polarization of the antennas (Hottinen et al., 2003) and

$$Z = \sqrt{\epsilon_r - \cos^2\theta} \tag{14.8}$$

for horizontal polarization, where ϵ_r is the relative permittivity of the ground. The ground permittivity is $\epsilon_r = 25$ for good ground and $\epsilon_r = 15$ for typical ground.

If the antennas are far from each other, d_0 is large, then $R \approx -1$, $\theta \approx 0$ and

$$\Delta\phi \approx \frac{4\pi h_t h_r}{\lambda} \tag{14.9}$$

These approximations are valid for angles smaller than 10 degrees and frequencies above 100 MHz.

Therefore,

$$P_r \approx g\left(\frac{h_t h_r}{d_0^2}\right)^2 \tag{14.10}$$

This approximation is valid after a given distance between the antennas. Before this critical distance, the power decay is proportional to the inverse of d_0^2.

In dB's, the expression above can be rewritten as

$$P_r \approx 10\log(g) + 20\log(h_t h_r) - 40\log(d_0) \tag{14.11}$$

It is worth recalling that the transmitter power is included in the value of g. The received power is usually measured in dBm, where the reference transmitted power is one milliwatt mW, or is measured in dBW, where the reference transmitted power is one watt W. For example, in dBm, the expression above should be rewritten as

$$P_r dBm \approx P_t dBm + 10\log(g_t) + 10\log(g_r) + 20\log(h_t h_r) - 40\log(d_0) \tag{14.12}$$

where $P_t dBm = 10\log(\frac{P_t}{0.001})$ with P_t denoting the transmitter power, g_t is the gain of the transmitter antenna, and g_r is the gain of the receiver antenna.

In practice, the attenuation model above applies to rural areas and highways. In suburban areas, a similar trend is followed by an additional loss. The model above can be modified depending on the region and the frequency employed, but the general trend is usually observed. In urban roads with high-rise buildings and antennas lower than the building tops, like in Manhattan, the decay is proportional to the inverse of d_0^2 due to the dominance of the building wall reflections.

The ray-tracing model can be generalized to multiple paths when the paths are fixed due to a stable environment. If the transmitter and receiver move, we can still employ ray tracing as long as the geometry of the environment and the location characteristics are known over time (Goldsmith, 2005). The statistical model should be used if the geometry is complex, with a large number of reflections whose surfaces are non-smooth.

14.4.2 *Shadowing*

Assuming an obstruction $\bar{l}$ has an attenuation constant $at_{\bar{l}}$ and a thickness $\Delta th_{\bar{l}}$, we can say that this obstruction attenuates the signal as follows:

$$s_{\bar{l}} \approx s_{\bar{l}-1} e^{-at_{\bar{l}}\Delta th_{\bar{l}}} \tag{14.13}$$

After several obstructions, the exponent becomes:

$$y = \sum_{\bar{l}=1}^{L} -at_{\bar{l}}\Delta th_{\bar{l}} \tag{14.14}$$

assuming large L we can say that y is approximately Gaussian and therefore s_L is log-normal. The quantity $10\log_{10} s_{\bar{l}}$ has an average μ_s and a variance σ_s^2 ranging from 4 to 12 dBs, according to some measurements.

This model results from averaging the effects of local multipath but felt at large distances from the transmitter, that is, the amplitude of the signal received at wide areas follows a log-normal distribution.

The varying behavior and complexity of radio channels require statistical models, as deterministic solutions are not always acceptably accurate.

14.5 Small-Scale Fading

The small-scale path loss characterizes the rapid fluctuation in signal strength over short-distance variations between the transmitter and receiver. The term small originates from the fact that the characterization entails small distance movements in the order of a few carrier wavelengths. Typically, the signal strength can vary a few orders of magnitude when the receiver moves by small distances in the order of fractions of the wavelength.

Small-scale fading originates from the summations of several reflections of the transmitted signal reaching the receiver at different times. As a result, the signal combination can increase or decrease the overall signal power at short distances.

These clusters of signals result from several multipath components reaching the receiver at different angles and with distinct amplitudes. In addition, the movement of the mobile and the objects surrounding the mobile will lead to small frequency shifts in the cluster signals.

The transmitted signal occupies a prescribed range of frequencies. Whenever this range is larger than the channel bandwidth, the received signal will reach the receiver in distinguishable multipath clusters. Several copies of the transmitted signal can be sensed at the receiver in the time domain, which may cause intersymbol interference.

At any rate, if the transmitted signal bandwidth (BW_s) is much smaller than the channel bandwidth (BW_c), a single cluster is perceived at the receiver. Otherwise, several clusters will be sensed.

As a result, if $BW_s < BW_c$, which is equivalent to having a cluster delay spread $<$ symbol delay spread, we will have a flat fading. Figure 14.4 illustrates the impulse response of the channel model, whereas Fig. 14.5 shows two distinct symbols to be transmitted. The received signal is depicted in Fig. 14.6, where it can be observed that minimal overlap exists among the transmitted symbols, making their respective detection easy. In the frequency domain, the flat fading corresponds to the case where the signal spectrum is narrow in comparison with the channel bandwidth, as illustrated in Fig. 14.7.

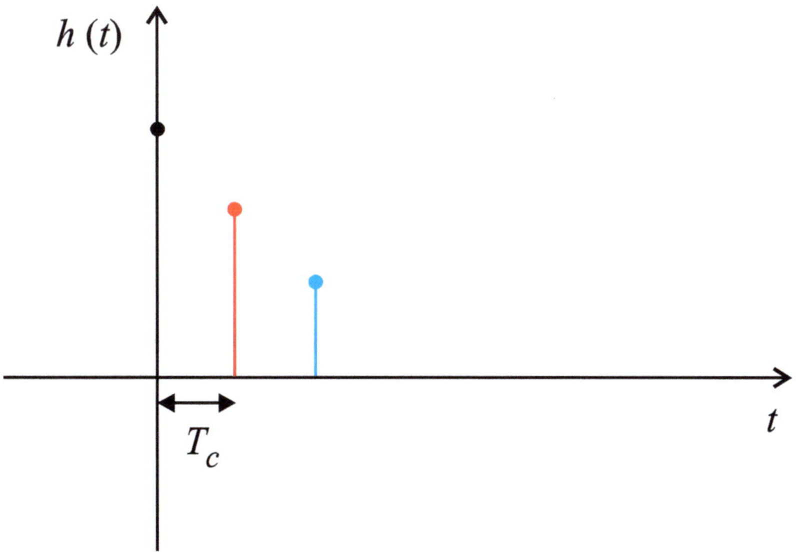

Fig. 14.4 Channel impulse response

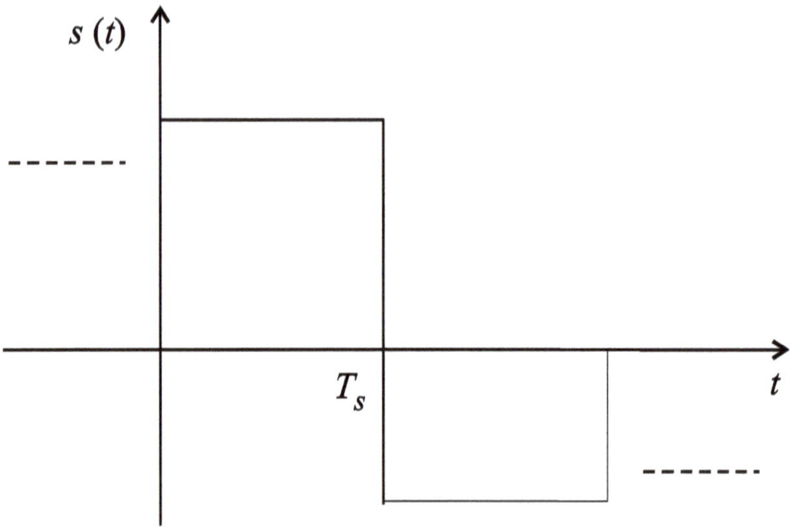

Fig. 14.5 Transmitted symbols

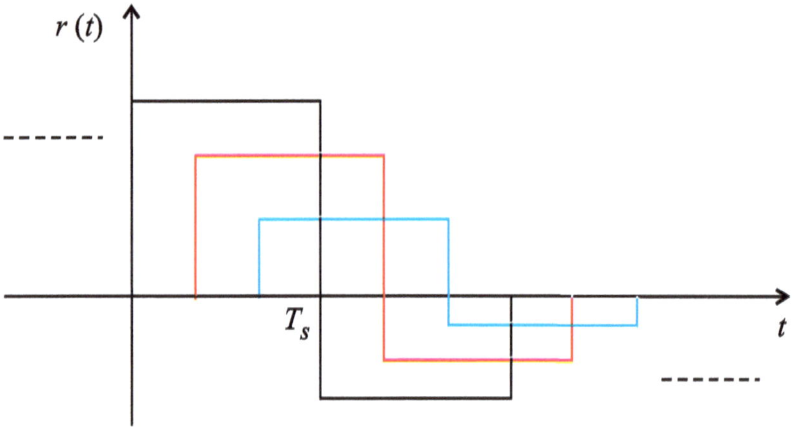

Fig. 14.6 Received symbols

For $BW_s > BW_c$, which is equivalent to cluster delay spread $>$ symbol delay spread, we will have a frequency-selective fading. For this case, a possible channel impulse response is as shown in Fig. 14.8, whereas, again, Fig. 14.5 shows two distinct symbols to be transmitted. The received signal for this frequency-selective channel is depicted in Fig. 14.9, where it is possible to verify that substantial overlap exists among the repetitions of transmitted symbols and other symbols (not explicitly shown), making an efficient equalization necessary. In the frequency domain, the frequency-selective fading corresponds to the case where the signal spectrum is broader than the channel bandwidth as illustrated in Fig. 14.10.

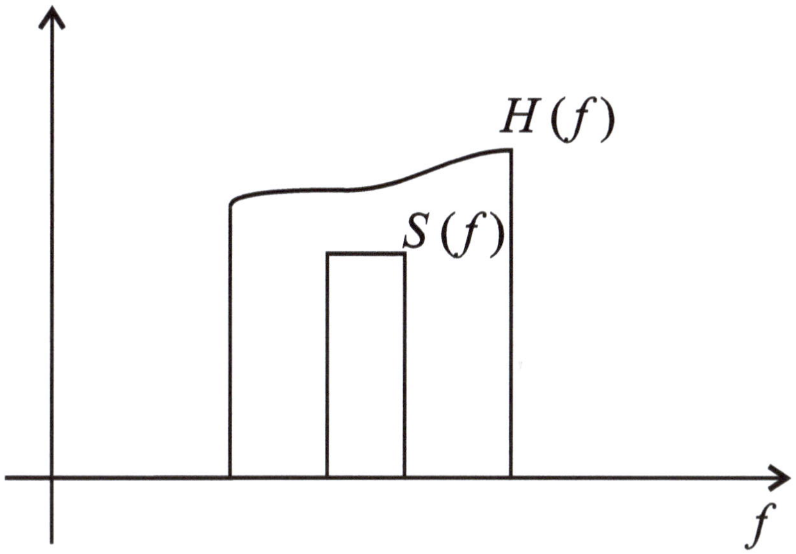

Fig. 14.7 Signal and channel bands

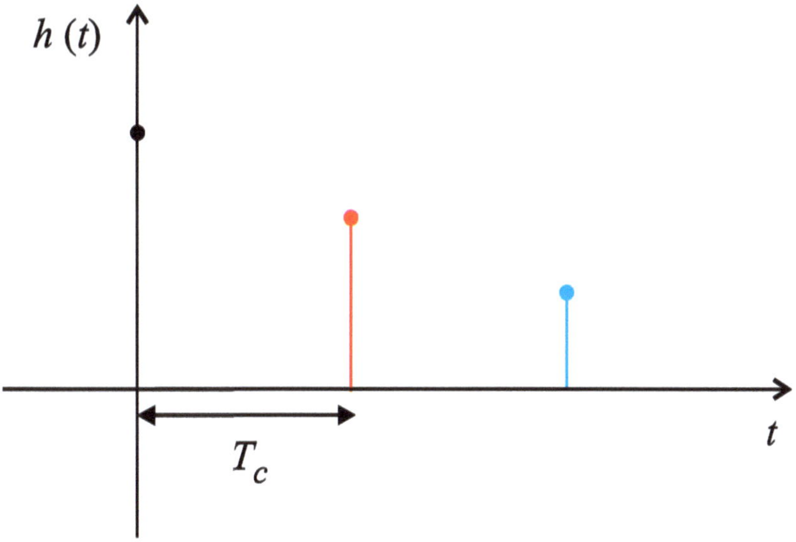

Fig. 14.8 Channel impulse response

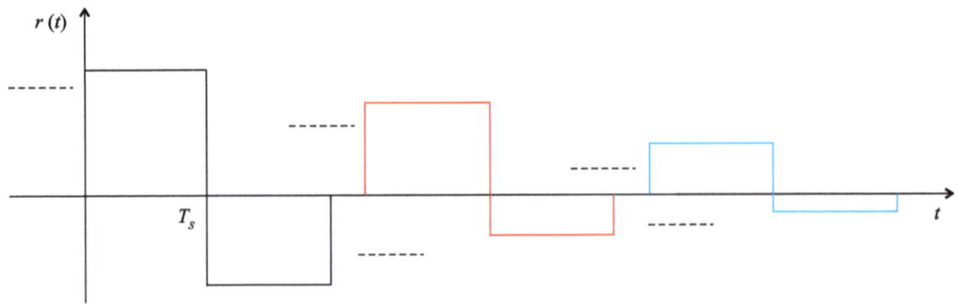

Fig. 14.9 Received symbols

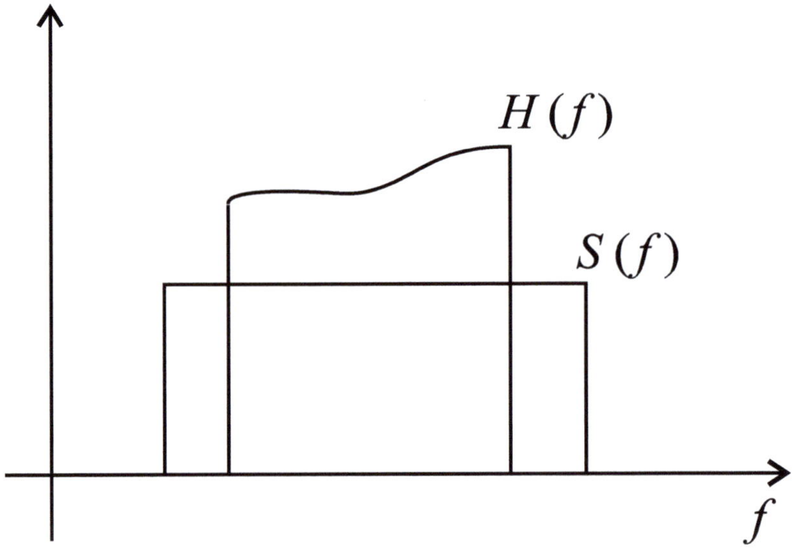

Fig. 14.10 Signal and channel bands. This is an exaggerated illustrative example where the data bandwidth exceeds the channel bandwidth

The condition in Fig. 14.10 is equivalent to have $|\tau_1 - \tau_2| \gg BW_s^{-1}$ where τ_1 and τ_2 are two multipath delays, indicating that the two multipath components are resolvable by the receiver. In this situation, the data stream in the channel will face severe intersymbol interference.

14.5.1 Multipath Fading

Typically, the transmitted signal is given by

$$s^c(t) = s(t)e^{j(2\pi f_c t + \theta)} \tag{14.15}$$

where $s^c(t)$ is a complex-valued signal and f_c is the carrier frequency and θ is the phase of the carrier. Due to multipath, the received signal is

$$r^c(t) = g\sum_{m=1}^{M} a_m s\left(t + \frac{f_c}{f_d}\cos\phi_m t - \tau_m\right) e^{j[2\pi((f_c+f_d\cos\phi_m)t - f_c\tau_m)+\theta]} \tag{14.16}$$

where it was assumed that the propagation distance was much larger than the propagation-distance spread, so that the attenuation due to distance is the same at each path. The parameter $\tau_m = \frac{d_m}{c}$ is the multipath delay where d_m is the propagation distance and c represents the speed of light, whereas the parameter a_m represents the attenuation on a given path. The frequency $f_d = \frac{v}{\lambda}$ is the maximum Doppler shift, with v being the mobile velocity, and ϕ_m is the direction of the mth scatterer with respect to the mobile speed vector.

14.5.2 Narrowband Case

Assuming the baseband signal $s(t)$ is a narrowband signal, then $s(t + \frac{f_d}{f_c}\cos\phi_m t - \tau_m) \approx s(t - \tau)$ for all m;[1] see the book (Tse and Viswanath, 2005) for alternative discussion on this matter on pages 20–25. From (14.16) we get:

$$\begin{aligned} r^c(t) &\approx g s(t-\tau)e^{j2\pi f_c t} \\ &\quad \times \left[\sum_{m=1}^{M} a_m e^{j[2\pi((f_d\cos\phi_m)t - f_c\tau_m)+\theta]}\right] \qquad (14.17) \\ &= g s(t-\tau)e^{j2\pi f_c t}\left[\sum_{m=1}^{M} a_m e^{j\psi_m(t)}\right] \\ &= g s(t-\tau)e^{j2\pi f_c t}\hat{r}(t)e^{j\psi(t)} \qquad (14.18) \end{aligned}$$

where $\tau_m \approx \tau_l$ for all m and l. From the equation above, we recognize that

$$h_l(t,\tau) = \delta(\tau - \tau_0)\hat{r}(t)e^{j\psi(t)} = \delta(\tau - \tau_0)h'(t) \tag{14.19}$$

[1] In a typical wireless connection, the term $\frac{f_d}{f_c}\cos\phi_m$ is considered small enough to be neglected, for a carrier frequency of $f_c = 1\,\text{GHz}$, a movement of $100\,\text{km/h}$, and a Doppler shift of $f_d = \frac{f_c}{c}v \approx \frac{1\text{Ghz}}{300000\,\text{km/s}}100\,\text{km/h} \approx 92.6\text{Hz}$, the neglected term is on the order of 92.5×10^{-9}. However, at the exponent, this term is multiplied by $2\pi f_c$ leading to an order of $2\pi 92.6$ resulting in a frequency change in the order of 92.6 Hz which sometimes should not be neglected.

The mean value of $h'(t)$ is zero. Assuming that a_m are uncorrelated and independent of $\psi_m(t)$, also $\psi_m(t)$ are assumed to be random variables and uniformly distributed between 0 and 2π. The autocorrelation of $h'(t)$ is

$$\begin{aligned}\rho_{h'}(\tau) =& \mathbb{E}[h'(t)h'^{*}(t+\tau)] \\ =& \sum_{m=1}^{M} \mathbb{E}[a_m^2]e^{-j2\pi f_d \cos(\phi_m)\tau}\end{aligned} \tag{14.20}$$

An alternative deduction for this expression employs (14.17)[2].

Assuming the baseband signal $s(t)$ is a narrowband signal, the probability distribution of $\hat{r}(t)$[3] is Rayleigh

$$f(x) = \frac{2x}{\sigma_r^2} e^{-\frac{x^2}{\sigma_r^2}} \tag{14.21}$$

where $x \geq 0$, assuming a_m has bounded variance as it is the usual case. Also σ_r^2 is the variance of $\hat{r}(t)$. This model is suitable in the case where there is no strong (no line of sight) received component, and rapid and small-scale amplitude fluctuation occurs.

For a direct line of sight with low-level scatterer paths, the probability distribution of $\hat{r}(t)$ is Rician

$$f(x) = \frac{2x}{\sigma_r^2} e^{-\frac{x^2+p^2}{\sigma_r^2}} I_0\left[2\frac{xp}{\sigma_r^2}\right] \tag{14.22}$$

where p^2 is the average signal power in the direct line of sight and I_0 is the modified zero-order Bessel function.[4]

The Nakagami distribution is better in urban areas and is given by

$$f(x) = \frac{2\kappa^{\kappa} x^{2\kappa-1}}{\Gamma(\kappa)\sigma_r^{2\kappa}} e^{-\frac{\kappa x^2}{\sigma_r^2}} \tag{14.23}$$

---

[2] Assuming that $h'(t)$ has the form in the square bracket of (14.17), we get $h'(t)h'^{*}(t+\tau) = \sum_{m=1}^{M} a_m e^{j[2\pi(f_d \cos\phi_m t - f_c\tau_m)+\theta]} \times \sum_{m=1}^{M} a_m e^{-j[2\pi f_d \cos\phi_m (t+\tau) - f_c\tau_m)+\theta]}$, which after applying the expected value and assuming that a_m are uncorrelated and independent of exponents, leading to $\sum_{m=1}^{M} \mathbb{E}[a_m^2]e^{-j2\pi f_d \cos(\phi_m)\tau}$.

[3] $\hat{r}(t)$ represents a single cluster of the received signal $r^c(t)$.

[4] $I_0 = \frac{1}{\pi}\int_0^{\pi} e^{x\cos\phi} d\phi$.

where $\kappa \geq \frac{1}{2}$, $\Gamma(\cdot)$ is the gamma function,[5] $\sigma_r^2 = \mathbb{E}[x^2]$, and $\kappa = \frac{\sigma_r^4}{\mathbb{E}[(x^2-\sigma_r^2)^2]}$. The Nakagami distribution is more general than the Rician and Rayleigh distributions, and does not require their assumptions. For example, $\kappa = 1$ leads to the Rayleigh distribution.

14.5.3 Wideband Case

If the signal bandwidth BW_s is large, we cannot assume that $s(t-\tau_m) \approx s(t-\tau)$ for all m. This approximation might be true for some clusters of signals coming from subpaths that are not resolvable by the receiver (this occurs when $|\tau_i - \tau_j| < \frac{1}{BW_s}$). Therefore,

$$r^c(t) = g \sum_{m=1}^{M} s(t-\tau_m)\hat{r}_m(t)e^{j\psi_m(t)} \tag{14.24}$$

where M, in this case, is the number of clusters, and $\hat{r}_m(t)$ represents the envelop of each cluster. From the equation above, we recognize that

$$h_l(t,\tau) = \sum_{m=1}^{M} \delta(\tau-\tau_m)\hat{r}_m(t)e^{j\psi_m(t)} \tag{14.25}$$

The model above is often known (Hashemi, 1979) as Turin's model for channel response, with wideband signals (Turin et al., 1972). As the mobile moves, all the parameters in the above equation change, including the number of clusters. Figure 14.11 illustrates the different impulse response profiles for the channel model perceived at distinct locations. Note that a similar aspect of the profile applies to the impulse responses measured at distinct times in the same location.

The appropriate model for the envelop correlation is (Naguib, 1996)

$$\rho_{\hat{r}}(\Delta\omega, \Delta t) = J_0^2(2\pi f_d \Delta t)|F_{\tau_m}(j\Delta\omega)|^2 \tag{14.26}$$

where J_0 is the zero-order Bessel function[6] and $F_{\tau_m}(j\Delta\omega)$ is the Fourier transform of the probability density function of τ_m, which is shown to approach a Poisson process or a modified Poisson process (especially when the signal arrives in clusters from reflections in buildings).

[5] $\Gamma(\kappa) = \int_0^\infty z^{\kappa-1}e^{-z}dz$.

[6] $J_0 = \frac{1}{\pi}\int_0^\pi e^{jx\cos\phi}d\phi$.

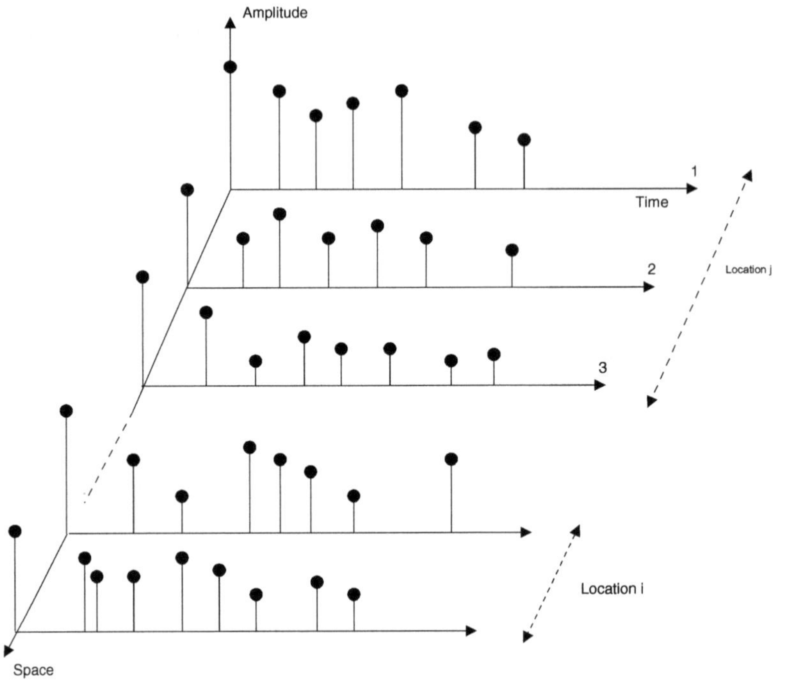

Fig. 14.11 Channel model profiles at different locations

14.5.4 Complex Channel Representation

From (14.16), it is straightforward to show that the received signal after removing the carrier, including its phase, is given by

$$r(t) = g \sum_{m=1}^{M} a_m s\left(t + \frac{f_c}{f_d} \cos \phi_m t - \tau_m\right) e^{j[2\pi f_d \cos \phi_m t - f_c \tau_m)]} \quad (14.27)$$

As can be observed, the complex signal representation is more straightforward than the real signal one discussed in the following section. However, since, in the actual communication system, we transmit real signals, it is worth discussing this case in some detail.

14.6 Real Model Representation

The modulated transmitted signal is real-valued in an actual implementation since the transmitting oscillator circuits do not generate complex exponentials. However, the baseband signal $s(t)$ can be complex-valued. The actual transmitted signal is given by

$$\begin{aligned} s^r(t) =& \text{Re}\left[s^c(t)\right] = \text{Re}\left[s(t)e^{j(2\pi f_c t+\theta)}\right] \\ =& \frac{1}{2}\left[s(t)e^{j(2\pi f_c t+\theta)}\right] + \frac{1}{2}\left[s^*(t)e^{-j(2\pi f_c t+\theta)}\right] \\ =& \text{Re}\left[s(t)\right]\cos(2\pi f_c t+\theta) - \text{Im}\left[s(t)\right]\sin(2\pi f_c t+\theta) \end{aligned} \tag{14.28}$$

In some textbooks, the transmitted signal has the form:

$$s^r(t) = \sqrt{2}\text{Re}\left[s(t)\right]\cos(2\pi f_c t+\theta) - \sqrt{2}\text{Im}\left[s(t)\right]\sin(2\pi f_c t+\theta) \tag{14.29}$$

where the term $\sqrt{2}$ guarantees that the power of the transmitted signal remains the same, i.e., it remains unchanged by the carrier.

If we imagine that the complex signal transmission is complex modulated, the real part of the received signal is then represented by

$$\begin{aligned} r^r(t) =& \text{Re}\left[g\sum_{m=1}^{M} a_m s(t+\frac{f_d}{f_c}\cos\phi_m t-\tau_m)e^{j[2\pi((f_c+f_d\cos\phi_m)t-f_c\tau_m)+\theta]}\right] \\ =& \frac{g}{2}\left[\left(\sum_{m=1}^{M} a_m s(t+\frac{f_d}{f_c}\cos\phi_m t-\tau_m)e^{j[2\pi((f_c+f_d\cos\phi_m)t-f_c\tau_m)+\theta]}\right)\right. \\ &\left.+\left(\sum_{m=1}^{M} a_m s^*(t+\frac{f_d}{f_c}\cos\phi_m t-\tau_m)e^{-j[2\pi((f_c+f_d\cos\phi_m)t-f_c\tau_m)+\theta]}\right)\right] \end{aligned} \tag{14.30}$$

If we remove the carrier from the received signal, assuming an accurate synchronization between transmitter and receiver, followed by low-pass filtering of the result, we obtain:

$$\begin{aligned} r(t) =& \text{LP}\left\{\frac{g}{2}\left[\left(\sum_{m=1}^{M} a_m s(t+\frac{f_d}{f_c}\cos\phi_m t-\tau_m)e^{j[2\pi((f_c+f_d\cos\phi_m)t-f_c\tau_m)+\theta]}\right)\right.\right. \\ &\left.+\left(\sum_{m=1}^{M} a_m s^*(t+\frac{f_d}{f_c}\cos\phi_m t-\tau_m)e^{-j[2\pi((f_c+f_d\cos\phi_m)t-f_c\tau_m)+\theta]}\right)\right] \\ &\left.\times\left[e^{-j[2\pi f_c(t-\tau)+\theta]}+e^{j[2\pi f_c(t-\tau)+\theta]}\right]\right\} \\ =& \text{LP}\left\{\frac{g}{2}\left[\left(\sum_{m=1}^{M} a_m s(t+\frac{f_d}{f_c}\cos\phi_m t-\tau_m)e^{j2\pi[f_d\cos\phi_m t+f_c(\tau-\tau_m)]}\right)\right.\right. \end{aligned}$$

$$+\left(\sum_{m=1}^{M} a_m s^*(t+\frac{f_d}{f_c}\cos\phi_m t-\tau_m)e^{-j2\pi[f_d\cos\phi_m)t+f_c(\tau-\tau_m)]}\right)$$
$$+\left(\sum_{m=1}^{M} a_m s(t+\frac{f_d}{f_c}\cos\phi_m t-\tau_m)e^{j[4\pi f_c(t-\tau_m)+2\pi f_d\cos\phi_m t+f_c(\tau-\tau_m)+2\theta]}\right)$$
$$+\left(\sum_{m=1}^{M} a_m s^*(t+\frac{f_d}{f_c}\cos\phi_m t-\tau_m)e^{-j[4\pi f_c(t-\tau_m)+2\pi f_d\cos\phi_m t+f_c(\tau-\tau_m)+2\theta]}\right)\Bigg]\Bigg\} \tag{14.31}$$

The last two terms in the above equation can be ignored due to the low-pass filter effect, leading to

$$r(t)=\frac{g}{2}\left[\left(\sum_{m=1}^{M} a_m s(t+\frac{f_d}{f_c}\cos\phi_m t-\tau_m)e^{j2\pi[f_d\cos\phi_m t+f_c(\tau-\tau_m)]}\right)\right.$$
$$\left.+\left(\sum_{m=1}^{M} a_m s^*(t+\frac{f_d}{f_c}\cos\phi_m t-\tau_m)e^{-j2\pi[f_d\cos\phi_m t+f_c(\tau-\tau_m)]}\right)\right] \tag{14.32}$$

The above expression can be rewritten in the form below if we recall that $f_c=\frac{c}{\lambda}$ and $f_d=\frac{v}{\lambda}$, so that $\frac{f_d}{f_c}=\frac{v}{c}$.

$$r(t)=\frac{g}{2}\left[\left(\sum_{m=1}^{M} a_m s(t+\frac{v}{c}\cos\phi_m t-\tau_m)e^{j2\pi[f_d\cos\phi_m t+f_c(\tau-\tau_m)]}\right)\right.$$
$$\left.+\left(\sum_{m=1}^{M} a_m s^*(t+\frac{v}{c}\cos\phi_m t-\tau_m)e^{-j2\pi[f_d\cos\phi_m t+f_c(\tau-\tau_m)]}\right)\right] \tag{14.33}$$

For a constant movement between transmitter and receiver, the distance the signal traverses, related to the mth path, is given by

$$d_m(t)=d_{m0}-v\cos\phi_m t \tag{14.34}$$

where d_{m0} is the initial distance between the transmitter and the receiver in the m-th path, so that

$$\frac{\partial d(t)}{\partial t}=v\sin\phi_m t \tag{14.35}$$

The overall delay is given by

$$\tau_m(t) = \frac{d_m(t)}{c} = \frac{d_{m0}}{c} - \frac{v}{c}\cos\phi_m t = \tau_m - \frac{v}{c}\cos\phi_m t \tag{14.36}$$

If there is no Doppler, i.e., $v = 0$, the received signal is given by

$$r(t) = g\mathrm{Re}\left(\sum_{m=1}^{M} a_m s(t-\tau_m)e^{j2\pi f_c(\tau-\tau_m)}\right) \tag{14.37}$$

In addition, if the carrier demodulation delay τ in (14.31) is equal to zero, the above equation becomes:

$$r(t) = g\mathrm{Re}\left(\sum_{m=1}^{M} a_m s(t-\tau_m)e^{-j2\pi f_c\tau_m}\right) \tag{14.38}$$

where

$$\hat{h}(t) = \sum_{m=1}^{M} a_m \delta(t-\tau_m)e^{-j2\pi f_c\tau_m} \tag{14.39}$$

is the baseband channel model corresponding to the passband channel model given by

$$h(t) = \sum_{m=1}^{M} a_m \delta(t-\tau_m) \tag{14.40}$$

As can be observed, the baseband channel model delays the baseband signal by the same amount of time as the passband signal faces the passband channel. However, each signal reflection faces a phase change given by the term $e^{-j2\pi f_c\tau_m}$ whose angle is a function of the delay τ_m and the carrier frequency f_c.

Note that in this section's discussion, we utilized the complex representation of the transmitted and received signals. For the sake of completion, we illustrate the full real-valued representation of the received signal.

14.6.1 *Real-Valued Demodulation*

For the case where the modulated transmitted signal is real-valued, and the demodulation is also a real cosine waveform, the actual received transmitted signal, followed by demodulation and low-pass filtering, is given by

$$\begin{aligned} r(t) =&\text{LP}\left\{g\left[\sum_{m=1}^{M} a_m \text{Re}\left[s(t+\frac{f_d}{f_c}\cos\phi_m t-\tau_m)\right]\right.\right.\\ &\cos\left(2\pi[(f_c+f_d\cos\phi_m)t-f_c\tau_m]+\theta\right)\\ &\left.\left.+\ \text{Im}\left[s(t+\frac{f_d}{f_c}\cos\phi_m t-\tau_m)\right]\sin\left(2\pi[(f_c+f_d\cos\phi_m)t-f_c\tau_m]+\theta\right)\right]\right\}\\ &\times 2\cos\left(2\pi f_c(t-\tau)+\theta\right) \end{aligned} \tag{14.41}$$

The gain of 2 multiplying the cosine leads to a more straightforward expression in the following discussion.[7] The terms at high frequencies in the vicinity of $4\pi f_c$[8] are eliminated by the low-pass filtering, resulting in

$$\begin{aligned} r(t) =&g\left[\sum_{m=1}^{M} a_m \text{Re}\left[s(t+\frac{f_d}{f_c}\cos\phi_m t-\tau_m)\right]\cos\left(2\pi(f_d\cos\phi_m t+f_c(\tau-\tau_m))\right)\right.\\ &\left.+\ \text{Im}\left[s(t+\frac{f_d}{f_c}\cos\phi_m t-\tau_m)\right]\sin\left(2\pi\left(f_d\cos\phi_m t+f_c(\tau-\tau_m)\right)\right)\right]\\ =&g\left[\sum_{m=1}^{M} a_m \text{Re}\left[s(t+\frac{v}{c}\cos\phi_m t-\tau_m)\right]\cos\left(2\pi(f_d\cos\phi_m t+f_c(\tau-\tau_m))\right)\right.\\ &\left.+\ \text{Im}\left[s(t+\frac{v}{c}\cos\phi_m t-\tau_m)\right]\sin\left(2\pi\left(f_d\cos\phi_m t+f_c(\tau-\tau_m)\right)\right)\right] \end{aligned} \tag{14.42}$$

Note that the real and imaginary parts of the complex signal can still be detected by adequately projecting $r(t)$ in the appropriate quadrature demodulation.

14.7 General Complex-Valued Signal and Carrier

For the sake of answering a common question of why not consider the signal and the carrier complex at the same time, we show how this case would look like without

[7] Usually, if the transmitted signal in the baseband is real and converted to the passband by a simple cosine, the carrier is $\sqrt{2}\cos(2\pi f_c t)$ so that the power of the baseband and passband signals are the same.

[8] We utilized the trigonometric equalities $\cos(a)\cos(b) = \frac{1}{2}[\cos(a-b)+\cos(a+b)]$ and $\sin(a)\cos(b) = \frac{1}{2}[\sin(a-b)+\sin(a+b)]$.

taking into account any channel model. The transmitted signal is given by

$$
\begin{aligned}
s^c(t) =& s(t)e^{j(2\pi f_c t+\theta(t))} \\
& \{\mathrm{Re}\,[s(t)] + j\mathrm{Im}\,[s(t)]\}\,\{\cos(2\pi f_c t + \theta(t)) + j\,\sin(2\pi f_c t + \theta(t))\} \\
=& \mathrm{Re}\,[s(t)]\cos(\theta(t))\cos(\pi f_c t) - \mathrm{Re}\,[s(t)]\sin(\theta(t))\sin(\pi f_c t) \\
& - \mathrm{Im}\,[s(t)]\sin(\pi f_c t)\cos(\theta(t)) - \mathrm{Im}\,[s(t)]\sin(\theta(t))\cos(\pi f_c t) \\
& j\,\{\mathrm{Im}\,[s(t)]\cos(\theta(t))\cos(\pi f_c t) - \mathrm{Im}\,[s(t)]\sin(\theta(t))\sin(\pi f_c t) \\
& + \mathrm{Re}\,[s(t)]\cos(\theta(t))\sin(\pi f_c t) + \mathrm{Re}\,[s(t)]\sin(\theta(t))\cos(\pi f_c t)\}
\end{aligned}
\tag{14.43}
$$

where $s(t)$ is a complex-valued signal and f_c is the carrier frequency and $\theta(t)$ is the phase of the carrier.

After a few manipulations, the real part of the above expression becomes:

$$
\begin{aligned}
\mathrm{Re}\left[s^c(t)\right] =& \{\mathrm{Re}\,[s(t)]\cos(\theta(t)) - \mathrm{Im}\,[s(t)]\sin(\theta(t))\}\cos(2\pi f_c t) \\
& - \{\mathrm{Re}\,[s(t)]\sin(\theta(t)) + \mathrm{Im}\,[s(t)]\cos(\theta(t))\}\sin(2\pi f_c t)
\end{aligned}
\tag{14.44}
$$

As can be observed, the transmitted signal can convey information in $s(t)$ without requiring a complex carrier since the second term is related to the $\sin(2\pi f_c t)$ would not be required to extract the terms $\mathrm{Re}\,[s(t)]$ and $\mathrm{Im}\,[s(t)]$.

It is worth mentioning that if the phase $\theta(t)$ conveys the data information, we are dealing with an FM-type transmission.

The distinct forms representing the transmitted signal and its received counterparts are intended to facilitate a quick understanding of the various descriptions available in the literature and simulation packages.

14.8 Remarks

Technological advancements have significantly influenced the field of channel modeling. Deterministic models, sometimes learned on the fly, replace traditional statistical models, eventually utilizing machine learning tools. There are many works on this subject that the interested reader can explore. Here are listed many references that treat several aspects of channel models spanning distinct presentations in terms of breadth and depth. Some examples are comprehensive textbooks on communication such as Lee (1998); Madhow (2008); Nguyen and Shwedyk (2009); Proakis and Salehi (1994); Haykin (2001); Barry (2003); Proakis (2001); Anderson (1999); Gallager (2008), and more focused books on multiuser detection, represented by the book (Verdu, 1998). There are also many textbooks covering more specific topics related to wireless communications, and some of those are Rappaport (2002); Tse and Viswanath (2005); Goldsmith (2005); Stüber (2011); Molisch (2005). More recent texts cover MIMO communication systems, such as Hottinen et al. (2003); Biglieri et al. (2007); Heath and Lozano (2019). The book by Berger presents essential topics related to rate distortion theory, providing some benchmarks to design communication systems (Berger, 1971). Of particular importance to the coverage of channel models are the papers Hashemi (1979, 1993) and the books Pätzold (2012); Rappaport (2002). In addition, the books Jeruchim et al. (2000); Tranter et al. (2004) provide helpful information for developing simulation examples.

The coverage of channel models here is superficial, focusing on a few cases of representations to help the reader quickly understand how they are used in actual simulations and deployment evaluations. In autonomous systems, a much wider range of channel models is required to address the fading faced by multiple-class signal sources. The channel model differs for each class. If the autonomous vehicle is immersed in an environment packed with wireless, lidar, and image sources, the channel models are likely embedded in an AI-support system to warrant robust and optimized use of the signal resources.

References

J.B. Anderson, *Digital Transmission Engineering* (IEEE Press, New York, 1999)

J.B. Barry, E.A. Lee, S. Messerschmitt, *Digital Communications*, 3rd edn. (Kluwer Academic Publishers, Norwell, MA, 2003)

T. Berger, *Rate Distortion Theory* (Prentice-Hall, Englewood Cliffs, 1971)

E. Biglieri, R. Calderbank, A. Constantinides, A. Goldsmith, A. Paulraj, H.V. Poor, *MIMO Wireless Communications* (Cambridge University, Cambridge, 2007)

R.G. Gallager, *Principles of Digital Communication* (Cambridge University, Cambridge, 2008)

A. Goldsmith, *Wireless Communications* (Cambridge University, Cambridge, 2005)

H. Hashemi, Simulation of the urban radio propagation channel. IEEE Trans. Veh. Technol. **VT-28**, 213–225 (1979)

H. Hashemi, The indoor radio propagation channel. Proc. IEEE **83**, 943–968 (1993)

S. Haykin, *Communication Systems*, 4th edn. (Wiley, New York, 2001)

R.W. Heath Jr., A. Lozano, *Foundations of MIMO Communication* (Cambridge University, Cambridge, 2019)

A. Hottinen, O. Tirkkonen, R. Wichman, *Multi-Antenna Transceiver Techniques for 3G and Beyond* (Wiley, New York, 2003)

M.C. Jeruchim, P. Balaban, S. Shanmugan, *Simulation of Communication System: Modeling, Methodology and Techniques*, 2nd edn. (Plenum Press, New York, 2000)

E.G. Larsson, P. Stoica, *Space-Time Block Coding for Wireless Communications* (Cambridge University, Cambridge, 2003)

W.C.Y. Lee, *Mobile Communications Engineering* (McGraw Hill, New York, 1998)

U. Madhow, *Fundamentals of Digital Communication* (Cambridge University, Cambridge, 2008)

T.L. Marzetta, E. Larsson, H. Yang, H.Q. Ngo, *Fundamentals of Massive MIMO* (Cambridge University, Cambridge, 2016)

A.F. Molisch, *Wireless Communications* (Wiley, Hoboken, 2005)

A.F. Naguib, Adaptive antennas for CDMA wireless networks, Ph.D. dissertation (Stanford University, Stanford,1996)

H. Nguyen, E. Shwedyk, *A first Course in Digital Communications* (Cambridge University, Cambridge, 2009)

M. Pätzold, *Mobile Fading Channels*, 2nd edn. (Wiley, Chischester, 2012)

A. Paulraj, R. Nabar, D. Gore, *Introduction to Space-Time Wireless Communications* (Cambridge University Press, Cambridge, 2003)

J.G. Proakis, *Digital Communications*, 4th edn. (McGraw Hill, New York, 2001)

J.G. Proakis, M. Salehi, *Communication Systems Engineering* (Prentice Hall, Upper Saddle River, NJ, 1994)

T.S. Rappaport, *Mobile Communications: Principles and Practice*, 2nd edn. (Prentice Hall, Upper Saddle River, 2002)

G. Stüber, *Principles of Mobile Communication*, 3rd edn. (Springer, New York, 2011)

W.H. Tranter, S. Shanmugan, T.S. Rappaport, and K.L. Kosbar, *Principles of Communication Systems Simulation with Wireless Applications* (Prentice Hall, Upper Saddle River, 2004)

D. Tse, P. Viswanath, *Fundamentals of Wireless Communications* (Cambridge University, Cambridge, 2005)

G.L. Turin, F.D. Clapp, T.L. Johnson, S.B. Fine, D. Lavry, A statistical model of urban multipath propagation. IEEE Trans. Veh. Technol. **VT-21**, 1–9 (1972)

H.L. Van Trees, *Optimum Array Processing: Detection, Estimation, and Modulation Theory, Part IV* (Wiley Interscience, New York, 2002)

S. Verdu, *Multiuser Detection* (Cambridge University, Cambridge, 1998)

A.J. Viterbi, *Principles of Spread Spectrum Communication* (Addison Wesley, Reading, MA, 1995)

Chapter 15
CDMA System

15.1 Introduction

In this chapter, we will discuss the basic concepts of CDMA, given its widespread use and its importance to modern wireless communication systems. Although CDMA is becoming less important with the advance of 5G systems, many related concepts form the basis of many building blocks in several autonomous vehicle solutions. The chapter discusses the basic concepts, channel capacity, spreading codes, and receiver design.

The spreading codes play a crucial role in satellite navigation systems, such as global positioning technology, which can offer precision, security, and efficiency. The Walsh-Hadamard sequence has been employed as part of orthogonal multiple access methods to allow users to share the same radio resources. These are just examples of the many uses of spreading codes discussed in this chapter.

15.2 Basic Concepts

Let us assume that in a CDMA system, the lth user bitstream is given by $b_l(kT_b)$ for $-\infty \leq k \leq \infty$. The parameter T_b represents the time allotted for a bit.

Each user bitstream is multiplied by its own signature waveform $c_l(t)$, with the following property:

$$\|c_l(t)\|^2 = \frac{1}{T_b}\int_0^{T_b} c_l^2(t)dt = 1 \tag{15.1}$$

where $c_l(t) = 0, 0 < t, t > T_b$.

The signature waveform (or code sequences) usually consists of a sequence of concatenated pulses, $p_l(t - iT_c)$, whose amplitudes are $+1$ or -1 according to

Y. C. Lim et al., *Signal Processing for Autonomous Systems*,
https://doi.org/10.1007/978-3-032-00675-2_15

the pseudorandom sequence of the user. The pulse duration is T_c with $T_c \ll T_b$. The inverse of T_c is called *chip rate*. With the spreading multiplication, the bandwidth of the baseband signal is much larger than that of the data sequence. Figure 15.1 illustrates a piece of a bitstream $(1, -1, 1)$ that multiplies a code sequence to generate the transmitted signal. As can be observed, the CDMA signal generation is quite simple.

The ratio

$$K = \frac{T_b}{T_c} \tag{15.2}$$

is known as processing gain. These definitions explain the term spread spectrum.

The signature waveform for the lth user is given by

$$c_l(t) = \sum_{i=1}^{K} a_{l,i}\, p_l(t - iT_c) \tag{15.3}$$

where the coefficients $a_{l,i} \in \{-1, +1\}$ for $i = 1, \ldots, K$ are the elements of the lth user chip code.

The system above is the so-called direct-sequence CDMA system (DS-CDMA). In practice, the concatenated pulses, $p_l(t - iT_c)$, need not be rectangular pulses. They can have other forms. Alternative pulse formats allow a reduction of the frequency content beyond the spectral null placed at $\frac{1}{T_c}$. The actual transmitted signal is the multiplication of the baseband signal with a carrier given by

$$\sqrt{2E}\cos(\omega_c t + \theta(l)) \tag{15.4}$$

where ω_c is the carrier frequency in rad/sec and $\theta(l)$ is the random phase of the lth user.

15.2.1 Transmitted Signal

The baseband signal of the lth user is given by

$$\begin{aligned} s_l(t) &= \sum_{n=-\infty}^{\infty} b_l(nT_b) c_l(t - nT_b) \\ &= \sum_{n=-\infty}^{\infty} b_l(nT_b) \sum_{i=0}^{K} a_{l,i}\, p_l(t - nT_b - iT_c) \end{aligned} \tag{15.5}$$

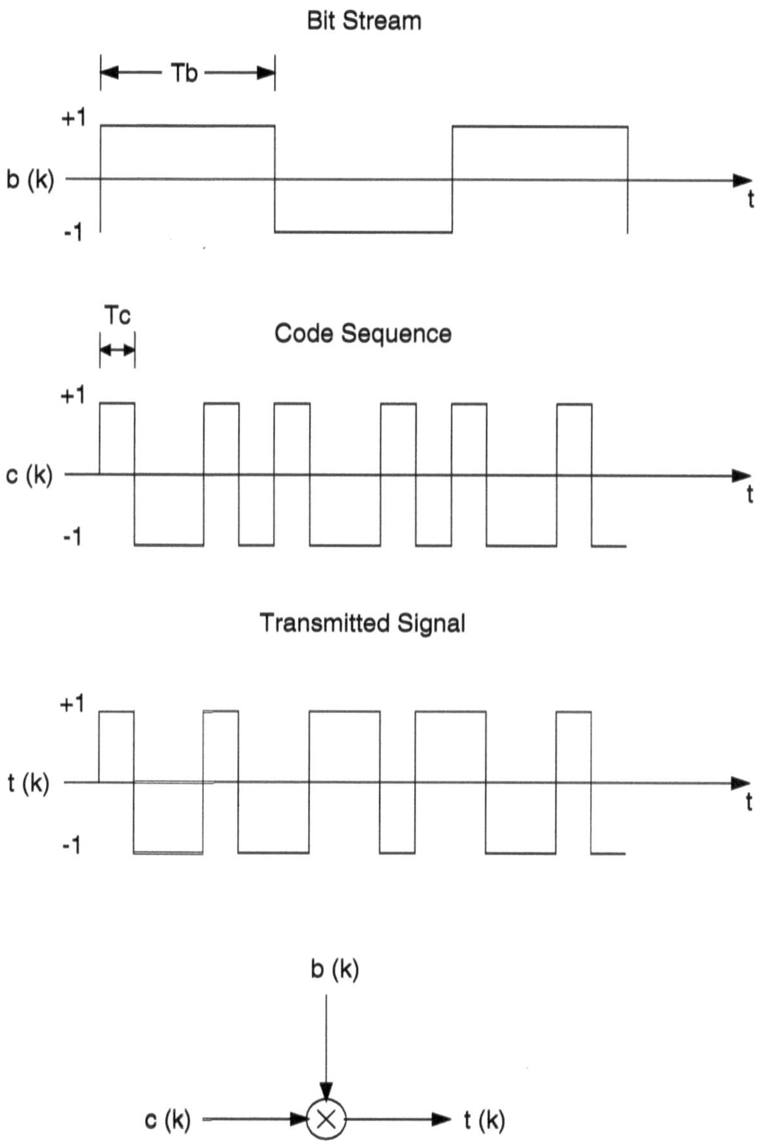

Fig. 15.1 CDMA signal generation

The actual transmitted signal is

$$m_l(t) = \sqrt{2E_l}[\sum_{n=-\infty}^{\infty} b_l(nT_b)c_l(t-nT_b)]\cos(\omega_c t + \theta(l))$$
$$= \sqrt{2E_l}[\sum_{n=-\infty}^{\infty} b_l(nT_b)\sum_{i=0}^{K} a_{l,i}p_l(t-nT_b-iT_c)]\cos(\omega_c t + \theta(l)) \quad (15.6)$$

15.2.2 *Received Signal*

The signal reaching the receiver is a combination of signals coming from different transmitters:

$$r(t) = \sum_{l=0}^{L}\sqrt{2E_l}[\sum_{n=-\infty}^{\infty} b_l(nT_b)c_l(t-nT_b-\tau_l)]\cos[\omega_c(t-\tau_l)+\theta(l)] + n(t)$$
$$= \sum_{l=0}^{L}\sqrt{2E_l}[\sum_{n=-\infty}^{\infty} b_l(nT_b)\sum_{i=0}^{K} a_{l,i}p_l(t-nT_b-iT_c-\tau_l)]$$
$$\times \cos[\omega_c(t-\tau_l)+\theta(l)] + n(t) \quad (15.7)$$

where $n(t)$ is the channel noise and τ_l is the relative delay of the lth user. The channel noise is usually considered a white Gaussian noise with variance σ_n^2.

For synchronous CDMA systems, τ_l is constant for all users, which means they are synchronized. On the other hand, for asynchronous CDMA, each user has a different value for the delay. As is usually the case, synchronous systems have better performance (concerning interference rejection); however, since it is sometimes difficult to control the delay introduced by the channel, asynchronous systems are more often used.

15.3 Spreading Codes

Spreading codes are sequences that allow multiple access allowind users to utilize the same band simultaneously, providing security and more robustness to jamming and inherent interferences. Their orthogonal versions help provide simultaneous services requiring distinct data rates, as long as they are synchronized.

The spreading codes $a_{l,i} \in \{-1, +1\}$ for $i = 0, 1, \ldots, K$ can be short such that $(K+1)T_c = T_b$, or long where $(K+1)T_c \gg T_b$. The UMTS 3G system uses long codes, and in addition, it is a wideband CDMA (WCDMA) system.

CDMA system using long codes was employed in the IS-95 system for the uplink connection (mobile to base station). This was also the choice for the third-generation mobile systems previously deployed.

Some issues related to the length of the spreading codes are worth mentioning. Short-code systems require good code planning, and assume strict power control; otherwise, the nonorthogonal signature of the interferer (with high power) has a strong projection on the desired user code direction. In a short-code system, since the interference remains unchanged from bit to bit, an unlucky user will have a lousy connection all the way. In addition, the received signal is cyclostationary for which powerful signal processing algorithms exist.

If power control is used in long-code systems, all users will be exposed to the same level of interference, and the initial delay estimation might take longer than for short-code systems. Figure 15.2 illustrates a simple case of long-code spreading. Each data bit multiples a different spreading code, unlike the short-code spreading of Fig. 15.1.

15.3.1 Finite-State Machine

Spreading codes are usually generated with shift registers (SRs) that use feedback logic which is a finite-state machine; see Fig. 15.3. A good set of codes is one in which their mean value is approximately zero, that is

$$\frac{1}{K+1}\sum_{i=0}^{K} a_{l,i} \approx 0 \tag{15.8}$$

and their autocorrelation is approximately given by

$$\frac{1}{K+1}\sum_{i=0}^{K} a_{l,i}a_{l,i+j} \approx \begin{cases} 1, \ j=0 \\ 0, \ j>0 \end{cases} \tag{15.9}$$

Although the conditions cannot be met in practice, they can be approximately achieved by maximum-length sequences. For these sequences

$$\sum_{i=0}^{K} a_{l,i} = \pm 1 \tag{15.10}$$

and as a consequence, we have:

$$\frac{1}{K+1}\sum_{i=0}^{K} a_{l,i}a_{l,i+j} \approx \begin{cases} 1, & j=0 \\ \frac{-1}{K+1}, & j>0 \end{cases} \tag{15.11}$$

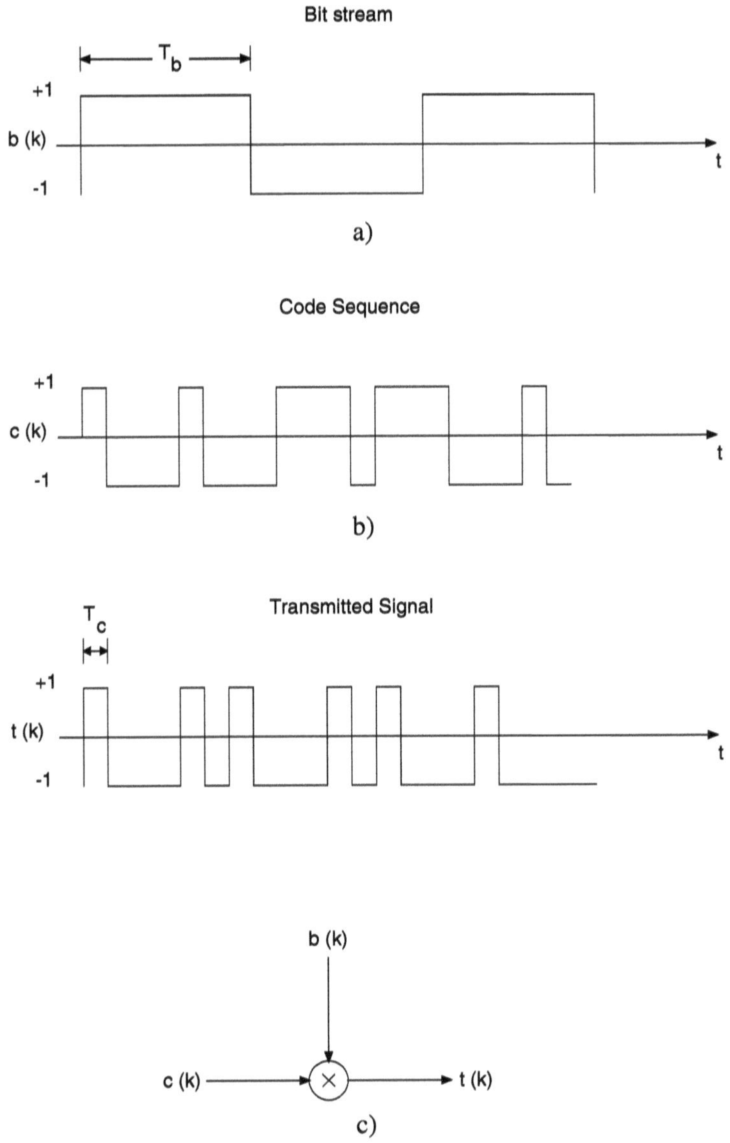

Fig. 15.2 Long-code example. (**a**) Original sequence of bits. (**b**) Long-code CDMA. (**c**) Long-coded signal

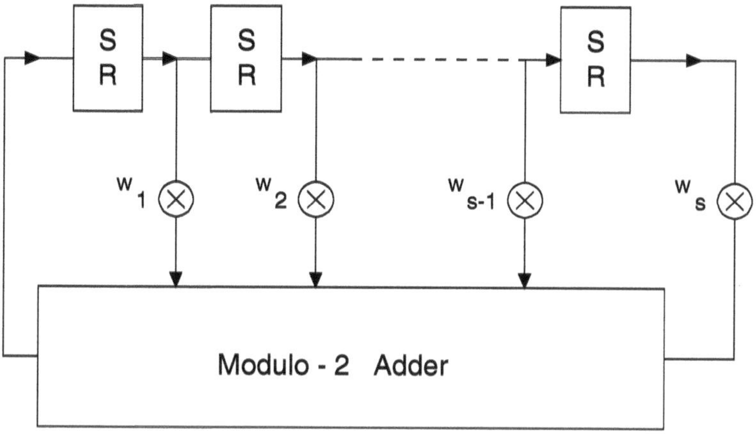

Fig. 15.3 Pseudo-noise (PN) generation

Also, for these sequences, the numbers of $+1$ and -1 differ by one.

A realization of a pseudo-code generator with s shift registers would lead to $(K+1) = 2^s - 1$ distinct states, excluding the undesirable zero state. Other alternatives for the maximum-length pseudo-noise are the Gold, Kasami, and Walsh-Hadamard sequences.

15.3.2 Wash-Hadamard Sequences

Widely used code sequences are the Walsh-Hadamard sequences that can be easily built. Starting with

$$\mathbf{H}_2 = \begin{bmatrix} 1 & 1 \\ 1 & -1 \end{bmatrix} \tag{15.12}$$

We can generate long orthogonal sequences as follows:

$$\mathbf{H}_{2L} = \begin{bmatrix} \mathbf{H}_L & \mathbf{H}_L \\ \mathbf{H}_L & -\mathbf{H}_L \end{bmatrix} \tag{15.13}$$

An inherent characteristic is that orthogonal codes are good only if the synchronization is accurate (within a small fraction of a chip). The cross-correlation between sequences with different shifts is high.

For example,

$$\mathbf{H}_4 = \begin{bmatrix} 1 & 1 & 1 & 1 \\ 1 & -1 & 1 & -1 \\ 1 & 1 & -1 & -1 \\ 1 & -1 & -1 & 1 \end{bmatrix} \tag{15.14}$$

This sequence is attractive for forward links in CDMA systems because they are orthogonal, and the signals leave the base station synchronized. Unfortunately, multipath fading can destroy their usefulness because of their poor autocorrelation properties. For example, Walsh-Hadamard sequences are used in the IS-95 and CDMA2000 wireless communications CDMA systems.

15.3.3 Variable-Length Codes

Variable-length orthogonal codes are used in mobile systems supporting variable data rates. An example is the wideband CDMA of the third- generation (3G) wireless communication systems.

Each user will have the same signal bandwidth. The user with the lowest bit rate will have its information spread by the longest spread sequence. A user with double the lowest bit rate will use a twice shorter spread sequence and so forth.

Suppose we start with $\mathbf{V}_1 = 1$ to generate

$$\mathbf{V}_2 = \begin{bmatrix} 1 & 1 \\ 1 & -1 \end{bmatrix} \tag{15.15}$$

Now, suppose the first row of $\mathbf{V}_2$ generates the first two rows of $\mathbf{V}_4$ and the second row of $\mathbf{V}_2$ generates the last two rows of $\mathbf{V}_4$, as follows:

$$\mathbf{V}_4 = \begin{bmatrix} 1 & 1 & 1 & 1 \\ 1 & 1 & -1 & -1 \\ 1 & -1 & 1 & -1 \\ 1 & -1 & -1 & 1 \end{bmatrix} \tag{15.16}$$

We can proceed as above to generate a larger orthogonal code.

Main features:

- At a given layer, the spreading codes are orthogonal.
- Any two codes of different layers are orthogonal if one is not the mother of the other (grandmother, great-grandmother, etc.).

15.3.3.1 Multiple Spreading Codes

Multiple spreading codes usually consist of two layers of spreading used to facilitate the deployment. One layer can maintain orthogonality among users of the same cell while maintaining mutual randomness among users of different cells. Some types of multiple spreading codes are channeling codes, where each user data is multiplied by a short spread sequence orthogonal to all other users in the cell (it could be a variable-length Walsh orthogonal code), and scrambling codes, where the resulting signal is multiplied by a long pseudorandom sequence belonging to all users of that particular cell in the downlink (forward link) and in the uplink (reverse link); each user has its own PN sequence.

15.3.3.2 Codes for IS-95 and WCDMA

IS-95 standard for 2G wireless communication used in the downlink a Walsh orthogonal sequence of length 64 followed by maximum-length sequences with $2^{15} - 1$ chips. Each cell uses the same PN sequence but with different offsets since the system works synchronously (Lee and Miller 1998). Each user employs a maximum-length sequence in the uplink with $2^{15} - 1$ chips. Also, each user employs a different offset.

WCDMA of 3G uses in the forward link a variable-length orthogonal sequence followed by ten milliseconds of $2^{18} - 1$ chip Gold sequence. Each cell uses the same PN sequence. Each user employs a variable-length orthogonal sequence in the reverse link, followed by a very large set of Kasami sequences or 10 milliseconds of a $2^{41} - 1$ Gold sequence.

Example 15.1 (Simple Example) In the following, we present a toy example to provide some insight into the CDMA implementation in simple environment.

Take as spreading-code sequences two Walsh-Hadamard sequences of length 8. Generate 4 data bits for each user by flipping a coin (± 1 values). (i) Transmit the user data in a noisy channel, where the noise is generated by flipping a coin eight times (the same noise is used on each bit frame) and using 0.25 for the gain. Check the detection for each user. Assume the received signal:

(a) synchronized with the transmitted signal,
(b) and delayed by one chip.
(c) Repeat the previous problem if the channel model is

$$\frac{1}{2}(1 + z^{-1}) \tag{15.17}$$

plus the additional noise.

Solution

(a)

1. $\mathbf{h}_8^1 = [1\ -1\ 1\ -1\ 1\ -1\ 1\ -1]^T$ and $\mathbf{h}_8^2 = [1\ 1\ -1\ -1\ 1\ 1\ -1\ -1]^T$ are selected as the spreading codes.
2. The user' bits are assembled in vectors as $\mathbf{c}_1 = [1\ -1\ 1\ -1]^T$ and $\mathbf{c}_2 = [-1\ -1\ 1\ 1]^T$, respectively.
3. The channel noise is assumed to be $\mathbf{n} = [0.25\ -0.25\ -0.25\ 0.25\ 0.25\ 0.25\ -0.25\ 0.25]^T$.
4. The spreading codes and the received bits are synchronized.
5. The integrated received bits, corresponding to the summation of the integration of the despread waveform within the time of support of each user data, are

$$\hat{\mathbf{c}}_1 = \left[7.5\ {-8.5}\ 7.5\ {-8.5}\right]^T,$$
$$\hat{\mathbf{c}}_2 = \left[-7.5\ {-7.5}\ 8.5\ 8.5\right]^T.$$

   It should be noted that no normalization was applied during the spreading and despreading process throughout this example.
6. The detection of the two users is

$$\hat{\mathbf{c}}_1 = \left[1\ {-1}\ 1\ {-1}\right]^T,$$
$$\hat{\mathbf{c}}_2 = \left[-1\ {-1}\ 1\ 1\right]^T.$$

7. The transferred bits are received correctly.

The signal waveforms are depicted in Fig. 15.4. The figure includes the spread information signals pertaining to each user, along with their combination according to the problem formulation. The figure also depicts the data signals of each user for proper comparison with the despread signals at the bottom. To detect the symbols of each user, we integrate the chips within each data symbol interval, and apply to the result a hard decision between the symbols 1 and -1.

(b)

1. The spreading codes and the received bits are not synchronized. The received bits are delayed by one chip period.
2. The integrated received bits are

$$\hat{\mathbf{c}}_1 = \left[-8.5\ 5.5\ {-4.5}\ 5.5\right]^T,$$

$$\hat{\mathbf{c}}_2 = \left[0.5\ {-1.5}\ 4.5\ {-1.5}\right]^T.$$

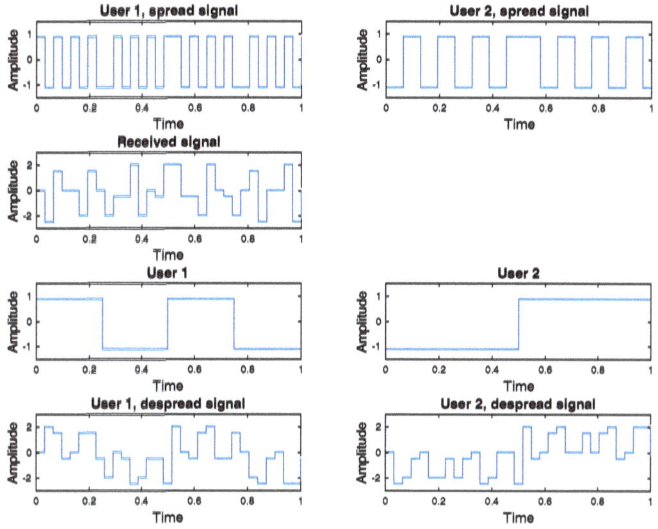

Fig. 15.4 Toy example sequences

3. The detection of the two users is

$$\hat{\mathbf{c}}_1 = \left[-1\ 1\ -1\ 1\right]^T,$$

$$\hat{\mathbf{c}}_2 = \left[1\ -1\ 1\ -1\right]^T.$$

4. The transferred bits are not received correctly. The second set of bits is lost because of the one-period delay of the received bits.

Figure 15.5 shows the relevant signals inherent to the problem. Their descriptions follow the previous items.

(c)

1. The spreading codes and the received bits are synchronized. The received bits face multipath.
2. The integrated received bits are

$$\hat{\mathbf{c}}_1 = \left[3.5\ -0.5\ -0.5\ 0.5\right]^T,$$

$$\hat{\mathbf{c}}_2 = \left[-3.5\ -7.5\ 0.5\ 8.5\right]^T.$$

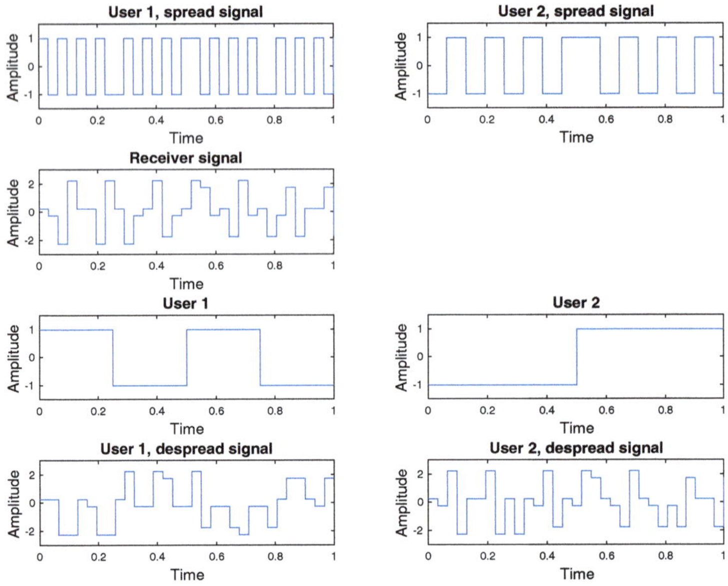

Fig. 15.5 Received and decoded sequences, including the received sequence; the decoded sequence I; and decoded sequence II; all were delayed by one chip

3. The detection of the two users is

$$\hat{\mathbf{c}}_1 = \begin{bmatrix} 1 & -1 & -1 & -1 \end{bmatrix}^T,$$

$$\hat{\mathbf{c}}_2 = \begin{bmatrix} -1 & -1 & 1 & 1 \end{bmatrix}^T.$$

4. The transferred bits are not received correctly. One bit of the first sequence is lost because of the multipath.

   Figure 15.6 depicts the relevant signals inherent to the problem.

15.3.4 Maximal-Length Sequences

The maximal-length sequences play a crucial role in generating spreading codes for several uses, particularly CDMA systems and wireless systems in general. The basic concepts used to design these sequences are outlined here. See (Dinan and Jabbari 1998; Sarwate and Pursley 1980) for details, and the textbooks on CDMA (Viterbi 1995) and (Verdu 1998).

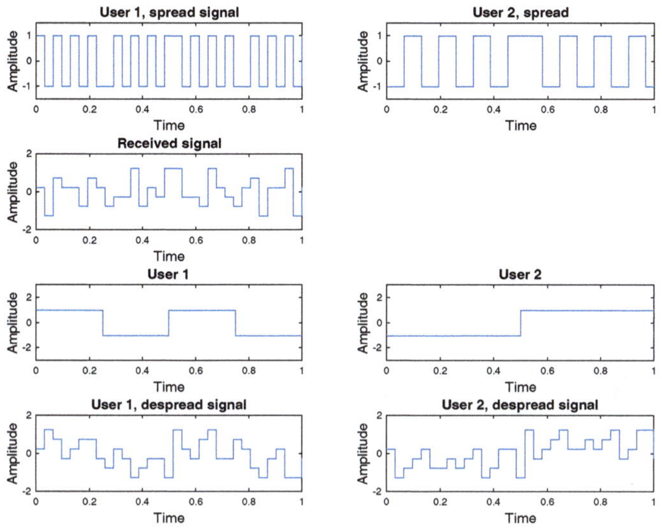

Fig. 15.6 Received and decoded sequences, including the received sequence; the decoded sequence I; and decoded sequence II; all frequency-selective channels

Assume that the generating function of the l-th user sequence is described as

$$F_l(z) = \sum_{i=0, Mod_2}^{\infty} a_{l,i} z^i \tag{15.18}$$

If the sequence $a_{l,i}$ is periodic with period $N = K + 1$, that is, $a_{l,i} = a_{l,i+N}$, the generating function can be written as

$$\begin{aligned} F_l(z) &= \sum_{i=0, Mod_2}^{\infty} z^{iN} (a_{l,0} \oplus a_{l,1} z^1 \oplus a_{l,2} z^2 \oplus \cdots \oplus a_{l,N-1} z^{N-1}) \\ &= \frac{a_{l,0} \oplus a_{l,1} z^1 \oplus a_{l,2} z^2 \oplus \cdots \oplus a_{l,N-1} z^{N-1}}{1 \oplus z^N} \end{aligned} \tag{15.19}$$

We can utilize the finite-state machine of Fig. 15.3 to generate the minimal-length sequence by starting with the following relation:

$$a_{l,0} = \sum_{j=1, Mod_2}^{S} a_{l,-j} w_j \tag{15.20}$$

or more generally

$$a_{l,i} = \sum_{j=1,Mod_2}^{S} a_{l,i-j} w_j \tag{15.21}$$

Combining (15.18) and (15.20), we get:

$$\begin{aligned}
F_l(z) &= \sum_{i=0,Mod_2}^{\infty} a_{l,i} z^i \\
&= \sum_{i=0,Mod_2}^{\infty} \sum_{j=1,Mod_2}^{S} a_{l,i-j} w_j z^i \\
&= \sum_{j=1,Mod_2}^{S} w_j z^j \sum_{i=0,Mod_2}^{\infty} a_{l,i-j} z^{i-j} \\
&= \sum_{j=1,Mod_2}^{S} w_j z^j (a_{l,-1} z^{-1} \oplus a_{l,-2} z^{-2} \oplus \cdots a_{l,-j} z^{-j} \oplus F_l(z))
\end{aligned} \tag{15.22}$$

so that

$$\begin{aligned}
F_l(z) &= \frac{\sum_{j=1,Mod_2}^{S} w_j z^j (a_{l,-1} z^{-1} \oplus a_{l,-2} z^{-2} \oplus \cdots a_{l,-j} z^{-j})}{1 \oplus \sum_{j=1,Mod_2}^{S} w_j z^j} \\
&= \frac{N_l(z)}{D(z)}
\end{aligned} \tag{15.23}$$

where $D(z)$ is the characteristic polynomial of the shift register circuit. $N_l(z)$ depends on the initial conditions of the SRs.

15.3.4.1 Properties of the Maximal-Length Sequences

Some properties characterize the maximum-length sequences as briefly described. The sequence generated has period N if $D(z)$ divides $1 \oplus z^{-N}$. See (15.19) and (15.23). To generate a maximum-length sequence, i.e., $N = 2^s - 1$, $D(z)$ must be irreducible, that is, $D(z) \neq D_1(z)D_2(z)$, this is a necessary but not sufficient condition.

An irreducible polynomial is called primitive (coefficients generated by $\frac{1}{D(z)}$ is of period $2^s - 1$), and the number of irreducible polynomials of degree s is given by

Dinan and Jabbari (1998)

$$N_P(s) = \frac{2^s - 1}{s} \prod_{i=1}^{k} \frac{P_i - 1}{P_i}$$

where P_i represents a decomposition of $2^s - 1$ with prime numbers.

The maximum-length sequences might have large cross-correlation, an undesirable feature. The cross-correlation satisfies the following property:

$$r_{l,j}(k) = \frac{1}{N} \sum a_{l,i} a_{j,i+k} \gg \sqrt{\frac{M-1}{MN-1}} \approx \frac{1}{\sqrt{N}}$$

However, this type of sequence has good autocorrelation; therefore, it is useful for synchronous systems. In asynchronous systems, low cross-correlations are of great importance.

The maximum-length sequences play a key role in the generation of other useful spread sequences.

15.3.5 Gold Sequences

Consider a maximum-length sequence $\mathbf{a}$ of length N. Generate a decimated sequence $\mathbf{a}_d$ of length N, where, in this case, copies of the original sequence can be thought to be concatenated before applying decimation. The resulting sequence will have a period N if and only if the greatest common divisor (gcd) between d and N is 1.

The sequences $\mathbf{a}$ and $\mathbf{a}_d$ are called *preferred sequences* for $N = 2^s - 1$, if:

1. s is either an odd number or if it is even, it is not a multiple of 4.
2. d must be odd and equal to $d = 2^i + 1$ or $d = 2^{2i} - 2^i + 1$.
3. Where i and s are either prime for s odd or their $gcd(s, i) = 2$ if s even (but not multiple of 4).

For odd s, the cross-correlation of the *preferred sequences* $\mathbf{a}$ and $\mathbf{a}_d$ can assume the values:

$$\frac{-1 - 2^{\frac{s+1}{2}}}{N}$$

$$\frac{-1}{N}$$

$$\frac{-1 + 2^{\frac{s+1}{2}}}{N}$$

and for s even:

$$\frac{-1 - 2^{\frac{s+2}{2}}}{N}$$
$$\frac{-1}{N}$$
$$\frac{-1 + 2^{\frac{s+2}{2}}}{N}$$

Once defined the preferred sequences, we can generate a family of sequences consisting of

$$\mathbf{a}$$
$$\mathbf{a}_d$$
$$\mathbf{a} \oplus \mathbf{a}_d$$
$$\mathbf{a} \oplus z^{-1}\mathbf{a}_d$$
$$\mathbf{a} \oplus z^{-2}\mathbf{a}_d$$
$$\vdots$$
$$\mathbf{a} \oplus z^{-N+1}\mathbf{a}_d$$

Any pair of these sequences of period N has the cross-correlation shown above. However, except for $\mathbf{a}$ and $\mathbf{a}_d$, the remaining sequences are not maximum length. Their autocorrelations assume the same values as the cross-correlations of the original sequences. The cross-correlation of a Gold sequence is bigger than the Welsh lower bound (a theoretical minimum for pseudorandom sequences) by two for s even and $\sqrt{2}$ for s odd.

15.3.6 Kasami Sequences

The family of Kasami sequences has very low cross-correlation.

For s even we can generate $2^{\frac{s}{2}}$ Kasami sequences of length $N = 2^s - 1$. Consider a maximum-length sequence $\mathbf{a}$ of length N. Generate a decimated sequence $\mathbf{a}_d$ of length $2^{\frac{s}{2}} - 1$ by employing a decimation factor $d = 2^{\frac{s}{2}} + 1$. Now add, modulo two, the bits of $\mathbf{a}$ with the bits of $\mathbf{a}_d$ and all its $2^{\frac{s}{2}} - 2$ cyclic shifts.

Including **a** we have $2^{\frac{s}{2}}$ sequences of length N. Their autocorrelation and crosscorrelations are

$$\frac{-1-2^{\frac{s}{2}}}{N}$$
$$\frac{-1}{N}$$
$$\frac{-1+2^{\frac{s}{2}}}{N}$$

These are called *small-set Kasami sequences*, which meet Welsh lower bounds.

We now discuss another class of Kasami sequences. Consider a maximum-length sequence **a** of length N. Generate a decimated sequence $\mathbf{a}_{d_1}$ by employing a decimation factor $d_1 = 2^{\frac{s}{2}} + 1$ (with the decimation sequence generated as in small-set Kasami) and another decimated sequence $\mathbf{a}_{d_2}$ by employing a decimation $d_2 = 2^{\frac{s+2}{2}} + 1$ (with the decimation sequence generated as in Gold). Add the sequence **a** with the decimated ones using all possible different shifts, where the decimated sequences are $\mathbf{a}_{d_1}$ and $\mathbf{a}_{d_2}$ (from Gold). The process is

$$\mathbf{a}$$
$$\mathbf{a}_{d_2}$$
$$\mathbf{a} \oplus z^{-m}\mathbf{a}_{d_2}$$
$$\mathbf{a} \oplus z^{-i}\mathbf{a}_{d_1}$$
$$\mathbf{a}_{d_2} \oplus z^{-i}\mathbf{a}_{d_1}$$
$$\mathbf{a} \oplus z^{-m}\mathbf{a}_{d_2} \oplus z^{-i}\mathbf{a}_{d_1}$$

If s is multiple of 4, we have $2^{\frac{3s}{2}}$ sequences; this case utilizes Gold-like sequences which are not really the ones presented here Sarwate and Pursley (1980). If s is even not multiple of 4, we have $2^{\frac{3s}{2}} + 2^{\frac{s}{2}}$ sequences, including the existing Gold sequence in this case. Their autocorrelation and cross-correlations are

$$\frac{-1 \pm 2^{\frac{s}{2}}}{N}$$
$$\frac{-1}{N}$$
$$\frac{-1 \pm 2^{\frac{s}{2}+1}}{N}$$

Table 15.1 Maximum-length sequences (generated with $p_1(z)$ and $p_2(z)$)

s_1	1	0	1	0	0	1	1
s_2	1	0	0	1	0	1	1

Table 15.2 Decimated sequences

Sequences		Decimated sequences						
$s_{1,d}$	$d = 3$	1	0	1	1	1	0	0
	$d = 5$	1	1	0	0	1	0	1
$s_{2,d}$	$d = 3$	1	1	1	0	1	0	0
	$d = 5$	1	1	1	0	1	0	0

Table 15.3 Gold sequences generated from sequence s_1 with decimation factor $d = 3$

Sequences	Gold sequences						
s_1	1	0	1	0	0	1	1
$s_{1,d}$	1	0	1	1	1	0	0
$s_1 \oplus s_{1,d}$	0	0	0	1	1	1	1
$s_1 \oplus z^{-1} s_{1,d}$	1	1	1	1	1	0	1
$s_1 \oplus z^{-2} s_{1,d}$	1	0	0	0	1	0	0
$s_1 \oplus z^{-3} s_{1,d}$	0	0	1	1	0	0	0
$s_1 \oplus z^{-4} s_{1,d}$	0	1	1	0	1	1	0
$s_1 \oplus z^{-5} s_{1,d}$	0	1	0	0	0	0	1
$s_1 \oplus z^{-6} s_{1,d}$	1	1	0	1	0	1	0

These are called *large-set Kasami sequences*, which do not meet Welsh lower bounds but are more efficient than Gold. The small-set Kasami and Gold sequences are special cases of the large-set Kasami sequences.

Example 15.2 (Spreading Codes Example) Generate Gold sequences with $N = 7$ $(s = 3)$.

Solution Possible generating polynomials: $p_1(z) = 1 + z^{-1} + z^{-3}$ and $p_2(z) = 1 + z^{-2} + z^{-3}$. In the sequel, we will use $p_1(z)$.

Table 15.1 shows the maximum-length sequences (generated with $p_1(z)$ and $p_2(z)$, respectively).

Possible decimation factors for $s = 3$: $d = \{3, 5\}$. Table 15.2 lists the decimated sequences.

Table 15.3 lists the Gold sequences generated from sequence s_1 with decimation factor $d = 3$.

Table 15.4 shows the Gold sequences generated from sequence s_2 with decimation factor $d = 3$:

Figure 15.7a, b illustrates the autocorrelation and cross-correlation inherent to the Gold sequences designed in this example, respectively. The relative comparison among the quantities in the vertical bar gives information regarding the effectiveness of the spreading code.

Table 15.4 Gold sequences generated from sequence s_2 with decimation factor $d = 3$

Sequences	Gold sequences						
s_2	1	0	0	1	0	1	1
$s_{2,d}$	1	1	1	0	1	0	0
$s_2 \oplus s_{2,d}$	0	1	1	1	1	1	1
$s_2 \oplus z^{-1}s_{2,d}$	1	1	1	0	0	0	1
$s_2 \oplus z^{-2}s_{2,d}$	1	0	1	0	1	1	0
$s_2 \oplus z^{-3}s_{2,d}$	0	0	0	0	1	0	1
$s_2 \oplus z^{-4}s_{2,d}$	1	1	0	1	1	0	0
$s_2 \oplus z^{-5}s_{2,d}$	0	0	1	1	0	0	0
$s_2 \oplus z^{-6}s_{2,d}$	0	1	0	0	0	1	0

15.4 DS-CDMA Receivers

15.4.1 Conventional Receiver

For the sake of simplifying, we will deal with real signals when discussing the receivers. The relation between real and complex-valued signals is given by

$$s_l(t)\cos(\omega_c t + \theta(l)) = \mathrm{Re}[s_l^c(t)e^{j(\omega_c t)}] \tag{15.24}$$

where

$$s_l^c(t) = s_l(t)e^{j(\theta(l))} \tag{15.25}$$

and Re[·] represents the real part of [·]. Please note a distinct notation from the previous chapter. Unlike (14.43), herein the above equation $\theta(l)$ represents the phase (constant) related to the lth user and $s_l(t)$ its symbol.

The detection of single-user bits is mathematically performed as

$$b_l'(kT_b) = \int_{kT_b+\tau_l}^{(k+1)T_b+\tau_l} c_l(t - kT_b - \tau_l)r(t)\cos[\omega_c(t - \tau_l) + \theta(l)]dt \tag{15.26}$$

where $r(t)$ represents the receiver signal, $\cos[\omega_c(t - \tau_l) + \theta(l)]$ is meant for the carrier removal with the appropriate delay and phase to match the part of the received signal pertaining to the l-th user, and $c_l(t - kT_b - \tau_l)$ is the user spread signal.

The detailed equation is

$$\begin{aligned} b_l'(kT_b) = \int_{kT_b+\tau_l}^{(k+1)T_b+\tau_l} & \{\sum_{m=0}^{L} \sqrt{2E_m}[\sum_{n=-\infty}^{\infty} b_m(nT_b)c_m(t - nT_b - \tau_m)] \\ & \times \cos[\omega_c(t - \tau_m) + \theta(m)] + n(t)\} \\ & \times c_l(t - kT_b - \tau_l)\cos[\omega_c(t - \tau_l) + \theta(l)]dt \end{aligned} \tag{15.27}$$

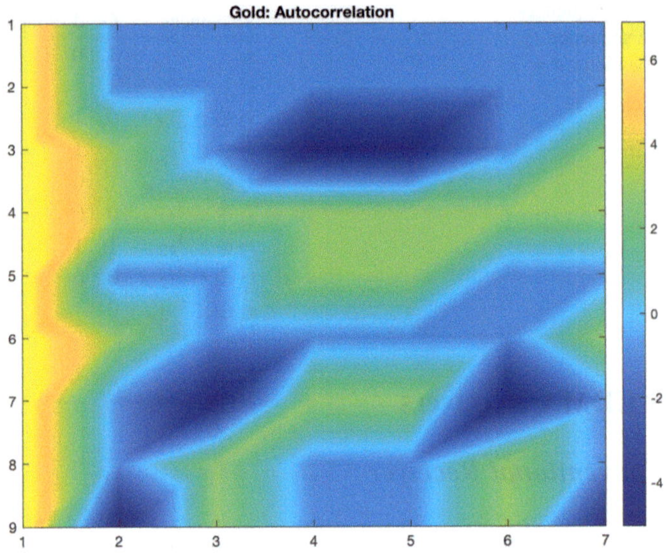

(a)

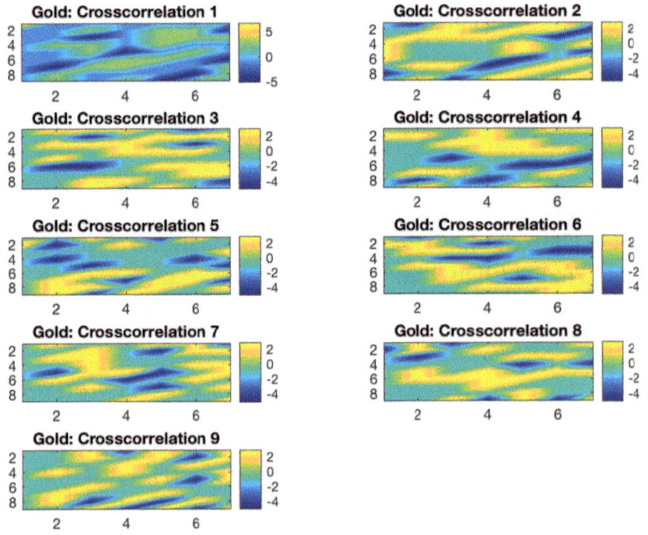

(b)

Fig. 15.7 Correlations of the Golden sequences. (**a**) Autocorrelation. (**b**) Crosscorrelation

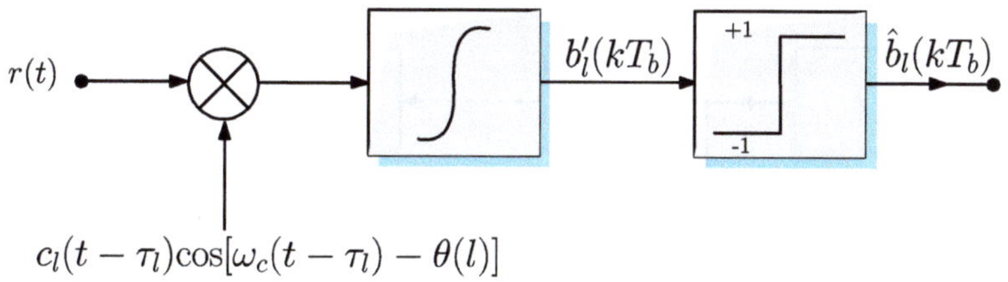

Fig. 15.8 Detection of a single user

An even more detailed form for the estimated symbol at time kT_b before a hard decision is

$$b_l'(kT_b) = \sqrt{2E_l}b_l(kT_b)\rho_{l,l} + \{ \sum_{m=0,m\neq l}^{L} \sqrt{2E_l} \int_{kT_b+\tau_l}^{(k+1)T_b+\tau_l}$$

$$\times \left[\sum_{n=-\infty}^{\infty} b_m(nT_b) \sum_{i=0}^{K} a_{m,i} p_m(t - nT_b - iT_c - \tau_m) \right] \cos[\omega_c(t-\tau_m) + \theta(m)]\}$$

$$\times \sum_{l=0}^{K} a_{l,i} p_l(t - kT_b - iT_c - \tau_l) \cos[\omega_c(t-\tau_l) + \theta(l)]dt + n_l(kT_b) \qquad (15.28)$$

where $n_l(kT_b)$ is the projected channel noise of the lth user.

If one applies the cosine $\cos a \cos b = \frac{1}{2}[\cos(a+b) + \cos(a-b)]$ to (15.27), it is easy to show that Fig. 15.8 represents the single-user receiver configuration. We consider one the baseband signal where the high-frequency term related to $2\omega_c$ is removed.

15.4.2 CDMA RAKE Receiver

The RAKE receiver is a special kind of receiver for CDMA systems. It explores the fact that the chip-shifted signal can be combined constructively.

We assume multipath versions of a user signal are uncorrelated and have useful information. The RAKE receivers attempt to combine the time-shifted versions of the desired signal constructively by exploiting the diversity provided by multipath signals. The receivers used in the IS-95 system employ three taps (fingers) by coherently combining the three most significant signal clusters perceived at the receiver.

The RAKE receiver consists of an FIR filter with chip period delays T_c whose coefficients are chosen such that, ideally, the filter is matched to the channel impulse response. The incoming signal is filtered by the FIR filter and the result filtered by

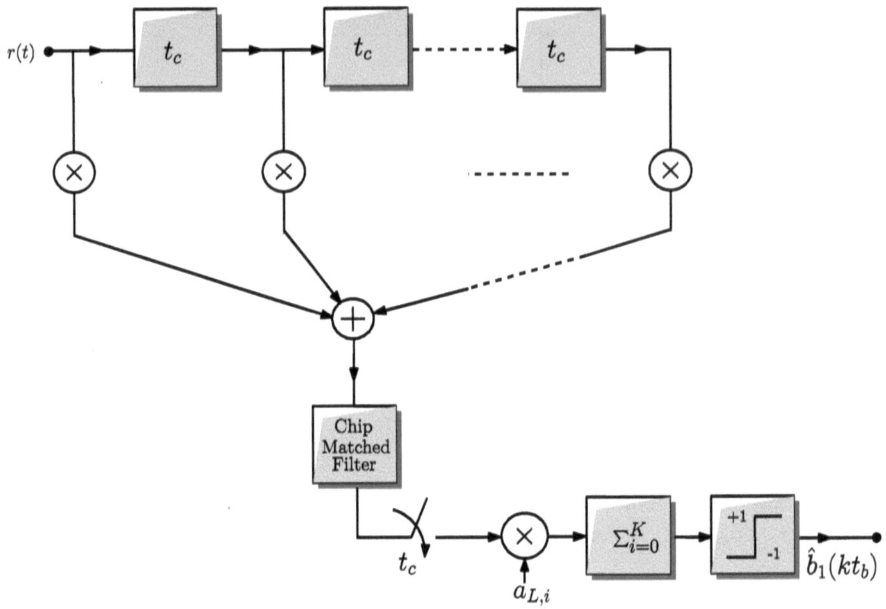

Fig. 15.9 RAKE receiver

the chip-matched filter. The main difficulty is finding accurate estimates for the FIR filter coefficients. Figure 15.9 details RAKE receiver configuration.

15.4.3 Matched Filter Receiver

Figure 15.10 shows the basic configuration of the standard matched filter receiver. The chip-matched filter refers to the pulse shape ($p_l(t - iT_c)$) of the chip.

Let us consider that a properly sampled version of the received signal containing a set of symbols sent by users, each utilizing its spread factor, can be represented as

$$\mathbf{r} = \mathbf{CAb} + \hat{\mathbf{n}}$$

where $\mathbf{r}$ represents a vector containing the samples of the received signal, $\mathbf{C}$ represents the matrix containing the spreading sequence of the lth user at its lth column, $\mathbf{A}$ is a diagonal matrix whose elements represent the power of the users, and $\hat{\mathbf{n}}$ is the environment noise. By applying the despreading matrix to the received

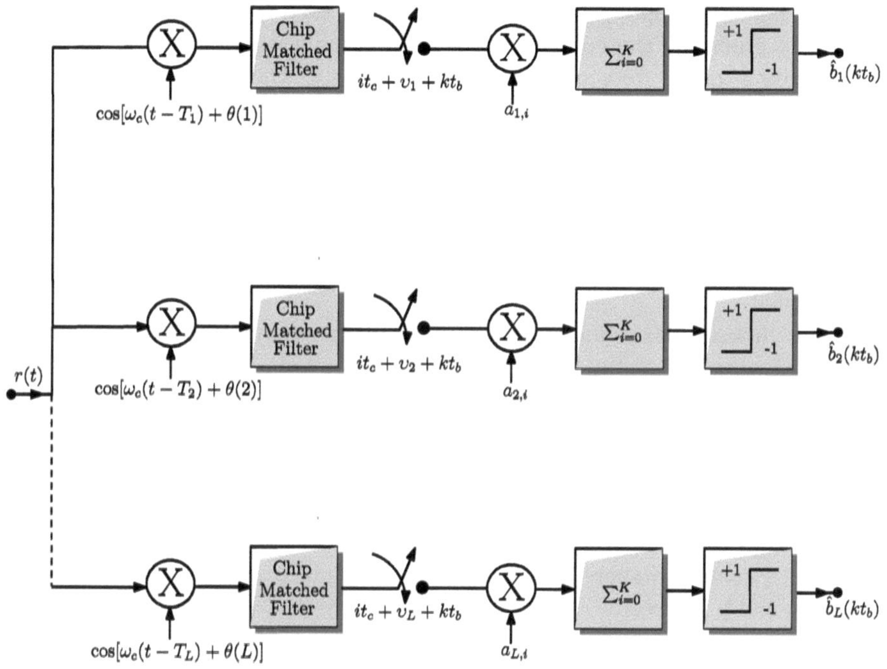

Fig. 15.10 Matched filter receiver

signal, we get:

$$
\begin{aligned}
\mathbf{y} =& \mathbf{C}^T \mathbf{r} = \mathbf{C}^T \mathbf{C} \mathbf{A} \mathbf{b} + \mathbf{C}^T \hat{\mathbf{n}} \\
=& \mathbf{R} \mathbf{A} \mathbf{b} + \mathbf{n}
\end{aligned}
$$

where $\mathbf{y} = \mathbf{C}^T \mathbf{r}$, $\mathbf{R} = \mathbf{C}^T \mathbf{C}$, and $\mathbf{n} = \mathbf{C}^T \hat{\mathbf{n}}$.

The matched filter receiver is optimum if white Gaussian noise is the only disturbance, but it is not optimum for non-Gaussian noise and interference from other users (multiple-access interference). This receiver is highly sensitive to the near-far effect. It works for a well-designed signal constellation; however, the implementation of the signal constellation is so complex that it is advantageous to use other multiple-access method.

15.4.4 Optimum Receiver

Is there an optimum demodulation scheme in the presence of white Gaussian background noise and multiple access interferences?

For coherent reception, optimum solutions exist, each specialized for the chosen optimality criterion. The computation of the optimum solution requires dynamic programming and is very computationally complex. The solution generated by the optimum demodulator does not have the performance disadvantages of the conventional demodulator. For a high-bit-to-noise ratio, the performance of the optimum demodulator approaches that of a single-user system.

The optimum receiver has some drawbacks, such as a much higher complexity than the competing solutions. Furthermore, it requires coupled processing (not a problem at the base station) and the knowledge of the amplitudes of the individual users' signal.

If one chooses maximum likelihood for detection, the objective is to maximize the a posteriori probability of $\mathbf{b}$ conditioned on $r(t)$. In the case all transmitted vectors are equally likely the a posteriori maximization is equivalent to the maximize the a priori one given by

$$P\{[r(t), t \in \mathbb{R}]|\mathbf{b}\} = Be^{\left(\frac{\Gamma(\mathbf{b})}{2\sigma_n^2}\right)} \tag{15.29}$$

where $P[\cdot]$ denotes probability of $[\cdot]$, B is a constant, $\mathbb{R}$ represents the set of real numbers, and $\mathbf{b}$ is a vector including the bits sent by all L users (Poor 1992; Verdu 1998). The parameter σ_n^2 is the measurement noise variance.

The function $\Gamma(\cdot)$ is defined as

$$\Gamma(\mathbf{b}) = 2\int_{-\infty}^{\infty} S(t, \mathbf{b})r(t)dt - \int_{-\infty}^{\infty} S^2(t, \mathbf{b})dt \tag{15.30}$$

where

$$S(t, \mathbf{b}) = \sum_{n=-\infty}^{\infty} \sum_{m=0}^{L} \sqrt{2E_m} b_m(nT_b) c_m(t - nT_b - \tau_m) \times \cos[\omega_c(t - \tau_m) + \theta(m)] \tag{15.31}$$

Note that $S(t, \mathbf{b})$ represents the signal prepared for transmission including the vector $\mathbf{b}$ we wish to detect. The optimum solution clearly requires the knowledge of all the information sent by all users all the time, along with their signature waveform. The asynchronous mode of operation allows each bit to overlap with two adjacent bits from each interfering user, turning the solution more complex.

Figure 15.11 depicts the optimum-receiver configuration.

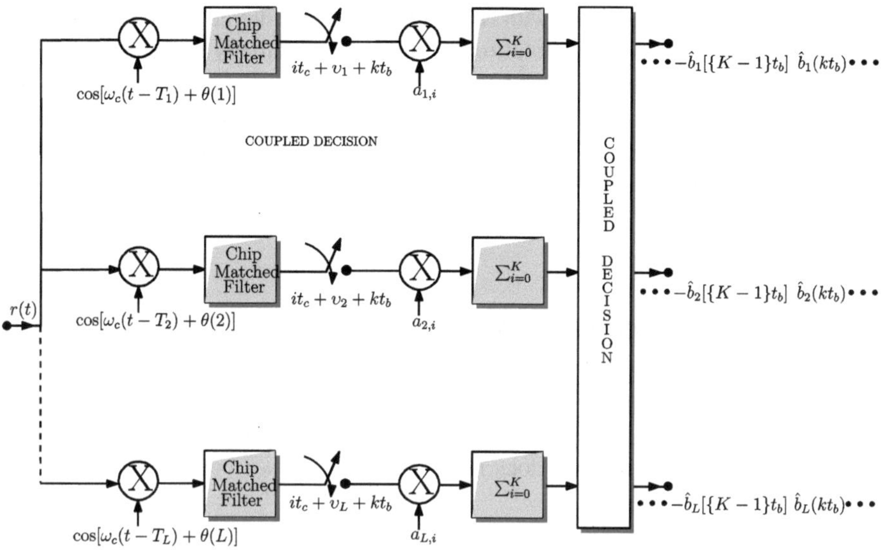

Fig. 15.11 Optimum receiver

15.4.5 Decorrelating Receiver

The decorrelating receiver applies a transformation to the matched filter output vector $\mathbf{y}$ in order to decouple the decisions utilizing the transformed vector:

$$\begin{aligned}\mathbf{R}^{-1}\mathbf{y} &= \mathbf{A}\mathbf{b} + \mathbf{R}^{-1}\mathbf{n} \\ &= \mathbf{A}\mathbf{b} + \mathbf{R}^{-1}\mathbf{C}^T\hat{\mathbf{n}}\end{aligned}$$

In the asynchronous case, the decorrelating receiver generalizes to an IIR filter.

The inverse of $\mathbf{R}$ must be computed or estimated for the decorrelating receiver. Also, the timing of all the users must be acquired. The receiver is coupled or centralized; the desired user is not the only one received, and the users' signatures must be known. In low SNR scenarios, it leads to severe noise enhancement. In addition, the decorrelating receiver performs worse than the matched filter receiver when all the interferers face poor performance in terms of bit error rate.

15.4.6 MMSE Receiver

The optimum linear minimum MSE receiver is designed as follows:

$$\min_{\mathbf{W}\in\mathcal{R}^{K\times K}} \mathbb{E}[\|\mathbf{b} - \mathbf{W}\mathbf{y}\|^2] \tag{15.32}$$

where the expectation is taken with respect to **b** and the filtered noise vector **n**.

Assuming **n** Gaussian, the solution for the MMSE actually replaces $\mathbf{R}^{-1}$ of the decorrelating receiver by $[\mathbf{R} + \sigma_n^2 \mathbf{A}^{-2}]^{-1}$.

It has the decorrelating receiver's features but requires the users' received amplitudes. If additional noise or interference is weak, the MMSE receiver approaches the single-user-matched filter detector. As $\sigma_n^2 \to 0$, the MMSE receiver tends to be the decorrelating receiver.

15.4.7 MMSE-Adaptive Receiver

The MMSE receiver can be made adaptive with an available reference signal. The desired user timing must be acquired if it is applied to asynchronous communication. The signature waveform does not need to be known but helps in algorithm initialization.

15.4.8 SIC and PIC Receivers

The SIC receiver detects the stronger user with a conventional matched filter detector and then subtracts the signal due to that user from the received waveform. We repeat the process to the resulting signal by removing the second strongest user. Note that the influence of the previous stronger user is not there if it was demodulated correctly. The method requires accurate user relative power estimation; otherwise, the decorrelating detector will be better.

The PIC receivers attempt to detect all users in parallel by mitigating the interference of all remaining users while detecting the desired user symbols.

15.4.9 Class of Receivers

The class of CDMA receivers discussed here is:

1. Conventional receiver with matched filter.
2. Optimum receiver with the coupled decision (also known as the optimum multiuser detector, exponentially complex with the number of users).
3. Optimum linear receiver (with a coupled decision).
4. Minimum MSE-adaptive receiver.
5. Decorrelating receiver.
6. Parallel interference canceller (PIC), serial interference canceller (SIC), and multistage receiver.
7. Blind receivers based on second- and higher-order statistics.

Table 15.5 Requirements of each approach

	Interf. codes	User code	Relative power	Timing user	Timing interference	Training sequence
1		*	*	*		
2	*	*	*	*	*	
3	*	*	*	*	*	
4				*		*
5	*	*		*	*	
6	*	*	*	*	*	
7		*		*		

Table 15.5 shows the information required by several types of receivers.

Example 15.3 (Simple Example Continued) The toy example described in Sect. 15.1 is revisited here in order to provide insight in the behavior of some receiver designs.

(a) The matched filter pertaining to each user is applied to the received signal, including noise, as described in the example description. We apply a hard limit output of the matched filter output at the chip level, representing the output with ± 1 for visualization purposes. The actual symbol detection requires the accumulation as depicted in Fig. 15.10. The signal waveforms are depicted in Fig. 15.12. As can be observed, the waveforms suggest that the correct decision is achieved.

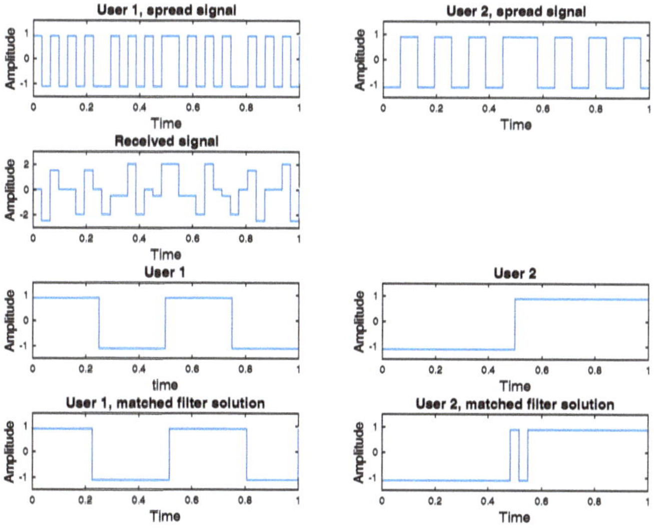

Fig. 15.12 Toy example sequences

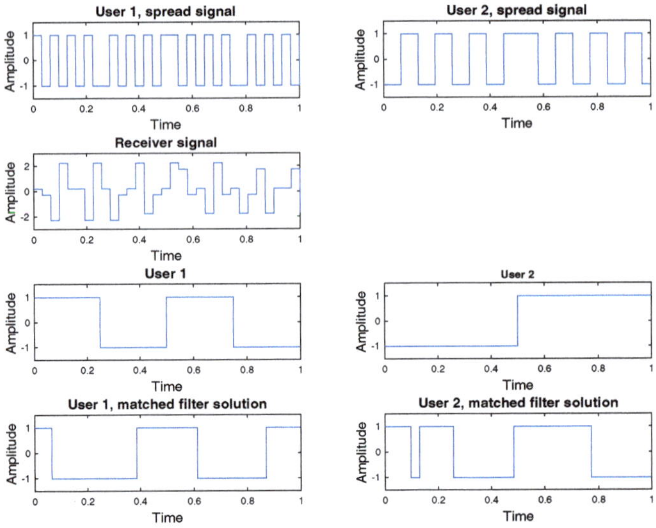

Fig. 15.13 Received and decoded sequences, including the received sequence; the decoded sequence I; and decoded sequence II; all were delayed by one chip

(**b**) In this case, the matched filters face a time shift of one chip, turning the CDMA signal detection more challenging. By observing the signal waveforms in Fig. 15.13, the detection of the received signals produces errors after chip integration within each symbol time frame.

(**c**) This example includes a frequency-selective channel, and the related waveforms are depicted in Fig. 15.14. It is possible to verify that the symbols are correctly detected after the accumulation of chips within each symbol time range. This example utilized a matched filter receiver.

Since this case considers a channel with memory, we can apply the decorrelating receiver. This time, we did not apply the hard limit to the chips, and the signal waveforms seen in Fig. 15.15 show a more reliable detection of the symbols, thanks to the decorrelation technique.

15.5 Frequency-Hopping CDMA

Each user employs a different frequency band at a given time frame. During transmission, the user frequency band hops to different bands according to a prescribed hopping pattern (code). The receiver hops synchronously with the transmitter with the knowledge of the code (Fig. 15.16).

The hopping can be slow, where hopping occurs at the bit rate, or fast, where more than one hop occurs during the bit duration. The latter case is more challenging to implement.

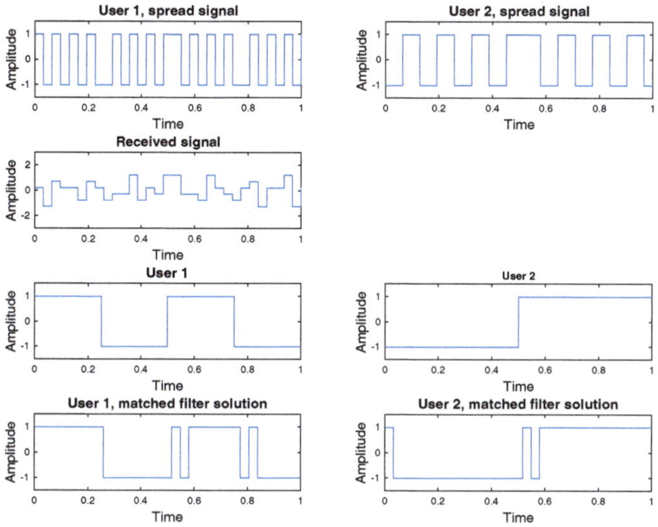

Fig. 15.14 Received and decoded sequences, including the received sequence; the decoded sequence I; and decoded sequence II; all frequency-selective channels

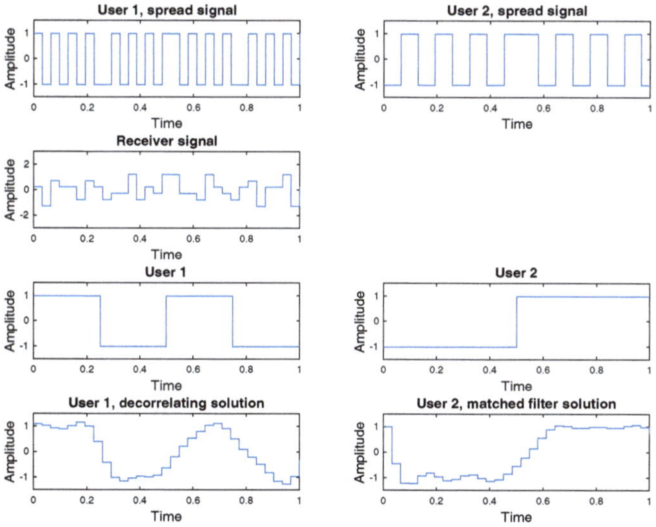

Fig. 15.15 Received and decoded sequences, including the received sequence; the decoded sequence I; and decoded sequence II; all frequency-selective channels

Example 15.4 (Example and Discussion) This example provides some possible mixtures of multiple-access methods as feasible solutions, but not necessarily adopted so far. For instance, the standard IS-95 utilizes a hybrid of FDMA and DS-CDMA.

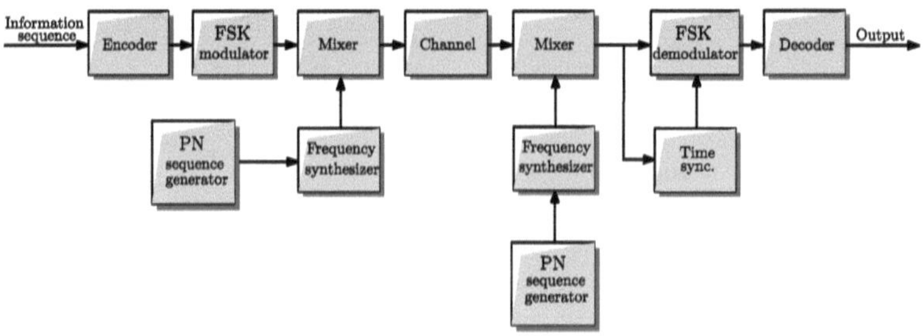

Fig. 15.16 Frequency-hopping CDMA

Let's consider the following abbreviations:

- FH—Frequency Hopping
- TH—Time Hopping
- DS—Direct Sequence
- MC—Multicarrier
- MT—Multitone

Then, it is possible to construct a myriad of MA systems using the discussed methods, such as

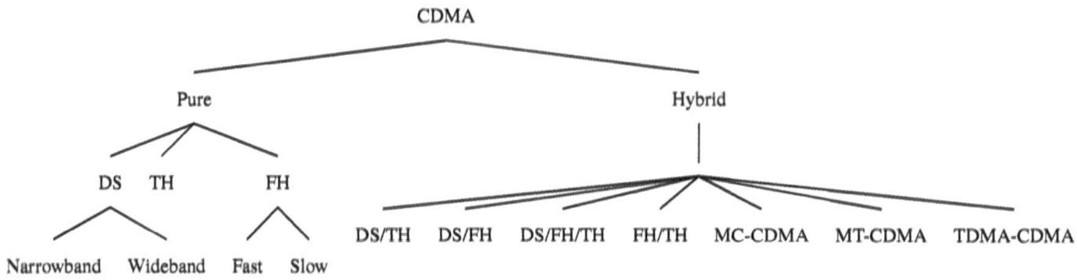

The main advantages of the CDMA systems are now discussed. CDMA is wideband compared to FDMA, and the use of a broader band in frequency-selective channels can be considered frequency diversity. The symbol time in CDMA is longer than in TDMA, so CDMA is less affected by systems with long impulse responses and requires simpler equalizers. Furthermore, the TDMA and FDMA have a fixed number of simultaneous users, and since CDMA does not have fixed numbers, if the number of users increases the service gracefully degrades.

When applied in cellular systems, CDMA allows soft handover between neighboring cells. As such, if the user equipment is close to the frontier where the signal is weaker, but since the user communicates with two base stations, the diversity helps compensate for the signal fading. In TDMA and FDMA, neighboring cells must use different frequencies to control interference based on spatial attenuation. In CDMA, all the cells use the same frequency range, eliminating the need for frequency use planning, whereas the TDMA and FDMA systems require adaptive frequency reallocation.

CDMA requires power control and base station synchronization. On the other hand, the receiver can suppress intercell interference.

The channel capacity of the CDMA system is crucial for understanding of how the CDMA system can achieve good performance in a multiple-access application.

15.6 Channel Capacity

This section discusses some properties pertaining to the capacity of multiple-access systems, to provide an idea of how these systems perform under the given assumptions. If we assume there is no additional noise, the SNR (signal-to-noise ratio) of the desired signal of a given user is

$$\mathrm{SNR} = \frac{P}{LP} = \frac{1}{L} \tag{15.33}$$

where $L + 1$ is the number of users in a cell, for example. P is the received signal power from each user (assumed to be the same at the receiver due to some power control). If we divide the desired user signal power by the bit rate R, and the interference power by the total bandwidth BW of the CDMA system, the SNR in terms of bit energy per interference density is given by

$$\frac{E_b}{I} = \frac{\frac{P}{R}}{L\frac{P}{BW}} = \frac{\frac{BW}{R}}{L} \tag{15.34}$$

where I is the overall interference spectral density, and E_b represents the energy per bit.

In case there is additional noise, we will have

$$\frac{E_b}{I} = \frac{\frac{P}{R}}{\frac{\sigma_n^2}{BW} + L\frac{P}{BW}} = \frac{\frac{BW}{R}}{L + \frac{\sigma_n^2}{P}} \tag{15.35}$$

Therefore, the total number of users that can access the system is

$$L = \frac{\frac{BW}{R}}{\frac{E_b}{I}} - \frac{\sigma_n^2}{P} \tag{15.36}$$

The way to increase the system capacity is to exploit the voice and data activity by reducing the power transmission (or rate transmission). This procedure achieves gains of $\approx$2.67 according to measurements. Furthermore, sectorized antennas bring about additional benefits; for a three-sector antenna, the gain is $\approx$2.4. The analysis above does not take into account that the other cells are also interfering, reducing the capacity by a factor of $\approx$1.6.

According to Viterbi (1995) if we employ error-correcting coding, dual-diversity antenna, and RAKE receiver, the CDMA system is expected to perform well for $\frac{E_b}{I} \approx 6\,\text{dB}$.

In a noiseless environment, by taking into consideration the gains above plus the acceptable bit energy to interference ratio, the number of users will be

$$L + 1 \approx \frac{BW}{R} \tag{15.37}$$

which is an order of magnitude greater than those based on TDMA and FDMA (Viterbi 1995) (note that this conclusion is not straightforward, especially for single-cell and fixed data rate users).

Example 15.5 For $BW = 5\,\text{MHz}$ and 25 Kbit/s, the potential number of users is $L + 1 \approx 200$.

15.6.1 FDMA

Channel capacity for $L + 1$ users:

$$(L + 1)C = \frac{1}{2} \log_2(1 + (L + 1)C\frac{E_b}{I_0}) \tag{15.38}$$

$\frac{1}{2}I_0$ is the noise spectral density of the additional noise and as usual E_b is the energy per bit.

The parameter C is the channel capacity per bandwidth in bits per transmission, where, in the expression above, we used the fact that

$$CE_b = \frac{P}{BW} \tag{15.39}$$

15.6.2 TDMA

Channel capacity for $L + 1$ users:

$$(L + 1)C = \frac{1}{2} \log_2(1 + (L + 1)C\frac{E_b}{I_0}) \tag{15.40}$$

However, in TDMA, each user has to transmit a power $(L + 1)P$ in its $\frac{1}{L+1}$th time slot per frame. This can be prohibitive for large L.

15.6.3 CDMA

If there is synchronization among the users, the total capacity is

$$C_{tot} = \sum_{l=0}^{L} C_l < \frac{1}{2}\log_2(1 + (L+1)C\frac{E_b}{I_0}) \tag{15.41}$$

whose upper bound is identical to the TDMA and FDMA cases. However, if the users require different rates, the CDMA system exceeds the capacity of the TDMA and FDMA.

Channel capacity for *two* users:

$$C_{tot} = \sum_{l=1}^{2} C_l = \frac{1}{2}\log_2(1 + \frac{P}{\sigma_n^2}) \tag{15.42}$$

where $P = P_2 + P_1$, then

$$C_2 = C_{tot} - C_1 = \frac{1}{2}\log_2(1 + \frac{P_2}{P_1 + \sigma_n^2}) \tag{15.43}$$

Using multiuser detection, we can optimize the capacity. For instance, when detecting user 2, we consider the signal from user one as background noise. Conversely, after detecting user 2, we can eliminate its effect when detecting user 1. That is why we considered only the effect of channel noise in user 1 detection.

15.6.4 Noise Floor

In a communication environment, the concept of noise floor relates to the minimum level at which the received signal can be detected while buried in the environment noise. The noise floor representing this quantity is given by

$$N_0 = 10\log_{10}(1000\kappa T) + N_f + 10\log_{10}(BW) \tag{15.44}$$

where $\kappa = 1.3810^{-23}$ represents the Boltzmann constant in Joules/Kelvin, BW represents the bandwidth in Hertz, T is the temperature in Kelvins, and N_f is the receiver noise figure in dB. The parameter N_f accounts for signal degradation in some devices or networks consisting of the ratio $\frac{SNR_{in}}{SNR_{out}}$, where SNR_{in} and SNR_{out} represent the SNR at the input and output of a network or device, respectively. The constant 1000 in (15.44) reflects the dBm measure. It is worth noting that the first and last terms in (15.44) relate to the thermal noise previously denoted as I_0.

Let's consider a case where temperature is 290[1] degrees in Kelvins, an amplifier reduces the SNR by 10 dB due to its internal noise, and the bandwidth is 20 MHz. In this instance, the noise floor is given by

$$\begin{aligned} N_0 =& 10\log_{10}(1000 \times 1.3810^{-23} \times 290) + 10 + 10\log_{10}(20.010^{6}) \\ =& -174 + 10 + 73.01 = -90.99\ \text{dBm} \end{aligned} \tag{15.45}$$

15.6.5 Capacity Examples

In this subsection, we present some capacity-based examples to illustrate how this tool plays an essential role in communication system designs. Let's consider:

$$\begin{aligned} C &= BW\log_2(1+\frac{P}{I}) = BW\log_2(1+\text{SNR}) \\ &= BW\log_2(1+\frac{E_b}{I_0}\frac{R}{BW}) \end{aligned} \tag{15.46}$$

where C in this case is the capacity per complex dimension.[2] The SNR is the signal-to-noise ratio. The received power can be expressed as

$$P = E_b R \tag{15.47}$$

where R is the data rate and E_b is the received energy per information bit. The quantity $I = I_0 BW$ where I_0 is the noise power spectral density measured in Watts/Hertz, usually measured at room temperature of 290 °K. The radio link bandwidth utilization is the ratio $\frac{R}{BW}$.

In the particular case of the long-term evolution (LTE), mobile communication systems, the following capacity-related information applies: the minimum bandwidth of a user is a single-resource block corresponding to $12 \times 15\,\text{kHz} = 180\,\text{kHz}$. The LTE system with a narrower bandwidth occupies 1.4 MHz and utilizes FFT[3] with a length of 128.

The LTE system with a broader bandwidth occupies 20 MHz and utilizes FFT with a length of 2048. In particular, the LTE employs 72 subcarriers for 1.4 MHz; 180 subcarriers for 3 MHz; 300 subcarriers for 5 MHz; 600 subcarriers for 10 MHz; and 1200 subcarriers for 20 MHz.

---

[1] Corresponding to 273 + 17 °C.

[2] $C = \frac{1}{2}BW\log_2(1+\frac{P}{I})$ for real dimension, where $\frac{I_0}{2}$ is noise variance per real symbol and power per real symbol is $\frac{P}{2BW}$.

[3] This corresponds to a discrete-Fourier transform (DFT) type of modulation called OFDM.

Recall that 64 QAM carries 6 bits, 16 QAM carries 4 bits, and 2 × 2 MIMO doubles the capacity. As such, a 64 QAM + 2 × 2 MIMO + 20 MHz → leads to a peak of ≈170 Mbp/s in terms of data rate.

Example 15.6 Assume the environment noise power spectral density in dBm is − 102 dBm. This means that

$$-102\ \text{dBm} = 10\log_{10}\left(\frac{I_0}{1\ \text{mW}}\right) \tag{15.48}$$

where $\frac{I_0}{2} = \frac{6.3110^{-14}}{2}$ Watts/Hertz is the power spectral density in Watts/Hertz, for a real signal. For a bandwidth of 20 MHz, the noise power will be

$$I = I_0 BW = 6.31 \times 20 \times 10^{-8}\ \text{W} = 1.26210^{-6}\ \text{W} \tag{15.49}$$

Now, if we assume the base station transmits with 40 W of power, then the maximum capacity for a single user is given by

$$\begin{aligned} C &= \frac{1}{2}BW\log_2(1+\frac{P}{I}) = \frac{1}{2}20\ \text{MHz}\log_2(1+\ \text{SNR}) \\ &= \frac{1}{2}20\ \text{MHz}\log_2(1+\frac{40}{1.26210^{-6}}) \\ &= 249.2\ \text{Mbps} \end{aligned} \tag{15.50}$$

For 15 kHz

$$I = 6.31 \times 15 \times 10^{-11}\ \text{W} = 9.46510^{-10}\ \text{W} \tag{15.51}$$

Then, the transmission power for this bandwidth in the LTE is

$$P = \frac{40}{1200}\ \text{W} \tag{15.52}$$

For the case of 1200 subcarriers, we assumed the 40 W is shared with them. Therefore,

$$C = \frac{1}{2}BW\log_2(1+\frac{P}{I}) = 188.02\ \text{kbps} \tag{15.53}$$

Since we have 1200 subcarriers with 15 kHz each, the total capacity can be estimated as 225.63 Mbps.

Concerning discrete-time channel capacity, an important parameter is the channel bandwidth available. Starting from this and assuming the channel is not frequency-selective, we face two situations: noiseless and noisy channels. In the noiseless case and real signals, the data rate is determined by the Nyquist rate as follows:

$$C = 2BW \log_2(2^b) \text{ bps} \tag{15.54}$$

where b is the number of bits per symbol. For the transmission of a binary signal, the capacity becomes $C = 2BW$, as widely employed in the literature, in the context of the classical sampling theorem, as can be seen in Molisch (2005), Diniz et al. (2010).

15.7 Current Applications

Although CDMA has decreased its importance in public wireless networks, when it was the main multiple-access system for the 3G systems, its related pseudorandom codes play a crucial role in satellite navigation systems and many wireless communications applications.

15.7.1 GPS

The Global Network Satellite System (GNSS) provides worldwide coverage through the use of a Medium Earth Orbit (MEO) constellation of satellites. There are a few GNSS systems where the Global Positioning System (GPS) is the most widely used.

Each satellite within a GNSS constellation has a unique pseudorandom (PR) code transmitted for coarse acquisition (CA) code, taking part in the navigation message. This PR CA code repeats itself every 1023 bits and modulates at a rate of 1 MHz, producing a CA code at 1.023 MHz. Then, it modulates an L1 carrier at 1575.42 MHz. It conveys a PR coding for timing and status messages, and a low-frequency signal giving information related to clock corrections, satellites' orbits, etc. It is employed for civilian purposes, and its simplified description is below.

The GPS satellites also transmit a second carrier frequency, namely, the L2 carrier, aiming for military purposes. The L2 carrier is 1227.60 MHz, employing a more precise PR code (Pcode). The Pcode has 10.23 MHz and repeats itself over 7 days.

A simple representation of the whole GPS can be described as

$$\begin{aligned}\text{L1 signal} &= \text{L1 carrier} \times (\text{CA} \oplus \text{Nav}) \times \text{Pcode}\\ \text{L2 signal} &= \text{L2 carrier} \times (\text{Pcode} \oplus \text{Nav})\end{aligned}$$

where the additions are module-two sums and NAV represents the navigation and system data. The Pcode can be encrypted for military use, being called Y code.

Both CA and Pcode are maximum-length sequences generated by their prescribed shift-register polynomial, and the latter is much longer.

The GPS receivers play a crucial role in location-based systems in cellular networks since they are integrated into smartphones; see Campos and Lovisolo (2015) for further details. The updating in GSP, as well as in the alternative systems such as the European Union Galileo, is continuous.

15.7.2 RFID

The frequency-hopping methods are used in RF IDentification (RFID) with a reader-tag communication technique, entailing frequency-hopping spread spectrum/listen before talk (FHSS/LBT) methods; see, for instance, https://rfid4u.com/rfid-regulations/.

15.7.3 Bluetooth

The Bluetooth wireless communication system belongs to the class of PANs; see Sect. 12.9.1. Bluetooth uses an adaptive frequency-hopping spread spectrum to cope with interference. In Bluetooth, data security and privacy are provided by a 128-bit random number, two keys, authentication (128 bits) and encryption (256 bits) (8 to 128 bits), and a device's 48-bit MAC address in security mode. The three modes of operation are nonsecure, service level, and link level.

These are only a few examples of CDMA and PR codes that are widely used in many current communication and network systems.

15.8 Remarks

Spreading codes, widely used in CDMA systems, play an essential role in positioning systems. They are crucial to distinguishing signals originating from different satellites, and their design is critical to minimizing interchannel interference in satellite navigation. The positioning systems are likely to continue to play a role in the implementation of autonomous vehicles.

References

R.S. Campos, L. Lovisolo, *RF Positioning: Fundamentals, Applications, and Tools (Gnss Technology and Applications)* (Artech House Publishers, Boston, 2015), ISBN: 978-1-60807-816-5

E.H. Dinan, B. Jabbari, Spreading codes for direct sequence CDMA and wideband CDMA cellular networks. IEEE Commun. Mag. **36**, 48–58 (1998)

P.S.R. Diniz, E.A.B. Silva, S.L. Netto, *Digital Signal Processing: System Analysis and Design*, 2nd edn. (Cambridge University Press, Cambridge, 2010)

S. L. Lee, L. E. Miller, *CDMA Systems Engineering Handbook* (Artech House Publishers, Boston, 1998). ISBN: 0-89006-990-5

A.F. Molisch, *Wireless Communications* (John Wiley & Sons, Hoboken, 2005)

H.V. Poor, Signal processing for wideband communications. IEEE Inform. Theory Soc. Newslett. **42**, 1–10 (1992)

D.V. Sarwate, M.B. Pursley, Crosscorrelation properties of pseudorandom and related sequences. Proc. IEEE **58**, 583–619 (1980)

S. Verdu, *Multiuser Detection* (Cambridge University Press, Cambridge, 1998)

A.J. Viterbi, *Principles of Spread Spectrum Communication* (Addison Wesley, Reading, 1995)

Chapter 16
Orthogonal Frequency Division Multiplex

16.1 Introduction

The main feature of the orthogonal frequency division multiplex (OFDM) transceivers is to avoid intersymbol interference (ISI), particularly when the information bandwidth is wider than the channel's coherence bandwidth. In OFDM and in more sophisticated multicarrier systems, the key idea is to decompose the channel into multiple parallel subchannels whose models are flat-fading Gaussian. This feature allows us to reach the channel capacity, particularly in the case the number of subchannels tends to infinity through proper power allocation (Hirt and Massey 1988). Also, the decoupling of the parallel subchannels allows their distribution among different users, giving rise to a multiple-access technique based on the multicarrier division. The more detailed presentation in Diniz et al. (2012) inspires the one here.

OFDM is a transmission technique that is currently used in several practical systems, including digital radio broadcasting or digital audio broadcasting (DAB); discrete multitone (DMT), which includes bit loading; wireless local area network (WLAN); wireless broadband links (WiMAX, IEEE 802.16); fourth- and fifth-generation cellular communications; digital broadcasting TV; and, more recently, radar systems. OFDM belongs to the family of multicarrier systems, which is briefly described in the following section.

16.2 Multicarrier Systems

Multicarrier transmissions are the standard in many communication systems being developed. The key role of the multicarrier modulation methods is to solve the equalization difficulties inherent to channels with moderate to severe intersymbol interference.

Y. C. Lim et al., *Signal Processing for Autonomous Systems*,
https://doi.org/10.1007/978-3-032-00675-2_16

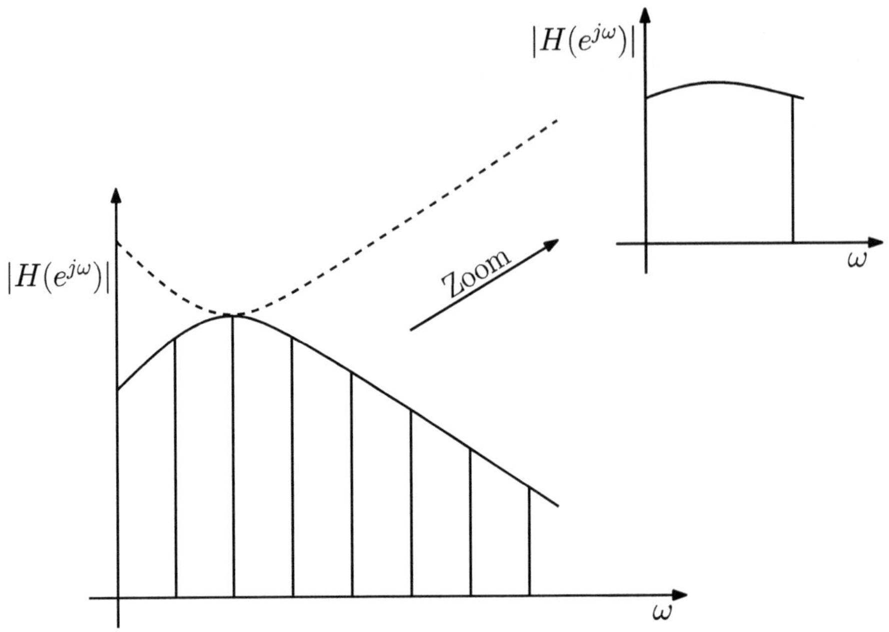

Fig. 16.1 Multicarrier system, flat subchannels

The OFDM is the most widely used solution among the multicarrier systems. Through OFDM, it is possible to perform reliable communication by exploiting wideband channels by simultaneously utilizing many narrowband subcarriers. The main advantage of the OFDM-based transceivers is their efficient implementation.

All multicarrier systems benefit from the partition of the physical channel into nonoverlapping narrowband subchannels whose adequate description is via a transmultiplexer structure. As illustrated, Fig. 16.1 shows how the wideband channel splitting enables the generation of flatter and frequency orthogonal subchannels. The narrow subchannels allow much simpler channel equalization so that we replace sophisticated equalizers by several trivial parallel subchannel equalizers, a tradeoff worth pursuing.

The subchannel division allows, whenever possible, the exploitation of signal-to-noise ratio (SNR) in the distinct subbands by managing their data load in each subchannel.

If knowledge about the signal-to-noise ratio at the channel output for each subcarrier is available to the transmitter, some loading scheme can be applied. Figure 16.2 illustrates a hypothetical multicarrier probing that would eventually lead to a data constellation choice depicted in Fig. 16.3.

In a general setup, the minimum redundancy required at the transmission in order to keep equalization as simple as possible is an important issue. In addition, several methods for jointly optimizing the transmitter and receiver FIR MIMO systems can be designed to combat near-end crosstalk and additive noise source.

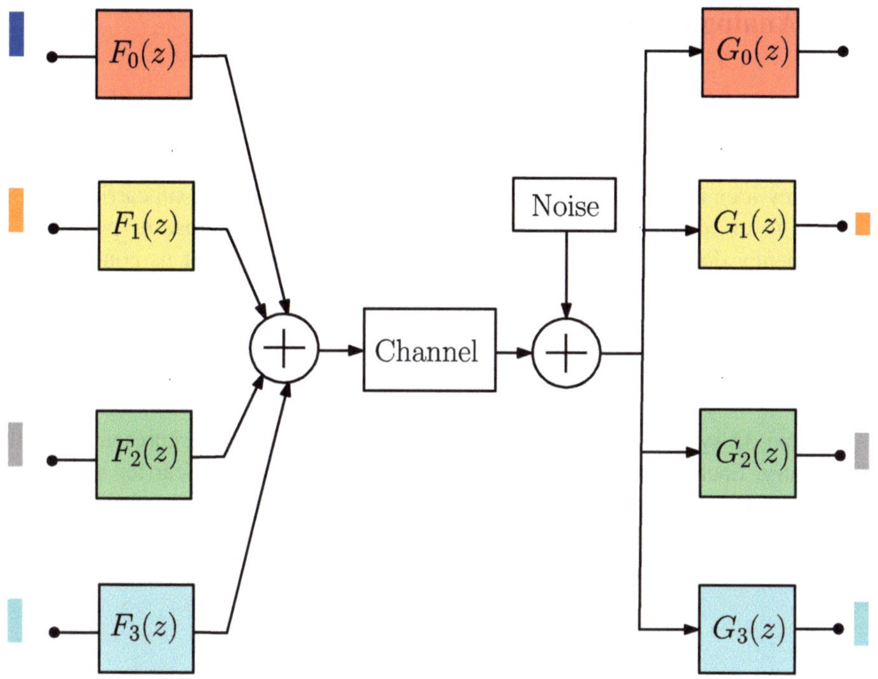

Fig. 16.2 Multicarrier system

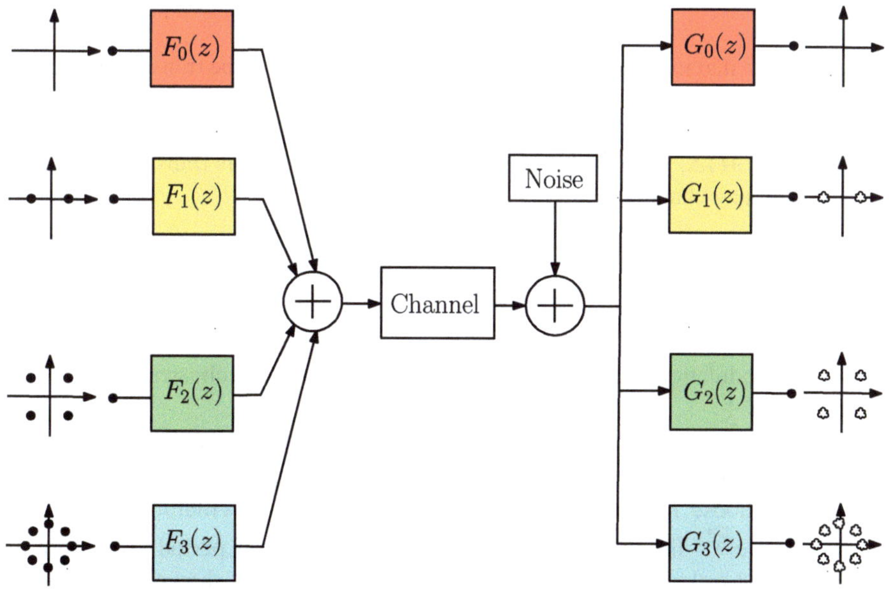

Fig. 16.3 Loading multicarrier system

16.3 Analog OFDM

The concept of OFDM started with its analog version in the 1960s, with the work Chang (1966) and some details are available in Diniz et al. (2012) which inspired the current presentation.

The key idea is to modulate each symbol in a data block by a subcarrier, which we refer to as tone in this section. These tones must be as close as possible in the frequency domain while maintaining their orthogonality. Let us consider two neighboring analog tones with frequency f_1 and f_2 as

$$f_1 - f_2 = \frac{1}{\Delta}. \tag{16.1}$$

Assume M symbols of a data block is transmitted, where b_i represents the ith block entry. Each symbol of the data block will modulate a subcarrier as per the following rule.

$$\begin{aligned} b_0(n\hat{T}) &\rightarrow f_0 = 0 \\ b_1(n\hat{T}) &\rightarrow f_1 = \frac{1}{\Delta} \\ b_k(n\hat{T}) &\rightarrow f_k = \frac{k}{\Delta} \\ b_{M-1}(n\hat{T}) &\rightarrow f_{M-1} = \frac{M-1}{\Delta}. \end{aligned}$$

If we assume that in case each complex symbol was meant to be transmitted serially and utilizing the minimum bandwidth, we can infer the Δ represents the time allotted for the symbol. Therefore, the continuous time span of the OFDM symbol is $0 \leq t \leq (M)\Delta$, so at each time slot the following signal is transmitted without considering any channel memory:

$$\sum_{l=0}^{M-1} b_l(n\hat{T})e^{\frac{\jmath 2\pi}{\Delta}lt}. \tag{16.2}$$

Since the modulation is complex, the sampling rate for each subcarrier must have a minimum of $\frac{1}{\Delta}$.

For memoryless channels, the time span of the OFDM systems $\hat{T}$ can be chosen so that the cross-correlation among all subcarriers is equal to zero $\forall l$. This way, we should choose $\hat{T}$ as a multiple of Δ so that the slowest oscillating and complex exponential $e^{\frac{\jmath 2\pi}{\Delta}t}$ fits in with a multiple of its fundamental period. Therefore, to guarantee the orthogonality among the subcarriers, it can be shown that $\hat{T}$ should be a multiple of Δ to guarantee the desired temporal zero cross-correlation among subcarriers within the OFDM symbol duration, with the simplest viable solution being $\Delta = \hat{T}$ (Diniz et al. 2012). This solution allows maximum data throughput.

The continuous-time OFDM description can be redefined in terms of the subcarrier functions as

$$\hat{\varphi}_l(t) = \mathrm{e}^{\frac{2\pi \mathrm{j} l t}{\hat{T}}}\, \hat{p}(t) \tag{16.3}$$

where

$$\hat{p}(t) = \frac{1}{\sqrt{\hat{T}}} \tag{16.4}$$

for $0 \leq t < \hat{T}$ and 0 otherwise; see Diniz et al. (2012). The transmitted signal is

$$s(t) = \sum_{n=-\infty}^{\infty} \sum_{l=0}^{M-1} b_l(n\hat{T})\hat{\varphi}_l(t - n\hat{T}) \tag{16.5}$$

where $b_l(n\hat{T})$ is the lth symbol in a block at a given instant $n\hat{T}$.

Figure 16.4 illustrates the pulse shape as defined in (16.4).

The choice of the pulse shape determines the subcarrier frequency spectrum, whereas for the case of the square pulse of Fig. 16.4, it corresponds to the spectrum of Fig. 16.5.

In the time domain, the OFDM signal resembles a random signal whose spectrum occupies a frequency band around the main carrier. This results from illuminating the channel with an additive combination of random symbols modulated by several subcarriers illustrated in Fig. 16.6.

In Fig. 16.7, it is possible to observe how the spectrum of neighboring subcarriers add up together. Note that the time-limited symbol shaping leads to cumulative

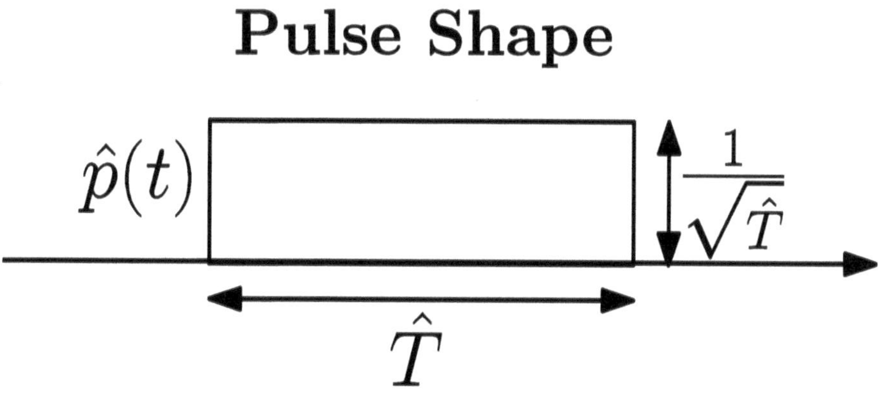

Fig. 16.4 OFDM pulse shape

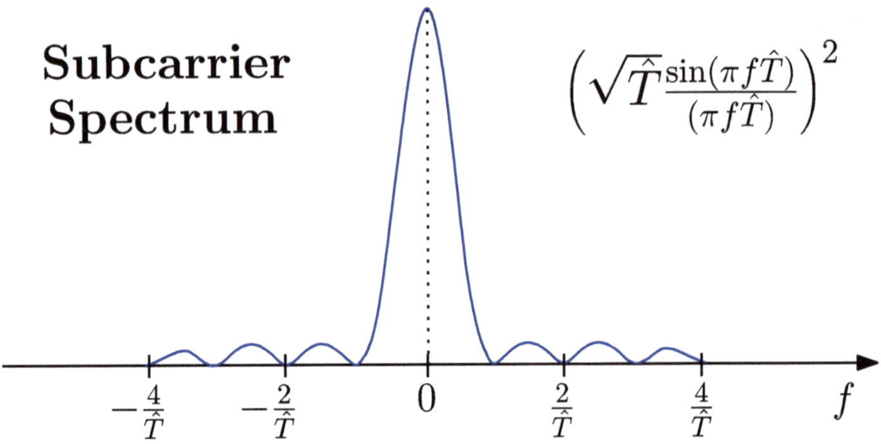

Fig. 16.5 OFDM subcarrier spectrum.

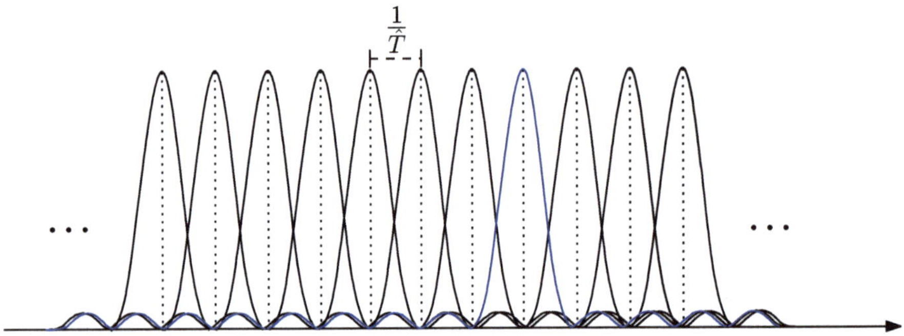

Fig. 16.6 OFDM subcarriers

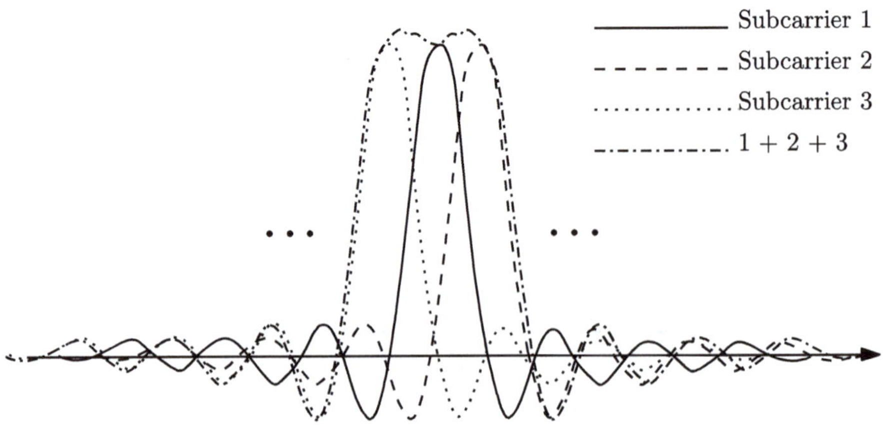

Fig. 16.7 OFDM subband frequency response

contribution from the secondary lobes in the frequency domain, which can be considered negligible after $\frac{M}{\hat{T}}$. Since a limited number of carriers on the edges of the spectrum are not used, we can use a sampled version of the signal to be transmitted for analysis.

Above, we discussed the OFDM transmitter aimed at generating orthogonal subcarriers before crossing the channel. The orthogonality feature is crucial at the receiver end; therefore, we should transmit OFDM symbols that maintain a decoupled detection among subcarriers' symbols when they reach the receiver end. The full task is to eliminate interference among transmitted constellation symbols and OFDM symbols. For channels with memory, it is necessary to include some sort of guard period between the block transmissions inherent to the OFDM symbol in order to avoid interblock interference (IBI). It is also desirable to maintain a decoupled detection of the symbols at the receiver, achieved by eliminating intersymbol interference (ISI).

16.3.1 Channel with Memory

In case there are multiple reflections, the choice of the sampling rate has to take into consideration the channel's time span τ. Instinctively, the complete set of subcarriers originally with time support equal to $\hat{T}$, should be stretched to cope with the multipath channel. Therefore, their time support should increase to T to address the IBI while keeping their frequency separation. In a way, ISI is avoided through a biorthogonality among the transmitter and receiver subcarriers' waveforms.

For that, it can be shown that the OFDM symbol should be extended to T such that $T - \tau = \hat{T}$, where τ represents the maximum channel reflection delay; see Diniz et al. (2012) for a thoughtful proof. In terms of resources, one can say that the budgetable data block spans the time $\hat{T}$.

In case $\tau \neq 0$, the solution is to use a large number of orthogonal carriers, which can be efficiently implemented, in which each subcarrier will be responsible for transporting a symbol during a time slot that takes into account for the longest path propagation time. Recalling that now the OFDM symbol has a duration of T, if the channel has memory, the orthogonality should be maintained within any time frame of the symbol of length $T - \tau$. The orthogonality can be certainly effected if we define M carriers separated in frequency by $\frac{1}{T-\tau}$ as

$$\varphi_l(t) = \mathrm{e}^{\frac{2\pi \mathrm{j} l t}{T-\tau}} p(t) \tag{16.6}$$

where

$$p(t) = \frac{1}{\sqrt{T - \tau}} \tag{16.7}$$

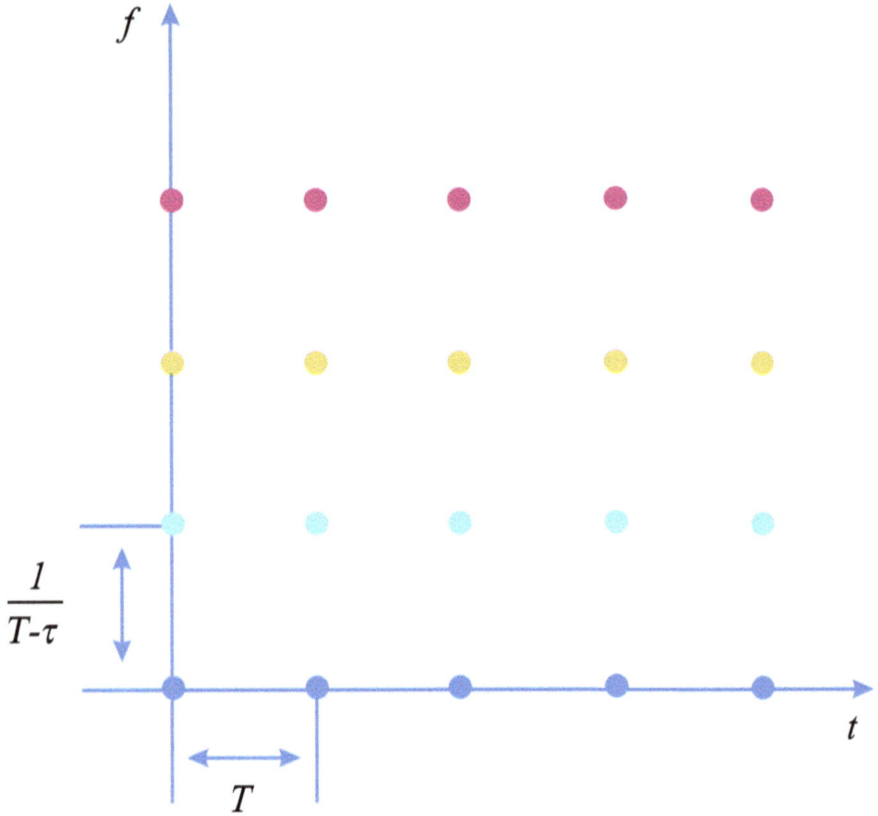

Fig. 16.8 Time-frequency map of OFDM signals

for $-\tau \leq t < T - \tau$ and 0 otherwise. As derived in Diniz et al. (2012), the starting of the pulse time support is an important issue for the proofs not included here. The transmitted signal is

$$s(t) = \sum_{n=-\infty}^{\infty} \sum_{l=0}^{M-1} b_l(nT)\varphi_l(t - nT) \tag{16.8}$$

where $b_l(nT)$ is the lth symbol in a block at a given instant nT.

In OFDM system analysis, it is instructional to depict the evolution of the signals communicated through each tone in the time domain, as illustrated in Fig. 16.8.

The interblock interference happens when the channel has memory and the OFDM symbols are transmitted without taking the multi-reflections into consideration. Using a guard period avoids interblock interference, as illustrated in Fig. 16.9. The choice of T should be much longer than the most prolonged multipath delay so that most of the time, profitable information occupies the channel. Also, the

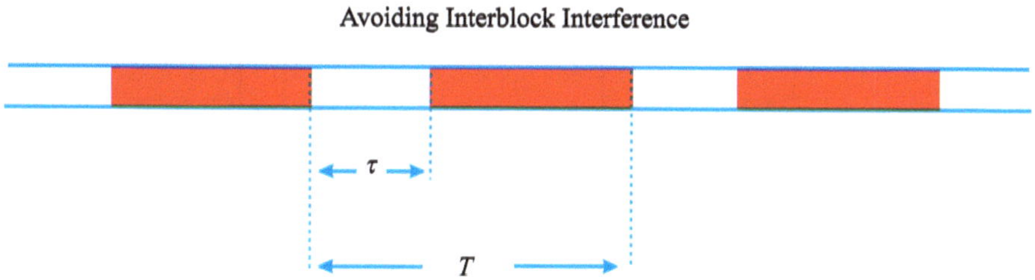

Fig. 16.9 Concatenated OFDM symbols with guard period

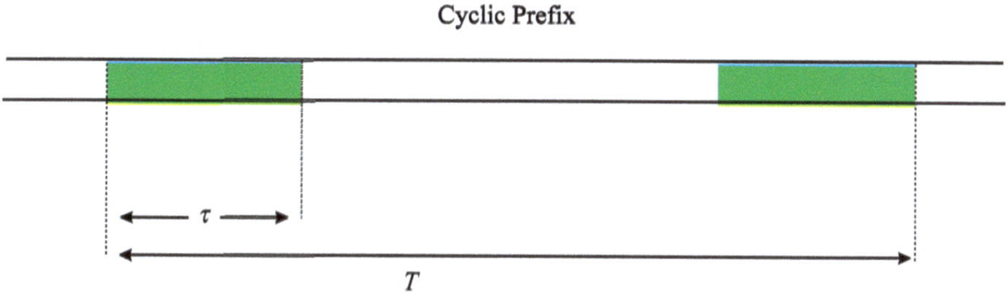

Fig. 16.10 Received OFDM symbol

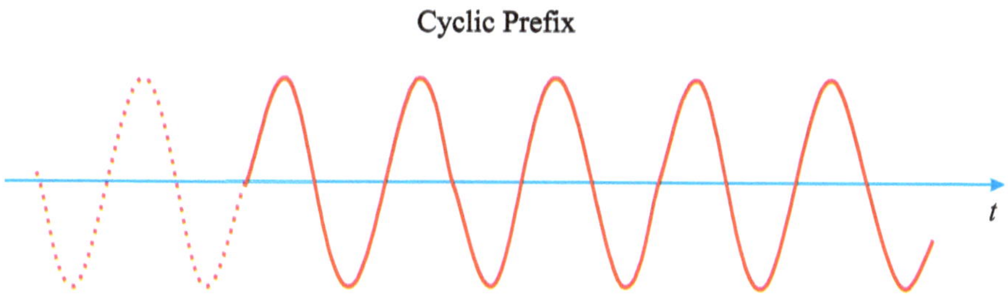

Fig. 16.11 Cyclic prefix for an OFDM carrier

guard interval τ is larger than the channel's longest multipath delay. The difference between T and τ is called valuable symbol time, whereas the latter is an overhead time used to avoid intersymbol interference.

Figure 16.10 depicts the delayed OFDM symbol at the receiver end. The darker part of the block represents the time frame where distinct OFDM blocks interfere with each other.

The critical issue is to keep the subcarriers orthogonal at the receiver when the OFDM signal crosses a channel with memory. Extending the support of the subcarriers in the time domain makes it possible to maintain orthogonality among them. The operation is related to introducing a cyclic prefix, as seen in Fig. 16.11. The extension of the subcarrier support in the continuous-time domain leads to an elegant proof for the cyclic-prefix extension in the discrete-time domain, which ultimately falls into efficient digital implementation as per Diniz et al. (2012).

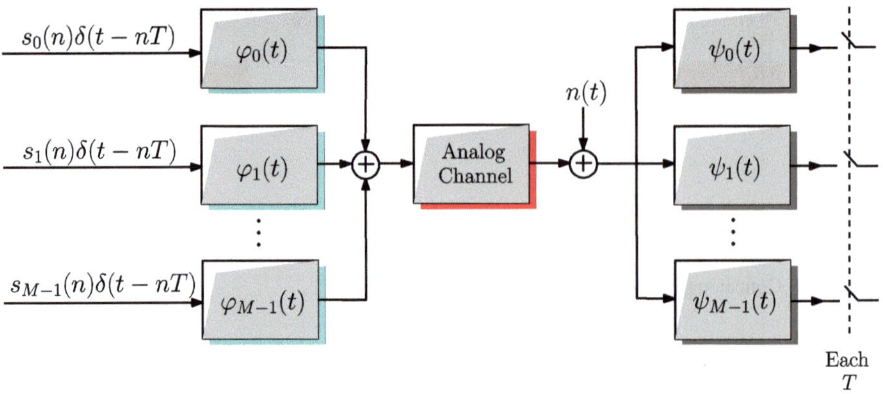

Fig. 16.12 OFDM continuous-time representation

The reader should note the differences between Eqs. (16.5) and (16.8), where the first case represents an OFDM transmission with no guard period. The received signal after crossing a multipath channel would not yield orthogonal subcarriers. Equation (16.8) represents the concatenation of infinitely many extended OFDM symbols using a cyclic prefix, ensuring that the orthogonality among subcarriers is recovered after passing the transient period. In Diniz et al. (2012), it is shown how the continuous-channel model affects each symbol transported by the subcarriers. The details are omitted here for the sake of brevity; however, the channel role at the detector will reappear in the sampled OFDM system description.

On a continuous-time implementation, the OFDM system is implemented as shown in Fig. 16.12, where the receiver employs matched filter functions:

$$\psi_l(t) = \mathrm{e}^{\frac{2\pi\,\mathrm{j}lt}{T-\tau}}\, p(t) \tag{16.9}$$

where

$$p(t) = \frac{1}{\sqrt{T-\tau}} \tag{16.10}$$

for $0 \leq t < T - \tau = \hat{T}$ and 0 otherwise.

If we compare the set of transmitter and receiver pulse waveforms of (16.6) and (16.9), it is possible to conclude that they form a biorthogonal set (Diniz et al. 2010). The continuous-time representation of the OFDM transceiver is shown in Fig. 16.12.

16.3.2 Spectral Efficiency

It is appropriate to briefly discuss how efficiently an OFDM system utilizes the spectrum available to service provision. Let us start by recalling that the bit rate for a constellation with 2^β points is

$$\text{BR} = \frac{M\beta}{T} \tag{16.11}$$

in bits per second.

If the spectrum of the symbols is truncated such that only parts of the spectrum below 20 dBs from the main lobe are removed (corresponding to 2 side lobes), the total bandwidth is

$$\text{BW} = \frac{(M-1)}{T-\tau} + 2\frac{3}{T} \tag{16.12}$$

The ratio

$$\frac{\text{BR}}{\text{BW}} = \frac{M\beta}{\frac{(M-1)T}{T-\tau} + 6} \tag{16.13}$$

represents the spectral efficiency, which tends to $\beta(1 - \frac{\tau}{T})$ for large $(M - 1)$. The latter expression shows that the OFDM is an optimal modulation in terms of spectral efficiency if $\frac{\tau}{T} << 1$. In conclusion, as long as the block duration is much longer than the channel time spread, the OFDM is very efficient regarding spectral utilization.

16.4 Discrete-Time Implementation

The issue now is how to implement the digital version of OFDM efficiently. The presentation here is simple, and there is no attempt to connect the analog and digital versions of the OFDM transceivers, as done in Diniz et al. (2012). In most applications, the discrete-time OFDM is the implementation choice.

If we sample the modulated signal (remember it is band-limited) with the sampling period $T_s = \frac{\hat{T}}{M} = \frac{T-\tau}{M}$, that is,

$$\begin{aligned} s(k\frac{T-\tau}{M}) &= \sum_{n=-\infty}^{\infty}\sum_{l=0}^{M-1} b_l(nT)\varphi_l(k\frac{T-\tau}{M} - nT) \\ &= \sum_{n=-\infty}^{\infty}\sum_{l=0}^{M-1} b_l(nT)e^{2\pi jl(k\frac{1}{M}-nT)}p(k\frac{T-\tau}{M} - nT). \end{aligned} \tag{16.14}$$

Without loss of generality, we can analyze each block of symbols separately and remove the time dependency of the above equation. That is, the signal sent at a given block slot nT is

$$\begin{aligned}\hat{s}(k\frac{T-\tau}{M}) &= \sum_{l=0}^{M-1} b_l \varphi_l(k\frac{T-\tau}{M}) \\ &= \frac{1}{\sqrt{T-\tau}} \sum_{l=0}^{M-1} b_l e^{2\pi \mathrm{j} l(\frac{k}{M})} \\ &= \sqrt{\frac{M}{T-\tau}} \mathrm{IDFT}[\mathbf{b}(nT)]_k. \end{aligned} \tag{16.15}$$

Note that the equation above is nothing but the inverse DFT[1] of b_l, where $\mathbf{b}(nT)$ is the data block.

From the above equation, we conclude that the receiver can recover the original block by

$$b_l = \sqrt{\frac{T-\tau}{M}} \mathrm{DFT}[\hat{\mathbf{s}}(k\frac{T-\tau}{M})]_l. \tag{16.16}$$

For $l = 0, 1, \cdots, M-1$, where $\hat{\mathbf{s}}(\cdot)$ represents a vector congregating the signals. As expected, with a memoryless channel inflicting no distortion in the received signal, the IDFT and DFT appear as natural transmitters and receivers for a discrete OFDM system. By taking into consideration their efficient implementation with the fast Fourier transform (FFT) algorithm, the discrete-time version of the OFDM promises to be implemented efficiently.

16.4.1 Channel with Memory

Note that, originally, each symbol had a time slot of $T_s < \frac{\hat{T}}{M} = \frac{T-\tau}{M}$ to be transmitted. The choice of τ is such that it is longer than the longest multipath delay.

In real practice, the received signal is

$$r(k) = \sum_{i=0}^{I} h_i s(k\frac{T-\tau}{M} - \tau_i) \tag{16.17}$$

[1] $\mathrm{IDFT}[\mathbf{a}] = \frac{1}{\sqrt{M}} \sum_{l=0}^{M-1} a_l e^{2\pi \mathrm{j} \frac{kl}{M}}$.

where τ_ls are the multipath delays, and

$$s(k\frac{T-\tau}{M}) = \sum_{n=-\infty}^{\infty} \sum_{l=0}^{M-1} b_l(nT)\varphi_l(k\frac{T-\tau}{M} - nT) \tag{16.18}$$

For the channel model $I = \lceil \frac{\tau_L}{M} \rceil$, where $\lceil \cdot \rceil$ represents the closest integer larger than $[\cdot]$. When choosing a prefix length K such that $\tau = KT_s$ and $K \geq I$, we know that the prefix is longer than the channel's memory (Diniz et al. 2012).

In general terms, the OFDM waveforms need to satisfy

$$\sum_{k=-\infty}^{\infty} \varphi_l(k\frac{T-\tau}{M} - nT)\varphi_j(k\frac{T-\tau}{M} - mT) = \delta(l-j)\delta(n-m) \tag{16.19}$$

which represents orthogonality.

The question is if orthogonality is sufficient to allow the detection of the transmitted symbols when the channel has memory. In case there is multipath propagation, we have

$$\begin{aligned}\hat{b}_l(nT) &= \text{DFT}[r(k)]_l = \text{DFT}\{\text{IDFT}[\mathbf{b}(nT)] * \mathbf{h}\}_l \\ &= g_l(n)b_l(nT) + e_l(n)\end{aligned} \tag{16.20}$$

where $*$ represents linear convolution, $\mathbf{h}$ is the channel impulse, $g_l(n)$ is the fading factor which can be compensated with an automatic gain control (one tap equalizer), and $e_l(n)$ is the error due to intersymbol interference, that is,

$$\begin{aligned}e_l(n) = \sum_{i=0}^{I} h_i \sum_{l=0}^{M-1} b_l(nT) \sum_{k=-\infty,(j,n)\neq(l,m)}^{\infty} \varphi_j(k\frac{T-\tau}{M} - nT - \tau_i) \\ \times \varphi_l(k\frac{T-\tau}{M} - mT).\end{aligned} \tag{16.21}$$

Note that due to the relative delays τ_l between the subcarriers, orthogonality is no longer sufficient to guarantee the correct recovery of the information.

Notice that the chosen basis function here satisfies the orthogonality condition and, as a byproduct, allows fast implementation via FFT. However, the error signal above is not negligible because multipath causes interference among the symbols of different OFDM channels.

An intelligent solution to this problem is to use basis sequences at the receiver, which is biorthogonal to those used at the transmitter, that is,

$$\sum_{k=-\infty}^{\infty} \varphi_l(k\frac{T-\tau}{M} - nT)\psi_j(k\frac{T-\tau}{M} - mT) = \delta(l-j)\delta(n-m) \tag{16.22}$$

which represents biorthogonality.

An appropriate choice for $\psi_l(t)$ is

$$\psi_l(t) = \frac{1}{\sqrt{T-\tau}} \mathrm{e}^{-\frac{2\pi\,\mathrm{j}lt}{T-\tau}}[u(t) - u(t-T+\tau)] \tag{16.23}$$

for $0 \leq l \leq M-1$. Note that, strictly speaking, $\psi_l(t) = \varphi_l(T-\tau)$ for $0 \leq t \leq T-\tau$. That means the basis used at the receiver is matched to the final part of the basis function at the transmitter, and the reader should bear in mind that a few steps omitted here can be found in Diniz et al. (2012).

The receptor should then estimate the received symbols as follows:

$$r_l(n) = \hat{b}_l(nT) = \frac{1}{\sqrt{T-\tau}} \sum_{k=0}^{M-1} r(k)\mathrm{e}^{-\frac{2\pi\,\mathrm{j}lk}{M}}. \tag{16.24}$$

Note that, considering the discrete case, $\psi_j(k)$ is orthogonal to shifted versions of the transmitter modulating waveforms $\varphi_l(k)$ for $l \neq j$, for any $\varphi_l(k\frac{T-\tau}{M} - nT - \tau_i)$ if the delay τ_i falls in the range $0 \leq \tau_i < \tau$, and intersymbol interference is removed if the delay spread does not exceed the guard time. This technique is called *cyclic prefix*. Figure 16.13 details the building blocks of an OFDM transceiver utilizing a cyclic prefix. Alternative OFDM-inspired transceivers will be presented in this chapter.

Notice that, sometimes, the role of $\psi_j(k)$ and $\varphi_j(k)$ are exchanged in the literature. They can be changed with minor implications, although it appears that the case previously shown seems more natural for transmission since the whole interval allotted for a block is utilized.

16.5 Block Multipath Model

We discuss the multipath model for an OFDM block transceiver in the following. Assume we have a multipath model where, at the receiver, there are delayed versions of the OFDM symbol as

$$r(t) = \sum_{i=0}^{I} h_i x(t-\tau_i) \tag{16.25}$$

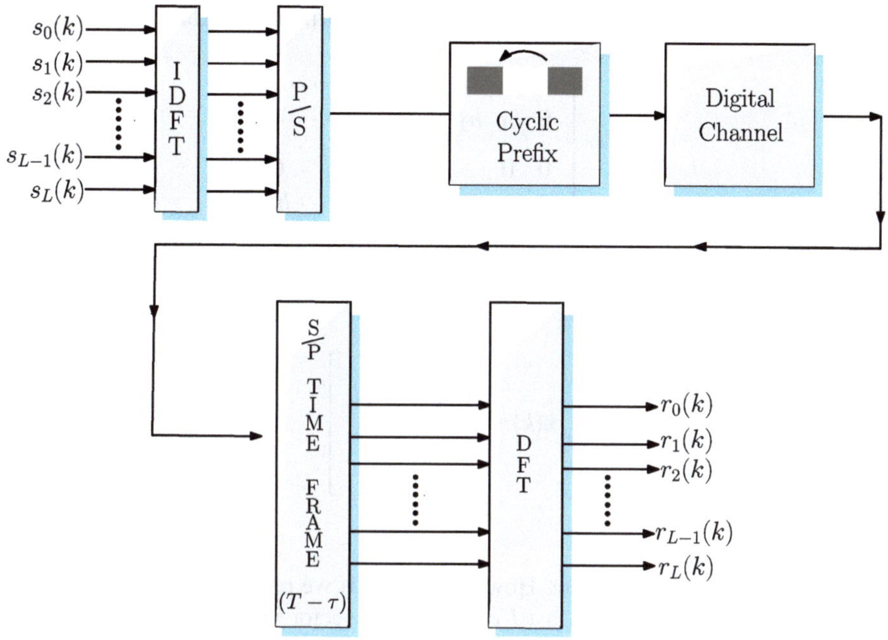

Fig. 16.13 OFDM digital transceiver

where h_i represents the channel gain related to a reflection cluster reaching the receiver with delay τ_i. Assume a time-series transmission in a streaming format and, after proper sampling, the received signal can be described in the discrete-time domain as

$$r(k) = \sum_{i=0}^{I} h_i(k-i) \tag{16.26}$$

whereas, by using a block description, we get

$$\mathbf{r}(k) = \begin{bmatrix} r(k) \\ r(k-1) \\ \vdots \\ r(k-M+1) \end{bmatrix} = \hat{\mathcal{H}}\mathbf{x}(k) + \mathbf{n}(k) \tag{16.27}$$

where $\hat{\mathcal{H}}$ represents the channel model, and its format here is more akin to the streaming data described in this section. At a later section, we will use the notation $\mathcal{H}$ for the channel matrix suitable for descriptions using data blocks.

Using the data-streaming format, the channel matrix has the following form

$$\hat{\boldsymbol{H}} = \begin{bmatrix} h_0 & h_1 & \cdots & h_I & 0 & 0 & \cdots & 0 \\ 0 & h_0 & h_1 & \cdots & h_I & 0 & \cdots & 0 \\ 0 & 0 & \ddots & \ddots & \ddots & \ddots & \ddots & 0 \\ 0 & \cdots & 0 & \cdots & 0 & 0 & \cdots & h_0 \end{bmatrix}$$

where $\hat{\boldsymbol{H}}$ is $M \times M$ and the input signal vector consists of a block of length M, that is

$$\mathbf{x}(k) = \begin{bmatrix} x(k) \\ x(k-1) \\ \vdots \\ x(k-M) \end{bmatrix} \tag{16.28}$$

where $I = \lceil \frac{\tau_L}{M} \rceil$.

However, $\hat{\boldsymbol{H}}$ is not circulant. How to force it? If we repeat the first I samples of the input-signal stream in the last I entries of the vector $\mathbf{x}(k)$, we obtain

$$\hat{\mathbf{x}}(k) = \begin{bmatrix} x(k) \\ x(k-1) \\ \vdots \\ x(k-M) \\ x(k) \\ x(k-1) \\ \vdots \\ x(k-M+I-1) \end{bmatrix} \tag{16.29}$$

Now the matrix $\hat{\boldsymbol{H}}$ has dimension $M \times (M+I-1)$. If we exploit the repetition in the entries of $\hat{\mathbf{x}}(k)$, the matrix representing the channel effect becomes

$$\boldsymbol{H} = \begin{bmatrix} h_0 & h_1 & \cdots & h_I & 0 & \cdots & 0 \\ 0 & h_0 & h_1 & \cdots & h_I & 0 & \vdots \\ 0 & 0 & \ddots & \ddots & \ddots & \ddots & \vdots \\ h_I & 0 & \cdots\cdots & h_0 & \cdots & & h_{I-1} \\ \vdots & \ddots & \ddots & \ddots & \ddots & \ddots & \vdots \\ h_1 & h_2 & \cdots & 0 & \cdots\cdots & & h_0 \end{bmatrix} \tag{16.30}$$

The resulting channel matrix $\boldsymbol{H}$ of dimension $M \times M$ is *circulant*. By applying the DFT to the resulting equation, it follows that

$$\begin{aligned}\mathrm{DFT}\left[\mathbf{r}(k)\right] &= \mathrm{DFT}\left[\mathcal{H}\mathbf{x}(k) + \mathbf{n}(k)\right] \\ &= \mathrm{DFT}\left[\mathcal{H}\ \mathrm{IDFT}[\mathbf{b}(k)]\right] + \mathbf{n}'(k)\end{aligned} \tag{16.31}$$

where

$$\mathbf{n}'(k) = \mathrm{DFT}[\mathbf{n}(k)].$$

If $\mathcal{H}$ is circulant, we have (and if the noise is white to guarantee the equations are really independent)

$$\mathfrak{r}(k) = \mathrm{DFT}\left[\mathbf{r}(k)\right] = \mathbf{\Lambda}\mathbf{b}(k) + \mathbf{n}'(k) \tag{16.32}$$

where $\mathbf{\Lambda}$ is the matrix with the eigenvalues of $\mathcal{H}$.

If the additional noise is not white, we can convert the channel into a white-noise equivalent by using $\mathbf{R}_n^{-\frac{1}{2}}$ as transformation matrix (guard period may need adjustment):

$$\mathfrak{r}_i(k) = \lambda_i b_i(k) + n_i'(k) \tag{16.33}$$

where $n_i'(k)$ is an element of $\mathbf{n}'(k)$.

For many carriers, that is, large blocks, the noise can be considered white. More details related to the channel gain will be discussed in this chapter.

Example 16.1 (OFDM Example) Design an OFDM communication system with four channels and a fixed channel with transfer function $H(z) = 1 + 0.1z^{-1}$.

$$\begin{bmatrix} r(k) \\ r(k-1) \\ r(k-2) \\ r(k-3) \end{bmatrix} = \begin{bmatrix} 1 & 0.1 & 0 & 0 & 0 \\ 0 & 1 & 0.1 & 0 & 0 \\ 0 & 0 & 1 & 0.1 & 0 \\ 0 & 0 & 0 & 0 & 0.1 \end{bmatrix} \begin{bmatrix} x(k) \\ x(k-1) \\ x(k-2) \\ x(k-3) \\ pref \end{bmatrix} \tag{16.34}$$

If a $pref = x(k)$ is used, then

$$\begin{bmatrix} r(k) \\ r(k-1) \\ r(k-2) \\ r(k-3) \end{bmatrix} = \begin{bmatrix} 1 & 0.1 & 0 & 0 \\ 0 & 1 & 0.1 & 0 \\ 0 & 0 & 1 & 0.1 \\ 0.1 & 0 & 0 & 1 \end{bmatrix} \begin{bmatrix} x(k) \\ x(k-1) \\ x(k-2) \\ x(k-3) \end{bmatrix} \tag{16.35}$$

Since $\mathbf{x}(k) = \mathcal{F}\mathbf{b}(k)$, where $\mathcal{F}$ is the DFT matrix, then

$$\mathcal{F}\begin{bmatrix} r(k) \\ r(k-1) \\ r(k-2) \\ r(k-3) \end{bmatrix} = \mathcal{F}\begin{bmatrix} 1 & 0.1 & 0 & 0 \\ 0 & 1 & 0.1 & 0 \\ 0 & 0 & 1 & 0.1 \\ 0.1 & 0 & 0 & 1 \end{bmatrix}\mathcal{F}^{-1}\begin{bmatrix} b(k) \\ b(k-1) \\ b(k-2) \\ b(k-3) \end{bmatrix}$$
$$= \begin{bmatrix} 1.1 & 0 & 0 & 0 \\ 0 & 1-0.1\mathrm{j} & 0 & 0 \\ 0 & 0 & 0.9 & 0 \\ 0 & 0 & 0 & 1+0.1\mathrm{j} \end{bmatrix}\begin{bmatrix} b(k) \\ b(k-1) \\ b(k-2) \\ b(k-3) \end{bmatrix} \tag{16.36}$$

The example shows that the transceiver operations lead to a decoupled reception of each symbol multiplied by a channel gain. This property allows simple equalization for proper detection of the transmitted symbols.

If a zero *pref* is used instead, then

$$\begin{bmatrix} r(k) \\ r(k-1) \\ r(k-2) \\ r(k-3) \end{bmatrix} = \begin{bmatrix} 1 & 0.1 & 0 & 0 \\ 0 & 1 & 0.1 & 0 \\ 0 & 0 & 1 & 0.1 \\ 0 & 0 & 0 & 1 \end{bmatrix}\begin{bmatrix} x(k) \\ x(k-1) \\ x(k-2) \\ x(k-3) \end{bmatrix} \tag{16.37}$$

Since $\mathbf{x}(k) = \mathcal{F}\mathbf{b}(k)$, where $\mathcal{F}$ is the DFT matrix, then

$$\mathcal{F}\begin{bmatrix} r(k) \\ r(k-1) \\ r(k-2) \\ r(k-3) \end{bmatrix} = \mathcal{F}\begin{bmatrix} 1 & 0.1 & 0 & 0 \\ 0 & 1 & 0.1 & 0 \\ 0 & 0 & 1 & 0.1 \\ 0 & 0 & 0 & 1 \end{bmatrix}\mathcal{F}^{-1}\begin{bmatrix} b(k) \\ b(k-1) \\ b(k-2) \\ b(k-3) \end{bmatrix}$$
$$= \begin{bmatrix} 1.075 & 0.025\mathrm{j} & 0.025 & -0.025\mathrm{j} \\ -0.025 & 1-0.075\mathrm{j} & 0.025 & -0.025\mathrm{j} \\ -0.025 & 0.025\mathrm{j} & 0.925 & -0.025\mathrm{j} \\ -0.025 & 0.025\mathrm{j} & 0.025 & 1+0.075\mathrm{j} \end{bmatrix} \cdot \tag{16.38}$$
$$\cdot \begin{bmatrix} b(k) \\ b(k-1) \\ b(k-2) \\ b(k-3) \end{bmatrix}$$

In this case, the transceiver offers coupled reception of each symbol. However, simple operations at the receiver end allow the desired decoupling, as will be addressed later.

16.6 DMT

From information theory (Cover and Thomas 2006), one can access the channel capacity (in bits/transmission or bits/real dimension, i.e., channel capacity normalized by the channel bandwidth), as defined by

$$C = \frac{1}{2}\log_2(1 + \text{SNR})$$
$$\text{SNR} = \frac{\epsilon}{\sigma_n^2} \tag{16.39}$$

where ϵ represented the energy per symbol (channel gain = 1), and σ_n^2 denotes noise variance.

Given that to transmit a bandwidth BW, we need a sample of at least $\frac{1}{T} = 2BW$, the number of bits per second corresponds to

$$\text{No of bits/s} = \frac{C\beta}{T} \tag{16.40}$$

where β is the number of bits per symbol or the dimension per symbol. T is the symbol period.

For a given small error probability P_e, the channel code requires infinite complexity. In practice, at a given fixed probability of symbol error, one can characterize the modulation/coding by a gap g that quantifies the effective loss in SNR with respect to capacity:

$$\overline{C} = \frac{1}{2}\log_2(1 + \frac{\text{SNR}}{g})$$
$$g = \frac{2^{2C} - 1}{2^{2\overline{C}} - 1}. \tag{16.41}$$

That means a smaller gap $\rightarrow$ the better the code. The constellations QAM and PAM achieve a gap of 9.8 dBs (Star et al. 1999), whereas good coding leads to gap in the range 3 to 5 dBs. Extremely powerful codes can reduce the gap to 1 to 2 dB, such as turbo codes, concatenated codes, and LDPC codes (Lin and Costello 2004).

16.6.1 Multichannel Transmission

Multichannel transmissions modulate the symbols with subcarriers, also called multitone or multicarrier system in a more straightforward case. For each subchannel

(N of those), each carrying

$$b_n = \frac{1}{2}\log_2(1 + \frac{\text{SNR}_n}{g}) \tag{16.42}$$

bits/dimension. Here, we are considering each subchannel carrying one bit (of N, and the dimension is one), and g is the channel gap, assuming they are equal for all subchannels.

The average number of bits is the sum of the number of bits carried on each channel divided by N (number of channels):

$$\begin{aligned}\overline{b} &= \frac{1}{N}\sum_{n=1}^{N}\frac{1}{2}\log_2(1 + \frac{\text{SNR}_n}{g}) \\ &= \frac{1}{2}\log_2\left[\prod_{n=1}^{N}(1 + \frac{\text{SNR}_n}{g})\right]^{\frac{1}{N}} \\ &= \frac{1}{2}\log_2(1 + \frac{\text{SNR geom.}}{g})\end{aligned} \tag{16.43}$$

where SNR geom. $\rightarrow$ geometric SNR, and the geometric SNR is defined as

$$\text{SNR geom.} = g\left[\prod_{n=1}^{N}(1 + \frac{\text{SNR}_n}{g})^{\frac{1}{N}} - 1\right]. \tag{16.44}$$

The parallel and independent channels together behave as a white Gaussian noise channel with geometric SNR, essentially equal to the geometric mean of the subchannels' SNRs (a good approximation for moderate to high SNR on all subchannels). In practical terms, the geometric SNR must be compared with the SNR of an equalized baseband system.

The geometric SNR can be improved considerably when the available energy is distributed nonuniformly over all or a subset of parallel channels, allowing high-performance multichannel systems (loading algorithms).

Consider a QAM transmission in each subchannel:

$$b_n = \frac{1}{2}\log_2\left(1 + \frac{\epsilon_n p_n^2}{g\sigma_n^2}\right) \tag{16.45}$$

where p_n is the channel gain (2 dimensions). The total number of bits is

$$b = \sum_{n=1}^{M} b_n \tag{16.46}$$

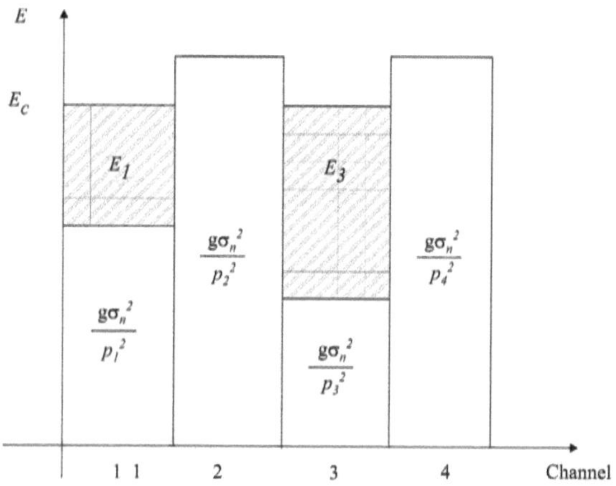

Fig. 16.14 Discrete water-filling for four subchannels

where M is the number of subchannels utilized. The number b must be maximized with the constraint that

$$\epsilon = \sum_{n=1}^{M} \epsilon_n \tag{16.47}$$

a procedure called water-filling.

A solution to the water-filling problem is Star et al. (1999)

$$\epsilon_n + \frac{g\sigma_n^2}{p_n^2} = \text{constant.} \tag{16.48}$$

Figure 16.14 illustrates a water-filling solution where we assign power to the subchannels less affected by the ratio between the noise and gain.

Example 16.2 (Water-Filling Example) Given a channel model with transfer function $H(z) = 1 - z^{-2}$, considering $\sigma_n^2 = 0.2$ and $g = 1$, compute the channel capacity and determine the power distribution for the case of 8 subchannels.

The values for λ_i^2 representing the channel gain at ith tone, $k_i = \frac{1}{\lambda_i^2}$, and σ_n^2/λ_i^2 are calculated as

	c_1	c_2	c_3	c_4	c_5	c_6	c_7	c_8
λ_i^2	0.00	2.00	4.00	2.00	0.00	2.00	4.00	2.00
k_i	∞	0.50	0.25	0.50	∞	0.50	0.25	0.50
σ_n^2/λ_i^2	∞	0.10	0.05	0.10	∞	0.10	0.05	0.10

Let us consider two different power constraints, $\epsilon = 0.08$ and $\epsilon = 1$.

(a) $\epsilon = 0.08$

The power distribution, the signal-to-noise ratio, and the capacity for each subchannel are summarized in the following table:

	c_1	c_2	c_3	c_4	c_5	c_6	c_7	c_8
ϵ_i	0.00	0.00	0.04	0.00	0.00	0.00	0.04	0.00
SNR_i	0.00	0.00	0.80	0.00	0.00	0.00	0.80	0.00
b_i	0.00	0.00	0.424	0.00	0.00	0.00	0.424	0.00

For this power distribution,

$$b = \sum_{n=1}^{M} b_n = 0.848 \text{ bits} \tag{16.49}$$

$$\text{SNR}_{\text{geom}} = 0.1583. \tag{16.50}$$

If the water-filling was not adopted, i.e., all channels were transmitted with $\epsilon_i = \epsilon/N = 0.01$, then

$$b_{\text{OFDM}} = 0.538 \text{ bits} \tag{16.51}$$

where b_{OFDM} represents the case with no bit loading (Fig. 16.15).

(b) $\epsilon = 1$
The power distribution, the signal-to-noise ratio, and the capacity for each subchannel are summarized in the following table:

	c_1	c_2	c_3	c_4	c_5	c_6	c_7	c_8
ϵ_i	0.00	0.15	0.20	0.15	0.00	0.15	0.20	0.15
SNR_i	0.00	1.50	4.00	1.50	0.00	1.50	4.00	1.50
b_i	0.00	0.661	1.161	0.661	0.00	0.661	1.161	0.661

For this power distribution (Fig. 16.16),

$$b = \sum_{n=1}^{M} b_n = 4.9658 \text{ bits} \tag{16.52}$$

$$\text{SNR}_{\text{geom}} = 1.3644 \tag{16.53}$$

$$b_{\text{OFDM}} = 4.1472 \text{ bits}. \tag{16.54}$$

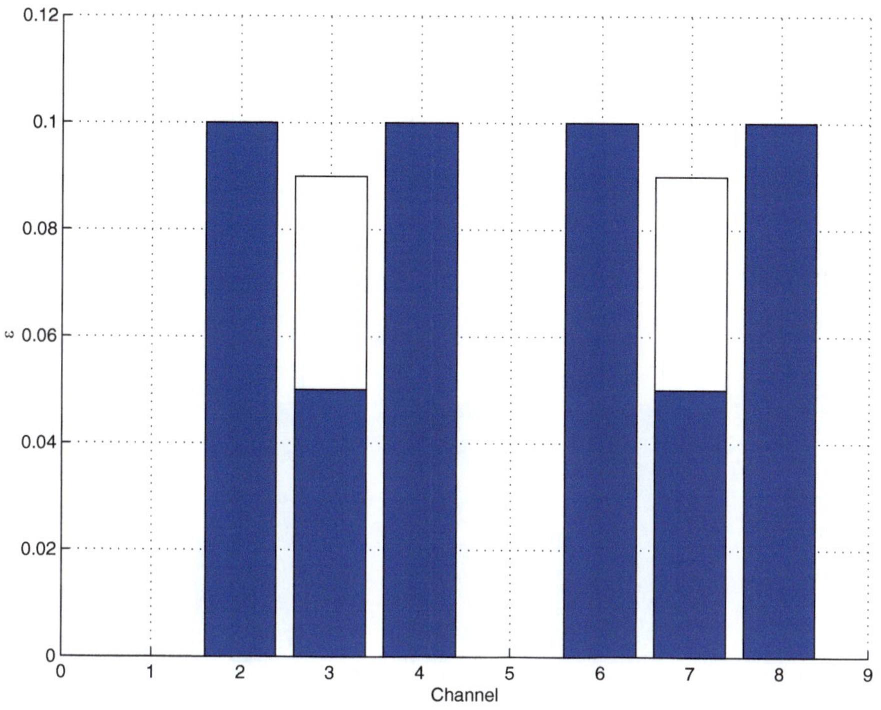

Fig. 16.15 Water-filling with $\epsilon = 0.08$

16.7 OFDM Systems

This section describes many options for configuring an OFDM-based transceiver. The presentation here follows the one in reference Diniz et al. (2012).

16.7.1 CP-OFDM

The cyclic prefix OFDM (CP-OFDM), previously illustrated in Fig. 16.13, inserts its redundancy in the cyclic format to induce a circulant format to the matrix-model channel representation.

The mathematical description of the CP-OFDM transceiver consists of the following transmitter and receiver matrices, respectively:

$$\mathcal{F} = \underbrace{\begin{bmatrix} \mathbf{0}_{K\times(M-K)} & \mathbf{I}_K \\ \multicolumn{2}{c}{\mathbf{I}_M} \end{bmatrix}}_{\mathbf{A}_{\mathrm{CP}}\in\mathbb{C}^{N\times M}} \mathbf{W}_M^H , \tag{16.55}$$

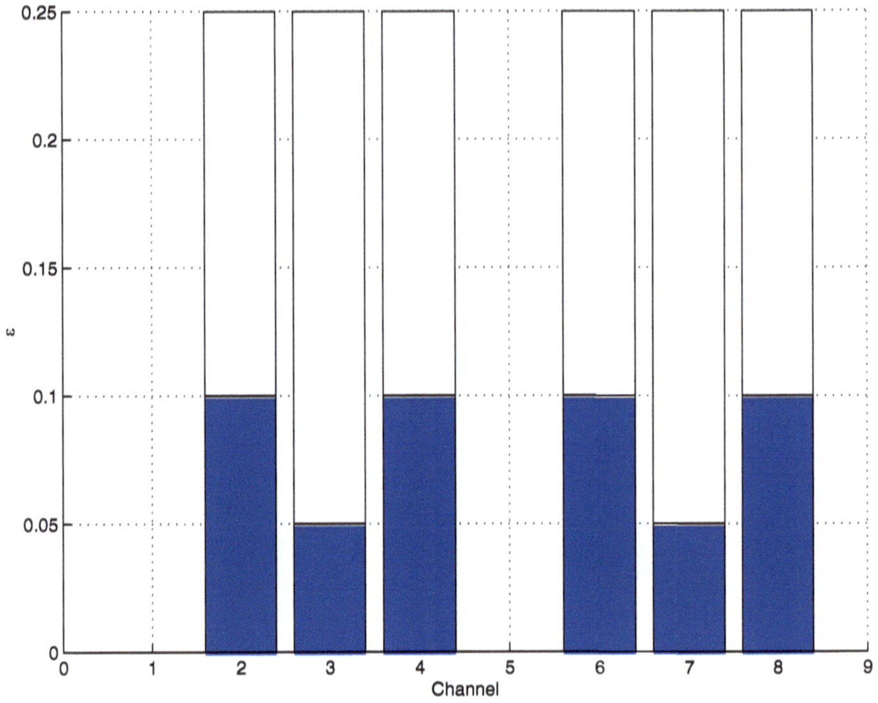

Fig. 16.16 Water-filling with $\epsilon = 1$

$$\mathcal{G} = \mathbf{E}\mathbf{W}_M \underbrace{\left[\mathbf{0}_{M\times K}\ \mathbf{I}_M\right]}_{\mathbf{R}_{\mathrm{CP}}\in\mathbb{C}^{M\times N}}, \tag{16.56}$$

where K represents the number of redundant elements, and $\mathbf{W}_M$ is the unitary $M \times M$ discrete Fourier transform (DFT) matrix, that is,

$$[\mathbf{W}_M]_{ml} = \frac{W_M^{ml}}{\sqrt{M}}, \tag{16.57}$$

with $(m, l) \in \mathcal{M}^2 = \mathcal{M} \times \mathcal{M}$ and

$$W_M = \mathrm{e}^{-\frac{2\pi}{M}}. \tag{16.58}$$

Note that $\mathbf{I}_M$ is the $M \times M$ identity matrix; $\mathbf{0}_{M\times N}$ is an $M \times N$ matrix, whose entries are zero; and $\mathbf{E}$ is an $M \times M$ diagonal equalizer matrix placed after scrapping the cyclic prefix and applying the DFT matrix.

Matrix $\mathbf{A}_{\text{CP}}$ adds and matrix $\mathbf{R}_{\text{CP}}$ removes the related cyclic prefix. It is possible to show that the product $\mathbf{R}_{\text{CP}}\mathcal{H}(z)\mathbf{A}_{\text{CP}} = \mathcal{H}_{\text{c}} \in \mathbb{C}^{M\times M}$ results in

$$\mathcal{H}_{\text{c}} = \begin{bmatrix} h(0) & 0 & \cdots & 0 & h(L) & \cdots & h(1) \\ h(1) & h(0) & \cdots & 0 & 0 & \ddots & \vdots \\ \vdots & \vdots & \ddots & & & & h(L) \\ h(L) & h(L-1) & & \ddots & \ddots & & 0 \\ 0 & h(L) & & & \ddots & & \vdots \\ \vdots & \ddots & \ddots & & & \ddots & 0 \\ 0 & \cdots & 0 & h(L) & \cdots & & h(0) \end{bmatrix}. \tag{16.59}$$

Note that the dependency on z disappears so that that $\mathbf{R}_{\text{CP}}$ eliminates the IBI, and matrix $\mathbf{A}_{\text{CP}}$ right-multiplies the resulting memoryless matrix $\mathbf{R}_{\text{CP}}\mathcal{H}(z) \in \mathbb{C}^{M\times N}$ inducing the overall matrix product to become a *circulant matrix* of dimension $M \times M$. As can be observed, each row of matrix $\mathcal{H}_{\text{c}}$ can be obtained by circular-shifting the corresponding previous row.

Finally, it is possible to infer that

$$\begin{aligned} \mathbf{\Lambda} &= \mathbf{W}_M \mathcal{H}_{\text{c}} \mathbf{W}_M^H \\ &= \mathbf{W}_M \mathbf{R}_{\text{CP}} \mathcal{H}(z) \mathbf{A}_{\text{CP}} \mathbf{W}_M^H \\ &= \operatorname{diag}\left\{ \sqrt{M} \mathbf{W}_M \begin{bmatrix} \mathbf{h} \\ \mathbf{0}_{(M-L-1)\times 1} \end{bmatrix} \right\} \\ &= \operatorname{diag}\{\lambda_m\}_{m\in\mathcal{M}}, \end{aligned} \tag{16.60}$$

in which $\mathbf{h} = [\, h(0) \;\; h(1) \;\; \cdots \;\; h(L)\,]^T$.

16.7.2 ZP-OFDM

Another popular form of OFDM-based system inserts zeros as redundancy, giving rise to the so-called zero-padding OFDM (ZP-OFDM). Figure 16.17 illustrates a possible and simplified configuration of ZP-OFDM. It can be shown that there are many variants of ZP-OFDM.

One of the alternatives is the ZP-OFDM-OLA (overlap-and-add), whose transmitter and receiver matrices are implemented as

$$\mathcal{F} = \underbrace{\begin{bmatrix} \mathbf{I}_M \\ \mathbf{0}_{K\times M} \end{bmatrix}}_{\mathbf{A}_{\text{ZP}}\in\mathbb{C}^{N\times M}} \mathbf{W}_M^H, \tag{16.61}$$

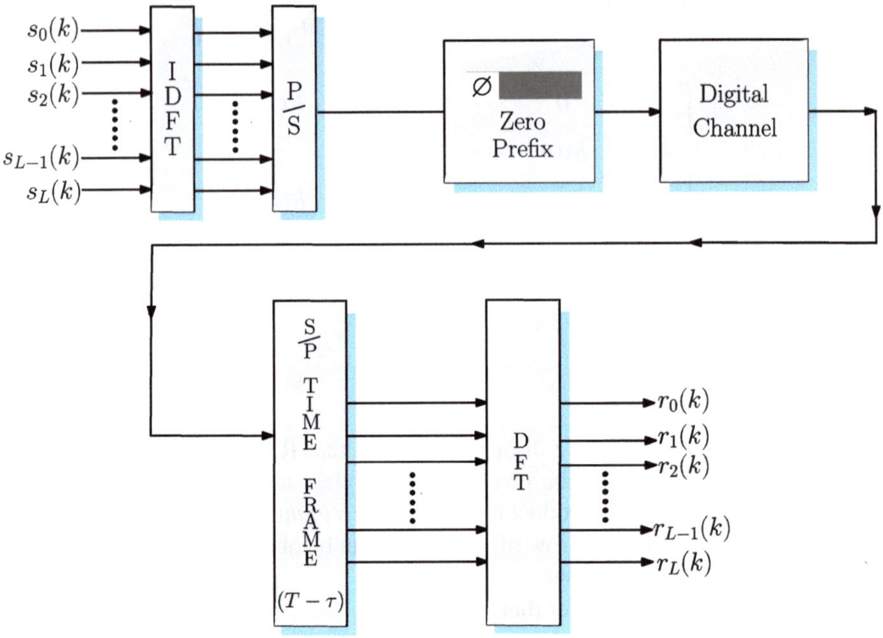

Fig. 16.17 Zero-prefix OFDM

$$\mathcal{G} = \mathbf{E}\mathbf{W}_M \underbrace{\begin{bmatrix} & \mathbf{I}_K \\ \mathbf{I}_M & \\ & \mathbf{0}_{(M-K)\times K} \end{bmatrix}}_{\mathbf{R}_{\mathrm{ZP}} \in \mathbb{C}^{M\times N}}, \tag{16.62}$$

where, as in the CP-OFDM case, $K \geq L$ elements are inserted as redundancy at the receiver end, and $N = M + K$. The OLA operation reestablishes the circulant property of the channel model.

The diagonalization of the overall communication of the data vector takes place because the product

$$\mathbf{R}_{\mathrm{ZP}}\mathcal{H}(z)\mathbf{A}_{\mathrm{ZP}} = \mathbf{R}_{\mathrm{CP}}\mathcal{H}(z)\mathbf{A}_{\mathrm{CP}} = \mathcal{H}_{\mathrm{c}} \in \mathbb{C}^{M\times M}. \tag{16.63}$$

since according to (16.59), the equivalent channel model is circulant. This property is the key for the DFT and IDFT matrices to decouple the entries of the information vector at the receiver. The price paid is that the matrix $\mathbf{R}_{\mathrm{ZP}}$ requires summations at the receiver end, also enhancing the channel noise.

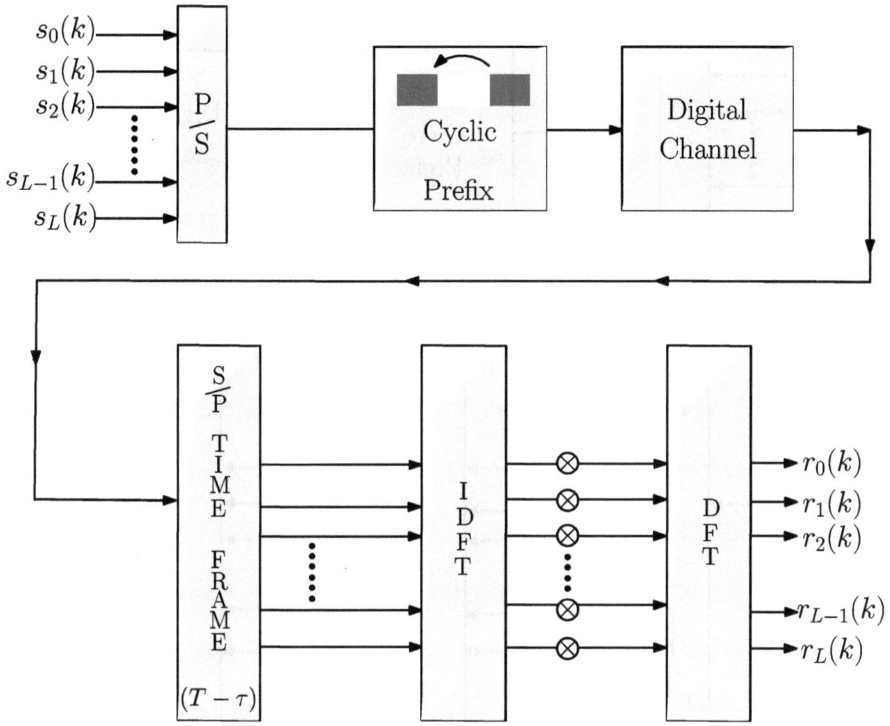

Fig. 16.18 Cyclic-prefix FD equalization

16.7.3 CP-SC-FD

The cyclic-prefix single-carrier with frequency-domain (CP-SC-FD) equalization transceiver uses cyclic prefix as redundancy, being closely related to the CP-OFDM transceiver. By extending the support of the subcarriers in the time domain, it is possible to maintain the orthogonality among them. The operation is related to introducing a cyclic prefix, as seen in Fig. 16.18.

The transmitter and receiver matrices related to the CP-SC-FD system are given by

$$\mathcal{F} = \begin{bmatrix} \mathbf{0}_{K\times(M-K)} \quad \mathbf{I}_K \\ \mathbf{I}_M \end{bmatrix}, \tag{16.64}$$

$$\mathcal{G} = \mathbf{W}_M^H \mathbf{E} \mathbf{W}_M \left[\mathbf{0}_{M\times K} \; \mathbf{I}_M \right]. \tag{16.65}$$

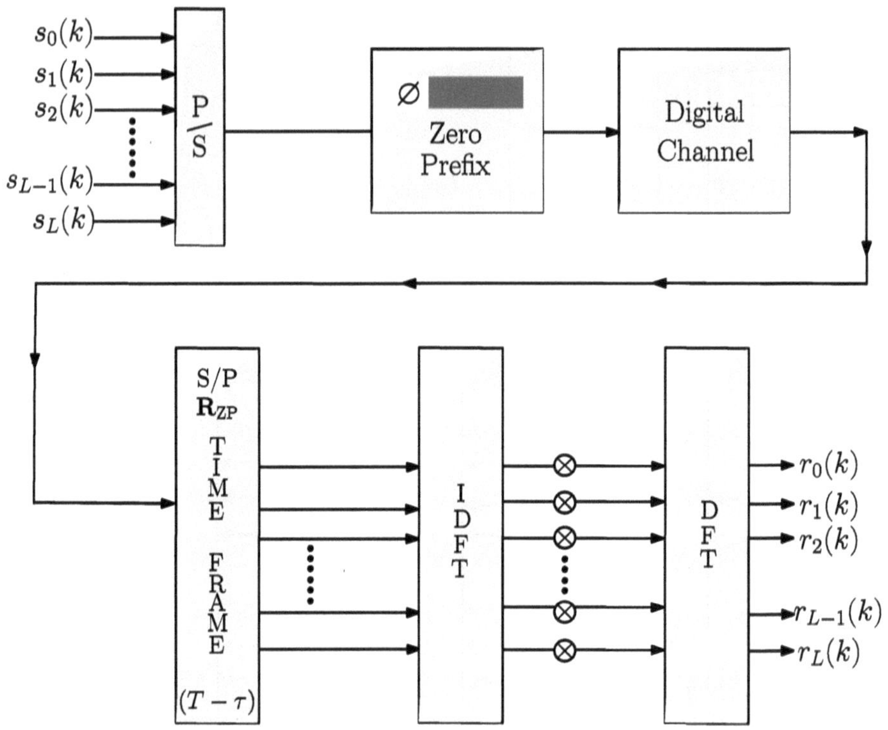

Fig. 16.19 Zero-prefix FD equalization

16.7.4 ZP-SC-FD

Yet another transceiver configuration is the so-called zero-padding single-carrier with frequency-domain (ZP-SC-FD) equalization transceiver. As in the ZP-OFDM transceiver case, it inserts redundant zeros into the block to transmit. Figure 16.19 illustrates the ZP-SC-FD transceiver where the equalization is performed in the frequency domain.

The FD versions of the transceivers are utilized in wireless systems with constraints to the power consumption related to the power amplifier due to the $PAPR$ (peak-to-average ratio) of the OFDM solutions. The uplink connection in the 4G wireless communication system is an example.

The overlap-and-add version of the ZP-SC-FD, namely, ZP-SC-FD-OLA, has the following transmitter and receiver matrix descriptions:

$$\mathcal{F} = \begin{bmatrix} \mathbf{I}_M \\ \mathbf{0}_{K\times M} \end{bmatrix}, \tag{16.66}$$

$$
\begin{aligned}
\mathcal{G} &= \mathbf{W}_M^H \mathbf{E} \mathbf{W}_M \begin{bmatrix} & \mathbf{I}_K \\ \mathbf{I}_M & \\ & \mathbf{0}_{(M-K)\times K} \end{bmatrix} \\
&= \mathbf{W}_M^H \mathbf{E} \mathbf{W}_M \begin{bmatrix} & \mathbf{I}_K \\ \mathbf{I}_M & \\ & \mathbf{0}_{(M-K)\times K} \end{bmatrix} \mathbf{R}_{\text{ZP}}. \qquad (16.67)
\end{aligned}
$$

Broadly, we can represent the observed vector at the receiver in a noiseless channel as $\mathbf{y} = \mathcal{H}_c \mathbf{s}$, where $\mathcal{H}_c$ is a circulant matrix which contains the channel coefficients. Moreover, we know from linear algebra that all circulant matrices may be diagonalized using the same set of orthonormal eigenvectors. These eigenvectors are the columns of the unitary DFT matrix, $\mathbf{W}$. In addition, the eigenvalues of circulant channel matrices are easily computed by means of the DFT of the first column of the circulant matrix (Diniz et al. 2012). Thus, we have $\mathbf{y} = \mathcal{H}_c \mathbf{s} = \mathbf{W}^H \boldsymbol{\Lambda} \mathbf{W} \mathbf{s}$, which implies that we can recover $\mathbf{s}$ by performing $\mathbf{s} = \mathcal{H}_c^{-1} \mathbf{y} = \mathbf{W}^H \boldsymbol{\Lambda}^{-1} \mathbf{W} \mathbf{y}$, considering that $\boldsymbol{\Lambda}^{-1}$ is computable, i.e., all eigenvalues of $\mathcal{H}_c$ are nonzero.

16.8 OFDM Equalization

To start the equalization process, it is worth mentioning that the matrix $\boldsymbol{\Lambda}$ congregates the distortion originating from the channel on each symbol of the data block. Therefore, the CP-OFDM transceiver model is described by

$$
\hat{\mathbf{s}} = \mathbf{E} \boldsymbol{\Lambda} \mathbf{s} + \mathbf{E} \mathbf{n}', \qquad (16.68)
$$

with $\mathbf{n}' = \mathbf{W}_M \mathbf{R}_{\text{CP}} \mathbf{n}$, where the time dependency of the expressions was omitted. As expected, the estimates of the transmitted symbols are uncoupled; that means each symbol can be estimated independently of any other symbol within the related block, avoiding intersymbol interference. A nice interpretation of this fact is that each symbol can be considered as being transmitted through a flat-fading subchannel.

The model described in (16.68) has a simple interpretation from a signal processing perspective. Indeed, the inclusion and removal of the cyclic prefix transform the linear convolution described in (17.72) into a *circular convolution*. Each subcarrier of the CP-OFDM system is weighted in the frequency domain by a constellation symbol. Then, the system applies an inverse discrete Fourier transformation, generating a vector in the time domain. The elements of this vector can be thought of as a periodic signal processed by the channel through a linear convolution. The result returns the signal to the original frequency domain. A fundamental property of digital signal processing is that the circular convolution of two signals can be implemented in the frequency domain by performing the product of the DFTs of the related signals. Therefore, the CP-OFDM system can be further simplified if we consider this fact. That means we can perform the entire processing in the frequency domain. The symbols that are loaded at each subcarrier can be directly mapped to the received signals at each subcarrier by performing the product with the frequency response of the channel, i.e., the DFT of the zero-padded impulse response of (16.60).

An equalizer design can have various solutions. For instance, the most popular forms of $\mathbf{E}$ are the *zero-forcing* (ZF) and the *minimum mean-square error* (MMSE) equalizers. The ZF solution entails compensation for the distortions introduced by the channel, disregarding the channel noise. As a result, if there is no noise, the ZF solution can perfectly recover the transmitted vector. The solution assumes that matrix $\boldsymbol{\Lambda}$ is invertible, thus yielding

$$\mathbf{E}_{\text{ZF}} = \boldsymbol{\Lambda}^{-1}. \tag{16.69}$$

In the MMSE solution, the invertibility of matrix $\boldsymbol{\Lambda}$ is not a requirement. Indeed, the linear MMSE equalizer matrix is the solution to the following optimization problem:

$$\mathbf{E}_{\text{MMSE}} = \arg\left\{\min_{\forall \mathbf{E}\in\mathbb{C}^{M\times M}} \xi(\mathbf{E})\right\}, \tag{16.70}$$

where ξ is a real-valued function of a complex-valued matrix argument defined as

$$\begin{aligned}
\xi(\mathbf{E}) &= \mathbb{E}\left[\|\mathbf{s} - \mathbf{E}(\boldsymbol{\Lambda}\mathbf{s} + \mathbf{n}')\|_2^2\right] \\
&= \mathbb{E}\left[\left(\mathbf{s} - \mathbf{E}\boldsymbol{\Lambda}\mathbf{s} - \mathbf{E}\mathbf{n}'\right)^H \left(\mathbf{s} - \mathbf{E}\boldsymbol{\Lambda}\mathbf{s} - \mathbf{E}\mathbf{n}'\right)\right] \\
&= \text{tr}\left\{\mathbb{E}\left[\left(\mathbf{s} - \mathbf{E}\boldsymbol{\Lambda}\mathbf{s} - \mathbf{E}\mathbf{n}'\right)\left(\mathbf{s} - \mathbf{E}\boldsymbol{\Lambda}\mathbf{s} - \mathbf{E}\mathbf{n}'\right)^H\right]\right\} \\
&= \text{tr}\left\{\sigma_s^2\mathbf{I}_M + \sigma_s^2\mathbf{E}\boldsymbol{\Lambda}\boldsymbol{\Lambda}^H\mathbf{E}^H - \sigma_s^2\mathbf{E}\boldsymbol{\Lambda} - \sigma_s^2\boldsymbol{\Lambda}^H\mathbf{E}^H + \sigma_v^2\mathbf{E}\mathbf{E}^H\right\},
\end{aligned} \tag{16.71}$$

where $\mathbb{E}[\cdot]$ and $\text{tr}\{\cdot\}$ are the expected value and trace operators, respectively. In the derivation, we assumed that the transmitted symbols and environment noise within a block are independent and identically distributed (i.i.d.), originating from wide-sense stationary (WSS) white random sequences with zero means and uncorrelated. The practical results are that $\mathbb{E}[\mathbf{s}\mathbf{v}'^H] = \mathbb{E}[\mathbf{s}]\mathbb{E}[\mathbf{v}']^H = \mathbf{0}_{M\times M} = \mathbb{E}[\mathbf{v}']\mathbb{E}[\mathbf{s}]^H = \mathbb{E}[\mathbf{v}'\mathbf{s}^H]$ and that $\mathbb{E}[\mathbf{s}\mathbf{s}^H] = \sigma_s^2\mathbf{I}_M$ and $\mathbb{E}[\mathbf{v}'\mathbf{v}'^H] = \sigma_v^2\mathbf{I}_M$, where the positive real numbers σ_s^2 and σ_v^2 are the variances of the related WSS random sequences.[2]

[2] In this book, we shall not employ distinct notations for deterministic and random variables.

Then, by employing the properties of derivatives of scalar functions of complex matrices (Vaidyanathan et al. 2010):

$$\frac{\partial \operatorname{tr}\left\{\mathbf{Z}\mathbf{A}\mathbf{Z}^H\right\}}{\partial \mathbf{Z}^*} = \mathbf{Z}\mathbf{A}, \tag{16.72}$$

$$\frac{\partial \operatorname{tr}\left\{\mathbf{A}\mathbf{Z}^H\right\}}{\partial \mathbf{Z}^*} = \mathbf{A}, \tag{16.73}$$

we have

$$\frac{\partial \xi(\mathbf{E})}{\partial \mathbf{E}^*} = \sigma_s^2 \mathbf{E}\boldsymbol{\Lambda}\boldsymbol{\Lambda}^H - \sigma_s^2 \boldsymbol{\Lambda}^H + \sigma_v^2 \mathbf{E}. \tag{16.74}$$

The optimal solution $\mathbf{E}_{\text{MMSE}}$ is such that $\frac{\partial \xi(\mathbf{E}_{\text{MMSE}})}{\partial \mathbf{E}^*} = \mathbf{0}_{M\times M}$, which implies that[3]

$$\begin{aligned}\mathbf{E}_{\text{MMSE}} &= \boldsymbol{\Lambda}^H \left(\boldsymbol{\Lambda}\boldsymbol{\Lambda}^H + \frac{\sigma_v^2}{\sigma_s^2}\mathbf{I}_M\right)^{-1} \\ &= \operatorname{diag}\left\{\frac{\lambda_m^*}{|\lambda_m|^2 + \frac{\sigma_v^2}{\sigma_s^2}}\right\}_{m\in\mathcal{M}}.\end{aligned} \tag{16.75}$$

If one assumes that the channel order L is less than the total length of the transmitted message N, then the received block $\mathbf{y}(n) \in \mathbb{C}^{N\times 1}$ will suffer from ISI and IBI effects according to the channel model described in Diniz et al. (2012), which can be rewritten as

$$\mathbf{y}(n) = \mathcal{H}_{\text{ISI}}\mathcal{F}\mathbf{s}(n) + \mathcal{H}_{\text{IBI}}\mathcal{F}\mathbf{s}(n-1) + \mathbf{n}(n), \tag{16.76}$$

where $\mathbf{n}(n) \in \mathbb{C}^{N\times 1}$ accounts for an additive noise at the receiver front end.

16.9 Coded OFDM

The usual complete coded OFDM system is depicted in Fig. 16.20 where two levels of coding and an interleaver are employed. These building blocks are required because:

[3] We encourage the reader to justify why this is actually the minimum solution of the objective function.

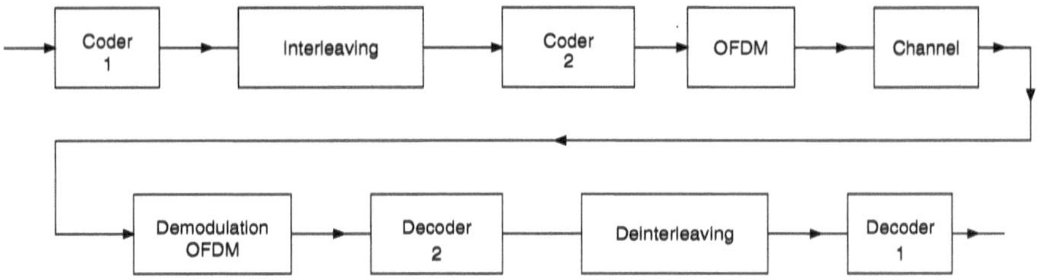

Fig. 16.20 Coded OFDM system

- Coder 1 is meant to insert redundancy in the bitstream (for example, using Reed-Solomon codes). This will increase the required bandwidth for transmission, but, on the other hand, it allows for a more reliable reception.

  Two approaches are to obtain block codes (code words) with increased Hamming distances or to encode the data continuously so that a correlation between the data is forced to help recognition at the receiver.
- Coder 2 is meant to increase the Euclidean distance among the symbols of the constellation. Usually using a trellis-coded modulation.

  The idea is to increase the number of symbols in comparison with the required number and subdivide them into subsets with greater Euclidean distances while maintaining the transmission energy.

  The subsets and the points in the subsets are tied using a trellis diagram derived from the convolutional code. A Viterbi decoder is used at the receiver (6 dB of improvement is expected).

  Error burst could occur if decoder 2 chooses the wrong decoding path which could then be corrected by the combination of the interleaver and decoder 1.

In the final analysis, the reason for coding is to provide a link among the symbols transmitted on different carriers such that fading carriers do not avoid the correct reception (which can be achieved with non-strongly faded subcarriers). The bottom line is that we are distributing responsibilities.

16.10 Multicarrier System

A generalization of the multicarrier communication systems utilizes sophisticated transmultiplexer design as depicted in Fig. 16.21. In this configuration, the transmitted signal block is filtered by a precoder with memory consisting of a MIMO FIR filter. The inherent memory at the transmitter can be viewed as a kind of redundancy, given that a signal block is transmitted more than once along with neighboring blocks. The more sophisticated transmitter requires a more complex receiver but might reduce the number of prefix signals to allow for the zero-forcing solution, for

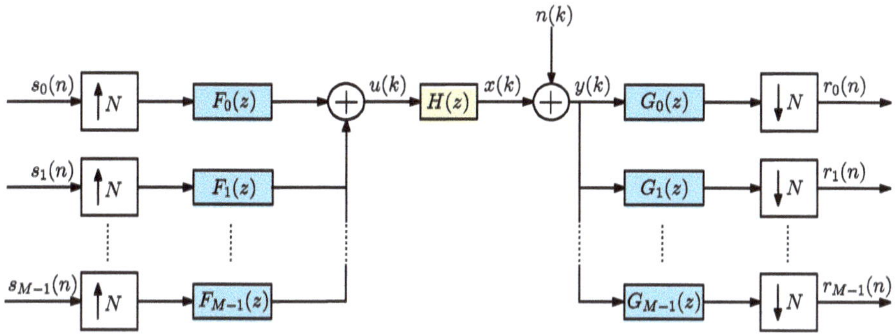

Fig. 16.21 TMUX multicarrier system

example. More information about multicarrier systems with longer precoders will be presented in Chap. 17.

16.11 Remarks

OFDM-based transceivers constitute the leading solution for dealing with frequency-selective channels. They also have the additional benefit of allowing flexibility to share the subcarriers among different users and services. In addition, these transceivers have very simple implementation by employing widely available signal processing tools. As a result, it is not surprising that they have been employed in many wireless communications and sensing solutions, taking part in autonomous vehicle deployments.

References

R.W. Chang, High-speed multichannel data transmission with bandlimited orthogonal signals. Bell. Sys. Tech. J. **45**, 1775–1796 (1966)

T.M. Cover, J.A. Thomas, *Elements of Information Theory*, 2nd edn. (John Wiley Interscience, New York, 2006)

P.S.R. Diniz, E.A. B. da Silva, S.L. Netto, *Digital Signal Processing: System Analysis and Design*, 2nd edn. (Cambridge University Press, Cambridge, 2010)

P.S.R. Diniz, W.A. Martins, M.V.S. Lima, *Block Transceivers: OFDM and Beyond* (Morgan & Claypool Publishers, Fort Collins, 2012)

W. Hirt, J.L. Massey, Capacity of the discrete-time Gaussian channel with intersymbol interference. IEEE Trans. Inform. Theory **34**, 380–388 (1988)

S. Lin, D.J. Costello, *Error Control Coding*, 2nd edn. (Prentice Hall, Upper Saddle River, 2004)

T. Star, J.M. Cioffi, P.J. Silverman, *Understanding Digital Subscriber Line Technology* (Prentice Hall, Upper Saddle River, 1999)

P.P. Vaidyanathan, S.M. Phoong, Y.P. Lin, *Signal Processing and Optimization for Transceiver Systems* (Cambridge University Press, Cambridge, 2010)

Chapter 17
MIMO Systems

17.1 Introduction

The multiple-input multiple-output (MIMO) model represents several communication systems encompassing: CDMA systems, multiple array systems, and multicarrier systems such as OFDM, DMT, filter-bank-based multicarrier systems, and many combinations of optimized MIMO systems. This chapter discusses many aspects of MIMO systems, starting with their general description, including several distinct configurations of communication systems. Then, a review of the basic concepts of statistical signal processing is introduced to provide tools to design some transceivers. These concepts are applied to design precoders and equalizers in the transceivers.

The chapter discusses a multiple-antenna configuration, where a singular value decomposition (SVD) formulation allows a precoder and postcoder design. In this context, we discuss how the channel capacity relates to the singular values of the channel matrix.

The MIMO transceivers with memory can be modeled using filter bank tools. Through the description of the transmitter and receiver using polyphase decomposition and the use of noble identities of multirate systems, it is possible to describe the channel as a MIMO system giving rise to a fully block-based formulation. The same model encompasses multiple transceivers' transmission with memory and redundancy operation on a channel described as a matrix including memory. The chapter includes a discussion on how to design a transceiver iteratively.

Using multiple antennas at the transmitter and receiver brings about three straightforward benefits: beamforming gain, spatial diversity, and spatial multiplexing. When incorporated into autonomous vehicle applications, these advantages bring flexibility and performance gains, where interference mitigation, direction of arrival, velocity, and positioning require robust solutions.

Y. C. Lim et al., *Signal Processing for Autonomous Systems*,
https://doi.org/10.1007/978-3-032-00675-2_17

17.2 Multi-Input Multi-Output Model

In communications, there are distinct situations where information is acquired from the environment, at different instances, and with distinct objectives. In some cases, pilot symbols are sent through the communication systems to estimate the channel model, a situation where the transmitted symbols are assumed to be known at the receiver side. In other cases, estimation relates to detecting the transmitted information symbols through the channel.

Let us consider a general model for the received information vector **r** certainly containing the information block of interest (Fig. 17.1):

$$\mathbf{r} = \mathcal{G}\mathcal{H}\mathcal{F}\,\mathbf{s} + \mathcal{G}\mathcal{H}_{\mathrm{mu}}\mathbf{s}_{\mathrm{mu}} + \mathcal{G}\mathbf{n} \tag{17.1}$$

where $\mathcal{F}$ represents the precoder matrix, $\mathcal{G}$ represents the receiver matrix, and $\mathcal{H}$ is the MIMO channel model. In some cases, the precoding process might include some redundancy to the transmitter block and process several blocks of data symbols. The MIMO channel model might also have memory, whose details are discussed in due course. The matrix $\mathcal{H}_{\mathrm{mu}}$ represents channels associated with the interference of other users in the environment and $\mathbf{s}_{\mathrm{mu}}$ represents the overall information sent to these users that are observed when trying to detect a block of the desired user. The vectors **n** represent the environment noise.

There are several ways to represent and configure the input and output signals and the channel in a communication system. By properly stacking the received information, the representation above is quite powerful for noisy channels. If we can consider that all nuisances originating from other users and the environmental noise are uncorrelated with the information of interest, the model above can be simplified as

$$\mathbf{r} = \mathcal{G}\mathcal{H}\mathcal{F}\,\mathbf{s} + \mathcal{G}\mathbf{n}_{\mathrm{t}} \tag{17.2}$$

where $\mathbf{n}_{\mathrm{t}}$ represents the total noise disturbance, including interference. We will use just the notation **n** in the following discussions except when we want the reader to consider other interference effects. In case the noise term includes more than the environmental noise, this will be clear from the context.

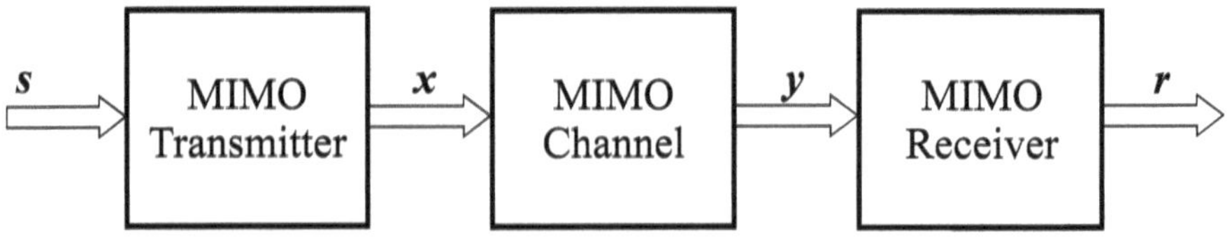

Fig. 17.1 MIMO model

The simplified formulation may result from a stacked vector representation of several distinct information origins. Examples include vectors representing data from different users transmitted through several transmitting antennas and acquired by several receiver antennas. It may also assemble data from several block instances, all depending on the problem at hand and whether some interference cancellation schemes should utilize part of the interfering information.

If the channel is noiseless, further simplification is possible, as in the following equation:

$$\mathbf{r} = \mathcal{G}\mathcal{H}\mathcal{F}\mathbf{s} \tag{17.3}$$

where again $\mathcal{F}$ represents the precoder matrix, $\mathcal{G}$ represents the receiver matrix, and $\mathcal{H}$ is the MIMO channel model.

17.3 Statistical Signal Processing Tools

Usually, the communication systems require the estimation of some parameters that, in the present discussion, are assembled in a vector denoted by $\mathbf{w}$. The entries of this vector can represent the channel model parameters, the symbols to be detected, etc.

From the statistical point of view, if the parameter vector is considered deterministic, the estimation method should fall in the family of classical estimation. On the other hand, Bayesian estimation methods are employed to estimate the representation of random vector $\mathbf{w}$ based on other observed random variables.

In the classical approach, the main information is modeled by the probability density function (PDF) of the random vector $\mathbf{y}$ that is dependent on deterministic vector $\mathbf{w}$ and denoted as $f(\mathbf{y};\mathbf{w})$.[1] This notation with a semicolon implies that the PDF is parameterized by $\mathbf{w}$, following the notation in Kay (1993). It should not be confused with the joint PDF denoted as $f(\mathbf{y},\mathbf{w})$ where both vectors in the argument are random.

The Bayesian estimation algorithms depend on the a posteriori distribution of the data $\mathbf{y}$ by relying on prior information about the PDF of $\mathbf{w}$, denoted as $f(\mathbf{w})$, reflecting some assumed knowledge. The a posteriori distribution employs the Bayes rule, as the name suggests, that is,

$$\begin{aligned} f(\mathbf{w}|\mathbf{y}) &= \frac{f(\mathbf{y}|\mathbf{w})\,f(\mathbf{w})}{\int f(\mathbf{y},\mathbf{w})\,f(\mathbf{w})\,d\mathbf{w}} \\ &= \frac{f(\mathbf{y}|\mathbf{w})\,f(\mathbf{w})}{f(\mathbf{y})} \end{aligned} \tag{17.4}$$

where the denominator of the first expression utilizes marginalization to compute $f(\mathbf{y})$. The estimation problem in statistical signal processing is grounded in the assumptions and definitions related to PDFs.

[1] This notation represents a family of PDFs, not a conditional PDF.

Let us consider a simplified linear model;

$$\mathbf{y} = \mathcal{H}\mathbf{w} + \mathbf{n} \tag{17.5}$$

where the vectors $\mathbf{y}$, $\mathbf{w}$, and $\mathbf{n}$ represent the observed signal, the unknown parameters, and the noise signal, of dimensions $M \times 1$, $N \times 1$, and $M \times 1$, respectively. The matrix $\mathcal{H}$ is part of a model and has dimension $M \times N$.

17.3.1 Classical Methods

Among the classical estimation methods, we can highlight the Cramer-Rao lower bound (CRLB) method, whose solution, when available, leads to the minimum variance unbiased (MVU) estimator (Kay 1993; Poor 1994). The presentation here follows Chaps. 2 and 3 of Kay (1993); see Chap. 2 of this book for details. Its general solution requires the knowledge $f(\mathbf{y}; \mathbf{w})$. This PDF might not be available, and sometimes, the MVU estimate cannot be found or does not exist, even when the PDF is available; see also Chap. 2 of Kay (1993). However, for a linear model as in (17.5) where $\mathcal{H}$ is considered known, and $\mathbf{n}$ is a Gaussian vector with zero mean and covariance $\mathbf{R_n}$, the PDF is

$$f(\mathbf{y}; \mathbf{w}) = \frac{1}{(2\pi)^{\frac{M}{2}} \det(\mathbf{R_n})} \mathrm{e}^{-\frac{1}{2}\left[(\mathbf{y}-\mathcal{H}\mathbf{w})^H \mathbf{R_n}^{-1} (\mathbf{y}-\mathcal{H}\mathbf{w})\right]} \tag{17.6}$$

where $\det(\cdot)$ denotes the determinant of $(\cdot)$. Then, the optimal estimate has a closed-form solution given by

$$\mathbf{w}_{\mathrm{o}} = \left(\mathcal{H}^H \mathbf{R_n}^{-1} \mathcal{H}\right)^{-1} \mathcal{H}^H \mathbf{R_n}^{-1} \mathbf{y} \tag{17.7}$$

calculated by equating $\frac{\partial f(\mathbf{y};\mathbf{w})}{\partial \mathbf{w}}$ to zero, which is an MVU estimator.

This same expression (17.7) also corresponds to the best linear unbiased estimator (BLUE), whose starting assumption is $\mathbb{E}[\mathbf{y}] = \mathcal{H}\mathbf{w}$ and its covariance is $\mathbf{R_y} = \mathbb{E}\left[(\mathbf{y} - \mathbb{E}[\mathbf{y}])^H (\mathbf{y} - \mathbb{E}[\mathbf{y}])\right]$,[2] that is, as long as the first- and second-order statistics of the data are available, we can compute the estimator.[3] In such a case, the model in (17.5) applies if $\mathbf{R_y} = \mathbf{R_n}$ and the environment noise is Gaussian and has zero mean. If the noise is non-Gaussian, the solution of (17.7), with $\mathbf{R_n}$ replaced

[2] The symbol $\mathbf{R_y}$ represents covariance here, but in many texts it denotes correlation. We expect no confusion here.

[3] The idea is to estimate $\mathbf{w}$ from a linear model of the observed data $\hat{\mathbf{z}} = \mathbf{A}\mathbf{y}$ where $\mathbf{z} = \mathbb{E}[\hat{\mathbf{z}}] = \mathbf{A}\mathbb{E}[\mathbf{y}]$, since $\mathbb{E}[\mathbf{y}] = \mathcal{H}\mathbf{w}$, the constraint $\mathbf{A}\mathcal{H} = \mathbf{I}$ should be satisfied while minimizing the estimated variance sitting at the diagonal entries of the matrix $\mathbf{A}^H \mathbf{R_y} \mathbf{A}$.

by $\mathbf{R_y}$, is still a BLUE estimation, but it is not an MVU estimation. The MVU, in this case, provides a better solution outside the domain of the BLUE estimator; that is, most likely, there is a better nonlinear estimator.

Another important estimator in the classical family is the maximum likelihood estimator (MLE), designed to maximize $f(\mathbf{y}; \mathbf{w})$ with respect to $\mathbf{w}$. In practice, the vector $\mathbf{y}$ corresponds to the observed data, and the MLE solution requires the PDF's knowledge. If there is an MVU solution, the MLE estimator finds it; if not, it provides an optimal solution for large data sets and is unbiased (Kay 1993, pp. 482). There are cases where a closed-form solution cannot be found, and the MLE problem might be solved iteratively; see, for instance, Kay (1993), Chaps. 7 and 14. If the PDF is assumed to be known, the model is linear for the case, and the noise follows a zero-mean Gaussian distribution, then the MLE solution is also given by (17.7). The MLE is the most widely used among the estimation methods; see Sect. 2.2.

17.3.2 Least Squares Estimator

The least squares estimator (LSE) assumes the model

$$\mathbf{y} = \mathfrak{F}(\mathbf{w}) + \mathbf{n} \tag{17.8}$$

where the vector $\mathfrak{F}(\cdot)$ represents a vector function. The LSE seeks the minimization of

$$\xi(\mathbf{w}) = (\mathbf{y} - \mathfrak{F}(\mathbf{w}))^H (\mathbf{y} - \mathfrak{F}(\mathbf{w})) \tag{17.9}$$

where this minimization does not necessarily mean a minimization of the parameter estimation error. For a zero-mean noise with Gaussian distribution, the LSE and MLE solutions are equivalent (Kay 1993). For a linear model, where $\mathfrak{F}(\mathbf{w}) = \mathcal{H}\mathbf{w}$, the LSE solution is

$$\mathbf{w}_o = \left(\mathcal{H}^H \mathcal{H}\right)^{-1} \mathcal{H}^H \mathbf{y} \tag{17.10}$$

where a closer comparison between Eqs. (17.10) and (17.7) show that for a diagonal noise covariance in the latter, the results are the same. That is, the LSE and MVU estimations coincide. Otherwise, we can modify the cost function in (17.9) by including a weight function as

$$\xi(\mathbf{w}) = (\mathbf{y} - \mathfrak{F}(\mathbf{w}))^H \mathcal{W} (\mathbf{y} - \mathfrak{F}(\mathbf{w})) \tag{17.11}$$

so that by making $\mathcal{W} = \mathbf{R_n^{-1}}$, the weighted LSE solution becomes equivalent to the MVU estimation, assuming $\mathbf{n}$ is a Gaussian vector with non-diagonal $\mathbf{R_n}$. In the latter case, the estimate has the form of (17.7). On the other hand, if the noise is non-Gaussian, the solution is also (17.7), but this time it is just the best unbiased linear solution, BLUE.

17.3.3 Bayesian Methods

We have some widely used approaches on the Bayesian estimation side, starting with the minimum mean squared error (MMSE) estimator; see Chap. 3. In the MMSE case, the joint PDF $f(\mathbf{y}, \mathbf{w})$ is considered known and the model coefficients assembled in $\mathbf{w}$ are assumed random. Usually, $f(\mathbf{y}|\mathbf{w})$ is the data model, and $f(\mathbf{w})$ is the prior PDF information about the model. With these definitions, we have $f(\mathbf{y}, \mathbf{w}) = f(\mathbf{y}|\mathbf{w})f(\mathbf{w}) = f(\mathbf{w}|\mathbf{y})f(\mathbf{y})$ so that the MMSE Bayesian estimator seeks to find

$$\mathbf{w}_\text{o} = \mathbb{E}_{f(\mathbf{w}|\mathbf{y})}[\mathbf{w}|\mathbf{y}] \tag{17.12}$$

where the expectation is calculated with respect to the posterior PDF $f(\mathbf{w}|\mathbf{y})$ and using (17.4).

A distinct feature of the Bayesian approach lies in the minimization of the MSE, taking into consideration the prior distribution of the parameters, defined as

$$\begin{aligned}
\text{MSE}_\text{B} &= \int\int (\mathbf{w} - \mathbb{E}[\mathbf{w}|\mathbf{y}])^H (\mathbf{w} - \mathbb{E}[\mathbf{w}|\mathbf{y}]) f(\mathbf{y}, \mathbf{w}) d\mathbf{y} d\mathbf{w} \\
&= \int\int (\mathbf{w} - \mathbb{E}[\mathbf{w}|\mathbf{y}])^H (\mathbf{w} - \mathbb{E}[\mathbf{w}|\mathbf{y}]) f(\mathbf{w}|\mathbf{y}) d\mathbf{w} f(\mathbf{y}) d\mathbf{y} \\
&= \int \text{var}\left[\mathbf{w} - \mathbb{E}[\mathbf{w}|\mathbf{y}]\right] f(\mathbf{y}) d\mathbf{y}
\end{aligned} \tag{17.13}$$

where var[·] denotes variance of [·]. The objective is to minimize $\mathbb{E}[(\mathbf{w} - \mathbb{E}[\mathbf{w}|\mathbf{y}])^H (\mathbf{w} - \mathbb{E}[\mathbf{w}|\mathbf{y}])]$ with respect to $f(\mathbf{y}, \mathbf{w})$.

In the classical non-Bayesian case, the MSE is calculated by

$$\text{MSE} = \int (\mathbf{w} - \mathbf{w}_\text{o})^H (\mathbf{w} - \mathbf{w}_\text{o}) f(\mathbf{y}; \mathbf{w}_\text{o}) d\mathbf{y}. \tag{17.14}$$

Although for some estimators, their expressions look the same, the Bayesian and classical principles are quite distinct. In (17.14), the PDF is a function of a fixed $\mathbf{w}_o$; as such, the MSE will depend on it. In the Bayesian case in (17.13), the simulation realization requires that we draw $\mathbf{w}_o$ according to a prior distribution.

For jointly Gaussian vectors $\mathbf{w}$ and $\mathbf{y}$, the Bayesian MMSE solution is

$$\mathbf{w}_\text{o} = \mathbb{E}[\mathbf{w}] + \mathbf{P}_{\mathbf{wy}} \mathbf{R}_\mathbf{y}^{-1} (\mathbf{x} - \mathbb{E}[\mathbf{y}]) \tag{17.15}$$

where $\mathbf{P}_{\mathbf{wy}} = \mathbb{E}[\mathbf{w}\mathbf{y}^H]$ is the cross-correlation between the vectors in the argument, and $\mathbf{R}_\mathbf{y}$ is the autocorrelation of the observed vector.

For a linear model, with a prior PDF of $\mathbf{w}$ being Gaussian with mean $\mathbf{w}_\mathrm{m}$ and covariance $\mathbf{R}_\mathbf{w}$, the MMSE solution becomes

$$\mathbf{w}_\mathrm{o} = \mathbf{w}_\mathrm{m} + \mathbf{R}_\mathbf{w}\mathcal{H}^H \left(\mathcal{H}\mathbf{R}_\mathbf{w}\mathcal{H}^H + \mathbf{R}_\mathbf{n}\right)^{-1} (\mathbf{y} - \mathcal{H}\mathbf{w}_\mathrm{m}) \tag{17.16}$$

where

$$\mathbf{R}_\mathbf{w} = \mathbb{E}\left[(\mathbf{w} - \mathbb{E}[\mathbf{w}])(\mathbf{w} - \mathbb{E}[\mathbf{w}])^H\right].$$

17.3.4 MAP Estimation

The maximum a posteriori (MAP) estimation is obtained by maximizing the a posteriori PDF $f(\mathbf{w}|\mathbf{y})$, that is,

$$\mathbf{w}_\mathrm{o} = \max_{\mathbf{w}} f(\mathbf{w}|\mathbf{y}) = \max_{\mathbf{w}} f(\mathbf{y}|\mathbf{w}) f(\mathbf{w}). \tag{17.17}$$

For the case $\mathbf{w}$ and $\mathbf{y}$ are jointly Gaussian, and the MAP and MMSE solutions are identical. For PDFs having the maxima at the same location as the mean, these estimators lead to the same result; see Kay (1993) pp. 488 and Sects. 3.3 and 3.4. Also, in the linear model case with the noise having a Gaussian PDF, the peak and the mean are at the same position, so that the solution in (17.15) is valid for the MAP estimator.

17.3.5 LMMSE Estimator

The last estimator discussed here of the Bayesian family is the linear minimum mean square error (LMMSE) estimator. The LMMSE estimator does not require knowledge of the joint PDF $f(\mathbf{y}, \mathbf{w})$. Basically, we assume that $\mathbf{w}$ is linearly estimated from the observation $\mathbf{y}$, relying on the correlation between them. If they are uncorrelated, $\mathbf{w}$ cannot be linearly estimated from $\mathbf{y}$. Also, the result is not optimal if the MMSE estimator is nonlinear. The solution assumes knowledge of the covariance matrix $\mathbf{R_y}$ and $\mathbf{P_{wy}}$, as well as $\mathbb{E}[\mathbf{w}]$ and $\mathbb{E}[\mathbf{y}]$, to derive the estimator as[4]

$$\mathbf{w}_o = \mathbb{E}[\mathbf{w}] + \mathbf{P_{wy}}\mathbf{R_y}^{-1}(\mathbf{x} - \mathbb{E}[\mathbf{y}]) \tag{17.18}$$

where

$$\mathbf{R_y} = \mathbb{E}\left[(\mathbf{y} - \mathbb{E}[\mathbf{y}])(\mathbf{y} - \mathbb{E}[\mathbf{y}])^H\right],$$

and

$$\mathbf{P_{wy}} = \mathbb{E}\left[(\mathbf{w} - \mathbb{E}[\mathbf{w}])(\mathbf{y} - \mathbb{E}[\mathbf{y}])^H\right].$$

The cost function given by (17.13) is minimized without any assumption about $f(\mathbf{y}, \mathbf{w})$, but assuming that the parameters can be estimated by a linear (or affine) function of the observation $\mathbf{y}$. Again, the case $\mathbf{w}$ and $\mathbf{y}$ are jointly Gaussian, and the LMMSE estimator, MAP, and MMSE solutions are identical (Kay 1993).

17.4 Precoder Design

The statistical signal processing theory sets the stage for producing the building blocks required to design a transceiver. The procedure generally involves designing or choosing a precoder, estimating the parameters of the channel model, and designing the postcoder, which is called the MIMO equalizer. In MIMO systems employing multiple antennas, the precoder maps the symbols to the transmit antennas. It should be designed appropriately, particularly in the downlink connections, leaving the main computational burden on the base station.

In many standards, the precoder is predetermined. However, it is possible to design the precoder in some cases, as long as the pertinent information is available, such as the channel model feedback to the receiver. An excellent general reference to the design of linear precoders can be found in Joham et al. (2005).

[4] The cost function is to minimize

$$\mathbb{E}\left[(\mathbf{w} - \mathbf{Ay})^H(\mathbf{w} - \mathbf{Ay})\right]$$

with respect to $\mathbf{A}$ that represents a linear weight to the vector $\mathbf{y}$.

17.4.1 Linear Zero-Forcing Precoder

The linear zero-forcing precoder aims at inverting the channel effect, subjected to constraining the transmission energy by E_t, so that $\mathbb{E}[\|\mathcal{F}\mathbf{s}\|^2] = \text{tr}\left[\mathcal{F}\mathbf{R}_s\mathcal{F}^H\right] = E_t$. The cost function consists of the minimization of $\mathbb{E}[\|\mathcal{F}\mathbf{s}\|^2]$, subject to the constraint that $\mathcal{G}\mathcal{H}\mathcal{F} = \mathbf{I}$.

The resulting precoder is

$$\mathcal{F} = \sqrt{\frac{E_t}{\text{tr}\left(\left(\mathcal{G}\mathcal{H}\mathcal{H}^H\mathcal{G}^H\right)^{-1}\mathbf{R}_s\right)}}\mathcal{H}^H\mathcal{G}^H\left(\mathcal{G}\mathcal{H}\mathcal{H}^H\mathcal{G}^H\right)^{-1}; \tag{17.19}$$

some attention should be given to the conditioning of the matrix $\mathcal{G}\mathcal{H}$. Section 17.10.1 presents a sketch of the proof.

17.4.2 Matched Filter Precoder

The matched filter precoder entails the maximization of the ratio between desired signal energy at the receiver with respect to the noise and constrained by the transmission energy E_t, that is,

$$\max_{\mathcal{F}} = \frac{\mathbb{E}\left[|\mathbf{s}^H\mathbf{r}|^2\right]}{\mathbb{E}\left[\|\mathcal{G}\mathbf{n}\|^2\right]} \tag{17.20}$$

$$\text{subject to}: \mathbb{E}[\|\mathcal{F}\mathbf{s}\|^2] = E_t \tag{17.21}$$

where $\mathbf{r}$ represents the receiver signal vector used to detect the transmitted symbols. Originally proposed to the CDMA system (Esmailzadeh and Nakagawa 1993), the matched filter precoder belongs to a class of precoder design, which includes the minimization of the bit error rate (Hjørungnes and Diniz 2005).

The solution is given by

$$\mathcal{F} = \sqrt{\frac{E_t}{\text{tr}\left(\mathcal{H}^H\mathcal{G}^H\mathbf{R}_s\mathcal{G}\mathcal{H}\right)}}\mathcal{H}^H\mathcal{G}^H \tag{17.22}$$

where E_t is the transmitter energy and $\mathbf{R}_s$ is the transmitter symbol autocovariance matrix. See Sect. 17.10.2 for a simplified proof.

17.4.3 Linear MMSE Precoder

In this linear MMSE precoder case, the cost function to be minimized is

$$\min_{\mathcal{F},\alpha} \frac{\mathbb{E}\left[\|\mathbf{s}-\alpha\mathbf{r}\|^2\right]}{\sigma_{\mathrm{n}}^2} \tag{17.23}$$

$$\text{subject to}: \mathbb{E}[\|\mathcal{F}\mathbf{s}\|^2] = E_{\mathrm{t}} \tag{17.24}$$

where α is a positive scalar.

Let us define the notation $\hat{\mathcal{H}} = \mathcal{G}\mathcal{H}$ and $\hat{\mathbf{R}}_{\mathbf{n}} = \mathcal{G}\mathbf{R}_{\mathbf{n}}\mathcal{G}^H$. After solving the optimization problem utilizing the Lagrange multiplier tool, Joham et al. (2005), the solution is

$$\mathcal{F} = \sqrt{\frac{E_{\mathrm{t}}}{\mathrm{tr}\left(\left(\hat{\mathcal{H}}\hat{\mathcal{H}}^H + \mathrm{tr}\left(\hat{\mathbf{R}}_{\mathbf{n}}\right)\right)^{-2}\left(\hat{\mathcal{H}}\mathbf{R}_{\mathbf{s}}\hat{\mathcal{H}}\right)\right)}}\left(\hat{\mathcal{H}}\hat{\mathcal{H}}^H + \mathrm{tr}\left(\hat{\mathbf{R}}_{\mathbf{n}}\right)\right)^{-1}\frac{\mathbf{I}_{N_{\mathrm{t}}}}{E_{\mathrm{t}}}\hat{\mathcal{H}}^H \tag{17.25}$$

where $\mathbf{I}_{N_{\mathrm{t}}}$ represents an identity matrix whose dimension N_{t} is the length of the transmitted vector. See Sect. 17.10.3 for a proof.

17.5 Equalizer Designs

The cases in which we seek a linear equalizer solution to take part in the receiver design are briefly shown, considering the linear channel model of (17.2).

17.5.1 Linear Zero-Forcing Equalizer

Equalizers are intended to minimize either the ISI or the mean square error (MSE) at the receiver. On the other hand, the zero-forcing (ZF) equalizer is thought to invert the channel effect at the receiver without taking into consideration the presence of channel noise.

In the zero-forcing case, the solution is

$$\begin{aligned}
\mathcal{G}_{\mathrm{ZF}} &= (\mathcal{H}\mathcal{F})^{\dagger} \\
&= [(\mathcal{H}\mathcal{F})^H(\mathcal{H}\mathcal{F})]^{-1}(\mathcal{H}\mathcal{F})^H \\
&= [\mathcal{F}^H\mathcal{H}^H\mathcal{H}\mathcal{F}]^{-1}\mathcal{F}^H\mathcal{H}^H
\end{aligned}$$

$$
\begin{aligned}
&= \mathcal{F}^{-1}(\mathcal{H}^H\mathcal{H})^{-1}\left(\mathcal{F}^H\right)^{-1}\mathcal{F}^H\mathcal{H}^H \\
&= \mathcal{F}^{-1}\underbrace{(\mathcal{H}^H\mathcal{H})^{-1}\mathcal{H}^H}_{\triangleq\mathcal{H}^\dagger} \\
&= \mathcal{F}^H\mathcal{H}^\dagger .
\end{aligned}
\tag{17.26}
$$

The given expression for $\mathcal{G}_{\text{ZF}}$ is valid in the case $\mathcal{H}$ has full rank. As a result, we are assuming that the FIR channel model leads to a rank-M matrix $\mathcal{H}$. The symbol $(\cdot)^\dagger$ denotes the Moore-Penrose pseudo-inverse of $(\cdot)$, and $\mathcal{F}$ was considered a unitary matrix but this is not a requirement.

By applying the solution to the received vector, we get

$$
\begin{aligned}
\mathcal{G}_{\text{ZF}}\mathbf{r} &= \mathcal{F}^{-1}(\mathcal{H}^H\mathcal{H})^{-1}\mathcal{H}^H\mathcal{H}\mathcal{F}\mathbf{s} + \mathcal{G}_{\text{ZF}}\mathbf{n}_{\text{t}} \\
&= \mathbf{s} + \mathcal{G}_{\text{ZF}}\mathbf{n}_{\text{t}};
\end{aligned}
\tag{17.27}
$$

in the last expression, the symbols $\mathbf{s}$ appear decoupled but very likely corrupted by a noise enhanced by the ZF equalizer, depending on its conditioning.

The ZF design does not require any estimate of the environment-noise variance and it is mainly discussed for theoretical purposes. This lack of information about the noise variance leads to poor performance when the SNR is low at the receiver end. A more informed ZF design consists of minimizing $\mathbb{E}[\|\mathcal{G}\mathbf{n}_{\text{t}}\|^2]$ with respect to $\mathcal{G}$, subjected to the constraint $\mathcal{G}\mathcal{H}\mathcal{F} = \mathbf{I}$, with the component matrices of appropriate dimension. In this case, the ZF solution is

$$
\mathcal{G}_{\text{ZF}} = \left(\mathcal{F}^H(\mathcal{H}^H\mathbf{R}_{\mathbf{n}_{\text{t}}}^{-1}\mathcal{H})\mathcal{F}\right)^{-1}\mathcal{F}^H\mathcal{H}^H\mathbf{R}_{\mathbf{n}_{\text{t}}}^{-1}.
\tag{17.28}
$$

Section 17.10.4 details a proof.

On the other hand, the practical minimum mean square error (MMSE) design is computationally more demanding than the ZF solution, but leads to more effective equalization.

17.5.2 Linear Minimum MSE Equalizer

The MMSE equalizer design consists of the minimization of the following cost function:

$$
\begin{aligned}
\xi_{\text{MSE}} &= \mathbb{E}\left\{\left[\mathcal{G}_{\text{MMSE}}\mathbf{r} - \mathbf{s}\right]^H\left[\mathcal{G}_{\text{MMSE}}\mathbf{r} - \mathbf{s}\right]\right\} \\
&= \mathbb{E}\left\{\left[\mathcal{G}_{\text{MMSE}}\left(\mathcal{F}^{-1}(\mathcal{H}^H\mathcal{H})^{-1}\mathcal{H}^H\mathcal{H}\mathcal{F}\mathbf{s} + \mathbf{n}_{\text{t}}\right)\right]^H\right.
\end{aligned}
$$

$$\times \left[\mathcal{G}_{\text{MMSE}}\left(\mathcal{F}^{-1}(\mathcal{H}^H\mathcal{H})^{-1}\mathcal{H}^H\mathcal{H}\mathcal{F}\mathbf{s}+\mathbf{n}_{\text{t}}\right)\right]\Big\} \tag{17.29}$$

where in the last expression, the symbols do not appear decoupled, but the noise term provides some regularization to the solution.

After a few derivations, the equalizer becomes

$$\begin{aligned}\mathcal{G}_{\text{MMSE}} &= \left[(\mathcal{H}\mathcal{F})^H\mathbf{R}_{\mathbf{n}_{\text{t}}}^{-1}\mathcal{H}\mathcal{F}+\mathbf{R}_{\text{s}}\right]^{-1}(\mathcal{H}\mathcal{F})^H\mathbf{R}_{\mathbf{n}_{\text{t}}}^{-1}\\ &= \left[\mathcal{F}^H\,\mathcal{H}^H\mathbf{R}_{\mathbf{n}_{\text{t}}}^{-1}\mathcal{H}\mathcal{F}+\mathbf{R}_{\text{s}}\right]^{-1}\mathcal{F}^H\,\mathcal{H}^H\mathbf{R}_{\mathbf{n}_{\text{t}}}^{-1}\end{aligned}$$

$$. \tag{17.30}$$

The optimal MMSE solution can be derived following the same procedure as Diniza et al. (2012). In addition, the vectors $\mathbf{s}$ and $\mathbf{n}_t$ are considered uncorrelated wide-sense stationary (WSS) random sequences with zero mean so that $\mathbb{E}[\mathbf{s}\mathbf{n}_{\text{t}}]^H = \mathbb{E}[\mathbf{s}]\mathbb{E}[\mathbf{n}_{\text{t}}]^H = \mathbb{E}[\mathbf{n}_{\text{t}}]\mathbb{E}[\mathbf{s}]^H = \mathbf{0}$. We also assume that $\mathbb{E}[\mathbf{s}\mathbf{s}^H] = \mathbf{R}_{\text{s}}$ and $\mathbb{E}[\mathbf{n}_{\text{t}}\,\mathbf{n}_{\text{t}}^H] = \mathbf{R}_{\mathbf{n}_{\text{t}}}$.

17.5.3 Maximum Likelihood Linear Equalizer

In the maximum likelihood receiver, the vector $\mathbf{r}$ is observed, and we have to choose one out of many possible transmitted vectors relying on a probabilistic argument. If the transmitted vector $\mathbf{s}$ belongs to a set $\mathcal{S}$, the objective is to maximize the probability that the chosen transmitted symbol induced the observed vector. This argument is equivalent to the maximum a posteriori (MAP) criterion, which maximizes $f(\mathbf{s}|\mathbf{r})$, for all $\mathbf{s} \in \mathcal{S}$, where $\mathcal{S}$ represents $\mathbb{C}^M$ the set of complex vectors with dimension M, in the present discussion.

The MAP solution is

$$\begin{aligned}\mathbf{s}^* &= \max_{\mathbf{s}\in\mathbb{C}^M}\{f(\mathbf{s}\text{ was sent}|\mathbf{r}\text{ is observed})\}\\ &= \max_{\mathbf{s}\in\mathbb{C}^M}\left\{\frac{f\,(\mathbf{r}\text{ is observed}|\mathbf{s}\text{ was sent})\,f\,(\mathbf{s}\text{ was sent})}{f\,(\mathbf{r}\text{ is observed})}\right\}\\ &= \max_{\mathbf{s}\in\mathbb{C}^M}\left\{\frac{f\,(\mathbf{r}|\mathbf{s})\,f\,(\mathbf{s})}{f\,(\mathbf{r})}\right\}.\end{aligned} \tag{17.31}$$

If the symbols are equiprobable, that is, $f(\mathbf{s})$ is the same for all symbol in $\mathcal{S}$, then maximizing $f(\mathbf{s}|\mathbf{r})$ is the same as maximizing $f(\mathbf{r}|\mathbf{s})$. That is,

$$\mathbf{s}^* = \max_{\mathbf{s}\in\mathbb{C}^M} \{f\ (\mathbf{r} \text{ is observed}|\mathbf{s} \text{ was sent})\} \tag{17.32}$$

considering that $f(\mathbf{r})$ is not a function of $\mathbf{s}$. The latter case corresponds to the maximum likelihood equalizer (MLE).[5]

Is there an equalization scheme that is optimum in the presence of white Gaussian background noise and multiple access interferences? In general, we can seek for optimum solutions, each specialized for the chosen optimality criterion. The computation of the optimum solution might be very computationally complex, but the optimum receiver serves as a performance reference for implementable systems. In general, it is expected that for high SNR, the performance of the simpler receivers is good enough for many applications.

The usual drawbacks of the optimal receivers relate to their much higher complexity and requirements of other users' information in multiuser environments.

If we choose maximum likelihood for detection, the objective is to maximize the a posteriori probability of $\mathbf{r}$ conditioned on $\mathbf{s}$. In the case all transmitted vectors are equally likely, the prior is non-informative (uniform), the a posteriori maximization is equivalent to maximizing the a priori one given by

$$f\{[\mathbf{r} \in \mathbb{C}^M]|\mathbf{s}\} = Be^{\left(\frac{\Gamma(\mathbf{s})}{2\sigma_n^2}\right)} \tag{17.33}$$

where, as before, $f[\cdot]$ denotes probability density of $[\cdot]$ governing the observation, B is a constant, $\mathbb{C}$ represents the set of complex numbers, and $\mathbf{s}$ is a vector including the M symbols sent (Van Trees 2013). Equation (17.33) represents the exponential likelihood function (Poor 1994). The parameter σ_n^2 is the variance of the measurement noise. If the noise is considered as white Gaussian, that is, the components of $\mathbf{r}$ are statistically independent, the function $\Gamma(\cdot)$ corresponds to

$$\Gamma(\mathbf{s}) = -\|\mathbf{r} - \mathbf{s}\|^2. \tag{17.34}$$

Note that $\mathbf{s}$ represents the signal prepared for transmission and the received vector $\mathbf{r}$ is the vector from which we should detect the transmitted signal. In the case there was interference from other users, the optimum solution requires the knowledge of all the information sent by all users.

According to the squared Euclidean distance (optimal for white Gaussian noise) to the received sequence, that is, the received vector $\mathbf{r}$ decodes symbol $\mathbf{s}$ that maximizes the likelihood function. In the case of white complex Gaussian

---

[5] The ML receiver can be considered as a special case of the MAP receiver since ML assumes all hypothesis are equality likely; see Sects. 2.2 and 3.4.

environment noise with zero mean, the decoding is equivalent to maximizing

$$\max\left\{2\mathrm{Re}\left[\mathbf{s}^H\mathbf{r}\right]-\mathbf{s}^H\mathbf{s}\right\}. \tag{17.35}$$

For Gaussian channels, the solution is equivalent to the matched filter, whereas for linear channels with memory, the maximum likelihood sequence estimator (MLSE) solution is also optimum (Goldsmith 2005). The MLSE equalization is complex to implement and might resource to the popular Viterbi algorithm (Viterbi 1995), wherein this case, we can benefit from the observation that a discrete-time channel resembles a convolution encoder, so that a trellis can be built, and the best path for the estimation can be found (Haykin 2001). Note that this equalizer solution does not employ an equalization filter; it just acts on the received signal, avoiding this way the resulting noise enhancement inherent to some equalization filters. Section 17.10.5 provides some details.

17.6 Multiple Antenna Case

In this section, we briefly discuss the mitigation of interference in a general multiple-antenna configuration.

17.6.1 Mitigating Interferences

There are many tools to provide enhancements in the communication performance by mitigating the interferences and distortions introduced at the physical layer. Some of these solutions include the use of the following techniques:

- Equalizers for MIMO systems
- Antenna arrays
- Optimized multicarrier systems

The primary sources of interference are the intersymbol interference due to multipath fading and co-channel interference due to multiple-user communications. In a wireless system, such as mobile radio, the fast time-varying fading calls for fast-tracking adaptive equalization techniques. Using multiple antennas at the transmitters and receivers is part of the solution to increase the data transfer in distinct scenarios. The following subsections briefly discuss some specific advantages of using multiple antennas.

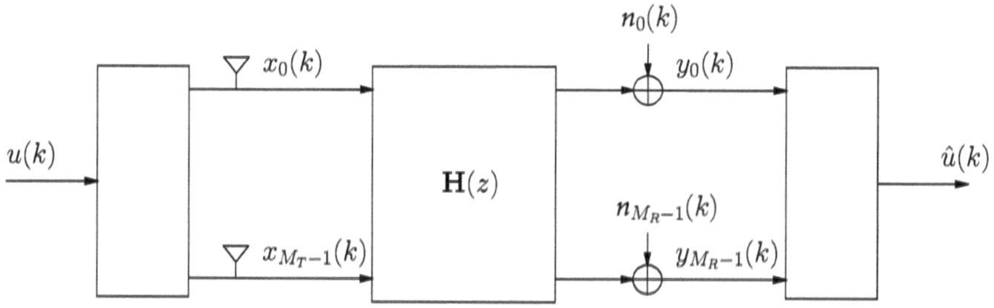

Fig. 17.2 Multiple antenna system

17.6.2 MIMO Systems

In the main MIMO systems, we use an antenna array at the transmitter and/or receiver in order to increase the communication capacity. The capacity gain of the MIMO systems with respect to the conventional SISO systems relies on:

- Number of antennas at the transmitter
- Number of antennas at the receiver
- Number of paths in the channel

The spatial diversity allows applications where we can implement spatial source separation, space-time coding, and MIMO equalization. Figure 17.2 illustrates a MIMO system employing multiple antennas.

As a rule, for multiple-input single-output (MISO) or single-input and multiple-output (SIMO) communication systems, as the SNR grows, the capacity gain with respect to the SISO ($\mathbf{H}_{l,j}$) approaches a constant. That means that the capacity curves of the MISO and SIMO solutions tend to be parallel to the corresponding curve of the SISO case for high SNR, obviously with higher values. Also, for MIMO communication systems, as the SNR grows, the capacity also increases at a rate $\rho > 1$ times the growth rate for MISO, SIMO, and SISO systems (Paulraj et al. 2003; Björnson and Demir 2003). The rate is given by

$$\rho \to \min(L + 1, M_R, M_T) \tag{17.36}$$

where $L + 1$ is the rank of the channel matrix, M_T is the number of transmitting antennas, and M_R is the number of receiving antennas.

The objective is to design a precoder (or the codeword to be sent) in order to maximize a performance criterion using the channel knowledge (when available). As long as there are multipaths, increasing the MIMO system is more advantageous since the channel with only a line of sight has rank one. In the latter case, all power gain originating from the MIMO connection is concentrated in the single eigenvector direction determined by the receiver array so that no spatial diversity is exploited. If the antennas are far apart or reflections and scatterings occur far away, there is a gain related to the direction of arrival (Tse and Viswanath 2005). However, if no multipath exists, the capacity curve as a function of the SNR resembles the results of a SISO channel obtained for high SNR and additive Gaussian white noise. The optimal precoder is computationally complex to obtain, requiring the calculation of singular value decomposition (SVD). The SVD is a powerful tool for addressing the capacity of MIMO channels.

For applications where the channel model is known or can be estimated, its model is represented by

$$\mathcal{H} = \mathcal{U}\boldsymbol{\Sigma}\mathcal{V}^H \tag{17.37}$$

where the channel matrix $\mathcal{H}$ has dimension $M_R \times M_T$ and has rank $L+1$. Matrix $\mathcal{U}$ contains the left singular vectors at its columns, whereas $\mathcal{V}$ has the right singular vectors at its columns.

Matrices $\mathcal{U}$ and $\mathcal{V}$ have dimensions $M_R \times M_R$ and $M_T \times M_T$, respectively, where

$$\mathcal{U}^H\mathcal{U} = \mathbf{I}_{M_R} \tag{17.38}$$

and

$$\mathcal{V}^H\mathcal{V} = \mathbf{I}_{M_T}. \tag{17.39}$$

Note that matrix $\mathcal{U}$ contains the eigenvectors of $\mathcal{H}\mathcal{H}^H$ and matrix $\mathcal{V}$ contains the eigenvectors of $\mathcal{H}^H\mathcal{H}$.

The matrix of the singular values is

$$\boldsymbol{\Sigma} = \begin{bmatrix} \sqrt{\lambda_0} & 0 & \cdots & 0 & \cdots 0 \\ 0 & \sqrt{\lambda_1} & \cdots & 0 & \cdots 0 \\ \mathbf{0} & \mathbf{0} & \ddots & \mathbf{0} & \mathbf{0}\ \mathbf{0} \\ 0 & 0 & \cdots & \sqrt{\lambda_L} & \cdots 0 \\ \mathbf{0} & \mathbf{0} & \ddots & \mathbf{0} & \ddots\ \mathbf{0} \end{bmatrix} \tag{17.40}$$

where $\sqrt{\lambda_0} \geq \sqrt{\lambda_1} \geq \cdots \sqrt{\lambda_L}$, $\boldsymbol{\Sigma}$ is an $M_R \times M_T$ matrix, and $(L+1) \leq \min[M_R, M_T]$.

Assume the original channel model is as follows:

$$\begin{aligned} \mathbf{y}(k) &= \mathcal{H}\mathbf{x}(k) + \mathbf{n}(k) \\ &= \mathcal{U}\boldsymbol{\Sigma}\mathcal{V}^H\,\mathbf{x}(k) + \mathbf{n}(k). \end{aligned} \tag{17.41}$$

By premultiplying the model by $\mathcal{U}^H$, we get

$$\begin{aligned}\mathcal{U}^H\mathbf{y}(k) &= \mathcal{U}^H\mathcal{U}\boldsymbol{\Sigma}\mathcal{V}^H\,\mathbf{x}(k) + \mathcal{U}^H\mathbf{n}(k)\\ &= \mathcal{U}^H\mathcal{U}\boldsymbol{\Sigma}\mathcal{V}^H\,\mathcal{V}\mathcal{V}^H\,\mathbf{x}(k) + \mathcal{U}^H\mathbf{n}(k)\\ &= \boldsymbol{\Sigma}\mathcal{V}^H\,\mathbf{x}(k) + \mathcal{U}^H\mathbf{n}(k).\end{aligned}$$

Then, by considering $\mathbf{s}(k) = \mathcal{V}^H\mathbf{x}(k)$ and $\mathbf{r}(k) = \mathcal{U}^H\mathbf{y}(k)$, we obtain

$$\mathbf{r}(k) = \boldsymbol{\Sigma}\,\mathbf{s}(k) + \mathcal{U}^H\mathbf{n}(k). \tag{17.42}$$

The second equality in (17.41) can be represented in a block form as illustrated in Fig. 17.3. Note that the resulting model works as if the transformed input vector is multiplied by $\mathcal{V}$ before transmission (precoder) and by $\mathcal{U}^H$ at the receiver.

The equality of (17.42) shows the decoupling effect of the SVD, allowing independent detection of transmitted data. After decomposition, the overall system consists of independent subchannels as shown in Fig. 17.4 where $\sqrt{\lambda_i}$ for $i = 0, \ldots, L$ represent the gains of the subchannels. Channel knowledge and SVD

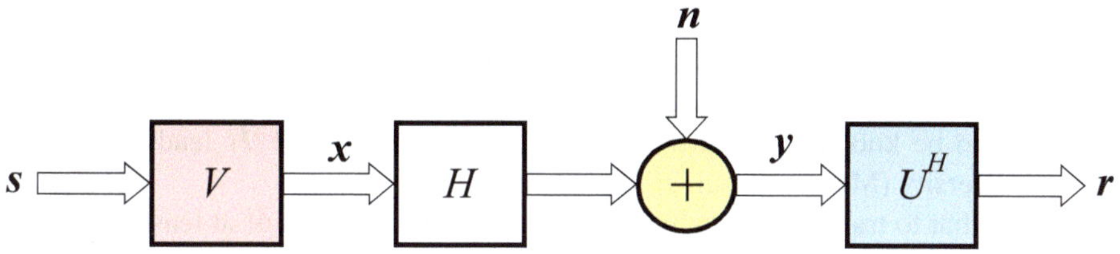

Fig. 17.3 Multiple antenna transceiver

Fig. 17.4 Decoupled multiple antenna system

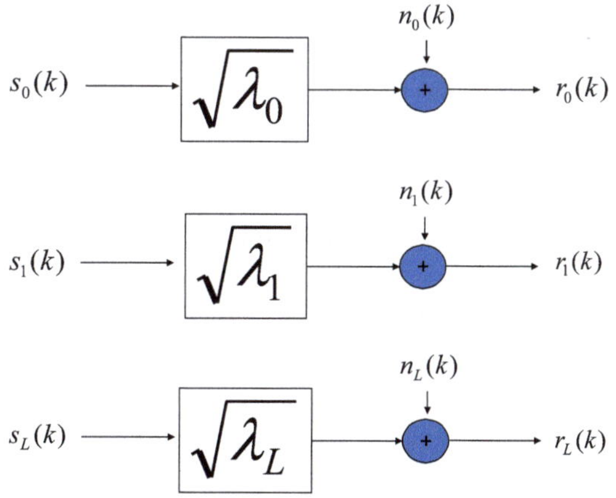

allow the allocation of variable energy among spatial subchannels, with the aim of maximizing the capacity similarly to the DMT bit loading described in Sect. 16.6.

17.6.3 Capacity Issues

Now, we discuss some capacity properties, taking into consideration the availability of the MIMO channel model at the transmitter. The channel model is always perfectly known or estimated and tracked at the receiver.

17.6.3.1 Known Channel

It is now possible to compute the channel capacity considering that the MIMO system consists of individual SISO subchannels connected in parallel. From information theory, we know that the channel capacity (in bits/transmission or bits/complex dimension, i.e., channel capacity normalized by the channel bandwidth) is given by

$$\mathrm{SNR}_i = \frac{\epsilon_i}{\sigma_{n_i}^2} \tag{17.43}$$

where ϵ_i represents energy per symbol (channel gain = $\sqrt{\lambda_i}$), $\sum_{i=0}^{L} \epsilon_i = M_T$, $\sigma_{n_i}^2$ denotes noise variance at each parallel subchannel, and the channel model is assumed to be known as seen in (17.42). The higher rank of $\mathcal{H}$ leads to greater spatial diversity (Marzetta et al. 2016).

Given that to transmit a bandwidth BW, we need a sample of at least $\frac{1}{T} = BW$ for complex symbols, and the number of bits per second corresponds to

$$\text{No of bits/s} = \frac{C\beta}{T} \tag{17.44}$$

where β is the number of bits per symbol or the dimension per symbol. T is the symbol period.

17.6.3.2 Unknown Channel

For applications where the channel model is unknown, we can describe the capacity of the MIMO channel through the following expression:

$$C = \max_{\mathrm{tr}[\mathbf{R_s}]=M_T} \log_2 \det\left(\mathbf{I}_{M_R} + \frac{\mathrm{SNR}}{M_T}\mathcal{H}\mathbf{R_s}\mathcal{H}^H\right) \tag{17.45}$$

with $\mathrm{SNR} = \frac{\epsilon}{\sigma_n^2}$, where ϵ represents the total energy of the transmit antennas.

If there is no prior information about the channel, there is no easier way to choose an efficient correlation matrix for the transmission vector $\mathbf{R_s}$ other than to assume that the elements of $\mathbf{s}$ have the same power and are independent. An instinctive solution is $\mathbf{R_s}$ being an identity matrix of rank M_T.

To derive an expression for the capacity, the signal and noise subspaces could be estimated using the decomposition of the eigenvalue to $\mathcal{H}\mathcal{H}^H$ or $\mathcal{H}^H\mathcal{H}$, choosing the smaller dimension one. Since $\mathcal{H}\mathcal{H}^H$ is an $M_R \times M_R$ positive semi-definite, whatever the dimension of $\mathcal{H}$, it follows that

$$\mathcal{H}\mathcal{H}^H = \mathcal{U}\Lambda_1\mathcal{U}^H = \mathcal{U}\Sigma\mathcal{V}^H\left(\mathcal{U}\Sigma\mathcal{V}^H\right)^H = \mathcal{U}\Sigma^2\mathcal{U}^H \tag{17.46}$$

where

$$\mathbf{\Lambda}_1 = \underbrace{\begin{bmatrix} \lambda_0 & 0 & \cdots & 0 & \cdots & 0 \\ 0 & \lambda_1 & \cdots & 0 & \cdots & 0 \\ \mathbf{0} & \mathbf{0} & \ddots & \mathbf{0} & \cdots & \mathbf{0} \\ 0 & 0 & \cdots & \lambda_L & \cdots & 0 \\ \mathbf{0} & \mathbf{0} & \ddots & \mathbf{0} & \cdots & \mathbf{0} \\ 0 & 0 & \cdots & \cdots & \cdots & 0 \end{bmatrix}}_{M_R} \tag{17.47}$$

where $\lambda_0 \geq \lambda_1 \geq \cdots \lambda_L$. Note that $\mathbf{\Lambda}_1$ is a diagonal matrix of dimension $M_R \times M_R$ and has null $M_R - L - 1$ eigenvalues.

Similarly,

$$\mathcal{H}^H\mathcal{H} = \mathcal{V}\Lambda_2\mathcal{V}^H = \left(\mathcal{U}\Sigma\mathcal{V}^H\right)^H \mathcal{U}\Sigma\mathcal{V}^H = \mathcal{V}\Sigma^2\mathcal{V}^H \tag{17.48}$$

where

$$\mathbf{\Lambda}_2 = \underbrace{\begin{bmatrix} \lambda_0 & 0 & \cdots & 0 & \cdots & 0 \\ 0 & \lambda_1 & \cdots & 0 & \cdots & 0 \\ \mathbf{0} & \mathbf{0} & \ddots & \mathbf{0} & \cdots & \mathbf{0} \\ 0 & 0 & \cdots & \lambda_L & \cdots & 0 \\ \mathbf{0} & \mathbf{0} & \ddots & \mathbf{0} & \cdots & \mathbf{0} \\ 0 & 0 & \cdots & \cdots & \cdots & 0 \end{bmatrix}}_{M_T} \tag{17.49}$$

where $\mathbf{\Lambda}_2$ is $M_T \times M_T$ and includes $M_T - L - 1$ null eigenvalues.

In this case, the channel capacity (in bits/transmission or bits/complex dimension, i.e., channel capacity normalized by the channel bandwidth) is

$$\begin{aligned} C &= \log_2 \det\left(\mathbf{I}_{M_R} + \frac{\text{SNR}}{M_T}\mathcal{H}\mathcal{H}^H\right) \\ &= \log_2 \det\left(\mathbf{I}_{M_R} + \frac{\text{SNR}}{M_T}\mathcal{U}\mathbf{\Lambda}_1\mathcal{U}^H\right) \\ &= \log_2 \det\left(\mathbf{I}_{M_R} + \frac{\text{SNR}}{M_T}\mathbf{\Lambda}_1\right) \\ &= \sum_{i=0}^{L} \log_2(1 + \frac{\text{SNR}}{M_T}\lambda_i) \\ \text{SNR} &= \frac{\epsilon}{\sigma_n^2} \end{aligned} \tag{17.50}$$

where ϵ is the total energy in the transmit antennas. Note that in this analysis, we could not benefit from the knowledge of the channel model at the transmitter and receiver, preventing the exploitation of the parallel SISO subchannels. In addition, by examining (17.50), it is possible to infer that using more than $L + 1$ receiving antennas brings no benefit to the capacity since only $L + 1$ entries of the receiver vector have useful information. Similarly, if $M_T > L + 1$, the capacity is reduced since only part of the transmitted information reaches the receiver, corresponding to the $L + 1$ modes. Note that M_T divides the SNR in (17.50).

17.7 Filter-Bank Transceivers

This section briefly describes the formulation of more general multicarrier transceivers borrowing from the development of filter banks widely utilized in signal analysis and compression (Vaidyanathan 1993). The presentation here follows closely Diniza et al. (2012), and reference Vaidyanathan et al. (2010) contains a lot of details in filter-bank-based transceivers.

Transmultiplexers, also known as *filter-bank transceivers*, are considered systems dual to the filter-bank configurations since the roles of analysis and synthesis filter banks are interchanged in transmultiplexers. Figure 17.5 depicts the transmultiplexer (TMUX) configuration including the channel model for our studies. Please note a slight change of notation in this section where the $x(k)$ here represents the signal after the channel and before the noise insertion; from now on in this chapter, the channel input is represented by $u(k)$, this facilitates the required block representation, and in the case the channel is memoryless with gain equal to one; both quantities coincide.

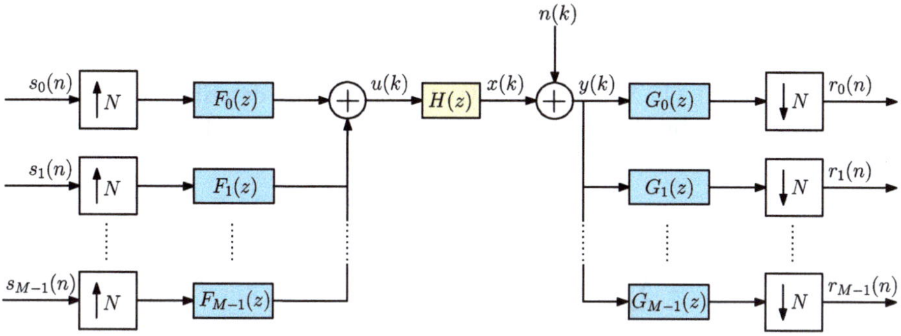

Fig. 17.5 TMUX multicarrier system

At the input of the TMUX, the data samples of each sequence $s_m(n)$ belong to a particular digital constellation $\mathcal{S} \subset \mathbb{C}$, such as PAM, QAM, or PSK. The sequence $s_m(n)$ represents the mth transceiver input, where $m \in \mathcal{M}$ and $n \in \mathbb{Z}$ represents the time index. The corresponding transceiver output is denoted as $r_m(n) = \hat{s}_m(n) \in \mathbb{C}$, which should be a reliable estimate of $s_m(n - \delta)$, where $\delta \in \mathbb{N}$ represents the delay introduced by the overall transmission/reception process.

A communication system can be designed by choosing the set of causal transmitter filters with impulse responses $\{f_m(k)\}_{m \in \mathcal{M}}$ and the set of causal receiver filters $\{g_m(k)\}_{m \in \mathcal{M}}$. The channel model can be represented by an FIR filter of order $L \in \mathbb{N}$ whose impulse response is $h(k) \in \mathbb{C}$.

We assume that the channel input signal is given by

$$u(k) = \sum_{(i,m) \in \mathbb{Z} \times \mathcal{M}} s_m(i) f_m(k - iN) \tag{17.51}$$

so that the relation between the input and output of the channel is described as

$$y(k) = \sum_{j \in \mathbb{Z}} h(j) u(k - j) + v(k). \tag{17.52}$$

On the right-hand side, the signal $y(k)$ is filtered in subbands to generate estimates of the transmitted data $s_m(n)$ through the following expression:

$$r_m(n) = \hat{s}_m(n) = \sum_{l \in \mathbb{Z}} g_m(l) y(nN - l). \tag{17.53}$$

17.7.1 Time-Domain Representation

From Eqs. (17.51), (17.52), and (17.53), it is possible to describe the relationship between the input signal $s_m(n)$ and its estimate $\hat{s}_m(n)$, as given by

$$\hat{s}_m(n) = \sum_{(i,j,l,m)\in\mathbb{Z}^3\times\mathcal{M}} g_m(l)h(j)s_m(i)f_m(nN-l-j-iN) + \sum_{l\in\mathbb{Z}} g_m(l)v(nN-l). \tag{17.54}$$

It is convenient to describe the transmitter and receiver filters by their *polyphase decompositions* of order N (Diniz et al. 2010). In the $\mathcal{Z}$-transform domain, the transmitter is described by

$$\begin{aligned}
F_m(z) &= \mathcal{Z}\{f_m(k)\} \\
&= \sum_{k\in\mathbb{Z}} f_m(k)z^{-k} \\
&= \sum_{j\in\mathbb{Z}} \Big[f_m(jN)z^{-jN} + f_m(jN+1)z^{-(jN+1)} + \cdots \\
&\quad + f_m(jN+N-1)z^{-(jN+N-1)} \Big] \\
&= \sum_{j\in\mathbb{Z}} \sum_{i\in\mathcal{N}} f_m(jN+i)z^{-(jN+i)} \\
&= \sum_{i\in\mathcal{N}} \sum_{j\in\mathbb{Z}} f_m(jN+i)z^{-jN}z^{-i} \\
&= \sum_{i\in\mathcal{N}} z^{-i} \underbrace{\sum_{j\in\mathbb{Z}} f_m(jN+i)z^{-jN}}_{\triangleq F_{i,m}(z^N)} \\
&= \sum_{i\in\mathcal{N}} z^{-i} F_{i,m}(z^N).
\end{aligned} \tag{17.55}$$

At the receiver end, the description follows another type of polyphase decomposition given by

$$\begin{aligned}
G_m(z) &= \mathcal{Z}\{g_m(k)\} \\
&= \sum_{k\in\mathbb{Z}} g_m(k)z^{-k}
\end{aligned}$$

$$
\begin{aligned}
&= \sum_{j\in\mathbb{Z}} \Big[g_m(jN)z^{-jN} + g_m(jN-1)z^{-(jN-1)} + \cdots \\
&\quad + g_m(jN-N+1)z^{-(jN-N+1)} \Big] \\
&= \sum_{j\in\mathbb{Z}} \sum_{i\in\mathcal{N}} g_m(jN-i)z^{-(jN-i)} \\
&= \sum_{i\in\mathcal{N}} \sum_{j\in\mathbb{Z}} g_m(jN-i)z^{-jN}z^{i} \\
&= \sum_{i\in\mathcal{N}} z^{i} \underbrace{\sum_{j\in\mathbb{Z}} g_m(jN-i)z^{-jN}}_{\triangleq G_{m,i}(z^N)} \\
&= \sum_{i\in\mathcal{N}} z^{i} G_{m,i}(z^N)\,,
\end{aligned}
\tag{17.56}
$$

where $m \in \mathcal{M}$, and $F_m(z)$ and $G_m(z)$ are the $\mathcal{Z}$-transforms of $f_m(k)$ and $g_m(k)$, respectively.

In the transmitter case, the transfer functions $F_{i,m}(z)$ are the type I polyphase components of order N associated with $F_m(z)$. In the receiver description, the transfer functions $G_{m,i}(z)$ are the type II polyphase components of order N associated with $G_m(z)$.

By using the noble identities of multirate systems to polyphase matrices (see Vaidyanathan 1993 and Diniz et al. 2010), it is possible to describe the transceiver in matrix form (Diniza et al. 2012):

$$
\begin{aligned}
\left[F_0(z) \cdots F_{M-1}(z)\right] &= \underbrace{\left[1\ z^{-1} \cdots z^{-(N-1)}\right]}_{\mathbf{d}^T(z)} \underbrace{\begin{bmatrix} F_{0,0}(z^N) & \cdots & F_{0,M-1}(z^N) \\ \vdots & \ddots & \vdots \\ F_{N-1,0}(z^N) & \cdots & F_{N-1,M-1}(z^N) \end{bmatrix}}_{\mathbf{F}(z^N)} \\
&= \mathbf{d}^T(z)\mathbf{F}(z^N),
\end{aligned}
\tag{17.57}
$$

$$
\begin{aligned}
\begin{bmatrix} G_0(z) \\ \vdots \\ G_{M-1}(z) \end{bmatrix} &= \underbrace{\begin{bmatrix} G_{0,0}(z^N) & \cdots & G_{0,N-1}(z^N) \\ \vdots & \ddots & \vdots \\ G_{M-1,0}(z^N) & \cdots & G_{M-1,N-1}(z^N) \end{bmatrix}}_{\mathbf{G}(z^N)} \underbrace{\begin{bmatrix} 1 \\ \vdots \\ z^{(N-1)} \end{bmatrix}}_{\mathbf{d}(z^{-1})} \\
&= \mathbf{G}(z^N)\mathbf{d}(z^{-1}).
\end{aligned}
\tag{17.58}
$$

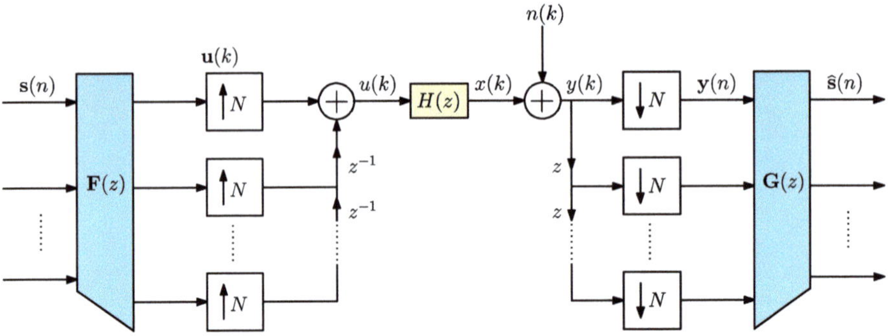

Fig. 17.6 TMUX multicarrier system: polyphase description

Defining $S_m(z) = \mathcal{Z}\{s_m(n)\}$, $U(z) = \mathcal{Z}\{u(k)\}$, $X(z) = \mathcal{Z}\{x(k)\}$, $V(z) = \mathcal{Z}\{v(k)\}$, $Y(z) = \mathcal{Z}\{y(k)\}$, and $\hat{S}_m(z) = \mathcal{Z}\{\hat{s}_m(n)\}$, we can write

$$U(z) = \mathbf{d}^T(z)\mathbf{F}(z^N)\underbrace{\begin{bmatrix} S_0(z^N) \\ \vdots \\ S_{M-1}(z^N) \end{bmatrix}}_{\triangleq \mathbf{s}(z)}, \tag{17.59}$$

$$X(z) = H(z)U(z), \tag{17.60}$$

$$Y(z) = X(z) + N(z), \tag{17.61}$$

$$\underbrace{\begin{bmatrix} \hat{S}_0(z) \\ \vdots \\ \hat{S}_{M-1}(z) \end{bmatrix}}_{\triangleq \hat{\mathbf{s}}(z)} = \Big[\mathbf{G}(z^N)\mathbf{d}(z^{-1})Y(z)\Big]_{\downarrow N}, \tag{17.62}$$

where $N(z)$ represents the noise in the $\mathcal{Z}$-domain.

It is now possible to illustrate the TMUX system of Fig. 17.5 in an equivalent representation employing polyphase decompositions as depicted in Fig. 17.6.

Figure 17.6 includes delays, forward delays, decimators, interpolators, and the SISO channel model can be represented by a *pseudo-circulant matrix* $\mathbf{H}(z)$ of dimension $N \times N$, given by

$$\mathbf{H}(z) = \begin{bmatrix} H_0(z) & z^{-1}H_{N-1}(z) & z^{-1}H_{N-2}(z) & \cdots & z^{-1}H_1(z) \\ H_1(z) & H_0(z) & z^{-1}H_{N-1}(z) & \cdots & z^{-1}H_2(z) \\ \vdots & \vdots & \ddots & \vdots & \vdots \\ H_{N-1}(z) & H_{N-2}(z) & H_{N-3}(z) & \cdots & H_0(z) \end{bmatrix}, \tag{17.63}$$

where

$$H(z) = \sum_{i \in \mathcal{N}} H_i(z^N)z^{-i} \quad \text{and} \quad H_i(z) = \sum_{\substack{j \in \mathbb{Z} \\ 0 \leq jN+i \leq L}} h(jN+i)z^{-j}. \tag{17.64}$$

By assuming that $n(k) = 0$, if $U_l(z)$ is the lth input at the transmitter end and $Y_m(z)$ is the mth output at the receiver end, then

$$\begin{aligned} Y_m(z) &= \left[z^{-l}H(z)U_l(z^N)z^m\right]_{\downarrow N} \\ &= \left[z^{m-l}H(z)U_l(z^N)\right]_{\downarrow N} \\ &= U_l(z)\left[z^{m-l}H(z)\right]_{\downarrow N}. \end{aligned} \tag{17.65}$$

We also know that $-(N-1) \leq m-l \leq N-1$, since m and l are within the set $\mathcal{N}$. Hence,

$$[\mathbf{H}(z)]_{ml} = \left[H_{(m-l)}(z^N)\right]_{\downarrow N} = H_{(m-l)}(z). \tag{17.66}$$

On the other hand, if $m - l < 0$, then

$$[\mathbf{H}(z)]_{ml} = \left[z^{-N}H_{(N-l+m)}(z^N)\right]_{\downarrow N} = z^{-1}H_{(N-l+m)}(z). \tag{17.67}$$

We will now describe the block-based transceivers in $\mathcal{Z}$-domain employing polyphase decompositions. Let us assume the transmitted and received vectors are denoted as

$$\mathbf{s}(n) = [\, s_0(n) \;\; s_1(n) \;\; \cdots \;\; s_{M-1}(n) \,]^T, \tag{17.68}$$

$$\hat{\mathbf{s}}(n) = [\, \hat{s}_0(n) \;\; \hat{s}_1(n) \;\; \cdots \;\; \hat{s}_{M-1}(n) \,]^T. \tag{17.69}$$

It is also possible to infer that the transfer matrix $\mathbf{T}(z)$ of the transceiver can be expressed as

$$\mathbf{T}(z) = \mathbf{G}(z)\mathbf{H}(z)\mathbf{F}(z), \tag{17.70}$$

where it was assumed that $n(k) = 0$ for all integer number k, giving rise to the *zero-forcing* (ZF) design. The zero-forcing condition means that $\mathbf{T}(z) = \alpha z^{-d}\mathbf{I}_M$, for some $\alpha \in \mathbb{C}$ and $d \in \mathbb{N}$. As a result, the zero-forcing solution produces a scaled and delayed version of all transmitted signals.

Let us assume that we choose $N > L$, i.e., the interpolation/decimation factor is greater than or equal to the channel order L, a desirable choice. We have that the only integer number j which satisfies the inequality constraint $0 \leq jN + i \leq L$ is $j = 0$, which leads us to conclude that $H_i(z) = h(i)$, for $\mathcal{N} \ni i \leq L$. On the other hand, if there exists $i > L$ within the set $\mathcal{N}$, then $H_i(z) = 0$ since there is no term to be added in order to form $H_i(z)$. We can say that, for $N > L$, each element of the matrix $\mathbf{H}(z)$ will consist of filters with a single coefficient, eventually a null coefficient.

The resulting pseudo-circulant channel matrix is represented by a first-order FIR matrix described as

$$\mathbf{H}(z) = \underbrace{\begin{bmatrix} h(0) & 0 & 0 & \cdots & 0 \\ h(1) & h(0) & 0 & \cdots & 0 \\ \vdots & \vdots & \vdots & \vdots & \vdots \\ h(L) & h(L-1) & \ddots & \cdots & 0 \\ 0 & h(L) & \cdots & \cdots & 0 \\ \vdots & \vdots & \vdots & \vdots & \vdots \\ 0 & 0 & h(L) & \cdots & h(0) \end{bmatrix}}_{\triangleq \mathbf{H}_{\mathrm{ISI}}} + z^{-1} \underbrace{\begin{bmatrix} 0 \cdots & 0 & h(L) & \cdots & h(1) \\ 0\ 0 & \cdots & 0 & \ddots & \vdots \\ \vdots\ \vdots & \vdots & \vdots & \vdots & h(L) \\ 0\ 0 & 0 & \cdots & 0 & 0 \\ 0\ 0 & 0 & \cdots & 0 & 0 \\ \vdots\ \vdots & \vdots & \vdots & \vdots & \vdots \\ 0\ 0 & 0 & 0 & \cdots & 0 \end{bmatrix}}_{\triangleq \mathbf{H}_{\mathrm{IBI}}}$$

$$= \mathbf{H}_{\mathrm{ISI}} + z^{-1}\mathbf{H}_{\mathrm{IBI}}. \tag{17.71}$$

The relation in the time domain

$$\begin{aligned} \mathbf{x}(k) &= \mathbf{H}_{\mathrm{ISI}}\mathbf{u}(k) + \mathbf{H}_{\mathrm{IBI}}\mathbf{u}(k-1) \\ &= \begin{bmatrix} \mathbf{H}_{\mathrm{IBI}} & \mathbf{H}_{\mathrm{ISI}} \end{bmatrix} \begin{bmatrix} \mathbf{u}(k-1) \\ \mathbf{u}(k) \end{bmatrix}, \end{aligned} \tag{17.72}$$

where

$$\mathbf{u}(k) = [\, u(kN - N + 1)\ \ u(kN - N + 2)\ \ \cdots\ \ u(kN)\,]^T, \tag{17.73}$$

$$\mathbf{x}(k) = [\, x(kN - N + 1)\ \ x(kN - N + 2)\ \ \cdots\ \ x(kN)\,]^T. \tag{17.74}$$

In current generations of wireless communications, multicarrier transmissions are the dominating standard. Multicarrier modulation methods play a crucial role in modern data transmission channels with severe and moderate intersymbol interference. The key idea behind the success of this technique is the partition of the physical channel into nonoverlapping narrowband subchannels through a transmultiplexer.

If the subchannels are narrow enough, the associated channel response in each subchannel frequency range appears to be flat, avoiding the use of sophisticated equalizers. The subchannel division allows, whenever possible, the exploitation of signal-to-noise ratio (SNR) in the different subchannels by managing the data load in each subchannel. A critical component of the multicarrier systems is the filter-bank-based transmultiplexer (TMUX).

We can design TMUXes with sharp transitions and a large number of sub-channels. However, the possible use of these TMUX-based transceivers for future-generation wireless systems and their current interest in developing local and broadband wireless systems is still being considered. A critical component of the multicarrier systems is the filter-bank-based TMUX.

The minimum data redundancy being required at the transmission in order to keep the equalization as simple as possible is another issue addressed in recent studies. Also, several methods for jointly optimizing the transmitter and receiver FIR MIMO filters can be developed in the presence of near-end cross-talk and additive noise source.

17.8 Filter-Bank-Based Multicarrier Systems

In this section, the multicarrier systems are revisited under the framework of the filter bank representation which provides powerful tools for transceiver designs.

As previously mentioned, some common multicarrier systems are OFDM, DMT, DWMT, CDMA, and TDMA. Filter-bank-based multicarrier techniques help to increase system capacity and provide multiple access. Usually, redundancy must be introduced in the transmitted block of symbols in order to achieve zero-forcing (ZF) equalization with FIR filter banks. For example, in OFDM and DMT systems, the minimum redundancy is equal to the channel order. If a criterion other than ZF is adopted, redundant symbols are unnecessary, even though they could improve system performance. We overview filter-bank-based multicarrier techniques and show how redundancy can be introduced in the transmitted block.

In a general multiplex description of a transceiver, we utilize the following definitions:

- For $N = M$, the system is called maximally decimated (or minimally interpolated).
- For $N < M$, equalization is not possible, and thus, this case will not be considered.
- Define the redundancy as $K = N - M$.

As previously shown, Fig. 17.5 depicts the classic multiplexer configuration. With the polyphase representation, the received signal is given by

$$\hat{\mathbf{s}}(z) = \mathbf{G}(z)\mathbf{H}(z)\mathbf{F}(z)\mathbf{s}(z). \tag{17.75}$$

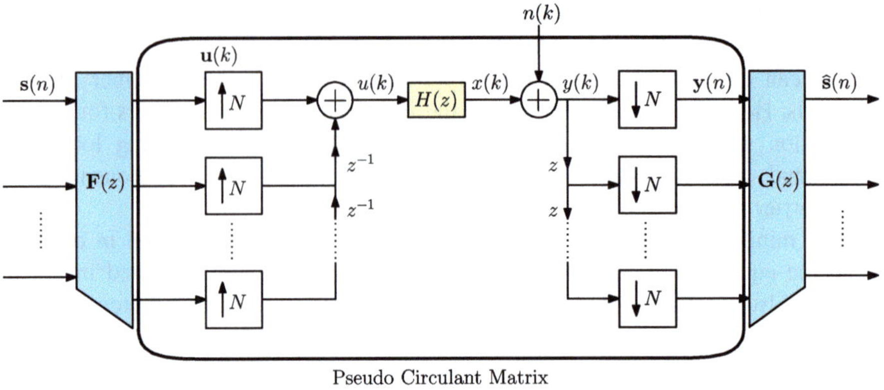

Fig. 17.7 TMUX multicarrier system: polyphase description

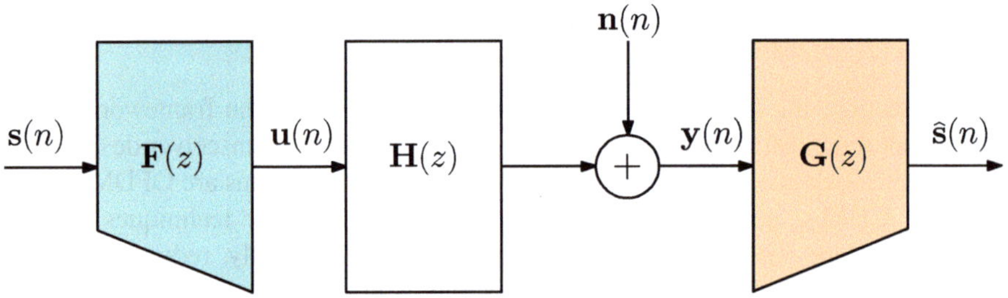

Fig. 17.8 Matrix representation of transceivers including channel-matrix model

The $N \times N$ channel matrix, $\mathbf{H}(z)$, is pseudocirculant:

$$\mathbf{H}(z) = \begin{bmatrix} H_0(z) & z^{-1}H_{N-1}(z) & \cdots & z^{-1}H_1(z) \\ H_1(z) & H_0(z) & \cdots & z^{-1}H_2(z) \\ \vdots & \vdots & \ddots & \vdots \\ H_{N-1}(z) & H_{N-2}(z) & \cdots & H_0(z) \end{bmatrix}. \tag{17.76}$$

The polyphase description of the TMUX enables a matrix formulation for the transceiver and channel interaction, which allows for a helpful matrix description of the channel. The matrix representing a linear-time invariant channel is a pseudo-circulant that is diagonalizable by DFT matrices. Figure 17.7 illustrates the matrix representation of the communication system.

A more compact representation is depicted in Fig. 17.8 where the transmitter, channel, and receiver are described by matrices with memory.

17.8.1 FIR MIMO Matrices of LTI Transceivers

Let us start with Fig. 17.8, which depicts the $\mathcal{Z}$-domain description of block-based LTI transceivers. In this figure, $\mathbf{F}(z)$ represents an FIR MIMO transmitter matrix whose entries are polynomials in the complex variable z^{-1}, as given in (17.77). If we assume that, among all entries of $\mathbf{F}(z)$, the highest polynomial order is $T-1$, with T being a given natural number,[6] then we can decompose $\mathbf{F}(z)$ as

$$\mathbf{F}(z) = \sum_{\tau \in \mathcal{T}} \mathbf{F}_{\tau} z^{-\tau}, \tag{17.77}$$

where $\mathcal{T} = \{0, \cdots, T-1\}$ is a set of integer indexes and $\mathbf{F}_{\tau}$ is an $N \times M$ matrix with complex-valued entries rather than polynomial entries. The natural numbers M and N are the number of symbols within a data block and the actual number of transmitted elements per block, respectively, which includes both data symbols and redundant elements. The entries of $\mathbf{F}_{\tau}$, for all $\tau \in \mathcal{T}$, can be determined directly from $\mathbf{F}(z)$ using (17.77).

A transmitter matrix with memory ($T > 1$) enables one to design high-selective transmitter filters, which may be very useful to avoid spectrum leakage (out-of-band transmissions), to reduce intercarrier interference (ICI) that occurs, especially when carrier-frequency offset (CFO) is significant, and to compensate for the channel distortions when channel-state information (CSI) is available at the transmitter side. Besides, having memory at the transmitter increases the transmitter diversity, which could enhance the performance of the related transceiver.

If we consider that the order of the FIR MIMO receiver matrix $\mathbf{G}(z)$ in Fig. 17.8 is $Q-1$, where Q is a natural number, we can decompose $\mathbf{G}(z)$ as

$$\mathbf{G}(z) = \sum_{q \in \mathcal{Q}} \mathbf{G}_{q} z^{-q}, \tag{17.78}$$

where $\mathbf{G}_q$ is an $M \times N$ complex-valued matrix, for each integer index q within the set $\mathcal{Q} = \{0, \cdots, Q-1\}$.

The entries of the matrix $\mathbf{G}_q$ are not important in our discussions here, but they belong to $\mathbf{G}(z)$ which in turn is defined in (17.78). It is worth mentioning that if the transmitted redundant elements are not able to eliminate the interblock interference (IBI) completely, then matrix $\mathbf{G}(z)$ must have memory, i.e., it is mandatory to have $Q > 1$ in some applications, as will be clearer soon.

---

[6] From now on, we will define the *order of a matrix* as the highest polynomial order among all its entries. For instance, the order of $\mathbf{F}(z)$ is $T-1$.

Once again, we will perform the same type of decomposition, but now for the FIR MIMO channel matrix $\mathbf{H}(z)$ in Fig. 17.8. Thus, we are interested in representing $\mathbf{H}(z)$ as

$$\mathbf{H}(z) = \sum_{b \in \mathcal{B}} \mathbf{H}_b z^{-b}, \tag{17.79}$$

where $\mathbf{H}_b$ is an $N \times N$ complex-valued Toeplitz matrix, for each integer index b within the set $\mathcal{B} = \{0, 1, \cdots, B\}$, where B stands for the order of the channel matrix $\mathbf{H}(z)$.

With those previous definitions, we can now derive a matrix description for the LTI transceiver of Fig. 17.8. To do that, let us first assume that the $\mathcal{Z}$-transform of a generic time-domain vector signal $\mathbf{y}(n)$ is denoted as

$$\mathbf{y}(z) = \mathcal{Z}\left\{\mathbf{y}(n)\right\}, \tag{17.80}$$

where $\mathbf{y}(n)$ has size $N_y \in \mathbb{N}$ and can be expressed as

$$\mathbf{y}(n) = [\, y(nN_y - N_y + 1) \;\; y(nN_y - N_y + 2) \;\; \cdots \;\; y(nN_y)\,]^T. \tag{17.81}$$

Considering a noiseless channel, i.e., $\mathbf{n}(n) = \mathbf{0}_{N\times 1}$ for all time index $n \in \mathbb{Z}$, the time-domain representations for the transmitted, received, and estimated vectors, are, respectively, given by

$$\begin{aligned}
\mathbf{u}(n) &= \mathcal{Z}^{-1}\left\{\mathbf{F}(z)\mathbf{s}(z)\right\} \\
&= \sum_{\tau \in \mathcal{T}} \mathbf{F}_\tau \mathcal{Z}^{-1}\left\{z^{-\tau}\mathbf{s}(z)\right\} \\
&= \sum_{\tau \in \mathcal{T}} \mathbf{F}_\tau \mathbf{s}(n-\tau),
\end{aligned} \tag{17.82}$$

$$\begin{aligned}
\mathbf{y}(n) &= \mathcal{Z}^{-1}\left\{\mathbf{H}(z)\mathbf{u}(z)\right\} \\
&= \sum_{b \in \mathcal{B}} \mathbf{H}_b \mathcal{Z}^{-1}\left\{z^{-b}\mathbf{u}(z)\right\} \\
&= \sum_{b \in \mathcal{B}} \mathbf{H}_b \mathbf{u}(n-b),
\end{aligned} \tag{17.83}$$

$$\begin{aligned}
\hat{\mathbf{s}}(n) &= \mathcal{Z}^{-1}\left\{\mathbf{G}(z)\mathbf{y}(z)\right\} \\
&= \sum_{q \in \mathcal{Q}} \mathbf{G}_q \mathcal{Z}^{-1}\left\{z^{-q}\mathbf{y}(z)\right\} \\
&= \sum_{q \in \mathcal{Q}} \mathbf{G}_q \mathbf{y}(n-q).
\end{aligned} \tag{17.84}$$

It is then possible to write the input-output relationship of FIR LTI transceivers in a compact form by conveniently stacking the variables in Eqs. (17.82), (17.83), and (17.84). Therefore, (17.84) can be rewritten as follows:

$$\begin{aligned}\hat{\mathbf{s}}(n) &= \underbrace{\left[\mathbf{G}_{Q-1} \cdots \mathbf{G}_0\right]}_{\triangleq \mathcal{G} \in \mathbb{C}^{M \times QN}} \underbrace{\begin{bmatrix} \mathbf{y}\left(n-(Q-1)\right) \\ \vdots \\ \mathbf{y}(n) \end{bmatrix}}_{\triangleq \bar{\mathbf{y}}(n) \in \mathbb{C}^{QN \times 1}} \\ &= \mathcal{G}\bar{\mathbf{y}}(n). \end{aligned} \tag{17.85}$$

Equation (17.85) requires the expressions for vectors $\mathbf{y}\left(n-Q+1\right), \cdots, \mathbf{y}(n)$, which derive from (17.83) as follows:

$$\begin{aligned}\bar{\mathbf{y}}(n) &= \underbrace{\begin{bmatrix} \mathbf{H}_B & \cdots & \mathbf{H}_0 & \mathbf{0} & \cdots & \mathbf{0} \\ \mathbf{0} & \mathbf{H}_B & \cdots & \mathbf{H}_0 & \cdots & \mathbf{0} \\ \vdots & \vdots & \ddots & & \ddots & \vdots \\ \mathbf{0} & \mathbf{0} & \cdots & \mathbf{H}_B & \cdots & \mathbf{H}_0 \end{bmatrix}}_{\triangleq \mathcal{H} \in \mathbb{C}^{QN \times (B+Q)N}} \underbrace{\begin{bmatrix} \mathbf{u}\left(n-(B+Q-1)\right) \\ \vdots \\ \mathbf{u}(n-1) \\ \mathbf{u}(n) \end{bmatrix}}_{\triangleq \bar{\mathbf{u}}(n) \in \mathbb{C}^{(B+Q)N \times 1}} \\ &= \mathcal{H}\bar{\mathbf{u}}(n). \end{aligned} \tag{17.86}$$

On the other hand, (17.86) depends on $\mathbf{u}\left(n-B-Q+1\right), \cdots, \mathbf{u}(n)$, and can be computed using (17.82) as follows:

$$\bar{\mathbf{u}}(n) = \underbrace{\begin{bmatrix} \mathbf{F}_{T-1} & \cdots & \mathbf{F}_0 & \mathbf{0} & \cdots & \mathbf{0} \\ \mathbf{0} & \mathbf{F}_{T-1} & \cdots & \mathbf{F}_0 & \cdots & \mathbf{0} \\ \vdots & \vdots & \ddots & & \ddots & \vdots \\ \mathbf{0} & \mathbf{0} & \cdots & \mathbf{F}_{T-1} & \cdots & \mathbf{F}_0 \end{bmatrix}}_{\triangleq \mathcal{F} \in \mathbb{C}^{(B+Q)N \times [B+Q+(T-1)]M}} \underbrace{\begin{bmatrix} \mathbf{s}\left(n-(B+Q+T-2)\right) \\ \vdots \\ \mathbf{s}(n-1) \\ \mathbf{s}(n) \end{bmatrix}}_{\triangleq \bar{\mathbf{s}}(n) \in \mathbb{C}^{[B+Q+(T-1)]M \times 1}} \tag{17.87}$$

$$= \mathcal{F}\bar{\mathbf{s}}(n). \tag{17.88}$$

From Eqs. (17.85), (17.86), and (17.87), it is possible to derive the overall input-output description of the block-based LTI transceiver of Fig. 17.8 given by

$$\hat{\mathbf{s}}(n) = \mathcal{G}\mathcal{H}\mathcal{F}\bar{\mathbf{s}}(n). \tag{17.89}$$

Note the above equation is the time-domain version of the following equation:

$$\hat{\mathbf{s}}(z) = \mathbf{G}(z)\mathbf{H}(z)\mathbf{F}(z)\mathbf{s}(z) \tag{17.90}$$

in the $\mathcal{Z}$-domain, where $\hat{\mathbf{s}}(z)$ is the last vector in the right-hand side of (17.88). The resulting transceiver is time-invariant since the transmitter and receiver matrices, represented by $\boldsymbol{\mathcal{F}}$ and $\boldsymbol{\mathcal{G}}$, respectively, are independent of the time index n. The matrix description is quite handy for the time-varying case, since we cannot apply the $\mathcal{Z}$-domain description of the TMUX of Fig. 17.6.

17.9 Transceiver Design Using Optimization

A more flexible way to design the transceivers is to jointly optimize the transmitter and receiver filters, leading to an optimal MMSE-based solution. If there are enough degrees of freedom, one can perform redundant-free transmissions, for instance, in the case the transceiver building blocks have memory. In this case, the extra parameters allow the joint design of the transmitter and receiver FIR MIMO filters, taking into account the near-end crosstalk and additive noise. Some optimization approaches have been described in Hjørungnes et al. (2004), Hjørungnes et al. (2005), Hjørungnes and Diniz (2005), Vaidyanathan et al. (2010).

The transceiver design starts by defining the matrices representing the transmitter and receiver, as previously described in Eqs. (17.86), (17.87), and (17.89). There are many design solutions for transmitter matrix $\boldsymbol{\mathcal{F}}(n)$, which has a block-Toeplitz structure as per (17.86)), and the receiver matrix $\boldsymbol{\mathcal{G}}(n)$.

A possible cost function consists of minimizing the MSE of symbols, $\xi_{\text{MSE}} \in \mathbb{R}_+$, leading to the minimum MSE (MMSE) design (Vaidyanathan et al. 2010). The fundamental consideration is that the FIR MIMO channel transfer matrices are known and that all the system input signals are uncorrelated. It also considered that the second-order statistics of all the input signals are known.

Assume that the transmitted vector $\bar{\mathbf{s}}(n)$ and the channel-noise vector $\bar{\mathbf{n}}(n)$ are zero-mean jointly wide-sense stationary (WSS) random sequences $\bar{\boldsymbol{s}}(n)$ and $\bar{\boldsymbol{n}}(n)$. Given that the estimated symbols are

$$\begin{aligned} \hat{\boldsymbol{s}}(n) &= \boldsymbol{\mathcal{G}}(n)\left[\boldsymbol{\mathcal{H}}\boldsymbol{\mathcal{F}}(n)\bar{\boldsymbol{s}}(n) + \bar{\boldsymbol{n}}(n)\right] \\ &= \boldsymbol{\mathcal{G}}(n)\boldsymbol{\mathcal{H}}\boldsymbol{\mathcal{F}}(n)\bar{\boldsymbol{s}}(n) + \boldsymbol{\mathcal{G}}(n)\bar{\boldsymbol{n}}(n), \end{aligned} \tag{17.91}$$

so that the symbols' MSE is given by

$$\begin{aligned} \xi_{\text{MSE}} &= \mathbb{E}\{\|\hat{\boldsymbol{s}}(n) - \boldsymbol{s}(n)\|_2^2\} \\ &= \mathbb{E}\{\left\|\boldsymbol{\mathcal{G}}(n)\boldsymbol{\mathcal{H}}\boldsymbol{\mathcal{F}}(n)\bar{\boldsymbol{s}}(n) + \boldsymbol{\mathcal{G}}(n)\bar{\boldsymbol{n}}(n) - [\,\mathbf{0} \quad \mathbf{I}_M\,]\,\bar{\boldsymbol{s}}(n)\right\|_2^2\} \\ &= \operatorname{tr}\left\{(\boldsymbol{\mathcal{G}}(n)\boldsymbol{\mathcal{H}}\boldsymbol{\mathcal{F}}(n) - [\,\mathbf{0} \quad \mathbf{I}_M\,])\mathbf{R}_{ss}(n)(\boldsymbol{\mathcal{G}}(n)\boldsymbol{\mathcal{H}}\boldsymbol{\mathcal{F}}(n) - [\,\mathbf{0} \quad \mathbf{I}_M\,])^H\right\} \\ &\quad + \operatorname{tr}\left\{\boldsymbol{\mathcal{G}}(n)\mathbf{R}_{\boldsymbol{nn}}(n)\boldsymbol{\mathcal{G}}^H(n)\right\}, \end{aligned} \tag{17.92}$$

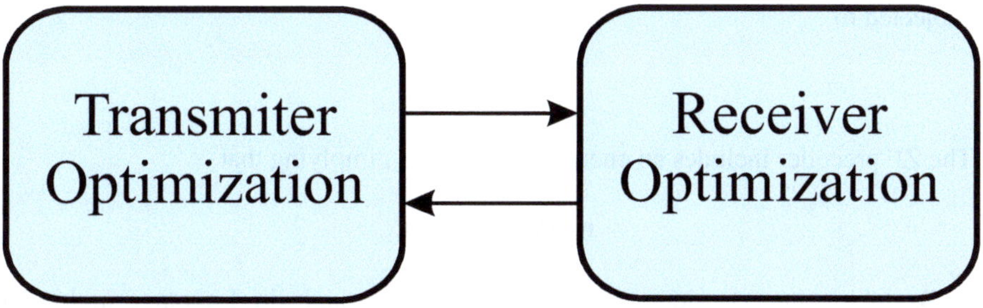

Fig. 17.9 Optimization strategy

in which $\mathbf{R}_{ss}(n) = \mathbb{E}\{\overline{s}(n)\overline{s}^H(n)\}$ and $\mathbf{R}_{nn}(n) = \mathbb{E}\{\overline{\boldsymbol{n}}(n)\overline{\boldsymbol{n}}^H(n)\}$. Note that we assumed that $\overline{s}(n)$ and $\overline{\boldsymbol{n}}(n)$ are uncorrelated, i.e.,

$$\mathbf{R}_{sn} = \mathbb{E}\{\overline{s}(n)\overline{\boldsymbol{n}}^H(n)\} = \mathbb{E}\{\overline{s}(n)\}\mathbb{E}\{\overline{\boldsymbol{n}}(n)\}^H = \mathbf{0}.$$

The equality $s(n) = [\,\mathbf{0} \quad \mathbf{I}_M\,]\,\overline{s}(n)$ originates from (17.88), where it has been assumed that a ZF solution was reached. The detailed derivation can be found in Diniza et al. (2012).

In practice, the MSE minimization requires the inclusion of power constraints. Reference Hjørungnes et al. (2004) proposed a solution through analytical expressions for the MSE criterion and power constraint. The answer involves deriving the Karush-Kuhn-Tucker (KKT) conditions for optimality and implementing an iterative numerical algorithm considering those necessary conditions. The approach consists of designing a transmitter assuming a known initial receiver and then designing a receiver considering the previously designed transmitter, as illustrated in Fig. 17.9. The minimization can be applied iteratively until no further reduction in the MSE is achieved.

17.10 Proof Details

This section presents the simplified details of the transceiver designs described in this chapter.

17.10.1 Proof of (17.19)

Minimize

$$\mathcal{F} = \min_{\mathcal{F},\kappa} \mathbb{E}[||\kappa\mathbf{r} - \mathbf{s}||^2] = \min_{\mathcal{F},\kappa} \mathbb{E}[||\kappa\mathbf{r} - \mathcal{G}\mathbf{n}||^2]$$

subjected to

$$\kappa \mathcal{G}\mathcal{H}\mathcal{F} = \mathbf{I}.$$

The ZF precoder includes an energy constraint E_t, implying that

$$\mathrm{tr}(\mathcal{F}\mathcal{R}_\mathrm{s}\mathcal{F}^H) = E_\mathrm{t}.$$

By taking into consideration the constraints through the Lagrange multiplier approach, the cost function becomes

$$\begin{aligned}\mathcal{L}(\mathcal{F}, \kappa, \mathbf{\Lambda}_1, \lambda_2) = \kappa^2 \mathrm{tr}(\mathcal{G}\mathcal{R}_n\mathcal{G}^H) + \mathrm{tr}(\mathbf{\Lambda}_1(\kappa\mathcal{G}\mathcal{H}\mathcal{F} - \mathbf{I}))\\ + \lambda_2(\mathrm{tr}(\mathcal{F}\mathcal{R}_\mathrm{s}\mathcal{F}^H) - E_\mathrm{t})\end{aligned}$$

where $\mathbf{\Lambda}_1$ and λ_2 represent the Lagrange multiplier matrix and scalar, respectively.

Since the cost function is real, i.e., $\mathcal{L}(\cdot) \in \mathbb{R}$, the properties $\mathrm{tr}(\mathcal{L}(\cdot)) = \mathrm{tr}(\mathcal{L}^*(\cdot)) = \mathrm{tr}(\mathcal{L}(\cdot)) \in \mathbb{R}$ and $\mathrm{tr}(\mathcal{L}(\cdot)) = \frac{1}{2}\mathrm{tr}(\mathcal{L}(\cdot)) + \frac{1}{2}\mathrm{tr}(\mathcal{L}^*(\cdot))$ hold. Therefore,

$$\begin{aligned}\mathcal{L}(\mathcal{F}, \kappa, \mathbf{\Lambda}_1, \lambda_2) = \kappa^2 \mathrm{tr}(\mathcal{G}\mathcal{R}_n\mathcal{G}^H) + \frac{1}{2}\mathrm{tr}(\mathbf{\Lambda}_1(\kappa\mathcal{G}\mathcal{H}\mathcal{F} - \mathbf{I}))\\ + \frac{1}{2}\mathrm{tr}((\kappa\mathcal{F}^H\mathcal{H}^H\mathcal{G}^H - \mathbf{I})\mathbf{\Lambda}_1^H) + \lambda_2(\mathrm{tr}(\mathcal{F}\mathcal{R}_\mathrm{s}\mathcal{F}^H) - E_\mathrm{t}).\end{aligned}$$

By taking the derivative of the cost function with respect to $\mathcal{F}^*$ and equating the the result to zero (see page 79 of Hjørungnes 2011 for details), it follows that

$$\frac{\partial \mathcal{L}(\mathcal{F}, \kappa, \mathbf{\Lambda}_1, \lambda_2)}{\partial \mathcal{F}^*} = \frac{\kappa}{2}\mathcal{H}^H\mathcal{G}^H\mathbf{\Lambda}_1^H + \lambda_2\mathcal{F}\mathcal{R}_\mathrm{s} = \mathbf{0}$$

so that

$$-\frac{\kappa}{2}\mathcal{H}^H\mathcal{G}^H\mathbf{\Lambda}_1^H = \lambda_2\mathcal{F}\mathcal{R}_\mathrm{s};$$

as a result

$$\mathcal{F} = -\frac{\kappa}{2\lambda_2}\mathcal{H}^H\mathcal{G}^H\mathbf{\Lambda}_1^H\mathcal{R}_\mathrm{s}^{-1}. \tag{17.93}$$

The next step is to apply the constraint $\kappa \mathcal{GHF} = \mathbf{I}$, to compute $\boldsymbol{\Lambda}_1^H$ as follows:

$$\mathbf{I} = \kappa \mathcal{GH}(-\frac{\kappa}{2\lambda_2}\mathcal{H}^H \mathcal{G}^H \boldsymbol{\Lambda}_1^H \mathcal{R}_\mathbf{s}^{-1})$$

$$\mathbf{I} = -\frac{\kappa^2}{2\lambda_2}(\mathcal{GHH}^H \mathcal{G}^H)\boldsymbol{\Lambda}_1^H \mathcal{R}_\mathbf{s}^{-1}$$

$$\boldsymbol{\Lambda}_1^H = -\frac{2\lambda_2}{\kappa^2}(\mathcal{GHH}^H \mathcal{G}^H)^{-1}\mathcal{R}_\mathbf{s}.$$

By replacing the result in (17.93), the outcome is

$$\mathcal{F} = -\frac{\kappa}{2\lambda_2}\mathcal{H}^H \mathcal{G}^H(-\frac{2\lambda_2}{\kappa^2}(\mathcal{GHH}^H \mathcal{G}^H)^{-1}\mathcal{R}_\mathbf{s})\mathcal{R}_\mathbf{s}^{-1} \tag{17.94}$$

$$= \overset{\frac{1}{\kappa}}{\cancel{-\frac{\kappa}{2\lambda_2}(-\frac{2\lambda_2}{\kappa^2})}}\mathcal{H}^H \mathcal{G}^H(\mathcal{GHH}^H \mathcal{G}^H)^{-1}\overset{\mathbf{I}}{\cancel{(\mathcal{R}_\mathbf{s}\mathcal{R}_\mathbf{s}^{-1})}} \tag{17.95}$$

$$= \frac{1}{\kappa}\mathcal{H}^H \mathcal{G}^H(\mathcal{GHH}^H \mathcal{G}^H)^{-1}. \tag{17.96}$$

The proper choice of a real and positive $\frac{1}{\kappa}$ enforces the energy constraint, as described in the following:

$$\begin{aligned}
E_\mathrm{t} &= \mathrm{tr}[\mathcal{F}\mathcal{R}_\mathbf{s}\mathcal{F}^H] \\
&= \mathrm{tr}[\frac{1}{\kappa}\mathcal{H}^H \mathcal{G}^H(\mathcal{GHH}^H \mathcal{G}^H)^{-1}\mathcal{R}_\mathbf{s}(\frac{1}{\kappa}\mathcal{H}^H \mathcal{G}^H(\mathcal{GHH}^H \mathcal{G}^H)^{-1})^H] \\
&= \left(\frac{1}{\kappa}\right)^2 \mathrm{tr}[\mathcal{K}^H(\mathcal{KK}^H)^{-1}\mathcal{R}_\mathbf{s}((\mathcal{KK}^H)^{-1})^H \mathcal{K}] \\
&= \left(\frac{1}{\kappa}\right)^2 \mathrm{tr}[\overset{\mathbf{I}}{\cancel{(\mathcal{KK}^H)(\mathcal{KK}^H)^{-1}}}\mathcal{R}_\mathbf{s}((\mathcal{KK}^H)^{-1})^H] \\
&= \left(\frac{1}{\kappa}\right)^2 \mathrm{tr}[\mathcal{R}_\mathbf{s}(\mathcal{KK}^H)^{-1}] = \left(\frac{1}{\kappa}\right)^2 \mathrm{tr}[(\mathcal{KK}^*)^{-1}\mathcal{R}_\mathbf{s}]
\end{aligned}$$

where $\mathcal{K} = \mathcal{GH}$ represents an auxiliary matrix to simplify the derivation. As a result,

$$\left(\frac{1}{\kappa}\right)^2 = \frac{E_\mathrm{t}}{\mathrm{tr}[(\mathcal{KK}^H)^{-1}\mathcal{R}_\mathbf{s}]} = \frac{E_\mathrm{t}}{\mathrm{tr}[(\mathcal{GHH}^H \mathcal{G}^H)^{-1}\mathcal{R}_\mathbf{s}]}$$

so that

$$\frac{1}{\kappa} = \sqrt{\frac{E_t}{\mathrm{tr}[(\mathcal{G}\mathcal{H}\mathcal{H}^H\mathcal{G}^H)^{-1}\mathcal{R}_s]}}$$

leading to the final expression:

$$\begin{aligned}\mathcal{F} &= \frac{1}{\kappa}\mathcal{H}^H\mathcal{G}^H(\mathcal{G}\mathcal{H}\mathcal{H}^H\mathcal{G}^H)^{-1} \\ &= \sqrt{\frac{E_t}{\mathrm{tr}[(\mathcal{G}\mathcal{H}\mathcal{H}^H\mathcal{G}^H)^{-1}\mathcal{R}_s]}}\mathcal{H}^H\mathcal{G}^H(\mathcal{G}\mathcal{H}\mathcal{H}^H\mathcal{G}^H)^{-1}.\end{aligned}$$

■

17.10.2 Proof of (17.22)

The objective is to maximize (17.20) taking into consideration the constraint of (17.21). The maximization is equivalent to

$$\mathcal{F}_{MF} = \min_{\mathcal{F}}\left[-\left|\mathbb{E}[\mathbf{s}^H\mathbf{r}]\right|^2\right] = \min_{\mathcal{F}}\left[-\left|\mathbb{E}[\mathbf{s}^H\mathcal{G}\mathcal{H}\mathcal{F}\mathbf{s}]\right|^2\right]$$

so that

$$\begin{aligned}\mathcal{L}(\mathcal{F},\lambda) &= -\left|\mathbb{E}[\mathbf{s}^H\mathcal{G}\mathcal{H}\mathcal{F}\mathbf{s}]\right|^2 + \lambda(\mathrm{tr}(\mathcal{F}\mathbf{R}_s\mathcal{F}^H) - E_t) \\ &= -\mathrm{tr}(\mathcal{G}\mathcal{H}\mathcal{F}\mathbf{R}_s)\mathrm{tr}(\mathbf{R}_s\mathcal{F}^H\mathcal{H}^H\mathcal{G}^H) + \lambda(\mathrm{tr}(\mathcal{F}\mathbf{R}_s\mathcal{F}^H) - E_t).\end{aligned}$$

By taking the derivative of $\mathcal{L}(\mathcal{F},\lambda)$ with respect to $\mathcal{F}^*$, while leaving $\mathcal{F}$ as independent variable, we get

$$\begin{aligned}\frac{\partial\mathcal{L}(\mathcal{F},\lambda)}{\partial\mathcal{F}^*} &= -\mathrm{tr}\,(\mathcal{G}\mathcal{H}\mathcal{F}\mathbf{R}_s)\,\mathcal{H}^H\mathcal{G}^H\mathbf{R}_s + \lambda\mathcal{F}\mathbf{R}_s = \mathbf{0} \\ &= -\mathrm{tr}(\mathcal{G}\mathcal{H}\mathcal{F}\mathbf{R}_s)\mathcal{H}^H\mathcal{G}^H\underbrace{\mathbf{R}_s\mathbf{R}_s^{-1}}_{\mathbf{I}} + \lambda\mathcal{F}\underbrace{\mathbf{R}_s\mathbf{R}_s^{-1}}_{\mathbf{I}}\end{aligned}$$

By equating the result to zero, the following expression arises:

$$\mathcal{F} = \frac{\mathrm{tr}(\mathcal{G}\mathcal{H}\mathcal{F}\mathbf{R}_s)}{\lambda}\mathcal{H}^H\mathcal{G}^H \tag{17.97}$$

$$= \kappa \mathcal{H}^H \mathcal{G}^H, \kappa \in \mathbb{R} \tag{17.98}$$

where $\kappa = \frac{\mathrm{tr}(\mathcal{G}\mathcal{H}\mathcal{F}\mathbf{R}_s)}{\lambda}$.

By replacing (17.98) in the constraint (17.21), we obtain

$$\begin{aligned}
E_t &= \mathrm{tr}(\mathcal{F}\mathbf{R}_s\mathcal{F}^H) \\
E_t &= \mathrm{tr}(\kappa \mathcal{H}^H \mathcal{G}^H \mathbf{R}_s \kappa \mathcal{G}\mathcal{H}) \\
E_t &= (\kappa)^2 \, \mathrm{tr}(\mathcal{H}^H \mathcal{G}^H \mathbf{R}_s \mathcal{G}\mathcal{H}) \\
(\kappa)^2 &= \frac{E_t}{\mathrm{tr}(\mathcal{H}^H \mathcal{G}^H \mathbf{R}_s \mathcal{G}\mathcal{H})}.
\end{aligned}$$

The value of κ is indeed real and positive, as below:

$$\kappa = \sqrt{\frac{E_t}{\mathrm{tr}(\mathcal{H}^H \mathcal{G}^H \mathbf{R}_s \mathcal{G}\mathcal{H})}} \tag{17.99}$$

so that by substituting (17.99) into (17.98), the desired result follows:

$$\mathcal{F} = \sqrt{\frac{E_t}{\mathrm{tr}(\mathcal{H}^H \mathcal{G}^H \mathbf{R}_s \mathcal{G}\mathcal{H})}} \mathcal{H}^H \mathcal{G}^H.$$

■

17.10.3 Proof of (17.25)

Let us start with the energy function

$$\mathbb{E}||\mathcal{F}\mathbf{s}||^2 = \mathrm{tr}\left(\mathcal{F}\mathbf{R}_s\mathcal{F}^H\right),$$

and expanding the cost function to be minimized, we get

$$\begin{aligned}
\mathbb{E}||\mathbf{s} - \alpha\mathbf{r}||^2 &= \mathbb{E}\left\{(\mathbf{s} - \alpha\mathbf{r})^H (\mathbf{s} - \alpha\mathbf{r})\right\} \\
&= \mathbb{E}\left\{\mathbf{s}^H\mathbf{s} - \mathbf{s}^H\alpha\mathbf{r} - \alpha\mathbf{r}^H\mathbf{s} + \alpha^2\mathbf{r}^H\mathbf{r}\right\} \\
&= \mathbb{E}\left\{\mathbf{s}^H\mathbf{s}\right\} - \alpha\mathbb{E}\left\{\mathbf{s}^H\mathbf{r}\right\} - \alpha\mathbb{E}\left\{\mathbf{r}^H\mathbf{s}\right\} + \alpha^2\mathbb{E}\left\{\mathbf{r}^H\mathbf{r}\right\}
\end{aligned}$$

.

Since the received vector is given by $\mathbf{r} = \mathcal{G}\mathcal{H}\mathcal{F}\mathbf{s} + \mathcal{G}\mathbf{n}$, it follows that

$$\mathbb{E}\left\{\mathbf{s}^H\mathbf{r}\right\} = \mathbb{E}\left\{\mathbf{s}^H\mathcal{G}\mathcal{H}\mathcal{F}\mathbf{s}\right\} + \cancelto{0}{\mathbb{E}\left\{\mathbf{s}^H\mathcal{G}\mathbf{n}\right\}}$$

$$\mathbb{E}\left\{\mathbf{r}^H\mathbf{s}\right\} = \mathbb{E}\left\{\mathbf{s}^H\mathcal{F}^H\mathcal{H}^H\mathcal{G}^H\mathbf{s}\right\} + \cancelto{0}{\mathbb{E}\left\{\mathbf{n}^H\mathcal{G}^H\mathbf{s}\right\}}$$

$$\mathbb{E}\left\{\mathbf{r}^H\mathbf{r}\right\} = \mathbb{E}\left\{\mathbf{s}^H\mathcal{F}^H\mathcal{H}^H\mathcal{G}^H\mathcal{G}\mathcal{H}\mathcal{F}\mathbf{s}\right\} + \mathbb{E}\left\{\mathbf{n}^H\mathcal{G}^H\mathcal{G}\mathbf{n}\right\}$$

so that an expression for $\mathbb{E}||\mathbf{s} - \alpha\mathbf{r}||^2$ takes the form

$$\begin{aligned}\mathbb{E}||\mathbf{s} - \alpha\mathbf{r}||^2 &= \text{tr}(\mathbf{R_s}) - \alpha\text{tr}(\mathcal{G}\mathcal{H}\mathcal{F}\mathbf{R_s}) - \alpha\text{tr}(\mathcal{H}^H\mathcal{G}^H\mathbf{R_s}\mathcal{F}^H)\\ &+ \alpha^2\text{tr}(\mathcal{H}^H\mathcal{G}^H\mathcal{G}\mathcal{H}\mathcal{F}\mathbf{R_s}\mathcal{F}^H) + \alpha^2\text{tr}(\mathcal{G}\mathbf{R_n}\mathcal{G}^H).\end{aligned}$$

Then, the unconstrained cost function using the Lagrange multiplier method becomes

$$\begin{aligned}\mathcal{L}(\mathcal{F}, \alpha, \lambda) &= \text{tr}(\mathbf{R_s}) - \alpha\text{tr}(\mathcal{G}\mathcal{H}\mathcal{F}\mathbf{R_s}) - \alpha\text{tr}(\mathcal{H}^H\mathcal{G}^H\mathbf{R_s}\mathcal{F}^H)\\ &+ \alpha^2\text{tr}(\mathcal{H}^H\mathcal{G}^H\mathcal{G}\mathcal{H}\mathcal{F}\mathbf{R_s}\mathcal{F}^H) + \alpha^2\text{tr}(\mathcal{G}\mathbf{R_n}\mathcal{G}^H)\\ &+ \lambda(\text{tr}(\mathcal{F}\mathbf{R_s}\mathcal{F}^H) - E_t).\end{aligned}$$

Now, we compute the derivative of the cost function with respect to $\mathcal{F}^*$ and equate the result to zero:

$$\frac{\partial\mathcal{L}(\mathcal{F}, \alpha, \lambda)}{\partial\mathcal{F}^*} = -\alpha\mathcal{H}^H\mathcal{G}^H\cancel{\mathbf{R_s}} + \alpha^2\mathcal{H}^H\mathcal{G}^H\mathcal{F}\cancel{\mathbf{R_s}} + \lambda\mathcal{F}\cancel{\mathbf{R_s}} = \mathbf{0}$$

leading to

$$\frac{1}{\alpha}\mathcal{H}^H\mathcal{G}^H = \left(\mathcal{H}^H\mathcal{G}^H\mathcal{G}\mathcal{H} + \frac{\lambda}{\alpha^2}\mathbf{I}_{Nt}\right)\mathcal{F} \tag{17.100}$$

and

$$\mathcal{F} = \frac{1}{\alpha}\left(\mathcal{H}^H\mathcal{G}^H\mathcal{G}\mathcal{H} + \frac{\lambda}{\alpha^2}\mathbf{I}_{Nt}\right)^{-1}\mathcal{H}^H\mathcal{G}^H.$$

Let us consider that α is the independent variable; taking the partial derivative with respect to α and equating the result to zero, we get

$$\begin{aligned}\frac{\partial\mathcal{L}(\mathcal{F}, \alpha, \lambda)}{\partial\alpha} &= -\text{tr}(\mathcal{G}\mathcal{H}\mathcal{F}\mathbf{R_s}) - \text{tr}(\mathcal{H}^H\mathcal{G}^H\mathbf{R_s}\mathcal{F}^H)\\ &+ 2\alpha\text{tr}(\mathcal{H}^H\mathcal{G}^H\mathcal{G}\mathcal{H}\mathcal{F}\mathbf{R_s}\mathcal{F}^H) + 2\alpha\text{tr}(\mathcal{G}\mathbf{R_n}\mathcal{G}^H)\end{aligned}$$

so that

$$2\text{tr}(\mathcal{H}^H \mathcal{G}^H \mathbf{R}_s \mathcal{F}^H) = 2\alpha \text{tr}(\mathcal{H}^H \mathcal{G}^H \mathcal{G}\mathcal{H}\mathcal{F}\mathbf{R}_s \mathcal{F}^H) + 2\alpha \text{tr}(\mathcal{G}\mathbf{R}_\mathbf{n}\mathcal{G}^H).$$

By multiplying both sides of (17.100) by $2\alpha \mathbf{R}_s \mathcal{F}^H$, it is possible to show that

$$\begin{aligned} 2\text{tr}[\mathcal{H}^H \mathcal{G}^H \mathbf{R}_s \mathcal{F}^H] &= 2\alpha \text{tr}[(\mathcal{H}^H \mathcal{G}^H \mathcal{G}\mathcal{H} + \frac{\lambda}{\alpha^2}\mathbf{I})\mathcal{F}\mathbf{R}_s \mathcal{F}^H] \\ &= 2\alpha \text{tr}[\mathcal{H}^H \mathcal{G}^H \mathcal{G}\mathcal{H}\mathcal{F}\mathbf{R}_s \mathcal{F}^H] + 2\frac{\lambda}{\alpha}\text{tr}[\mathcal{F}\mathbf{R}_s \mathcal{F}^H]. \end{aligned}$$

Then, by inspecting the last two equations above, we are able to perform the following simplification:

$$\begin{aligned} &\cancel{2\alpha \text{tr}[\mathcal{H}^H \mathcal{G}^H \mathcal{G}\mathcal{H}\mathcal{F}\mathbf{R}_s \mathcal{F}^H]} + \cancel{2\alpha}\text{tr}[\mathcal{G}\mathbf{R}_\mathbf{n}\mathcal{G}^H] \\ &= \cancel{2\alpha \text{tr}[\mathcal{H}^H \mathcal{G}^H \mathcal{G}\mathcal{H}\mathcal{F}\mathbf{R}_s \mathcal{F}^H]} + \cancel{2\alpha}\frac{\lambda}{\alpha^2}\text{tr}[\mathcal{F}\mathbf{R}_s \mathcal{F}^H] \end{aligned}$$

so that

$$\frac{1}{\alpha^2} = \frac{\text{tr}[\mathcal{G}\mathbf{R}_s\mathcal{G}^H]}{\text{tr}[\mathcal{F}\mathbf{R}_s\mathcal{F}^H]} = \frac{\text{tr}[\mathcal{G}\mathbf{R}_\mathbf{n}\mathcal{G}^H]}{E_t}.$$

Therefore, the desired intermediate results follow:

$$\begin{aligned} \mathcal{F} &= \frac{1}{\alpha}\left(\mathcal{H}^H \mathcal{G}^H \mathcal{G}\mathcal{H} + \frac{\text{tr}[\mathcal{G}\mathbf{R}_\mathbf{n}\mathcal{G}^H]}{E_t}\mathbf{I}\right)\mathcal{H}^H \mathcal{G}^H \\ E_t &= \frac{1}{\alpha^2}\text{tr}\left[\mathcal{H}^H \mathcal{G}^H \mathcal{G}\mathcal{H} + \frac{\text{tr}[\mathcal{G}\mathbf{R}_\mathbf{n}\mathcal{G}^H]}{E_t}\mathbf{I}\right]\mathcal{H}^H \mathcal{G}^H \mathbf{R}_s \mathcal{G}\mathcal{H} \\ \alpha &= \sqrt{\frac{\text{tr}\left[\mathcal{H}^H \mathcal{G}^H \mathcal{G}\mathcal{H} + \frac{\text{tr}[\mathcal{G}\mathbf{R}_\mathbf{n}\mathcal{G}^H]}{E_t}\mathbf{I}\right]\mathcal{H}^H \mathcal{G}^H \mathbf{R}_s \mathcal{G}\mathcal{H}}{E_t}}. \end{aligned}$$

By employing the derived expressions, the desired result of (17.25) follows:

$$\mathcal{F} = \sqrt{\frac{E_t}{\text{tr}\left(\left(\hat{\mathcal{H}}\hat{\mathcal{H}}^H + \text{tr}\left(\hat{\mathbf{R}}_\mathbf{n}\right)^{-2}\right)\left(\hat{\mathcal{H}}\mathbf{R}_s\hat{\mathcal{H}}\right)\right)}}\left(\hat{\mathcal{H}}\hat{\mathcal{H}}^H + \text{tr}\left(\hat{\mathbf{R}}_\mathbf{n}\right)\right)^{-1}\frac{\mathbf{I}_{Nt}}{E_t}\hat{\mathcal{H}}^H.$$

■

17.10.4 Proof of (17.28)

Cost function

$$\begin{aligned}\mathcal{L}(\mathcal{G},\boldsymbol{\lambda}) &= \operatorname{tr}\left(\mathcal{G}\mathcal{R}_{\mathbf{n}_t}\mathcal{G}^H\right) + \operatorname{tr}\left(\boldsymbol{\lambda}\left(\mathbf{I}-\mathcal{G}\mathcal{H}\mathcal{F}\right)\right)\\ &= \operatorname{tr}\left(\mathcal{G}\mathcal{R}_{\mathbf{n}_t}\mathcal{G}^H\right) + \operatorname{tr}\left(\boldsymbol{\lambda}\mathbf{I}\right) + \operatorname{tr}\left(\boldsymbol{\lambda}\mathcal{G}\mathcal{H}\mathcal{F}\right).\end{aligned}$$

Derivation with respect to $\mathcal{G}$ equals zero to derive an expression for $\mathcal{G}$ as a function of $\boldsymbol{\lambda}$:

$$\begin{aligned}\frac{\partial\mathcal{L}(\mathcal{G},\boldsymbol{\lambda})}{\partial\mathcal{G}} &= \mathcal{R}_{\mathbf{n}_t}\mathcal{G}^H - \mathcal{H}\mathcal{F}\boldsymbol{\lambda} = \mathbf{0}\\ \mathcal{R}_{\mathbf{n}_t}\mathcal{G}^H &= \mathcal{H}\mathcal{F}\boldsymbol{\lambda}\\ \mathcal{G}^H &= \mathcal{R}_{\mathbf{n}_t}^{-1}\mathcal{H}\mathcal{F}\boldsymbol{\lambda}.\end{aligned}$$

Replace $\mathcal{G}^H$ in the constraints and compute $\boldsymbol{\lambda}$:

$$\begin{aligned}\mathbf{I} &= \mathcal{G}\mathcal{H}\mathcal{F} = \mathcal{F}^H\mathcal{H}^H\mathcal{G}^H\\ \mathbf{I} &= \mathcal{F}^H\mathcal{H}^H\mathcal{R}_{\mathbf{n}_t}^{-1}\mathcal{H}\mathcal{F}\boldsymbol{\lambda}\\ \boldsymbol{\lambda} &= \left(\mathcal{F}^H\mathcal{H}^H\mathcal{R}_{\mathbf{n}_t}^{-1}\mathcal{H}\mathcal{F}\right)^{-1}.\end{aligned}$$

It is now possible to get the desired result.

$$\begin{aligned}\mathcal{G}^H &= \mathcal{R}_{\mathbf{n}_t}^{-1}\mathcal{H}\mathcal{F}\boldsymbol{\lambda}\\ \mathcal{G}^H &= \mathcal{R}_{\mathbf{n}_t}^{-1}\mathcal{H}\mathcal{F}\left(\mathcal{F}^H\mathcal{H}^H\mathcal{R}_{\mathbf{n}_t}^{-1}\mathcal{H}\mathcal{F}\right)^{-1}\\ \mathcal{G} &= \left(\mathcal{F}^H\left(\mathcal{H}^H\mathcal{R}_{\mathbf{n}_t}^{-1}\mathcal{H}\right)\mathcal{F}\right)^{-1}\mathcal{F}^H\mathcal{H}^H\mathcal{R}_{\mathbf{n}_t}^{-1}.\end{aligned}$$

■

17.10.5 Proof of (17.35)

We know that

$$\max_{\mathbf{r}}\left(F\{[\mathbf{r}\in\mathbf{C}]|\mathbf{s}\}\right) = \max_{\mathbf{r}}\left(-||\mathbf{r}-\mathbf{s}||^2\right)$$

that should be equivalent to

$$\begin{aligned} f_1(\mathbf{r}, \mathbf{s}) &= -||\mathbf{r} - \mathbf{s}||^2 \\ &= -(\mathbf{r} - \mathbf{s})(\mathbf{r} - \mathbf{s})^H \\ &= -(\mathbf{r} - \mathbf{s})(\mathbf{r}^H - \mathbf{s}^H) \\ &= -(\mathbf{r}\mathbf{r}^H - \mathbf{r}\mathbf{s}^H - \mathbf{s}\mathbf{r}^H + \mathbf{s}\mathbf{s}^H) \\ &= -\mathbf{r}\mathbf{r}^H + \mathbf{r}\mathbf{s}^H + \mathbf{s}\mathbf{r}^H - \mathbf{s}\mathbf{s}^H, \end{aligned}$$

since

$$z + z^H = \text{Re}\{z\} + \text{Im}\{z\} + \text{Re}\{z\} - \text{Im}\{z\} = 2\text{Re}\{z\}.$$

The argument we will manipulate is f_2:

$$\begin{aligned} f_2 &= 2\text{Re}\{\mathbf{s}^H\mathbf{r}\} - \mathbf{s}^H\mathbf{s} \\ &= \mathbf{s}^H\mathbf{r} + (\mathbf{s}^H\mathbf{r})^H - \mathbf{s}^H\mathbf{s} \\ &= \mathbf{s}^H\mathbf{r} + \mathbf{r}^H\mathbf{s} - \mathbf{s}^H\mathbf{s}. \end{aligned}$$

By using f_2 to express f_1

$$\begin{aligned} f_1 &= -\mathbf{r}\mathbf{r}^H + \mathbf{r}\mathbf{s}^H + \mathbf{s}\mathbf{r}^H - \mathbf{s}\mathbf{s}^H \\ &= -\mathbf{r}\mathbf{r}^H + f_2^T. \end{aligned}$$

Then, by calculating the derivative of f_1 with respect to $\mathbf{r}^H$ and equating the result to zero, we get that $\mathbf{r} = \mathbf{s}$:

$$\begin{aligned} \frac{\partial f_1}{\partial \mathbf{r}^H} &= -\mathbf{r} + \frac{\partial f_2^T}{\partial \mathbf{r}^H} = \mathbf{0} \\ \mathbf{r} &= \frac{\partial}{\partial \mathbf{r}^H}\left(\mathbf{s}^H\mathbf{r} + \mathbf{r}^H\mathbf{s} - \mathbf{s}^H\mathbf{s}\right) \\ &= \cancelto{0}{\frac{\partial}{\partial \mathbf{r}^H}\left(\mathbf{s}^H\mathbf{r}\right)} + \frac{\partial}{\partial \mathbf{r}^H}\left(\mathbf{r}^H\mathbf{s}\right) + \cancelto{0}{\frac{\partial}{\partial \mathbf{r}^H}\left(-\mathbf{s}^H\mathbf{s}\right)} \\ \mathbf{r} &= \mathbf{s}. \end{aligned}$$

Since only f_1 is a function of $\mathbf{r}$, we reach the desired maximization. ■

17.11 Remarks

This chapter introduced some tools for designing transceivers whose data information is assembled in blocks. This framework encompasses many practical communication configurations, including those considered for autonomous vehicle applications.

References

E. Björnson, O.T. Demir, *Introduction to Multiple Antenna Communications and Reconfigurable Arrays* (Now Publishers, Delft, 2003)

P.S.R. Diniz, E.A.B. da Silva, S.L. Netto, *Digital Signal Processing: System Analysis and Design*, 2nd edn. (Cambridge University Press, Cambridge, 2010)

P.S.R. Diniz, W.A. Martins, M.V.S. Lima, *Block Transceivers: OFDM and Beyond* (Morgan & Claypool Publishers, Fort Collins, 2012)

R. Esmailzadeh, M. Nakagawa, Pre-rake diversity combination for direct sequence spread spectrum communications systems, in *Proceedings of an International Conference on Communications* (Geneva, 1993), pp. 463–467

A. Goldsmith, *Wireless Communications* (Cambridge University Press, Cambridge, 2005)

S. Haykin, *Communication Systems*, 4th edn. (John Wiley, New York, 2001)

A. Hjørungnes, *Complex-Valued Matrix Derivatives* (Cambridge University Press, Cambridge, 2011)

A. Hjørungnes, P.S.R. Diniz, Minimum BER fir prefilter transform for communications systems with binary signaling and known fir mimo channel. IEEE Signal Process. Lett. **12**, 234–237 (2005)

A. Hjørungnes, P.S.R. Diniz, M.C.R. de Campos, Jointly minimum BER fir mimo transmitter and receiver filters for binary signal vectors. IEEE Trans. Signal Process. **52**, 1021–1036 (2004)

A. Hjørungnes, M.C.R. de Campos, P.S.R. Diniz, Jointly optimized transmitter and receiver filters fir mimo filters in the presence of near-end crosstalk. IEEE Trans. Signal Process. **53**, 346–359 (2005)

M. Joham, W. Utschick, J.A. Nossek, Linear transmit processing in mimo communications systems. IEEE Trans. Signal Process. **53**, 2700–2712 (2005)

S.M. Kay, *Fundamentals of Statistical Signal Processing: Estimation Theory* (Prentice Hall, Upper Saddle River, 1993)

T.L. Marzetta, E. Larsson, H. Yang, H.Q. Ngo, *Fundamentals of Massive MIMO* (Cambridge University Press, Cambridge, 2016)

A. Paulraj, R. Nabar, D. Gore, *Introduction to Space-Time Wireless Communications* (Cambridge University Press, Cambridge, 2003)

H.V. Poor, *An Introduction to Signal Detection and Estimation*, 2nd edn. (Springer, New York, 1994)

H. Van Trees, K.L. Bell, Z. Tian, *Detection, Estimation, and Modulation Theory Part I*, 2nd edn. (John Wiley, New York, 2013)

D. Tse, P. Viswanath, *Fundamentals of Wireless Communications* (Cambridge University Press, Cambridge, 2005)

P.P. Vaidyanathan, *Multirate Systems and Filter Banks* (Prentice Hall, Upper Saddle River, 1993)
P.P. Vaidyanathan, S.M. Phoong, Y.P. Lin, *Signal Processing and Optimization for Transceiver Systems* (Cambridge University Press, Cambridge, 2010)
A.J. Viterbi, *Principles of Spread Spectrum Communication* (Addison Wesley, Reading, 1995)

Chapter 18
Radar Signal Processing

18.1 Introduction

With the advent of the Internet-of-Things (IoT), 5G mobile communications, and autonomous vehicle communications, there is a growing interest in developing technologies that allow high data throughput at short distances. A way to achieve this goal is to exploit solutions operating at gigahertz (GHz) carrier frequencies along with wide bandwidth (BW). Here, we briefly discuss the application of radar[1] solutions in situations where millimeter waves (mm waves) are employed in radar systems, bringing about the benefit of using small antennas. In fact, a set of antennas can fit easily in small spaces forming reconfigurable arrays (Balanis and Ioannides 2007). As will be seen, the wide bandwidth allows a high resolution in the radar signal delay estimates, the target range, and the Doppler effects originating from the relative speed between the target and the radar transmitter.

In particular, vehicles include a growing number of sensors to acquire environmental data, which can be fused to allow safer travel. Several cars in proximity, along with fixed infrastructure such as wireless communications, might form a network, enabling somewhat accurate information about the environment. The overall system requires signal processing, communication, computational resources, and even some machine learning tools.

This chapter complements Chap. 6 and exploits how radar systems work and how they can be integrated with wireless communication systems. The radar systems are meant to detect targets through the reflected transmitted signal while using the electromagnetic spectrum, also employed by wireless communication systems. However, the latter aims to transport information from the transmitter to the receiver in a protected manner to allow its correct detection. Both systems might occupy spectra closer to each other in the vehicle network environment, providing the stage

[1] Acronym for RAdio Detection and Ranging.

Y. C. Lim et al., *Signal Processing for Autonomous Systems*,
https://doi.org/10.1007/978-3-032-00675-2_18

to discuss their joint design despite their distinct waveforms. But when they operate in the millimeter wave at close carrier frequencies, it is viable that at least one system might aid the other with helpful information. For instance, a radar at a cellular base station can provide proximity and speed estimates to be transferred to the nearby vehicles to improve their information about the current environmental situation.

18.2 Motivation

Vehicles can also incorporate other technologies to acquire information about the surroundings, such as by scanning the field of view through laser beams operating at near-infrared wavelength (see Li and Ibanez-Guzman 2020 and Rapp et al. 2020), known as light detectors and ranging (lidar) systems. Video cameras and Global Positioning Systems (GPS), along with laser sensors, are additional acquisition devices that can incorporate further information about the environment where a vehicle is immersed.

Many autonomous vehicles incorporate radar, lidar, and cameras, forming a set of source information, each with its own attractive feature. Radars have a low cost but provide low resolution related to depth and might confuse objects that are close to each other. On the other hand, the cameras generate high-resolution images but do not stand out in recognizing object depth information. Cameras also face difficulties in low light situations and shadows. Constructively, lidar sensors have features not found in cameras and radars since they can produce high-resolution images with accurate depth information and independent of lighting conditions. Figure 18.1 illustrates a vehicle incorporating lidar, camera, communications, and radar. If properly designed, the infrastructure enables the autonomous vehicles to have an accurate situational 3D representation of all objects and obstacles around them.

The challenge is to fuse a large amount of data, sometimes coming from different sensing systems, to access the situation awareness better. In the long run, machine learning tools and distributed cloud computations will be increasingly incorporated.

In the remaining part of this chapter, we briefly show how the radar systems work and discuss some common and complementary features with the standard wireless communication systems. The joint use of radars and wireless communication systems brings about advantages in terms of robustness and cost if they operate in the same frequency range (Ma et al. 2020). Traditional radar and communication systems commonly utilize electromagnetic signals to perform their task. The radar probes the environment by sending an electromagnetic waveform and analyzing its reflection at the receiver located at the same site. In wireless communications, the aim is to exchange information among devices and distinct types of infrastructure. The devices installed in a vehicle communicate with traffic infrastructure, another vehicle, a wireless network, a computer server, a pedestrian, etc. The communication links are collectively called V2X communications.

Fig. 18.1 Autonomous vehicle including lidar, radars, camera, and wireless connections

18.3 Doppler Effects

This section discusses a few pieces of information about the Doppler effects originating from moving targets (Richards 2022; Mahafza 2013).

Let us assume a target is moving towards[2] the radar transmitter and the radar is sending a waveform of time length T_{ch}, where the subscript denotes the chirp to be discussed. When the leading edge to the waveform reaches the target and is reflected, the trailing edge is at $d_{\text{tr}} = T_{\text{ch}}c$ away from the target. By the time the waveform trailing edge reaches the target, the latter has moved a distance of

[2] Or away from for negative velocity. It is also assumed that the velocity is constant. Otherwise, we have a time-varying change of $d(t)$.

$\Delta d = v\Delta t$, whereas the leading edge has traveled the distance $d' = c\Delta t$ back in the radar direction. At this point, the pulse width has changed to T'_{ch} so that $cT'_{\text{ch}} = c\Delta t - v\Delta t$. On the other hand, the trailing edge had to travel $d_{\text{tr}} = c\Delta t$, and the target itself had traveled $v\Delta t$ so that by adding these distances, the result is cT_{ch}. Therefore,

$$\begin{aligned} cT_{\text{ch}} &= c\Delta t + v\Delta t \\ cT'_{\text{ch}} &= c\Delta t - v\Delta t \end{aligned} \tag{18.1}$$

The time dilation factor is given by

$$T'_{\text{ch}} = \frac{c - v}{c + v} T_{\text{ch}}. \tag{18.2}$$

In case the chirp pulses are sent periodically at a frequency $\frac{1}{T} = f_0$, for $T_{\text{ch}} \le T$, the perceived frequency at the receiver will be

$$f'_0 = \frac{c + v}{c - v} f_0 \tag{18.3}$$

so that the Doppler shift is

$$f_{\text{D}} = f'_0 - f_0 = \frac{c + v}{c - v} f_0 - f_0 = \frac{2v}{c - v} f_0 \tag{18.4}$$

where for $c \gg v$, $f_{\text{D}} \approx \frac{2v}{c} f_0$.

18.4 Radar Waveforms

This section discusses some issues related to the radar waveforms (Richards 2022). Among the waveforms discussed for radar transmissions is the sinusoid representing the carrier multiplied by a square pulse, as described in the following:

$$x(t) = \mathrm{e}^{\mathrm{j}2\pi f_c t} \left[u(t) - u\left(t - T\right)\right] = \mathrm{e}^{\mathrm{j}2\pi f_c t} p(t) \tag{18.5}$$

where f_c, $p(t)$, and $u(t)$ represent the carrier frequency and the pulse shaping the unitary step function, respectively. We will concentrate on this waveform due to its simplicity in the following.

Let us consider that the radar transmits a single pulse $x(t)$ of duration T. The received signal reflected in a moving target is

$$\begin{aligned} r(t) &= gx(t-\tau) = gp(t-\tau)e^{j2\pi f_c(t-\tau)} \\ &= gp(t-\tau)e^{j2\pi \frac{c}{\lambda_c}(t-\tau)} = gp(t-\tau)e^{j2\pi f_c(t-\frac{2v}{c}t+\frac{2d}{c})} \\ &= gp(t-\tau)e^{j2\pi f_c(t-\tau_0)}e^{j2\pi f_D t} \end{aligned} \tag{18.6}$$

where $\lambda_c = \frac{c}{f_c}$ is the carrier wavelength, v is the relative velocity between the radar and the target, τ is the return delay, and g is the channel gain. The delay is

$$\tau = \frac{2d}{c} - \frac{2v}{c}t = \tau_0 - \frac{2v}{c}t \tag{18.7}$$

and d is the distance in meters between the radar and the target when the waveform first reaches the latter. The symbol g represents the returned signal gain, f_c is the carrier frequency, and $f_D \approx f_c\frac{2v}{c}$ is the Doppler frequency for $v \ll c$, as per (18.4). Note that for positive v, meaning the target is moving in the radar direction, the corresponding f_D is positive in our convention. If we had the knowledge of the Doppler frequency and delay, the appropriate matched filter to detect the radar signal would have the following impulse response:

$$h(t) = \frac{1}{g}r^*(-t)e^{j2\pi f_D(t)} = p(-\tau-t)e^{-j2\pi f_c(-t-\tau_0)}e^{j2\pi f_D(-t)}e^{j2\pi f_D(t)} \tag{18.8}$$

whose Fourier transform is

$$H(f) = R^*(f - f_D) \tag{18.9}$$

where $r^*(\cdot)$ represents the complex conjugate of $r(\cdot)$. The frequency representation shows that the ideal matched filter takes into account the frequency and time shift of the reflected signal.

Consider we know the delay τ_0 for simplicity and that the peak of the matched filter is at $t = 0$, and the matched filter does have the correct information about the Doppler shift. The output of the matched filter can then be calculated as

$$\begin{aligned} y(t) &= \int_{-\infty}^{\infty} r(\nu)h(t-\nu)d\nu \\ &= g\int_{-\infty}^{\infty} p(\nu-\tau)e^{j2\pi f_c(\nu-\tau_0)}e^{j2\pi f_D \nu}p(-t+\nu-\tau) \\ &\quad \times e^{-j2\pi f_c(-t+\nu-\tau_0)}e^{j2\pi f_D(t-\nu)}d\nu \\ &= ge^{j2\pi(f_c+f_D)t}\int_{-\infty}^{\infty} p(\nu-\tau)p(-t+\nu-\tau)d\nu \end{aligned} \tag{18.10}$$

where we considered the square pulse $p(t) = [u(t) - u(t-T)]$ to make the expression more complete, with more detail later in the text. Then

$$y(t) = g e^{j2\pi(f_c + f_D)t} \int_{t+\tau}^{T+\tau} dv \text{ for } t \geq 0 + \int_{\tau}^{T+\tau+t} dv \text{ for } t < 0$$
$$= g e^{j2\pi(f_c + f_D)t} (T - |t|) \tag{18.11}$$

for $-T \leq t \leq T$ and the integral limits depend on the value of t. For instance, considering $\tau = 0$, for $t \geq 0$ the range is t to T, whereas for $t \leq 0$ it is 0 to $T - |t|$.

For the pulse waveform of (18.5) if we assume that the frequency shift is not the exact value (represented by f_D'), with a simple derivation it is possible to show that

$$|y(0)| = \left| \frac{2g \sin\left[(f_D - f_D')\frac{T}{2})\right]}{f_D - f_D'} \right| \tag{18.12}$$

where $y(t)$ represents the matched filter output. The above expression shows that if $\frac{1}{T} \gg (f_D - f_D')$ the peak of the matched filter will not be reduced substantially due to Doppler mismatch. The derivation of this expression is simple and can be found in Richards (2022, pp. 176).

For another pedagogic analysis, let us assume the input signal to the matched filter consists of the transmitted signal affected by a Doppler shift. Then, the matched filter output with its peak center at $t = 0$, for simplification, is described by

$$y(t, f_D) = g \int_{-\infty}^{\infty} x(v) e^{j2\pi f_D v} x^*(v - t) dv \tag{18.13}$$

which is usually called a complex ambiguity function. An alternative way to calculate the complex ambiguity function in the frequency domain is

$$y(t, f_D) = g \int_{-\infty}^{\infty} X^*(f) X(f - f_D) e^{j2\pi f t} df \tag{18.14}$$

where the latter expression is derived by recognizing that (18.13) is a linear convolution and by utilizing the properties that the Fourier transforms of $x^*(-t)$ and $x(t)e^{j2\pi f_D t}$ are $X^*(f)$ and $X(f - f_D)$, respectively.

Following Richards (2022, pp. 178), we define the actual ambiguity function as

$$A(t, f_D) = |y(t, f_D)|\,. \tag{18.15}$$

The ambiguity function is a handy tool for qualifying and quantifying the performance of radar waveforms when subjected to Doppler shift and its combined effect on delay uncertainty.

The ambiguity function of the carrier sinusoid multiplied by a square pulse is given by

$$A(t, f_{\mathrm{D}}) = g\left(1 - \frac{|t|}{T}\right)\left|\frac{\sin\left[\pi f_{\mathrm{D}}(T - |t|)\right]}{\pi f_{\mathrm{D}}(T - |t|)}\right| \tag{18.16}$$

for $-T \leq t \leq T$ and the matched filter peak is at $t = 0$. This expression is a special case of the ambiguity function discussed in the following section.

With the aid of the ambiguity function, it is possible to access information related to the ambiguities of the range and the Doppler effect, the resolution in their estimate as well as how the waveform design connects them.

18.5 Frequency-Modulated Continuous-Wave Radar

A popular waveform for pulsed radar relies on linear frequency modulated pulse. This section briefly describes the radar system based on frequency modulation. See the textbooks Richards (2022) and Mahafza (2013) and the tutorials Aydogdu et al. (2020) and Sun et al. (2020) for details. The frequency-modulated waveform is described as

$$\begin{aligned} x(t) &= \mathrm{e}^{\mathrm{j}\left[2\pi\left(f_{\mathrm{c}}t + \frac{1}{2}at^2\right)\right]}\left[u(t) - u\left(t - T\right)\right] \\ &= \mathrm{e}^{\mathrm{j}2\pi f_{\mathrm{c}}t}\mathrm{e}^{\mathrm{j}\left[\pi a t^2\right]}\left[u(t) - u\left(t - T\right)\right] \end{aligned} \tag{18.17}$$

where a represents the frequency slope to be discussed. This waveform gives rise to the frequency-modulated continuous-wave (FMCW) radar.

The ambiguity function related to the frequency-modulated waveform is

$$A(t, f_{\mathrm{D}}) = g\left(1 - \frac{|t|}{T}\right)\left|\frac{\sin\left[\pi\left(f_{\mathrm{D}} + \frac{at}{T}\right)(T - |t|)\right]}{\pi\left(f_{\mathrm{D}} + \frac{at}{T}\right)(T - |t|)}\right| \tag{18.18}$$

for $-T \leq t \leq T$, where, in the case of zero Doppler, the matched filter peak is centered at $t = 0$. For a broader discussion related to the ambiguity function, see Abatzoglou and Gheen (1998). The maximum value of the ambiguity function happens at $t = -\frac{f_{\mathrm{D}}T}{a}$ and not at $t = 0$. This time shift leads to an error in the range induced by the Doppler effect. In this case, $2\Delta\tilde{d} = c\Delta t = c(t - 0)$ since the radio signal travels at the speed of light. The range error is given by

$$\Delta\tilde{d} = -\frac{cf_{\mathrm{D}}T}{2a}. \tag{18.19}$$

It also reduces ambiguity amplitude by a factor $(1 - \frac{|t|}{T})$, in this case equal to $(1 - \frac{f_{\mathrm{D}}}{a})$ since $t = -\frac{f_{\mathrm{D}}T}{a}$. Despite the FMCW shift-range value, this waveform enables the detection of Doppler shifts in a much broader range in comparison with simpler waveforms such as sinusoid and square pulse.

This resolution relates to the frequency sweep bandwidth determined by a, a flexibility introduced by the frequency-modulated waveform. Suppose one compares the ambiguity functions of Eqs. (18.16) and (18.18). In that case, the latter one is a skewed version of the former in the plane formed by the delay $\times$ Doppler axes, and the interested reader should consult (Richards 2022, pages 181 and 203). An FMCW pulse with the same duration as the square or sinusoidal pulses gives rise to matched filter output with significant power for a broader range of Doppler shifts (Richards 2022).

The presence of environment noise induces variances in the Doppler shift and in the range that are obviously inversely proportional to the SNR. In the case of the range, it is also inversely proportional to the bandwidth squared of the radar (Mahafza 2013). In contrast, the Doppler shift is inversely proportional to the waveform's effective duration; see Mahafza (2013, pp. 157–158).[3]

Given the carrier frequency f_c, we can define the waveform utilized in the FMCW radar with multiple chirps as follows:

$$\begin{aligned} s(t) &= \sum_{m=0}^{M-1} x(t - mT) \\ &= \sum_{m=0}^{M-1} \mathrm{e}^{\mathrm{j}\left[2\pi\left(f_c(t-mT)+\frac{1}{2}a(t-mT)^2\right)\right]} \left[u(t-mT) - u\left(t-(m+1)T\right)\right] \end{aligned} \tag{18.20}$$

whereby considering BW the bandwidth of the radar system, the parameter $a = \frac{BW}{T}$ represents the chirp slope. The term $[u(t - mT) - u\left(t - (m+1)T\right)]$ represents a pulse of width T, distinguishing each of the transmission chirp, where $0 \le t \le T$.

In many situations, the chirp duration within a pulse repetition interval is T_{ch}. In this case, the FMWC waveform is[4]

$$s(t) = \sum_{m=0}^{M-1} \mathrm{e}^{\mathrm{j}\left[2\pi\left(f_c(t-mT)+\frac{1}{2}a(t-mT)^2\right)\right]} \left[u(t-mT) - u\left(t-mT-T_{\text{ch}}\right)\right] \tag{18.21}$$

[3] The variance of the delay, quantifying the uncertainty, is proportional to $1/(BW^2 SNR)$, whereas the Doppler estimation variance is proportional to $1/(\tau_{\max}^2 SNR)$, where the $SNR = \frac{2E_r}{N_0}$ with E_r represents the transmitted signal energy and $N_0/2 = I_0$ is the power spectrum density of a band-limited white noise.

[4] The time support of a single chirp is T_{ch}, and the time between two pulses is the so-called pulse repetition interval (PRI) in the literature (Richards 2022), usually denoted as T_{PRI}. In this chapter, instead of T_{PRI} we use T as notation, and in any situation $T \ge T_{\text{ch}}$.

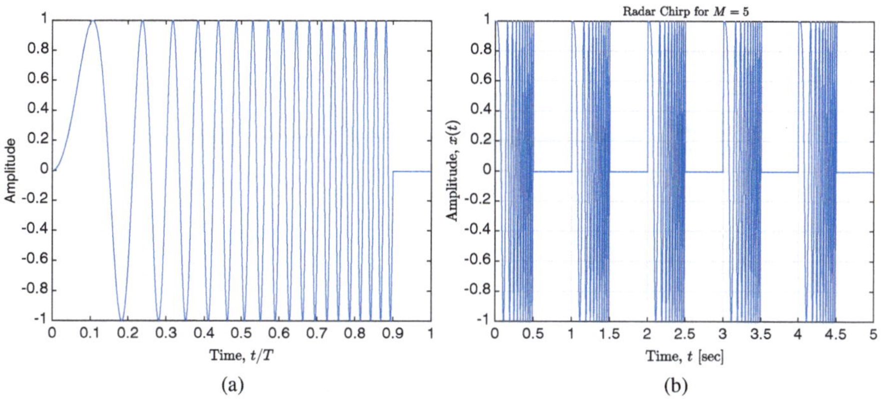

Fig. 18.2 Real part of an FMCW waveform with up-chirp characteristics: single-chirp pulse and multiple-chirp pulse. (**a**)Single chirp. (**b**) Multiple chirps

whereby the chirp slope, in this case, is $a = \frac{BW}{T_{\text{ch}}}$, and each chirp has support at $t \in [mT, mT + T_{\text{ch}}]$.

Figure 18.2 shows the real part of a single-chirp radar waveform with $aT_{\text{ch}} = 36$, $T_{\text{ch}} = 0.9T$. The figure also shows a multiple-chirp waveform in Fig. 18.2b with $M = 5$, $T_{\text{ch}} = 0.5T$, and $aT_{\text{ch}} = 10$.

In the FMCW implementation, the chirp signal illuminates the antenna and, at the same time, feeds the local correlator; more details are given in Fig. 18.3. The latter will mix the chirp signal with the signal reflected in the target. The received reflection usually has low power, requiring amplification via a low-noise amplifier. Then, the better-conditioned signal is mixed with the stored chirp signal, an operation called dechirping.[5] If the transmitter and the target are moving, the round-trip delay affects the reflected signal, and the Doppler effect is observed as a carrier's frequency shift.

As per the simplified description of the pulsed radar of Fig. 18.3, the baseband waveform is generated in the digital domain and then applied to a D/A converter. This signal traverses a mixer followed by a duplexer that, in turn, feeds the power amplifier, conditioning the signal to be sent through the transmitting antenna. The duplex also feeds a mixer whose other input originates from a linear RF amplifier whose output consists of an amplified version of the receiver's antenna signal. The mixer output feeds a low-pass filter, followed by a DC trimming to avoid saturation in the A/D converter while reducing the transmitter signal's leakage to the receiver. Note that the low-pass filter smooths the mixer output signal (it is actually a bandpass filter where the low frequencies are also attenuated) and transforms it to the digital domain. One should note that the A/D converter operates at a much lower frequency than the sweep signal. Therefore, the overall resulting baseband signal is

[5] In here, we call the mixing operation of the received signal with the transmit signal as dechirping.

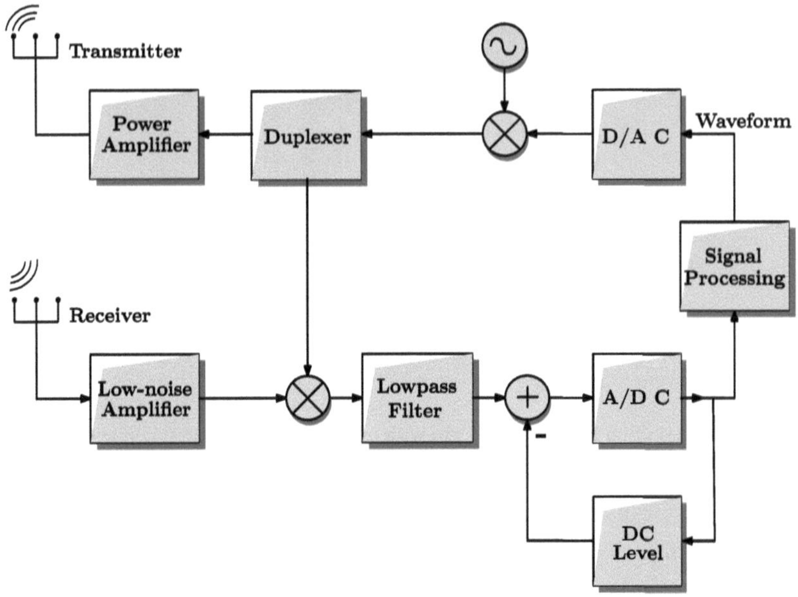

Fig. 18.3 Simplified FMCW radar system

analyzed in the digital domain to estimate the target range and Doppler. The radar configuration presented in Fig. 18.3 is quite simplified. For instance, some radars use superheterodyne configuration at the receiver where a cascade of intermediate frequency (IF) filters replaces the low-pass filter before reaching the baseband level.

Example 18.1 For a radar system show:

1. For a chirp supported for a normalized period of time, $T_{\text{ch}} = 1 = T$:
   - Plot the real part of the radar chirp for $aT = 20$.
   - Plot the ambiguity function.
   - Plot the spectrum of the radar signal.
   - Repeat the above plots for the sinusoid wave pulse with period $\frac{1}{10}$.
2. Show the real part of a radar chirp waveform with $aT = 20$, $T = 2T_{\text{ch}}$, and $M = 5$ for illustration.
3. Plot the ambiguity function for a normalized Doppler shift given by $f_{\text{D}}T_{\text{ch}}$ ranging from -10 to 10 and a normalized delay $\frac{t}{T}$. Also, plot the spectrum of the radar signal.
4. Repeat the previous item for $aT = 5$.
5. Try different values of M ranging from 1 to 10.

Solution

1. Figure 18.4 shows the real part of the chirp waveform along with its ambiguity function and the corresponding spectral content of the radar signal, typical of

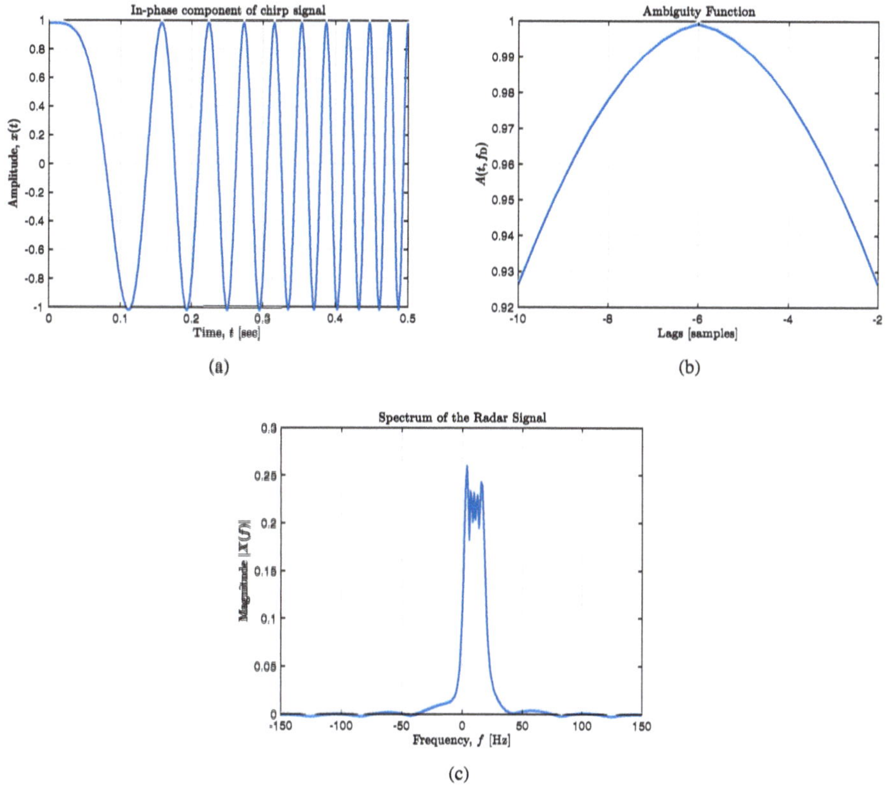

Fig. 18.4 Chirp waveform, ambiguity function, and spectrum. (**a**) Chirp waveform. (**b**) Ambiguity function. (**c**) Spectrum

FMCW radars. Figure 18.5 depicts the sinusoidal pulse waveform, its ambiguity function, and the corresponding spectral content of the radar signal.

2. Figure 18.2b shows the real part of a radar chirp waveform with $aT = 20$, $T = 2T_{\mathrm{ch}}$, and $M = 5$.
3. In Fig. 18.6, we show, for the FMCW waveform, the ambiguity function for a normalized Doppler shift given by $f_{\mathrm{D}}T_{\mathrm{ch}}$ ranging from -10 to 10 and a normalized delay $\frac{t}{\tau}$. The same figure plots the spectrum of the radar signal. This figure allows an exciting interpretation related to the ambiguity function in case there is no Doppler mismatch; the ambiguity has a triangle shape in the delay direction at $F_D\tau = 0$. In this case, if the Doppler slightly changes, the peak related to the range estimate does not change drastically (not shown here). On the other direction, by choosing $t = 0$ in (18.18) the ambiguity function becomes a sinc-like function:

$$A(0, f_{\mathrm{D}}) = g\left|\frac{\sin\left[\pi f_{\mathrm{D}}T)\right]}{\pi f_{\mathrm{D}}T}\right| \tag{18.22}$$

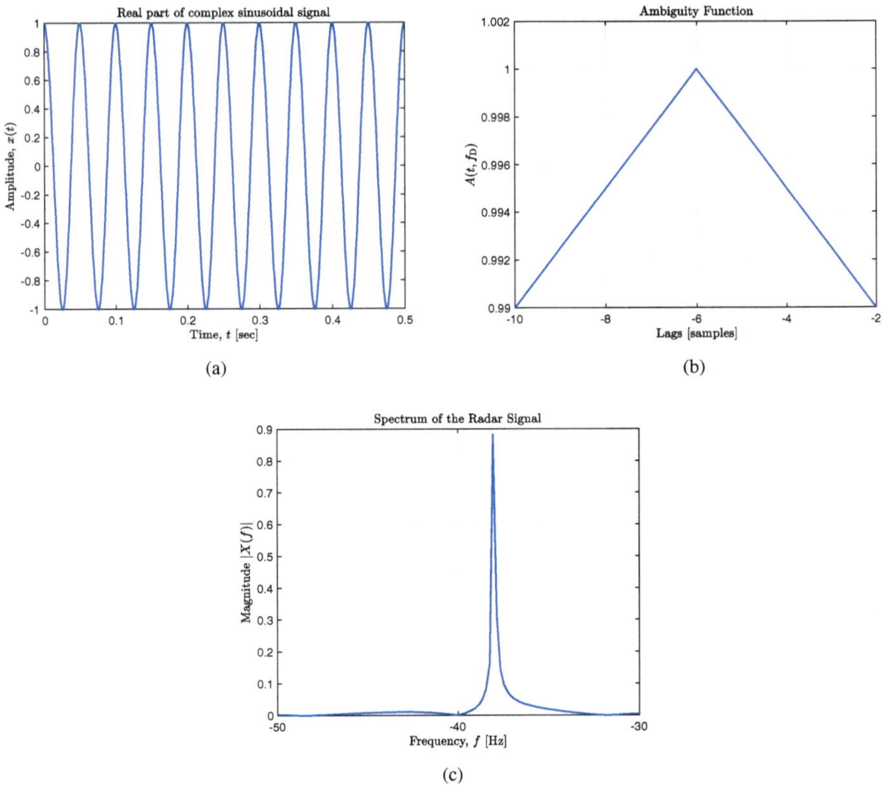

Fig. 18.5 Sinusoid waveform, ambiguity function, and spectrum. (**a**) Sinusoid waveform. (**b**) Ambiguity function. (**c**) Spectrum

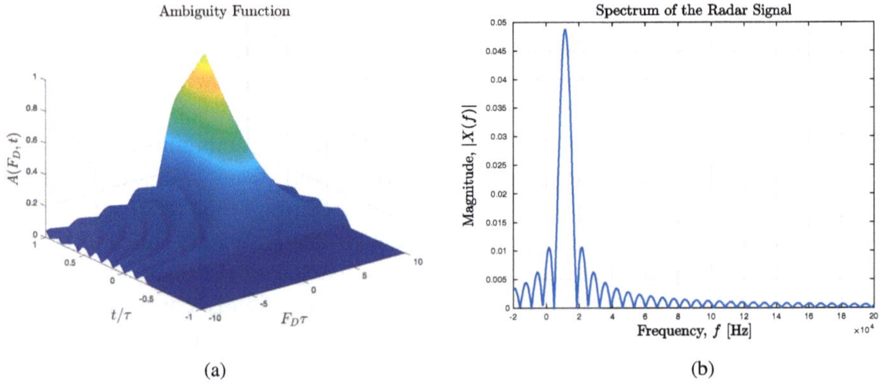

Fig. 18.6 Ambiguity function and spectrum, $aT = 20$. (**a**) Ambiguity function, $A(F_D, t) \times t/\tau \times F_D\tau$. (**b**) Spectrum

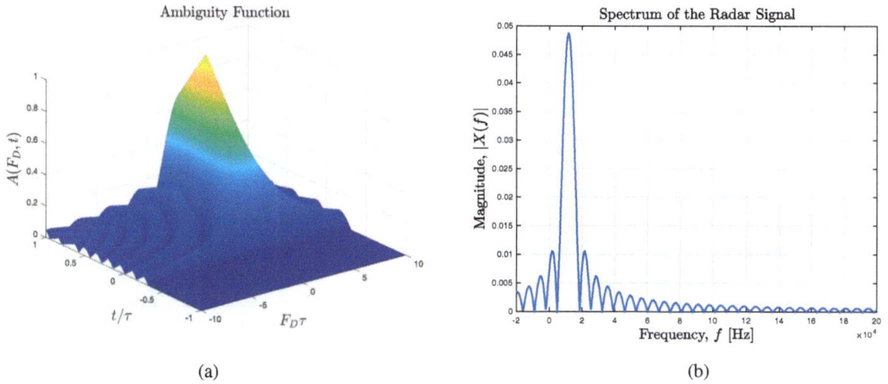

Fig. 18.7 Ambiguity function and spectrum, $aT = 5$. (**a**) Ambiguity function, $A(F_D, t) \times t/\tau \times F_D\tau$. (**b**) Spectrum

which shows compatible Doppler resolution with the simple sinusoid or pulse cases. The ambiguity function crosses zero at $f_\text{D} = \frac{1}{T}$ leading to a resolution in the velocity of $\frac{\lambda_\text{c}}{2T}$ m/s. A Doppler mismatch between the reflected received signal and the matched filter around this value leads to substantial peak reduction. Larger pulses lead to a better resolution in velocity. In fact, the plot of Fig. 18.6 consists of the cross correlation between a transmitted radar signal with the signal reflected in a target at 120 km/h at a range of 15 km and with negligible background noise. At the peak of the ambiguity function, the estimated velocity was 120.19 km/h, and the range estimation is 14953.24 m.

4. Figure 18.7 illustrates the case where $aT = 5$. For this case, the peak of the ambiguity function leads to a velocity of 120.19 km/h and a range of 14953.24 m as in the previous case, also with very-low-power background noise.
5. Figures 18.8 and 18.9 plot the ambiguity function for values of $M = 3, 6, 10$, respectively. In particular, Fig. 18.8 also shows the corresponding signal spectrum for $M = 3$. In this item, the plot of Figs. 18.8 and 18.9 represent the cross correlation between a transmitted radar signal with the signal reflected in a target at 120 km/h at a range of 15 km. A random noise was added to the reflected signal, explaining the bumpy behavior of the plots for an SNR of -5 dBs. For high SNR, the range and velocity estimates are pretty accurate for this example, irrespective of the value of M; these are 14953.24 m and 120.19 km/h, respectively. For SNR $= -5$ dBs, the range and velocity tend to vary with M, but the variations are random around the nominal values.

End

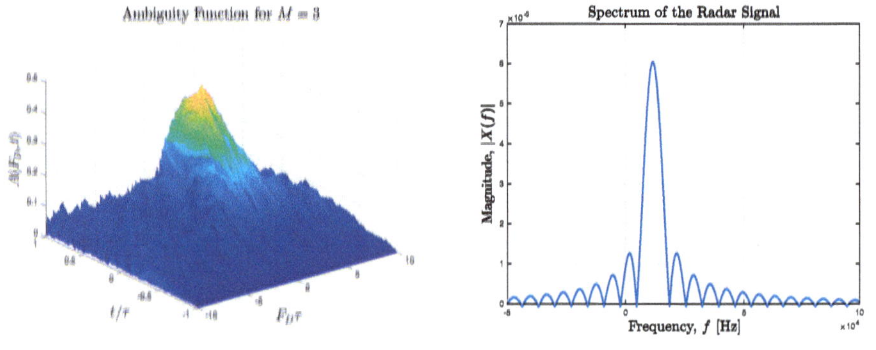

(a) (b)

Fig. 18.8 Ambiguity function and spectrum, including environment noise. (**a**) Ambiguity function $M = 3$, SNR $= -5$ dB, $A(F_D, t) \times t/\tau \times F_D\tau$. (**b**) Spectrum

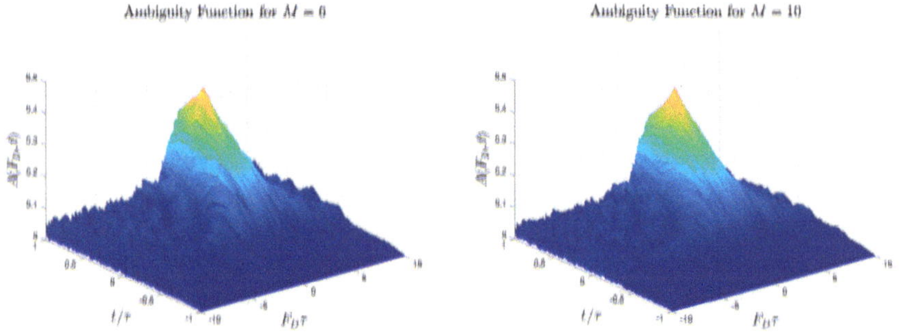

(a) (b)

Fig. 18.9 Ambiguity function for distinct values of M. (**a**) Ambiguity function $M = 6$, SNR $= -5$ dB, $A(F_D, t) \times t/\tau \times F_D\tau$. (**b**) Ambiguity function $M = 10$, SNR $= -5$ dB, $A(F_D, t) \times t/\tau \times F_D\tau$

18.5.1 Data Organization

Typically, the propagation delay, considering the round-trip and no relative velocity, is $\tau_0 = \frac{2d}{c}$. The Doppler shift is $f_\mathrm{D} = 2\frac{f_\mathrm{c}}{c}v$ where v represents here the relative radial velocity between the target and the radar, whereas c is the speed of light. The factor of 2 in the propagation delay and Doppler shift account for the round-trip condition. In the literature, it is common to define and use the relative Doppler shift as $\vartheta = \frac{f_\mathrm{D}}{f_\mathrm{c}} = 2\frac{v}{c}$.

As per (18.7), by taking into consideration the relative velocity, the actual return delay is $\tau = \frac{2(d+vt)}{c}$. By employing this relationship, the received signal related to the mth chirp is

$$
\begin{aligned}
r_m(t) &= g_m x(t - mT - \tau) + n_m(t) \\
&= g_m \mathrm{e}^{\mathrm{j}\left[2\pi\left(f_\mathrm{c}(t-mT-\tau)+\frac{1}{2}a(t-mT-\tau)^2\right)\right]} \\
&\times [u(t - mT - \tau) - u(t - mT - T - \tau)] \\
&+ n_m(t) \\
&= g_m \mathrm{e}^{\mathrm{j}\left[2\pi\left(f_\mathrm{c}\left(t-mT-\frac{2(d+v(t-mT))}{c}\right)+\frac{1}{2}a\left(t-mT-\frac{2(d+v(t-mT))}{c}\right)^2\right)\right]} \\
&[u(t - mT - \tau) - u(t - mT - T - \tau)] \\
&+ n_m(t) \\
&= g_m \mathrm{e}^{\mathrm{j}\left[2\pi\left(f_\mathrm{c}\left(t-mT-\frac{2d}{c}-\vartheta(t-mT)\right)\right)+\frac{1}{2}a2\pi\left(t-mT-\frac{2d}{c}-\frac{2v}{c}(t-\tau)\right)^2\right]} \\
&[u(t - mT - \tau) - u(t - mT - T - \tau)] \\
&+ n_m(t) \\
&= g_m \mathrm{e}^{\mathrm{j}[2\pi(f_\mathrm{c}(t-mT-\tau_0)-f_\mathrm{D}(t-mT))+a\pi(t-mT-\tau_0-\vartheta(t-mT))^2]} \\
&[u(t - mT - \tau) - u(t - mT - T - \tau)] \\
&+ n_m(t)
\end{aligned}
\tag{18.23}
$$

where $0 \leq t \leq T$ and g_m is a complex parameter representing the path loss, antenna gain, and radar cross section. The signal $n_m(t)$ accounts for the measurement noise. Note that for each chirp, the received signal might be analyzed separately to avoid the use of the time shift $t - mT$ in the expression, except when discussing the Doppler effect term. This model requires a quasi-stationary assumption where the instantaneous velocity, $\frac{\partial d(t)}{\partial t}$, of a target moving at a given time t at a range $d(t)$ is much smaller than c.[6]

Figure 18.10 shows the plotted frequency versus time where the transmitted chirps repeat every period T by starting from the carrier frequency f_c to the upper frequency $f_\mathrm{c} + BW$ where BW denotes the radar bandwidth. We observe the reflected chirp signal measured at the receiver antenna in the same plot. The bottom part of Fig. 18.10 shows the organization of the received data by assembling the tiles for visual purposes, where each tile represents a single sample in the baseband. The fast-time direction represents the L samples acquired at each chirp period, utilized

[6] For time-varying range, the expression describing the received signal is $r_m(t) = -g_m(1 - \frac{\partial q(t)}{\partial t})x(2q(t) - t)$, where $q(t) + \frac{d(q(t))}{c} = t$, and $q(t)$ represents the time the waveform should be sent to reach the moving target at t, at a range $d(t)$; see [https://radarsp.weebly.com/uploads/2/1/4/7/21471216/doppler_shift.pdf].

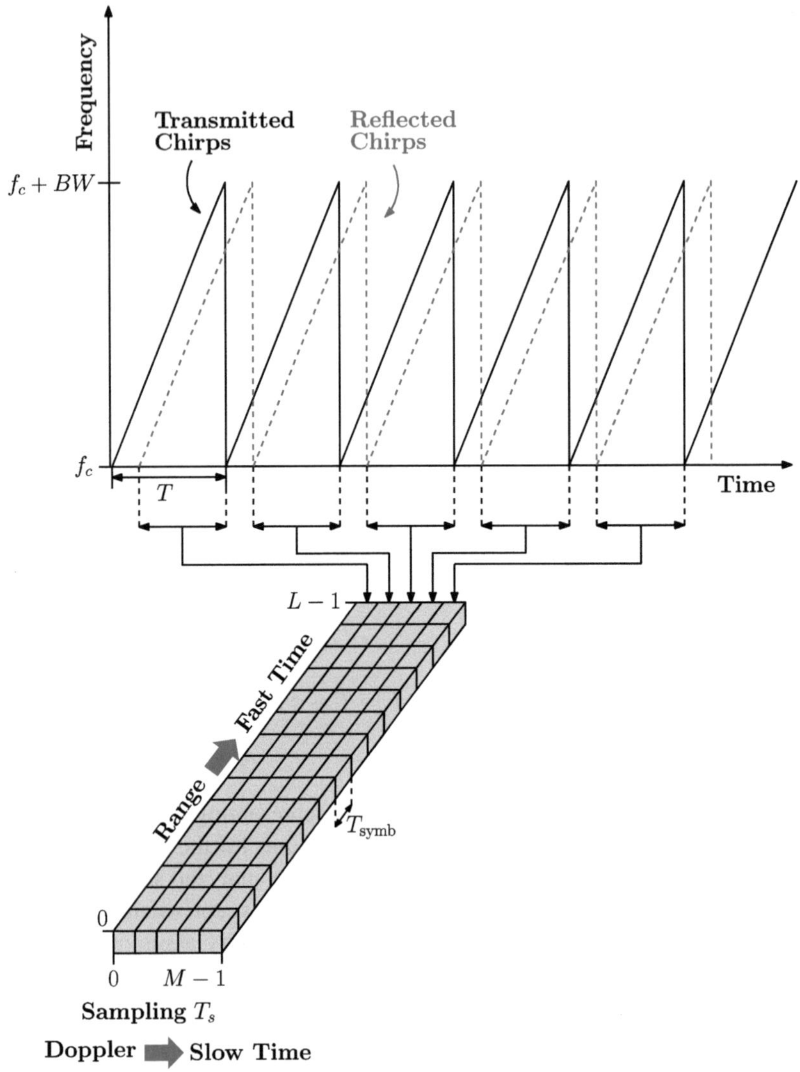

Fig. 18.10 Transmitted and received chirps and their sample organization method

for the range estimation.[7] The time frame related to one fast-time tile is $T_{\text{symb}} < \frac{1}{BW}$ to allow some margin over the Nyquist sampling rate. The slow-time direction corresponds to the samples acquired during one chirp period whose sampling period is T_{ch} in this representation, where T_{ch} is the chirp duration and $T_{\text{ch}} \leq T$. The slow-time axis' sampling (observation) period is $T_{\text{s}} = T$. The slow-time information

[7] The range resolution is roughly $\frac{c}{2BW}$ meters, as it will be detailed in (18.27).

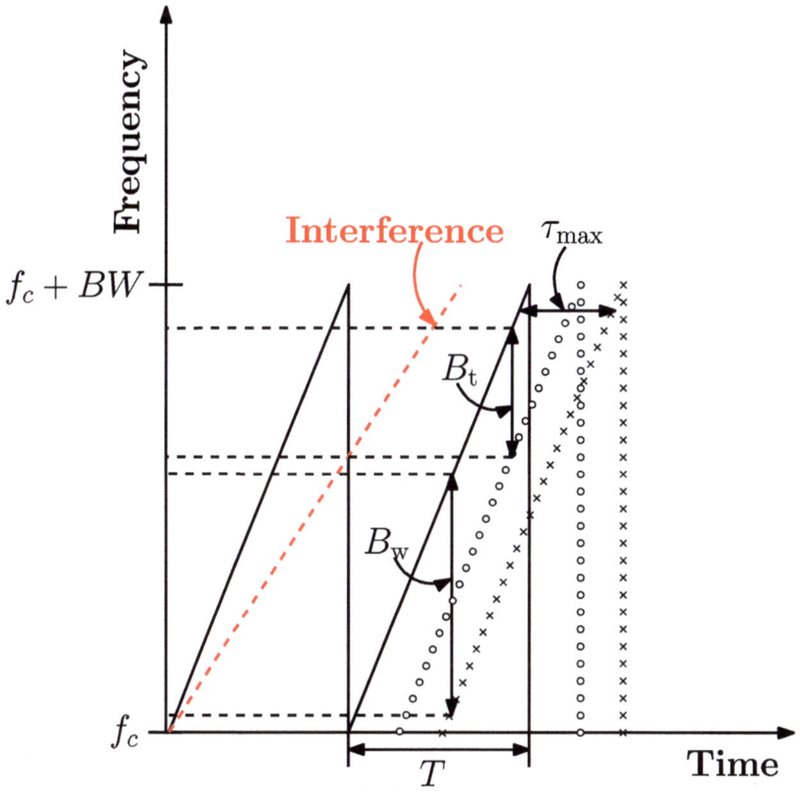

Fig. 18.11 Transmitted and received chirps and their sample organization method. The bandwidth B_t would be enough to detect the first reflection, whereas, for the detection of both reflections, the requirement is, at least, B_w

is used to estimate the Doppler shift and the related Doppler ambiguity. As such, assuming a given Doppler shift of f_D, the sampling frequency in the slow-time axis should be $f_{PRI} = \frac{1}{T} \geq 2f_D$, so that a frequency shift can be observed in the slow-time direction.

Figure 18.11 illustrates the frequency versus time plot of two chirps, including the effect of interference in the first chirp and the effects of two targets' reflections related to the second chirp to discuss the required sampling periods and bandwidth (Aydogdu et al. 2020). In the second chirp, the last and second reflections in the figure determine the minimum required bandwidth B_w to observe the two reflections. Besides, Fig. 18.11 indicates the maximum time delay between the reflections, denoted as $\tau_{\max}$, and the minimum bandwidth necessary to resolve the Doppler shifts, represented by B_w. For a given maximum delay $\tau_{\max} = \frac{B_w}{a} = \frac{B_w T}{BW}$, so that, as will be discussed, it determines the sampling period of the dechirped waveform and the minimum bandwidth of the analog-to-digital (A/D) converter. As such, B_w is called radar bandwidth of interest, which in turn is upper-bounded by the bandwidth of the utilized A/D converter.

18.5.2 Dechirped Signal

Each chirp reflection is mixed in the dechirping process by

$$
\begin{aligned}
\bar{x}_m(t) &= r_m(t)x^*(t - mT - \tau') \\
&= g_m x(t - mT - \tau)x^*(t - mT - \tau') + n_m(t)x^*(t - mT - \tau') \\
&= g_m e^{j\left[2\pi\left(f_c(t-mT-\tau)+\frac{1}{2}a(t-mT-\tau)^2\right)\right]} \\
&\times [u(t - mT - \tau) - u(t - mT - T - \tau)] \\
&\times e^{-j[2\pi(f_c(t-mT-\tau')+\frac{a}{2}(t-mT-\tau')^2)]} \\
&\times \left[u(t - mT - \tau') - u\left(t - mT - T - \tau'\right)\right] \\
&+ \bar{n}_m(t)
\end{aligned} \quad (18.24)
$$

where $0 < t-\tau \leq T$, and $\bar{n}_m(t) = n_m(t)x^*(t-\tau')$ represents the environment noise after dechirping. The model also ignores the possibility of a large phase increase during the processing interval of the M chirps (Richards 2022), related to the term mT in each chirp given by $e^{j2\pi(f_D\vartheta mT)}$, in (18.23).

By detailing the dechirping expression of (18.24), it follows that

$$
\begin{aligned}
\bar{x}_m(t) &= g_m e^{j[2\pi f_c(\tau'-\tau)]} e^{j\pi a[(t-mT-\tau)^2-(t-mT-\tau')^2]} \\
&\times [u(t - mT - \tau_1) - u(t - mT - T - \tau_2)] + \bar{n}_m(t) \\
&= g_m e^{j[2\pi f_c(\tau'-\tau)]} e^{j\pi a[(2(t-mT)(\tau'-\tau)+(\tau^2-(\tau')^2)]} \\
&\times [u(t - mT - \tau_1) - u(t - mT - T - \tau_2)] + \bar{n}_m(t)
\end{aligned} \quad (18.25)
$$

where $\tau_1 = \tau'$ and $\tau_2 = \tau$, if $\tau' > \tau$. Otherwise, $\tau_1 = \tau$ and $\tau_2 = \tau'$, if $\tau' < \tau$. We assumed that the matched filter delay was not aligned with receiving filter. If they are, that is, $\tau' = \tau$, we have

$$
\bar{x}_m(t) = g_m [u(t - mT - \tau) - u(t - mT - T - \tau)] + \bar{n}_m(t). \quad (18.26)
$$

Note that by integrating $\bar{x}_m(t)$ up to $t = \tau + T$, we have the best moment to detect the radar signal in the presence of noise.

A comment is worthwhile at this point, assuming we know the relative velocity or, considering it negligible, the expression $2\pi f_c(\tau' - \tau)$ might also reflect the distance between a reference target and the actual one so that

$$2\pi f_c(\tau' - \tau) = 2\pi f_c \frac{2(d' - d)}{c} = 2\pi f_c \frac{2(\Delta d)}{c} \approx \pi f_c \frac{1}{BW} \tag{18.27}$$

where d' represents the reference target distance; note that BW represents the total bandwidth required to estimate any target in the designated range. We will assume $d' = 0$ representing no target reference so that the target initial distance appears directly.

Suppose the target is moving with speed v; we know from (18.7) that from an estimate of reflection delay τ, it is possible to estimate an approximate value of the initial range d or Δd if we have a reference of the target, i.e., the relative target radial distance when the waveform hits it. In FMCW transmission, while the reflected signal faces a return delay τ, the transmitted signal increases its frequency by

$$a\tau = \frac{2a(d + vt)}{c}. \tag{18.28}$$

For simplification, let us consider a single chirp, i.e., $m = 0$; by observing the first exponent of (18.25) and the relation in (18.27), it is possible to infer the effect of the information delay in the dechirped signal, also known as the beat signal. The discussed frequency content sits at[8]

$$f_t = f_r + f_D = \frac{2BWd}{Tc} + \frac{2f_c}{c}v = \frac{2d}{c}a + \frac{2f_c}{c}v \tag{18.29}$$

with f_t representing the total dechirped signal frequency contributed by the range plus the Doppler frequencies, and $a = \frac{BW}{T}$ in this discussion. If one assumes, like in automotive applications, that $f_r \gg f_D$ by measuring f_t through the peak of the modulus of a proper DFT, applied to the sampled beat signal along the fast time, one can estimate the distance of a target by $d = \frac{Tcf_t}{2BW} \approx \frac{Tcf_r}{2BW}$. Also, after discretizing the dechirped signal, it remains to estimate the Doppler shift.

Estimating the Doppler shift accurately for small T and a single pulse $m = 0$ is difficult. Then, we need more flexible solutions for the distance and Doppler shift joint estimation. In the following, we provide more details of the case where $m > 0$.

The dechirping described here consists of a pulse-by-pulse-based, to be utilized in two-dimension form as depicted in Fig. 18.12. In reality, the matched filter should involve the full signal burst transmitted by the radar. However, this strategy is not computationally efficient. The best solution is to match the reflected signal with a single pulse and integrate the slow-time samples' resulting tiles after acquiring the single-pulse matched filter in the fast-time direction (Richards 2022).

If the peak of the matched filtering processing occurs within T seconds, at let us say τ, and there is a single target, slow-time integration using the dark tiles

[8] Note that we considered only the effect of the term $-a\tau t$, for small avt in (18.28), originating from the second exponent in (18.23) since the remaining terms lead to negligible contribution.

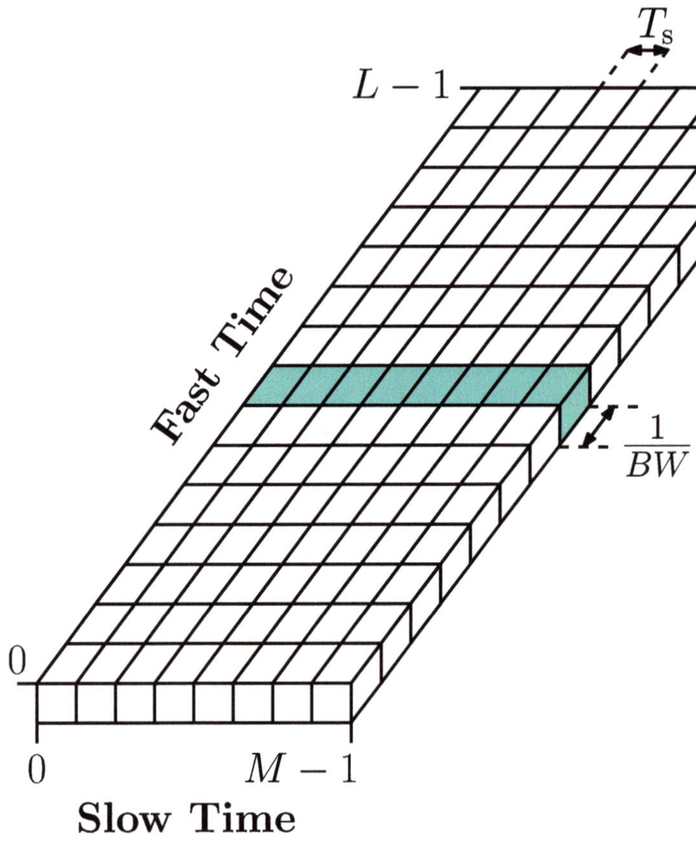

Fig. 18.12 Tile representation of the radar system where the shaded tiles should be added to represent the matched filter related to the full radar burst signal

results in Fig. 18.12 will lead to the maximum SNR. The reflection brings energy to the first tile in the slow-time direction, assuming no noise or other interference. However, if the reflections originate from a target more than T seconds away in terms of delay, at $\tau + T$, the received waveform remains the same except for a possibly reduced amplitude (Richards 2022); therefore, including the first tile to estimate the range might not offer the best SNR condition. Another interpretation is that the amplitude reduction occurs because the burst pulse waveform is being correlated with a reflection beyond T seconds and some delayed matching takes place. We integrate the tiles in the slow-time direction starting from the second tile, as illustrated in Fig. 18.13, avoiding the effect of a noninformative tile in the integration process. A similar reduction occurs when the matched filtering peak is missed for some reason, and if the SNR is high enough, one can mistake a target with a delay $\tau + T$ with one at τ, for instance.

In addition, assume we have a target at delay $\tau < T$; another target at a range between T and $2T$ seconds will generate ambiguous reflections between $2T$ and

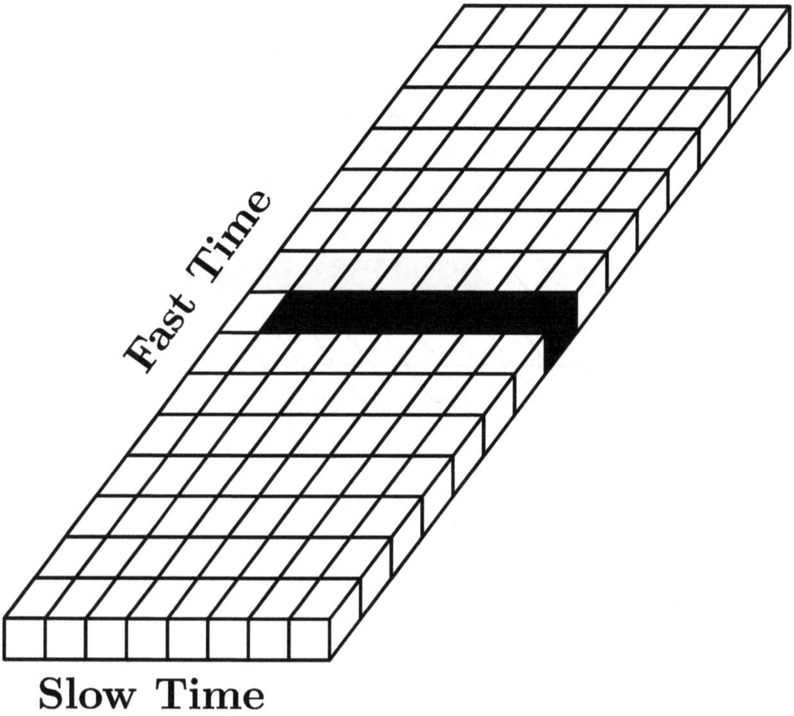

Fig. 18.13 Tile representation of the radar system with an unused tile. Then, we should add shaded tiles to represent the matched filter related to the full radar burst signal

$3T$, for instance. Indeed, the number of unambiguous ranges relates to the number of samples in the fast-time direction. If more of the target's reflections arrive with delays equal to the main target delay plus a multiple of T seconds, we will have a situation of multiple ambiguous ranges.

In the case there is more than one ambiguity, we can extend the transmitted waveform, including additional chirps corresponding to the range of ambiguous reflections. Figure 18.14 shows a tile representation example whose target reflection can be placed up to $3T$ seconds. This strategy allows the detection of targets whose reflection delays $\tau + iT$, for $i = 0, 1, 2$, since the matched filter will offer full SNR levels for detection. In these cases, the slow-time information includes targets and clutter contribution of all intervals of T seconds only after the transient.

18.5.3 Sampling Issues

We now discuss the sampling of the dechirped signal in order to fill the entries of the fast-time columns. By properly choosing the sampling period T_r, we can generate a

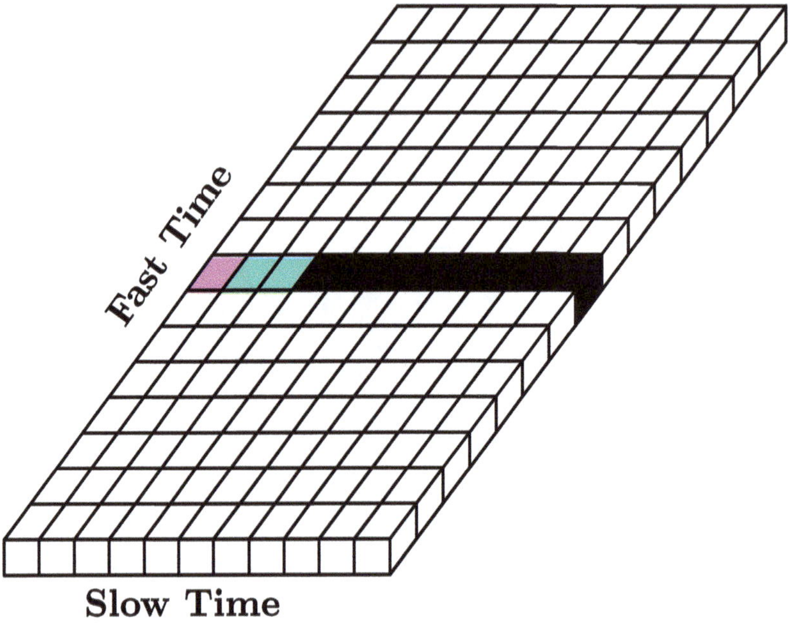

Fig. 18.14 Tile representation of the radar system with three unused tiles, where the colored tiles should be added to represent the matched filter related to the full radar burst signal

discrete-time version of the reflected signal as

$$
\begin{aligned}
\bar{x}_m(n) &= gx(nT_r - mT - \tau)x^*(nT_r - mT - \tau') + n_m(nT_r)x^*(nT_r - mT - \tau') \\
&= \mathrm{e}^{\mathrm{j}\left[2\pi\left(f_c(nT_r - mT - \tau) + \frac{1}{2}a(nT_r - mT - \tau)^2\right)\right]} \\
&\times \left[u(nT_r - mT - \tau) - u\left(nT_r - mT - T - \tau\right)\right] \\
&\times \mathrm{e}^{-\mathrm{j}\left[2\pi\left(f_c(nT_r - mT - \tau') + \frac{1}{2}a(nT_r - mT - \tau')^2\right)\right]} \\
&\times \left[u(nT_r - mT - \tau') - u\left(nT_r - mT - T - \tau'\right)\right] + \bar{n}_m(t) \\
&= \mathrm{e}^{\mathrm{j}\left[2\pi f_c(\tau' - \tau)\right]}\mathrm{e}^{\mathrm{j}\frac{1}{2}a\left[2(nT_r - mT)(\tau' - \tau) + (\tau^2 - (\tau')^2)\right]} \\
&\times \left[u(nT_r - mT - \tau_1) - u\left(nT_r - mT - T - \tau_2\right)\right] + \bar{n}_m(t)
\end{aligned}
\qquad (18.30)
$$

where $m = 0, \ldots, M - 1$, and τ' is a delay applied to the transmitted signal that could either represent a reference distance or an initial range estimate. It is worth mentioning that in some radar systems, the receiver is turned off during transmission to avoid damage to the reception circuitry originating from the relatively high power of the transmitter.

The sampling times of interest must start after the maximum reflection delay, which we assume can be accurately detected, ending at the chirp period T, that is, in the time frame $\tau_{\max} \leq t \leq T$. The value of the maximum range is

$$d_{\max} = \min(\frac{cT}{2}, \frac{cT}{2}\frac{1}{aT_r}) \tag{18.31}$$

where the first relates to the sweeping repetition interval and the last term reflects the sampling rate of the fast time, and it usually results in a small value. The value of $\tau_{\max}$ indicates the time interval for unambiguous target range detection.

The range of values of n can be prescribed to start at

$$n_0 = \lfloor\frac{\tau_{\min}}{T_r}\rfloor \tag{18.32}$$

and ends at

$$L - 1 + n_0 = N \tag{18.33}$$

where T_r was defined before as the sampling period, and $\tau_{\min}$ is a prescribed value related to the defined minimum range, and n_0 takes into consideration how many samples should be discarded before the chirp reflected signal reaches the radar. In the following, we denote the range n_0 to N as 0 to $L - 1$.[9]

The range swath represents the difference between the maximum range that can be detected unambiguously and the minimum range the radar is able to detect a target.

$$d_{\mathrm{w}} = d_{\max} - d_{\min} \tag{18.34}$$

where $d_{\min}$ leads to low beat frequency so that the noise would likely define the minimum range if it is too small. In autonomous vehicles, the prescribed value for the minimum range could be as small as one meter. On the other hand, the range in the forward and sideway directions should be around 300 and 70 meters, respectively. The minimum range is also related to the pulse duration T when the radar receiver has to be isolated from antennas to avoid circuitry damages (Richards 2022).

The sampling in the fast time should lie in the range

$$t_1 = \frac{2d_{\min}}{c} \to t_2 = \frac{2d_{\max}}{c} + T_{\mathrm{ch}} \tag{18.35}$$

[9] As previously mentioned, the range resolution is roughly $\Delta d = \frac{c}{2BW}$ meters. The value of N is upperbounded by $\lfloor\frac{T}{T_r}\rfloor$. If there is a range swath d_{w}, i.e., a strip of the surface from which data is collected, then the value of L is $L = \frac{d_{\mathrm{w}}}{\Delta d}$. This way, we will have enough bins in the fast time to observe the range swath. The value of d_{w} depends on the application at hand.

with the rates ranging from hundreds of KHz to a few tens of MHz (Richards 2022). Usually, the fast time ranges from tens to hundreds μs.

18.5.4 Quantities and Parameters of Interest

In the following, we describe some quantities and parameters related to the FMCW radar:

- The chirp period should be chosen such that

$$T >> |\frac{1}{2f_{\mathrm{D}}}| \tag{18.36}$$

  so that the Doppler shift can be properly observed.
- In the pulse repetition, the starting point of the $M-1$th pulse, for $M \geq 1$, the chirp received signal starts at

$$d_m = d_0 + v(m-1)T \tag{18.37}$$

  where the expression above considers the vehicle's movement along the previous $m-1$ pulse periods with constant velocity v.
- The corresponding frequency faced by the transmitting chirp will be

$$f_{\mathrm{shift}} = a\frac{2d_m}{c} + \frac{2v}{\lambda_c}. \tag{18.38}$$

  This is exactly the frequency shift[10] measured after dechirping.
- The maximum velocity denoted by $v_{\max}$ relates to chirp repetition as follows:

$$\frac{1}{T} \geq \frac{4v_{\max}}{\lambda_c} \tag{18.39}$$

  where the carrier wavelength is $\lambda_c = \frac{c}{f_c}$.[11]
- The range resolution Δd is $\frac{c}{2BW}$ meters, and the maximum range that unambiguous detected is

$$d_{\max} = \frac{cT}{2}. \tag{18.40}$$

---

[10] Frequently defined as beat frequency measuring the difference between the received and transmitted frequencies.

[11] The maximum Doppler frequency $f_{\mathrm{D}_{\max}} = \frac{2f_c v_{\max}}{c} = \frac{2v_{\max}}{\lambda_c}$, from (18.36), we can consider $T \leq |\frac{\lambda_c}{4v_{\max}}|$ satisfying the Nyquist condition, so that $v_{\max} \leq \pm\frac{\lambda_c}{4T}$.

- The sampling frequency at the fast time should be chosen such that

$$\frac{1}{T_r} \geq 2f_t = 2(a\frac{2d_{\max}}{c} + 2\frac{v_{\max}}{\lambda_c}). \tag{18.41}$$

  This choice warrants that the sampling frequency at the fast time is high enough to represent the dechirped signal.

Example 18.2 In automotive communications, the carrier frequency range lies between 76 and 81 GHz. Assume a bandwidth of $BW = 500$ MHz, a carrier of $f_c = 80$ GHz, a pulse duration of $T = 0.04$ ms, and a range of detection of $d = 300$ m. Assume the target is moving at $v = 300$ km/h. Consider that we need to cover a range swath of $d_w = 200$ m so that you can choose L. Choose $T_{ch} = 0.8$T. In this case, the Doppler effect originates a frequency shift of $f_D = f_c\frac{2v}{c} = 44444.44$, Hz.

Use a pulse burst FMCW with one and four pulses and estimate the range and the Doppler shift in both cases. Then, use a DFT of the fast-time axis to estimate the range where the peak identifies f_t; see (18.29). Also, estimate the Doppler shift from the data delay for $M = 2$ and by exploring the slow-time information for $M = 4$.

You can automatically detect a single target's delay by comparing the outcome of the matched filter with some prescribed threshold.

Solution Some quantities related to the given problem are listed below:

- For each 10 m of distance, the round-trip delay is $\frac{20}{c} = 66.666$ ns.
- In 300 m, the reflected signal will have an offset frequency with respect to the current transmitted signal by BW in T_{ch}s, and in 2 μs (corresponding to 300 m) the offset is $\frac{BW \times 2\mu s}{T_{ch}} = 31.25$ MHz.
- The frequency increases at a rate of $\frac{BW}{T_{ch}} = 15.6250$ kHz/ns.
- With the receiver sampling at T_r collects in T_{ch}, $\frac{T_{ch}}{T_r} = 288000$ samples. Using an FFT with 1024 points, the frequency resolution is $\frac{287800}{1024} = 281.250$ Hz.
- The chirp period should be chosen such that $T >> |\frac{1}{2f_D}| = \frac{1}{88888.88} = 1.124250^{-5}$.

We start by analyzing the given data in order to infer what we expect from the simulation. The expected reflection delay should be

$$\tau = \frac{2d}{c} = \frac{600.0}{3.0\,10^8} = 2\,\mu\text{s}$$

when not considering the relative velocity. The minimum range resolution is

$$\Delta d = \frac{c}{2BW}\frac{T}{T - \tau_{\max}} = \frac{3.0\,10^8}{10^9}1.1111 = 0.3333\,\text{m}$$

where the ratio $\frac{T}{T-\tau_{\max}}$ takes into consideration the minimum reflection delay, whereas the number of samples in the fast time axis should be

$$L = \frac{d_{\rm w}}{\Delta d} = \frac{600}{0.3333} \approx 1800.$$

According to Eqs. (18.32) and (18.33), we choose $L = 1800$, and $T_r = 1.111110^{-10}$ s. In this case $n_0 = \lfloor \frac{\tau_{\min}}{T_r} \rfloor = 120$, leading to a minimum $N = 1921$. The choice led to more samples than the minimum required.

The carrier wavelength is

$$\lambda_{\rm c} = \frac{c}{f_{\rm c}} = \frac{3.0\,10^8}{80.0\,10^9} = \frac{15}{4}\,10^{-3}\,\text{m} = 3.75\,\text{mm}$$

allowing us to compute the Doppler resolution for this experiment.

$$\begin{aligned}\Delta v &= \lambda_{\rm c} \frac{1}{2MT} \\ &= \frac{15}{4}\,10^{-3} \frac{1}{0.8\,10^{-4} M} \\ &= \frac{15}{2.4M}\,\text{m/s}\end{aligned}$$

where $T = 0.04$ ms. For $M = 1$, we have $\Delta v = 46.8426$ m/s $= 13.0118$ Km/h. The range and Doppler frequencies are, respectively, given by

$$\begin{aligned}f_{\rm r} &= \frac{2BWd}{Tc} = \frac{1.0\,10^9\,300.0}{40.0\,10^{-6}\,300.0\,10^6} = 25\,\text{MHz} \\ f_{\rm D} &= \frac{2f_{\rm c}}{c} v = \frac{160.0\,10^9}{300.0\,10^6} \frac{300.0\,10^3}{3600} = 44444.44\,\text{Hz}.\end{aligned}$$

Figure 18.15 shows the DFT of the fast-time vector and the resulting in dechirped signal values for distinct choices of M (not clearly distinctive). From these simulations, we estimate the range and the velocity of the target, whose results are listed below.

The distance (range) estimation by locating the maximum value in the dechirped signal matrix was:

- $M = 0 :$ 281.7277 and 303.3991 m
- $M = 1 :$ 303.3991, 303.3991 m.
- $M = 10 :$ 303.3991, 303.3991 m.

for SNR $= -5$ and 10 dBs, respectively.

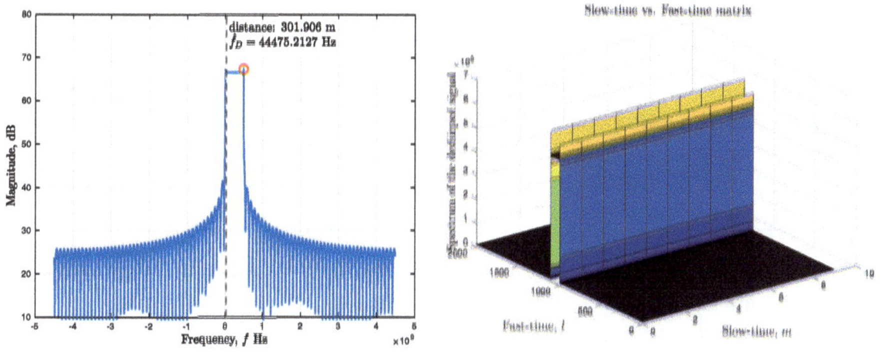

(a) (b)

Fig. 18.15 Fast-time and dechirped signal information. (**a**) DFT fast time, for $m = 10$. (**b**) Acquired dechirped signal

The velocity estimation is performed by examining the evolution of the range in the slow-time direction. The resulting estimates for SNR $= -5$ and 10 dBs, were:

- $M = 0$: 292.9695 and 299.3258 km/h.
- $M = 1$: 292.9695 and 298.7480 km/h
- $M = 10$: 298.1701 and 299.9037 km/h

respectively.

End

18.5.5 DFT Range-Doppler Estimation

It is possible to apply a two-dimension DFT to the signal $\bar{x}_m(l)$ of (18.30) as follows:

$$\bar{tf}_x(\bar{\tau}, \bar{\vartheta}) = \sum_{m=0}^{M-1} \sum_{l=0}^{L-1} \bar{x}_m(l) e^{j2\pi(a\bar{\tau}(n_0+l)T_r)} e^{-j2\pi(f_c\bar{\vartheta}mT)}. \tag{18.42}$$

This quantity represents a delay-Doppler (or range-Doppler) spectrum from where we can recover information related to the Doppler effect and receiver carrier delay.

For a single target if one computes $|\bar{tf}_x(\bar{\tau}, \bar{\vartheta})|$, its maximum value should be placed at $\bar{\tau} = \tau - \frac{f_c\vartheta}{a}$ and $\bar{\vartheta} = \vartheta$. This maximum value represents a processing gain with value $g = ML$, as can be inferred when the exponential product equals one in (18.42).

In practical situations, one must consider the interferer's signals to quantify their influence in the target detection, on top of the environment's thermal noise.

18.5.6 Interference Considerations

The interference of other radar users in the neighboring location of the user of interest affects the parameter's estimates and should be taken into consideration. Some observations related to interferences are following described Hourani et al. (2018), Aydogdu et al. (2020):

- The term $\frac{f_c \vartheta}{a}$ relates to the range-Doppler coupling.
- For $n_0 \ll L - 1 + n_0$, the velocity and range resolutions are $\Delta v = \frac{\lambda_c}{2MT}$, for $M > 0$, and $\Delta d = \frac{c}{2BW}$, where λ_c represents the carrier wavelength.
- Desired signal gain can be shown to be

$$|g|^2 = P_g \frac{\sigma_{rcs} \lambda_c^2}{(4\pi)^3 d^4} = P_g \frac{\sigma_{rcs} f_c^2}{(4\pi)^3 c^2 d^4} \tag{18.43}$$

  where σ_{rcs} is the area related to the radar cross section of the target, and d is the distance between the target and the radar; see equations (6)–(8) of Hourani et al. (2018) for details. P_g includes the transmitter and receiver antennas' gains and the transmission power (Balanis and Ioannides 2007).
- Interferer signal gain is estimated as

$$|g_{int}|^2 = P_g \frac{\lambda_c^2}{(4\pi)^2 r^2} \tag{18.44}$$

  where λ_c is the operation wavelength, and r is the distance between the victim and interferer radar. It is possible to have distinct wavelengths among the victim radar and the interferers.

If one examines Eqs. (18.43) and (18.44), it is likely that the interferer has more power than the reflected radar signal originating from the target. This phenomenon happens, for example, when r and d are comparable, and σ_{rcs} is a few squared meters. In case there are more interferers, the situation worsens when their contributions are added, reducing the signal-to-interference ratio. A more accurate interference model should take into consideration the level of coherence between the victim radar signal and each interference signal. The combined effect of the interferers can be represented by an average interference probability p_{int} giving rise to an average interference power $p_{int}|g_{int}|^2 P_{g_r}$ where P_{g_r} represents the radar processing gain. This interference power reduces the signal-to-noise ratio, decreasing correct radar signal detection ability.

The value of p_{int} can reach one when an interfering radar operates in the same carrier frequency f_c as the victim radar and the chirp period T, and the chirp slopes $a = \frac{BW}{T}$ are the same. By adequately choosing these parameters in distinct radars, it is possible to reduce p_{int} and the interference power. Whenever these parameters' choice turns the interfering radar incoherent, their influence is perceived as an increase in the environment noise level. This background noise increase might reduce the detection of targets with low radar cross sections, such as pedestrians, animals, and cyclists.

In actual road environments with several lanes and vehicles, the effectiveness of radar monitoring requires joint configuration of the parameters in the form of ad hoc network. Besides, a wireless communication network can coordinate the radar operation allowing more features and flexibility to enhance security.

A more thorough analysis, as per Aydogdu et al. (2020), shows that the signal-to-interference ratio can be approximated as

$$\text{SIR} = \frac{P_g |g|^2}{P_{g_r} p_{\text{int}} |g_{\text{int}}|^2} \tag{18.45}$$

where $P_g \geq P_{g_r}$, with $P_g = P_{g_r}$ in coherent transmission, and $P_{g_r} = 1$ for incoherent interference.

There are reactive methods to reduce interference. They act as some narrow-band or when short-lived interferences are active. However, more sophisticated approaches to mitigate interference are required for general cases.

As discussed in Aydogdu et al. (2020), asymptotically, the interference probability can be modeled as

$$p_{\text{int}} = \frac{a\tau_{\max} d_c}{BW} = \frac{a\tau_{\max} T_{\text{ch}}}{TBW} \tag{18.46}$$

in a network environment consisting of several vehicles. The parameter $d_c = \frac{T_{\text{ch}}}{T}$ is the radar duty cycle. This expression shows that reducing the value of a and increasing the chirp duration are ways to reduce the interference probability.

Another way to reduce p_{int} is to employ quasi-orthogonal waveforms. For a given chirp, one can measure the leakage caused by its replica being delayed by $\Delta\tau$ as follows Aydogdu et al. (2020):

$$\begin{aligned} l(\Delta\tau) &= \left| \frac{1}{T} \int_0^T x(t) x^*(t - \Delta\tau) dt \right|^2 \\ &= \left| \frac{1}{T} \int_0^T \mathrm{e}^{\mathrm{j}\left[2\pi\left(f_c t + \frac{1}{2} a(t)^2\right)\right]} \mathrm{e}^{-\mathrm{j}\left[2\pi\left(f_c(t-\Delta\tau) + \frac{1}{2} a(t-\Delta\tau)^2\right)\right]} dt \right|^2 \\ &\leq \frac{1}{\pi^2 BW^2 (\Delta\tau)^2}. \end{aligned} \tag{18.47}$$

Suppose we have K interfering signals with the same T and duty cycle so that two chirps delayed by $\Delta\tau$ generate a leakage as described below. By defining $\frac{T}{K} = \Delta\tau$, the requirement for achieving higher radar signal power than interference is

$$\begin{aligned}|g|^2 &= P_g \frac{\sigma_{\text{rcs}}\lambda_{\text{c}}^2}{(4\pi)^3 d^4} \geq P_g l(\Delta\tau)\frac{\lambda_{\text{c}}^2}{(4\pi)^2 r^2} \\ &= P_g \frac{K^2}{\pi^2 BW^2 T^2}\frac{\lambda_{\text{c}}^2}{(4\pi)^2 r^2}\end{aligned} \tag{18.48}$$

according to Eqs. (18.43), (18.44), and (18.47). With the expression above, and assuming r and d comparable, one can estimate how many intervals of $\Delta\tau$ fits the inequality requirement, that is,

$$K \leq TBW\frac{\sqrt{\sigma_{\text{rcs}}}}{r} \tag{18.49}$$

so that the overall interference is acceptable. The ratio $\frac{BW\Delta\tau}{T}$ represents the instantaneous frequency difference between chirps. In a way, the interest is knowing how many interferers are bearable in detecting the reflected signal. In this case, the set of waveforms is considered quasi-orthogonal.

Example 18.3 In automotive communications, assume a bandwidth of BW = 500 MHz, a carrier of f_{c} = 80 GHz, a pulse duration of T = 40 ms, and a range of detection of d = 300 m. Assume the target is moving at v = 150 km/h.

Plot (18.47) to locate the Doppler shift and range and determine how many radars can operate simultaneously and be considered quasi-orthogonal. Simulate the situation with the largest number of radars and report the results.

Solution Figure 18.16a shows the upper bounds predicted by (18.47), including the maximum leakage and the estimated leakage as a function of $\Delta\tau$. This analysis provides information related to the interference of other users to be taken into consideration in simulation tools for radar systems. Figure 18.16b depicts the leakage versus the target cross section and the ranges. The last plot illustrates the interrelations among the radar cross section, the range, and the maximum number of radars according to (18.49). These figures illustrate how the delay $\Delta\tau$ affects the leakage more drastically for small values, as expected. Similarly, combining a large cross section and short range increases the leakage, ultimately determining the number of radar signals that are bearable for proper detection of the desired target.

End

It is important to consider the power required to allow the target's detection in practical applications. The following expression defines the minimum required signal-to-noise ratio (SNR) so that the radar can detect a target placed at the maximum range (Xu et al. 2019).

$$d_{max} = \left(\frac{MTg_t g_r \lambda^2 \sigma_{\text{rcs}}}{(4\pi)^3 \kappa K N_{\text{F}} \text{SIR}_{\text{min}}}\right)^{\frac{1}{4}} \tag{18.50}$$

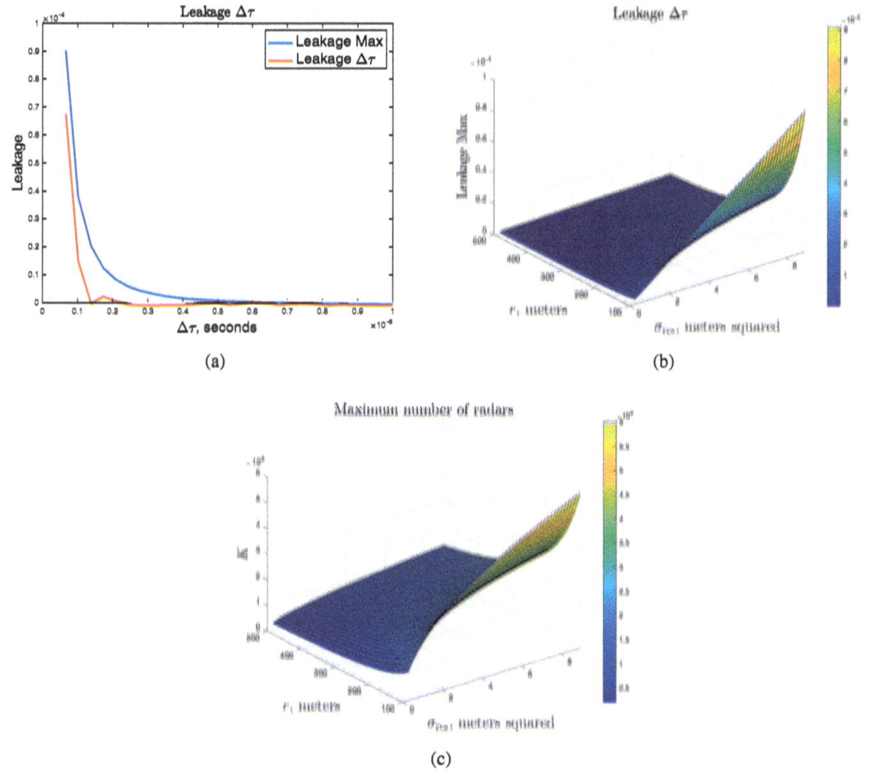

Fig. 18.16 Interference estimation in the multi-radar environment. (**a**) Leakage and its upper bound. (**b**) Leakage as function of target cross section and range. (**c**) Maximum number of radars as a function of target cross section and range

where $\kappa = 8.617333262 \times 10^{-5}$ eV/K in electronvolts per kelvin, K is the temperature in Kelvins, N_F is the noise figure at the receiver, and g_t and g_r are the antenna gains at the transmitter and receiver, respectively. The quantity SNR_{min} represents the minimum SNR. The N_F measures how much a device degrades the SNR; in our discussion, it is the ratio between SNR at the receiver input and its output.

18.5.7 Dechirping and Ambiguity Connection

The sampled version of the dechirped signal in (18.30) might lead to an inaccurate estimate of Doppler frequency shift. In this case, there is an ambiguous relation between τ related to the target delay (range) and the velocity (speed) v between the subjects, particularly for small values of T. A possible solution is returning to the dechirped signal in (18.24) and getting

$$
\begin{aligned}
\bar{x}_m(t) &= r_m(t)x^*(t - mT - \tau') \\
&= gx(t - mT - \tau)x^*(t - mT - \tau') + n_m(t)x^*(t - mT - \tau') \\
&= g\mathrm{e}^{\mathrm{j}\left[2\pi\left(f_c(t-mT-\tau)+\frac{1}{2}a(t-mT-\tau)^2\right)\right]} \\
&\times [u(t - mT - \tau) - u(t - (m+1)T - \tau)] \\
&\times \mathrm{e}^{-\mathrm{j}\left[2\pi\left(f_c(t-mT-\tau')+\frac{1}{2}a(t-mT-\tau')^2\right)\right]} \\
&\times \left[u(t - mT - \tau') - u\left(t - (m+1)T - \tau'\right)\right] \\
&+ \bar{n}_m(t).
\end{aligned}
\tag{18.51}
$$

As can be observed, $\bar{x}_m(t)$ relates to the argument of the matched-filter integral of (18.13), this time including the chirp pulse index m. The integral of $\bar{x}_m(t)$ with respect to t, spanning all chirps' periods in the range $m = 0, \ldots, M - 1$, gives rise to the matched filter output as a function of f_D and delay mismatch $\tau - \tau'$. The modulus of the resulting function is the ambiguity function.

The maximum of the modulus squared of the Fourier transform of $\bar{x}_m(t)$ produces an estimate of $f_c\vartheta - a\tau$, where a is the chirp slope. This Fourier transform of $\bar{x}_m(t)$ can be employed to estimate the velocity numerically by analyzing the angle of the transform, which should increase with the velocity (Aydogdu et al. 2020). This strategy works well for a single target.

This approach can be compared with the solution where we use the slow-time and fast-time analysis of the FMCW radar, where data acquired in the chirp period is stored in the columns of a matrix, whereas in the rows of the same matrix we store samples acquired along with several chirps. In the latter case, the maximum of the modulus of a Fourier function representing the Doppler spread leads to the speed information, and for sure, only a single target is related to a single beat frequency. In the slow-time and fast-time formulation, it is also possible to get several targets in a range and Doppler resolution cell (Aydogdu et al. 2020).

18.6 Combined Radar and Wireless Communication

18.6.1 Radar Coordination

A way to control the interference among radars operating in the vicinity is to coordinate their transmissions with the aid of wireless communication infrastructure. One possible configuration is to create a joint radar wireless communication (WC)[12] unit, where the WC can be either time-multiplexed with the radars or utilized to coordinate the operation of the radars to reduce their mutual interference by turning their transmission quasi-orthogonal.

The WCs in different vehicles operate in broadcast mode, informing the frequency-time resources and their starting chirp times. All the information related to the joint radar WC unit (JRWCU) on a given vehicle is broadcasted simultaneously in the WC channel. Each vehicle has its own JRWCU that might include several radar subunits. The choice of some parameters is to produce near orthogonality among the radars, where they either operate in distinct frequency ranges or start transmissions at different times in the same frequency range.

18.6.2 Joint Radar and Wireless Communication

A possible alternative is to perform joint radar and wireless communications employing multicarrier transmission, such as those used in current wireless[13] and Wi-Fi communications based on OFDM (Braun 2014; Lellough et al. 2015; Schweizer et al. 2018).

The OFDM signal itself, utilized as radar waveform, faces the difficulty of requiring much narrower bandwidth per carrier than the plain FMCW radar waveform, requiring narrowband A/D converter. This issue affects the range resolution of the OFDM radar system. A possible alternative is to use stepped-frequency OFDM, where the vehicle radars transmit at a set of orthogonal resources separated in the frequency domain, alleviating the burden on the A/D converter bandwidth. It basically consists of frequency hopping of each radar frame into narrower bands than the OFDM radar system. Figure 18.17 depicts a radio resource allocation in a time-frequency plane utilizing the stepped-frequency allocation where the same color indicated the frequency range used by a user in the respective time range.

One can also employ a narrowband OFDM system to share the available bandwidth among the radars without employing frequency-hopping, but this would certainly be less robust to jamming and eventual interference in a particular band. Indeed, the stepped-frequency OFDM brings about more accurate ranging estimates while reducing the Doppler shift estimates' accuracy and accommodating a larger number of vehicles within the same spectral resources. Figure 18.18 illustrates the allocation of radio resources in the narrowband OFDM on the time-frequency plane, where the same color indicates the frequency range used by a user throughout the time range. The wideband OFDM can attend even more vehicles as a function of

[12] By WC, we mean systems that implement cellular standards that can be used for vehicle-to-everything (V2X) or adequate IEEE 802.11 type of standards.

[13] Such as 4G and 5G systems.

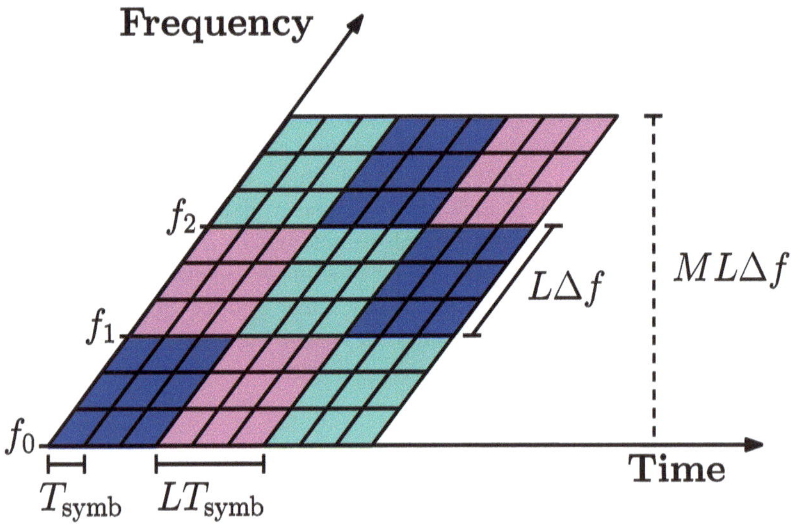

Fig. 18.17 Stepped-frequency OFDM time-frequency resource allocation

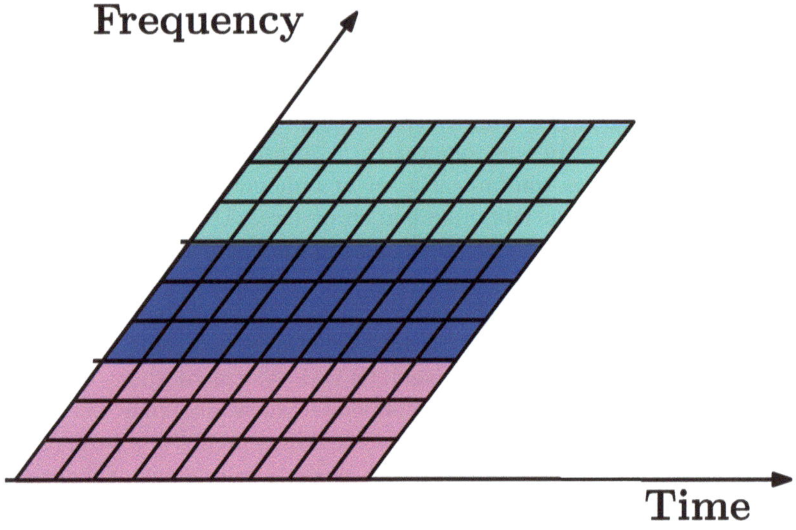

Fig. 18.18 Narrowband OFDM time-frequency resource allocation

subcarrier SNR, but the A/D converter requires a much higher sampling rate, and the velocity accuracy is limited, according to Aydogdu et al. (2020).

In typical OFDM configuration with appropriate cyclic prefix and small Doppler shift assumption, the received symbol at the mth subcarrier of the lth symbol of the

nth OFDM frame can be described as

$$\bar{x}_{m,l,n} = g\, d(m,l,n)\mathrm{e}^{-\mathrm{j}[2\pi(f_m+n\Delta f)]\tau_0}\mathrm{e}^{\mathrm{j}[2\pi(f_0(mL+l+1))T_{\mathrm{sym}}\vartheta]} + \bar{n}(m,l,n) \tag{18.52}$$

where $d(m,l,n)$ is the complex-valued transmitted data. By appropriately selecting the set of time-frequency tiles, we can separate the information from different vehicles, for example.

Example 18.4 In automotive communications, assume a bandwidth of $BW = 600\,\text{MHz}$, a carrier of $f_c = 80\,\text{GHz}$, a pulse duration of $T = 40\,\mu\text{s}$, and a detection range of $d = 300$ m. Assume that the target is moving at $v = 60\,\text{km/h}$. Consider that we need to cover a range swath of 600 m by properly choosing L. Also, use $M = 4$.

Divide the bandwidth among three vehicles, each using its own OFDM waveform to estimate its relative distances and Doppler shifts. Plot the normalized dechirped information using the range-velocity profile and discuss the results. Use SNR of 50 dB and 10 dB, respectively.

Solution We considered $\tau_{\min} \ll T_r$ and may choose $L = 801 > \frac{d_w}{\Delta d} = 800.5538$. The bandwidth allowed to each user is 200 MHz since there are three vehicles sharing the environment.

The related quantities are $f_{D_{\max}} \leq \frac{1}{2T} = 1250$ Hz; $f_D = \frac{2f_c v}{c} = 8895\,\text{Hz}$; $v_{\max} = 234213$; $T = 40 \times 10^{-6}$.

The 3D plot of the output signal of the matched filter as a function of time and frequency, depicted in Fig. 18.19, shows a peak at a delay estimate corresponding to the range of 299.200 m of 374 seconds. In contrast, in the frequency axis, it corresponds to a Doppler frequency of 9030 Hz of $v = (cf_D)/(2f_c)$, which is equivalent to a velocity of 60.1903 km/h. These results refer to the case of SNR $=$ 50 dB. For SNR $= 10$ dB, we get 8982 Hz of Doppler estimation corresponding to 60.5866, with a noisier 3D plot as depicted in Fig. 18.20.

End

18.7 MIMO Radar

The classical radar systems utilized in vehicles utilize FMCW waveforms received in a single antenna. The radar system has to estimate the range and the target velocity with the acquired information. The idea of MIMO radar lies in the wish to determine the angle of direction of the target. The classical requirement is to have a set of transmitting and receiving antennas whose signal composition forms an equivalent of a uniform linear array. The transmit antennas are configured to send orthogonal FMCW waveforms so that the receive antennas can identify the signals from each antenna participating in the transmission. For that, the MIMO radar system might

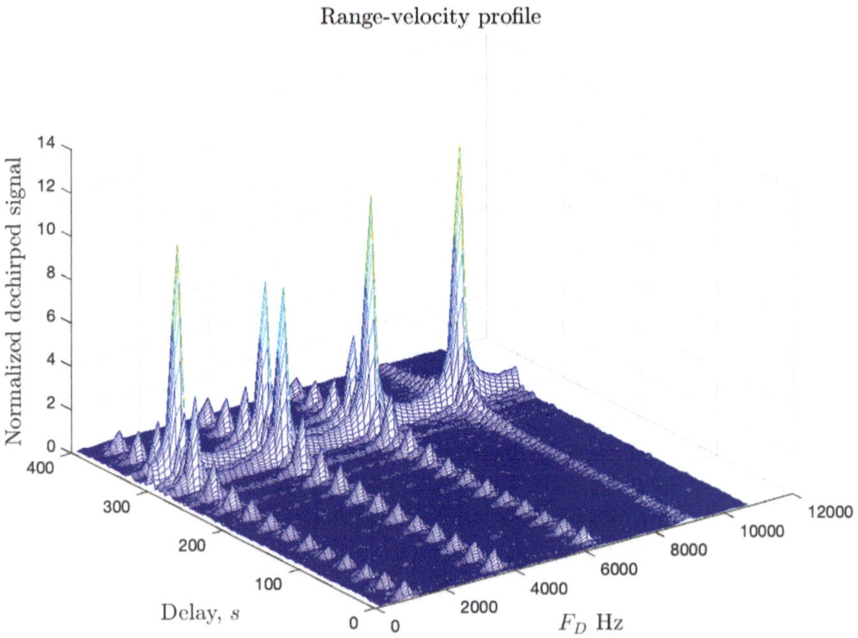

Fig. 18.19 Range and velocity estimation using time and frequency analysis, SNR $= 50\,\text{dB}$

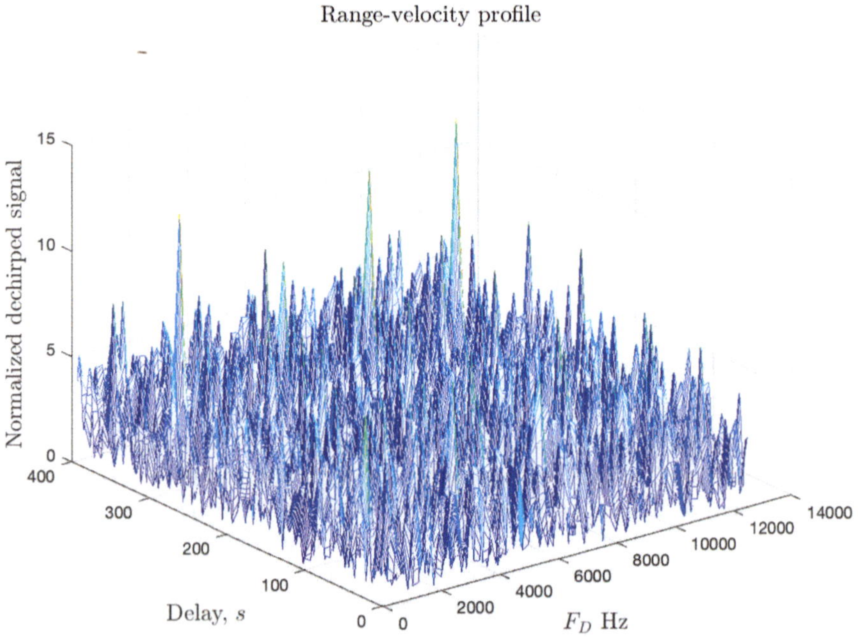

Fig. 18.20 Range and velocity estimation using time and frequency analysis, SNR $= 10\,\text{dB}$

utilize the concept of virtual array consisting of M_{t} transmit antennas and M_{r} receive antennas.

18.7.1 Virtual Array

Radar systems are very efficient in providing good estimates of Doppler effects and range. There is a need to increase useful angular information with an array of sensors with high aperture lengths. It is then possible to get increased angular resolution by utilizing a MIMO system that implements many virtual antenna elements. The way to construct these virtual antenna arrays in the radar context follows the works of Zoeke and Ziroff (2015) and Wang (2012). In addition, other virtual antenna array methods employ relays more akin to wireless communication setup (Dohler 2003). A major issue in virtual antennas is the assumption that the channel is stationary during the measurement time of a cycle of the array configuration, representing the MIMO switching strategy. This cycle corresponds to the overall channel switching, which affects the angle resolution due to the target movement.

Let us assume the transmit antennas form a linear separated by distance denoted as d_{t}, whereas the receive antennas form the same pattern whose elements are d_{r} apart.

The phase progression vector for a target in the far field is

$$\mathbf{a}_{\mathrm{t}} = \begin{bmatrix} e^{j\phi_{1,\mathrm{t}}} \\ e^{j\phi_{2,\mathrm{t}}} \\ \vdots \\ e^{j\phi_{N_{\mathrm{t}}-1,\mathrm{t}}} \\ e^{j\phi_{N_{\mathrm{t}},\mathrm{t}}} \end{bmatrix} \tag{18.53}$$

with

$$\phi_{m,\mathrm{t}} = \frac{2\pi}{\lambda_{\mathrm{c}}}(m - 1)d_{\mathrm{t}} \sin\theta_{\mathrm{t}} \tag{18.54}$$

where θ_{t} is the transmit angle to a target in the far field.

The incident wavefront at the receiving antennas is represented by

$$\mathbf{a}_{\mathrm{r}} = \begin{bmatrix} e^{j\phi_{1,\mathrm{r}}} \\ e^{j\phi_{2,\mathrm{r}}} \\ \vdots \\ e^{j\phi_{N_{\mathrm{r}}-1,\mathrm{r}}} \\ e^{j\phi_{N_{\mathrm{r}},\mathrm{r}}} \end{bmatrix} \tag{18.55}$$

with

$$\phi_{n,\mathrm{r}} = \frac{2\pi}{\lambda_\mathrm{c}}(n-1)d_\mathrm{r}\sin\theta_\mathrm{r} \tag{18.56}$$

where θ_r is the receiving angle of a target in the far field. One should bear in mind that the entries of the vector above should have different gains since each antenna receives a reflected signal. However, we assume the gains are very close and equal to one for simplicity.

It is possible to synthesize an equivalent virtual uniform linear array (ULA) with $M_\mathrm{t}M_\mathrm{r}$ elements. Indeed, the ULA will have as characteristic an equivalent array described by

$$\begin{aligned}\mathbf{A} &= \mathbf{a}_\mathrm{r} \otimes \mathbf{a}_\mathrm{t} \\ &= \begin{bmatrix} e^{j(\phi_{1,\mathrm{r}}+\phi_{1,\mathrm{t}})} & \dots & e^{j(\phi_{1,\mathrm{r}}+\phi_{M_\mathrm{t},\mathrm{t}})} \\ \vdots & \ddots & \vdots \\ e^{j(\phi_{M_\mathrm{r},\mathrm{r}}+\phi_{1,\mathrm{t}})} & \dots & e^{j(\phi_{M_\mathrm{r},\mathrm{r}}+\phi_{M_\mathrm{t},\mathrm{t}})} \end{bmatrix}\end{aligned} \tag{18.57}$$

where $\otimes$ represents the Kronecker product defined as the multiplication of the vectors entry by entry.

When the antennas' separation is substantially smaller than the target range, the receiving and transmitting angles are close, that is, $\theta_\mathrm{r} \approx \theta_\mathrm{t}$, then it is possible to form virtual arrays through orthogonal transmissions. In this case, the exponents are

$$\phi_{n_\mathrm{r},\mathrm{r}} + \phi_{n_\mathrm{t},\mathrm{t}} = \frac{2\pi}{\lambda_\mathrm{c}}[(n-1)d_\mathrm{r} + (m-1)d_\mathrm{t}]\sin\theta_\mathrm{t} \tag{18.58}$$

for n_r and n_t range from 1 to M_r and 1 to M_t, respectively.

In the case the target is moving, the phase added by the channel when the transmitted and reflected signal reaches back the receiver is

$$\begin{aligned}\phi_\mathrm{ch}(\Delta t) &= 2\pi f_\mathrm{c}\tau + \frac{4\pi v\Delta t}{\lambda_\mathrm{c}} \\ &= \frac{4\pi d}{\lambda_\mathrm{c}} + \frac{4\pi v\Delta t}{\lambda_\mathrm{c}}\end{aligned} \tag{18.59}$$

where $\tau_0 = \frac{2d}{c}$, $\lambda_\mathrm{c} = \frac{c}{f_\mathrm{c}}$, v is the radial velocity, and d is the initial target distance. Then, the channel matrix of (18.57) becomes

$$\mathcal{A}(\Delta t) = \mathbf{a}_\mathrm{r} \otimes \mathbf{a}_\mathrm{t} e^{j\phi_\mathrm{ch}(\Delta t)} \tag{18.60}$$

affecting all the entries of the original matrix.

By recalling that the maximum velocity that can be unambiguously detected in FMCW radars is

$$v_{\max} = \frac{\lambda_c}{4T}, \tag{18.61}$$

it is possible to observe the role of the period T of a single chirp in this velocity. For a target moving at $v_{\max}$ and the time-division multiplex with period exactly $\Delta t = T$, from the second term of (18.59), the added phase becomes π. That means, adding π to $\phi_{n,r}$ is equivalent to add a phase shift of $\frac{\pi}{2}$ in θ_r of (18.56) when $d_r = \frac{\lambda_c}{2}$.

Generalizing (18.59) for any mth switched transmission of non neighboring antennas, it follows that

$$\begin{aligned}\phi_{\text{ch}}(t) &= \frac{4\pi r}{\lambda_c} + (m-1)\frac{4\pi v \Delta t}{\lambda_c} \\ &= \frac{4\pi r}{\lambda_c} + (m-1)\frac{\pi v}{v_{\max}}.\end{aligned} \tag{18.62}$$

The second term in the last equation should be compensated for every target in the environment in an individual way. It can also be noted that if v is an even multiple of $v_{\max}$, the phase shift is not changed from chirp to chirp as long as the target is located in a single radar range (Zoeke and Ziroff 2015).

These basic discussions allow the introduction of virtual arrays.

18.7.2 Virtual Array with TDM

There are several strategies to promote orthogonality among the signals transmitted on radar systems (Sun et al. 2020). The first classical method is the time-division multiplex (TDM), where each antenna has a transmission time slot, whereas the others remain inactive. The active antennas transmit an FMCW chirp. At the receiver processing, each antenna output computes an FFT in the range domain of length L for each chirp, while the FFTs of M chirps compose N_t matrices representing each transmitted chirp through the radar antennas.

Assume for the time being that the target is not moving with respect to the radar system, and the transmitting antennas are time-multiplexed. There are some ways to form a virtual uniform linear array, as depicted in Fig. 18.21. In the first case, the transmitting antennas are separated by $M_r d_r$, whereas the M_r receiving antennas are mutually separated by d_r. As can be observed, each transmitting antenna in TDM mode illuminates the entire receiving array in a different block time range leading to a virtual ULA whose elements are separated by d_r. For instance, since $d_t = M_r d_r$ and $d_r = \frac{\lambda_c}{2}$, the resulting virtual ULA has element separation equal to $\frac{\lambda_c}{2}$.

In the second case, the two transmitting antennas are separated by d_t, whereas the M_r receiving antennas are mutually separated by $2d_t$. Similarly, each transmitting

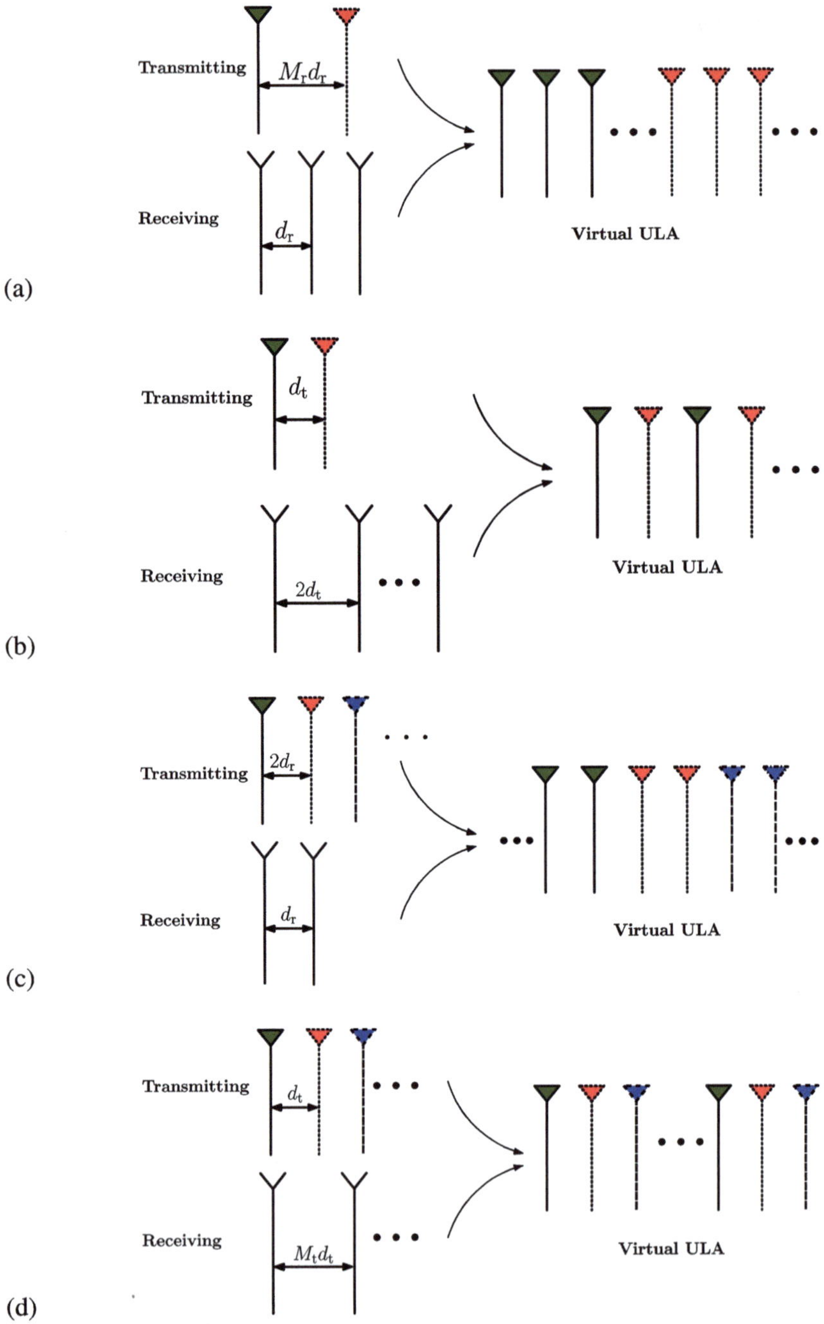

Fig. 18.21 Virtual arrays: (**a**) stacked $2 \times M_r$, (**b**) interleaved $2 \times M_r$, (**c**) stacked multi-transmitter $M_t \times 2$, and (**d**) mixed $M_t \times M_r$

antenna in TDM mode illuminates the entire receiving array in an alternating fashion, leading to a virtual ULA. Their elements are separated by d_t. For instance, if $d_t = \frac{\lambda_c}{2}$ and $d_r = \lambda_c$, that is, $d_t = \frac{d_r}{2}$, the resulting virtual ULA has element separation equal to $\frac{\lambda_c}{2}$.

In the third case, the M_t transmitting antennas are separated by $2d_r$, whereas the two receiving antennas are mutually separated by d_r, giving rise to a virtual ULA with the same separation of the elements equal to d_r.

In the fourth case, M_t transmitting antennas are separated by d_t, whereas the M_r receiving antennas are mutually separated by $M_t d_t$, giving rise to a virtual ULA with the same separation of the elements of d_t.

In all virtual array cases discussed, any nonstationary behavior of the target will destroy the ULA emulation. As a result, we require a strategy to reverse the effect of the phase distortion $\phi_{ch}(\Delta t)$ in (18.60). The trivial solution is to reduce Δt, which is not always viable since it entails reducing the frequency increase period as discussed in Zoeke and Ziroff (2015). The resulting effect is an increase in the demand for A/D converters, raising the costs.

Example 18.5 Investigate the phase migration due to target movement and array alignment discussed in Zoeke and Ziroff (2015) and reproduce their example here. The motivation is to discuss the effect of the target movement on distinct antenna arrangements. The nonstationarity of the antennas at the target will result in a departure from the linear array behavior.

According to the example in Zoeke and Ziroff (2015), the target starts 10 m from the source with 0 m of azimuth. The examples consist of the transmit antennas moving at $-0, 5v_{max}$ in the first experiment and the receiving antennas moving at $-v_{max}$, respectively.

Solution Let us start with the stationary case. Figure 18.22 shows the expected result of the virtual array in a stationarity environment for comparison with the forthcoming illustrations.

Figure 18.23 illustrates the effects of nonstationarity of the environment on the performance of the virtual array for moving transmit antennas. Unless some preprocessing takes place, the results might be unsatisfactory. Figure 18.24 illustrates the effects of nonstationarity of the environment in the performance of the virtual array for moving receiving antennas. As before, some preprocessing is required to recover the uniform array behavior.

End

A possible way to reduce the phase migration is to reduce the chirp ramp, which would increase the A/D converter's costs. Also, after the phase correction, some procedures are required to mitigate possible angular measurement errors and distortions originating from the relative temporal alignment of virtual elements (Zoeke and Ziroff 2015).

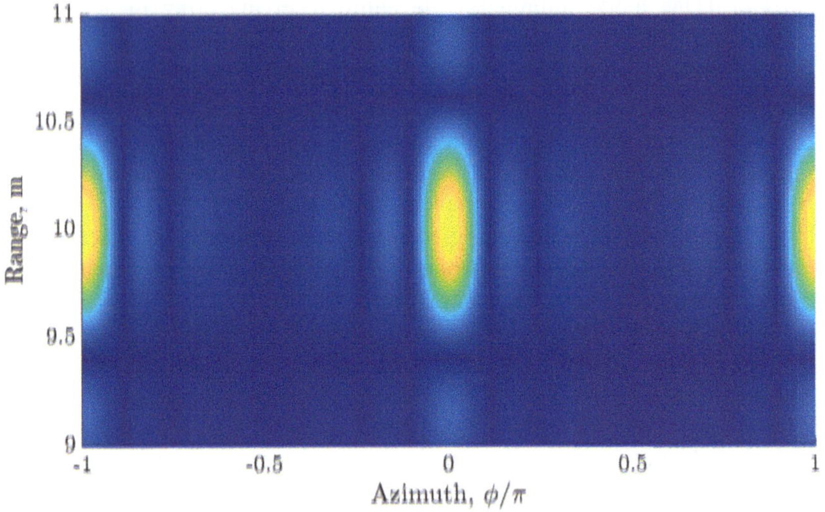

Fig. 18.22 Range × azimuth plot of the virtual array; stationary environment, interleaved array

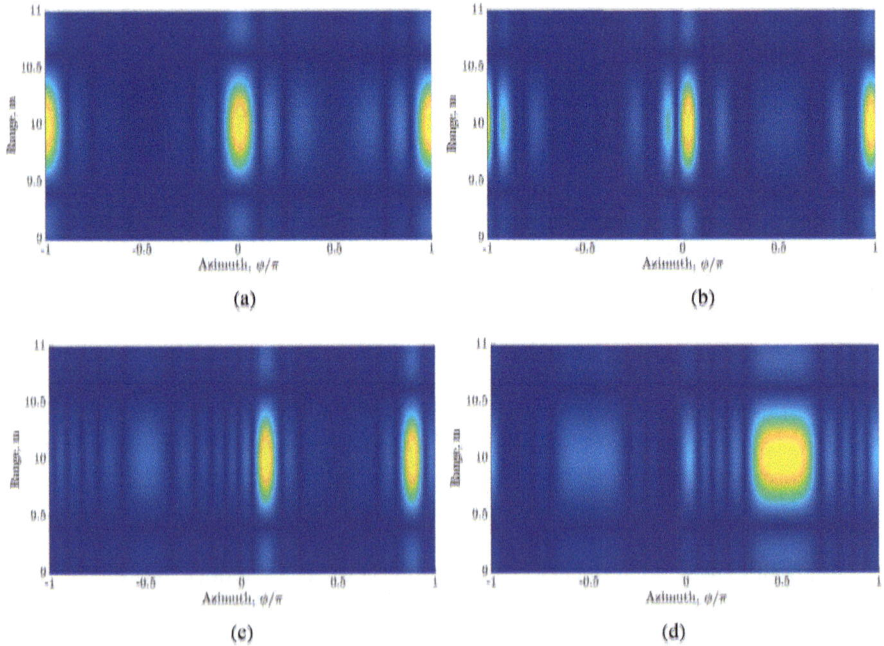

Fig. 18.23 Effect of phase errors in a virtual array, transmitter movement. (**a**) Interleaved configuration. (**b**) Stacked configuration. (**c**) Stacked multi-transmitter. (**d**) Mixed array

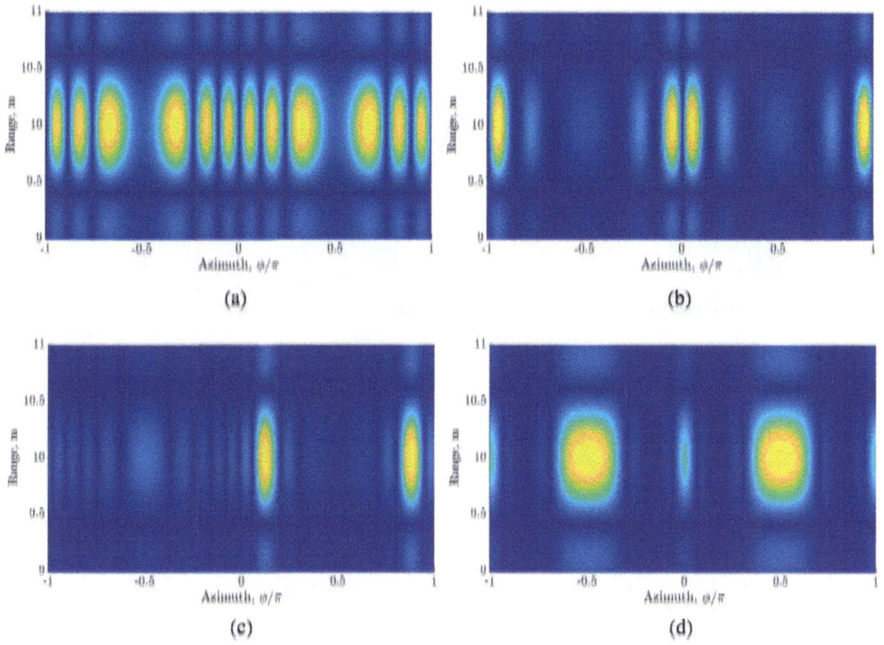

Fig. 18.24 Effect of phase errors in a virtual array, receiver movement. (**a**) Interleaved configuration. (**b**) Stacked configuration. (**c**) Stacked multi-transmitter. (**d**) Mixed array

18.7.3 TDM Illustration

Assume we have a TDM MIMO radar where there are two transmit antennas, each scheduled to communicate at a different time slot, as depicted in Fig. 18.25. At each receiver antenna, the so-called chirp range FFT of length L is applied to each individual chirp. The FFT's outputs are collected in two sets of tiles, as illustrated at the bottom of Fig. 18.25. The chirps sent by distinct transmitter antennas can be interpreted as two subarrays forming a virtual array at the receiver. Depending on the configuration, this example might be a stacked $2 \times M_\mathrm{r}$ or an interleaved $2 \times M_\mathrm{r}$ virtual array.

Assuming the transmitter antennas are switched to transmit a chirp T seconds apart, it is possible to observe a phase variation originated from a target velocity v between two consecutive chirps as

$$\Delta\phi = 2\pi f_\mathrm{c} T \vartheta = 4\pi f_\mathrm{c} T \frac{v}{c} = 4\pi T \frac{v}{\lambda_\mathrm{c}} = 2\pi T f_\mathrm{D}. \tag{18.63}$$

The expression above indicates how much phase correction should be applied to the signal collected by the virtual array to reduce the beam pattern distortion caused by the target movements. If there are many targets, phase compensation is required for

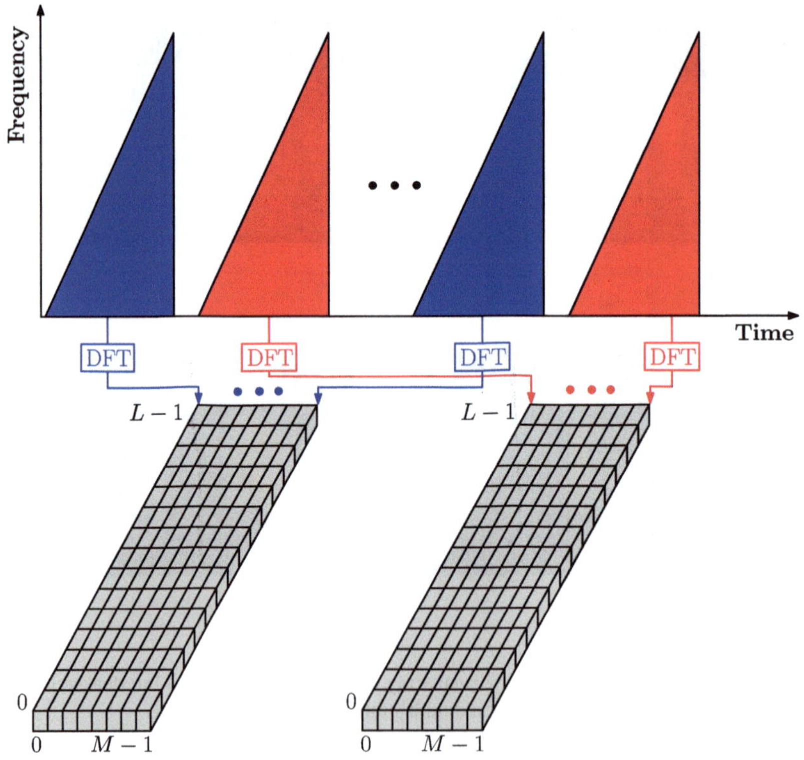

Fig. 18.25 TDM MIMO radar configuration with two transmitter antennas using FMCW

each target. In all cases, the respective velocities have to be previously estimated through the data corresponding to one receiving antenna, represented by the 2D FFT or some form combination of the 2D FFTs of the same subarray; see Zoeke and Ziroff (2015) and Sun et al. (2020).

Another way to enforce orthogonality is to transmit the chirps at distinct carrier frequencies, where each FMCW is transmitted through its separate carrier. The carrier frequency offset has to be larger than the antialiasing bandpass filter related to the ability to detect the Doppler offset and the range unambiguously.

Yet another way to achieve orthogonality is to perform the so-called Doppler division multiplex (DDM) where for M transmitted chirps, each repeated every T seconds, the transmit antennas send the same signal, but each antenna multiplies it by a complex function defined as

$$g_m(n) = \mathrm{e}^{-\mathrm{j}[2\pi(\phi_m(n))]} \tag{18.64}$$

where $\phi_m(n)$ is a phase sequence that should be properly chosen to promote orthogonality. Typical forms of the phase sequence are

$$\phi_m(n) = \phi_0 n \tag{18.65}$$

or

$$\phi_m(n) = \mathrm{j}\left(\frac{\pi}{M_p}\right) m(n+1)(n) \tag{18.66}$$

for $n = 1, \ldots, M_p$ and $m = 1, \ldots, M_t$ and for (18.66) M_p should be a prime number, and M_t represents the number of transmitting antennas. The sequence

$$g_m(n) = \mathrm{e}^{\mathrm{j}\left(\frac{\pi}{M_p}\right)m(n+1)(n)} \tag{18.67}$$

forms the so-called Zadoff-Chu sequence (Chu 1972).

One should bear in mind that when a MIMO radar uses isotropic antennas and orthogonal transmissions, the array beam pattern becomes omnidirectional. As such, its array gain is not as good as the case of phased-array radar beam formers at the transmitter, but this drawback is acceptable in automotive radar.

Example 18.6 In automotive communications, assume a bandwidth of $BW = 500\,\text{MHz}$, a carrier of $f_c = 80\,\text{GHz}$, a pulse duration of $T = 40\,\mu\text{s}$, and a range of detection of $d = 30$ m. Assume the target is moving at $v = 50\,\text{km/h}$ in the radial direction and the angle of movement is 15.0000°. We have two transmitter antennas and two receiving antennas. Estimate the range and the velocity.

Solution Each antenna transmits the chirp waveform in a time-division duplex. Figure 18.26a shows the chirp signal applied to each antenna in a TDD fashion, along with the ambiguity function and the receiver signal spectrum. The corresponding ambiguity functions at the antennas are depicted in Fig. 18.26b, c.

In the first attempt, the range estimation using DFT is performed for each transmission antenna by searching the maximum value of the information matrix. We observed this maximum value for different values of the slow-time axis, denoted here as M, as listed below:

- For $M = 4$, the estimated ranges at antennas 1 and 2 are 30.3297 m and 30.2586 m, respectively. The estimated angle of arrival is 13.9872°, whereas the estimated velocity is 0.0000 km/h. The estimated distance projected into the vehicle direction is 30.2942 m. Note that it is not possible to obtain a reasonable estimate of the velocity in this case.
- For $M = 64$, the estimated ranges at antennas 1 and 2 are 30.3653 m and 30.2942 m, respectively. The estimated angle of arrival is 13.9869°, whereas the estimated velocity is 48.5756 km/h. The estimated distance projected into the vehicle direction is 29.4650 m.

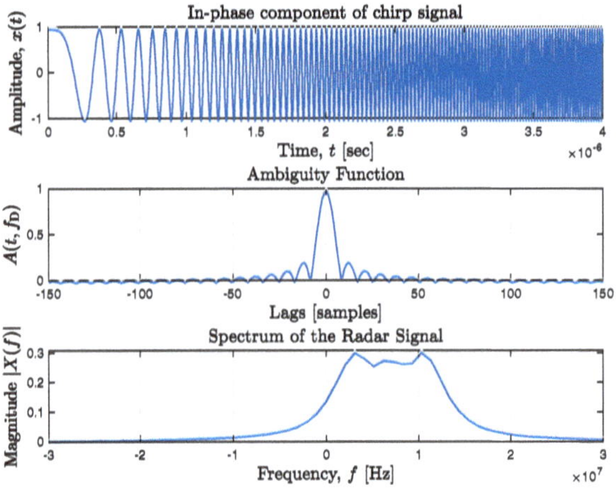

(a)

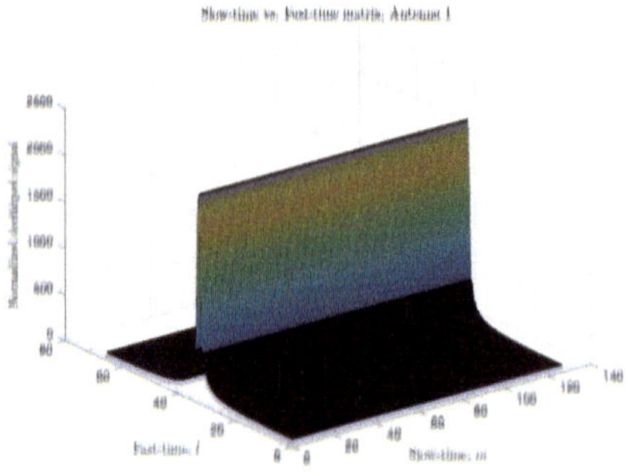

(b)

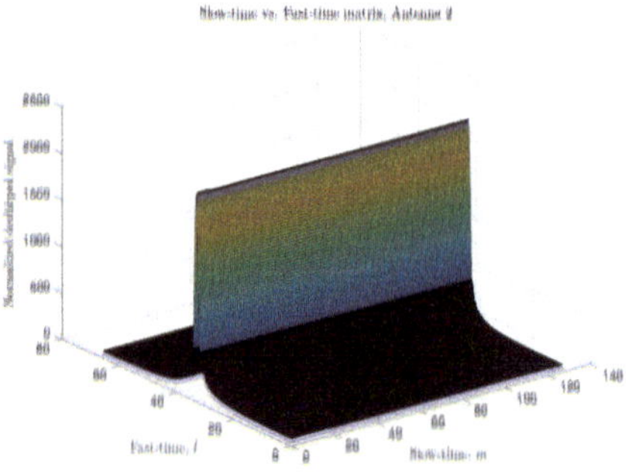

(c)

Fig. 18.26 Two-antenna radar system. (**a**) Chirp signal, ambiguity, and spectrum. (**b**) Ambiguity of antenna 1, 2D. (**c**) Ambiguity of antenna 2, 2D

- For $M = 128$, the estimated ranges at antennas 1 and 2 are 30.4009 m and 30.3297 m, respectively. The estimated angle of arrival is 13.9865°, whereas the estimated velocity is 48.5756 km/h. The estimated distance projected into the vehicle direction is 29.4996 m.

Alternatively, the estimation of the range and angle of arrival was performed in the time-domain dechirped signal with the following results.

- For $M = 4$, the initial range (distance) is 30.0017 m, reaching 31.0599 m from antenna 1 and 30.9836 from antenna 2. The angle of arrival is estimated at 14.9992°.

- For $M = 64$, the initial range (distance) is 30.0350 m, reaching 31.0921 m from antenna 1 and 31.0159 from antenna 2. The estimated angle of arrival is 14.9833°.
- For $M = 128$, the initial range (distance) is 30.0706 m, reaching 31.1264 m from antenna 1 and 31.0503 from antenna 2. The estimated angle of arrival is 14.9664°.

End

18.7.4 Frequency-Time-Angle Information

With a virtual ULA, it is possible to estimate the angle of arrival through beamforming where we organize the information acquired in several combinations of the virtual antenna in a set of time-frequency planes, each representing the channel information as shown in Fig. 18.27 (Sun et al. 2020). The same information might be organized in a tile format in Fig. 18.28 (Richards 2022), also known as a data cube. Note that the amount of channel information depends on the number of antennas in the array.

Beamforming consists of applying the FFT in the antennas' information direction, forming another tile formatting where the number of antennas axis, also called angle axis, as illustrated Fig. 18.29. We should bear in mind that beamforming does not provide high-resolution angle detection without high cost. For an order of a degree, one might require over a hundred antennas separated by half-wavelength (Richards 2022). This resolution can be costly. However, as pointed out in Sun et al. (2020), in a massive MIMO implementation, it is possible to address the complexity issue by employing sparse or nonuniform linear arrays.

Once a ULA is formed, it is possible to apply the knowledge of Chap. 9 to determine the azimuth angle of arrival. The beamformer width of an antenna array, delimited by -3 dB attenuation, is given by

$$BW_{\mathrm{a}} = 2 \arcsin \frac{1.4\lambda_{\mathrm{c}}}{O\pi} \tag{18.68}$$

where O is the aperture size; in the present case of ULA, it relates to the number of array elements and their separation.

If the elevation angle is also desired, we need a 2D array, as previously discussed. For the time being, only the ULA is discussed.

The data cube presented in Fig. 18.29 enables the representation of the data source for many signal processing operations that can be applied to acquire information about the targets. The possibilities rely on the selection of the tile set from the data cube. For instance, a vector inner product between a weight vector and the data from antenna elements forms a beamformer. A two-dimensional matched filtering in the fast- and slow-time plane produces synthetic aperture imaging.

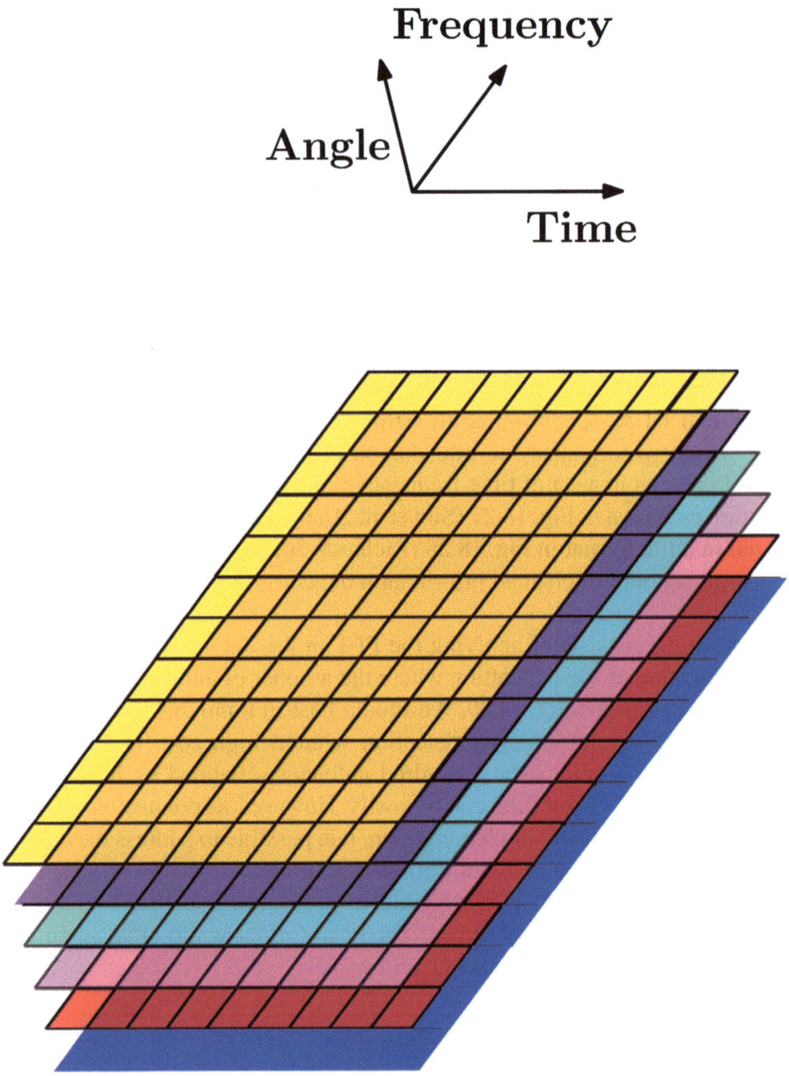

Fig. 18.27 Time-frequency-channel information MIMO radar configuration

A two-dimensional space-time adaptive filtering is possible using the slow-time antenna-element plane.

If one draws a vector along the fast-time axis with a single tile from the other domains, it is possible to apply one-dimension filtering for pulse compression purposes. One vector along the slow-time axis with a single tile from the other domains allows for a one-dimension filtering or spectrum analysis in pursuit of Doppler information processing. Mathematical tools using tensor methods can also

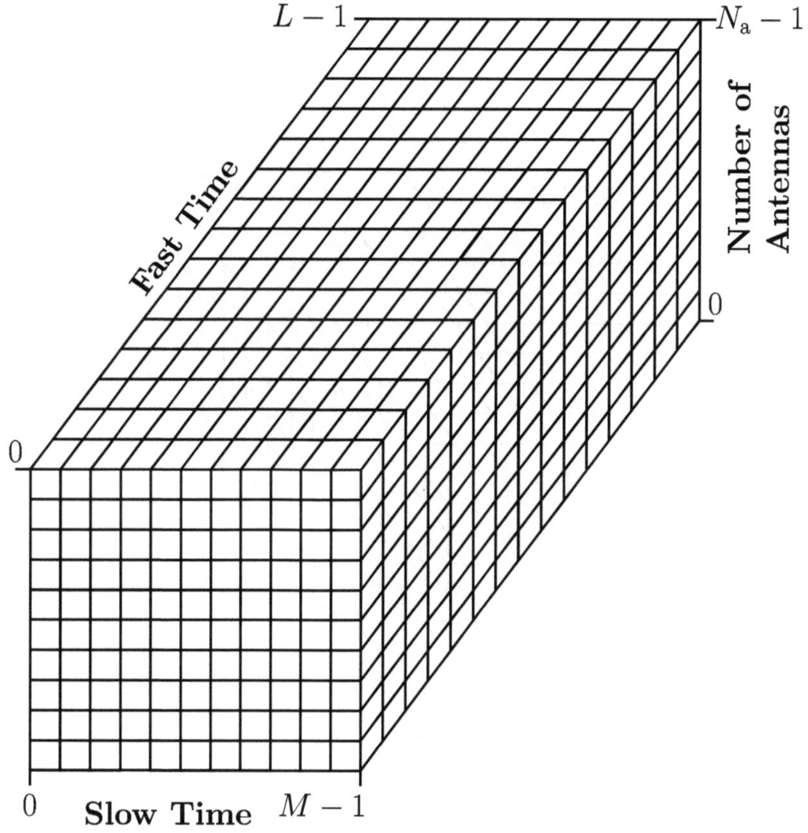

Fig. 18.28 Time-frequency-channel in a tile-format representation called data cube

exploit the tile array multidimensionality, enabling more accurate velocity, range, and direction estimation.

18.7.5 Direction of Arrival

Let us describe the virtual antenna matrix of (18.57) as a function of the angle of arrival θ_r as $\mathbf{A}(\theta_r)$, so the model of the received signal is described as

$$\begin{aligned}\mathbf{r} &= \mathbf{A}(\theta_r)\,\mathbf{s} + \mathbf{n} \\ &= \mathbf{A}(\theta_r)\begin{bmatrix} s_1 \\ \vdots \\ s_N \end{bmatrix} + \mathbf{n}\end{aligned} \tag{18.69}$$

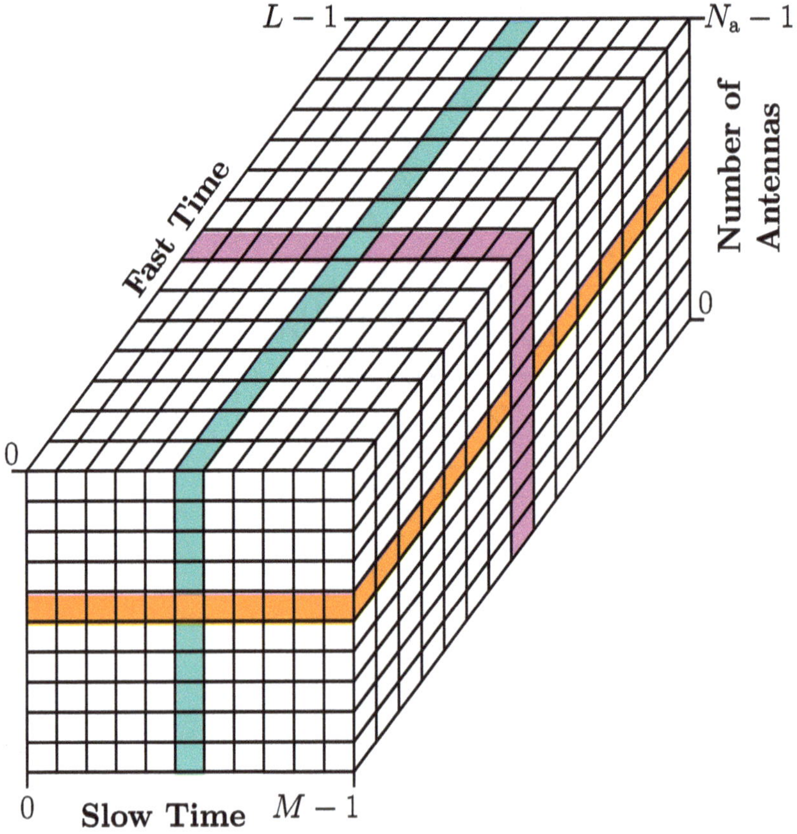

Fig. 18.29 Time-frequency-angle information MIMO radar configuration

where in the present discussion $N = M_t M_r$, $\mathbf{s}$ represents the target reflection coefficient vector where s_i is the coefficient related to the ith target. Some snapshots of the array response can be used to estimate the target range and Doppler shift by using the information from all virtual array receivers located in the same bin. However, if the environment is highly dynamic, the number of snapshots should be small. Chapters 9 and 10 of this book present several methods to design beamformers.

Example 18.7 In automotive communications, a vehicle is located at the front view and using a virtual array of the radar with 20 elements in place, implementing a beamformer aimed at fifteen degrees from the source vehicle movement direction corresponding to the target position.

Assume a bandwidth of $BW = 500\,\text{MHz}$, a carrier of $f_c = 80\,\text{GHz}$, a pulse duration of $T = 40\,\mu\text{s}$, and a range of detection of $d = 100$ m. Assume the target is not moving. We must cover a range swath of 600 m by choosing L accordingly. Use $M = 10$.

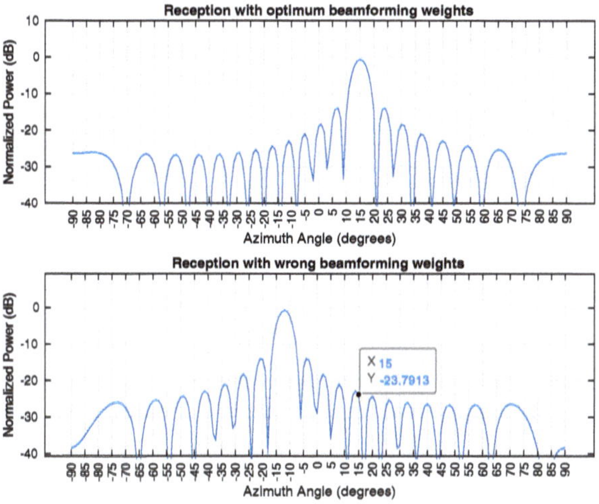

(a)

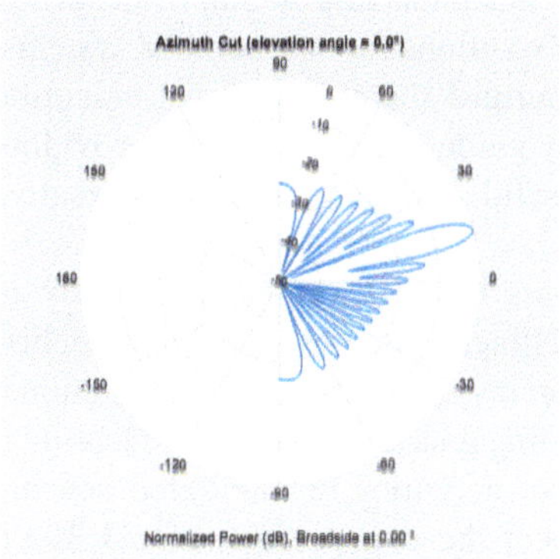

(b)

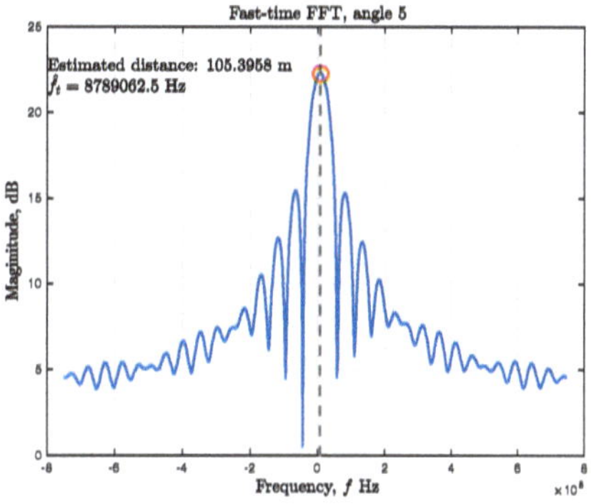

(c)

Fig. 18.30 MIMO radar. (**a**) Beamformers. (**b**) Angle of arrival. (**c**) Spectrum

Solution The value of L in this example is 600. In Fig. 18.30a, it is possible to observe the optimal beamformer whose main lobe sits in the incoming signal direction. Assume the beamformer changes its angle to -12 degrees, with the source vehicle maintaining its direction. In such a case, it is possible to observe that the received signal gain will reduce by -23.7913 dBs, undoubtedly affecting the range and Doppler shift estimation. It is also possible to verify the lobe distribution of the array in the polar description in Fig. 18.30b. The last subplot shows the DFT of the fast-time axis from where it is possible to estimate the range at 105.3958 meters approximately.

End

Many challenging issues are related to the automotive use of radar systems, including extending their reliability and accuracy in the measured signals required to infer situational position. Some issues are discussed in Engels et al. (2021) and the references therein. An available data set for automotive applications is described in Schumann et sl. (2021). In Parker (2017), some practical discussions are related to automotive radar implementation.

18.8 Remarks

Autonomous systems, and vehicles in particular, require a multitude of signal processing tools for successful and reliable implementations. The present chapter addresses several methods related to radar signal processing and its possible integration with some communication methods. The integration of sensors and communications plays a central role in this area, which is still witnessing fast development.

It remains to be seen whether the proper fusion of signals originating from several types of sources ultimately results in ultrareliable solutions for autonomous systems.

It is important that the regulatory bodies nurture the international scientific community and develop technical standards to guide the industry in providing competitive solutions. This will ensure successful and economically attractive products related to autonomous systems.

Deploying robust autonomous vehicle systems is challenging, particularly in urban areas. Possible solutions involve fusing distinct sensing sources and sophisticated environment modeling. The practical systems might rely on computational intelligence, which requires large amounts of training data and some online learning.

At the time of this writing, there are a lot of activities in integrated sensing and communication (ISAC), for instance, to meet the expectations of 6G 3GPP standards 3GPP23 (2023) for localization, intrusion detection, unmanned aerial vehicle control, etc. The design of ISAC systems aims at optimization of the spectral resources while orchestrating the communication and sensing tasks Gonzalez-Prelcic et al. (2024).

References

T.J. Abatzoglou, G.O. Gheen, Range, radial velocity, and acceleration MLE using radar LFM pulse train. IEEE Trans. Aerospace Electron. Syst. **34**, 1070–1084 (1998)

C. Aydogdu, M.F. Keskin, G.K. Carvajal, O. Eriksson, H. Hellsten, E. Nilsson, M. Rydström, K. Vanäs, H. Wymeersch, Radar interference mitigation for automated driving. IEEE Signal Process. Mag. **37**, 72–84 (2020)

C.A. Balanis, P.I. Ioannides, *Introduction to Smart Antennas* (Morgan & Claypool Publishers, Fort Collins, 2007)

M. Braun, Radar algorithms in mobile communication networks. Ph.D. Dissertation, Karlsruher Instituts für Technologie, Karlsruhe, Germany, 2014

D. Chu, Polyphase codes with good periodic correlation properties. IEEE Trans. Inform. Theory **18**, 531–532 (1972)

M. Dohler, *Virtual Antenna Arrays*. King's College, University of London, London, UK: Ph.D. Thesis, 2003

F. Engels, P. Heidenreich, M. Wintermantel, L. Stäker, M. Al Kad, A.M. Zoubir, Automotive radar signal processing: Research directions and practical challenges. IEEE J. Sel. Topics Signal Process. **15**, 865–878 (2021)

A. Al-Hourani, R.J. Evans, S. Kandeepan, B. Moran, H. Eltom, Stochastic geometry methods for modeling automotive radar interference. IEEE Trans. Intell. Transp. Syst. **19**, 333–344 (2018)

N. Gonzalez-Prelcic, M.F. Keskin, O. Kaltiokallio, M. Valkama, D. Dardari, X. Shen, Y. Shen, M. Bayraktar, H. Wymeersch, The integrated sensing and communication revolution for 6G: Vision, techniques, and applications. Proc. IEEE **112**, 676–723 (2024)

G. Lellough, A.K. Mishra, M. Inggs, Stepped-carrier OFDM-radar processing scheme to retrieve high-resolution range-velocity profile at low sampling rate. IEEE Trans. Aerospace Electron. Syst. **51**, 937–950 (2015)

Y. Li, J. Ibanez-Guzman, Lidar for autonomous driving. IEEE Signal Process. Mag. **37**, 50–61 (2020)

D. Ma, N. Shlezinger, T. Huang, Y. Liu, Y.C. Eldar, Joint radar-communications strategies for autonomous vehicles. IEEE Signal Process. Mag. **37**, 85–97 (2020)

B.R. Mahafza, *Radar Systems Analysis and Design*, 3rd edn. (CRC Press, Boca Raton, 2013)

M. Parker, *Digital Signal Processing 101*, 2nd edn. (Elsevier, Oxford, 2017)

J. Rapp, J. Tachella, Y. Altmann, S. McLaughlin, V.K. Goyal, Advances in single-photon lidar for autonomous vehicles. IEEE Signal Process. Mag. **37**, 62–71 (2020)

M.A. Richards, *Fundamentals of Radar Signal Processing*, 3rd edn. (McGraw-Hill, New York, 2022)

O. Schumann, M. Hahn, N. Scheiner, F. Weishaupt, J.F. Tilly, J. Dickmann, C. Wöhler, Radar-science: a real-world radar point cloud data set for automotive applications (2021). https://arxiv.org/abs/2104.02493

B. Schweizer, C. Knill, D. Schindler, C. Waldschmidt, Stepped OFDM radar technique to resolve range and doppler simultaneously. IEEE Trans. Microwave Theory Tech. **66**, 1610–1618 (2018)

S. Sun, A.P. Petropulu, H.V. Poor, Mimo radar for advanced driver-assistance systems and autonomous driving. IEEE Signal Process. Mag. **37**, 98–117 (2020)

W.Q. Wang, Virtual antenna array analysis for mimo synthetic aperture radars. Int. J. Antennas Propag. **2012**, 1–10 (2012)

Z. Xu, C.J. Baker, S. Pooni, Range and doppler cell migration in wideband automotive radar. IEEE Trans. Veh. Technol. **68**, 5527–5536 (2019)

D. Zoeke, A. Ziroff, Phase migration effects in moving target localization using switched mimo arrays, in *Proceedings of 12th European Radar Conference* (France, 2015), pp. 85–88

3GPP TR 22.837 V19.0.0, "Feasibility study on integrated sensing and communication," Jun. 2023

Index

Y. C. Lim et al., *Signal Processing for Autonomous Systems*,
https://doi.org/10.1007/978-3-032-00675-2

V

W

Z

MIX
Papier aus verantwortungsvollen Quellen
Paper from responsible sources
FSC® C105338

If you have any concerns about our products,
you can contact us on
ProductSafety@springernature.com

In case Publisher is established outside the EU,
the EU authorized representative is:
Springer Nature Customer Service Center GmbH
Europaplatz 3, 69115 Heidelberg, Germany

Printed by Libri Plureos GmbH
in Hamburg, Germany